Birkhäuser

Operator Theory: Advances and Applications

Volume 217

Founded in 1979 by Israel Gohberg

Yuri Arlinskii
Sergey Belyi
Eduard Tsekanovskii

Conservative Realizations of Herglotz–Nevanlinna Functions

Yuri Arlinskii
Department of Mathematics
East Ukrainian National University
Kvartal Molodizhny, 20-A
91034 Lugansk, Ukraine
yury_arlinskii@yahoo.com

Sergey Belyi
Department of Mathematics
Troy University
Troy, AL 36082, USA
sbelyi@troy.edu

Eduard Tsekanovskii
Department of Mathematics
Niagara University,
NY 14109, USA
tsekanov@niagara.edu

2010 Mathematics Subject Classification: 47A, 47B

ISBN 978-3-7643-9995-5 e-ISBN 978-3-7643-9996-2
DOI 10.1007/978-3-7643-9996-2

Library of Congress Control Number: 2011930752

Cover design: deblik, Berlin

Printed on acid-free paper

Springer Basel AG is part of Springer Science+Business Media

www.birkhauser-science.com

To the memory of Moshe Livšic, a Teacher
and a great mathematician

Preface

Consider the system of equations

$$\begin{cases} i\frac{d\chi}{dt} + T\chi(t) = KJ\psi_-(t), \\ \chi(0) = x \in \mathcal{H}, \\ \psi_+ = \psi_- - 2iK^*\chi(t), \end{cases} \tag{0.1}$$

where T is a bounded linear operator from a Hilbert space $\mathcal{H}$ into itself, K is a bounded linear operator from a Hilbert space E ($\dim E < \infty$) into $\mathcal{H}$, $J = J^* = J^{-1}$ maps E into itself, and $\operatorname{Im} T = KJK^*$. If for a given continuous in E function $\psi_-(t) \in L^2_{[0,\tau_0]}(E)$ we have that $\chi(t) \in \mathcal{H}$ and $\psi_+(t) \in L^2_{[0,\tau_0]}(E)$ satisfy the system (0.1), then the following metric conservation law holds:

$$2\|\chi(\tau)\|^2 - 2\|\chi(0)\|^2 = \int_0^\tau (J\psi_-, \psi_-)_E\, dt - \int_0^\tau (J\psi_+, \psi_+)_E\, dt, \quad \tau \in [0, \tau_0]. \tag{0.2}$$

Given an input vector $\psi_- = \varphi_- e^{izt} \in E$, we seek solutions to the system (0.1) as an output vector $\psi_+ = \varphi_+ e^{izt} \in E$ and a state-space vector $\chi(t) = xe^{izt}$ in $\mathcal{H}$, ($z \in \mathbb{C}$). Substituting the expressions for $\psi_\pm(t)$ and $\chi(t)$ in (0.1) allows us to cancel exponential terms and convert the system to stationary algebraic format

$$\begin{cases} (T - zI)x = KJ\varphi_-, \\ \varphi_+ = \varphi_- - 2iK^*x; \end{cases} \qquad \operatorname{Im} T = KJK^*, \quad z \in \rho(T), \tag{0.3}$$

where $\rho(T)$ is the set of regular points of the operator T. The type of an open system in (0.3) was introduced and studied by Livšic [191]. A brief form of an open system (0.3) can be written as a rectangular array known in operator theory as an operator colligation [89]

$$\Theta = \begin{pmatrix} T & K & J \\ \mathcal{H} & & E \end{pmatrix}, \quad \operatorname{Im} T = KJK^*. \tag{0.4}$$

The transfer function of the system Θ of the form (0.3)-(0.4) is given by

$$W_\Theta(z) = I - 2iK^*(T - zI)^{-1}KJ, \tag{0.5}$$

and satisfies, for $z \in \rho(T)$,

$$\begin{aligned} W^*_\Theta(z) J W_\Theta(z) &\geq J, \quad (\mathrm{Im}\ z > 0), \\ W^*_\Theta(z) J W_\Theta(z) &= J, \quad (\mathrm{Im}\ z = 0), \\ W^*_\Theta(z) J W_\Theta(z) &\leq J, \quad (\mathrm{Im}\ z < 0). \end{aligned} \tag{0.6}$$

We call the function

$$V_\Theta(z) = K^*(\mathrm{Re}\, T - zI)^{-1} K = i[W_\Theta(z) + I]^{-1}[W_\Theta(z) - I]J, \tag{0.7}$$

the impedance function of the system Θ. This function $V_\Theta(z)$ is a Herglotz-Nevanlinna function in E. The condition $\mathrm{Im}\, T = KJK^*$ plays a crucial role in determining the analytical properties of the functions $W_\Theta(z)$ and $V_\Theta(z)$. The open system Θ in (0.3)-(0.4) has the property that its transfer function $W_\Theta(z)$ becomes a J-unitary operator for real $z \in \rho(T)$, i.e.,

$$[\varphi_+, \varphi_+] = [W_\Theta(z)\varphi_-, W_\Theta(z)\varphi_-],$$

where $[\cdot,\cdot] = (J\cdot,\cdot)_E$, and $(\cdot,\cdot)_E$ is an inner product in E.

Let us look at a simple but motivating example leading to a system of the form (0.3)-(0.4). Consider a four-terminal electrical circuit in Figure 1. Let C denote the capacity of the capacitor and let L represent the inductance of an induction coil. Given a harmonic input

$$\psi_- = \varphi_- e^{i\omega t}, \quad \varphi_- = \begin{pmatrix} \sqrt{2}\, I^- \\ \sqrt{2}\, U^- \end{pmatrix},$$

where I^- is the current and U^- is the voltage, we are trying to find the harmonic output

$$\psi_+ = \varphi_+ e^{i\omega t}, \quad \varphi_+ = \begin{pmatrix} I^+ \\ U^+ \end{pmatrix},$$

and also describe the state of the capacitor and the induction coil

$$\chi = x e^{i\omega t}, \quad x = \begin{pmatrix} \sqrt{L}\, I \\ \sqrt{C}\, U \end{pmatrix}.$$

Here I is the current on the induction coil, U is the voltage on the capacitor, I^+ and U^+ are the output current and voltage, respectively. Using electrical circuit equations

$$L\frac{dI}{dt} = U^-(t), \qquad C\frac{dU}{dt} = -I(t) + I^-(t),$$

we can obtain system (0.1), separate variables, and arrive at the system

$$\begin{cases} (T - \omega I)x = KJ\varphi_-, \\ \varphi_+ = \varphi_- - 2iK^*x, \end{cases}, \quad \omega \in \rho(T),$$

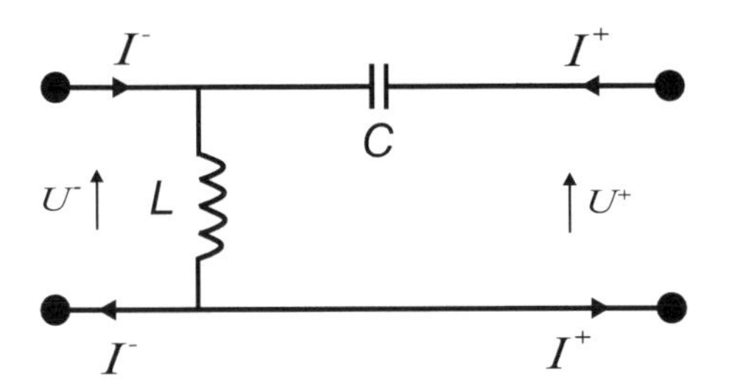

Figure 1: Four-terminal circuit

where

$$T = \begin{pmatrix} 0 & 0 \\ \frac{i}{\sqrt{LC}} & 0 \end{pmatrix}, \quad K = \begin{pmatrix} \frac{i}{\sqrt{2L}} & 0 \\ 0 & \frac{i}{\sqrt{2C}} \end{pmatrix}, \quad J = \begin{pmatrix} 0 & 1 \\ 1 & 0 \end{pmatrix}, \tag{0.8}$$

and x, $\varphi_\pm$ are defined above. By a routine argument one obtains $\text{Im}\, T = KJK^*$. This open system can be re-written in the form (0.4)

$$\Theta = \begin{pmatrix} T & K & J \\ \mathbb{C}^2 & & \mathbb{C}^2 \end{pmatrix}, \quad \text{Im}T = KJK^*,$$

whose transfer function is of the form (0.5) and actually reads

$$W_\Theta(\omega) = I - 2iK^*(T - \omega I)^{-1}KJ = \begin{pmatrix} 1 & \frac{i}{\omega L} \\ \frac{i}{\omega C} & 1 - \frac{1}{\omega LC} \end{pmatrix},$$

where T, K, and J are defined in (0.8). It is easy to see that $W_\Theta(\omega)$ satisfies the conditions (0.6) with $z = \omega \in \rho(T)$.

When a physical system (for instance a lengthy line) has distributed parameters, the state-space operator T of the system becomes unbounded. As a result, the above mentioned system Θ does not (as an algebraic structure) have any meaning, since the imaginary part of an unbounded operator T may not be defined properly because the domains of T and T^* may not coincide. However, some examples [191] of systems with unbounded operators show that the ranges of the channel operators K belong to some triplets of Hilbert spaces $\mathcal{H}_+ \subset \mathcal{H} \subset \mathcal{H}_-$ while not being a part of $\mathcal{H}$. In the 1960s Livšic formulated a problem [191] of developing a theory of open systems and their transfer functions that would involve unbounded operators and at the same time preserve the algebraic structure existing in the case when the state-space operator of the system is bounded. The importance of a problem of using generalized functions in system theory (especially in systems with distributed parameters) was pointed out in the 1970s independently by Helton [155]. The solution to this difficult problem is the main subject of the current book. The monograph also covers the research in this area for the last three decades. Different approaches to realization problems of various types of systems with continuous time and conservativity condition (or without) have been considered by Arov-Dym [56], [57], [58], [119], Ball-Staffans [69], Staffans [235], [237], Bart-Gohberg-Kaashoek-Ran [70], [71], and others.

Below we provide a brief description of the results considered in the current text.

- In Chapter 1 we consider some basic facts related to the theory of extensions of linear symmetric operators. In particular, we study the parameterizations of the domains of all self-adjoint extensions in dense and non-dense domain cases. This includes the von Neumann and Krasnoselkiĭ decomposition and parametrization formulas.

- In Chapter 2 we study the extensions of symmetric non-densely defined operators in triplets of rigged Hilbert spaces. The Krasnoselskiĭ formulas discussed in Chapter 1 are based on indirect decomposition, where linear manifolds such as the domain of the symmetric operator and its deficiency subspaces may be linearly dependent. Introduction of rigged Hilbert spaces allows us to obtain direct decomposition for the domain of the adjoint operator and parametrization of all self-adjoint extensions. This direct decomposition is written in terms of the semi-deficiency subspaces and is an analogue of the classical von Neumann formulas.

- Chapter 3 is dedicated to the development of a new extension theory of symmetric operators in triplets of Hilbert spaces, the so-called *bi-extension theory* that will be put to extensive use later in the text.

- Chapter 4 studies quasi-self-adjoint extensions of symmetric operators and contains the definition of the so-called $(*)$-extension. The $(*)$-extensions will be used later in the book in the definition of L-systems. We also present an analysis of these extensions together with their description and parametrization.

- Chapter 5 contains the main concepts and ideas of the classical theory of the Livšic canonical systems (operator colligations) with bounded operators. A comprehensive study of operator colligations of the form (0.4) in operator theory was developed by Brodskiĭ and Livšic [89], [91]. We provide a collection of known results for such a type of systems in terms of transfer functions and their linear-fractional transformations. These results include couplings of these systems and multiplication and factorization theorems for the transfer functions.

- In Chapter 6, we introduce rigged canonical systems with an unbounded operator T, $(\rho(T) \neq \emptyset)$, where $T \supset \dot{A}$, $T^* \supset \dot{A}$, $\dot{A}$ is a symmetric operator, $\mathcal{H}_+ \subset \mathcal{H} \subset \mathcal{H}_-$ is a rigged Hilbert space generated by $\dot{A}$. If we consider system (0.1) and its stationary version (0.3) with an unbounded operator T such that $\operatorname{Ran}(K) \subset \mathcal{H}_-$, we will run into substantial difficulties even at the stage of defining the solution of the system. This happens because, for a given input $\varphi_- \in E$, the first equation of system (0.3) does not have regular solutions $x \in \mathrm{Dom}(T)$. In order to treat this case adequately, we need to perform a certain regularization of the system that is based on the bi-extension and $(*)$-extension theory developed in Chapters 3 and 4. This regularization also allows us to determine the imaginary part $\mathrm{Im}\, T$ of the unbounded operator T. Let $\mathbb{A}, \mathbb{A}^* \in [\mathcal{H}_+, \mathcal{H}_-]$ be $(*)$-extensions of T and consider the system

$$\begin{cases} i\frac{d\chi}{dt} + \mathbb{A}\chi(t) = KJ\psi_-(t), \\ \chi(0) = x \in \mathcal{H}_+, \\ \psi_+ = \psi_- - 2iK^*\chi(t), \end{cases} \tag{0.9}$$

where $K \in [E, \mathcal{H}_-]$, $K^* \in [\mathcal{H}_+, E]$, J is a self-adjoint and unitary operator in E, and $\mathrm{Im}\,\mathbb{A} = KJK^*$. If for a given continuous in E function $\psi_-(t) \in L^2_{[0,\tau_0]}(E)$ we have that a continuous in $\mathcal{H}_+$ and strongly differentiable in $\mathcal{H}$ function $\chi(t) \in \mathcal{H}_+$ and a function $\psi_+(t) \in L^2_{[0,\tau_0]}(E)$ satisfy the system (0.9), then the metric conservation law (0.2) also holds. In line with the approach of the bounded case, we look for stationary solutions and convert our system to the algebraic form

$$\begin{cases} (\mathbb{A} - zI)x = KJ\varphi_-, \\ \varphi_+ = \varphi_- - 2iK^*x, \end{cases} \qquad z \in \rho(T), \tag{0.10}$$

where φ_- and φ_+ are input and output vectors in E, respectively and vector $x \in \mathcal{H}_+$ is a vector of the state space. The system (0.10) above is called *the Livšic rigged, canonical system* or *L-system* and can be written as an array

$$\Theta = \begin{pmatrix} \mathbb{A} & K & J \\ \mathcal{H}_+ \subset \mathcal{H} \subset \mathcal{H}_- & & E \end{pmatrix}, \quad z \in \rho(T), \tag{0.11}$$

where $\mathbb{A}$ is a $(*)$-extension of T such that

$$\mathrm{Im}\, T = \mathrm{Im}\,\mathbb{A} = KJK^*, \quad \mathrm{Ran}(\mathrm{Im}\,\mathbb{A}) = \mathrm{Ran}(K). \tag{0.12}$$

The transfer function of the system Θ has the form

$$W_\Theta(z) = I - 2iK^*(\mathbb{A} - zI)^{-1}KJ, \quad z \in \rho(T),$$

and satisfies analytical conditions (0.6). The impedance function of the system Θ is

$$V_\Theta(z) = K^*(\mathrm{Re}\,\mathbb{A} - zI)^{-1}K = i[W_\Theta(z) + I]^{-1}[W_\Theta(z) - I]J.$$

Clearly, systems of the form (0.11)-(0.12) have the same algebraic structure as systems (0.3)-(0.4) with bounded operators. Therefore we can refer to the canonical systems in Chapter 5 as L-systems as well. We recall that, for a holomorphic function that maps the open upper half-plane into itself, one can find the names Herglotz [137], Nevanlinna [42], and R-functions [159] (sometimes depending on the geographical origin of authors). In this text we adopt the term Herglotz-Nevanlinna function and extend it to both scalar and operator-valued cases. An important criteria is obtained for the class of Herglotz-Nevanlinna functions in Hilbert space E ($\dim E < \infty$) of the form

$$V(z) = Q + Lz + \int_{\mathbb{R}} \left(\frac{1}{t-z} - \frac{t}{1+t^2} \right) d\Sigma(t)$$

where $Q = Q^*$, $L \geq 0$, and $\int_{\mathbb{R}} \frac{(d\Sigma(t)x,x)_E}{1+t^2} < \infty$ for all $x \in E$ that can be realized as impedance functions of some scattering ($J = I$) L-system. The proof of this criteria relies on the constructed minimal L-system involving an operator of multiplication by the independent variable in the model functional Hilbert space. This model L-system is extensively used in the following chapters.

- In Chapter 7 we introduce three distinct subclasses of the class of all realizable Herglotz-Nevanlinna operator-valued functions. Complete proofs of direct and inverse realization theorems are given for each subclass. We also provide several multiplication theorems related to this class partition.

- In Chapter 8 we give the definition and describe the properties of normalized canonical L-systems. We also prove an important theorem about the constant J-unitary factor. This theorem states that if an operator-valued function $W(z)$ is realizable as a transfer function of an L-system Θ, then for an arbitrary constant J-unitary operator B the functions $W(z)B$ and $BW(z)$ can be realized as transfer functions of the same type of L-system that contains the same unbounded operator T as Θ but a different channel operator.

- In Chapter 9 we present the solution to the restricted Phillips-Kato extension problem on the existence and description of all proper accretive and sectorial maximal extensions of a given densely defined non-negative symmetric operator. This description (parametrization) is presented in terms of contractive extensions of a given symmetric contraction that are linear fractional transformations of the corresponding accretive operators. The established parametrization strengthens the classical Krein theorem on self-adjoint contractive extensions of symmetric contractions. On the basis of this new parametrization we establish a criterion in terms of the impedance function $V_\Theta(z)$ of an L-system Θ when the state-space operator T of this system is a contraction or so-called an α-co-sectorial contraction. Also in this chapter we establish the criteria for a given Stieltjes or inverse Stieltjes function to be realized as an impedance function of some L-system Θ whose state-space operator T is maximal accretive or α-sectorial.

- Chapter 10 is dedicated to an important application: it provides the descriptions of all accretive and sectorial boundary value problems T_h, where T_h is a Schrödinger operator in $L_2[a, +\infty)$ with a complex boundary parameter h. We also provide a complete description of all L-systems with Schrödinger operator T_h. Moreover, we describe the class of scalar Stieltjes (inverse Stieltjes) like functions that can be realized as impedance functions of L-systems with Schrödinger operator T_h. It is shown that the Schrödinger operator T_h of an L-system is accretive if and only if the impedance function of this L-system is either a Stieltjes or inverse Stieltjes function. We derive the formulas that restore an L-system uniquely from a given Stieltjes (inverse Stieltjes) like function to become the impedance function of this L-system. These formulas allow us to solve the inverse problem and find the exact value of the parameter h in the definition of T_h, as well as a real parameter μ that appears in the construction of the elements of the realizing L-system. An elaborate investigation of these formulas shows the dynamics of the restored parameters h and μ in terms of the changing constant term from the integral representation of the realizable function. We also point out an important connection between the impedance functions of L-systems with Schrödinger operator T_h and the Krein-von Neumann and Friedrichs extensions of a minimal non-negative Schrödinger operator.
- In Chapter 11 we consider a new type of solutions of Nevanlinna-Pick interpolation problems for the class of scalar Herglotz-Nevanlinna functions. These are explicit system solutions that are impedance functions of some L-systems with bounded operators. The conditions for the existence and uniqueness of solutions are presented in terms of interpolation data. We also find new properties of the classical Pick matrices. The exact formula for the angle of sectoriality of the corresponding state-space operator in the explicit system solution is derived. The criterion for this operator to be accretive, but not α-sectorial for any angle $\alpha \in (0, \pi/2)$, is obtained in terms of interpolation data and classic Pick matrices. We find conditions on interpolation data under which the explicit system solution of a scalar Nevanlinna-Pick interpolation problem is generated by the dissipative L-system whose state-space operator is a non-self-adjoint, prime dissipative Jacobi matrix with a rank-one imaginary part. In order to obtain these results, we establish a new model for prime, bounded, dissipative operators with rank-one imaginary part and show that a semi-infinite (finite) bounded Jacobi matrix is a new model. In addition, an inverse spectral problem for finite non-self-adjoint Jacobi matrices with rank-one imaginary part is solved. It is shown that any finite sequence of non-real numbers in the open upper half-plane is the set of eigenvalues (counting multiplicities) of some dissipative non-self-adjoint Jacobi matrix with rank-one imaginary part. The algorithm of reconstruction of the unique Jacobi matrix from its non-real eigenvalues is presented.
- In Chapter 12 we consider non-canonical rigged systems and show that the metric conservation law holds for them as well. Moreover, it is easily seen

that, in the special case when a non-canonical system becomes canonical, the metric conservation law for the non-canonical system matches its canonical version. Later on in this chapter we utilize non-canonical systems to present the solution of the general realization problem for an arbitrary Herglotz-Nevanlinna function. In particular it is shown that an arbitrary Herglotz-Nevanlinna operator-valued function can be realized as the transfer function of the corresponding non-canonical impedance system or NCI-system. The conditions on an arbitrary Herglotz-Nevanlinna function to be an impedance function of a non-canonical L-system (or NCL-system) are also provided.

Over the last several decades many books and papers have been dedicated to the analysis of infinite-dimensional systems and realization problems for different function classes. The literature on this subject is too extensive to be discussed exhaustively but we refer in this matter to [1]–[274] and the literature therein. In this text we propose a comprehensive analysis of the above mentioned L-systems with, generally speaking, unbounded operators that satisfy the *metric conservation law*. We also treat realization problems for Herglotz-Nevanlinna functions and their various subclasses when members of these subclass are realized as impedance functions of L-systems. This type of realizations is called *conservative*. The detailed study provided relies on a new method involving extension theory of linear operators with the exit into rigged triplets of Hilbert spaces. In particular, it is possible to set a one-to-one correspondence between the impedance of L-systems and related $(*)$-extensions of unbounded operators. The theory of *singular systems* developed in the current monograph leads to several useful and important applications including systems with non-self-adjoint Schrödinger operator, non-self-adjoint Jacobi matrices, and system interpolation. In summary, we hope that this book contains new developments and will be of value and interest to researchers in the field of operator theory, spectral analysis of differential operators, and system theory. We also think that this text may be used to teach a graduate level special topics course on this subject.

Acknowledgements

We are very grateful to Harm Bart and Olof Staffans for their reviews and valuable comments. We are also indebted to Vladimir Peller and Peter Kuchment for their helpful suggestions. Special thanks goes to our co-authors and colleagues Vladimir Derkach, Seppo Hassi, Fritz Gesztesy, Konstantin Makarov, Mark Malamud, and Henk de Snoo for their valuable input and fruitful discussions.

Lugansk, Troy, Niagara Falls,

Yury Arlinskii
Sergey Belyi
Eduard Tsekanovskii

October 2010

Contents

Chapter 1

Extensions of Symmetric Operators

In this chapter we deal with extensions of densely and non-densely defined symmetric operators. The parametrization of the domains of all self-adjoint extensions in both dense and non-dense cases is given in terms of von Neumann's and Krasnoselskiĭ's formulas, respectively. The so-called admissible unitary operators serve as parameters in Krasnoselskiĭ's formulas.

1.1 Deficiency indices of symmetric operators

Let $\mathcal{H}$ be a complex separable Hilbert space with inner product $(\cdot, \cdot)$ and the identity operator I. A linear operator $\dot{A}$ in a Hilbert space $\mathcal{H}$ is called **symmetric** if

$$(\dot{A}x, y) = (x, \dot{A}y), \qquad \forall x, y \in \mathrm{Dom}(\dot{A}).$$

If in this case also $\mathrm{Dom}(\dot{A})$ is dense in $\mathcal{H}$ and $\dot{A} = \dot{A}^*$, then the operator $\dot{A}$ is **self-adjoint**. An operator $\dot{A}$ is **closed** if the relations $\{x_n\} \subset \mathrm{Dom}(\dot{A})$, $x_n \to x$, $\dot{A}x_n \to y$ imply that $x \in \mathrm{Dom}(\dot{A})$ and $\dot{A}x = y$.

Let $\dot{A}$ be a closed linear operator whose domain $\mathrm{Dom}(\dot{A})$ is dense in $\mathcal{H}$. A number $\lambda \in \mathbb{C}$ is called a **regular point** of the operator $\dot{A}$ if the operator $(\dot{A} - \lambda I)^{-1}$ exists, is bounded, and is defined on the entire space $\mathcal{H}$. The set of all regular points of the operator $\dot{A}$ forms a **resolvent set** and is denoted by $\rho(\dot{A})$. We call a complex number λ a point of **regular type** of the operator $\dot{A}$ if there exists $k = k(\lambda) > 0$ such that, for all $x \in \mathrm{Dom}(\dot{A})$,

$$\|(\dot{A} - \lambda I)x\| \geq k\|x\|.$$

It follows from this definition that λ is a point of regular type of the operator $\dot{A}$ if and only if $(\dot{A} - \lambda I)^{-1}$ exists and is bounded. The set $\pi(\dot{A})$ of all points of regular type is called the **field of regularity** of the operator $\dot{A}$. It is easy to see that $\pi(\dot{A})$ is an open set and $\rho(\dot{A}) \subset \pi(\dot{A})$. By $\sigma(\dot{A}) = \mathbb{C} \setminus \rho(\dot{A})$ we denote the **spectrum** of the operator $\dot{A}$ and by $\sigma_p(\dot{A})$ its **point spectrum**. A complex number λ belongs to the point spectrum of $\dot{A}$ if λ is an eigenvalue of $\dot{A}$. If $\dot{A}$ is a symmetric operator and $z = x + iy$, $(y \neq 0)$, then for any $f \in \mathrm{Dom}(\dot{A})$,

$$\|(\dot{A} - zI)f\|^2 = \|(\dot{A} - xI)f\|^2 + y^2\|f\|^2 \geq y^2\|f\|^2,$$

which implies that both the upper and the lower half-planes are connected components of the field of regularity of any symmetric operator.

For a closed symmetric operator $\dot{A}$ we consider the set

$$\mathfrak{M}_\lambda = (\dot{A} - \lambda I)\mathrm{Dom}(\dot{A}).$$

Then the subspace

$$\mathfrak{N}_\lambda = \mathcal{H} \ominus \mathfrak{M}_{\bar{\lambda}}$$

is called the **deficiency subspace** of the operator $\dot{A}$. Obviously, $\mathfrak{M}_\lambda$ is closed and makes a subspace in $\mathcal{H}$ for $\lambda \in \pi(\dot{A})$.

Lemma 1.1.1. *Let $\dot{A}$ be a closed symmetric operator and $\lambda_0 \in \pi(\dot{A})$. Then the dimension of the deficiency subspaces $\mathfrak{N}_\lambda$ is the same for each λ in some neighborhood of λ_0.*

Proof. Let $\mathcal{H}_1$ and $\mathcal{H}_2$ be subspaces of a Hilbert space $\mathcal{H}$ and $\dim \mathcal{H}_2 > \dim \mathcal{H}_1$. Let also $\mathcal{G} \subset \mathcal{H}_2$ be the set of all vectors in $\mathcal{H}_2$ that are orthogonal to $\mathcal{H}_1$. We are going to show that $\mathcal{G} \neq \{0\}$. Indeed, let P_1 be an orthoprojection operator in $\mathcal{H}$ onto $\mathcal{H}_1$ and let $T = P_1 \restriction \mathcal{H}_2$. Suppose that $\mathcal{G} = \{0\}$. We will show that in this case $\mathrm{Ker}\,(T) = \{0\}$. Indeed, let $x \in \mathrm{Ker}\,(T)$, i.e., $x \in \mathcal{H}_2$ and $Tx = P_1 x = 0$. The latter means that $x \perp \mathcal{H}_1$ and hence $x \in \mathcal{G} = \{0\}$ implying $x = 0$. Thus, $\mathrm{Ker}\,(T) = \{0\}$. Taking into account the invertibility of T we get $\dim \mathcal{H}_2 = \dim T\mathcal{H}_2 \leq \dim \mathcal{H}_1$ and arrive at a contradiction. Therefore, $\mathcal{G} \neq \{0\}$.

Now we get back to the proof of the lemma. Since $\lambda_0 \in \pi(\dot{A})$ then $\bar{\lambda}_0$ is also a point of regular type of the operator $\dot{A}$, and hence, there exists a number $k_0 = k(\bar{\lambda}_0) > 0$ such that

$$\|(\dot{A} - \bar{\lambda}_0 I)g\| \geq k_0\|g\|, \qquad \forall g \in \mathrm{Dom}(\dot{A}). \tag{1.1}$$

We have already mentioned that $\pi(\dot{A})$ is an open set. Hence $\pi(\dot{A})$ contains some ε-neighborhood of $\bar{\lambda}_0$. Without loss of generality we can assume that $\varepsilon \leq k_0$. Let $|\lambda - \lambda_0| < \varepsilon \leq k_0$. Assume that

$$\dim \mathfrak{N}_\lambda > \dim \mathfrak{N}_{\lambda_0}. \tag{1.2}$$

Then, according to the above, the set of vectors $\mathcal{G} \subset \mathfrak{N}_\lambda$ orthogonal to $\mathfrak{N}_{\lambda_0}$ is nontrivial, and thus there exists an element $h \in \mathfrak{N}_\lambda$ $(h \neq 0)$ such that $h \perp \mathfrak{N}_{\lambda_0}$ and therefore $h \in \mathfrak{M}_{\bar{\lambda}_0}$. Consequently, $h = (\dot{A} - \bar{\lambda}_0 I)g$, $(g \neq 0)$. But $h = f + (\bar{\lambda} - \bar{\lambda}_0)g$, where $f = (A - \bar{\lambda} I)g \in \mathfrak{M}_{\bar{\lambda}}$, and hence $f \perp h$, $h \in \mathfrak{N}_\lambda$. Then

$$|\lambda - \lambda_0|^2 \|g\|^2 = \|h - f\|^2 = \|h\|^2 + \|f\|^2, \tag{1.3}$$

where by (1.1)

$$\|h\| \geq k_0 \|g\|, \quad \|f\| \geq k\|g\|, \qquad (k = k(\lambda) > 0). \tag{1.4}$$

Substituting (1.4) into (1.3) and canceling $\|g\|^2$ we get

$$|\lambda - \lambda_0|^2 \geq k_0^2 + k^2,$$

which contradicts $|\lambda - \lambda_0| < k_0$. Thus (1.2) yields a contradiction. Similarly we conclude that the opposite to (1.3) inequality does not take place. Therefore

$$\dim \mathfrak{N}_\lambda = \dim \mathfrak{N}_{\lambda_0}, \qquad (|\lambda - \lambda_0| < \varepsilon).$$ □

Theorem 1.1.2. *The dimension of a deficiency subspace $\mathfrak{N}_\lambda$ of a closed symmetric operator $\dot{A}$ is the same for each λ in the open upper (lower) half-plane.*

Proof. It is easy to see that all non-real points are points of regular type of a symmetric operator $\dot{A}$. Let λ and μ be two arbitrary fixed points from upper (lower) half-plane. We connect them with a line segment. For each point of this segment there is a neighborhood such that dimensions of deficiency subspaces of the operator $\dot{A}$ are the same for every λ from this neighborhood. Thus we get an open cover of a line segment that contains a finite subcover. As a result we conclude that dimensions of $\mathfrak{N}_\lambda$ and $\mathfrak{N}_\mu$ are the same. □

Let us note that neither Lemma 1.1.1 nor Theorem 1.1.2 assume that operator $\dot{A}$ is densely defined.

For a closed symmetric operator $\dot{A}$ with the deficiency subspaces $\mathfrak{N}_\lambda$ the numbers

$$n_+ = \dim \mathfrak{N}_i, \qquad n_- = \dim \mathfrak{N}_{-i}$$

are called **deficiency indices (numbers)** of the operator $\dot{A}$. They are also written in the form of an ordered pair (n_+, n_-). It follows from Theorem 1.1.2 that

$$\dim \mathfrak{N}_\lambda = \begin{cases} n_+, & \mathrm{Im}\lambda > 0, \\ n_-, & \mathrm{Im}\lambda < 0. \end{cases}$$

We note that deficiency indices of a symmetric operator are not necessarily equal and can be finite or infinite.

Theorem 1.1.3. *If a symmetric operator $\dot{A}$ has a real point of a regular type, then its deficiency indices are equal.*

Proof. If $\lambda_0 \in \pi(\dot{A})$ and $\lambda_0 \in \mathbb{R}$, then in some neighborhood the dimensions of deficiency subspaces of $\dot{A}$ are the same. Then by Theorem 1.1.2, $n_+ = n_-$. □

1.2 The first von Neumann formula in the dense case

In this section we assume that our symmetric operator $\dot{A}$ has a dense domain, i.e., $\overline{\mathrm{Dom}(\dot{A})} = \mathcal{H}$. In what follows, for a linear set $\mathcal{B}$ in a Hilbert space $\mathcal{H}$ by $\overline{\mathcal{B}}$ we denote the closure of $\mathcal{B}$ in $\mathcal{H}$. As we have already mentioned above for a closed symmetric operator $\dot{A}$ we have

$$(\dot{A}f, g) = (f, \dot{A}g), \qquad \forall f, g \in \mathrm{Dom}(\dot{A}).$$

This implies that if $g \in \mathrm{Dom}(\dot{A})$, then $g \in \mathrm{Dom}(\dot{A}^*)$ and $\dot{A}^* g = \dot{A}g$. Thus $\dot{A}^*$ is an extension of $\dot{A}$. Let us consider a deficiency subspace $\mathfrak{N}_\lambda$ of the operator $\dot{A}$ and $\varphi \in \mathfrak{N}_\lambda$. For any $g \in \mathrm{Dom}(\dot{A})$ we have $((\dot{A} - \bar{\lambda}I)g, \varphi) = 0$ or

$$(\dot{A}g, \varphi) = (g, \lambda\,\varphi), \qquad \forall g \in \mathrm{Dom}(\dot{A}). \tag{1.5}$$

That means $\varphi \in \mathrm{Dom}(\dot{A}^*)$ and

$$\dot{A}^*\varphi = \lambda\varphi, \qquad \varphi \in \mathfrak{N}_\lambda$$

Evidently, the converse statement is also true, i.e., $\dot{A}^* f = \lambda f$ implies $f \in \mathfrak{N}_\lambda$. Thus the deficiency subspace of the operator $\dot{A}$ is a kernel of the operator $\dot{A}^* - \lambda I$, or

$$\mathfrak{N}_\lambda = \mathrm{Ker}\,(\dot{A}^* - \lambda I). \tag{1.6}$$

Let us agree to call the linear manifolds L_1, $L_2, \ldots$, L_n $(n < \infty)$ **linearly independent** if the equation

$$x_1 + x_2 + \cdots + x_n = 0, \qquad x_k \in L_k, \quad k = 1, 2, \ldots, n,$$

implies that $x_1 = x_2 = \cdots = x_n = 0$. It is easy to see that two linear manifolds L and M are linearly independent if and only if $L \cap M = \{0\}$.

The following result is known as **the first von Neumann's formula**.

Theorem 1.2.1. *Let $\dot{A}$ be an arbitrary closed symmetric operator and* $\mathrm{Im}\,\lambda \neq 0$. *Then the linear manifolds* $\mathrm{Dom}(\dot{A})$, $\mathfrak{N}_\lambda$, *and* $\mathfrak{N}_{\bar{\lambda}}$ *are linearly independent. Moreover,*

$$\mathrm{Dom}(\dot{A}^*) = \mathrm{Dom}(\dot{A}) \dotplus \mathfrak{N}_\lambda \dotplus \mathfrak{N}_{\bar{\lambda}}. \tag{1.7}$$

Proof. Let $g \in \mathrm{Dom}(\dot{A})$, $\varphi \in \mathfrak{N}_\lambda$, and $\psi \in \mathfrak{N}_{\bar{\lambda}}$. Suppose that

$$g + \varphi + \psi = 0. \tag{1.8}$$

Applying $\dot{A}^*$ to both sides of this equation and taking into account that $\dot{A}^* g = \dot{A}g$, $\dot{A}^*\varphi = \lambda\varphi$, and $\dot{A}^*\psi = \bar{\lambda}\psi$, we get

$$\dot{A}g + \lambda\varphi + \bar{\lambda}\psi = 0. \tag{1.9}$$

Multiplying both sides of (1.8) by $\bar{\lambda}$ and subtracting the result from (1.9) yields

$$(\dot{A} - \bar{\lambda}I)g + (\lambda - \bar{\lambda})\varphi = 0.$$

Consequently, $(\dot{A}-\bar{\lambda}I)g=0$ and $\varphi=0$ because the vectors $(\dot{A}-\bar{\lambda}I)g\in\mathfrak{M}_{\bar{\lambda}}$ and $\varphi\in\mathfrak{N}_\lambda$ are orthogonal. But $\lambda\in\pi(\dot{A})$, $(\operatorname{Im}\lambda\neq 0)$ and so $g=0$. Hence (1.8) can only take place if all the terms in the left-hand side are zeros. This proves linear independence of $\operatorname{Dom}(\dot{A})$, $\mathfrak{N}_\lambda$, and $\mathfrak{N}_{\bar{\lambda}}$.

Now let us consider the linear manifold $L=\operatorname{Dom}(\dot{A})\dotplus\mathfrak{N}_\lambda\dotplus\mathfrak{N}_{\bar{\lambda}}$. Since all the sets $\operatorname{Dom}(\dot{A})$, $\mathfrak{N}_\lambda$, and $\mathfrak{N}_{\bar{\lambda}}$ are contained in $\operatorname{Dom}(\dot{A}^*)$, we have $L\subset\operatorname{Dom}(\dot{A}^*)$. Then all we have to show is that $\operatorname{Dom}(\dot{A}^*)\subset L$.

Let $f\in\operatorname{Dom}(\dot{A}^*)$ and $f_1=(\dot{A}^*-\bar{\lambda}I)f$. The Hilbert space $\mathcal{H}$ can be written as $\mathcal{H}=\mathfrak{M}_{\bar{\lambda}}\oplus\mathfrak{N}_\lambda$. Then $f_1=g_1+\varphi_1$, where $g_1\in\mathfrak{M}_{\bar{\lambda}}$, $\varphi_1\in\mathfrak{N}_\lambda$. Therefore,

$$g_1=(\dot{A}-\bar{\lambda}I)g=(\dot{A}^*-\bar{\lambda}I)g,$$

where $g\in\operatorname{Dom}(\dot{A})$. Consider the vector

$$\varphi=\frac{1}{\lambda-\bar{\lambda}}\varphi_1.$$

Following (1.5) we get

$$\varphi_1=(\lambda-\bar{\lambda})\varphi=(\dot{A}^*-\bar{\lambda}I)\varphi.$$

Using this we can rewrite $f_1=g_1+\varphi_1$ in the form

$$(\dot{A}^*-\bar{\lambda}I)f=(\dot{A}^*-\bar{\lambda}I)g+(\dot{A}^*-\bar{\lambda}I)\varphi,$$

or

$$(\dot{A}^*-\bar{\lambda}I)(f-g-\varphi)=0.$$

The latter means that the vector $\psi=f-g-\varphi\in\mathfrak{N}_{\bar{\lambda}}$, and thus, $f=g+\varphi+\psi\in L$. This proves $\operatorname{Dom}(\dot{A}^*)\subset L$ and (1.7). □

Corollary 1.2.2. *A closed symmetric operator is self-adjoint if and only if both its deficiency indices equal zero.*

Proof. If $\dot{A}$ is self-adjoint, then $\operatorname{Dom}(\dot{A})=\operatorname{Dom}(\dot{A}^*)$ and by (1.7), $\mathfrak{N}_\lambda=\mathfrak{N}_{\bar{\lambda}}=\{0\}$ implying $n_+=n_-=0$. Conversely, if $n_+=n_-=0$, then $\operatorname{Dom}(\dot{A})=\operatorname{Dom}(\dot{A}^*)$ and $\dot{A}$ is self-adjoint. □

1.3 Parametrization of symmetric and self-adjoint extensions. The second von Neumann formula

Let A be a closed symmetric or self-adjoint extension of a densely defined symmetric operator $\dot{A}$. Since $\dot{A}\subset A$ and $A\subset A^*$, we have that A^* is also an extension of $\dot{A}$, i.e., $\dot{A}\subset A^*$. But then $A^*\subset\dot{A}^*$ and $A\subset\dot{A}^*$. Thus for any symmetric extension A of a symmetric operator $\dot{A}$ we have

$$\dot{A}\subset A\subset\dot{A}^*,\qquad \dot{A}\subset A^*\subset\dot{A}^*. \tag{1.10}$$

Consequently, for any $f_1, f_2 \in \mathrm{Dom}(A)$,

$$f_1 = g_1 + \varphi_1 + \psi_1, \quad f_2 = g_2 + \varphi_2 + \psi_2, \tag{1.11}$$

where $g_1, g_2 \in \mathrm{Dom}(\dot{A})$, $\varphi_1, \varphi_2 \in \mathfrak{N}_\lambda$, $\psi_1, \psi_2 \in \mathfrak{N}_{\bar{\lambda}}$. Also, because $A \subset \dot{A}^*$ we have

$$Af_1 = \dot{A}g_1 + \lambda\varphi_1 + \bar{\lambda}\psi_1, \quad Af_2 = \dot{A}g_2 + \lambda\varphi_2 + \bar{\lambda}\psi_2.$$

Using the relations

$$\begin{aligned} (\dot{A}g_1, \varphi_2) &= \bar{\lambda}(g_1, \varphi_2), \quad (\dot{A}g_1, \psi_2) = \lambda(g_1, \psi_2), \\ (\varphi_1, \dot{A}g_2) &= \lambda(\varphi_1, g_2), \quad (\psi_1, \dot{A}g_2) = \bar{\lambda}(\psi_1, g_2), \end{aligned}$$

we get

$$0 = (Af_1, f_2) - (f_1, Af_2) = (\bar{\lambda} - \lambda)\big[(\psi_1, \psi_2) - (\varphi_1, \varphi_2)\big].$$

Therefore, if $f_1, f_2 \in \mathrm{Dom}(A)$, where both f_1 and f_2 are defined by (1.11), then

$$(\psi_1, \psi_2) = (\varphi_1, \varphi_2).$$

In particular, if $f_1 = f_2$, then we have

$$\|\psi_1\| = \|\varphi_1\|. \tag{1.12}$$

Recall that an operator U acting from a Hilbert space $\mathcal{H}_1$ into a Hilbert space $\mathcal{H}_2$ is called an **isometric operator** or an **isometry** if for any $x, y \in \mathrm{Dom}(U)$ we have that $(Ux, Uy) = (x, y)$ or, equivalently, $\|Ux\| = \|x\|$ for all $x \in \mathrm{Dom}(U)$. In particular, if $\mathrm{Dom}(U) = \mathcal{H}_1$ and $\mathrm{Ran}(U) = \mathcal{H}_2$, the isometric operator U is called **unitary**.

Theorem 1.3.1 (von Neumann). *Let $\dot{A}$ be a closed symmetric operator and let $\mathfrak{N}_\lambda$ and $\mathfrak{N}_{\bar{\lambda}}$ (Im $\lambda \neq 0$) be any pair of its deficiency subspaces. Let also A be a symmetric extension of $\dot{A}$. Then there exists an isometric operator U from $\mathrm{Dom}(U) \subset \mathfrak{N}_\lambda$ into $\mathfrak{N}_{\bar{\lambda}}$ such that the following representation for $f \in \mathrm{Dom}(A)$, $\varphi \in \mathrm{Dom}(U)$, and Af takes place:*

$$\begin{aligned} f &= g + \varphi - U\varphi, \\ Af &= \dot{A}g + \lambda\varphi - \bar{\lambda}U\varphi. \end{aligned} \tag{1.13}$$

Conversely, if U is an isometric operator from $\mathrm{Dom}(U) \subset \mathfrak{N}_\lambda$ into $\mathfrak{N}_{\bar{\lambda}}$, then the operator A defined by (1.13) is a symmetric extension of the operator $\dot{A}$.

Proof. Since any element $f \in \mathrm{Dom}(A)$ is uniquely represented in the form

$$f = g + \varphi + \psi, \qquad \Big(g \in \mathrm{Dom}(\dot{A}),\ \varphi \in \mathfrak{N}_\lambda,\ \psi \in \mathfrak{N}_{\bar{\lambda}}\Big), \tag{1.14}$$

we can consider operator $P_\lambda : \mathrm{Dom}(A) \to \mathfrak{N}_\lambda$ defined by the rule $P_\lambda f = -\varphi$. This operator is linear and its range $\mathrm{Ran}(P_\lambda)$ is a linear manifold in $\mathfrak{N}_\lambda$. Similarly, if $P_{\bar{\lambda}} f = \psi$, then $\mathrm{Ran}(P_{\bar{\lambda}})$ is a linear manifold in $\mathfrak{N}_{\bar{\lambda}}$.

Now consider operator $U : \mathrm{Ran}(P_\lambda) \to \mathrm{Ran}(P_{\bar\lambda})$ defined by $U\varphi = \psi$, where the vectors φ and ψ are defined in (1.14) (and thus $\varphi + \psi \in \mathrm{Dom}(A)$). Then by (1.12) we have that

$$\|U\varphi\| = \|\varphi\|,$$

and U is an isometric operator from $\mathfrak{N}_\lambda$ into $\mathfrak{N}_{\bar\lambda}$. We should also mention that both f and Af are given in the form (1.13).

Now let U be an isometric operator acting from $\mathfrak{N}_\lambda$ into $\mathfrak{N}_{\bar\lambda}$ and A be defined by (1.13). It is clear that $\dot A \subset A$ and all we have to check is that A is symmetric. Repeating the argument we used in the proof of (1.12) we get that, for any f_1 and f_2 in $\mathrm{Dom}(A)$,

$$(Af_1, f_2) - (f_1, Af_2) = (\bar\lambda - \lambda)\left[(U\varphi_1, U\varphi_2) - (\varphi_1, \varphi_2)\right] = 0,$$

and hence the operator A is symmetric. □

Remark 1.3.2. If in the formulas (1.13) the isometric operator U is replaced by $\mathcal{U} = -U$ then the von Neumann formulas take the form

$$\begin{aligned} f &= g + \varphi + \mathcal{U}\varphi, \\ Af &= \dot A g + \lambda\varphi + \bar\lambda \mathcal{U}\varphi, \end{aligned} \tag{1.15}$$

that is often seen in some literature on the subject.

Corollary 1.3.3. *If one of the deficiency indices of a closed symmetric operator $\dot A$ is zero, then $\dot A$ does not have symmetric extensions.*

Proof. Indeed, if one of the deficiency numbers is zero we can not possibly construct an isometric operator U with non-zero domain acting from one deficiency subspace to the other. □

A symmetric operator $\dot A$ is called **maximal** if it does not have a non-trivial symmetric extension acting in the same space as $\dot A$. Using the above argument we conclude that a closed symmetric operator is maximal if and only if at least one of its deficiency indices is zero.

Theorem 1.3.4. *A symmetric operator $\dot A$ in a Hilbert space $\mathcal{H}$ has self-adjoint extensions in the same space if and only if its deficiency indices are equal.*

Proof. Let A be a non-trivial symmetric extension of the operator $\dot A$ defined by (1.13). In order to find the deficiency subspaces of $\dot A$ we use (1.13) to see that

$$(A - \bar\lambda I)f = (\dot A - \bar\lambda I)g + (\lambda - \bar\lambda)\psi \in \mathfrak{M}_{\bar\lambda} \oplus \mathrm{Dom}(U),$$

where $\mathrm{Dom}(U) \subset \mathfrak{N}_\lambda$. On the other hand, if $h \in \mathfrak{M}_{\bar\lambda} \oplus \mathrm{Dom}(U)$, then $h = (A - \lambda I)f$, where f is a vector from $\mathrm{Dom}(A)$. Thus,

$$(A - \lambda I)\mathrm{Dom}(A) = \mathfrak{M}_{\bar\lambda} \oplus \mathrm{Dom}(U),$$

and deficiency subspace N_λ of the operator A is defined as

$$N_\lambda = \mathcal{H} \ominus (A - \bar{\lambda}I)\mathrm{Dom}(A) = \mathfrak{N}_\lambda \ominus \mathrm{Dom}(U).$$

Similarly we find that $N_{\bar{\lambda}} = \mathfrak{N}_{\bar{\lambda}} \ominus \mathrm{Ran}(U)$, where $\mathrm{Ran}(U)$ is the range of the operator U.

Consequently, if $\mathrm{Dom}(U) = \mathfrak{N}_\lambda$ (or $\mathrm{Ran}(U) = \mathfrak{N}_{\bar{\lambda}}$), then $N_\lambda = \{0\}$ (or $N_{\bar{\lambda}} = \{0\}$), and one of the deficiency indices of A is zero. This means that under the above conditions our operator A is maximal. Moreover, if

$$\mathrm{Dom}(U) = \mathfrak{N}_\lambda, \quad \mathrm{Ran}(U) = \mathfrak{N}_{\bar{\lambda}}, \tag{1.16}$$

then both deficiency indices of the operator A are zeros implying that A is a self-adjoint operator. It also means that in this case operator U is unitary.

Therefore, an extension A of a symmetric operator $\dot{A}$ defined by (1.13) is self-adjoint if and only if operator U in (1.13) is a unitary operator mapping the deficiency subspace $\mathfrak{N}_\lambda$ onto the deficiency subspace $\mathfrak{N}_{\bar{\lambda}}$. Hence the dimensions of $\mathfrak{N}_\lambda$ and $\mathfrak{N}_{\bar{\lambda}}$ are the same and the deficiency indices are equal. □

Remark 1.3.5. Combining Theorems 1.3.4 and 1.3.1 one can see that formulas (1.13) set up a one-to-one correspondence between all the unitary operators U satisfying (1.16) and the set of all self-adjoint extensions A of the symmetric operator $\dot{A}$.

1.4 The Cayley transform

Let $\dot{A}$ be a symmetric operator in $\mathcal{H}$ and λ be such that $\mathrm{Im}\,\lambda \neq 0$ and $h \in \mathrm{Dom}(\dot{A})$. We set

$$(\dot{A} - \bar{\lambda}I)h = f, \tag{1.17}$$
$$(\dot{A} - \lambda I)h = g, \tag{1.18}$$

where $f \in \mathfrak{M}_{\bar{\lambda}}$ and $g \in \mathfrak{M}_\lambda$, respectively. It is easy to see that (1.17) sets a one-to-one mapping from $\mathrm{Dom}(\dot{A})$ onto $\mathfrak{M}_{\bar{\lambda}}$, while (1.18) defines a one-to-one mapping from $\mathrm{Dom}(\dot{A})$ onto $\mathfrak{M}_\lambda$. Thus for every $f \in \mathfrak{M}_{\bar{\lambda}}$ there is a unique element $h \in \mathrm{Dom}(\dot{A})$ satisfying (1.17). Once we have defined this element h, we find $g \in \mathfrak{M}_\lambda$ using (1.18). Therefore we can define a linear operator $U_\lambda(\dot{A})$ with the domain $\mathrm{Dom}(U_\lambda(\dot{A})) = \mathfrak{M}_{\bar{\lambda}}$ and the range $\mathrm{Ran}(U_\lambda(\dot{A})) = \mathfrak{M}_\lambda$ such that

$$g = U_\lambda(\dot{A})f,$$

where

$$U_\lambda(\dot{A}) = (\dot{A} - \lambda I)(\dot{A} - \bar{\lambda}I)^{-1}. \tag{1.19}$$

It is easy to see that $U_\lambda(\dot{A})$ is an isometric operator. It is linear and

$$\|f\|^2 = \|(\dot{A} - (\mathrm{Re}\lambda)I)h\|^2 + (\mathrm{Im}\lambda)^2\|h\|^2 = \|g\|^2.$$

This operator $U_\lambda(\dot{A})$ is called a **Cayley transform** of a symmetric operator $\dot{A}$. Solving (1.19) for $\dot{A}$ we get

$$\dot{A} = \big(\bar{\lambda}U_\lambda(\dot{A}) - \lambda I\big)\big(U_\lambda(\dot{A}) - I\big)^{-1}. \tag{1.20}$$

We note that the Cayley transform of a self-adjoint operator is a unitary operator.

It is very important that the deficiency indices (n_+, n_-) of the operator $\dot{A}$ are the same as the deficiency indices of the operator $U_\lambda(\dot{A})$. Indeed

$$n_+ = \dim(\mathcal{H} \ominus \mathfrak{M}_{\bar{\lambda}}), \ \operatorname{Im}\lambda > 0,$$

but $\operatorname{Dom}(U_\lambda(\dot{A})) = \mathfrak{M}_{\bar{\lambda}}$, so $n_+ = \operatorname{def}(\operatorname{Dom}(U_\lambda(\dot{A})))$. Similarly, if $\operatorname{Im}\lambda < 0$, then $n_- = \operatorname{def}(\operatorname{Ran}(U_\lambda(\dot{A})))$. Here by "def" we denote the deficiency of a set, that is the dimension of its orthogonal complement.

Theorem 1.4.1. *If U is an isometric operator such that $\operatorname{Ran}(U - I)$ is dense in $\mathcal{H}$, then the operator $\dot{A}$ which is defined by the formula*

$$\dot{A} = (\bar{\lambda}U - \lambda I)(U - I)^{-1}$$

is symmetric, densely-defined operator, and the operator U is its Cayley transform, i.e., $U = U_\lambda(\dot{A})$.

Proof. Since $\operatorname{Ran}(U - I)$ is dense in $\mathcal{H}$, the inverse operator $(U - I)^{-1}$ exists, and therefore, the operator

$$\dot{A} = (\bar{\lambda}U - \lambda I)(U - I)^{-1}$$

exists and its domain is dense in $\mathcal{H}$. We show that this operator is symmetric. Let f and g be arbitrary elements of $\operatorname{Dom}(\dot{A}) = \operatorname{Ran}(U - I)$ so that

$$f = U\varphi - \varphi, \quad g = U\psi - \psi, \qquad \varphi, \psi \in \operatorname{Dom}(U).$$

Then

$$\dot{A}f = (\bar{\lambda}U - \lambda I)\varphi = \bar{\lambda}U\varphi - \lambda\varphi,$$
$$\dot{A}g = (\bar{\lambda}U - \lambda I)\psi = \bar{\lambda}U\psi - \lambda\psi.$$

Therefore,

$$(\dot{A}f, g) = (\bar{\lambda}U\varphi - \lambda\varphi, U\psi - \psi) = (\bar{\lambda} + \lambda)(\varphi, \psi) - \bar{\lambda}(U\varphi, \psi) - \lambda(\varphi, U\psi),$$

and

$$(f, \dot{A}g) = (U\varphi - \varphi, \bar{\lambda}U\psi - \lambda\psi) = (\lambda + \bar{\lambda})(\varphi, \psi) - \bar{\lambda}(U\varphi, \psi) - \lambda(\varphi, U\psi),$$

so that $(\dot{A}f, g) = (f, \dot{A}g)$. The proof of the relation

$$U = (\dot{A} - \lambda I)(\dot{A} - \bar{\lambda}I)^{-1}$$

is not difficult. Thus, the operator U is the Cayley transform of the operator $\dot{A}$, i.e., $U = U_\lambda(\dot{A})$. □

The next theorem immediately follows from the proposition proved above.

Theorem 1.4.2. *Let A_1 and A_2 be symmetric operators and let $U_\lambda(A_1)$ and $U_\lambda(A_2)$ be their Cayley transforms. Then A_2 is an extension of A_1 if and only if $U_\lambda(A_2)$ is an extension of $U_\lambda(A_1)$.*

The theorem above allows us reduce the problem of finding extensions of a symmetric operator to the problem of finding isometric extensions of its Cayley transform.

Since the closed linear manifolds F and G can be the domain and range of an isometric operator if and only if their dimensions coincide, we can construct isometric extensions of Cayley transform $U_\lambda(\dot{A})$ as follows. In the deficiency spaces $\mathcal{H}\ominus\mathrm{Dom}(U_\lambda(\dot{A}))$ and $\mathcal{H}\ominus\mathrm{Ran}(U_\lambda(\dot{A}))$ we choose two subspaces of equal dimensions F and G and construct an arbitrary isometric operator W defined on F with G as its range. Further, we define a linear operator U on $\mathrm{Dom}(U) = \mathrm{Dom}(U_\lambda(\dot{A})) \oplus F$ and $\mathrm{Ran}(U) = \mathrm{Ran}(U_\lambda(\dot{A})) \oplus G$ by the formula

$$Uf = \begin{cases} U_\lambda(\dot{A})f, & \text{for } f \in \mathrm{Dom}(U_\lambda(\dot{A})); \\ Wf, & \text{for } f \in F. \end{cases}$$

Obviously, U is an isometric extension of $U_\lambda(\dot{A})$ and also changing F, G, and W we will obtain all isometric extensions U of $U_\lambda(\dot{A})$. In order to find a symmetric extension A of $\dot{A}$ we need to take the Cayley transform of $\dot{A}$, extend as prescribed above, and finally invert the resulting extensions U to get A.

1.5 Non-densely defined symmetric operators and semi-deficiency subspaces

In the present section we will describe the deficiency structure of a symmetric operator $\dot{A}$ with non-dense domain. According to [3] the **aperture of two linear manifolds** M_1 and M_2 in $\mathcal{H}$ is denoted by $\Theta(M_1, M_2)$ and defined as

$$\Theta(M_1, M_2) := \|P_1 - P_2\|,$$

where P_1, P_2 are the orthogonal projection operators on the subspaces $\overline{M_1}$ and $\overline{M_2}$, respectively. It follows from the definition of aperture that

$$\Theta(M_1, M_2) = \Theta(\overline{M_1}, \overline{M_2}) = \Theta(\mathcal{H} \ominus M_1, \mathcal{H} \ominus M_2).$$

Applying the relation

$$P_2 - P_1 = P_2(I - P_1) - (I - P_2)P_1,$$

to an element $h \in \mathcal{H}$ yields

$$(P_2 - P_1)h = P_2(I - P_1)h - (I - P_2)P_1 h.$$

Then the orthogonality of $P_2(I-P_1)h$ and $(I-P_2)P_1h$ implies

$$\begin{aligned}\|(P_2-P_1)h\|^2 &= \|P_2(I-P_1)h\|^2 + \|(I-P_2)P_1h\|^2\\ &\le \|(I-P_1)h\|^2 + \|P_1h\|^2 = \|h\|^2.\end{aligned}$$

The last inequality shows that the aperture of two linear manifolds does not exceed 1, i.e.,

$$\Theta(M_1, M_2) \le 1.$$

Moreover, one can see that the aperture always equals 1 if one of the manifolds contains a non-zero vector orthogonal to the other manifold. It is known that the relation

$$\Theta(M_1, M_2) = \max\left\{\sup_{f\in\overline{M_1},||f||=1} ||(I-P_1)f||, \sup_{f\in\overline{M_2},||f||=1} ||(I-P_2)f||\right\} \tag{1.21}$$

holds.

In what follows we denote by $[\mathcal{H}_1, \mathcal{H}_2]$ the **class of all bounded linear operators** from $\mathcal{H}_1$ into $\mathcal{H}_2$.

Lemma 1.5.1. *Let M_1 and M_2 be subspaces of a Hilbert space $\mathcal{H}$. Then the following conditions are equivalent*

(i) $P_1M_2 = M_1$ *and* $M_2 \cap M_1^\perp = \{0\}$,

(ii) $P_2M_1 = M_2$ *and* $M_1 \cap M_2^\perp = \{0\}$,

(iii) $\Theta(M_1, M_2) < 1$,

where $M_k^\perp := \mathcal{H} \ominus M_k$, $k = 1, 2$.

Proof. Suppose $P_1M_2 = M_1$ and $M_2 \cap M_1^\perp = \{0\}$. Then $\ker(P_1 \upharpoonright M_2) = \{0\}$ and by Banach's inverse mapping theorem there exists a number $c > 0$ such that

$$||P_1h|| \ge c||h|| \quad \text{for all} \quad h \in M_2.$$

Let the operator $S \in [M_2, M_2]$ be defined as

$$S := P_2P_1 \upharpoonright M_2.$$

Then

$$(Sh, h) = (P_2P_1h, h) = ||P_1h||^2 \ge c^2||h||^2,\ h \in M_2.$$

It follows that $||Sh|| \ge c^2||h||$ for all $h \in M_2$. Hence

$$||P_2P_1h|| \ge c||h|| \ge c^2||P_1h||,\ h \in M_2 \Rightarrow ||P_2g|| \ge c^2||g||,\ g \in M_1.$$

It follows that $M_1 \cap M_2^\perp = \{0\}$. Since $M_1 \cap M_2^\perp = \{0\}$ we get $P_2M_1 = M_2$. In addition, the right-hand side of (1.21) is less then 1. So, (i)$\Rightarrow$(ii) and (i)$\Rightarrow$(iii). Similarly, (ii)$\Rightarrow$(i) and (ii)$\Rightarrow$(iii).

Now suppose $\Theta(M_1, M_2) = \gamma < 1$. Then from (1.21) we get that

$$||P_1 h||^2 \geq (1-\gamma^2)||h||^2 \quad \text{for all} \quad h \in M_2,$$
$$||P_2 g||^2 \geq (1-\gamma^2)||g||^2 \quad \text{for all} \quad g \in M_1.$$

Hence $M_2 \cap M_1^\perp = \{0\}$, $M_1 \cap M_2^\perp = \{0\}$, and $P_1 M_2 = M_1$, $P_2 M_1 = M_2$. □

Let $\mathcal{H}_0 = \overline{\mathrm{Dom}(\dot{A})}$, and $\dot{A}^*$ be the adjoint to the operator $\dot{A}$ (we consider $\dot{A}$ as acting from $\mathcal{H}_0$ into $\mathcal{H}$). It is easy to see that for the symmetric operator $\dot{A}$, $\mathrm{Dom}(\dot{A}) \subset \mathrm{Dom}(\dot{A}^*)$, and

$$\dot{A}^* g = P\dot{A}g, \qquad g \in \mathrm{Dom}(\dot{A}),$$

where P is an orthogonal projection of $\mathcal{H}$ onto $\mathcal{H}_0$. Following our notation for the dense case we set

$$\mathfrak{L} := \mathcal{H} \ominus \mathcal{H}_0, \quad \mathfrak{M}_\lambda := (\dot{A} - \lambda I)\mathrm{Dom}(\dot{A}), \quad \mathfrak{N}_\lambda := (\mathfrak{M}_{\bar{\lambda}})^\perp.$$

Lemma 1.5.2. *If* $\mathrm{Im}\,\lambda \neq 0$, *then*

$$\mathfrak{L} \cap \mathfrak{M}_\lambda = \{0\}.$$

Proof. Let $h = \dot{A}g - \bar{\lambda}g \in \mathfrak{L} \cap \mathfrak{M}_\lambda$, $g \in \mathrm{Dom}(\dot{A})$. Then $\dot{A}g = h + \bar{\lambda}g$ and, since $(h, g) = 0$,

$$\mathrm{Im}(g, \dot{A}g) = \mathrm{Im}\lambda \cdot (g, g).$$

But for symmetric operators

$$\mathrm{Im}(\dot{A}g, g) = 0,$$

and therefore $g = 0$ and $h = 0$. □

Let $P_{\mathfrak{N}_\lambda}$ be the orthogonal projection operator onto $\mathfrak{N}_\lambda$. We set

$$\mathfrak{B}_\lambda = P_{\mathfrak{N}_\lambda}\mathfrak{L}, \tag{1.22}$$

$$\mathfrak{N}'_\lambda = \mathfrak{N}_\lambda \ominus \overline{\mathfrak{B}_\lambda}. \tag{1.23}$$

The subspace $\mathfrak{N}'_\lambda$ is called the **semi-deficiency subspace** of the operator $\dot{A}$.

Recall that P is the orthogonal projection in $\mathcal{H}$ onto $\mathcal{H}_0$. We define the operator $\dot{A}_0$ by the formula

$$\dot{A}_0 = P\dot{A}, \quad \mathrm{Dom}(\dot{A}_0) = \mathrm{Dom}(\dot{A}). \tag{1.24}$$

Then $\dot{A}_0$ is a densely defined symmetric operator in the Hilbert space $\mathcal{H}_0$. Indeed, for $f, g \in \overline{\mathrm{Dom}(\dot{A})}$ we have

$$(\dot{A}_0 f, g) = (P\dot{A}f, g) = (\dot{A}f, g) = (f, \dot{A}f) = (f, P\dot{A}g) = (f, \dot{A}_0 g).$$

Theorem 1.5.3. *The semi-deficiency subspace $\mathfrak{N}'_\lambda$ of a symmetric operator $\dot{A}$ is the defect subspace $\mathfrak{N}_\lambda(\dot{A}_0)$ of the operator $\dot{A}_0$ of the form* (1.24). *Therefore the dimensions of semi-deficiency subspaces of a symmetric operator $\dot{A}$ are the same for all values of λ from the open upper (resp. lower) half-plane.*

Proof. Let $\varphi \in \mathfrak{N}'_\lambda$. Then φ is orthogonal to $\mathfrak{L}$ and therefore $\varphi \in \overline{\mathrm{Dom}(\dot{A})}$. Thus $P\varphi = \varphi$ and

$$(\varphi, P\dot{A}g - \bar{\lambda}g) = (\varphi, \dot{A}g - \bar{\lambda}g) = 0, \qquad g \in \mathrm{Dom}(\dot{A}),$$

and $\varphi \in \mathfrak{N}_\lambda(\dot{A}_0)$. Thus,

$$\mathfrak{N}'_\lambda \subset \mathfrak{N}_\lambda(\dot{A}_0).$$

Conversely, if $\psi \in \mathfrak{N}_\lambda(\dot{A}_0)$, then $\psi = P\psi$ and hence

$$(\psi, \dot{A}g - \bar{\lambda}g) = (\psi, P\dot{A}g - \bar{\lambda}g) = 0, \qquad g \in \mathrm{Dom}(\dot{A}).$$

Since $\psi \in \mathfrak{N}'_\lambda$ and $\psi \in \mathcal{H}_0$, we get $\psi \perp \mathfrak{B}_\lambda$ and therefore

$$\mathfrak{N}_\lambda(\dot{A}_0) \subset \mathfrak{N}'_\lambda.$$

□

The numbers $\dim\mathfrak{N}'_\lambda$ and $\dim\mathfrak{N}'_{\bar{\lambda}}$ $(\mathrm{Im}\,\lambda \neq 0)$ are called the **semi-defect numbers** or the **semi-deficiency indices** of the operator $\dot{A}$.

Theorem 1.5.4. *The following statements are equivalent:*

(i) $\mathfrak{B}_\lambda$ *is a subspace at least for one* λ, $\mathrm{Im}\,\lambda \neq 0$,

(ii) $\Theta(\mathfrak{L}, \mathfrak{B}_\lambda) < 1$ *at least for one* λ, $\mathrm{Im}\,\lambda \neq 0$,

(iii) *the operator $\dot{A}_0$ is closed.*

Proof. (i)⇒(ii)⇒(iii). Suppose $\mathfrak{B}_\lambda$ is a subspace for some λ, $\mathrm{Im}\,\lambda \neq 0$. From Lemma 1.5.1 and Lemma 1.5.2 we get that $\Theta(\mathfrak{L}, \mathfrak{B}_\lambda) < 1$. Hence

$$\Theta(\mathcal{H}_0, \mathcal{H} \ominus \mathfrak{B}_\lambda) < 1.$$

Again by Lemma 1.5.1 we get

$$P\,(\mathcal{H} \ominus \mathfrak{B}_\lambda) = \mathcal{H}_0.$$

We have the equality

$$\mathcal{H} \ominus \mathfrak{B}_\lambda = (\dot{A} - \bar{\lambda}I)\mathrm{Dom}(\dot{A}_0) \oplus \mathfrak{N}'_\lambda. \tag{1.25}$$

Since $\mathfrak{N}'_\lambda \subset \mathcal{H}_0$, and

$$P\left((\dot{A} - \bar{\lambda}I)\mathrm{Dom}(\dot{A}_0)\right) = (\dot{A}_0 - \bar{\lambda}I)\mathrm{Dom}(\dot{A}_0),$$

we get that the linear manifold $(\dot{A}_0 - \bar{\lambda}I)\mathrm{Dom}(\dot{A}_0)$ is a subspace (in $\mathcal{H}_0$). Hence the operator $\dot{A}_0$ is closed.

(iii)⇒(ii). Let $\dot{A}_0$ be a closed operator. Then the linear manifold $(\dot{A}_0 - \bar{\lambda}I)\mathrm{Dom}(\dot{A}_0)$ is a subspace for all non-real λ. Since

$$P\left((\dot{A} - \bar{\lambda}I)\mathrm{Dom}(\dot{A}_0)\right) = (\dot{A}_0 - \bar{\lambda}I)\mathrm{Dom}(\dot{A}_0),$$

and $(\dot{A}_0 - \bar{\lambda}I)\mathrm{Dom}(\dot{A}_0) \oplus \mathfrak{N}'_\lambda = \mathcal{H}_0$, we get from (1.25) that

$$P\left(\mathcal{H} \ominus \mathfrak{B}_\lambda\right) = \mathcal{H}_0,$$

and

$$(\mathcal{H} \ominus \mathfrak{B}_\lambda) \cap \mathfrak{L} = \{0\}.$$

Now Lemma 1.5.1 yields $\Theta(\mathcal{H}_0, \mathcal{H} \ominus \mathfrak{B}_\lambda) < 1$. Hence $\Theta(\mathfrak{L}, \mathfrak{B}_\lambda) < 1$.

(ii)⇒(i). Let $\Theta(\mathfrak{L}, \mathfrak{B}_\lambda) < 1$. Then $\Theta(\mathfrak{L}, \overline{\mathfrak{B}_\lambda}) < 1$. Let $P_{\overline{\mathfrak{B}_\lambda}}$ be the orthogonal projection in $\mathcal{H}$ onto $\overline{\mathfrak{B}}_\lambda$. From Lemma 1.5.1 we get $P_{\overline{\mathfrak{B}_\lambda}}\mathfrak{L} = \overline{\mathfrak{B}_\lambda}$. On the other hand, from (1.22) and the inclusion $\overline{\mathfrak{B}_\lambda} \subseteq \mathfrak{N}_\lambda$ we have

$$P_{\overline{\mathfrak{B}_\lambda}}\mathfrak{L} = \mathfrak{B}_\lambda.$$

Therefore, $\mathfrak{B}_\lambda$ is a subspace. □

Corollary 1.5.5. *The linear sets $\mathfrak{B}_\lambda$ for any λ $(\mathrm{Im}\,\lambda \neq 0)$ are either subspaces or non-closed linear sets in the Hilbert space $\mathcal{H}$.*

1.6 Symmetric extensions of a non-densely defined symmetric operator

The Cayley transform of a symmetric non-densely defined operator $\dot{A}$ is defined via (1.19), that is

$$U_\lambda(\dot{A}) = (\dot{A} - \lambda I)(\dot{A} - \bar{\lambda}I)^{-1},$$

for $(\mathrm{Im}\lambda \neq 0)$. Equivalently,

$$\begin{aligned} f &= \dot{A}g - \bar{\lambda}g, \\ U_\lambda(\dot{A})f &= \dot{A}g - \lambda g, \end{aligned} \qquad \left(f \in \mathrm{Dom}(U_\lambda(\dot{A})),\ g \in \mathrm{Dom}(\dot{A})\right).$$

Since both λ and $\bar{\lambda}$, $(\mathrm{Im}\,\lambda \neq 0)$ are the points of a regular type for $\dot{A}$, then the domain $\mathrm{Dom}(U_\lambda(\dot{A})) = \mathrm{Ran}(\dot{A} - \bar{\lambda}I) = \mathfrak{M}_{\bar{\lambda}}$ and the range $\mathrm{Ran}(U_\lambda(\dot{A})) = \mathrm{Ran}(\dot{A} - \lambda I) = \mathfrak{M}_\lambda$ are the subspaces. Similarly to the dense case in the Section 1.4 one can show that the operator $U_\lambda(\dot{A})$ is an isometry.

Lemma 1.6.1. *For any element $h \in \mathfrak{L}$ we have*

$$U_\lambda(\dot{A})P_{\mathfrak{M}_{\bar{\lambda}}}h = P_{\mathfrak{M}_\lambda}h, \quad (\mathrm{Im}\,\lambda \neq 0).$$

Proof. Since $U_\lambda(\dot{A})\mathfrak{M}_{\bar\lambda} = \mathfrak{M}_\lambda$, we only need to show that the element $h - U_\lambda(\dot{A})P_{\mathfrak{M}_{\bar\lambda}}h$ is orthogonal to the subspace $\mathfrak{M}_\lambda$. Indeed, if $\psi \in \mathfrak{M}_\lambda$, then $\psi = U_\lambda(\dot{A})\varphi$, where $\varphi \in \mathfrak{M}_{\bar\lambda}$. Thus,

$$\left(h - U_\lambda(\dot{A})P_{\mathfrak{M}_{\bar\lambda}}h, \psi\right) = \left(h, U_\lambda(\dot{A})\varphi - \varphi\right),$$

but by (1.20)

$$U_\lambda(\dot{A})\varphi - \varphi = g \in \mathrm{Dom}(\dot{A}),$$

and hence

$$\left(h - U_\lambda(\dot{A})P_{\mathfrak{M}_{\bar\lambda}}h, \psi\right) = (h, g) = 0. \qquad \square$$

We introduce a new linear operator V_λ defined by the formula

$$V_\lambda P_{\mathfrak{N}_\lambda} h = P_{\mathfrak{N}_{\bar\lambda}} h, \qquad h \in \mathfrak{L}, \ \mathrm{Im}\,\lambda \neq 0. \tag{1.26}$$

By Lemma 1.5.2 this operator V_λ is well defined. It is called the **exclusion operator** and plays an important role in the extension theory of operators with non-dense domain. The domain and the range of the operator V_λ are the sets $\mathfrak{B}_\lambda$ and $\mathfrak{B}_{\bar\lambda}$ respectively. Lemma 1.6.1 implies that

$$\|V_\lambda P_{\mathfrak{N}_{\bar\lambda}} h\|^2 = \|P_{\mathfrak{N}_{\bar\lambda}} h\|^2 = \|h\|^2 - \|P_{\mathfrak{M}_{\bar\lambda}} h\|^2 = \|h\|^2 - \|P_{\mathfrak{M}_\lambda} h\|^2 = \|P_{\mathfrak{N}_\lambda} h\|^2,$$

and operator V_λ is an isometry. This means that the dimensions of the subspaces $\overline{\mathfrak{B}_\lambda}$ and $\overline{\mathfrak{B}_{\bar\lambda}}$ are the same. Now we can conclude that the equality of the semi-deficiency numbers of a symmetric operator $\dot{A}$ implies the equality of its deficiency numbers. The inverse statement is not true when $\dim \mathfrak{L} = \infty$, i.e., one can easily construct a non-densely defined symmetric operator with equal deficiency indices but with nonequal semi-deficiency indices.

We should note several basic properties of the operator V_λ. According to Theorem 1.5.4, the operator V_λ is closed if and only if

$$\Theta(\mathfrak{L}, \mathfrak{B}_\lambda) < 1.$$

The above inequality also implies that the operator V_λ is closed if the symmetric operator $\dot{A}$ is bounded.

Let $\mathfrak{F}$ be the orthogonal complement of the linear span of the sets $\mathrm{Dom}(\dot{A})$ and $\mathrm{Ran}(\dot{A})$ in $\mathcal{H}$. One may notice that $\mathfrak{F}$ is the set of elements $\varphi \in \mathrm{Dom}(V_\lambda)$ for which $V_\lambda\varphi = \varphi$. Indeed, if $\varphi \in \mathfrak{F}$, then $(\varphi, \dot{A}\varphi - \bar\lambda\varphi) = 0$, and hence $\varphi \in \mathfrak{N}_\lambda$, $\varphi \in \mathfrak{N}_{\bar\lambda}$, and $V_\lambda\varphi = \varphi$. Conversely, if $V_\lambda\varphi = \varphi$ then $\varphi \in \mathfrak{N}_\lambda$, $\varphi \in \mathfrak{N}_{\bar\lambda}$, and

$$(\varphi, \dot{A}g - \bar\lambda g) = 0, \qquad (\varphi, \dot{A}g - \lambda g) = 0, \qquad g \in \mathrm{Dom}(\dot{A}).$$

The latter implies that $\varphi \in \mathfrak{F}$.

An isometric extension $\widetilde{U}_\lambda$ of the Cayley transform $U_\lambda(\dot{A})$ of the symmetric operator $\dot{A}$ is called an **admissible extension** if

$$\widetilde{U}_\lambda f = f, \qquad f \in \mathrm{Dom}(\widetilde{U}_\lambda),$$

implies that $f = 0$.

Lemma 1.6.2. *If $\widetilde{U}_\lambda$ is an admissible extension of $U_\lambda(\dot{A})$, then the operator*

$$A = (\lambda \widetilde{U}_\lambda - \bar{\lambda} I)(\widetilde{U}_\lambda - I)^{-1}, \tag{1.27}$$

is a symmetric extension of the symmetric operator $\dot{A}$.

Proof. It is obvious that A is an extension of $\dot{A}$. All we have to show is that A is symmetric. Note that operator A is defined on the elements g such that

$$g = \widetilde{U}_\lambda f - f, \qquad f \in \operatorname{Dom}(\widetilde{U}_\lambda), \tag{1.28}$$

on which it takes the values

$$Ag = \lambda \widetilde{U}_\lambda f - \bar{\lambda} f. \tag{1.29}$$

Consequently,

$$\begin{aligned} \operatorname{Im}(Ag, g) &= \operatorname{Im}(\lambda \widetilde{U}_\lambda f - \bar{\lambda} f, \widetilde{U}_\lambda f - f) \\ &= \operatorname{Im}\left[(\lambda + \bar{\lambda})(f, f) - \lambda(\widetilde{U}_\lambda f, f) - \bar{\lambda}(f, \widetilde{U}_\lambda f)\right] = 0, \end{aligned}$$

which proves the lemma. □

It is easy to see that the inverse statement takes place as well: for every symmetric extension A of operator $\dot{A}$ there exists an admissible isometric extension $\widetilde{U}_\lambda = U_\lambda(A)$ (the Cayley transform of A) of the isometric operator $U_\lambda(\dot{A})$. Therefore, there is a one-to-one correspondence between symmetric extensions of operator $\dot{A}$ and admissible isometric extensions of the operator $U_\lambda(\dot{A})$. In particular, in order to obtain the *self-adjoint* extensions of $\dot{A}$ we need to construct admissible *unitary* extensions of $U_\lambda(\dot{A})$. In other words we should have

$$\operatorname{Dom}(U_\lambda(A)) = \operatorname{Ran}(A - \bar{\lambda} I) = \mathcal{H}, \ \operatorname{Ran}(U_\lambda(A)) = \operatorname{Ran}(A - \lambda I) = \mathcal{H}. \tag{1.30}$$

We can summarize this as follows: An isometric (unitary) extension $\widetilde{U}_\lambda$ of the Cayley transform $U_\lambda(\dot{A})$ of symmetric operator $\dot{A}$ corresponds to the symmetric (self-adjoint) extension A of the symmetric operator $\dot{A}$ if and only if the operator $\widetilde{U}_\lambda$ is an admissible isometric (unitary) extension of the operator $U_\lambda(\dot{A})$.

Now we focus on the properties of non-admissible extensions of $U_\lambda(\dot{A})$.

Lemma 1.6.3. *If for an isometric non-admissible extension $\widetilde{U}_\lambda$ of the operator $U_\lambda(\dot{A})$ we have that, for some element $h \in \operatorname{Dom}(\widetilde{U}_\lambda)$,*

$$\widetilde{U}_\lambda h = h,$$

then $h \in \mathfrak{L}$.

Proof. If $\widetilde{U}_\lambda h = h$, then for all $f \in \operatorname{Dom}(\widetilde{U}_\lambda)$,

$$\left(\widetilde{U}_\lambda h - h, U_\lambda(\dot{A})f\right) = 0.$$

Since $\widetilde{U}_\lambda f = U_\lambda(\dot{A})f$ the last equality implies

$$\left(h, f - U_\lambda(\dot{A})f\right) = 0.$$

By (1.28) we have $(h, g) = 0$ for $g \in \operatorname{Dom}(\dot{A})$. □

Corollary 1.6.4. *If operator $\dot{A}$ is densely defined, then all isometric extensions of the operator $U_\lambda(\dot{A})$ are admissible.*

Theorem 1.6.5. *If for some isometric non-admissible extension $\widetilde{U}_\lambda$ of the operator $U_\lambda(\dot{A})$ and for some element $h \in \operatorname{Dom}(\widetilde{U}_\lambda)$ we have that*

$$\widetilde{U}_\lambda h = h,$$

then

$$\widetilde{U}_\lambda P_{\mathfrak{N}_\lambda} h = V_\lambda P_{\mathfrak{N}_\lambda} h = P_{\mathfrak{N}_{\bar{\lambda}}} h.$$

Proof. It follows from Lemma 1.6.3 that $h \in \mathfrak{L}$. Since the operator $\widetilde{U}_\lambda$ is an extension of $U_\lambda(\dot{A})$ we have

$$\widetilde{U}_\lambda P_{\mathfrak{M}_{\bar{\lambda}}} h = U_\lambda(\dot{A}) P_{\mathfrak{M}_{\bar{\lambda}}} h,$$

and thus by Lemma 1.6.1

$$\widetilde{U}_\lambda P_{\mathfrak{M}_{\bar{\lambda}}} h = P_{\mathfrak{M}_\lambda} h.$$

Then

$$\widetilde{U}_\lambda P_{\mathfrak{N}_\lambda} h = \widetilde{U}_\lambda (h - P_{\mathfrak{M}_{\bar{\lambda}}} h) = h - P_{\mathfrak{M}_\lambda} h = P_{\mathfrak{N}_{\bar{\lambda}}} h.$$ □

The last theorem allows us to describe the subclass of all admissible extensions in the class of all isometric extensions of the operator $\dot{A}$. Evidently, the admissible extensions are the ones that do not take the same value as operator V_λ on any element of $\mathfrak{B}_\lambda$. Thus an isometric extension $\widetilde{U}_\lambda$ of the Cayley transform $U_\lambda(\dot{A})$ coincides with the Cayley transform $U_\lambda(A)$ of a symmetric extension A of the operator $\dot{A}$ if and only if the equality $\widetilde{U}_\lambda \varphi = V_\lambda \varphi$, where V_λ is the exclusion operator (1.26), implies $\varphi = 0$.

It follows from the definition of the operator V_λ and Lemma 1.6.1 that operator $U_\lambda(\dot{A}) \dot{+} V_\lambda$ is isometric [1] and leaves every element of $\mathfrak{L}$ invariant,

$$\left(U_\lambda(\dot{A}) \dot{+} V_\lambda\right) h = h, \qquad h \in \mathfrak{L}. \tag{1.31}$$

The operator $U_\lambda(\dot{A}) \dot{+} V_\lambda$ is the minimal isometric extension of $U_\lambda(\dot{A})$ leaving every element of $\mathfrak{L}$ invariant.

[1] Here we denote by $\dot{+}$ a **direct sum of two operators** [169]. Let S and T be linear operators with $\operatorname{Dom}(S) \cap \operatorname{Dom}(T) = \{0\}$. Then $(D \dot{+} T)(\varphi + \psi) = S\varphi + T\psi$, $\varphi \in \operatorname{Dom}(S)$, $\psi \in \operatorname{Dom}(T)$.

1.7 Indirect decomposition and the Krasnoselskiĭ formulas

A process of construction of an isometric extension $\widetilde{U}_\lambda$ of the operator $U_\lambda(\dot{A})$ is equivalent to the construction of an isometric operator U with domain $\mathrm{Dom}(U) \subset \mathfrak{N}_\lambda$ and range $\mathrm{Ran}(U) \in \mathfrak{N}_{\bar{\lambda}}$ so that

$$\widetilde{U}_\lambda = U_\lambda(\dot{A}) \dotplus U.$$

We are particularly interested in a possibility of constructing admissible unitary extensions $\widetilde{U}_\lambda$ of the operator $U_\lambda(\dot{A})$. Obviously, these extensions exist only if our operator $\dot{A}$ has equal deficiency indices. Moreover, the equality of deficiency indices is also a sufficient condition of the existence of unitary admissible extensions of the Cayley transform of the operator $\dot{A}$.

Let us assume first that semi-deficiency indices of $\dot{A}$ are zeros. This means

$$\overline{\mathfrak{B}_\lambda} = \mathfrak{N}_\lambda, \qquad \overline{\mathfrak{B}_{\bar{\lambda}}} = \mathfrak{N}_{\bar{\lambda}}.$$

In this case every admissible extension $\widetilde{U}_\lambda$ of $U_\lambda(\dot{A})$ is determined by the formula

$$\widetilde{U}_\lambda = U_\lambda(\dot{A}) \dotplus \overline{U},$$

where U is an isometric operator defined on $\mathfrak{B}_\lambda$ with the range in $\mathfrak{N}_{\bar{\lambda}}$ that does not take the same value as V_λ on any element of $\mathfrak{B}_\lambda$ and $\overline{U}$ is the closure of operator U in $\mathcal{H}$. For example, one can use $U = -V_\lambda$.

Now let us assume that our operator $\dot{A}$ has equal deficiency indices. Let also U' be some, generally speaking non-admissible, unitary extension of $U_\lambda(\dot{A})$. We will show that this extension can be corrected, i.e., one can construct an admissible extension $\widetilde{U}_\lambda$ using U'. Let $\mathfrak{L}'$ be a subspace invariant under U'. By Lemma 1.6.3,

$$\mathfrak{L}' \subset \mathfrak{L}.$$

Here, by $\mathfrak{B}'_\lambda$ we denote the projection of the subspace $\mathfrak{L}'$ onto the deficiency subspace $\mathfrak{N}_\lambda$. Consider the part U^0 of the operator U' defined on $\mathcal{H} \ominus \overline{\mathfrak{B}_\lambda}$. The operator U^0 is a closed admissible extension of $U_\lambda(\dot{A})$. By Lemma 1.6.2 and (1.27) there exists a closed symmetric extension A^0 of $\dot{A}$. Also we should note that

$$\overline{\mathrm{Dom}(A^0)} \oplus \mathfrak{L}' = \mathcal{H}. \tag{1.32}$$

Indeed, let V'_λ be a part of the operator V_λ defined on $\mathfrak{B}'_\lambda$. The operator U^0 has an isometric extension $\tilde{U}^0$ leaving every element of $\mathcal{H} \ominus \overline{\mathrm{Dom}(A^0)}$ invariant. To construct such an extension one needs to construct an operator V_λ first and then use (1.31). The operator $\tilde{U}^0$ will be an isometric extension of the operator $U^0 \dotplus V'_\lambda$. Since U' is the unitary closure of the operator $U^0 \dotplus V'_\lambda$, then U' is an extension of $\tilde{U}^0$. But the subspace $\mathfrak{L}'$ is only invariant with respect to U' which implies (1.32).

According to (1.32) the operator U^0 is a Cayley transform of the operator A^0 whose semi-deficiency indices are zeros. In this case, as it was shown above, the operator U^0 has admissible unitary extensions $\tilde{U}$. Evidently these extensions are admissible unitary extensions of $U_\lambda(\dot{A})$.

If we summarize the above reasoning and the formulas (1.28) and (1.29), we get the Krasnoselskiĭ formulas that generalize the von Neumann Theorem 1.3.1. However, unlike the von Neumann formulas (1.13) for the dense case, in the representation (1.33) below one must choose an admissible (not arbitrary) isometric operator U.

Theorem 1.7.1. *A closed symmetric operator $\dot{A}$ acting in the Hilbert space $\mathcal{H}$ has self-adjoint extensions if and only if its deficiency indices are equal. In this case a self-adjoint extension A of the operator $\dot{A}$ is defined by the formula*

$$\begin{aligned} f &= g + \varphi - U\varphi, \\ Af &= Ag + \lambda\varphi - \bar{\lambda}U\varphi, \end{aligned} \tag{1.33}$$

where $g \in \mathrm{Dom}(\dot{A})$, $\varphi \in \mathfrak{N}_\lambda$, and U is an admissible isometric operator with

$$\mathrm{Dom}(U) = \mathfrak{N}_\lambda, \quad \mathrm{Ran}(U) = \mathfrak{N}_{\bar{\lambda}}.$$

Similar argument brings us to the generalization of the second von Neumann Theorem 1.3.1.

Theorem 1.7.2. *Any closed symmetric operator $\dot{A}$ in the Hilbert space $\mathcal{H}$ has maximal symmetric extensions in $\mathcal{H}$.*

This theorem can be proved even faster if we first extend the operator $U_\lambda(\dot{A})$ to U' in a way that the symmetric operator

$$A' = (\lambda U' - \bar{\lambda}I)(U' - I)^{-1}$$

has a dense in $\mathcal{H}$ domain. To do this all we need is

$$U' = U_\lambda(\dot{A}) \dotplus (-V_\lambda).$$

Then all isometric extensions of U' are going to be admissible.

Now we consider symmetric extensions A of a symmetric operator $\dot{A}$ with non-equal deficiency indices that exit to the space $\mathcal{H} \oplus \mathcal{H}'$. Choosing the space $\mathcal{H}'$ of sufficiently high dimension we can achieve the equality of the deficiency indices of $\dot{A}$ by considering it as acting in a wider space $\mathcal{H} \oplus \mathcal{H}'$. This follows from the fact that $\mathcal{H}'$ is included in both deficiency subspaces of $\dot{A}$ (acting in $\mathcal{H} \oplus \mathcal{H}'$). Thus we can now construct self-adjoint extensions of $\dot{A}$ in $\mathcal{H} \oplus \mathcal{H}'$.

Theorem 1.7.3. *Every closed symmetric operator has self-adjoint extensions (possibly with exit to the wider space).*

Now we raise the question of the linear dependence of the sets $\mathfrak{N}_\lambda$, $\mathfrak{N}_{\bar\lambda}$, and $\mathrm{Dom}(\dot A)$ as defined in Section 1.2. The following theorems by M. Naimark partially answer it.

Theorem 1.7.4. *The linear sets $\mathfrak{N}_\lambda$, $\mathfrak{N}_{\bar\lambda}$, and $\mathrm{Dom}(\dot A)$ are linearly dependent if and only if $\mathrm{Dom}(\dot A)$ is not dense in $\mathcal{H}$. In this case*

$$\varphi+\psi+g=0,\quad (\varphi\in\mathfrak{N}_\lambda,\ \psi\in\mathfrak{N}_{\bar\lambda},\ g\in\mathrm{Dom}(\dot A)),$$

holds if and only if there exists an element $h\in\mathfrak{L}$ such that

$$\begin{aligned}&(1)\ P_{\mathfrak{M}_{\bar\lambda}}h=\frac{\dot Ag-\bar\lambda g}{\lambda-\bar\lambda},\\ &(2)\ \varphi=P_{\mathfrak{N}_\lambda}h,\\ &(3)\ \psi=-P_{\mathfrak{N}_{\bar\lambda}}h=-V_\lambda\varphi.\end{aligned}$$

Proof. Suppose $\overline{\mathrm{Dom}(\dot A)}\neq\mathcal{H}$. Then according to Lemma 1.6.1 and the definition of $U_\lambda(\dot A)$ for every element $h\in\mathfrak{L}$ there are elements $\varphi\in\mathfrak{N}_\lambda$, $\psi\in\mathfrak{N}_{\bar\lambda}$, and $g\in\mathrm{Dom}(\dot A)$ such that (1), (2), and (3) hold and

$$\begin{aligned}h&=\varphi+\frac{\dot Ag-\bar\lambda g}{\lambda-\bar\lambda},\\ h&=-\psi+\frac{\dot Ag-\lambda g}{\lambda-\bar\lambda}.\end{aligned}$$

This implies that

$$\varphi+\psi+g=0,$$

i.e., the linear sets $\mathfrak{N}_\lambda$, $\mathfrak{N}_{\bar\lambda}$, and $\mathrm{Dom}(\dot A)$ are linearly dependent.

To show the necessity of (1), (2), and (3) we note that

$$\varphi+\psi+g=0,\quad (\varphi\in\mathfrak{N}_\lambda,\ \psi\in\mathfrak{N}_{\bar\lambda},\ g\in\mathrm{Dom}(\dot A)),$$

is equivalent to

$$\frac{\dot Ag-\bar\lambda g}{\lambda-\bar\lambda}+\varphi=\frac{\dot Ag-\lambda g}{\lambda-\bar\lambda}-\psi. \tag{1.34}$$

Since $(\dot Ag-\bar\lambda g,\varphi)=0$, $(\dot Ag-\lambda g,\psi)=0$, and $\|\dot Ag-\bar\lambda g\|=\|\dot Ag-\lambda g\|$, we have

$$\|\varphi\|=\|\psi\|. \tag{1.35}$$

Let W be a linear operator defined on elements $\{k\varphi,\ k\in\mathbb{C}\}$ by the formula

$$W(k\varphi)=-k\psi.$$

According to (1.35) the operator $U_\lambda(\dot A)\dotplus W$ is an isometric extension of $U_\lambda(\dot A)$. Also $U_\lambda(\dot A)\dotplus W$ is not an admissible extension of $U_\lambda(\dot A)$ because

$$(U_\lambda(\dot A)\dotplus W)\left(\frac{\dot Ag-\bar\lambda g}{\lambda-\bar\lambda}+\varphi\right)=\frac{\dot Ag-\lambda g}{\lambda-\bar\lambda}-\psi,$$

and, as it follows from (1.34), the element

$$h = \frac{\dot{A}g - \bar{\lambda}g}{\lambda - \bar{\lambda}} + \varphi$$

is invariant for the operator $U_\lambda(\dot{A}) \dotplus W$. From Lemma 1.6.3 it follows then that $\overline{\mathrm{Dom}(\dot{A})} \neq \mathcal{H}$. □

We should also note that the element $h \in \mathfrak{L}$ in the conditions of Theorem 1.7.4 can be defined uniquely.

Theorem 1.7.5. *Let f be an element in the domain $\mathrm{Dom}(A)$ of a symmetric extension A of the symmetric operator $\dot{A}$. Then according to Theorem 1.7.1 it has a representation*

$$f = g + \varphi - U\varphi, \quad g \in \mathrm{Dom}(\dot{A}),\ \varphi \in \mathrm{Dom}(U) \subset \mathfrak{N}_\lambda,\ U\varphi \in \mathfrak{N}_{\bar{\lambda}},$$

and uniquely defines the elements $g \in \mathrm{Dom}(\dot{A})$ and $\varphi \in \mathfrak{N}_\lambda$.

Proof. Suppose f has two representations

$$\begin{aligned} f &= g_1 + \varphi_1 - U\varphi_1, \\ f &= g_1 + \varphi_2 - U\varphi_2, \end{aligned}$$

where g_1, $g_2 \in \mathrm{Dom}(\dot{A})$ and φ_1, $\varphi_2 \in \mathfrak{N}_\lambda$. Then

$$(g_1 - g_2) + (\varphi_1 - \varphi_2) - U(\varphi_1 - \varphi_2) = 0,$$

and the previous theorem implies that

$$U(\varphi_1 - \varphi_2) = V(\varphi_1 - \varphi_2).$$

By Theorem 1.7.1 we have $\varphi_1 = \varphi_2$ and thus $g_1 = g_2$. □

Chapter 2

Geometry of Rigged Hilbert Spaces

In this chapter we study extensions of symmetric non-densely defined operators in the triplets $\mathcal{H}_+ \subset \mathcal{H} \subset \mathcal{H}_-$ of rigged Hilbert spaces. The Krasnoselskiĭ formulas discussed in Section 1.7 are based upon the indirect decomposition (1.33), where deficiency subspaces and the domain of symmetric operator may be linearly dependent. Introduction of the rigged Hilbert spaces allows us to obtain the direct decomposition and parameterization for the domain of the adjoint operator. This direct decomposition is written in terms of the semi-deficiency subspaces and is an analogue of the von Neumann formulas (1.7) and (1.13) for the case of the symmetric operator $\dot{A}$ whose domain is not dense in $\mathcal{H}$.

2.1 The Riesz-Berezansky operator

In this section we are going to equip our Hilbert space $\mathcal{H}$ with spaces $\mathcal{H}_+$ and $\mathcal{H}_-$, called spaces with positive and negative norms, respectively.

We start with a Hilbert space $\mathcal{H}$ with inner product (x, y) and norm $\|\cdot\|$. Let $\mathcal{H}_+$ be a dense in $\mathcal{H}$ linear set that is a Hilbert space itself with respect to another inner product $(x, y)_+$ generating the norm $\|\cdot\|_+$. We assume that $\|x\| \leq \|x\|_+$, $(x \in \mathcal{H}_+)$, i.e., the norm $\|\cdot\|_+$ generates a stronger than $\|\cdot\|$ topology in $\mathcal{H}_+$. The space $\mathcal{H}_+$ is called the **space with positive norm**.

Now let $\mathcal{H}_-$ be a space dual to $\mathcal{H}_+$. It means that $\mathcal{H}_-$ is a space of linear functionals defined on $\mathcal{H}_+$ and continuous with respect to $\|\cdot\|_+$. By the $\|\cdot\|_-$ we denote the norm in $\mathcal{H}_-$ that has a form

$$\|h\|_- = \sup_{u \in \mathcal{H}_+} \frac{|(h, u)|}{\|u\|_+}, \ h \in \mathcal{H}.$$

The value of a functional $f \in \mathcal{H}_-$ on a vector $u \in \mathcal{H}_+$ is denoted by (u, f). The space $\mathcal{H}_-$ is called a **space with negative norm**.

Further on in this chapter we will need to consider an embedding operator $\sigma : \mathcal{H}_+ \mapsto \mathcal{H}$ that embeds $\mathcal{H}_+$ into $\mathcal{H}$. Since $\|\sigma f\| \leq \|f\|_+$ for all $f \in \mathcal{H}_+$, then $\sigma \in [\mathcal{H}_+, \mathcal{H}]$. The adjoint operator σ^* maps $\mathcal{H}$ into $\mathcal{H}_-$ and satisfies the condition $\|\sigma^* f\|_- \leq \|f\|$ for all $f \in \mathcal{H}$. Since σ is a monomorphism with a $(\cdot)$-dense range, then σ^* is a monomorphism with $(-)$-dense range. By identifying $\sigma^* f$ with f ($f \in \mathcal{H}$) we can consider $\mathcal{H}$ embedded in $\mathcal{H}_-$ as a $(-)$-dense set and $\|f\|_- \leq \|f\|$. Also, the relation

$$(\sigma f, h) = (f, \sigma^* h), \qquad f \in \mathcal{H}_+,\ h \in \mathcal{H},$$

implies that the value of the functional $\sigma^* h \in \mathcal{H}$ calculated at a vector $f \in \mathcal{H}_+$ as $(f, \sigma^* h)$ corresponds to the value (f, h) in the space $\mathcal{H}$.

It follows from the Riesz representation theorem that there exists an isometric operator $\mathcal{R}$ which maps $\mathcal{H}_-$ onto $\mathcal{H}_+$ such that $(f, g) = (f, \mathcal{R}g)_+$ ($\forall f \in \mathcal{H}_+$, $g \in \mathcal{H}_-$) and $\|\mathcal{R}g\|_+ = \|g\|_-$. Now we can turn $\mathcal{H}_-$ into a Hilbert space by introducing $(f, g)_- = (\mathcal{R}f, \mathcal{R}g)_+$. Thus,

$$\begin{aligned} (f, g)_- &= (f, \mathcal{R}g) = (\mathcal{R}f, g) = (\mathcal{R}f, \mathcal{R}g)_+, && (f, g \in \mathcal{H}_-), \\ (u, v)_+ &= (u, \mathcal{R}^{-1}v) = (\mathcal{R}^{-1}u, v) = (\mathcal{R}^{-1}u, \mathcal{R}^{-1}v)_-, && (u, v \in \mathcal{H}_+). \end{aligned} \tag{2.1}$$

The operator $\mathcal{R}$ (or $\mathcal{R}^{-1}$) will be called the **Riesz-Berezansky operator**. Applying the above reasoning, we define a triplet $\mathcal{H}_+ \subset \mathcal{H} \subset \mathcal{H}_-$ to be called the **rigged Hilbert space**.

In what follows we use symbols $(+)$, $(\cdot)$, and $(-)$ to indicate the norms $\|\cdot\|_+$, $\|\cdot\|$, and $\|\cdot\|_-$ by which geometrical and topological concepts are defined in $\mathcal{H}_+$, $\mathcal{H}$, and $\mathcal{H}_-$. When considering continuity or closeness of an operator, we first indicate the topology of its domain and then the topology of the range. For instance, an operator B is called $(-, \cdot)$-continuous, if

$$\mathrm{Dom}(B) \subset \mathcal{H}_-, \quad \mathrm{Ran}(B) \subset \mathcal{H}, \qquad \sup_{h \in \mathrm{Dom}(B)} \frac{\|Bh\|}{\|h\|_-} < \infty.$$

Similarly, the closure of a set $\mathcal{L}$ using the norms $\|\cdot\|_+$, $\|\cdot\|$, and $\|\cdot\|_-$ is denoted by $\overline{\mathcal{L}}^{(+)}$, $\overline{\mathcal{L}}$, and $\overline{\mathcal{L}}^{(-)}$, respectively. If $\mathcal{L}$ is a subset from $\mathcal{H}_+$, then its orthogonal complement $\mathcal{L}^\perp$ will be a set of those functionals from $\mathcal{H}_-$ that annihilate $\mathcal{L}$. Thus, $\mathcal{L}^\perp \subset \mathcal{H}_-$. Likewise, if $\mathcal{L} \subset \mathcal{H}_-$ then its orthogonal complement $\mathcal{L}^\perp \subset \mathcal{H}_+$ is a set of those elements $x \in \mathcal{H}_+$ that annihilate all the functionals from $\mathcal{L}$. If $\mathcal{L} \subset \mathcal{H}$, then $\mathcal{L}^\perp \subset \mathcal{H}$ and is defined as a set of elements from $\mathcal{H}$ that are $(\cdot)$-orthogonal to $\mathcal{L}$.

The following theorem establishes the relationship between the orthogonal complements of the set $\mathcal{L}$.

Theorem 2.1.1. 1. *If $\mathcal{L}$ is a subspace in $\mathcal{H}$, then*

$$\left(\overline{\mathcal{L}}^{(-)}\right)^{\perp} = \mathcal{L}^{\perp} \cap \mathcal{H}_+, \quad (\mathcal{L} \cap \mathcal{H}_+)^{\perp} = \overline{\mathcal{L}^{\perp}}^{(-)}. \tag{2.2}$$

2. *If $\mathcal{L}$ is a subspace in $\mathcal{H}_+$, then*

$$\left(\overline{\mathcal{L}}\right)^{\perp} = \mathcal{L}^{\perp} \cap \mathcal{H}. \tag{2.3}$$

3. *If $\mathcal{L}$ is a subspace in $\mathcal{H}_-$, then*

$$(\mathcal{L} \cap \mathcal{H})^{\perp} = \overline{\mathcal{L}^{\perp}}. \tag{2.4}$$

Proof. Let $f \in \overline{\mathcal{L}}^{(-)}$. Then there is a sequence of elements $\{f_n\} \subset \mathcal{L}$ such that $f_n \xrightarrow{(-)} f$. Therefore $(h, f_n) \to (h, f)$ for any $h \in \mathcal{H}_+$. In particular, if $h \in \mathcal{L}^{\perp} \cap \mathcal{H}_+$, then $(h, f_n) = 0$, and consequently, $(h, f) = 0$. Thus, $\mathcal{L}^{\perp} \cap \mathcal{H}_+ \subset (\overline{\mathcal{L}}^{(-)})^{\perp}$.

Conversely, let $h \in (\overline{\mathcal{L}}^{(-)})^{\perp}$. Then $h \in \mathcal{H}_+$ and $(h, f) = 0$ for any $f \in \overline{\mathcal{L}}^{(-)}$, and, in particular, for any $f \in \mathcal{L}$. Hence $h \in \mathcal{L}^{\perp}$ and $h \in \mathcal{L}^{\perp} \cap \mathcal{H}_+$. This proves the first part of (2.2). The second part is being proved similarly by substituting $\mathcal{L}^{\perp}$ for $\mathcal{L}$.

Statements (2.3) and (2.4) can be proved in a similar way. We just note that one can obtain (2.4) from (2.3) by the respective substitution of $\mathcal{L}(\subset \mathcal{H}_+)$ for $\mathcal{L}^{\perp}(\subset \mathcal{H}_-)$. □

Theorem 2.1.1 can be interpreted as asserting the commutativity of the following diagram:

$$\begin{array}{ccc} \mathfrak{N} & \overset{\longleftarrow}{\longrightarrow} & \mathfrak{N} \\ \downarrow\Big\uparrow & & \Big\downarrow\uparrow \\ \mathfrak{N}_+ & \overset{\longrightarrow}{\longleftarrow} & \mathfrak{N}_- \end{array}$$

Here $\mathfrak{N}_+$, $\mathfrak{N}$, and $\mathfrak{N}_-$ are the classes of all $(+)$-closed, $(\cdot)$-closed, and $(-)$-closed linear manifolds in $\mathcal{H}_+$, $\mathcal{H}$, and $\mathcal{H}_-$, respectively. The horizontal arrows denote the passage to the orthogonal complement. The short down arrow $\downarrow$ denotes intersection with $\mathcal{H}_+$, and the long down arrow stands for the $(-)$-closure. The long up arrow on the left represents $(\cdot)$-closure whenever the short up arrow $\uparrow$ on the right denotes intersection with $\mathcal{H}$.

If B is an operator in the class $[\mathcal{H}_+, \mathcal{H}_-]$, then its **adjoint operator** B^* is defined by the formula $(Bf, g) = (f, B^*g)$ $(\forall f, g \in \mathcal{H}_+)$. This operator B^* acts from $\mathcal{H}_+$ into $\mathcal{H}_-$, is bounded, and therefore $B^* \in [\mathcal{H}_+, \mathcal{H}_-]$ as well. Thus, the class $[\mathcal{H}_+, \mathcal{H}_-]$ is invariant under taking adjoint. The class $[\mathcal{H}_-, \mathcal{H}_+]$ has a similar property. The concept of a bounded **self-adjoint operator** is, therefore, well defined in both of these classes. For instance, for the class $[\mathcal{H}_+, \mathcal{H}_-]$ such an operator is characterized by the quadratic functional (Bf, f) $(f \in \mathcal{H}_+)$ taking real values only. If $(Bf,f) \geq 0$ for all $f \in \mathcal{H}_+$, then B is called **non-negative**. For an operator $B \in [\mathcal{H}_+, \mathcal{H}_-]$ we introduce a new operator

$$\hat{B} = B \upharpoonright \mathrm{Dom}(\hat{B}), \quad \mathrm{Dom}(\hat{B}) = \{f \in \mathcal{H}_+ : Bf \in \mathcal{H}\}. \tag{2.5}$$

This operator $\hat{B}$ is called a **quasi-kernel** of the operator B.

For the remainder of this text we will need the following theorem.

Theorem 2.1.2. *Let $\mathcal{H}_1$, $\mathcal{H}_2$, and $\mathcal{H}$ be Hilbert spaces and let B and C be operators in $[\mathcal{H}_1, \mathcal{H}]$ and $[\mathcal{H}_2, \mathcal{H}]$, respectively. The following conditions are equivalent:*

(i) $\mathrm{Ran}(B) \subset \mathrm{Ran}(C)$;

(ii) $\ker(C^*) \subset \ker(B^*)$ *and*

$$\sup_{h \in \mathcal{H}, h \notin \ker(C^*)} \frac{\|B^* h\|}{\|C^* h\|} < \infty;$$

(iii) *there exists an operator $W \in [\mathcal{H}_1, \mathcal{H}_2]$ such that $B = CW$.*

Proof. First we show that (i)$\Rightarrow$(ii). The first part of condition (ii) can be derived from (i) by passing to the orthogonal complements. Now let us assume that the second part of the condition (ii) is not true. Then there exists a sequence of $f_n \in \mathcal{H}$ such that $\|B^* f_n\| \to \infty$ and $\|C^* f_n\| \to 0$. By condition (i) for any $h_1 \in \mathcal{H}_1$ there is an element $h_2 \in \mathcal{H}_2$ such that $Bh_1 = Ch_2$. We have

$$(B^* f_n, h_1) = (f_n, Bh_1) = (f_n, Ch_2) = (C^* f_n, h_2) \to 0.$$

Therefore, $B^* f_n$ converges weakly to zero which contradicts $\|B^* f_n\| \to \infty$.

(ii)$\Rightarrow$(iii). To every vector $\varphi = C^* f \in \mathrm{Ran}(C^*) \subset \mathcal{H}_2$ we assign a vector $\psi \in B^* f \in \mathrm{Ran}(B^*) \subset \mathcal{H}_1$. According to condition (ii), the operator $\psi = U' \varphi$ is well defined and bounded. We extend U' onto $\mathcal{H}_2$ to an operator $U \in [\mathcal{H}_2, \mathcal{H}_1]$ for which $B^* = UC^*$. Then, $B = CU^*$ and we can defined $W = U^*$.

It is very easy to see that (iii)$\Rightarrow$(i). □

Remark 2.1.3. Theorem 2.1.2 can be stated equivalently in the form: For every $A, B \in [\mathcal{H}, \mathcal{H}]$ the following statements are equivalent:

(i) $\mathrm{Ran}(A) \subset \mathrm{Ran}(B)$;

(ii) $A = BC$ for some $C \in [\mathcal{H}, \mathcal{H}]$;

(iii) $AA^* \leq \lambda BB^*$ for some $\lambda \geq 0$.

In this case there is a unique C satisfying $\|C\|^2 = \inf\{\,\lambda : \ AA^* \leq \lambda BB^*\,\}$ and $\mathrm{Ran}(C) \subset \overline{\mathrm{Ran}(B^*)}$, in which case $\ker C = \ker A$.

The following three results are based upon Theorem 2.1.2.

Theorem 2.1.4. *Let $\mathcal{H}_+ \subset \mathcal{H} \subset \mathcal{H}_-$ be a rigged Hilbert space and $\mathcal{G}$ be a Hilbert space. If $B \in [\mathcal{G}, \mathcal{H}_-]$ and $\mathrm{Ran}(B) \subset \mathcal{H}$, then $B \in [\mathcal{G}, \mathcal{H}]$. Moreover, if $B \in [\mathcal{G}, \mathcal{H}_-]$ or $B \in [\mathcal{G}, \mathcal{H}]$ and $\mathrm{Ran}(B) \subset \mathcal{H}_+$, then $B \in [\mathcal{G}, \mathcal{H}_+]$.*

Proof. For the embedding operator σ defined in the beginning of this section we have that $\sigma^* \in [\mathcal{H}, \mathcal{H}_-]$ and $\mathrm{Ran}(B) \subset \mathrm{Ran}(\sigma)$. According to Theorem 2.1.2 there exists an operator $W \in [\mathcal{G}, \mathcal{H}]$ such that $B = \sigma^* W$. Since σ^* is an embedding of $\mathcal{H}$ into $\mathcal{H}_-$, then $B \in [\mathcal{G}, \mathcal{H}]$. The proof of the second statement is similar. □

Theorem 2.1.5. *Let $C \in [\mathcal{H}_+, \mathcal{H}_-]$. Then C is a monomorphism and C^{-1} is $(-,\cdot)$-continuous if and only if $\mathrm{Ran}(C^*) \supset \mathcal{H}$.*

Proof. The existence and $(-,\cdot)$-continuity of operator C^{-1} are equivalent to

$$\inf_{h \in \mathcal{H}_+} \frac{\|Ch\|_-}{\|h\|} > 0,$$

i.e., $\ker(C) = \{0\}$ and $\sup \frac{\|\sigma h\|}{\|Ch\|_-} < \infty$. Applying Theorem 2.1.2 we see that the latter is equivalent to $\mathrm{Ran}(\sigma^*) \subset \mathrm{Ran}(C^*)$ that means $\mathcal{H} \subset \mathrm{Ran}(C^*)$. □

The following theorem can be proven similarly.

Theorem 2.1.6. *Let $C \in [\mathcal{H}_+, \mathcal{H}_-]$. Then C is a monomorphism and C^{-1} is $(-,-)$-continuous if and only if $\mathrm{Ran}(C^*) \supset \mathcal{H}_+$.*

2.2 Construction of the operator generated rigging

Let now $\dot{A}$ be a closed symmetric operator whose domain $\mathrm{Dom}(\dot{A})$ is not assumed to be dense in $\mathcal{H}$. Setting $\overline{\mathrm{Dom}(\dot{A})} = \mathcal{H}_0$, we can consider $\dot{A}$ as a densely defined operator from $\mathcal{H}_0$ into $\mathcal{H}$. Clearly, $\mathrm{Dom}(\dot{A}^*)$ is dense in $\mathcal{H}$ and $\mathrm{Ran}(\dot{A}^*) \subset \mathcal{H}_0$.

We introduce a new Hilbert space $\mathcal{H}_+ = \mathrm{Dom}(\dot{A}^*)$ with inner product

$$(f, g)_+ = (f, g) + (\dot{A}^* f, \dot{A}^* g), \qquad (f, g \in \mathcal{H}_+), \tag{2.6}$$

and then construct the rigged Hilbert space $\mathcal{H}_+ \subset \mathcal{H} \subset \mathcal{H}_-$.

Theorem 2.2.1. *Let $\dot{A}$ be a closed symmetric operator in $\mathcal{H}$. Then*

1. *The operator $\dot{A}$ is $(\cdot,-)$-continuous.*
2. *If $\overline{\dot{A}}$ is an extension of $\dot{A}$ by $(\cdot,-)$-continuity to $\mathcal{H}_0$, then the Riesz-Berezansky operator is given by the formula*

$$\mathcal{R}^{-1} = I + \overline{\dot{A}} \dot{A}^*. \tag{2.7}$$

3. $\mathcal{R}\mathcal{H} = \mathrm{Dom}(\dot{A}\dot{A}^*)$.

Proof. (1) Since $\|\dot{A}^*h\| \le \|h\|_+$, $(\forall h \in \mathcal{H}_+)$, then

$$\|\dot{A}g\|_- = \sup_{h\in\mathcal{H}_+} \frac{|(\dot{A}g,h)|}{\|h\|_+} = \sup_{h\in\mathcal{H}_+} \frac{|(g,\dot{A}^*h)|}{\|h\|_+} \le \sup_{h\in\mathcal{H}_+} \frac{\|g\|\cdot\|\dot{A}^*h\|}{\|h\|_+} \le \|g\|,$$

for all $g \in \operatorname{Dom}(\dot{A})$. This yields the $(\cdot,-)$-continuity of $\dot{A}$. Now let $\bar{\dot{A}}$ be an extension of $\dot{A}$ onto $\overline{\operatorname{Dom}(\dot{A})} = \mathcal{H}_0$ using $(\cdot,-)$-continuity. We will show that

$$(\bar{\dot{A}}g, f) = (g, \dot{A}^*f), \qquad (g \in \mathcal{H}_0,\ f \in \mathcal{H}_+). \tag{2.8}$$

Let $g \in \mathcal{H}_0$. Then there is a sequence $\{g_n\} \subset \operatorname{Dom}(\dot{A})$ such that $g_n \to g$ in $(\cdot)$-metric. Hence $\dot{A}g_n \to \bar{\dot{A}}g$ in $(-)$-metric. Letting $n \to \infty$ in

$$(\dot{A}g_n, f) = (g_n, \dot{A}^*f),$$

we get (2.8). We should note that (2.8) indicates that $\bar{\dot{A}} \in [\mathcal{H}_0, \mathcal{H}_-]$ is the adjoint to $\dot{A}^* \in [\mathcal{H}_+, \mathcal{H}_0]$ operator.

The condition that $\bar{\dot{A}}g \in \mathcal{H}$ for some $g \in \mathcal{H}_0$ implies $g \in \operatorname{Dom}(\dot{A})$. Indeed, it follows from (2.8) that for an arbitrary $f \in \mathcal{H}_+$ we have $g \in \operatorname{Dom}(\dot{A}^{**})$ and $\bar{\dot{A}}g = \dot{A}^{**}g$. Since $\dot{A} = \dot{A}^{**}$ we have that $g \in \operatorname{Dom}(\dot{A})$.

(2) For any $g, f \in \mathcal{H}_+$,

$$\begin{aligned}(\mathcal{R}^{-1}g, f) &= (g,f)_+ = (g,f) + (\dot{A}^*g, \dot{A}^*f)\\ &= (g,f) + (\bar{\dot{A}}\dot{A}^*g, f) = ((I + \bar{\dot{A}}\dot{A}^*)g, f),\end{aligned}$$

which implies (2.7).

(3) Obviously, for $g \in \operatorname{Dom}(\dot{A}\dot{A}^*)$ $\mathcal{R}^{-1}g = (I + \bar{\dot{A}}\dot{A}^*)g \in \mathcal{H}$. Conversely, if $g \in \mathcal{H}_+$, $\mathcal{R}^{-1}g \in \mathcal{H}$. Then $\bar{\dot{A}}\dot{A}^*g = \mathcal{R}^{-1}g - g \in \mathcal{H}$. As we have shown above $\dot{A}^*g \in \operatorname{Dom}(\dot{A})$ and thus $g \in \operatorname{Dom}(\dot{A}\dot{A}^*)$. So, the conditions $g \in \operatorname{Dom}(\dot{A}\dot{A}^*)$ and $\mathcal{R}^{-1}g \in \mathcal{H}$ are equivalent. □

Theorem 2.2.2. *Let $f \in \mathcal{H}_0$ and $\bar{\dot{A}}$ be an extension of $\dot{A}$ by $(\cdot,-)$-continuity to $\mathcal{H}_0$. Then $\bar{\dot{A}}f$ belongs to $\mathcal{H}$ if and only if $f \in \operatorname{Dom}(\dot{A})$.*

Proof. The sufficiency part is obvious because $\bar{\dot{A}}f = \dot{A}f \in \mathcal{H}$ for $f \in \operatorname{Dom}(\dot{A})$. Assume that for a $g \in \mathcal{H}_+ = \operatorname{Dom}(\dot{A}^*)$, $(\dot{A}^*g, f) = (g, \bar{\dot{A}}f)$. Since $\dot{A}$ is closed and $\dot{A}^{**} = \dot{A}$, we have that $f \in \operatorname{Dom}(\dot{A})$ and $\dot{A}f = \bar{\dot{A}}f$. □

2.3 Direct decomposition and an analogue of the first von Neumann's formula

We call an operator $\dot{A}$ **regular**, if $P\dot{A}$ is a closed operator in $\mathcal{H}_0$. Here P is an orthogonal projection in $\mathcal{H}$ onto $\mathcal{H}_0 = \overline{\operatorname{Dom}(\dot{A})}$. Obviously, any densely defined closed symmetric operator is regular. For a regular operator $\dot{A}$ we construct a rigged Hilbert space $\mathcal{H}_+ \subset \mathcal{H} \subset \mathcal{H}_-$ using the technique from the previous section. If $\dot{A}$ is densely defined, then by the first von Neumann formula (1.7) we have

$$\mathcal{H}_+ = \mathrm{Dom}(\dot{A}) \dotplus \mathfrak{N}_\lambda \dotplus \mathfrak{N}_{\bar{\lambda}}. \tag{2.9}$$

This decomposition is (+)-orthogonal for $\lambda = \pm i$. When the domain of $\dot{A}$ is not dense in $\mathcal{H}$, Theorem 1.7.1 implies an indirect decomposition

$$\mathcal{H}_+ = \mathrm{Dom}(\dot{A}) + \mathfrak{N}_\lambda + \mathfrak{N}_{\bar{\lambda}}, \tag{2.10}$$

where $\mathrm{Dom}(\dot{A})$, $\mathfrak{N}_\lambda$, and $\mathfrak{N}_{\bar{\lambda}}$ may be linearly dependent (see Theorem 1.7.4). Now we are going to derive an analogue of the first von Neumann's formula that has a direct decomposition of the involved linear manifolds. Define two subspaces of $\mathcal{H}_+$:

$$\mathcal{D}^* := \mathcal{H}_+ \cap \mathcal{H}_0, \quad \mathcal{D} := \overline{\mathrm{Dom}(\dot{A})}^{(+)}. \tag{2.11}$$

Clearly, $\mathcal{D}$ is the domain of the closure of a densely defined in $\mathcal{H}_0$ symmetric operator $P\dot{A}$ and $\mathcal{D}^* = \mathrm{Dom}((P\dot{A})^*)$. Hence

$$\mathcal{D}^* = \mathcal{D} \dotplus \mathfrak{N}'_\lambda \dotplus \mathfrak{N}'_{\bar{\lambda}}, \quad \mathrm{Im}\,\lambda \neq 0,$$

where $\mathfrak{N}'_\lambda$ and $\mathfrak{N}'_{\bar{\lambda}}$ are defined by (1.23). If $\dot{A}$ is regular, then $\mathcal{D} = \mathrm{Dom}(\dot{A})$. According to Theorem 2.1.1 the orthogonal complement of the subspace $\mathcal{D}^*$ is $\overline{\mathfrak{L}}^{(-)}$ where

$$\mathfrak{L} = \mathcal{H} \ominus \mathcal{H}_0. \tag{2.12}$$

This makes

$$\mathfrak{N} = \mathcal{R}\overline{\mathfrak{L}}^{(-)} \tag{2.13}$$

a (+)-orthogonal complement of $\mathcal{D}^*$. Thus we have

$$\mathcal{H}_+ = \mathcal{D} \dotplus \mathfrak{N}'_\lambda \dotplus \mathfrak{N}'_{\bar{\lambda}} \dotplus \mathfrak{N}, \qquad (\mathrm{Im}\lambda \neq 0). \tag{2.14}$$

This is a generalization of the first von Neumann's formula. For $\lambda = \pm i$ we obtain the (+)-orthogonal decomposition

$$\mathcal{H}_+ = \mathcal{D} \oplus \mathfrak{N}'_i \oplus \mathfrak{N}'_{-i} \oplus \mathfrak{N}. \tag{2.15}$$

Let

$$\mathfrak{M} = \mathfrak{N}'_i \oplus \mathfrak{N}'_{-i} \oplus \mathfrak{N}, \tag{2.16}$$

and let $\mathcal{F}$ ($\subset \mathcal{H}_-$) be the ($\cdot$)-orthogonal complement of $\mathrm{Dom}(\dot{A})$ ($\subset \mathcal{H}_+$), i.e.,

$$\mathcal{F} = \left\{\varphi \in \mathcal{H}_- : (\varphi, f) = 0 \quad \text{for all} \quad f \in \mathrm{Dom}(\dot{A})\right\}. \tag{2.17}$$

It is clear that

$$\mathcal{F} = \mathcal{R}^{-1}\mathfrak{M} = \mathcal{R}^{-1}\mathfrak{N}'_i \oplus \mathcal{R}^{-1}\mathfrak{N}'_{-i} \oplus \overline{\mathfrak{L}}^{(-)}, \tag{2.18}$$

and the last decomposition is $(-)$-orthogonal.

Here and below by $P^+_{\mathcal{G}}$ we denote the orthogonal projection in $\mathcal{H}_+$ onto a subspace $\mathcal{G}$ of $\mathcal{H}_+$. Respectively, $P_{\mathcal{G}}$ would represent the orthogonal projection in $\mathcal{H}$ onto a subspace $\mathcal{G}$ of $\mathcal{H}$.

Theorem 2.3.1. 1. *The operator* $\dot{A}^*\upharpoonright\mathfrak{N}$ *is symmetric in* $\mathcal{H}$ *and* $\dot{A}^*\mathfrak{N}\subset\mathcal{D}$.

2. *The operator* $\dot{A}^*\pm iI$ *maps* $\mathfrak{N}$ $(+,\cdot)$*-isometrically on* $\overline{\mathfrak{B}}_{\pm i}$*, where* $\mathfrak{B}_\lambda$ *is defined in* (1.22).

3. *The operator* $\dot{B}$ *given by the relations*

$$\mathrm{Dom}(\dot{B})=\mathrm{Dom}(\dot{A})\oplus\mathfrak{N},\quad \dot{B}=\dot{A}P^+_{\mathrm{Dom}(\dot{A})}+\dot{A}^*P^+_{\mathfrak{N}},\tag{2.19}$$

is closed, densely defined and symmetric in $\mathcal{H}$*, and*

$$\mathrm{Dom}(\dot{B}^*)=\mathrm{Dom}(\dot{B})\oplus\mathfrak{N}'_i\oplus\mathfrak{N}'_{-i},\ \dot{B}^*=AP^+_{\mathrm{Dom}(\dot{A})}+\dot{A}^*P^+_{\mathfrak{M}}.\tag{2.20}$$

Proof. (1) Let $g\in\mathrm{Dom}(\dot{A})$ and $f\in\mathfrak{M}$. Then

$$0=(g,f)_+=(g,f)+(\dot{A}^*g,\dot{A}^*f)=(g,f)+(P\dot{A}g,\dot{A}^*f)=(g,f)+(\dot{A}g,\dot{A}^*f).$$

Hence,

$$(\dot{A}g,\dot{A}^*f)=(g,-Pf),\qquad\forall g\in\mathrm{Dom}(\dot{A}).$$

Consequently, $\dot{A}^*f\in\mathrm{Dom}(\dot{A}^*)=\mathcal{H}_+$ and $(\dot{A}^*)^2f=-Pf$. On the other hand, since $\mathrm{Ran}(\dot{A}^*)\subset\mathcal{H}_0$, we get $\dot{A}^*f\in\mathcal{D}^*$, i.e., $\dot{A}^*\mathfrak{M}\subset\mathcal{D}^*$.

Let $\psi,f\in\mathfrak{M}$. Then

$$\begin{aligned}(\psi,\dot{A}^*f)_+&=(\psi,\dot{A}^*f)+(\dot{A}^*\psi,(\dot{A}^*)^2f)=(\psi,\dot{A}^*f)+(\dot{A}^*\psi,-Pf)\\&=(\psi,\dot{A}^*f)-(\dot{A}^*\psi,f).\end{aligned}$$

In particular, if $\psi\in\mathfrak{N}$, then ψ is $(+)$-orthogonal to $\dot{A}^*f\in\mathcal{D}^*$. This implies that

$$(\psi,\dot{A}^*f)=(\dot{A}^*\psi,f),\qquad\forall\psi\in\mathfrak{N},\ \forall f\in\mathfrak{M}.$$

Therefore, the operator $\dot{A}^*\upharpoonright\mathfrak{N}$ is symmetric.

Now let $\psi\in\mathfrak{N}$, $f\in\mathfrak{N}'_{\pm i}$. Then

$$(\dot{A}^*\psi,f)_+=(\dot{A}^*\psi,f)+(-P\psi,\dot{A}^*f)=(\dot{A}^*\psi,f)-(\psi,\dot{A}^*f)=0.$$

Thus, $\dot{A}^*\mathfrak{N}$ is $(+)$-orthogonal to $\mathfrak{N}'_i$ and $\mathfrak{N}'_{-i}$. It follows from (2.15) that $\dot{A}^*\mathfrak{N}\subset\mathcal{D}$.

(2) Using the symmetric property of $\dot{A}^*\upharpoonright\mathfrak{N}$ for $\psi\in\mathfrak{N}$ we get

$$\|(\dot{A}^*\pm iI)\psi\|^2=\|\dot{A}^*\psi\|^2+\|\psi\|^2=\|\psi\|^2_+.$$

This implies that $(\dot{A}^* \pm iI)\restriction\mathfrak{N}$ is an $(+,\cdot)$-isometry. Letting $f \in \mathcal{H}_+$ and $g \in \operatorname{Dom}(\dot{A})$ we have

$$\begin{aligned}\Big((\dot{A}^* + iI)f, (\dot{A} + iI)g\Big) &= (\dot{A}^* f, \dot{A}g) + i(f, \dot{A}g) - i(\dot{A}^* f, g) + (f, g)\\ &= (\dot{A}^* f, P\dot{A}g) + (f, g) = (f, g)_+.\end{aligned}$$

In particular, if $f \in \mathfrak{M}$, then $\Big((\dot{A}^* + iI)f, (\dot{A} + iI)g\Big) = 0$ and hence

$$(\dot{A}^* + iI)\mathfrak{M} \subset \mathfrak{N}_i.$$

Now we will show that for any $\phi \in \mathfrak{L}$,

$$P_i\phi = -i(\dot{A}^* + iI)\mathcal{R}\phi, \tag{2.21}$$

where $\mathcal{R}$ is the Riesz-Berezansky operator and

$$\|P_i\phi\| = \|\phi\|_-. \tag{2.22}$$

Here and below

$$P_\lambda = P_{\mathfrak{N}_\lambda}.$$

Since $\mathcal{R}\phi \in \mathcal{R}\mathfrak{L} \subset \mathfrak{M}$, $(\dot{A}^* + iI)\mathcal{R}\phi \in \mathfrak{N}_i$. On the other hand, by Theorem 2.2.1, we have $\mathcal{R}\phi \in \operatorname{Dom}(\dot{A}\dot{A}^*)$ and

$$\begin{aligned}\phi = (I + \dot{A}\dot{A}^*)\mathcal{R}\phi &= \Big[(\dot{A} + iI)\dot{A}^* - i(\dot{A}^* + iI)\Big]\mathcal{R}\phi\\ &= (\dot{A} + iI)\dot{A}^*\mathcal{R}\phi - i(\dot{A}^* + iI)\mathcal{R}\phi,\end{aligned}$$

so $\phi + i(\dot{A}^* + iI)\mathcal{R}\phi \in \mathfrak{M}_{-i}$. This proves (2.21). The latter also implies via (2.1) that

$$\|P_i\phi\| = \|(\dot{A} + iI)\mathcal{R}\phi\| = \|\mathcal{R}\phi\|_+ = \|\phi\|_-, \quad (\phi \in \mathfrak{L}).$$

Thus, the operator $(\dot{A}^* + iI)$ maps a linear $(+)$-dense in $\mathfrak{N}$ set $\mathcal{R}\mathfrak{L}$ $(+,\cdot)$-isometrically onto $P_i\mathfrak{L} = \mathfrak{B}_i$. That is why

$$(\dot{A}^* + iI)\mathfrak{N} = \overline{\mathfrak{B}_i}.$$

Similarly we can show the same for $(\dot{A}^* - iI)$. Also

$$P_{-i}\phi = i(\dot{A}^* - iI)\mathcal{R}\phi, \qquad \|P_{-i}\phi\| = \|\phi\|_-, \quad (\phi \in \mathfrak{L}), \tag{2.23}$$

so that $\|P_i\phi\| = \|P_{-i}\phi\|$.

(3) Let the operator $\dot{B}$ be given by (2.19). If a vector $h \in \mathcal{H}$ is orthogonal to $\operatorname{Dom}(\dot{B})$, then $h \in \mathfrak{L}$ and $(h, f) = 0$ for all $f \in \mathfrak{N}$. From definition (2.13) of $\mathfrak{N}$ we get

$$0 = (h, \mathcal{R}h) = \|h\|_-^2.$$

Therefore, $h = 0$ and $\mathrm{Dom}(\dot{B})$ is dense in $\mathcal{H}$.

Since $\dot{A}^* \upharpoonright \mathfrak{N}$ is symmetric in $\mathcal{H}$, the operator $\dot{B}$ is also symmetric in $\mathcal{H}$. The relations $(\dot{A}^* \pm iI)\mathfrak{N} = \overline{\mathfrak{B}_{\pm i}}$ yield

$$(\dot{B} \pm iI)\mathrm{Dom}(\dot{B}) = (\dot{A} \pm iI)\mathrm{Dom}(\dot{A}) \oplus \overline{\mathfrak{B}_{\pm i}}.$$

Hence, the linear manifolds $(\dot{B} \pm iI)\mathrm{Dom}(\dot{B})$ are closed in $\mathcal{H}$. It follows that the operator $\dot{B}$ is closed, and

$$\mathcal{H} \ominus \left((\dot{B} \pm iI)\mathrm{Dom}(\dot{B})\right) = \mathfrak{N}'_{\pm i}.$$

Thus, semi-deficiency subspace $\mathfrak{N}'_{\pm i}$ of the operator $\dot{A}$ coincides with the deficiency subspace of $\dot{B}$ corresponding to the number $\pm i$. In accordance with the first von Neumann formula for $\dot{B}^*$ we get (2.20). □

Notice that the operator $\dot{B}$ is a symmetric extension of the operator $\dot{A}$ and $\dot{B}$ is self-adjoint if and only if the semi-deficiency indices of $\dot{A}$ are zero.

Corollary 2.3.2. *The operator $\dot{A}^* \pm iI$ maps $\mathfrak{N}'_{\pm i} \oplus \mathfrak{N}$ with $(+)$-metric homeomorphically onto the subspace $\mathfrak{N}_{\pm i}$ with either $(\cdot)$- or $(+)$-metric. Moreover,*

$$(\dot{A}^* \pm iI)\mathfrak{M} = \mathfrak{N}_{\pm i}.$$

Proof. Indeed, using (1.23) we get

$$(\dot{A}^* \pm iI)(\mathfrak{N}'_{\pm i} \oplus \mathfrak{N}) = \mathfrak{N}'_{\pm i} \dotplus (\dot{A}^* \pm iI)\mathfrak{N} = \mathfrak{N}'_{\pm i} \dotplus \overline{\mathfrak{B}_{\pm i}} = \mathfrak{N}_{\pm i}.$$

$(+,\cdot)$-continuity of $(\dot{A}^* \pm iI)$ follows from

$$\|\dot{A}^* h \pm ih\| \le \|\dot{A}^* h\| + \|h\| \le 2\|h\|_+, \quad h \in \mathcal{H}_+.$$

Let $\phi \in \mathfrak{N}'_{\pm i} \oplus \mathfrak{N}$, i.e., $\phi = \varphi + \psi$, where $\varphi \in \mathfrak{N}'_{\pm i}$ and $\psi \in \mathfrak{N}$. Then $\|\phi\|_+^2 = \|\varphi\|_+^2 + \|\psi\|_+^2$. Further,

$$(\dot{A}^* \pm iI)\phi = 2i\varphi + (\dot{A}^* \pm iI)\psi,$$

and the terms $\pm 2i\varphi \in \mathfrak{N}'_{\pm i}$ and $(\dot{A}^* \pm iI)\psi \in \overline{\mathfrak{B}_{\pm i}}$ are $(\cdot)$-orthogonal. Consequently,

$$\|(\dot{A}^* \pm iI)\phi\|^2 = 4\|\varphi\|^2 + \|(\dot{A}^* \pm iI)\psi\|^2 = 2\|\varphi\|_+^2 + \|\psi\|_+^2 \ge \|\phi\|_+^2,$$

which implies the continuity of the inverse mapping. □

We should note that the operator $\dot{A}^* \pm iI$ maps $\mathfrak{N}_{\mp i}$ to zero and acts like $\pm 2iI$ on $\mathfrak{N}'_{\pm i}$. Therefore, it maps $\mathfrak{M}$ onto $\mathfrak{N}_{\pm i}$, the mapping is one-to-one, and mutually $(+,+)$-continuous on $\mathfrak{N}'_{\pm i} \oplus \mathfrak{N}$.

Following Section 1.6 we use (1.26) to introduce an isometric exclusion operator $V = V_i : \mathfrak{B}_i \to \mathfrak{B}_{-i}$ defined by the formula

$$VP_i f = P_{-i} f, \qquad f \in \mathfrak{L}, \;\; P_{\pm i} = P_{\mathfrak{N}_{\pm i}}. \tag{2.24}$$

Its closure $\overline{V}$ maps $\overline{\mathfrak{B}_i}$ isometrically onto $\overline{\mathfrak{B}_{-i}}$. It follows from (2.21) and (2.23) that

$$V(\dot{A}^* + iI)\mathcal{R}f = -(\dot{A}^* - iI)\mathcal{R}f, \qquad (f \in \mathfrak{L}),$$

and hence

$$\overline{V}(\dot{A}^* + iI)\mathcal{R}f = -(\dot{A}^* - iI)\mathcal{R}f,$$

for all $f \in \overline{\mathfrak{L}}^{(-)}$. Thus

$$\overline{V}(\dot{A}^* + iI)\psi = -(\dot{A}^* - iI)\psi, \qquad \psi \in \mathfrak{N}.$$

Theorem 2.3.3. *The operator $P^+_{\mathfrak{M}}$ is a bijection and a homeomorphism of $\mathfrak{N}_{\pm i}$ with the $(\cdot)$-metric onto $(\mathfrak{N}'_{\pm i} \oplus \mathfrak{N})$ with the $(+)$-metric.*

Proof. Let $\phi \in \mathfrak{N}_i$. Then $\phi = \varphi + \psi$, where $\varphi \in \mathfrak{N}'_i$ and $\psi \in \overline{\mathfrak{B}_i}$. According to Theorem 2.3.1, there exists such an element $h \in \mathfrak{N}$ that $\dot{A}^*h + ih = \psi$ and $\|\psi\| = \|h\|_+$. Since $\dot{A}^*h \in \operatorname{Dom}(\dot{A})$, then $P^+_{\mathfrak{M}}\psi = ih$, and hence, $P^+_{\mathfrak{M}}\phi = \varphi + ih$. Thus $P^+_{\mathfrak{M}}(\mathfrak{N}_i) = \mathfrak{N}'_{\pm i} \oplus \mathfrak{N}$.

Furthermore,

$$\|\phi\|^2_+ \geq \|P^+_{\mathfrak{M}}\phi\|^2_+ = \|\varphi\|^2_+ + \|h\|^2_+ = 2\|\varphi\|^2 + \|\psi\|^2 \geq \|\varphi\|^2 + \|\psi\|^2 = \|\phi\|^2,$$

implies the conclusion of the theorem for $\mathfrak{N}_i$. The proof of the theorem for $\mathfrak{N}_{-i}$ is similar. □

Let us now denote by $P^+_{\mathfrak{N}}$, the orthogonal projection operator from $\mathcal{H}_+$ onto $\mathfrak{N}$. We introduce a new inner product $(\cdot,\cdot)_1$ defined by

$$(f,g)_1 = (f,g)_+ + (P^+_{\mathfrak{N}}f, P^+_{\mathfrak{N}}g)_+ \tag{2.25}$$

for all $f, g \in \mathcal{H}_+$. The obvious inequality

$$\|f\|^2_+ \leq \|f\|^2_1 \leq 2\|f\|^2_+$$

shows that the norms $\|\cdot\|_+$ and $\|\cdot\|_1$ are topologically equivalent. It is easy to see that the spaces $\operatorname{Dom}(\dot{A})$, $\mathfrak{N}'_i$, $\mathfrak{N}'_{-i}$, $\mathfrak{N}$ are (1)-orthogonal. We write $\mathfrak{M}_1$ for the Hilbert space $\mathfrak{M} = \mathfrak{N}'_i \oplus \mathfrak{N}'_{-i} \oplus \mathfrak{N}$ with inner product $(f,g)_1$. We denote by $\mathcal{H}_{+1}$ the space $\mathcal{H}_+$ with norm $\|\cdot\|_1$, and by $\boldsymbol{\mathcal{R}}$ the corresponding Riesz-Berezansky operator related to the triplet $\mathcal{H}_{+1} \subset \mathcal{H} \subset \mathcal{H}_{-1}$. Both operators $\mathcal{R}$ and $\boldsymbol{\mathcal{R}}$ act according to Fig. 2.1 below.

One can also see that

$$\boldsymbol{\mathcal{R}}^{-1} = \mathcal{R}^{-1}(I + P^+_{\mathfrak{N}}), \quad \boldsymbol{\mathcal{R}} = (I - \frac{1}{2}P^+_{\mathfrak{N}})\mathcal{R}. \tag{2.26}$$

Note also that the operators $2^{-\frac{1}{2}}P^+_{\mathfrak{M}} \upharpoonright \mathfrak{N}_i$ and $2^{-\frac{1}{2}}P^+_{\mathfrak{M}} \upharpoonright \mathfrak{N}_{-i}$ mentioned in Theorem 2.3.3 are $(\cdot, 1)$-isometries from $\mathfrak{N}_i$ and $\mathfrak{N}_{-i}$ onto $\mathfrak{N}'_i \oplus \mathfrak{N}$ and $\mathfrak{N}'_{-i} \oplus \mathfrak{N}$, respectively. It follows from the proof of Theorem 2.3.1 that the explicit expression for $P^+_{\mathfrak{M}} \upharpoonright \mathfrak{N}_{\pm i}$ is of the form

$$P^+_{\mathfrak{M}}\phi = \pm i(2P^+_{\mathfrak{N}'_{\pm i}} + P^+_{\mathfrak{N}})(\dot{A}^* \pm iI)^{-1}\phi, \quad \phi \in \mathfrak{N}_{\pm i}. \tag{2.27}$$

Figure 2.1: Operators $\mathcal{R}$ and $\mathfrak{R}$

2.4 Regular and singular symmetric operators

At the beginning of Section 2.3 we introduced the definition of a regular operator $\dot{A}$. In this section we will provide a criteria for an operator $\dot{A}$ to be regular. It was shown in Corollary 1.5.5 that for all λ ($\text{Im}\,\lambda \neq 0$) the manifolds $\mathfrak{B}_\lambda$ are either all closed or all non-closed. We will show that operator $\dot{A}$ is regular in the first case and is called **singular** in the second.

Theorem 2.4.1. *The following statements are equivalent for a closed symmetric operator $\dot{A}$:*

(1) *The manifolds $\mathfrak{B}_\lambda$ are $(\cdot)$-closed for all λ $(\text{Im}\lambda \neq 0)$.*

(2) $\text{Dom}(\dot{A})$ *is $(+)$-closed.*

(3) *$\dot{A}$ is regular ($P\dot{A}$ is closed).*

(4) *$\dot{A}$ is $(+,\cdot)$-bounded.*

(5) *$\mathfrak{L}$ is $(-)$-closed.*

(6) *$\mathfrak{N}$ is $(\cdot)$-closed.*

(7) $\text{Dom}(\dot{B}^*) = \text{Dom}(\dot{A}^*)$, *where $\dot{B}$ is defined by* (2.19).

Proof. The equivalence (1) $\iff$ (3) is already proved (see Theorem 1.5.4 and Corollary 1.5.5). The equivalence (2) $\iff$ (3) follows from the definition of $(+)$-norm.

(2) $\Rightarrow$ (4) follows from the Closed Graph Theorem and the inequality $\|f\| \leq \|f\|_+$, $f \in \mathcal{H}_+$. Since $\dot{A}$ is closed we get (4)$\Rightarrow$ (2).

(1) $\iff$ (5). Because of (2.22) the operator $P_i = P_{\mathfrak{N}_i}$ maps the set $\mathfrak{L}$ isometrically onto the set $\mathfrak{B}_i$. Thus $(-)$-closure of $\mathfrak{L}$ is equivalent to the $(\cdot)$-closure of $\mathfrak{B}_i$.

(5) $\iff$ (6). Since $\|f\|_-^2 = (\mathcal{R}f, f) = \|\mathcal{R}f\|_+^2$ for $f \in \mathcal{H}_-$, and hence $\|f\|_-^2 \leq \|\mathcal{R}f\|\,\|f\|$ for $f \in \mathcal{H}$, we get

$$\begin{aligned} &\gamma\|f\| \leq \|f\|_- \leq \|f\| \quad \text{for all} \quad f \in \mathcal{L} \quad \text{and for some} \quad \gamma \in (0,1) \\ &\iff \|\mathcal{R}f\| \geq \gamma\|\mathcal{R}f\|_+. \end{aligned}$$

Comparing (2.15) with (2.19) and (2.20) we get that the equivalence (2) and (7) holds true. □

This theorem immediately implies the following independently sufficient conditions for a closed symmetric operator $\dot{A}$ to be regular:

- $\mathrm{Dom}(\dot{A})$ has a finite codimension ($\dim \mathfrak{L} < \infty$);
- $\mathfrak{L} \subset \mathcal{H}_+$, where $\mathfrak{L}$ is defined in (2.12).

Proposition 2.4.2. *If $\dot{A}$ is a regular symmetric operator, then the direct decomposition*

$$\mathcal{H} = \mathcal{H}_0 \dot{+} \mathfrak{N}$$

holds.

Proof. Since $\dot{A}$ is regular, by Theorem 2.4.1 and Theorem 1.5.4 the linear manifold $\mathfrak{B}_i$ is $(\cdot)$-closed and $\Theta(\mathfrak{L}, \mathfrak{B}_i) < 1$. Now from Lemma 1.5.1 we get the equality

$$P_{\mathfrak{L}} \mathfrak{B}_i = \mathfrak{L}.$$

On the other hand, from Theorem 2.3.1 we have that $\mathfrak{B}_i = (\dot{A}^* + iI)\mathfrak{N}$. Hence, $P_{\mathfrak{L}}(\dot{A}^* + iI)\mathfrak{N} = \mathfrak{L}$. But $\dot{A}^*\mathfrak{N} \subset \mathrm{Dom}(\dot{A}) \subset \mathcal{H}_0$. Therefore,

$$P_{\mathfrak{L}} \mathfrak{N} = \mathfrak{L}.$$

Taking into account that $\mathfrak{N} \cap \mathcal{H}_0 = \{0\}$ and $\mathcal{H} = \mathcal{H}_0 \oplus \mathfrak{L}$, we get the equality $\mathcal{H} = \mathcal{H}_0 \dot{+} \mathfrak{N}$. □

A closed symmetric operator $\dot{A}$ is said to be an **O-operator** if both its semi-deficiency indices equal zero. For such an operator $\mathfrak{M} = \mathfrak{N}$.

Theorem 2.4.3. *A closed symmetric operator $\dot{A}$ is a regular O-operator if and only if $\mathcal{F} \subset \mathcal{H}$, where $\mathcal{F}$ is of the form* (2.17).

The proof easily follows from the definition of a regular O-operator.

2.5 Closed symmetric extensions

Let A be a closed symmetric extension of a symmetric operator $\dot{A}$. Then

$$(Af, g) = (f, Ag), \qquad (\forall f, g \in \mathrm{Dom}(A)),$$

and, in particular

$$(\dot{A}f, g) = (f, Ag) = (f, PAg), \qquad (\forall f \in \mathrm{Dom}(\dot{A}),\ g \in \mathrm{Dom}(A)).$$

It follows then that $g \in \mathrm{Dom}(\dot{A}^*)$ and $PAg = \dot{A}^* g$, and thus $\mathrm{Dom}(A) \subset \mathcal{H}_+$ and

$$PAf = \dot{A}^* f,\ f \in \mathrm{Dom}(A), \tag{2.28}$$

where P is an orthogonal projection operator in $\mathcal{H}$ onto $\mathcal{H}_0 = \overline{\mathrm{Dom}(\dot{A})}$. The next theorem is an immediate consequence of (2.28) and the Closed Graph Theorem.

Theorem 2.5.1. *For a closed symmetric extension A of an operator $\dot{A}$ the following conditions are equivalent:*

(1) *The set* $\mathrm{Dom}(A)$ *is* $(+)$*-closed.*

(2) *The operator A is* $(+,\cdot)$*-bounded.*

(3) *The operator* $\dot{A}^*\upharpoonright \mathrm{Dom}(A)$ *is* $(\cdot,\cdot)$*-closed.*

Moreover, under the conditions (1)–(3) *the operator $\dot{A}$ is regular.*

A closed symmetric extension A of a closed symmetric operator $\dot{A}$ satisfying the conditions (1)–(3) of Theorem 2.5.1 is called a **regular symmetric extension**.

Let us recall some aspects of the theory of extensions of closed symmetric operators $\dot{A}$ with non-dense domain developed in Chapter 1. Once again we denote the exclusion operator (2.24) by V. Following Section 1.6 we call an isometric operator U ($\mathrm{Dom}(U)=\overline{\mathrm{Dom}(U)}\subset\mathfrak{N}_i$, $\mathrm{Ran}(U)\subset\mathfrak{N}_{-i}$) an **admissible operator** if $Ux=Vx$ only for $x=0$. The general form of a closed symmetric extension A of an operator $\dot{A}$ follows from (1.33) and is given by formulas

$$\begin{aligned} \mathrm{Dom}(A) &= \mathrm{Dom}(\dot{A}) \dot{+} \mathrm{Ran}(I-U),\\ A(g+\varphi-U\varphi) &= \dot{A}g+i(\varphi+U\varphi),\\ g\in \mathrm{Dom}(\dot{A}),&\ \varphi\in\mathrm{Dom}(U), \end{aligned} \tag{2.29}$$

where U is an admissible operator.

The operator A is self-adjoint if and only if $\mathrm{Dom}(U)=\mathfrak{N}_i$ and $\mathrm{Ran}(U)=\mathfrak{N}_{-i}$. Also, if $\varphi\in\mathfrak{B}_i$ then $\varphi-V\varphi\in\mathrm{Dom}(\dot{A})$. Conversely, if $\varphi-\psi\in\mathrm{Dom}(\dot{A})$, where $\varphi\in\mathfrak{N}_i$, $\psi\in\mathfrak{N}_{-i}$, then $\varphi\in\mathfrak{B}_i$, and $\psi=V\varphi\in\mathfrak{B}_{-i}$. Hence, in particular, for admissible U the equality $x=Ux$ holds only for $x=0$. Thus, the operator $I-U$ is injective and $(\cdot,+)$-continuous.

Now we are going to prove an auxiliary result of a general geometrical nature. First let us recall the notion of the **minimal angle** $\alpha(\mathcal{L},\mathcal{M})$ **between subspaces** $\mathcal{L}$ and $\mathcal{M}$ of a Hilbert space $\mathcal{H}$:

$$\cos\alpha(\mathcal{L},\mathcal{M}) := \sup_{\substack{x\in\mathcal{L},x_2\in\mathcal{M},\\ \|x\|=\|y\|=1}} |(x,y)| \tag{2.30}$$

Definition (2.30) yields the equalities

$$\cos\alpha(\mathcal{L},\mathcal{M}) = ||P_{\mathcal{L}}\upharpoonright\mathcal{M}|| = ||P_{\mathcal{M}}\upharpoonright\mathcal{L}||, \tag{2.31}$$

where $P_{\mathcal{L}}$ and $P_{\mathcal{M}}$ are orthogonal projections onto $\mathcal{L}$ and $\mathcal{M}$, respectively.

Lemma 2.5.2. *Let $\mathcal{L}_1$ be a subspace of a Hilbert space $\mathcal{H}$ and $\mathcal{L}_2$ be a linear manifold in $\mathcal{H}$ that is a range of a bounded operator T mapping a Banach space into $\mathcal{H}$. Let also $P_{\mathcal{L}_1}$ be an orthogonal projection from $\mathcal{H}$ onto $\mathcal{L}_1$ and $P_{\mathcal{L}_1^\perp}=I-P_{\mathcal{L}_1}$. Then the following statements are equivalent:*

(1) $\mathcal{L}_1\cap\mathcal{L}_2=\{0\}$ *and the linear set* $\mathcal{L}_1\dot{+}\mathcal{L}_2$ *is closed.*

(2) $\mathcal{L}_1 \cap \mathcal{L}_2 = \{0\}$ *and the linear set* $P_{\mathcal{L}_1^\perp}\mathcal{L}_2$ *is closed.*

(3) $\mathcal{L}_2$ *is closed and the minimal angle between spaces* $\mathcal{L}_1$ *and* $\mathcal{L}_2$ *is positive.*

(4) $\mathcal{L}_2$ *is closed and* $P_{\mathcal{L}_1^\perp} \upharpoonright \mathcal{L}_2$ *is a homeomorphism.*

Proof. (1) $\Leftrightarrow$ (2). It is clear that

$$\mathcal{L}_1 \dot{+} \mathcal{L}_2 = \mathcal{L}_1 \oplus P_{\mathcal{L}_1^\perp}\mathcal{L}_2,$$

and hence, the linear manifolds $\mathcal{L}_1 \dot{+} \mathcal{L}_2$ and $P_{\mathcal{L}_1^\perp}\mathcal{L}_2$ are closed simultaneously.

(1) $\Leftrightarrow$ (3). Assume that $\mathcal{L}_2 = T\mathcal{L}_0$, where $\mathcal{L}_0$ is a Banach space, $T : \mathcal{L}_0 \to \mathcal{H}$ is bounded and injective. Consider a Banach space $\mathcal{M} = \mathcal{L}_1 \times \mathcal{L}_0$ with the norm

$$|| \langle f, g\rangle ||_{\mathcal{M}} := ||f||_{\mathcal{L}_1} + ||g||_{\mathcal{L}_0}, \; f \in \mathcal{L}_1, \; g \in \mathcal{L}_0.$$

Then the mapping $\mathcal{Z} : \mathcal{M} \to \mathcal{L}_1 \dot{+} \mathcal{L}_2$ given by

$$\mathcal{Z} \langle f, g\rangle = f + Tg$$

is continuous and bijective. Applying the Banach inverse mapping theorem, we get that $\mathcal{L}_2 = \mathcal{Z} \langle 0 \times \mathcal{L}_0 \rangle$ is closed in $\mathcal{H}$. Let us show that the minimal angle between $\mathcal{L}_1$ and $\mathcal{L}_2$ is positive, i.e., $\cos\alpha(\mathcal{L}_1, \mathcal{L}_2) < 1$. Since $\mathcal{L}_2$ is closed, $\mathcal{L}_1 \cap \mathcal{L}_2 = \{0\}$, and $P_{\mathcal{L}_1^\perp}\mathcal{L}_2$ is closed, from Banach's inverse mapping theorem we get that there exists a constant $\gamma \in (0, 1)$ such that

$$||P_{\mathcal{L}_1^\perp} h|| \geq \gamma ||h|| \quad \text{for all} \quad h \in \mathcal{L}_2. \tag{2.32}$$

It follows that $||P_{\mathcal{L}_1} \upharpoonright \mathcal{L}_2|| < 1$. This means $\alpha(\mathcal{L}_1, \mathcal{L}_2) > 0$. Thus (1)$\Rightarrow$(3).

Now suppose that (3) holds. Due to (2.31) we have

$$\cos\alpha(\mathcal{L}_1, \mathcal{L}_2) < 1 \iff (2.32).$$

Therefore, the operator $P_{\mathcal{L}_1^\perp} \upharpoonright \mathcal{L}_2$ is a homeomorphism. Taking into account the relation

$$||f + h||^2 = ||f + P_{\mathcal{L}_1} h||^2 + ||P_{\mathcal{L}_1^\perp} h||^2, \; f \in \mathcal{L}_1, \; h \in \mathcal{L}_2,$$

we get that $\mathcal{L}_1 \dot{+} \mathcal{L}_2$ is closed. So, (3) $\iff$ (4) and (3)$\Rightarrow$(1). $\square$

Now let $\dot{A}$ be a closed symmetric operator and A be its closed symmetric extension with the corresponding admissible isometric operator U. Applying Lemma 2.5.2 to $\mathcal{H} = \mathcal{H}_+$, $\mathcal{L}_1 = \mathrm{Dom}(\dot{A})$, and $\mathcal{L}_2 = \mathrm{Dom}(U)$ and taking into account that $\mathrm{Dom}(U)$ is the range of a $(\cdot, +)$-continuous operator $I - U$, we obtain the following theorem:

Theorem 2.5.3. *The following statements are equivalent:*

(1) A *is a regular symmetric extension of* $\dot{A}$ *(that is,* $\mathrm{Dom}(A)$ *is a* $(+)$*-closed set).*

(2) $P^+_{\mathfrak{M}}(\operatorname{Ran}(I-U))$ *is a* $(+)$*-closed set.*

(3) $\operatorname{Ran}(I-U)$ *is* $(+)$*-closed, and the minimal angle between* $\operatorname{Dom}(\dot{A})$ *and* $\operatorname{Ran}(I-U)$ *is positive (with respect to the* $(+)$*-metric).*

(4) $\operatorname{Ran}(I-U)$ *is* $(+)$*-closed, and* $P^+_{\mathfrak{M}}\restriction \operatorname{Ran}(I-U)$ *is a homeomorphism.*

We note that under the conditions of Theorem 2.5.3 there is a constant $c>0$ such that

$$\|g+\varphi-U\varphi\|_+ \geq c\|\varphi\|, \quad (\forall g\in \operatorname{Dom}(\dot{A}),\ \forall\varphi\in\operatorname{Dom}(U)). \tag{2.33}$$

Indeed, the positivity of the minimal angle between $\operatorname{Dom}(\dot{A})$ and $\operatorname{Dom}(U)$ implies that orthoprojection on $\operatorname{Dom}(U)$ parallel to $\operatorname{Dom}(\dot{A})$ is $(+)$-continuous, i.e.,

$$c_1\|\varphi-U\varphi\|_+ \leq \|g+\varphi-U\varphi\|_+,$$

for some $c_1>0$. On the other hand, it follows from the Closed Graph Theorem that

$$\|\varphi-U\varphi\|_+ \geq c_2\|\varphi\|, \quad (\varphi\in\operatorname{Dom}(U)),$$

for some $c_2>0$. This proves (2.33).

Theorem 2.5.4. *If A is a regular symmetric extension of a regular closed symmetric operator $\dot{A}$, then there is a constant $c>0$ such that*

$$\|V\varphi-U\varphi\|_+ \geq c\|\varphi\|, \quad \forall\varphi\in\mathfrak{B}_i\cap\operatorname{Dom}(U), \tag{2.34}$$

where V is of the form (2.24). The converse is valid if $\operatorname{Dom}(U)\supset\mathfrak{B}_i$*, in particular, for self-adjoint A.*

Proof. For $\varphi\in\mathfrak{B}_i\cap\operatorname{Dom}(U)$ we have

$$V\varphi-U\varphi = -(\varphi-V\varphi)+(\varphi-U\varphi).$$

Since $-(\varphi-V\varphi)\in\operatorname{Dom}(\dot{A})$, then Theorem 2.5.3 yields (2.34).

Conversely, let $\operatorname{Dom}(U)\supset\mathfrak{B}_i$ and for all $\varphi\in\mathfrak{B}_i$

$$\|V\varphi-U\varphi\|_+ \geq c\|\varphi\|,$$

for some $c>0$. We will show that $P^+_{\mathfrak{M}}\operatorname{Dom}(U)$ is $(+)$-closed. Let

$$P^+_{\mathfrak{M}}(g_n-Ug_n)\xrightarrow{(+)}0,$$

for $g_n=g_n'+g_n''$, $(g_n'\in\mathfrak{N}_i',\ g_n''\in\mathfrak{B}_i)$. Then

$$g_n'+P^+_{\mathfrak{M}}(g_n''-Ug_n)\xrightarrow{(+)}0.$$

It was shown in Theorem 2.3.3 that

$$P^+_{\mathfrak{M}}\mathfrak{B}_i=\mathfrak{N},\quad P^+_{\mathfrak{M}}\mathfrak{N}_{-i}=\mathfrak{N}\oplus\mathfrak{N}_i'.$$

Therefore, $P^+_{\mathfrak{M}}(g_n - Ug_n) \subset \mathfrak{N} \oplus \mathfrak{N}'_i$, i.e., the elements g'_n and $P^+_{\mathfrak{M}}(g''_n - Ug_n)$ are (+)-orthogonal. Hence, $g'_n \xrightarrow{(+)} 0$, and thus $P^+_{\mathfrak{M}}(g''_n - Ug_n) \xrightarrow{(+)} 0$.

Furthermore,

$$g''_n - Ug''_n = g''_n - Vg''_n + Vg''_n - Ug''_n.$$

Since the vector $g''_n - Vg''_n \in \mathrm{Dom}(\dot{A})$ is (+)-orthogonal to $\mathfrak{M}$, then

$$P^+_{\mathfrak{M}}(Vg''_n - Ug''_n) \xrightarrow{(+)} 0.$$

According to Theorem 2.3.3 we have $Vg''_n - Ug''_n \xrightarrow{(+)} 0$. It follows from (2.34) that $g''_n \xrightarrow{(+)} 0$. Therefore, $g_n \xrightarrow{(+)} 0$ which implies that $P^+_{\mathfrak{M}}(\mathrm{Dom}(U))$ is (+)-closed. Then, by Theorem 2.5.3, A is a regular closed symmetric extension of the operator $\dot{A}$. □

In the case of a regular closed symmetric operator $\dot{A}$ we can describe its symmetric (in particular, self-adjoint) extensions in terms other than those in Chapter 1. The following theorem gives a characterization of the regular extensions for a regular closed symmetric operator $\dot{A}$.

Theorem 2.5.5.

I. *For each closed symmetric extension A of a regular operator $\dot{A}$ there exists a* (1)*-isometric* (*see formula* (2.25) *for the definition of* (1)*-metric*) *operator $\mathcal{U} = \mathcal{U}(A)$ on $\mathfrak{M}_1$ with the properties:*

   (a) $\mathrm{Dom}(\mathcal{U})$ *is* (+)*-closed and belongs to* $\mathfrak{N} \oplus \mathfrak{N}'_i$, $\mathrm{Ran}(\mathcal{U}) \subset \mathfrak{N} \oplus \mathfrak{N}'_{-i}$;

   (b) $\mathcal{U}\psi = \psi$ *only for* $\psi = 0$, *and*

$$\begin{aligned} \mathrm{Dom}(A) &= \mathrm{Dom}(\dot{A}) \oplus (I - \mathcal{U})\mathrm{Dom}(\mathcal{U}), \\ A(g + (I - \mathcal{U})\psi)) &= \dot{A}g + \dot{A}^*(I - \mathcal{U})\psi + i(\dot{A}\dot{A}^* + I)P^+_{\mathfrak{N}}(I + \mathcal{U})\psi, \end{aligned} \tag{2.35}$$

   where $g \in \mathrm{Dom}(\dot{A})$, $\psi \in \mathrm{Dom}(\mathcal{U})$.

   Conversely, for each (1)*-isometric operator $\mathcal{U}$ with the properties* (a) *and* (b) *the operator A defined by* (2.35) *is a closed symmetric extension of $\dot{A}$.*

II. *The extension A is regular if and only if the manifold* $\mathrm{Ran}(I - \mathcal{U})$ *is* (1)*-closed.*

III. *The operator A is self-adjoint if and only if* $\mathrm{Dom}(\mathcal{U}) = \mathfrak{N} \oplus \mathfrak{N}'_i$, $\mathrm{Ran}(\mathcal{U}) = \mathfrak{N} \oplus \mathfrak{N}'_{-i}$.

Proof. Let $\dot{A}$ be a regular closed symmetric operator and A be its closed symmetric extension whose domain is defined by (2.29) via the corresponding operator U. Besides, $\mathrm{Dom}(A)$ admits (+)-orthogonal decomposition

$$\mathrm{Dom}(A) = \mathrm{Dom}(\dot{A}) \oplus P^+_{\mathfrak{M}}(\mathrm{Ran}(I - U)).$$

We introduce a new operator $\mathcal{U} = \mathcal{U}(A)$ by the formula

$$\mathcal{U} P^+_{\mathfrak{M}} \varphi = P^+_{\mathfrak{M}} U \varphi, \qquad \varphi \in \operatorname{Dom}(U). \tag{2.36}$$

Therefore,

$$\begin{aligned} \operatorname{Dom}(\mathcal{U}) &= P^+_{\mathfrak{M}} \operatorname{Dom}(U) \subset \mathfrak{N} \oplus \mathfrak{N}'_i, \\ \operatorname{Ran}(\mathcal{U}) &= P^+_{\mathfrak{M}} (\operatorname{Ran}(U)) \subset \mathfrak{N} \oplus \mathfrak{N}'_{-i}. \end{aligned}$$

It follows from Theorem 2.3.3 that $\operatorname{Dom}(\mathcal{U})$ is $(+)$-closed and thus the definition (2.36) makes sense. Since the operator $2^{-\frac{1}{2}} P^+_{\mathfrak{M}} \upharpoonright \mathfrak{N}_{\pm i}$ is a $(\cdot,1)$-isometry and U is a $(\cdot,\cdot)$-isometry, the operator $\mathcal{U}$ is isometric with respect to $\|\cdot\|_1$ metric, i.e., (1,1)-isometry. Let us assume that $\mathcal{U}\psi = \psi$ for some $\psi \in \operatorname{Dom}(\mathcal{U})$. Then $P^+_{\mathfrak{M}}\varphi - P^+_{\mathfrak{M}} U\varphi = 0$ for some $\varphi \in \operatorname{Dom}(U)$. This implies that $\varphi - U\varphi \in \operatorname{Dom}(\dot{A})$. Consequently, $\varphi = 0$ and $\psi = 0$.

From (2.27) we get the parametric expression for $\mathcal{U}$

$$\begin{cases} \psi = i(2P^+_{\mathfrak{N}'_i} + P^+_{\mathfrak{N}})(\dot{A}^* + iI)^{-1}\varphi \\ \mathcal{U}\psi = -i(2P^+_{\mathfrak{N}'_{-i}} + P^+_{\mathfrak{N}})(\dot{A}^* - iI)^{-1} U g \end{cases}, \ \varphi \in \operatorname{Dom}(U). \tag{2.37}$$

It follows that

$$\begin{cases} \varphi = -i(\dot{A}^* + iI)(\frac{1}{2}P^+_{\mathfrak{N}'_i} + P^+_{\mathfrak{N}})\psi \\ U\varphi = i(\dot{A}^* - iI)(\frac{1}{2}P^+_{\mathfrak{N}'_{-i}} + P^+_{\mathfrak{N}})\mathcal{U}\psi \end{cases}, \ \psi \in \operatorname{Dom}(\mathcal{U}). \tag{2.38}$$

Hence,

$$\begin{aligned} \varphi - U\varphi &= \psi - \mathcal{U}\psi - i\dot{A}^* P^+_{\mathfrak{N}}(I + \mathcal{U})\psi, \\ i(\varphi + U\varphi) &= i(I + \mathcal{U})\psi + \dot{A}^* P^+_{\mathfrak{N}}(I - \mathcal{U})\psi. \end{aligned}$$

Let $g \in \operatorname{Dom}(\dot{A})$, $\varphi \in \operatorname{Dom}(U)$. Then

$$A(g + (I - U)\varphi)) = \dot{A}g + i(I + U)\varphi.$$

Therefore

$$\begin{aligned} A(g + (I - \mathcal{U})\psi)) &= A(g + i\dot{A}^* P^+_{\mathfrak{N}}(I + \mathcal{U})\psi + \varphi - Ug) \\ &= \dot{A}g + i\dot{A}\dot{A}^* P^+_{\mathfrak{N}}(I + \mathcal{U})\psi + i(I + U)\varphi \\ &= \dot{A}g + i\dot{A}\dot{A}^* P^+_{\mathfrak{N}}(I + \mathcal{U})\psi + i(I + \mathcal{U})\psi + \dot{A}^* P^+_{\mathfrak{N}}(I - \mathcal{U})\psi \\ &= \dot{A}g + \dot{A}^*(I - \mathcal{U})\psi + i(\dot{A}\dot{A}^* + I)P^+_{\mathfrak{N}}(I + \mathcal{U})\psi. \end{aligned}$$

Conversely, let $\mathcal{U}$ be a (1,1)-isometry in the subspace $\mathfrak{M}$ with (1)-closed domain $\operatorname{Dom}(\mathcal{U}) \subset \mathfrak{N} \oplus \mathfrak{N}'_i$, and $\operatorname{Ran}(\mathcal{U}) \subset \mathfrak{N} \oplus \mathfrak{N}'_{-i}$ such that $\mathcal{U}\psi = \psi$ only for $\psi = 0$. Let the operator A be defined by (2.35). Then A is an extension of $\dot{A}$ and by direct calculations one can check that A is symmetric. Relations (2.38) define

$(\cdot,\cdot)$-isometry U with $\mathrm{Dom}(U) = \overline{\mathrm{Dom}(U)} \subset \mathfrak{N}_i$ and $\mathrm{Ran}(U) \subset \mathfrak{N}_{-i}$ such that (2.36) and consequently (2.29) hold.

It remains to show that U is an admissible operator. Suppose $U\varphi = V\varphi$ for some $\varphi \in \mathfrak{B}_i \cap \mathrm{Dom}(U)$. Since the element $\varphi - V\varphi \in \mathrm{Dom}(\dot{A})$ is $(+)$-orthogonal to $\mathfrak{M}$, then $P^+_{\mathfrak{M}}\varphi = P^+_{\mathfrak{M}}V\varphi$ and thus $P^+_{\mathfrak{M}}U\varphi = P^+_{\mathfrak{M}}\varphi$, and $\mathcal{U}(P^+_{\mathfrak{M}}\varphi) = P^+_{\mathfrak{M}}\varphi$. Therefore, $P^+_{\mathfrak{M}}\varphi = 0$ and $\varphi = 0$. □

Theorem 2.5.6. *A regular closed symmetric operator $\dot{A}$ has a regular self-adjoint extension if and only if its semi-deficiency indices are equal.*

Proof. Let A be a regular self-adjoint extension of $\dot{A}$. Then by Theorem 2.5.5 the corresponding operator $\mathcal{U} = \mathcal{U}(A)$ maps $\mathfrak{N} \oplus \mathfrak{N}'_i$ (1)-isometrically onto $\mathfrak{N} \oplus \mathfrak{N}'_{-i}$. Hence it is clear that the dimensions $\dim \mathfrak{N}'_{-i}$ and $\dim \mathfrak{N}'_i$ are the same and the semi-deficiency indices of $\dot{A}$ are equal.

Conversely, suppose $\dim \mathfrak{N}'_{-i} = \dim \mathfrak{N}'_i$. It is not hard to construct a (1)-isometry $\mathcal{U}$ in $\mathfrak{M}_1$ that maps $\mathfrak{N} \oplus \mathfrak{N}'_i$ onto $\mathfrak{N} \oplus \mathfrak{N}'_{-i}$ with number 1 as a regular point. For example, we can take $\mathcal{U} \upharpoonright \mathfrak{N}'_i$ to be arbitrary (1)-isometric operator with the range $\mathfrak{N}'_{-i}$ and set $\mathcal{U}h = \varepsilon h$, $h \in \mathfrak{N}$, where $|\varepsilon| = 1$, $\varepsilon \neq 1$. Then the corresponding operator A will become a regular self-adjoint extension of the operator $\dot{A}$. □

Theorem 2.5.7. *Let A be a regular self-adjoint extension of a regular symmetric operator $\dot{A}$. The following statements are valid:*

1. *the operator $PA \upharpoonright (\mathrm{Dom}(A) \cap \mathcal{H}_0)$ is self-adjoint in $\mathcal{H}_0$;*
2. *if $U \in [\mathfrak{N}_i, \mathfrak{N}_{-i}]$ is an admissible operator that determines A by formulas (2.29) and if*
$$\tilde{\mathfrak{N}}_i := \{\varphi \in \mathfrak{N}_i, (U - I)\varphi \in \mathcal{H}_0\}, \tag{2.39}$$
then
$$\mathcal{H}_+ = \mathrm{Dom}(A) \dotplus (U + I)\tilde{\mathfrak{N}}_i. \tag{2.40}$$

Proof. Let us set
$$\mathfrak{N}_i^0 := (A - iI)^{-1}\mathfrak{L}.$$
Then $\mathfrak{N}_i^0 \subset \mathfrak{N}_i$. Indeed, if $f \in \mathfrak{L}$ and $g \in \mathfrak{M}_{-i} = (\dot{A} + iI)\mathrm{Dom}(\dot{A})$, then
$$(g, (A - iI)^{-1}f) = ((A + iI)^{-1}g, f) = ((\dot{A} + iI)^{-1}g, f) = 0,$$
Since A is a regular self-adjoint extension of $\dot{A}$, then the operator A is $(+,\cdot)$-bounded and therefore the resolvent $(A - iI)^{-1}$ has the estimate
$$c_1||f|| \le ||(A - iI)^{-1}f||_+ \le c_2||f|| \quad \text{for all} \quad f \in \mathcal{H},$$
with $c > 0$. Therefore $\mathfrak{N}_i^0$ is a subspace in $\mathcal{H}_+$ (and simultaneously in $\mathcal{H}$).

Let $\tilde{\mathfrak{N}}_i$ be defined by (2.39). We are going to prove that
$$\mathfrak{N}_i^0 \oplus \tilde{\mathfrak{N}}_i = \mathfrak{N}_i, \tag{2.41}$$

where the sum is $(\cdot)$-orthogonal. By (2.29) we have

$$(A+iI)(U-I)\varphi = -i\varphi - iU\varphi + iU\varphi - i\varphi = -2i\varphi, \quad \varphi \in \mathfrak{N}_i,$$

and hence

$$(A+iI)^{-1}\varphi = -\frac{1}{2i}(U-I)\varphi.$$

This yields the equivalence relations

$$(U-I)\varphi \in \mathcal{H}_0 \iff (A+iI)^{-1}\varphi \perp \mathfrak{L} \iff \varphi \perp \mathfrak{N}_i^0,$$

and thus the decomposition (2.41) holds true. Thus, it is proved that

$$\mathrm{Dom}(A) \cap \mathcal{H}_0 = \mathrm{Dom}(\dot{A}) \dot{+} (U-I)\tilde{\mathfrak{N}}_i.$$

Suppose the vector $\psi \in \mathcal{H}_0$ is $(\cdot)$-orthogonal to $(PA+iI)(\mathrm{Dom}(A)\cap\mathcal{H}_0)$. Then $\psi \in \mathfrak{N}_i' \cap \mathfrak{N}_i^0$. Since $\mathfrak{N}_i^0 \subset \mathrm{Dom}(A)$, we get that $\psi \in \mathrm{Dom}(A)$ and (see (2.28)) $PA\psi = \dot{A}^*\psi = i\psi$. Because $PA\restriction(\mathrm{Dom}(A)\cap\mathcal{H}_0$ is symmetric, we get that $\psi = 0$. This means $(PA+iI)(\mathrm{Dom}(A)\cap\mathcal{H}_0)$ is dense in $\mathcal{H}_0$. Similarly it can be proved that $(PA-iI)(\mathrm{Dom}(A)\cap\mathcal{H}_0)$ is dense in $\mathcal{H}_0$. Thus the operator $(PA+iI)(\mathrm{Dom}(A)\cap\mathcal{H}_0)$ is essentially self-adjoint in $\mathcal{H}_0$. On the other hand, since A is a regular extension, Theorem 2.5.1 yields that the set $\mathrm{Dom}(A)\cap\mathcal{H}_0$ is $(+)$-closed. Consequently, the operator $(PA+iI)(\mathrm{Dom}(A)\cap\mathcal{H}_0)$ is $(\cdot)$-closed and hence is self-adjoint in $\mathcal{H}_0$. The first statement is proved.

Since for all $\phi \in \tilde{\mathfrak{N}}_i$, $(U+I)\phi = (U-I)\phi + 2\phi$, then $(U+I)\phi \in \mathrm{Dom}(A)$ implies $\phi \in \mathrm{Dom}(A)$. But then $PA\phi = \dot{A}^*\phi = iP\phi$ and thus $P(A-iI)\phi = 0$ or equivalently $(A-iI)\phi \in \mathfrak{L}$. Therefore, $\phi = 0$ and hence $\mathrm{Dom}(A)\cap(U+I)\tilde{\mathfrak{N}}_i = 0$. It is easy to see that $(U+I)\mathfrak{N}_i^0 \subset \mathrm{Dom}(A)$.

According to (2.10) $\mathrm{Dom}(\dot{A}^*) = \mathrm{Dom}(A) + \mathfrak{N}_i + \mathfrak{N}_{-i}$. Let

$$f = g + \varphi + \psi,$$

where $g \in \mathrm{Dom}(\dot{A})$, $\varphi \in \mathfrak{N}_i$, and $\psi \in \mathfrak{N}_{-i}$. Then

$$f = g + \frac{1}{2}(U-I)(U^{-1}\psi - \varphi) + \frac{1}{2}(U+I)(U^{-1}\varphi + \varphi) = x + (U+I)y,$$

where $x = g + \frac{1}{2}(U-I)(U^{-1}\psi - \varphi)$ and $y = \frac{1}{2}(U+I)(U^{-1}\varphi + \varphi)$. This implies that $\mathrm{Dom}(\dot{A}^*) \subseteq \mathrm{Dom}(A) + (U+I)\mathfrak{N}_i$. Since the inverse inclusion is obvious, we conclude that

$$\mathrm{Dom}(\dot{A}^*) = \mathrm{Dom}(A) + (U+I)\mathfrak{N}_i. \tag{2.42}$$

Combining (2.42), $\mathrm{Dom}(A)\cap(U+I)\tilde{\mathfrak{N}}_i = 0$, and $(U+I)\mathfrak{N}_i^0 \subset \mathrm{Dom}(A)$ we obtain

$$\mathrm{Dom}(\dot{A}^*) = \mathrm{Dom}(A) \dot{+} (U+I)\tilde{\mathfrak{N}}_i.$$

This completes the proof of statement 2. □

Theorem 2.5.8. *Let $\dot{A}$ be a regular O-operator. Then all its regular self-adjoint extensions are of the form*

$$\mathrm{Dom}(A) = \mathcal{H}_+, \quad A = \dot{A}P^+_{\mathrm{Dom}(\dot{A})} + (\dot{A}^* + (\dot{A}\dot{A}^* + I)^{-1}S)P^+_{\mathfrak{N}},$$

where S is an arbitrary $(+)$-self-adjoint and $(+)$-bounded operator in $\mathfrak{N}$.

Proof. We apply Theorem 2.5.5 for the case $\mathfrak{N}'_i = \mathfrak{N}'_{-i} = \{0\}$. Then there is a one-to-one correspondence between all (1)-unitary operators $\mathcal{U}$ in $\mathfrak{N}$ having the number 1 as a regular point and all regular self-adjoint extensions A of $\dot{A}$, which take the form (2.35). Let

$$S = i(I + \mathcal{U})(I - \mathcal{U})^{-1}.$$

Then $S \in [\mathfrak{N}, \mathfrak{N}]$, $S = S^*$, and (2.35) can be re-written as

$$\mathrm{Dom}(A) = \mathrm{Dom}(\dot{A}) \oplus \mathfrak{N} = \mathcal{H}_+, \; A = \dot{A}P^+_{\mathrm{Dom}(\dot{A})} + (\dot{A}^* + (\dot{A}\dot{A}^* + I)^{-1}S)P^+_{\mathfrak{N}}. \qquad \Box$$

Chapter 3

Bi-extensions of Closed Symmetric Operators

In this chapter we introduce and consider a new type of extensions of a given symmetric operator with its exit into the triplets of Hilbert spaces. These new extensions are called *bi-extensions*. We present a complete description and parameterizations of these bi-extensions. Special attention is paid to so-called twice self-adjoint (*t-self-adjoint*) bi-extensions that will play an important role in the remaining part of this text.

3.1 Bi-extensions

Definition 3.1.1. An operator $\mathbb{A} \in [\mathcal{H}_+, \mathcal{H}_-]$ is called a **bi-extension** of a symmetric operator $\dot{A}$ if $\mathbb{A} \supset \dot{A}$ and $\mathbb{A}^* \supset \dot{A}$. If, in addition, $\mathbb{A} = \mathbb{A}^*$, then the bi-extension $\mathbb{A}$ is called self-adjoint.

The class of all bi-extensions of $\dot{A}$ is denoted by $\mathcal{E}(\dot{A})$. By definition if $\mathbb{A} \in \mathcal{E}(\dot{A})$, then $\mathbb{A}^* \in \mathcal{E}(\dot{A})$. It is easy to see that $\mathcal{E}(\dot{A})$ is a convex set[1] in $[\mathcal{H}_+, \mathcal{H}_-]$. In particular, if $\mathbb{A} \in \mathcal{E}(\dot{A})$, then its real part $\operatorname{Re}\mathbb{A} = (1/2)(\mathbb{A} + \mathbb{A}^*)$ also belongs to $\mathcal{E}(\dot{A})$.

Let $\mathbb{A} \in \mathcal{E}(\dot{A})$ and $\mathbb{A}' \in [\mathcal{H}_+, \mathcal{H}_-]$. Then $\mathbb{A}'$ belongs to $\mathcal{E}(\dot{A})$ if and only if

$$\operatorname{Dom}(\dot{A}) \subset \ker(\mathbb{A}' - \mathbb{A}) \quad \text{and} \quad \operatorname{Dom}(\dot{A}) \subset \ker(\mathbb{A}'^* - \mathbb{A}^*).$$

The last inclusion is equivalent to

$$\operatorname{Ran}(\mathbb{A}' - \mathbb{A}) \subseteq \mathcal{F},$$

[1] That means for any $\mathbb{A}_1, \mathbb{A}_2 \in \mathcal{E}(\dot{A})$ all the operators of the form $\lambda\mathbb{A}_1 + (1-\lambda)\mathbb{A}_2$ belong to $\mathcal{E}(\dot{A})$ for any $\lambda \in \mathbb{C}$.

where $\mathcal{F}(\subset \mathcal{H}_-)$ is the $(\cdot)$-orthogonal complement of $\mathrm{Dom}(\dot{A})$ given by (2.17) and (2.18). Hence, in particular, the imaginary part $\mathrm{Im}\,\mathbb{A} = (1/2i)(\mathbb{A} - \mathbb{A}^*)$ of an operator $\mathbb{A} \in \mathcal{E}(\dot{A})$ satisfies the condition

$$\mathrm{Dom}(\dot{A}) \subset \ker(\mathrm{Im}\,\mathbb{A}) \iff \mathrm{Ran}(\mathrm{Im}\,\mathbb{A}) \subset \mathcal{F}. \tag{3.1}$$

Let $\mathbb{A}$, $\mathbb{A}' \in \mathcal{E}(\dot{A})$. The operator $(\mathbb{A}' - \mathbb{A})\upharpoonright \mathfrak{M}$ is $(+,-)$-continuous and its range belongs to $\mathcal{F}$. Hence the operator

$$S = \mathcal{R}(\mathbb{A}' - \mathbb{A})\upharpoonright \mathfrak{M}, \tag{3.2}$$

is $(+,+)$-continuous as acting from $\mathfrak{M}$ into $\mathfrak{M}$, i.e., $S \in [\mathfrak{M}, \mathfrak{M}]$. Here $\mathcal{R}$ is the Riesz-Berezansky operator of the triplet $\mathcal{H}_+ \subset \mathcal{H} \subset \mathcal{H}_-$. Let $P^+_{\mathfrak{M}}$ be the $(+)$-orthogonal projection in $\mathcal{H}_+$ onto $\mathfrak{M}$ described by (2.16). We know that $\mathrm{Dom}(\dot{A})$ is $(+)$-orthogonal to $\mathfrak{M}$, and thus

$$\mathbb{A}' - \mathbb{A} = \mathcal{R}^{-1} S P^+_{\mathfrak{M}}.$$

It is easy to check that the inverse statement is also true. That is, if S in (3.2) belongs to $[\mathfrak{M}, \mathfrak{M}]$, then $\mathbb{A} + \mathcal{R}^{-1} S P^+_{\mathfrak{M}} \in \mathcal{E}(\dot{A})$. Consequently, in order to describe the entire class $\mathcal{E}(\dot{A})$ we need only one representative operator of this class. Below we construct two "particular" operators of the class $\mathcal{E}(\dot{A})$.

Let us recall the spaces $\mathcal{D}$ and $\mathcal{D}^*$ introduced in (2.11)

$$\mathcal{D} = \overline{\mathrm{Dom}(\dot{A})}^{(+)}, \quad \mathcal{D}^* = \mathcal{H}_+ \cap \mathcal{H}_0, \quad (\mathcal{H}_0 = \overline{\mathrm{Dom}(\dot{A})}),$$

and $\bar{\dot{A}}$ be the $(\cdot,-)$-continuity extension of the operator $\dot{A}$ onto $\mathcal{H}_0$ described in Theorem 2.2.1. Let us show that

$$(\bar{\dot{A}} f, g) = (f, \bar{\dot{A}} g), \qquad (f \in \mathcal{D},\ g \in \mathcal{D}^*). \tag{3.3}$$

There exists a sequence $\{f_n\} \subset \mathrm{Dom}(\dot{A})$ such that $f_n \xrightarrow{(+)} f$, or $f_n \xrightarrow{(\cdot)} f$ and $\dot{A}^* f_n \xrightarrow{(\cdot)} \dot{A}^* f$. We have that $(\dot{A} f_n, g) = (f_n, \dot{A}^* g)$ and since $g \in \mathcal{H}_0$,

$$(f_n, \dot{A}^* g) = (\dot{A} f_n, g) = (P\dot{A} f_n, g) = (\dot{A}^* f_n, g).$$

Letting $n \to \infty$ we get $(f, \dot{A}^* g) = (\dot{A}^* f, g)$. Applying (2.8) we obtain (3.3).

Now we use formulas (2.9) and (2.10) with $\lambda = i$. Following our convention above we will denote by $P^+_{\mathcal{G}}$ a $(+)$-orthoprojection operator in $\mathcal{H}_+$ onto a subspace $\mathcal{G}$ of $\mathcal{H}_+$. We set

$$\mathbb{A}_0 = \bar{\dot{A}} P^+_{\mathcal{D}} + \dot{A}^* P^+_{\mathfrak{M}}, \qquad \mathbb{A}_1 = \bar{\dot{A}} P^+_{\mathcal{D}^*} + \dot{A}^* P^+_{\mathfrak{N}}. \tag{3.4}$$

Since the operator $\bar{\dot{A}}$ is $(\cdot,-)$-continuous while $\dot{A}^*$ is $(+,\cdot)$-continuous, then both $\mathbb{A}_0$ and $\mathbb{A}_1$ belong to $[\mathcal{H}_+, \mathcal{H}_-]$. It is clear that $\mathbb{A}_0 \supset \dot{A}$ and $\mathbb{A}_1 \supset \dot{A}$. Relying on

(3.4) and using (2.8) and (3.3) we have, for arbitrary $f, g \in \mathcal{H}_+$,

$$\begin{aligned}(\mathbb{A}_0 f, g) &= (\overline{\dot{A}} P^+_{\mathcal{D}} f + \dot{A}^* P^+_{\mathfrak{M}} f, P^+_{\mathcal{D}^*} g + P^+_{\mathfrak{N}} g)\\ &= (\overline{\dot{A}} P^+_{\mathcal{D}} f, P^+_{\mathcal{D}^*} g) + (\overline{\dot{A}} P^+_{\mathcal{D}} f, P^+_{\mathfrak{N}} g) + (\dot{A}^* P^+_{\mathfrak{M}} f, P^+_{\mathcal{D}_*} g) + (\dot{A}^* P^+_{\mathfrak{M}} f, P^+_{\mathfrak{N}} g)\\ &= (P^+_{\mathcal{D}} f, \overline{\dot{A}} P^+_{\mathcal{D}^*} g) + (P^+_{\mathcal{D}} f, \dot{A}^* P^+_{\mathfrak{N}} g) + (P^+_{\mathfrak{M}} f, \overline{\dot{A}} P^+_{\mathcal{D}^*} g) + (P^+_{\mathfrak{M}} f, \dot{A}^* P^+_{\mathfrak{N}} g)\\ &= (P^+_{\mathcal{D}} f + P^+_{\mathfrak{M}} f, \overline{\dot{A}} P^+_{\mathcal{D}^*} g + \dot{A}^* P^+_{\mathfrak{N}} g) = (f, \mathbb{A}_1 g).\end{aligned}$$

This implies the equality

$$\mathbb{A}_0^* = \mathbb{A}_1, \tag{3.5}$$

hence, $\mathbb{A}_0, \mathbb{A}_1 \in \mathcal{E}(\dot{A})$. All the above reasoning can be summarized in the following theorem:

Theorem 3.1.2. *Let $\dot{A}$ be a symmetric operator in $\mathcal{H}$. The relations*

$$\mathbb{A} = \mathbb{A}_0 + \mathcal{R}^{-1} S_{\mathbb{A}} P^+_{\mathfrak{M}}, \quad S_{\mathbb{A}} = \mathcal{R}(\mathbb{A} - \mathbb{A}_0), \tag{3.6}$$

set a bijective mapping $\mathbb{A} \leftrightarrow S_{\mathbb{A}}$ between the classes $\mathcal{E}(\dot{A})$ and $[\mathfrak{M}, \mathfrak{M}]$. The adjoint operator $\mathbb{A}^*$ is given by

$$\mathbb{A}^* = \mathbb{A}_1 + \mathcal{R}^{-1} S^*_{\mathbb{A}} P^+_{\mathfrak{M}}.$$

We recall from Chapter 2 that $\mathfrak{L} = \mathcal{H} \ominus \mathcal{H}_0$ (see (2.12)), where $\mathcal{H}_0 = \overline{\mathrm{Dom}(\dot{A})}$ and $\mathcal{F}$ ($\subset \mathcal{H}_-$) is the $(\cdot)$-orthogonal complement of $\mathrm{Dom}(\dot{A})$ ($\subset \mathcal{H}_+$) given by (2.18). It follows from Theorem 2.1.1 that $\mathcal{F} \cap \mathcal{H} = \mathfrak{L}$, and hence, $\mathcal{F} \cap \mathcal{H}_0 = \{0\}$. The linear set

$$\mathcal{B} = \mathcal{H} + \mathcal{F} = \mathcal{H}_0 \dot{+} \mathcal{F}, \tag{3.7}$$

is $(-)$-dense in $\mathcal{H}_-$. Let Π be a projection operator in $\mathcal{B}$ onto $\mathcal{H}_0$ parallel to $\mathcal{F}$. It is easy to see that $\Pi \restriction \mathcal{H}$ coincides with P, the orthogonal projection operator from $\mathcal{H}$ onto $\mathcal{H}_0$. One can also show that Π is a $(-,\cdot)$-closed operator.

Lemma 3.1.3. 1. *A vector $f \in \mathcal{H}_-$ belongs to the set $\mathcal{B}$ of the form (3.7) if and only if the functional $F_f(\varphi) = (\varphi, f)$ is $(\cdot)$-continuous on* $\mathrm{Dom}(\dot{A})$.

2. *Let $f \in \mathcal{H}_0$. Then an element $\overline{\dot{A}} f$ belongs to the set $\mathcal{B}$ if and only if $f \in \mathcal{D}^* = \mathcal{H}_+ \cap \mathcal{H}_0$. Moreover,*

$$\Pi \overline{\dot{A}} f = \dot{A}^* f.$$

Proof. 1. If $f \in \mathcal{B}$, $f = f_1 + f_2$, where $f_1 = \Pi f \in \mathcal{H}_0$ and $f_2 \in \mathcal{F}$, then $(\varphi, f) = (\varphi, f_1)$ is a $(\cdot)$-continuous functional with respect to $\varphi \in \mathrm{Dom}(\dot{A})$.

Conversely, let $F_f(\varphi) = (\varphi, f)$ be a $(\cdot)$-continuous on $\mathrm{Dom}(\dot{A})$ functional. Then there exists an element $f_1 \in \overline{\mathrm{Dom}(\dot{A})} = \mathcal{H}_0$ such that $F_f(\varphi) = (\varphi, f_1)$. Thus, $f - f_1 \in \mathcal{F}$ and $f = f_1 + f_2 \in \mathcal{B}$.

2. According to statement 1. the condition $\overline{\dot{A}} f \in \mathcal{B}$ is equivalent to $(\overline{\dot{A}} f, \varphi)$ being a $(\cdot)$-continuous with respect to $\varphi \in \mathrm{Dom}(\dot{A})$ functional. Since

$$(\overline{\dot{A}} f, \varphi) = (f, \dot{A}^* \varphi) = (f, A^* \varphi) = (f, \Pi \dot{A} \varphi) = (f, \dot{A} \varphi),$$

then the $(\cdot)$-continuity of this functional is equivalent to $f \in \mathcal{H}_+$. Consequently, $(f, \dot{A}\varphi) = (\dot{A}^* f, \varphi)$. Thus, $\overline{\dot{A}} f \in \mathcal{B}$ if and only if $f \in \mathcal{H}_+ \cap \mathcal{H}_0$ and besides $(\overline{\dot{A}} f, \varphi) = (\dot{A}^* f, \varphi)$ for all $\varphi \in \mathrm{Dom}(\dot{A})$. That is why $\overline{\dot{A}} f - \dot{A}^* f \in \mathcal{F}$ and $\Pi \overline{\dot{A}} f = \dot{A}^* f$. $\square$

Theorem 3.1.4. *If* $\mathbb{A}$ *belongs to* $\mathcal{E}(\dot{A})$, *then* $\mathrm{Ran}(\mathbb{A}) \subset \mathcal{B}$ *and* $\Pi\mathbb{A} = \dot{A}^*$. *Conversely, if* $\mathbb{A} \in [\mathcal{H}_+, \mathcal{H}_-]$, $\mathrm{Ran}(\mathbb{A}) \subset \mathcal{B}$, $\mathrm{Ran}(\mathbb{A}^*) \subset \mathcal{B}$, *and* $\Pi\mathbb{A} = \Pi\mathbb{A}^* = \dot{A}^*$, *then* $\mathbb{A} \in \mathcal{E}(\dot{A})$.

Proof. By Lemma 3.1.3 we have $A\mathcal{D}_* \subset \mathcal{B}$ and $\Pi \overline{\dot{A}} f = \dot{A}^* f$ for all $f \in \mathcal{D}_*$. Taking into account that $\mathrm{Ran}(\dot{A}^*) \subset \mathcal{H}_0$ and relying on (3.4) we get

$$\mathbb{A}_1 f = \overline{\dot{A}} P^+_{\mathcal{D}^*} f + \dot{A}^* P^+_{\mathfrak{N}} f \in \mathcal{B},$$

and

$$\Pi\mathbb{A}_1 f = \Pi\overline{\dot{A}} P^+_{\mathcal{D}^*} f + \dot{A}^* P^+_{\mathfrak{N}} f = \dot{A}^*(P^+_{\mathcal{D}^*} f + P^+_{\mathfrak{N}} f) = \dot{A}^* f.$$

Now let $\mathbb{A}$ be an arbitrary operator in $\mathcal{E}(\dot{A})$. The inclusion $\mathrm{Ran}(\mathbb{A} - \mathbb{A}_1) \subset \mathcal{F}$ (see (3.1)) and the equality $\Pi\mathcal{F} = 0$ imply that

$$\mathrm{Ran}(\mathbb{A}) \subset \mathcal{B} + \mathcal{F} = \mathcal{B},$$

and

$$\Pi\mathbb{A} f = \Pi\mathbb{A}_1 f = \dot{A}^* f, \ f \in \mathcal{H}_+.$$

Conversely, let $\mathbb{A} \in [\mathcal{H}_+, \mathcal{H}_-]$, $\mathrm{Ran}(\mathbb{A}) \subset \mathcal{B}$, and $\Pi\mathbb{A} = \dot{A}^*$. Then $\mathbb{A}f - \dot{A}^* f \in \mathcal{F}$ for all $f \in \mathcal{H}_+$. If $g \in \mathrm{Dom}(\dot{A})$, then

$$(\mathbb{A}f, g) = (\dot{A}^* f, g) = (f, \dot{A}g).$$

This implies that $\mathbb{A}^* g = \dot{A}g$, i.e., $\mathbb{A}^* \supset \dot{A}$. Similarly conditions $\mathrm{Ran}(\mathbb{A}) \subset \mathcal{B}$ and $\Pi\mathbb{A} = \dot{A}^*$ imply $\mathbb{A} \supset \dot{A}$. That is why, if $\mathrm{Ran}(\mathbb{A})$, $\mathrm{Ran}(\mathbb{A}^*) \subset \mathcal{B}$ and $\Pi\mathbb{A} = \Pi\mathbb{A}^* = \dot{A}^*$ then $\mathbb{A}^* \supset \dot{A}$, $\mathbb{A} \supset \dot{A}$, and thus $\mathbb{A} \in \mathcal{E}(\dot{A})$. $\square$

Let $\mathbb{A} \in [\mathcal{H}_+, \mathcal{H}_-]$. We recall (see (2.5)) that the operator

$$\mathrm{Dom}(\hat{A}) = \{f \in \mathcal{H}_+ : \mathbb{A}f \in \mathcal{H}\}, \ \hat{A} = \mathbb{A}\restriction \mathrm{Dom}(\hat{A})$$

is called the quasi-kernel of $\mathbb{A}$.

Theorem 3.1.5. *If* $\mathbb{A}$ *belongs to the class* $\mathcal{E}(\dot{A})$, *then the quasi-kernel* $\hat{A}$ *of the operator* $\mathbb{A}$ *is* $(\cdot,\cdot)$*-closed.*

Proof. Let $\{f_n\} \subset \mathrm{Dom}(\hat{A})$ and $f_n \xrightarrow{(\cdot)} f$, while

$$\mathbb{A}f_n = \hat{A}f_n \xrightarrow{(\cdot)} g.$$

We recall that $\Pi\restriction\mathcal{H} = P$. According to Theorem 3.1.4 $\dot{A}^* f_n \to Pg$. Since $\dot{A}^*$ is $(\cdot,\cdot)$-closed, $f \in \mathcal{H}_+$ and $Py = \dot{A}^* f$. Therefore, $f_n \xrightarrow{(+)} f$ and hence $\mathbb{A}f_n \xrightarrow{(-)} \mathbb{A}f$. On the other hand $\mathbb{A}f_n = \hat{A}f_n \xrightarrow{(\cdot)} g \in \mathcal{H}$ and so $\mathbb{A}f = g \in \mathcal{H}$, $f \in \mathrm{Dom}(\hat{A})$, and $g = \hat{A}f$. This proves that the operator $\hat{A}$ is closed. $\square$

Now we will study continuity of operators from the class $\mathcal{E}(\dot{A})$ in $(+,\cdot)$- and $(\cdot,-)$-topologies.

Theorem 3.1.6. *The following statements about a closed symmetric operator $\dot{A}$ and its bi-extensions $\mathbb{A}_0$ and $\mathbb{A}_1$ are equivalent:*

(i) *$\dot{A}$ is a regular operator;*

(ii) *$\mathbb{A}_0$ is $(+,\cdot)$-continuous;*

(iii) *$\mathbb{A}_1$ is $(\cdot,-)$-continuous.*

Proof. First we will show that (i) implies (ii). For a regular operator $\dot{A}$ we have $\mathcal{D} = \mathrm{Dom}(\dot{A})$ and thus in the formula (3.4) for $\mathbb{A}_0$ one can replace $\overline{\dot{A}}$ by $\dot{A}$. Then $\mathrm{Ran}(\mathbb{A}_0) \subset \mathcal{H}$ and hence $\mathbb{A}_0 \in [\mathcal{H}_+,\mathcal{H}]$. Taking into account that $\mathbb{A}_0 \supset \dot{A}$ we conclude that (ii) implies (i). The equivalence of (ii) and (iii) follows from the fact that $\mathbb{A}_0 = \mathbb{A}_1^*$. □

We note that if $\dot{A}$ is regular, then $\mathbb{A}_0 = \dot{B}^*$ (see (2.20)), where $\dot{B}$ is a closed symmetric operator with dense domain in $\mathcal{H}$ defined by (2.19). The following theorem refines the result of Theorem 3.1.6.

Theorem 3.1.7. *If $\dot{A}$ is densely defined in $\mathcal{H}$, i.e., $\overline{\mathrm{Dom}(\dot{A})} = \mathcal{H}$, then $\mathcal{E}(\dot{A})$ contains a unique $(+,\cdot)$-continuous operator $\mathbb{A}_0 = \dot{A}^*$ and a unique $(\cdot,-)$-continuous operator*

$$\mathbb{A}_1 = \overline{\dot{A}}\restriction \mathcal{D}^* = \overline{\dot{A}}\restriction \mathcal{H}_+.$$

Proof. According to Theorem 2.4.1 a closed densely defined symmetric operator $\dot{A}$ is automatically regular, $\mathcal{D} = \mathrm{Dom}(\dot{A})$, $\mathfrak{N} = 0$, $\mathcal{D}^* = \mathcal{H}_+$, and from (3.5)

$$\mathbb{A}_0 = \dot{A}P^+_{\mathcal{D}} + \dot{A}^* P^+_{\mathfrak{M}} = \dot{A}^*, \qquad \mathbb{A}_1 = \mathbb{A}_0^* = \overline{\dot{A}}\restriction \mathcal{D}^* = \overline{\dot{A}}\restriction \mathcal{H}_+.$$

It follows that $\mathbb{A}_0$ is $(+,\cdot)$-continuous while $\mathbb{A}_1$ is $(\cdot,-)$-continuous.

If an operator $\mathbb{A}$ from the class $\mathcal{E}(\dot{A})$ is $(\cdot,-)$-continuous, then, being an extension of $\dot{A}$, it coincides with $\overline{\dot{A}}$ on $\mathcal{H}_+$, i.e., $\mathbb{A} = \mathbb{A}_1$. Consequently, $\mathbb{A}_1$ is the unique $(\cdot,-)$-continuous operator in the class $\mathcal{E}(\dot{A})$. If an operator $\mathbb{A} \in \mathcal{E}(\dot{A})$ is $(+,\cdot)$-continuous, then $\mathbb{A}^* \in \mathcal{E}(\dot{A})$ is $(\cdot,-)$-continuous. Then $\mathbb{A}^* = \mathbb{A}_1$ and $\mathbb{A} = \mathbb{A}_0$, i.e., $\mathbb{A}_0$ is unique $(+,\cdot)$-continuous of the class $\mathcal{E}(\dot{A})$. □

We should also note that if a closed symmetric operator $\dot{A}$ is not self-adjoint, then $\mathbb{A}_0$ and $\mathbb{A}_1$ are not self-adjoint in $[\mathcal{H}_+,\mathcal{H}_-]$. Thus, in the class of all bi-extensions of the operator $\dot{A}$ that are self-adjoint operators in $[\mathcal{H}_+,\mathcal{H}_-]$ there are no $(+,\cdot)$- and $(\cdot,-)$-continuous operators.

Theorem 3.1.8. *Let $\dot{A}$ be a regular closed symmetric operator, $\mathbb{A} \in \mathcal{E}(\dot{A})$, and $\hat{A}$ is a quasi-kernel of $\mathbb{A}$. Then $\hat{A}$ is $(+,\cdot)$-continuous and its domain $\mathrm{Dom}(\hat{A})$ is $(+)$-closed.*

Proof. According to Theorem 3.1.2 the operator $\mathbb{A}$ admits the representation (3.6) for some $S_{\mathbb{A}} \in [\mathfrak{M}, \mathfrak{M}]$. If $\dot{A}$ is a regular operator then, according to Theorem 3.1.6, $\mathrm{Ran}(\mathbb{A}_0) \subset \mathcal{H}$. That is why the expression $\mathbb{A}f \in \mathcal{H}$ is equivalent to $\mathcal{R}^{-1}S_{\mathbb{A}}P^+_{\mathfrak{M}}f \in \mathcal{H}$. Since $\mathcal{R}^{-1}\mathfrak{M} \cap \mathcal{H} = \mathcal{F} \cap \mathcal{H}$ we have that

$$\mathcal{R}^{-1}S_{\mathbb{A}}P^+_{\mathfrak{M}}f \in \mathfrak{L}, \qquad S_{\mathbb{A}}P^+_{\mathfrak{M}}f \in \mathcal{R}\mathfrak{L},\ f \in \mathcal{H}_+.$$

In the case of a regular operator $\dot{A}$ we have $\mathcal{R}\mathfrak{L} = \mathfrak{N}$. Therefore, $\mathbb{A}f \in \mathcal{H}$ if and only if $P^+_{\mathfrak{M}}f \in S^{-1}_{\mathbb{A}}(\mathfrak{N}) = G$, where $G = \{f \in \mathfrak{M} \mid S_{\mathbb{A}}f \in \mathfrak{N}\}$. The latter linear manifold is (+)-closed as a complete pre-image of the (+)-closed linear manifold $\mathfrak{N}$. Thus the set of such $f \in \mathcal{H}_+$ that $\mathbb{A}f \in \mathcal{H}$ coincides with the (+)-closed linear manifold $\mathrm{Dom}(\dot{A}) \oplus G$. Using Theorem 3.1.5 we get that the operator $\hat{A}$ is $(\cdot,\cdot)$-closed and hence $(+,\cdot)$-closed. But it is defined on a (+)-closed linear manifold and hence $(+,\cdot)$-continuous. □

Theorem 3.1.9. *The formula*

$$\mathbb{A} = \overline{\dot{A}}P^+_{\mathcal{D}} + (\dot{A}^* + \mathcal{R}^{-1}S)P^+_{\mathfrak{M}} - \frac{i}{2}\mathcal{R}^{-1}P^+_{\mathfrak{N}'_i} + \frac{i}{2}\mathcal{R}^{-1}P^+_{\mathfrak{N}'_{-i}}, \tag{3.8}$$

establishes a bijective correspondence between all operators $\mathbb{A} \in \mathcal{E}(\dot{A})$ and all operators $S \in [\mathfrak{M}, \mathfrak{M}]$. The adjoint operator $\mathbb{A}^*$ is given by

$$\mathbb{A}^* = \overline{\dot{A}}P^+_{\mathcal{D}} + (\dot{A}^* + \mathcal{R}^{-1}S^*)P^+_{\mathfrak{M}} - \frac{i}{2}\mathcal{R}^{-1}P^+_{\mathfrak{N}'_i} + \frac{i}{2}\mathcal{R}^{-1}P^+_{\mathfrak{N}'_{-i}},$$

where $S^* \in [\mathfrak{M}, \mathfrak{M}]$ is (+)-adjoint to S. In particular, the operator $\mathbb{A} \in \mathcal{E}(\dot{A})$ is self-adjoint if and only if $S = S^*$.

Proof. From (3.4) and (3.5) we have

$$\mathbb{A}_1 - \mathbb{A}_0 = \overline{\dot{A}}(P^+_{\mathcal{D}_*} - P^+_{\mathcal{D}}) - \dot{A}^*(P^+_{\mathfrak{M}} - P^+_{\mathfrak{N}}) = (\overline{\dot{A}} - \dot{A}^*)(P^+_{\mathfrak{N}'_i} + P^+_{\mathfrak{N}'_{-i}}).$$

By virtue of Theorem 2.2.1, if $\varphi \in \mathfrak{N}'_{\pm i}$, then

$$\mathcal{R}^{-1}\varphi = \varphi + \overline{\dot{A}}\dot{A}^*\varphi = \varphi \pm i\overline{\dot{A}}\varphi = \pm i(\overline{\dot{A}}\varphi \mp i\varphi) = \pm i(\overline{\dot{A}}\varphi - \dot{A}^*\varphi).$$

Therefore

$$\mathbb{A}_1 - \mathbb{A}_0 = -i\mathcal{R}^{-1}(P^+_{\mathfrak{N}'_i} - P^+_{\mathfrak{N}'_{-i}}). \tag{3.9}$$

Now we will establish a connection between $S_{\mathbb{A}}$ and $S_{\mathbb{A}^*}$ given by (3.6). Let S^* be the operator adjoint to $S \in [\mathfrak{M}, \mathfrak{M}]$ with respect to the (+)-metric. Let $\mathfrak{J}$ be the operator in $\mathfrak{M}$ defined by the formula

$$\mathfrak{J} = (P^+_{\mathfrak{N}'_i} - P^+_{\mathfrak{N}'_{-i}}) \upharpoonright \mathfrak{M}, \tag{3.10}$$

Notice that $\mathfrak{J}$ is (+)-self-adjoint and $\mathfrak{J}^2 = P^+_{\mathfrak{N}'_i} + P^+_{\mathfrak{N}'_{-i}}$. By Theorem 3.1.2 each operator $\mathbb{A} \in \mathcal{E}(\dot{A})$ takes the form

$$\mathbb{A} = \mathbb{A}_0 + \mathcal{R}^{-1} S_{\mathbb{A}} P^+_{\mathfrak{M}},$$

where $S \in [\mathfrak{M}, \mathfrak{M}]$ and $\mathbb{A}^* = \mathbb{A}_1 + \mathcal{R}^{-1} S^*_{\mathbb{A}} P^+_{\mathfrak{M}}$. Define the operator $S \in [\mathfrak{M}, \mathfrak{M}]$:

$$S = S_{\mathbb{A}} + \frac{i}{2}\mathfrak{J},$$

where $\mathfrak{J}$ is defined by (3.10). Then $\mathbb{A} = \mathbb{A}_0 + \mathcal{R}^{-1}(S - \frac{i}{2}\mathfrak{J})P^+_{\mathfrak{M}}$. Using (3.9) one has

$$\mathbb{A}^* = \mathbb{A}_0 + \mathcal{R}^{-1}(S^* - \frac{i}{2}\mathfrak{J})P^+_{\mathfrak{M}}.$$

Clearly, $S_{\mathbb{A}} = S - \frac{i}{2}\mathfrak{J}$, $S_{\mathbb{A}^*} = S^* - \frac{i}{2}\mathfrak{J}$, and $\mathbb{A}^* = \mathbb{A}$ if and only if $S = S^*$. □

Observe that relation (3.9) for the case of a densely defined operator $\dot{A}$ can be rewritten as

$$(\overline{\dot{A}} - \dot{A}^*)f = -\mathcal{R}^{-1}\dot{A}^* P^+_{\mathfrak{M}} f, \; f \in \mathcal{H}_+.$$

Due to Theorem 2.4.1 in the case of a regular operator $\dot{A}$ we can replace $\overline{\dot{A}}$ by $\dot{A}$ in the formula (3.8) and obtain

$$\mathbb{A} = \dot{A}P^+_{\mathrm{Dom}(\dot{A})} + (\dot{A}^* + \mathcal{R}^{-1}S)P^+_{\mathfrak{M}} - \frac{i}{2}\mathcal{R}^{-1}P^+_{\mathfrak{N}'_i} + \frac{i}{2}\mathcal{R}^{-1}P^+_{\mathfrak{N}'_{-i}}. \tag{3.11}$$

In addition, using the operator $\dot{B}$ and its adjoint $\dot{B}^*$ given by (2.19) and (2.20) we can rewrite the right-hand side of (3.11) as

$$\mathbb{A} = \dot{B}^* + \mathcal{R}^{-1}\left(S - \frac{i}{2}P^+_{\mathfrak{N}'_i} + \frac{i}{2}P^+_{\mathfrak{N}'_{-i}}\right)P^+_{\mathfrak{M}}. \tag{3.12}$$

Remark 3.1.10. If the (+)-inner product in $\mathcal{H}_+$ is replaced by the inner product $(f,g)_1 = ((I + P^+_{\mathfrak{N}})f, g)_+$, then the operator $\mathcal{R}$ must be replaced by the operator $\boldsymbol{\mathcal{R}} = (I - \frac{1}{2}P^+_{\mathfrak{N}})\mathcal{R}$ (see figure 2.1, (2.25) and (2.26)). It follows that if $\mathbb{A}$ is given by (3.8), then one has to replace $\mathcal{R}^{-1}$ by $\boldsymbol{\mathcal{R}}^{-1}$ and the operator S in $\mathfrak{M}$ by the operator $S_1 = (I - P^+_{\mathfrak{N}}/2)S$. Therefore $\mathbb{A}$ and $\mathbb{A}^*$ are of the form

$$\begin{aligned}\mathbb{A} &= \overline{\dot{A}}P^+_{\mathcal{D}} + (\dot{A}^* + \boldsymbol{\mathcal{R}}^{-1}S_1)P^+_{\mathfrak{M}} - \tfrac{i}{2}\boldsymbol{\mathcal{R}}^{-1}P^+_{\mathfrak{N}'_i} + \tfrac{i}{2}\boldsymbol{\mathcal{R}}^{-1}P^+_{\mathfrak{N}'_{-i}},\\ \mathbb{A}^* &= \overline{\dot{A}}P^+_{\mathcal{D}} + (\dot{A}^* + \boldsymbol{\mathcal{R}}^{-1}S_1^*)P^+_{\mathfrak{M}} - \tfrac{i}{2}\boldsymbol{\mathcal{R}}^{-1}P^+_{\mathfrak{N}'_i} + \tfrac{i}{2}\boldsymbol{\mathcal{R}}^{-1}P^+_{\mathfrak{N}'_{-i}},\end{aligned}$$

where $S_1^* \in [\mathfrak{M}, \mathfrak{M}]$ is (1)-adjoint to S_1. Moreover, $\mathbb{A} = \mathbb{A}^* \iff S_1 = S_1^*$.

In sequel we will use the parametrization of all bi-extensions of a regular symmetric operator $\dot{A}$ with non-zero semi-deficiency numbers in the form

$$\mathbb{A} = \dot{A}P^+_{\mathrm{Dom}(\dot{A})} + \left(\dot{A}^* + \boldsymbol{\mathcal{R}}^{-1}(S - \frac{i}{2}\mathfrak{J})\right)P^+_{\mathfrak{M}}, \tag{3.13}$$

where $\mathfrak{J}$ is given by (3.10). Clearly, if $\dot{A}$ is densely defined, then $\mathfrak{N} = \{0\}$ and $\mathcal{R} = \mathcal{R}$.

Proposition 3.1.11. *Let $\dot{A}$ be a regular closed symmetric operator and let $\mathbb{A}$ be a bi-extension of $\dot{A}$ given by (3.13). Define a subspace G_S in $\mathfrak{M}$ as*

$$G_S = \ker\left((P^+_{\mathfrak{N}'_i} + P^+_{\mathfrak{N}'_{-i}})S - \frac{i}{2}\mathfrak{J}\right). \tag{3.14}$$

Then the domain of the quasi-kernel $\hat{A}$ of $\mathbb{A}$ can be described as

$$\begin{aligned} &\mathrm{Dom}(\hat{A}) = \mathrm{Dom}(\dot{A}) \oplus G_S, \\ &\hat{A}(g+f) = \dot{A}g + \dot{A}^*f + \mathcal{R}^{-1}P^+_{\mathfrak{N}}Sf, \\ &g \in \mathrm{Dom}(\dot{A}), f \in G_S. \end{aligned} \tag{3.15}$$

Proof. From (3.13) it follows that if $f \in \mathfrak{M}$, then

$$\mathbb{A}f \in \mathcal{H} \iff \mathcal{R}^{-1}\left(S - \frac{i}{2}\mathfrak{J}\right)f \in \mathcal{H}.$$

On the other hand, every vector from $\mathcal{F} = \mathcal{R}^{-1}(\mathfrak{M})$ is $(\cdot)$-orthogonal to $\mathrm{Dom}(\dot{A})$. Taking into account (2.18) and the equality $\overline{\mathfrak{L}}^{(-)} = \mathfrak{L}$ (see Theorem 2.4.1) we have $\mathcal{F} \cap \mathcal{H} = \mathfrak{L}$ and therefore

$$\mathcal{R}^{-1}\left(S - \frac{i}{2}\mathfrak{J}\right)f \in \mathcal{H} \iff (S - \frac{i}{2}\mathfrak{J})f \in \mathfrak{N} \iff f \in G_S. \qquad \square$$

3.2 Bi-extensions of O-operators

Now we will consider the bi-extensions of O-operators. If $\dot{A}$ is a regular O-operator, then all its bi-extensions take the form

$$\mathbb{A} = \dot{A}P^+_{\mathrm{Dom}(\dot{A})} + (\dot{A}^* + (\dot{A}\dot{A}^* + I)^{-1}S)P^+_{\mathfrak{N}},$$

where $S \in [\mathfrak{N}, \mathfrak{N}]$ and $\mathbb{A} = \mathbb{A}^*$ iff $S = S^*$. Then it follows that $\mathrm{Ran}(\mathbb{A}) \subset \mathcal{H}$.

Theorem 3.2.1. *If $\dot{A}$ is a regular O-operator, then all its bi-extensions are $(+,\cdot)$-continuous and $(\cdot,-)$-continuous. Moreover, if a closed symmetric operator $\dot{A}$ has a $(+,\cdot)$-continuous or $(\cdot,-)$-continuous self-adjoint bi-extension $\mathbb{A}$, then $\dot{A}$ is a regular O-operator.*

Proof. As mentioned above, an arbitrary bi-extension $\mathbb{A}$ of a regular O-operator $\dot{A}$ satisfies the condition $\mathrm{Ran}(\mathbb{A}) \subset \mathcal{H}$. This implies $(+,\cdot)$-continuity of operators $\mathbb{A}$ and $\mathbb{A}^*$. It follows that $\mathbb{A}$ and $\mathbb{A}^*$ are $(+,\cdot)$-continuous.

In order to prove the second statement of the theorem we note that an operator $B \in [\mathcal{H}_+, \mathcal{H}_-]$ is $(+,\cdot)$-continuous if and only if B^* is $(\cdot,-)$-continuous.

Thus we can consider a bi-extension of $\dot{A}$ that is a self-adjoint operator in $[\mathcal{H}_+, \mathcal{H}_-]$ and both $(+,\cdot)$- and $(\cdot,-)$-continuous. Since $\dot{A}$ is a restriction of $\mathbb{A}$, then the operator $\dot{A}$ is $(+,\cdot)$-continuous, i.e., it is regular. We prove that $\dot{A}$ is also an O-operator. Using the fact that $\mathbb{A} \supset \dot{A}$ and $(\cdot,-)$-continuity of both $\mathbb{A}$ and $\dot{A}$ we apply Theorem 2.2.1 and obtain

$$\mathbb{A}f = \overline{\dot{A}}f, \quad (\mathbb{A}f, f) = (\overline{\dot{A}}f, f) = (f, \dot{A}^* f), \quad \forall f \in \mathcal{H}_+ \cap \mathcal{H}_0,$$

where $\overline{\dot{A}}$ is the $(\cdot,-)$-continuity extension of the operator $\dot{A}$ onto $\mathcal{H}_0$. It is proved that $f \in \mathfrak{N}'_{\pm i}$. Then

$$\dot{A}^* f = \pm i f, \qquad (\mathbb{A}f, f) = \mp i(f, f).$$

Taking into account that $(\dot{A}f, f)$ is real we get $f = 0$. Therefore, $\mathfrak{N}'_i = \mathfrak{N}'_{-i} = 0$ and $\dot{A}$ is an O-operator. □

Now let $\dot{A}$ be a regular O-operator and $\mathbb{A} \in \mathcal{E}(\dot{A})$. One can consider operator $\mathbb{A}$ as an operator $B : \mathcal{H} \to \mathcal{H}$ with the domain $\mathrm{Dom}(B) = \mathcal{H}_+$ that is dense on $\mathcal{H}$. Similarly, by C we denote the operator $\mathbb{A}^*$ as acting from $\mathcal{H}$ into $\mathcal{H}$ and such that $\mathrm{Dom}(C) = \mathcal{H}_+$. Obviously, $B \supset \dot{A}$, $C \supset \dot{A}$ and hence $\mathrm{Dom}(B^*) \subset \mathrm{Dom}(\dot{A}^*) = \mathcal{H}_+$, and $\mathrm{Dom}(C^*) \subset \mathrm{Dom}(\dot{A}^*) = \mathcal{H}_+$. We will show that $B = C^*$ and the operators B and C are $(\cdot)$-closed. Indeed, for $f, g \in \mathcal{H}_+$ we have

$$(Bf, g) = (\mathbb{A}f, g) = (f, \mathbb{A}^* g) = (f, Cg),$$

which implies that $B \subset C^*$. Since $\mathrm{Dom}(B) \subset \mathrm{Dom}(C^*) \subset \mathcal{H}_+$, then $\mathrm{Dom}(B) = \mathrm{Dom}(C^*)$ and $B = C^*$. In particular, if $\mathbb{A}$ $(\mathbb{A} = \mathbb{A}^*)$ is a bi-extension of $\dot{A}$, then considering $\mathbb{A}$ as an operator in $\mathcal{H}$ we obtain a usual self-adjoint extension of $\dot{A}$.

Theorem 3.2.2. *If $\dot{A}$ is a regular O-operator, then any of its bi-extensions $\mathbb{A}$ $(\mathbb{A} = \mathbb{A}^*)$ admits a $(-,-)$-closure. Moreover, if some bi-extension $\mathbb{A}$ $(\mathbb{A} = \mathbb{A}^*)$ of a closed symmetric operator $\dot{A}$ admits a $(-,-)$-closure, then $\dot{A}$ is an O-operator.*

Proof. As we mentioned above, $\mathbb{A}$ can be considered as a usual self-adjoint extension of $\dot{A}$ and in this case $\mathbb{A}$ is $(+,\cdot)$-continuous. It is known that $(\mathbb{A} + iI)$ maps $\mathrm{Dom}(\mathbb{A}) = \mathcal{H}_+$ onto $\mathcal{H}$ bijectively. According to the Banach inverse mapping theorem the operator $(\mathbb{A} + iI)^{-1}$ maps $\mathcal{H}$ onto $\mathcal{H}_+$ bijectively and $(\cdot,-)$-continuously. Consequently, this operator maps the $(\cdot)$-dense in $\mathcal{H}$ linear manifold $\mathcal{H}_+$ onto some $(+)$-dense in $\mathcal{H}_+$ linear manifold $\mathcal{L}_0$. Consider an operator

$$B_0 = \mathbb{A} \restriction \mathcal{L}_0$$

acting in $\mathcal{H}_+$ and $(+)$-densely defined in $\mathcal{H}_+$. Its adjoint B_0^* operates in $\mathcal{H}_-$ and is $(-,-)$-closed. Since $\mathbb{A} \supset B_0$, then $\mathbb{A} \subset B_0^*$, i.e., B_0^* is a $(-,-)$-closed extension of $\mathbb{A}$.

The second part of the theorem is proved by repeating the argument of Theorem 3.2.1. □

3.3 Self-adjoint and t-self-adjoint bi-extensions

In this sections we will study in detail self-adjoint bi-extensions of a regular closed symmetric operator $\dot{A}$. First, let us note that if G_S is defined by (3.14), then

$$G_S = \ker\left(P^+_{\mathfrak{N}'_i}(S - \frac{i}{2}I)\right) \cap \ker\left((P^+_{\mathfrak{N}'_{-i}}(S + \frac{i}{2}I)\right). \tag{3.16}$$

Let S be a (1)-self-adjoint operator in $\mathfrak{M}$ and let

$$\mathcal{V} = \left(S - \frac{i}{2}I\right)\left(S + \frac{i}{2}I\right)^{-1} \tag{3.17}$$

be the Cayley transform of S. Then $\mathcal{V}$ is (1)-unitary in $\mathfrak{M}$. In what follows the operator $\mathcal{V}$ will play an important role. Define

$$\mathfrak{L}_{\mathcal{V}} = \{h \in \mathfrak{N}'_i \oplus \mathfrak{N} : \mathcal{V}h \in \mathfrak{N}'_{-i} \oplus \mathfrak{N}\}. \tag{3.18}$$

It follows from (3.16) that

$$f \in G_S \iff \begin{cases} h = (S + \frac{i}{2}I)f \in \mathfrak{L}_{\mathcal{V}} \\ \mathcal{V}h = (S - \frac{i}{2}I)f \end{cases}.$$

Hence

$$G_S = (I - \mathcal{V})\mathfrak{L}_{\mathcal{V}}. \tag{3.19}$$

The next statement is well known.

Proposition 3.3.1. *Let F be a bounded linear operator in a Hilbert space $\mathcal{K}$ defined on a subspace $\mathcal{K}_1$ of $\mathcal{K}$. If F is symmetric, then there exists a bounded self-adjoint extension of F in $\mathcal{K}$.*

Proof. Let $\mathcal{K}_2 = \mathcal{K} \ominus \mathcal{K}_1$ and let $F_{11} = P_{\mathcal{K}_1}F$, $F_{21} = P_{\mathcal{K}_2}F$. According to the decomposition $\mathcal{K} = \mathcal{K}_1 \oplus \mathcal{K}_2$, the operator $\tilde{F} \in [\mathcal{K}, \mathcal{K}]$ given by the block-operator matrix

$$\tilde{F} = \begin{pmatrix} F_{11} & F^*_{21} \\ F_{21} & F_{22} \end{pmatrix}$$

is a bounded extension of F. Then $\tilde{F}$ is self-adjoint iff $F_{22} = F^*_{22}$. □

Theorem 3.3.2. *If the semi-deficiency indices of a regular closed symmetric operator $\dot{A}$ with non-dense domain are infinite, then $\dot{A}$ admits a self-adjoint bi-extension $\mathbb{A}$ with the quasi-kernel $\hat{A} = \dot{A}$.*

Proof. Let $\dim \mathfrak{N}'_i = \dim \mathfrak{N}'_{-i} = \infty$ and U be a (+)-isometric operator from $\mathfrak{N}'_i$ onto $\mathfrak{N}'_{-i}$. Then $(U + I)\mathfrak{N}'_i$ is a (1)-subspace in $\mathfrak{M}$. Since $\mathcal{H}$ is separable then $\mathcal{H}_+$ is separable as well (see [81]). Consequently,

$$\dim(\mathfrak{N}'_i \oplus \mathfrak{N}) = \dim(U + I)\mathfrak{N}'_i.$$

Let W be a (1)-isometric operator from $\mathfrak{N}'_i \oplus \mathfrak{N}$ onto $(U+I)\mathfrak{N}'_i$. Then for $\varphi \in \mathfrak{N}'_i \oplus \mathfrak{N}$ there exists a unique $\psi \in \mathfrak{N}'_i$ such that $W\varphi = U\psi + \psi$. Furthermore,

$$\begin{aligned}\|W\varphi - \varphi\|_1^2 &= \|U\psi + \psi - \varphi\|_1^2 = \|U\psi\|_1^2 + \|P^+_{\mathfrak{N}}\varphi\|_1^2 + \|\psi - P^+_{\mathfrak{N}'_i}\varphi\|_1^2 \\ &= \|\psi\|_1^2 + \|P^+_{\mathfrak{N}}\varphi\|_1^2 + \|\psi - P^+_{\mathfrak{N}'_i}\varphi\|_1^2 \\ &= \|P^+_{\mathfrak{N}}\varphi\|_1^2 + \frac{1}{2}\left(\|P^+_{\mathfrak{N}'_i}\varphi\|_1^2 + \|2\psi - P^+_{\mathfrak{N}'_i}\varphi\|_1^2\right) \\ &\geq \frac{1}{2}\left(\|P^+_{\mathfrak{N}'_i}\varphi\|_1^2 + \|2\psi - P^+_{\mathfrak{N}'_i}\varphi\|_1^2\right) = \frac{1}{2}\|\varphi\|_1^2.\end{aligned}$$

The last estimate implies that $(I - W)(\mathfrak{N}'_i \oplus \mathfrak{N})$ is a subspace in $\mathfrak{M}$. We define a linear operator S' in $\mathfrak{M}$ as follows:

$$\operatorname{Dom}(S') = \operatorname{Ran}(I - W), \quad S'(I - W)\varphi = \frac{i}{2}(W + I)\varphi.$$

Then $\operatorname{Dom}(S')$ is a subspace in $\mathfrak{M}$ and therefore S' is a bounded operator. In addition S' is (1)-symmetric. Then by Proposition 3.3.1 S' admits a bounded (1)-self-adjoint extension S in $\mathfrak{M}$. Let $\mathcal{V}$ be the Cayley transform (3.17) of S. Then $\mathcal{V}$ is a (1)-unitary extension of W. Let $\mathbb{A}$ be a self adjoint bi-extension of $\mathbb{A}$ determined by S, i.e.,

$$\mathbb{A} = \dot{A}P^+_{\operatorname{Dom}(\dot{A})} + \left(\dot{A}^* + \mathcal{R}^{-1}(S - \frac{i}{2}\mathfrak{J})\right)P^+_{\mathfrak{M}}.$$

Let us show that its quasi-kernel $\hat{A}$ coincides with $\dot{A}$. Indeed, if there is a vector $h \in \mathfrak{M}$ such that

$$\left((P^+_{\mathfrak{N}'_i} + P^+_{\mathfrak{N}'_{-i}})S - \frac{i}{2}\mathfrak{J}\right)h = 0,$$

then,

$$\left(S - \frac{i}{2}I\right)h \in \mathfrak{N}'_{-i} \oplus \mathfrak{N}, \ \left(S + \frac{i}{2}I\right)h \in \mathfrak{N}'_i \oplus \mathfrak{N}.$$

Because

$$\left(S - \frac{i}{2}I\right)h = \mathcal{V}\left(S + \frac{i}{2}I\right)h,$$

and $\varphi = \left(S + \frac{i}{2}I\right)h \in \mathfrak{N}'_i \oplus \mathfrak{N}$, we get $\mathcal{V}\varphi = W\varphi \in \mathfrak{N}'_{-i} \oplus \mathfrak{N}$. By construction $\operatorname{Ran}(W) \cap (\mathfrak{N}_{-i} \oplus \mathfrak{N}) = \{0\}$. It follows that $\varphi = 0$ and $h = 0$. Thus

$$\ker\left((P^+_{\mathfrak{N}'_i} + P^+_{\mathfrak{N}'_{-i}})S - \frac{i}{2}\mathfrak{J}\right) = \{0\}.$$

Then Proposition 3.1.11 yields $\hat{A} = \dot{A}$. □

Theorem 3.3.3. *If one of the semi-deficiency indices of a regular closed symmetric non-densely defined operator $\dot{A}$ is finite, then $\dot{A}$ does not admit self-adjoint bi-extensions $\mathbb{A}$ with $\hat{A} = \dot{A}$.*

Proof. Let $p = \dim \mathfrak{N}'_i < \infty$. Assume that $\dot{A}$ admits a bi-extension $\mathbb{A}$ defined by the formula (3.13) that is a self-adjoint operator in $[\mathcal{H}_+, \mathcal{H}_-]$ and with $\dot{A} = \hat{A}$. According to Proposition 3.1.11 we have $G_S = \{0\}$. Hence for every $\phi \neq 0$, $\phi \in \mathfrak{M}$ at least one of the vectors $P^+_{\mathfrak{N}'_i}\left(S - \frac{i}{2}I\right)\phi$ and $P^+_{\mathfrak{N}'_{-i}}\left(S + \frac{i}{2}I\right)\phi$ is different from zero. Put

$$\mathfrak{M}_0 := \ker\left(P^+_{\mathfrak{N}'_{-i}}(S + \frac{i}{2}I)\right).$$

Then

$$\left(S + \frac{i}{2}I\right)\phi \in \mathfrak{N}'_i \oplus \mathfrak{N}, \quad (\forall \phi \in \mathfrak{M}_0).$$

Besides, for every $\phi \neq 0$, $\phi \in \mathfrak{M}_0$ we have $P^+_{\mathfrak{N}'_i}\left(S - \frac{i}{2}I\right)\phi \neq 0$. Hence,

$$\left(S - \frac{i}{2}I\right)\phi \notin \mathfrak{N}'_{-i} \oplus \mathfrak{N}, \qquad (\phi \neq 0,\ \phi \in \mathfrak{M}_0).$$

Let $\mathcal{V}$ be the Cayley transform (3.17) of S. It follows from the above reasoning that

$$(\mathfrak{N}'_{-i} \oplus \mathfrak{N}) \cap \mathcal{V}(\mathfrak{N}'_i \oplus \mathfrak{N}) = \{0\}.$$

Let $\psi \in \mathcal{V}(\mathfrak{N}'_i \oplus \mathfrak{N})$ and $\psi \neq 0$. Then $P^+_{\mathfrak{N}'_i}\psi \neq 0$ or otherwise the equality $P^+_{\mathfrak{N}'_i}\psi = 0$ would imply $\psi = P^+_{\mathfrak{N}'_{-i}\oplus\mathfrak{N}}\psi$ which means that $\psi \in \mathfrak{N}'_{-i} \oplus \mathfrak{N}$. The latter is impossible and we get a contradiction. Therefore, the operator $P^+_{\mathfrak{N}'_i}$ maps the subspace $\mathcal{V}(\mathfrak{N}'_i \oplus \mathfrak{N})$ bijectively onto some subspace of $\mathfrak{N}'_i$. Because $\dim \mathfrak{N}'_i = p$, we get $\dim \mathcal{V}(\mathfrak{N}'_i \oplus \mathfrak{N}) \leq p$. On the other hand,

$$\dim \mathcal{V}(\mathfrak{N}'_i \oplus \mathfrak{N}) = \dim(\mathfrak{N}'_i \oplus \mathfrak{N}) > p.$$

The contradiction has arrived. This proves the theorem. □

Proposition 3.3.4. *Let S be a (1)-self-adjoint and bounded operator in $\mathfrak{M}$. Define the subspace*

$$\tilde{G}_S := \ker\left(S(P^+_{\mathfrak{N}'_i} + P^+_{\mathfrak{N}'_{-i}}) + \frac{i}{2}\mathfrak{J}\right). \tag{3.20}$$

Then the operator

$$\tilde{A} =: \left(\dot{A}P^+_{\mathrm{Dom}(\dot{A})} + \dot{A}^* P^+_{\mathfrak{M}}\right) \upharpoonright (\mathrm{Dom}(\dot{A}) \oplus \tilde{G}_S) \tag{3.21}$$

is a closed symmetric extension of the operator $\dot{B}$ defined by (2.19).

Proof. Recall that

$$\mathrm{Dom}(\dot{B}) = \mathrm{Dom}(\dot{A}) \oplus \mathfrak{N}, \quad \dot{B} = \dot{A}P^+_{\mathrm{Dom}(\dot{A})} + \dot{A}^* P^+_{\mathfrak{N}},$$

$\dot{B}$ is closed, densely defined and symmetric in $\mathcal{H}$, $\dot{B} \supset \dot{A}$, and the subspaces $\mathfrak{N}'_i$, $\mathfrak{N}'_{-i}$ are deficiency subspaces of $\dot{B}$, corresponding to the numbers i and $-i$, respectively. Clearly, $\tilde{G}_S \supseteq \mathfrak{N}$. Let $f \in \tilde{G}_S \ominus \mathfrak{N}$. Then the relation

$$(S + \frac{i}{2}I)P^+_{\mathfrak{N}'_i} f + (S - \frac{i}{2}I)P^+_{\mathfrak{N}'_{-i}} f = 0$$

is equivalent to $P^+_{\mathfrak{N}'_i} f = -\mathcal{V}P^+_{\mathfrak{N}'_{-i}} f$. Consequently, $\mathcal{V}P^+_{\mathfrak{N}'_{-i}} \tilde{G}_S = P^+_{\mathfrak{N}'_i} \tilde{G}_S$ and

$$\tilde{G}_S \ominus \mathfrak{N} = (I - \mathcal{V})(P^+_{\mathfrak{N}'_{-i}} \tilde{G}_S). \tag{3.22}$$

Let the operator U be given by

$$\mathrm{Dom}(U) = P^+_{\mathfrak{N}'_i}(\tilde{G}_S \ominus \mathfrak{N}) \subseteq \mathfrak{N}'_i, \quad U = \mathcal{V}^{-1} \upharpoonright \mathrm{Dom}(U).$$

Then U is (1)- and $(\cdot)$-isometric and $\tilde{G}_S = \mathfrak{N} \oplus (I - U)\mathrm{Dom}(U)$. Because $\tilde{G}_S$ is a subspace in $\mathfrak{M}$, the linear manifold $\mathrm{Dom}(U)$ is closed in $\mathfrak{N}'_i$. Therefore, the operator $\tilde{A}$ given by (3.21) coincides with $\dot{B}$ or is its closed symmetric extension. □

Definition 3.3.5. A self-adjoint bi-extension $\mathbb{A}$ of a regular symmetric operator $\dot{A}$ is called **twice-self-adjoint** (**t-self-adjoint**) if its quasi-kernel $\hat{A}$ is a self-adjoint operator in $\mathcal{H}$.

Theorem 3.3.6. *Let*

$$\mathbb{A} = \dot{A}P^+_{\mathrm{Dom}(\dot{A})} + \left(\dot{A}^* + \mathcal{R}^{-1}(S - \frac{i}{2}\mathfrak{J})\right) P^+_{\mathfrak{M}}$$

be a self-adjoint bi-extension of a regular symmetric operator $\dot{A}$ with equal semi-deficiency numbers. Then the following statements are equivalent:

1) *the operator $\mathbb{A}$ is t-self-adjoint;*
2) *the operator $\tilde{A}$ given by (3.21) is a self-adjoint extension of $\dot{B}$;*
3) $G_S \dot{+} (\tilde{G}_S \ominus \mathfrak{N}) = \mathfrak{M}$;
4) *the Cayley transform $\mathcal{V} = (S - \frac{i}{2}I)(S + \frac{i}{2}I)^{-1}$ possesses the property*

$$\mathcal{V}(\mathfrak{N}'_{-i}) = \mathfrak{N}'_i.$$

Proof. 1)⇒ 2), 1)⇒ 3), and 1)⇒ 4). Suppose the quasi-kernel $\hat{A}$ of $\mathbb{A}$ is a self-adjoint operator. By Theorem 2.5.5 and Proposition 3.1.11 we have

$$\mathfrak{L}_{\mathcal{V}} = \mathfrak{N}'_i \oplus \mathfrak{N} = (S + \tfrac{i}{2}I)G_S, \quad \mathcal{V}\mathfrak{L}_{\mathcal{V}} = \mathfrak{N}'_{-i} \oplus \mathfrak{N} = (S - \tfrac{i}{2}I)G_S,$$
$$G_S = (I - \mathcal{V})\mathfrak{L}_{\mathcal{V}},$$

and $\mathrm{Dom}(\hat{A}) = \mathrm{Dom}(\dot{A}) \oplus (I - \mathcal{V})\mathfrak{L}_{\mathcal{V}}$, where $\mathcal{V}$ and $\mathfrak{L}_{\mathcal{V}}$ are defined by (3.17) and (3.18). Since $\mathcal{V}$ is (1)-unitary in $\mathfrak{M}$ and maps $\mathfrak{N}'_i \oplus \mathfrak{N}$ onto $\mathfrak{N}'_{-i} \oplus \mathfrak{N}$, the inverse operator $\mathcal{V}^{-1}$ maps $\mathfrak{N}'_i$ onto $\mathfrak{N}'_{-i}$. Therefore $\mathcal{V}(\mathfrak{N}'_{-i}) = \mathfrak{N}'_i$. Let $U = \mathcal{V}^{-1}\upharpoonright\mathfrak{N}'_i$. Since

$$S(I - \mathcal{V}^{-1})f = -\frac{i}{2}(I + \mathcal{V}^{-1})f, \quad f \in \mathfrak{M},$$

we have $\left(S + \frac{i}{2}\mathfrak{J}\right)(I - U)\varphi = 0$, for all $\varphi \in \mathfrak{N}'_i$. Hence, for the subspace $\tilde{G}_S$, defined by (3.20), we get

$$\tilde{G}_S = (I - U)\mathfrak{N}'_i \oplus \mathfrak{N}$$

and, as a result, the operator $\tilde{A}$ is a self-adjoint extension of $\dot{B}$. Since $(I - U)\mathfrak{N}'_i = (I - \mathcal{V})\mathfrak{N}'_{-i}$ and $\mathrm{Ran}(I - \mathcal{V}) = \mathfrak{M}$, we get $G_S \dot{+} (\tilde{G}_S \ominus \mathfrak{N}) = \mathfrak{M}$.

2)⇒1). Suppose $\tilde{A}$ is a self-adjoint extension of $\dot{B}$. Then there exists a (1)-isometric operator U such that

$$\mathrm{Dom}(U) = \mathfrak{N}'_i, \quad \mathrm{Ran}(U) = \mathfrak{N}'_{-i}, \quad \tilde{G}_S = \mathfrak{N} \oplus (I - U)\mathfrak{N}'_i.$$

It follows that

$$\overline{\mathrm{Ran}\left((P^+_{\mathfrak{N}'_i} + P^+_{\mathfrak{N}'_{-i}})S - \frac{i}{2}\mathfrak{J}\right)} = (I + U)\mathfrak{N}'_i,$$

where the bar stands for (+)-closure. In particular,

$$||P^+_{\mathfrak{N}'_{-i}}(S + \frac{i}{2}I)f||_1 = ||P^+_{\mathfrak{N}'_i}(S - \frac{i}{2}I)f||_1 \quad \text{for all} \quad f \in \mathfrak{M}. \tag{3.23}$$

Let G_S be defined by (3.14). By Proposition 3.1.11 we have $\mathrm{Dom}(\hat{A}) = \mathrm{Dom}(\dot{A}) \oplus G_S$ and

$$(S + \frac{i}{2}I)G_S \subseteq \mathfrak{N}'_i \oplus \mathfrak{N}, \quad (S - \frac{i}{2}I)G_S \subseteq \mathfrak{N}'_{-i} \oplus \mathfrak{N}.$$

Assume $(S + \frac{i}{2}I)G_S \neq \mathfrak{N}'_i \oplus \mathfrak{N}$. Then there exists a non-zero vector $h \notin G_S$ such that

$$(S + \frac{i}{2}I)h \in \mathfrak{N}'_i \oplus \mathfrak{N}.$$

Then

$$P^+_{\mathfrak{N}'_{-i}}(S + \frac{i}{2}I)h = 0.$$

It follows from (3.23) that $P^+_{\mathfrak{N}'_i}(S - \frac{i}{2}I)h = 0$. Therefore $h \in G_S$ and we have a contradiction. Hence,

$$(S + \frac{i}{2}I)G_S = \mathfrak{N}'_i \oplus \mathfrak{N}.$$

Similarly $(S - \frac{i}{2}I)G_S = \mathfrak{N}'_{-i} \oplus \mathfrak{N}$. From Theorem 2.5.5 and relations (3.17), (3.18), and (3.19) we get that $\hat{A}$ is a self-adjoint extension of $\dot{A}$, i.e., $\mathbb{A}$ is a t-self-adjoint bi-extension of $\dot{A}$.

3)⇒1). Suppose $G_S \dot{+} (\tilde{G}_S \ominus \mathfrak{N}) = \mathfrak{M}$. Then from (3.22)

$$(I-\mathcal{V})\mathfrak{L}_{\mathcal{V}} \dot{+} (I-\mathcal{V})(P^+_{\mathfrak{N}'_{-i}}\tilde{G}_S) = \mathfrak{M}.$$

Since the number 1 is a regular point for $\mathcal{V}$, we get $\mathfrak{L}_{\mathcal{V}} \oplus (P^+_{\mathfrak{N}'_{-i}}\tilde{G}_S) = \mathfrak{M}$. This yields the equalities

$$\mathfrak{L}_{\mathcal{V}} = \mathfrak{N}'_i \oplus \mathfrak{N}, \quad P^+_{\mathfrak{N}'_{-i}}\tilde{G}_S = \mathfrak{N}'_{-i}.$$

Because $\mathcal{V}$ is (1)-unitary, we have

$$(I-\mathcal{V}^{-1})(\mathcal{V}\mathfrak{L}_{\mathcal{V}}) \dot{+} (I-\mathcal{V}^{-1})(\mathcal{V}P^+_{\mathfrak{N}'_{-i}}\tilde{G}_S) = \mathfrak{M}.$$

Similarly, we have that the equalities $\mathcal{V}\mathfrak{L}_{\mathcal{V}} = \mathfrak{N}'_{-i} \oplus \mathfrak{N}$ and $\mathcal{V}P^+_{\mathfrak{N}'_{-i}}\tilde{G}_S = \mathfrak{N}'_i$ hold. Therefore, the operator $\hat{A}$ is a self-adjoint extension of $\dot{A}$.

4)⇒1). Let $\mathcal{V}(\mathfrak{N}'_{-i}) = \mathfrak{N}'_i$. Since $\mathcal{V}$ is (1)-unitary, the operator $\mathcal{V}$ maps $\mathfrak{N}'_i \oplus \mathfrak{N}$ onto $\mathfrak{N}'_{-i} \oplus \mathfrak{N}$. Define $W = \mathcal{V}\upharpoonright(\mathfrak{N}'_i \oplus \mathfrak{N})$. Then

$$S(I-W)f = \frac{i}{2}(I+W)f, \quad f \in \mathfrak{N}'_i \oplus \mathfrak{N}.$$

Therefore $G_S = (I-W)(\mathfrak{N}'_i \oplus \mathfrak{N})$ and the quasi-kernel $\hat{A}$ of $\mathbb{A}$ is a self-adjoint extension of $\dot{A}$, i.e., $\mathbb{A}$ is a t-self-adjoint extension of $\dot{A}$. □

The next statement easily follows from Proposition 3.3.4 and Theorem 3.3.6.

Corollary 3.3.7. *If*

$$\mathbb{A} = \dot{A}P^+_{\mathrm{Dom}(\dot{A})} + \left(\dot{A}^* + \mathcal{R}^{-1}(S - \frac{i}{2}\mathfrak{J})\right)P^+_{\mathfrak{M}},$$

is a t-self-adjoint bi-extension of a regular symmetric operator $\dot{A}$ with equal non-zero semi-deficiency numbers and if $\hat{A}$ is the quasi-kernel of $\mathbb{A}$, then

$$\mathrm{Dom}(\hat{A}) + (\mathrm{Dom}(\tilde{A}) \cap \mathcal{H}_0) = \mathcal{H}_+,$$

where $\tilde{A}$ is given by (3.21).

Note that by Theorem 2.5.7 the operator $P\tilde{A}\upharpoonright(\mathrm{Dom}(\tilde{A}) \cap \mathcal{H}_0)$ is a self-adjoint extension of the operator $P\dot{A}$ in the Hilbert space $\mathcal{H}_0$.

Now we give a parametrization of all t-self-adjoint bi-extensions whose quasi-kernel is a fixed regular self-adjoint extension $\hat{A}$ of a regular symmetric operator with equal semi-deficiency numbers.

Theorem 3.3.8. *Let $\dot{A}$ be a regular symmetric operator in $\mathcal{H}$ with non-zero equal semi-deficiency numbers and let $\hat{A}$ be a regular self-adjoint extension of $\dot{A}$. Then* *there exists a t-self adjoint bi-extension $\mathbb{A}$ of $\dot{A}$ whose quasi-kernel is $\hat{A}$. Moreover, if* $\mathrm{Dom}(\hat{A})$ *is given by*

$$\mathrm{Dom}(\hat{A}) = \mathrm{Dom}(\dot{A}) \oplus (I - \mathcal{U})(\mathfrak{N}'_i \oplus \mathfrak{N}),$$

where $\mathcal{U} \in [\mathfrak{N}'_i \oplus \mathfrak{N}, \mathfrak{N}'_{-i} \oplus \mathfrak{N}]$ *is* (1)*-isometric operator,* $\ker(I - \mathcal{U}) = \{0\}$, $\mathrm{Ran}(\mathcal{U}) = \mathfrak{N}'_{-i} \oplus \mathfrak{N}$, *and* $\mathrm{Ran}(I - \mathcal{U})$ *is a subspace in* $\mathfrak{M}$ *(see Theorem 2.5.5), then there is a bijective correspondence between all t-self-adjoint bi-extensions* $\mathbb{A}$ *with quasi-kernel* $\hat{A}$ *and all* (1)*-isometric operators* U *mapping* $\mathfrak{N}'_i$ *onto* $\mathfrak{N}'_{-i}$*, satisfying the condition*

$$(I - \mathcal{U})(\mathfrak{N}'_i \oplus \mathfrak{N}) \dot{+} (I - U)\mathfrak{N}'_i = \mathfrak{M}. \tag{3.24}$$

This correspondence is given by the formula

$$\mathbb{A} = \dot{A} P^+_{\mathrm{Dom}(\dot{A})} + \left(\dot{A}^* + \mathcal{R}^{-1}(S - \frac{i}{2}\mathfrak{J})\right) P^+_{\mathfrak{M}},$$

where S *is* (1)*-self-adjoint operator in* $\mathfrak{M}$ *of the form*

$$\begin{aligned} &S\left((I - \mathcal{U})f + (I - U)\varphi\right) \\ &= \frac{i}{2}(I + \mathcal{U})f - \frac{i}{2}(I + U)\varphi, \; f \in \mathfrak{N}'_i \oplus \mathfrak{N}, \; \varphi \in \mathfrak{N}'_i. \end{aligned} \tag{3.25}$$

Proof. Define an operator S' by the formula

$$\mathrm{Dom}(S') = (I - \mathcal{U})(\mathfrak{N}'_i \oplus \mathfrak{N}), \quad \begin{cases} f = (I - \mathcal{U})\varphi, \\ S' f = \frac{i}{2}(I + \mathcal{U})\varphi, \end{cases} \quad \varphi \in \mathfrak{N}'_i \oplus \mathfrak{N}.$$

Then S' is (1)-symmetric, (1)-bounded, closed, and non-densely defined in $\mathfrak{M}$. Moreover,

$$\left((P^+_{\mathfrak{N}'_i} + P^+_{\mathfrak{N}'_{-i}})S' - \frac{i}{2}\mathfrak{J}\right) f = 0, \; f \in \mathrm{Dom}(S').$$

Let $S \in [\mathfrak{M}, \mathfrak{M}]$ be a bounded (1)-self-adjoint extension of S' and let

$$\mathbb{A} = \dot{A} P^+_{\mathrm{Dom}(\dot{A})} + \left(\dot{A}^* + \mathcal{R}^{-1}(S - \frac{i}{2}\mathfrak{J})\right) P^+_{\mathfrak{M}}$$

be a self-adjoint bi-extension of $\dot{A}$. Applying Proposition 3.1.11 we get that $\hat{A}$ is the quasi-kernel of $\mathbb{A}$. On the other hand, the Cayley transform

$$\mathcal{V} = (S - \frac{i}{2}I)(S + \frac{i}{2}I)^{-1},$$

is a (1)-unitary extension of $\mathcal{U}$ on $\mathfrak{M}$ and $\mathrm{Ran}(I - \mathcal{V}) = \mathfrak{M}$. It follows that $\mathcal{V}^{-1}$ (1)-unitarily maps $\mathfrak{N}'_{-i}$ onto $\mathfrak{N}'_i$. Define $U := \mathcal{V}^{-1} \restriction \mathfrak{N}'_i$. Since

$$(I - U)\mathfrak{N}'_i = (I - \mathcal{V}^{-1})\mathfrak{N}'_i = (I - \mathcal{V})\mathfrak{N}'_{-i},$$

relation (3.24) holds. Because $\mathcal{V}$ is the Cayley transform of S, we have (3.25).

If $U \in [\mathfrak{N}'_i, \mathfrak{N}'_{-i}]$ is (1)-isometric, $\mathrm{Ran}(U) = \mathfrak{N}'_{-i}$, and (3.24) holds, then one can verify that S defined by (3.25) is (1)-self-adjoint in $\mathfrak{M}$, $G_S = (I - \mathcal{U})(\mathfrak{N}'_i \oplus \mathfrak{N})$, and $\tilde{G}_S = (I - U)\mathfrak{N}'_i \oplus \mathfrak{N}$. Hence, $\hat{A}$ is a quasi-kernel of $\mathbb{A}$. □

Notice that equality (3.24) is equivalent to

$$\mathrm{Dom}(\hat{A}) \dot{+} (I - U)\mathfrak{N}'_i = \mathrm{Dom}(\hat{A}) \dot{+} (\mathrm{Dom}(\tilde{A}) \ominus \mathrm{Dom}(\dot{B})) = \mathcal{H}_+,$$

where $\tilde{A}$ is defined by (3.21).

3.4 The case of a densely defined symmetric operator

If $\dot{A}$ is a densely defined closed symmetric operator, then $\mathfrak{N} = \{0\}$, the semi-deficiency subspaces $\mathfrak{N}'_\lambda$ become the deficiency subspaces $\mathfrak{N}_\lambda$, $\mathfrak{M} = \mathfrak{N}_i \oplus \mathfrak{N}_{-i}$, the operator $\mathfrak{J}$ defined by (3.10) becomes

$$\mathfrak{J} = P^+_{\mathfrak{N}_i} - P^+_{\mathfrak{N}_{-i}},$$

and Theorem 3.1.9 takes the following form:

Theorem 3.4.1. *Let $\dot{A}$ be a densely defined closed symmetric operator in the Hilbert space $\mathcal{H}$. Then the formula*

$$\mathbb{A} = \dot{A}^* + \mathcal{R}^{-1}(S - \frac{i}{2}\mathfrak{J})P^+_{\mathfrak{M}}, \tag{3.26}$$

establishes a bijective correspondence between all bi-extensions of $\dot{A}$ and all operators $S \in [\mathfrak{M}, \mathfrak{M}]$. The adjoint operator $\mathbb{A}^*$ is of the form

$$\mathbb{A}^* = \dot{A}^* + \mathcal{R}^{-1}(S^* - \frac{i}{2}\mathfrak{J})P^+_{\mathfrak{M}},$$

and $\mathbb{A}$ is self-adjoint ($\mathbb{A} = \mathbb{A}^*$) if and only if $S = S^*$.

Notice that the subspace G_S defined by (3.14) for a densely defined $\dot{A}$ takes the form

$$G_S = \ker\left(S - \frac{i}{2}\mathfrak{J}\right),$$

and $\tilde{G}_S$ (3.20) becomes G_{-S}. It is easy to see that $\dot{A}$ admits self-adjoint bi-extensions with quasi-kernel $\dot{A}$. Actually, if $S = 0$, then the operator

$$\mathbb{A} = \dot{A}^* - \frac{i}{2}\mathcal{R}^{-1}\mathfrak{J}P^+_{\mathfrak{M}},$$

is self-adjoint and $G_0 = \ker(-\frac{i}{2}\mathfrak{J}) = \{0\}$. Hence, $\hat{A} = \dot{A}$.

Theorem 3.4.2. *Let $\hat{A}$ be an arbitrary closed densely defined symmetric operator extending $\dot{A}$. Then there is a self-adjoint bi-extension $\mathbb{A}$ of $\dot{A}$ whose quasi-kernel is $\hat{A}$.*

Proof. According to the von Neumann Theorem 1.3.1 we have

$$\mathrm{Dom}(\hat{A}) = \mathrm{Dom}(\dot{A}) \oplus (I - U)\mathrm{Dom}(U), \tag{3.27}$$

where

$$\mathfrak{N}_i \supseteq \mathrm{Dom}(U) \overset{U}{\mapsto} \mathrm{Ran}(U) \subseteq \mathfrak{N}_{-i}, \tag{3.28}$$

is a $(\cdot)$ (and $(+)$)-isometric operator, $\mathrm{Dom}(U)$ is $(\cdot)$ (and $(+)$)-closed. Define

$$G := (I - U)\mathrm{Dom}(U), \quad G^\perp := \mathfrak{M} \ominus G.$$

Since U is $(+)$-isometric, the inclusion $(I + U)\mathrm{Dom}(U) \subseteq G^\perp$. Let S' be defined as

$$\mathrm{Dom}(S') = G, \quad S'(I - U)f = \frac{i}{2}(I + U)f, \quad f \in \mathrm{Dom}(U).$$

Then S' is $(+)$-symmetric, $(+)$-bounded, and non-densely defined in $\mathfrak{M}$. In addition, $\mathrm{Ran}(S) \subset G^\perp$ and $\left(S' - \frac{i}{2}\mathfrak{J}\right) f = 0$ for $f \in \mathrm{Dom}(S') = G$. It follows from Proposition 3.3.1 that each $(+)$-bounded and $(+)$-self-adjoint extension S of S' on $\mathfrak{M}$ with respect to the decomposition $\mathfrak{M} = G \oplus G^\perp$ takes the block-operator matrix form

$$S = \begin{pmatrix} 0 & -\frac{i}{2}P_G^+\mathfrak{J}\restriction G^\perp \\ \frac{i}{2}\mathfrak{J}\restriction G & L \end{pmatrix}, \tag{3.29}$$

where $\mathfrak{J} = P_{\mathfrak{N}_i}^+ - P_{\mathfrak{N}_{-i}}^+$, P_G^+ is $(+)$-orthogonal projection onto G in $\mathfrak{M}$, and L is an arbitrary $(+)$-bounded and $(+)$-self-adjoint operator in $G^\perp$. Clearly

$$G_S = G \iff \left(L - \frac{i}{2}P_G^+\mathfrak{J} - \frac{i}{2}\mathfrak{J}\right) f \neq 0 \quad \text{for all} \quad f \in G^\perp \setminus \{0\}.$$

Let L be a zero operator in $G^\perp$. If

$$\left(-\frac{i}{2}P_G^+\mathfrak{J} - \frac{i}{2}\mathfrak{J}\right) f = 0,$$

for some $f \in G^\perp$, then $\mathfrak{J}f = 0$. Hence, $f = 0$. Thus, if

$$S_0 = \begin{pmatrix} 0 & -\frac{i}{2}P_G^+\mathfrak{J}\restriction G^\perp \\ \frac{i}{2}\mathfrak{J}\restriction G & 0 \end{pmatrix},$$

then $G_{S_0} = G$. Thus the quasi-kernel of the self-adjoint bi-extension

$$\mathbb{A}_0 = \dot{A}^* + \mathcal{R}^{-1}(S_0 - \frac{i}{2}\mathfrak{J})P^+_{\mathfrak{M}},$$

coincides with $\hat{A}$. □

Below we establish in terms of aperture a criterion for a self-adjoint bi-extension $\mathbb{A}$ of a closed densely defined symmetric operator to be t-self-adjoint.

Theorem 3.4.3. *Let $\mathbb{A} = \dot{A}^* + \mathcal{R}^{-1}(S - \frac{i}{2}\mathfrak{J})P^+_{\mathfrak{M}}$ be a self-adjoint bi-extension of a closed densely defined symmetric operator $\dot{A}$ with equal deficiency numbers. Then $\mathbb{A}$ is t-self-adjoint if and only if the apertures below satisfy*

$$\Theta\{G_S, \mathfrak{N}_i\} < 1 \quad \textit{and} \quad \Theta\{G_S, \mathfrak{N}_{-i}\} < 1. \tag{3.30}$$

Proof. Let $\hat{A}$ be the quasi-kernel of $\mathbb{A}$. Then by Proposition 3.1.11,

$$\operatorname{Dom}(\hat{A}) = \operatorname{Dom}(\dot{A}) \oplus G_S,$$

(3.27), and (3.28) hold. Moreover, $\hat{A}$ is a self-adjoint extension of $\dot{A}$ if and only if $\operatorname{Dom}(U) = \mathfrak{N}_i$ and $\operatorname{Ran}(U) = \mathfrak{N}_{-i}$. By Lemma 1.5.1 conditions (3.30) are equivalent to

$$P^+_{\mathfrak{N}_i} G_S = \mathfrak{N}_i \quad \text{and} \quad P^+_{\mathfrak{N}_{-i}} G_S = \mathfrak{N}_{-i}.$$

The latter are equivalent to $\operatorname{Dom}(U) = \mathfrak{N}_i$ and $\operatorname{Ran}(U) = \mathfrak{N}_{-i}$, i.e., to the operator $\hat{A}$ being a self-adjoint extension of $\dot{A}$. □

Notice that when $\hat{A}$ is self-adjoint, we have $G_S = (I - U)\mathfrak{N}_i$ and hence,

$$\|P^+_{\mathfrak{N}_i} f\|^2_+ = \|P^+_{\mathfrak{N}_{-i}} f\|^2_2 = \frac{1}{2}\|f\|^2_+, \; f \in G_S.$$

Since $\mathfrak{M} \ominus G_S = (I + U)\mathfrak{N}_i$,

$$\varphi = \frac{1}{2}(I - U)\varphi + \frac{1}{2}(I + U)\varphi, \; \varphi \in \mathfrak{N}_i,$$

and

$$\psi = -\frac{1}{2}(I - U)U^{-1}\psi + \frac{1}{2}(I + U)U^{-1}\psi, \; \psi \in \mathfrak{N}_{-i},$$

we get

$$\begin{aligned} \|(I - P^+_{G_S})\varphi\|^2_+ &= \frac{1}{2}\|\varphi\|^2_+, \; \varphi \in \mathfrak{N}_i, \\ \|(I - P^+_{G_S})\psi\|^2_+ &= \frac{1}{2}\|\psi\|^2_+, \; \psi \in \mathfrak{N}_{-i}. \end{aligned}$$

It follows from (1.21) that

$$\Theta\{G_S, \mathfrak{N}_i\} = \Theta\{G_S, \mathfrak{N}_{-i}\} = \frac{1}{\sqrt{2}}.$$

Therefore Theorem 3.4.3 yields

Corollary 3.4.4. *Under the conditions of Theorem* 3.4.3 *formula* (3.30) *implies*

$$\Theta\{G_S, \mathfrak{N}_i\} = \Theta\{G_S, \mathfrak{N}_{-i}\} = \frac{1}{\sqrt{2}}.$$

The next theorem is an analogue of Theorem 3.3.6.

Theorem 3.4.5. *Let $\dot{A}$ be a densely defined closed symmetric operator with equal deficiency numbers. Then the following statements are equivalent:*

1) *the operator*
$$\mathbb{A} = \dot{A}^* + \mathcal{R}^{-1}(S - \frac{i}{2}\mathfrak{J})P^+_{\mathfrak{M}}$$
is a t-self-adjoint bi-extension of $\dot{A}$;

2) *the operator*
$$\tilde{\mathbb{A}} = \dot{A}^* + \mathcal{R}^{-1}(-S - \frac{i}{2}\mathfrak{J})P^+_{\mathfrak{M}}$$
is a t-self-adjoint bi-extension of $\dot{A}$;

3) $G_S \dot{+} G_{-S} = \mathfrak{M}$;

4) *the Cayley transform $\mathcal{V} = (S - \frac{i}{2}I)(S + \frac{i}{2}I)^{-1}$ possesses the property*
$$\mathcal{V}(\mathfrak{N}_{-i}) = \mathfrak{N}_i.$$

It should be mentioned that the operators $\mathbb{A}$ and $\tilde{\mathbb{A}}$ in Theorem 3.4.5 are connected by the relation

$$\tilde{\mathbb{A}} + \mathbb{A} = 2\dot{A}^* + \mathcal{R}^{-1}\left(-iP^+_{\mathfrak{N}_i} + iP^+_{\mathfrak{N}_{-i}}\right) = 2\dot{A}^* - \mathcal{R}^{-1}\dot{A}^* P^+_{\mathfrak{M}}.$$

The latter is equivalent to the equality

$$((\tilde{\mathbb{A}} + \mathbb{A})f, g) = (\dot{A}^* f, g) + (f, \dot{A}^* g), \;\; f, g \in \mathcal{H}_+. \tag{3.31}$$

Suppose that a (+)-self-adjoint operator $S \in [\mathfrak{M}, \mathfrak{M}]$ satisfies condition 3) of Theorem 3.4.5 and let $\mathcal{V}$ be the Cayley transform (3.17) of S. Then the quasi-kernels $\hat{A}$ and $\tilde{A}$ of self-adjoint-bi-extensions $\mathbb{A}$ and $\tilde{\mathbb{A}}$ are self-adjoint extensions of $\dot{A}$ and posses the properties (see proof of Theorem 3.3.6 and Corollary 3.3.7):

$$\begin{aligned}
\mathrm{Dom}(\hat{A}) &= \mathrm{Dom}(\dot{A}) \oplus (I - \mathcal{V})\mathfrak{N}_i,\\
\mathrm{Dom}(\tilde{A}) &= \mathrm{Dom}(\dot{A}) \oplus (I - \mathcal{V})\mathfrak{N}_{-i} = \mathrm{Dom}(\dot{A}) \oplus (I - \mathcal{V}^{-1})\mathfrak{N}_i,\\
\mathrm{Dom}(\hat{A}) &+ \mathrm{Dom}(\tilde{A}) = \mathrm{Dom}(\dot{A}^*) = \mathcal{H}_+.
\end{aligned}$$

Now we reformulate Theorem 3.3.8 for the case of a densely defined symmetric operator.

Theorem 3.4.6. *Let $\dot{A}$ be a densely defined closed symmetric operator in $\mathcal{H}$ with equal deficiency numbers and let $\hat{A}$ be a self-adjoint extension of $\dot{A}$ given by the von Neumann formula*

$$\mathrm{Dom}(\hat{A}) = \mathrm{Dom}(\dot{A}) \oplus (I-U)\mathfrak{N}_i.$$

Then there is a bijective correspondence between all $(\cdot)$-isometric operators W mapping $\mathfrak{N}_i$ onto $\mathfrak{N}_{-i}$, satisfying the condition

$$\mathrm{Ran}(I - W^{-1}U) = \mathfrak{N}_i, \tag{3.32}$$

and all t-self-adjoint bi-extensions $\mathbb{A}$ with quasi-kernel $\hat{A}$. This correspondence is given by the formula

$$\mathbb{A} = \dot{A}^* + \mathcal{R}^{-1}(S - \frac{i}{2}\mathfrak{J})P^+_{\mathfrak{M}},$$

where S is $(+)$-self-adjoint operator in $\mathfrak{M}$ that is the inverse Cayley transform of the form

$$S = \frac{i}{2}\left(I_{\mathfrak{M}} + UP^+_{\mathfrak{N}_i} + W^{-1}P^+_{\mathfrak{N}_{-i}}\right)\left(I_{\mathfrak{M}} - UP^+_{\mathfrak{N}_i} - W^{-1}P^+_{\mathfrak{N}_{-i}}\right)^{-1}.$$

Notice that condition (3.32) is equivalent to

$$(I-U)\mathfrak{N}_i \dot{+} (I-W)\mathfrak{N}_i = \mathfrak{M},$$

and

$$S\left((I-U)f + (I-W)\varphi\right) = \frac{i}{2}(I+U)f - \frac{i}{2}(I+W)\varphi,\ f \in \mathfrak{N}_i,\ \varphi \in \mathfrak{N}_i. \tag{3.33}$$

Hence

$$\mathrm{Ran}(\mathbb{A} - \dot{A}^*) = \mathcal{R}^{-1}(I+W)\mathfrak{N}_{-i}. \tag{3.34}$$

Now we give a parametrization in a slightly different form of all t-self-adjoint bi-extension whose quasi-kernel is a fixed self-adjoint extension $\hat{A}$. We need a notion of mutually transversal self-adjoint extensions of a given densely defined symmetric operator.

Definition 3.4.7. Let $\dot{A}$ be a closed densely defined symmetric operator with equal defect numbers. Two self-adjoint extensions A_1 and A_2 of $\dot{A}$ are called **relatively prime (disjoint)** if $\mathrm{Dom}(A_1) \cap \mathrm{Dom}(A_2) = \mathrm{Dom}(\dot{A})$ and **transversal** if

$$\mathrm{Dom}(A_1) + \mathrm{Dom}(A_2) = \mathrm{Dom}(\dot{A}^*).$$

Proposition 3.4.8. *Let A_1 and A_2 be self-adjoint extensions of a closed densely defined symmetric operator $\dot{A}$. Then the following statements are equivalent:*

1) *A_1 and A_2 are transversal;*

2) $\operatorname{Ran}\left((A_1-\lambda I)^{-1}-(A_2-\lambda I)^{-1}\right)=\mathfrak{N}_\lambda$ *for some (and then for all) non-real* λ.

Proof. Clearly, for all λ, $\operatorname{Im}\lambda\neq 0$,

$$\operatorname{Ran}\left((A_1-\lambda I)^{-1}-(A_2-\lambda I)^{-1}\right)\subseteq\mathfrak{N}_\lambda.$$

If $\operatorname{Ran}\left((A_1-\lambda I)^{-1}-(A_2-\lambda I)^{-1}\right)=\mathfrak{N}_\lambda$, then $\mathfrak{N}_\lambda\subset\operatorname{Dom}(A_1)+\operatorname{Dom}(A_2)$. This yields

$$\operatorname{Dom}(A_1)+\operatorname{Dom}(A_2)=\operatorname{Dom}(\dot{A}^*).$$

Conversely, if A_1 and A_2 are transversal, then every vector $f\in\mathfrak{N}_\lambda$ can be represented as

$$f=f_1+f_2,\ f_1\in\operatorname{Dom}(A_1),\ f_2\in\operatorname{Dom}(A_2).$$

Then it follows that

$$\left((A_1-\lambda I)^{-1}-(A_2-\lambda I)^{-1}\right)(A_1-\lambda I)f_1=f.$$

Therefore, $\operatorname{Ran}\left((A_1-\lambda I)^{-1}-(A_2-\lambda I)^{-1}\right)=\mathfrak{N}_\lambda$. □

Let A_1 and A_2 be self-adjoint extensions of $\dot{A}$ and let

$$\operatorname{Dom}(A_k)=\operatorname{Dom}(\dot{A})\oplus(I-U_k)\mathfrak{N}_i,\quad k=1,2. \tag{3.35}$$

Since $U_k=(A_k-iI)(A_k+iI)^{-1}\upharpoonright\mathfrak{N}_i$ and

$$U_1-U_2=2i\left(A_2+iI)^{-1}-(A_1+iI)^{-1}\right)\upharpoonright\mathfrak{N}_i,$$

it follows from Proposition 3.4.8 that A_1 and A_2 are mutually transversal if and only if

$$\begin{aligned}\operatorname{Ran}(U_1-U_2)=\mathfrak{N}_{-i}&\iff\operatorname{Ran}(I-U_1^{-1}U_2)=\mathfrak{N}_i\\&\iff(I-U_1)\mathfrak{N}_i\dot{+}(I-U_2)\mathfrak{N}_i=\mathfrak{N}_i\oplus\mathfrak{N}_{-i}=\mathfrak{M}.\end{aligned}$$

Besides, two transversal self-adjoint extensions are relatively prime. The transversality of A_1 and A_2 yields the direct decompositions of $\mathcal{H}_+$,

$$\mathcal{H}_+=\operatorname{Dom}(A_1)\dot{+}(I-U_2)\mathfrak{N}_i=\operatorname{Dom}(A_2)\dot{+}(I-U_1)\mathfrak{N}_i. \tag{3.36}$$

Set

$$\mathfrak{M}_{A_k}:=(I-U_k)\mathfrak{N}_i=\operatorname{Dom}(A_k)\ominus\operatorname{Dom}(\dot{A}),\quad k=1,2,$$

Then the following conditions are equivalent:

1. A_1 and A_2 are transversal self-adjoint extensions of $\dot{A}$;
2. $\mathcal{H}_+=\operatorname{Dom}(A_1)\dot{+}\mathfrak{M}_{A_2}$;
3. $\mathcal{H}_+=\operatorname{Dom}(A_2)\dot{+}\mathfrak{M}_{A_1}$.

Theorem 3.4.9. *Let $\dot{A}$ be a densely defined closed symmetric operator in $\mathcal{H}$ with equal deficiency numbers and let $\hat{A}$ be a self-adjoint extension of $\dot{A}$. Then there is a bijective correspondence between all transversal to $\hat{A}$ self-adjoint extensions $\tilde{A}$ and all t-self-adjoint bi-extensions of $\dot{A}$ whose quasi-kernel is $\hat{A}$. This correspondence is given by the formula*

$$\mathbb{A} = \dot{A}^* - \mathcal{R}^{-1}\dot{A}^*(I - \mathcal{P}_{\hat{A}\tilde{A}}),$$

where $\mathcal{P}_{\hat{A}\tilde{A}}$ is a projector in $\mathcal{H}_+$ onto $\mathrm{Dom}(\hat{A})$ *corresponding to the decomposition*

$$\mathcal{H}_+ = \mathrm{Dom}(\hat{A}) \dot{+} \mathfrak{M}_{\tilde{A}}.$$

Proof. Let

$$\mathrm{Dom}(\hat{A}) = \mathrm{Dom}(\dot{A}) \oplus (I - W)\mathfrak{N}_i = \mathrm{Dom}(\dot{A}) \oplus \mathfrak{M}_{\hat{A}}.$$

By Theorem 3.4.6 there is a bijective correspondence between all t-self-adjoint bi-extensions with quasi-kernel $\hat{A}$ and all $(\cdot)$-isometric operators U, $\mathrm{Dom}(U) = \mathfrak{N}_i$, $\mathrm{Ran}(U) = \mathfrak{N}_{-i}$ such that

$$\mathfrak{M}_{\hat{A}} \dot{+} (I - U)\mathfrak{N}_i = \mathfrak{M}.$$

Define $\tilde{A}$ by

$$\mathrm{Dom}(\tilde{A}) = \mathrm{Dom}(\dot{A}) \oplus (I - U)\mathfrak{N}_i, \quad \tilde{A} = \dot{A}^* \upharpoonright \mathrm{Dom}(\tilde{A}).$$

Then $\tilde{A}$ is a self-adjoint extension of $\dot{A}$ transversal to $\hat{A}$ and $\mathfrak{M}_{\tilde{A}} = (I - U)\mathfrak{N}_i$. Hence $\mathcal{H}_+ = \mathrm{Dom}(\hat{A}) \dot{+} \mathfrak{M}_{\tilde{A}}$.

Define a $(+)$-self-adjoint operator S in $\mathfrak{M}$ by (3.33). Then

$$(S - \frac{i}{2}\mathfrak{J})(I - U)\varphi = -i(I + U)\varphi = -\dot{A}^*(I - U)\varphi,$$

for all $\varphi \in \mathfrak{N}_i$ and

$$(S - \frac{i}{2}\mathfrak{J})(I - W)f = 0, \; f \in \mathfrak{N}_i.$$

Let $\mathbb{A} = \dot{A}^* + \mathcal{R}^{-1}(S - \frac{i}{2}\mathfrak{J})P^+_{\mathfrak{M}}$. Then

$$\mathbb{A} = \dot{A}^* - \mathcal{R}^{-1}\dot{A}^*(I - \mathcal{P}_{\hat{A}\tilde{A}}).$$ □

Let $\mathfrak{N}_1 = \mathfrak{N}_i$, $\mathfrak{N}_2 = \mathfrak{N}_{-i}$, and let P_k $(k = 1, 2)$ be a $(+)$-orthogonal projection operator from $\mathfrak{M}$ onto $\mathfrak{N}_k$. If $\mathbb{A}$ is a self-adjoint bi-extension of a closed symmetric operator $\dot{A}$ then it takes a form of (3.26), where Q is of the form

$$Q = S - \frac{i}{2}\mathfrak{J}.$$

It is easy to see that Q can be written in the form

$$Q = \begin{pmatrix} Q_{11} & Q_{12} \\ Q_{21} & Q_{22} \end{pmatrix},$$

where Q_{jk} operates from $P_k\mathfrak{M}$ into $P_j\mathfrak{M}$.

Theorem 3.4.10. *Let A be a self-adjoint extension of a closed symmetric operator $\dot{A}$ with* $\mathrm{Dom}(A)$ *defined by von Neumann's formulas* (1.15) *using isometric operator $\mathcal{U}$. Then A is a quasi-kernel of a bi-extension $\mathbb{A}$ ($\mathbb{A} = \mathbb{A}^*$) if and only if*

$$\begin{aligned} Q_{11} + Q_{12}\mathcal{U} &= 0, \\ Q_{21} + Q_{22}\mathcal{U} &= 0. \end{aligned} \tag{3.37}$$

The proof of this theorem immediately follows from Theorem 3.4.1.

Let now $\dot{A}$ be a closed densely defined symmetric operator with deficiency indices (n, n). It is not hard to see that according to Theorem 3.4.1 any self-adjoint bi-extension $\mathbb{A}$ of $\dot{A}$ takes a form

$$\begin{aligned} \mathbb{A} = \dot{A}^* &+ \sum_{k=1}^{n} \left[\sum_{j=1}^{n} a_{kj}(\cdot, \alpha_j) + \sum_{j=1}^{n} b_{kj}(\cdot, \beta_j) \right] \alpha_k \\ &+ \sum_{k=1}^{n} \left[\sum_{j=1}^{n} c_{kj}(\cdot, \alpha_j) + \sum_{j=1}^{n} d_{kj}(\cdot, \beta_j) \right] \beta_k, \end{aligned} \tag{3.38}$$

where the coefficient block-matrix $\begin{pmatrix} \mathcal{A} & \mathcal{B} \\ \mathcal{C} & \mathcal{D} \end{pmatrix}$ satisfies the relations

$$\begin{aligned} a_{kj} &= \overline{a_{jk}}, \quad d_{kj} = \overline{d_{jk}}, \ (k \neq j), \quad \mathcal{C} = \mathcal{B}^*, \\ \mathrm{Im}\, a_{jj} &= -\frac{1}{2}, \ \mathrm{Im}\, d_{jj} = \frac{1}{2}, \quad (k, j = 1, \dots, n), \end{aligned}$$

and $\{\alpha_j\}$, $\{\beta_j\}$, $(j = 1, \dots, n)$ are $(-)$-orthonormal bases in subspaces $\mathcal{R}^{-1}(\mathfrak{N}_i)$ and $\mathcal{R}^{-1}(\mathfrak{N}_{-i})$, respectively.

Theorem 3.4.11. *Let $\mathbb{A}$ be a self-adjoint bi-extension of a closed symmetric densely defined operator with deficiency indices (n, n) given by* (3.38)*. Then $\mathbb{A}$ is a t-self-adjoint bi-extension of $\dot{A}$ if and only if*

$$\mathrm{rank} \begin{pmatrix} \mathcal{A} & \mathcal{B} \\ \mathcal{C} & \mathcal{D} \end{pmatrix} = n. \tag{3.39}$$

Theorem 3.4.11 implies that if (3.39) holds, then any self-adjoint extension A of $\dot{A}$ is described by the formula $Ah = \dot{A}^* h$, where $h \in \mathcal{H}_+$ satisfies the conditions

$$\begin{aligned} \sum_{j=1}^{n} a_{kj}(h, \alpha_j) + \sum_{j=1}^{n} b_{kj}(h, \beta_j) &= 0, \\ \sum_{j=1}^{n} c_{kj}(h, \alpha_j) + \sum_{j=1}^{n} d_{kj}(h, \beta_j) &= 0. \end{aligned} \tag{3.40}$$

Formula (3.40) above provides the abstract set of boundary conditions for a self-adjoint extension A of $\dot{A}$ under assumption (3.39).

Chapter 4

Quasi-self-adjoint Extensions

In this chapter we consider quasi-self-adjoint extensions of, generally speaking, non-densely defined symmetric operators and establish analogues of von Neumann's and Krasnoselkiĭ's formulas in cases of direct and indirect decompositions of their domains. The quasi-self-adjoint bi-extensions and the so-called $(*)$-extensions (with exit into triplets of rigged Hilbert spaces) of symmetric operators will be introduced. We also present an analysis of these extensions together with their description and parametrization.

4.1 Quasi-self-adjoint extensions of symmetric operators

Once again, let $\dot{A}$ be a closed symmetric operator in a Hilbert space $\mathcal{H}$ and let $\mathcal{H}_+ \subset \mathcal{H} \subset \mathcal{H}_-$ be a rigged Hilbert space generated by $\dot{A}$ with $\mathcal{H}_+ = \mathrm{Dom}(\dot{A}^*)$ (see Section 2.2). A closed densely defined linear operator T acting on a Hilbert space $\mathcal{H}$ is called a **quasi-self-adjoint extension** of a closed symmetric operator $\dot{A}$ if

$$T \supset \dot{A} \quad \text{and} \quad T^* \supset \dot{A}.$$

For such operators

$$\dot{A}^* \supset PT, \qquad \dot{A}^* \supset PT^*, \tag{4.1}$$

where P is the $(\cdot)$-orthogonal projection of $\mathcal{H}$ onto $\mathcal{H}_0 = \overline{\mathrm{Dom}(\dot{A})}$. Therefore, $\mathrm{Dom}(T) \subset \mathcal{H}_+$, $\mathrm{Dom}(T^*) \subset \mathcal{H}_+$. A quasi-self-adjoint extension T of a closed symmetric operator $\dot{A}$ is called **regular** if PT and PT^* are closed linear operators in $\mathcal{H}$.

Definition 4.1.1. A quasi-self-adjoint extension T of a closed symmetric operator $\dot{A}$ belongs to the **class** $\Omega(\dot{A})$ if the resolvent set $\rho(T)$ is not empty.

Clearly, self-adjoint extensions of $\dot{A}$ (if they exist) belong to the class $\Omega(\dot{A})$ and $T \in \Omega(\dot{A})$ if and only if $T^* \in \Omega(\dot{A})$.

Theorem 4.1.2. *The following statements about a quasi-self-adjoint extension T of a closed symmetric operator $\dot{A}$ are equivalent:*

(i) *The linear manifolds* $\mathrm{Dom}(T)$ *and* $\mathrm{Dom}(T^*)$ *are $(+)$-closed.*

(ii) *The operators T and T^* are $(+,\cdot)$-bounded.*

(iii) *T is a regular quasi-self-adjoint extension of $\dot{A}$.*

Proof. First we prove that statement (i) is equivalent to the statement (ii). Both operators T and T^* are closed and thus $(+,\cdot)$-closed. By the Closed Graph Theorem the operator T (T^*, respectively) is $(+,\cdot)$-bounded if and only if $\mathrm{Dom}(T)$ ($\mathrm{Dom}(T^*)$, respectively) is $(+)$-closed.

Now let us show that (ii) $\Leftrightarrow$ (iii). It follows from (4.1) that densely defined in $\mathcal{H}$ operators PT and PT^* admit closure in $\mathcal{H}$, that we will denote by $\overline{PT}$ and $\overline{PT^*}$, respectively. It is enough to show that $\mathrm{Dom}(\overline{PT})$ ($\mathrm{Dom}(\overline{PT^*})$) coincides with the $(+)$-closure of $\mathrm{Dom}(T)$ ($\mathrm{Dom}(T^*)$). An element $f \in \mathcal{H}$ belongs to $\mathrm{Dom}(\overline{PT})$ if and only if there exists such a sequence $\{f_n\} \subset \mathrm{Dom}(PT) = \mathrm{Dom}(T)$ that $f_n \xrightarrow{(\cdot)} f$,

$$\dot{A}^* f_n = PTf_n \xrightarrow{(\cdot)} \overline{PT} = \dot{A}^* f,$$

or $f_n \xrightarrow{(+)} f$. The latter is equivalent to that f belongs to the $(+)$-closure of $\mathrm{Dom}(T)$. Similar reasoning can be used to show that the same is true about $\mathrm{Dom}(T^*)$. □

Thus, a necessary condition for the existence of regular quasi-self-adjoint extensions of an operator $\dot{A}$ is that it is $(+,\cdot)$-bounded, that is, regular. It is obvious that in the case $\overline{\mathrm{Dom}(\dot{A})} = \mathcal{H}$ every quasi-self-adjoint extension T of $\dot{A}$ is regular, since $\dot{A}^* f = Tf$, $\|Tf\| = \|\dot{A}^* f\| \leq \|f\|_+$ for all $f \in \mathrm{Dom}(T)$, and, similarly, $\|T^* f\| \leq \|f\|_+$ $(\forall f \in \mathrm{Dom}(T^*)$. The following proposition is more general.

Theorem 4.1.3. *If* $\operatorname{codim} \overline{\mathrm{Dom}(\dot{A})} < \infty$, *then every quasi-self-adjoint extension T of $\dot{A}$ is regular.*

Proof. It is sufficient to show that T is $(+,\cdot)$-bounded. Since

$$\|Tf\|^2 < \|f\|^2 + \|Tf\|^2 = \|(I + T^*T)^{1/2} f\|, \quad (f \neq 0),$$

then the operator $\Gamma = T(I + T^*T)^{-\frac{1}{2}}$ satisfies the condition

$$\|\Gamma f\| < \|f\|, \qquad (f \in \mathcal{H}, \quad f \neq 0).$$

Let P be, as above, the orthoprojection operator in $\mathcal{H}$ onto $\overline{\mathrm{Dom}(\dot{A})}$ and $Q = I - P$. The operator $Q\Gamma$ then is finite-dimensional and $\|Q\Gamma f\| < \|f\|$ $(f \in \mathcal{H})$ and hence $\|Q\Gamma\| = q < 1$. Consequently, we have that

$$\|Q\Gamma f\| \leq q\|f\|, \quad f \in \mathcal{H}.$$

Letting $g = (I + T^*T)^{-\frac{1}{2}} f$ we have that $\Gamma f = Tg$,

$$\|f\|^2 = \|(I + T^*T)^{\frac{1}{2}} g\|^2 = \|g\|^2 + \|Tg\|^2,$$

and hence

$$\|QTg\|^2 \leq q^2(\|g\|^2 + \|Tg\|^2).$$

Since $\dot{A}^* g = PTg$, $(g \in \mathrm{Dom}(T))$, then

$$\begin{aligned} \|g\|_+^2 &= \|g\|^2 + \|\dot{A}^* g\|^2 = \|g\|^2 + \|Tg\|^2 - \|QTg\|^2 \\ &\geq (1 - q^2)(\|g\|^2 + \|Tg\|^2) > (1 - q^2)\|Tg\|^2. \end{aligned}$$

□

Following Section 1.7 we call the operators $\mathcal{U} \in [\mathfrak{N}_i, \mathfrak{N}_{-i}]$ and $\mathcal{W} \in [\mathfrak{N}_{-i}, \mathfrak{N}_i]$ **admissible** if

$$\mathcal{U}\phi = V\phi,\ \phi \in \mathfrak{B}_i, \qquad \text{and} \qquad \mathcal{W}\phi = V^{-1}\phi,\ \phi \in \mathfrak{B}_{-i},$$

respectively, imply that $\phi = 0$, where V is the exclusion operator defined in Section 1.7 by (2.24). The admissibility of $\mathcal{U}$ and $\mathcal{W}$ is equivalent to

$$\mathrm{Dom}(\dot{A}) \cap (\mathcal{U} - I)\mathfrak{N}_i = \{0\} \quad \text{and} \quad \mathrm{Dom}(\dot{A}) \cap (\mathcal{W} - I)\mathfrak{N}_{-i} = \{0\},$$

respectively (see Theorem 1.7.4). The lemma below is the analogue of the Krasnoselskiĭ formulas in the case of indirect decomposition of the domains of quasi-self-adjoint extensions.

Lemma 4.1.4. *If $T \in \Omega(\dot{A})$ and $-i \in \rho(T)$, then*

$$\begin{aligned} \mathrm{Dom}(T) &= \mathrm{Dom}(\dot{A}) \dotplus (I - \mathcal{U})\mathfrak{N}_i, \\ \mathrm{Dom}(T^*) &= \mathrm{Dom}(\dot{A}) \dotplus (\mathcal{U}^* - I)\mathfrak{N}_{-i}, \end{aligned} \tag{4.2}$$

where $\mathcal{U}$ and $\mathcal{U}^*$ are admissible operators in $[\mathfrak{N}_i, \mathfrak{N}_{-i}]$ and $[\mathfrak{N}_{-i}, \mathfrak{N}_i]$, respectively. Further, if $f \in \mathrm{Dom}(T)$ or $f \in \mathrm{Dom}(T^*)$, then

$$\left\{\begin{array}{l} f = g + \varphi - \mathcal{U}\varphi, \\ Tf = \dot{A}g + i\varphi + i\mathcal{U}\varphi, \\ g \in \mathrm{Dom}(\dot{A}), \varphi \in \mathfrak{N}_i, \\ f \in \mathrm{Dom}(T), \end{array}\right. \quad \text{or} \quad \left\{\begin{array}{l} f = g + \mathcal{U}^*\psi - \psi, \\ T^*f = \dot{A}g + i\mathcal{U}^*\psi + i\psi, \\ g \in \mathrm{Dom}(\dot{A}), \psi \in \mathfrak{N}_{-i}, \\ f \in \mathrm{Dom}(T^*). \end{array}\right. \tag{4.3}$$

Moreover, the formulas (4.2) and (4.3) establish a bijection between the set of all quasi-self-adjoint extensions $T \in \Omega(\dot{A})$ with $-i \in \rho(T)$ and the set of all admissible linear operators $\mathcal{U} \in [\mathfrak{N}_i, \mathfrak{N}_{-i}]$ and $\mathcal{U}^* \in [\mathfrak{N}_{-i}, \mathfrak{N}_i]$.

Proof. Let us consider the Cayley transform (1.19) of the operator T

$$U = (T - iI)(T + iI)^{-1} = I - 2i(T + iI)^{-1}.$$

Then

$$U^* = (T^* + iI)(T^* - iI)^{-1}.$$

Recall the notation $\mathfrak{M}_\lambda = (\dot{A} - \lambda I)\mathrm{Dom}(\dot{A})$ (see Chapter 1). Clearly,

$$U : \mathfrak{M}_{-i} \to \mathfrak{M}_i, \; U^* : \mathfrak{M}_i \to \mathfrak{M}_{-i},$$

$$(U - I)\mathfrak{M}_{-i} = (U^* - I)\mathfrak{M}_i = \mathrm{Dom}(\dot{A}).$$

It follows from the orthogonal decompositions

$$\mathcal{H} = \mathfrak{M}_{-i} \oplus \mathfrak{N}_i = \mathfrak{M}_i \oplus \mathfrak{N}_{-i}$$

that $U : \mathfrak{N}_i \to \mathfrak{N}_{-i}$ and $U^* : \mathfrak{N}_{-i} \to \mathfrak{N}_i$. Let

$$\mathcal{U} = U \restriction \mathfrak{N}_i.$$

Then $\mathcal{U} \in [\mathfrak{N}_i, \mathfrak{N}_{-i}]$ and for $\mathcal{U}^* \in [\mathfrak{N}_{-i}, \mathfrak{N}_i]$ one has $\mathcal{U}^* = U^* \restriction \mathfrak{N}_{-i}$. Since

$$\begin{cases} f = i(U - I)x \\ Tf = (U + I)x \end{cases}, \quad \begin{cases} h = i(I - U^*)x \\ T^*h = (U^* + I)x \end{cases}, \quad x \in \mathcal{H},$$

and because $\mathrm{Dom}(T)$ and $\mathrm{Dom}(T^*)$ are dense in $\mathcal{H}$, we get the equalities

$$\ker(U - I) = \ker(U^* - I) = \{0\},$$

$$\mathrm{Dom}(\dot{A}) \cap (U - I)\mathfrak{N}_i = \{0\}, \; \mathrm{Dom}(\dot{A}) \cap (U^* - I)\mathfrak{N}_i = \{0\},$$

and relations (4.2), (4.3).

Now suppose that $\mathcal{U} \in [\mathfrak{N}_i, \mathfrak{N}_{-i}]$ and $\mathcal{U}^* \in [\mathfrak{N}_{-i}, \mathfrak{N}_i]$ are both admissible operators. Using formulas (4.2) we define the operator T. Then $T \supset \dot{A}$ and

$$\ker(T + iI) = \{0\}, \quad \mathrm{Ran}(T + iI) = \mathcal{H}.$$

Moreover, formulas (4.3) yield that T is a closed operator in $\mathcal{H}$. Hence $-i$ is a regular point of T. Let us show that $\mathrm{Dom}(T)$ is dense in $\mathcal{H}$. If $x \in \mathcal{H}$ is orthogonal to $\mathrm{Dom}(T)$, then $x \in \mathfrak{L} = \mathcal{H} \ominus \mathrm{Dom}(\dot{A})$ and

$$(x, \varphi - \mathcal{U}\varphi) = 0,$$

for all $\varphi \in \mathfrak{N}_i$. It follows that $P_{\mathfrak{N}_i}x = \mathcal{U}^* P_{\mathfrak{N}_{-i}}x$. Since $\mathcal{U}^*$ is admissible, we get that $P_{\mathfrak{N}_i}x = P_{\mathfrak{N}_{-i}}x = 0 \iff x = 0$.

Let $U = (T - iI)(T + iI)^{-1}$ and $U_i(\dot{A}) = (\dot{A} - iI)(\dot{A} + iI)^{-1}$ be the Cayley transforms of T and $\dot{A}$, respectively. Then, clearly,

$$U = U_i(\dot{A})P_{\mathfrak{M}_{-i}} + \mathcal{U}P_{\mathfrak{N}_i}.$$

Hence

$$U^* = U_i^{-1}(\dot{A})P_{\mathfrak{M}_i} + \mathcal{U}^* P_{\mathfrak{N}_{-i}},$$

where $\mathcal{U}^* \in [\mathfrak{N}_{-i}, \mathfrak{N}_i]$ is the adjoint operator to $\mathcal{U} \in [\mathfrak{N}_i, \mathfrak{N}_{-i}]$. Let T^* be the adjoint to T. Because $U^* = (T^* + iI)(T^* - iI)^{-1}$, we get that

$$\mathrm{Dom}(T^*) = \mathrm{Dom}(\dot{A}) \dot{+} (\mathcal{U}^* - I)\mathfrak{N}_{-i},$$

and

$$T^*(g + \mathcal{U}^*\psi - \psi) = \dot{A}g + i\mathcal{U}^*\psi + i\psi,$$

for all $g \in \mathrm{Dom}(\dot{A})$, $\psi \in \mathfrak{N}_{-i}$. Thus $T^* \supset \dot{A}$. □

Theorem 4.1.5. *The following statements are equivalent for $T \in \Omega(\dot{A})$, $-i \in \rho(T)$:*

1. *T is a regular extension of $\dot{A}$;*
2. *$(I-\mathcal{U})\mathfrak{N}_i$ and $(\mathcal{U}^*-I)\mathfrak{N}_{-i}$ are $(+)$-closed, and the minimal angles (in the $(+)$-metric) between $\mathrm{Dom}(\dot{A})$ and $(I-\mathcal{U})\mathfrak{N}_i$ and between $\mathrm{Dom}(\dot{A})$ and $(\mathcal{U}^*-I)\mathfrak{N}_{-i}$ are positive;*
3. *$P^+_{\mathfrak{M}}((I-\mathcal{U})\mathfrak{N}_i)$ and $P^+_{\mathfrak{M}}((\mathcal{U}^*-I)\mathfrak{N}_{-i})$ are $(+)$-closed linear manifolds;*
4. *$(I-\mathcal{U})\mathfrak{N}_i$ and $(\mathcal{U}^*-I)\mathfrak{N}_{-i}$ are $(+)$-closed, and $P^+_{\mathfrak{M}} \upharpoonright (I-\mathcal{U})\mathfrak{N}_i$ and $P^+_{\mathfrak{M}} \upharpoonright (\mathcal{U}^*-I)\mathfrak{N}_{-i}$ are homeomorphisms.*

Proof. The proof of the theorem immediately follows from Lemma 2.5.2 if one sets $H = \mathcal{H}_+$, $\mathcal{L}_1 = \mathrm{Dom}(\dot{A})$, and $\mathcal{L}_2 = (I - \mathcal{U})\mathfrak{N}_i$ and Theorem 4.1.2. □

We note that when the conditions (1)–(4) of the theorem are satisfied, there are constants $d_1 > 0$ and $d_2 > 0$ such that for all $\varphi \in \mathfrak{N}_i$, $\psi \in \mathfrak{N}_{-i}$, $g \in \mathrm{Dom}(\dot{A})$ we have

$$\|g + \varphi - \mathcal{U}\varphi\|_+ \geq d_1\|\varphi\|, \qquad \|g + \mathcal{U}^*\psi - \psi\|_+ \geq d_2\|\psi\|.$$

Remark 4.1.6. Let $T \in \Omega(\dot{A})$ and $\lambda \in \rho(T)$. Then using the Cayley transform $(T - \bar{\lambda}I)(T - \lambda I)^{-1}$ and its restriction

$$M_\lambda = (T - \bar{\lambda}I)(T - \lambda I)^{-1} \upharpoonright \mathfrak{N}_{\bar{\lambda}} : \mathfrak{N}_{\bar{\lambda}} \to \mathfrak{N}_\lambda,$$

one can prove that the relations

$$\begin{aligned} \mathrm{Dom}(T) &= \mathrm{Dom}(\dot{A}) \dot{+} (I - M_\lambda)\mathfrak{N}_{\bar{\lambda}}, \\ \mathrm{Dom}(T^*) &= \mathrm{Dom}(\dot{A}) \dot{+} (I - M^*_\lambda)\mathfrak{N}_\lambda, \end{aligned} \tag{4.4}$$

and

$$\begin{aligned} T(g + (I - M_\lambda)\varphi) &= \dot{A}g + \bar{\lambda}\varphi - \lambda M_\lambda\varphi, \; g \in \mathrm{Dom}(\dot{A}), \; \varphi \in \mathfrak{N}_{\bar{\lambda}}, \\ T^*(g + (I - M^*_\lambda)\psi) &= \dot{A}g + \lambda\psi - \bar{\lambda}M^*_\lambda\psi, \; g \in \mathrm{Dom}(\dot{A}), \; \psi \in \mathfrak{N}_\lambda \end{aligned} \tag{4.5}$$

hold true.

Recall that the operator T in the Hilbert space $\mathcal{H}$ is called **dissipative** if $\operatorname{Im}(Tf, f) \geq 0$ for all $f \in \operatorname{Dom}(T)$. A dissipative operator T is called **maximal dissipative (m-dissipative)** if one of the equivalent conditions is satisfied:

- T has no dissipative extensions in $\mathcal{H}$;
- the resolvent set $\rho(T)$ contains a point from the open lower half-plane;
- T is densely defined, closed, and $-T^*$ is a dissipative operator.

A closed dissipative operator T possess the properties:

1. the field of regularity of T contains the open lower half-plane;
2. the numerical range
$$W(T) = \{(Tf, f), \ ||f|| = 1\}$$
of T is contained in the closed upper half-plane.

The resolvent set $\rho(T)$ of m-dissipative operator T contains the open lower half-plane and
$$||(T - \lambda I)^{-1}|| \leq \frac{1}{|\operatorname{Im}\lambda|}, \quad \operatorname{Im}\lambda < 0.$$
The Cayley transform $U(T) = (T - iI)(T + iI)^{-1}$ of m-dissipative operator T is a contraction and if T is an extension of a closed symmetric operator $\dot{A}$, then $U(T)$ is a contractive extension of the Cayley transform of A, which is an isometry from $\mathfrak{M}_{-i}$ onto $\mathfrak{M}_i$. The next statement is well known [219].

Proposition 4.1.7. *Let $\dot{A}$ be a closed symmetric operator in $\mathcal{H}$. If T is an m-dissipative extension of $\dot{A}$, then T^* is also an extension of $\dot{A}$. Moreover, if $\dot{A}$ is densely defined, then each m-dissipative extension of $\dot{A}$ is a restriction of $\dot{A}^*$.*

We complete Lemma 4.1.4 by the following assertion.

Lemma 4.1.8. *Each closed symmetric operator $\dot{A}$ admits m-dissipative extensions. Moreover, there is a one-to-one correspondence between all admissible contractions $\mathcal{U} : \mathfrak{N}_i \to \mathfrak{N}_{-i}$ and all m-dissipative extensions of $\dot{A}$.*

It should be noted that in view of Proposition 4.1.7, if $\mathcal{U} : \mathfrak{N}_i \to \mathfrak{N}_{-i}$ is an admissible contraction, then $\mathcal{U}^* : \mathfrak{N}_{-i} \to \mathfrak{N}_i$ is automatically admissible.

We recall that in Section 2.3 we introduced an inner product $(\cdot, \cdot)_1$ on $\mathcal{H}_+$ defined by (2.25). Then the (1)-orthogonal decomposition
$$\mathcal{H}_+ = \operatorname{Dom}(\dot{A}) \oplus \mathfrak{N}'_i \oplus \mathfrak{N}'_{-i} \oplus \mathfrak{N},$$
holds true. The following theorem is a generalization of the von Neumann formulas to the case of quasi-self-adjoint extensions and direct decompositions of their domains.

Theorem 4.1.9. *Let $\dot{A}$ be a regular closed symmetric operator.*

I. *To each $T \in \Omega(\dot{A})$, $-i \in \rho(T)$, there corresponds an operator $M \in [\mathfrak{N}'_i \oplus \mathfrak{N}, \mathfrak{N}'_{-i} \oplus \mathfrak{N}]$ such that*

$$\ker(M + I) = \{0\}, \quad \ker(M^* + I) = \{0\}, \tag{4.6}$$

and

$$\begin{aligned} &\mathrm{Dom}(T) = \mathrm{Dom}(\dot{A}) \oplus (M + I)(\mathfrak{N}'_i \oplus \mathfrak{N}), \\ &T(g + (M + I)\varphi) = \dot{A}g + \dot{A}^*(I + M)\varphi + i(\dot{A}\dot{A}^* + I)P^+_{\mathfrak{N}}(I - M)\varphi, \\ &\qquad g \in \mathrm{Dom}(\dot{A}),\ \varphi \in \mathfrak{N}'_i \oplus \mathfrak{N}, \end{aligned} \tag{4.7}$$

$$\begin{aligned} &\mathrm{Dom}(T^*) = \mathrm{Dom}(\dot{A}) \oplus (M^* + I)(\mathfrak{N}'_{-i} \oplus \mathfrak{N}), \\ &T^*(g + (M^* + I)\psi) = \dot{A}g + \dot{A}^*(I + M^*)\psi + i(\dot{A}\dot{A}^* + I)P^+_{\mathfrak{N}}(M^* - I)\psi, \\ &\qquad g \in \mathrm{Dom}(\dot{A}),\ \psi \in \mathfrak{N}'_{-i} \oplus \mathfrak{N}. \end{aligned} \tag{4.8}$$

II. *Conversely, for each operator $M \in [\mathfrak{N}'_i \oplus \mathfrak{N}, \mathfrak{N}'_{-i} \oplus \mathfrak{N}]$ satisfying (4.6), formulas (4.7) and (4.8) determine an operator $T \in \Omega(\dot{A})$, $-i \in \rho(T)$, and its adjoint T^*.*

III. *An operator T is a regular extension of $\dot{A}$ if and only if the linear manifolds* $\mathrm{Ran}(M + I)$ *and* $\mathrm{Ran}(M^* + I)$ *are (+)-closed.*

IV. *The operator T is an m-dissipative extension of $\dot{A}$ if and only if the operator M is a (1)-contraction.*

Proof. Define the operator $M \in [\mathfrak{N}'_i \oplus \mathfrak{N}, \mathfrak{N}'_{-i} \oplus \mathfrak{N}]$ as

$$\begin{cases} h = P^+_{\mathfrak{M}}\varphi, \\ Mh = -P^+_{\mathfrak{M}}\mathcal{U}\varphi, \end{cases} \quad \varphi \in \mathfrak{N}_i, \tag{4.9}$$

where $\mathcal{U}$ is an admissible operator defining T via formulas (4.2) and (4.3). By (2.27) we have

$$\begin{cases} h = i(2P^+_{\mathfrak{N}'_i} + P^+_{\mathfrak{N}})(\dot{A}^* + iI)^{-1}\varphi, \\ Mh = i(2P^+_{\mathfrak{N}'_{-i}} + P^+_{\mathfrak{N}})(\dot{A}^* - iI)^{-1}\mathcal{U}\varphi, \end{cases} \quad \varphi \in \mathfrak{N}_i.$$

Because

$$\begin{cases} \varphi = -i\left(\dot{A}^* + iI\right)\left(\frac{1}{2}P^+_{\mathfrak{N}'_i} + P^+_{\mathfrak{N}}\right)h, \\ \mathcal{U}\varphi = -i\left(\dot{A}^* - iI\right)\left(\frac{1}{2}P^+_{\mathfrak{N}'_{-i}} + P^+_{\mathfrak{N}}\right)Mh, \end{cases} \tag{4.10}$$

we obtain

$$(I - \mathcal{U})\varphi = (I + M)h - i\dot{A}^* P^+_{\mathfrak{N}}(I - M)h.$$

Since $\dot{A}^* P^+_{\mathfrak{N}}(h - Mh) \in \mathrm{Dom}(\dot{A})$ (see Theorem 2.3.1), it follows that $\ker(I + M) = \{0\}$ and

$$\mathrm{Dom}(T) = \mathrm{Dom}(\dot{A}) \oplus (I + M)(\mathfrak{N}'_i \oplus \mathfrak{N}).$$

From (4.10) we also have $i(I+\mathcal{U})\varphi = \dot{A}^*(I+M)h + iP^+_{\mathfrak{N}}(I-M)h$.

Now let $f \in \mathrm{Dom}(T)$. Then $f = g + (I+M)h$, where $g \in \mathrm{Dom}(\dot{A})$, $h \in (\mathfrak{N}'_i \oplus \mathfrak{N})$. From (4.3) we obtain

$$\begin{aligned} Tf &= T(g+(I+M)h) = T\left((g+i\dot{A}^*P^+_{\mathfrak{N}}(I-M)h)+(I-\mathcal{U})\varphi\right) \\ &= \dot{A}\left(g+i\dot{A}^*P^+_{\mathfrak{N}}(I-M)h\right)+i(I+\mathcal{U})\varphi \\ &= \dot{A}g+\dot{A}^*(I+M)h+i(\dot{A}\dot{A}^*+I)P^+_{\mathfrak{N}}(I-M)h. \end{aligned}$$

Thus, (4.7) is proved.

Recall that the operators $2^{-\frac{1}{2}}P^+_{\mathfrak{M}}\restriction\mathfrak{N}_i$ and $2^{-\frac{1}{2}}P^+_{\mathfrak{M}}\restriction\mathfrak{N}_{-i}$ are $(\cdot,1)$-isometries from $\mathfrak{N}_i$ and $\mathfrak{N}_{-i}$ onto $\mathfrak{N}'_i\oplus\mathfrak{N}$ and $\mathfrak{N}'_{-i}\oplus\mathfrak{N}$, respectively. It follows from (4.9) that the (1)-adjoint operator $M^* \in [\mathfrak{N}'_{-i}\oplus\mathfrak{N}, \mathfrak{N}'_i\oplus\mathfrak{N}]$ is given by

$$\begin{cases} x = P^+_{\mathfrak{M}}\psi \\ M^*x = -P^+_{\mathfrak{M}}\mathcal{U}^*\psi \end{cases}, \quad \psi\in\mathfrak{N}_{-i}.$$

Arguing as above we obtain the equality $\ker(M^*+I)=\{0\}$ and relations (4.7).

Let now $M \in [\mathfrak{N}'_i\oplus\mathfrak{N}, \mathfrak{N}'_{-i}\oplus\mathfrak{N}]$ satisfy conditions (4.6). Define the operator T by (4.7). Using (4.10) one can check that T is of the form (4.3). Hence, T is a quasi-self-adjoint extension of $\dot{A}$ and for its adjoint T^* relations (4.8) are valid.

Part III of the theorem follows directly from Theorem 4.1.2 and formulas (4.7). Part IV is a consequence of Lemma 4.1.8. □

Let us find the (1)-orthogonal complements

$$[\mathrm{Ran}(M+I)]^{\perp} = \mathfrak{M}\ominus\mathrm{Ran}(M+I),$$

and

$$[\mathrm{Ran}(M^*+I)]^{\perp} = \mathfrak{M}\ominus\mathrm{Ran}(M^*+I),$$

where $M\in[\mathfrak{N}'_i\oplus\mathfrak{N}, \mathfrak{N}'_{-i}\oplus\mathfrak{N}]$. Suppose $\psi\in[\mathrm{Ran}(M^*+I)]^{\perp}$. Then

$$((M^*+I)\varphi,\psi)_1 = 0,\ \varphi\in\mathfrak{N}'_{-i}\oplus\mathfrak{N}.$$

Furthermore, using the (1)-orthogonality relation one can show that

$$\begin{aligned} 0 = ((M^*+I)\varphi,\psi)_1 &= \left((M^*+I)\varphi, P^+_{\mathfrak{N}'_i}\psi + P^+_{\mathfrak{N}'_{-i}}\psi + P^+_{\mathfrak{N}}\psi\right)_1 \\ &= \left(\varphi, M(P^+_{\mathfrak{N}'_i}+P^+_{\mathfrak{N}})\psi\right)_1 + \left(\varphi,(P^+_{\mathfrak{N}'_{-i}}+P^+_{\mathfrak{N}})\psi\right)_1. \end{aligned}$$

Therefore, we have that

$$M(P^+_{\mathfrak{N}'_i}+P^+_{\mathfrak{N}})\psi = -(P^+_{\mathfrak{N}'_{-i}}+P^+_{\mathfrak{N}})\psi. \tag{4.11}$$

Let us set $\phi = (P^+_{\mathfrak{N}'_i} + P^+_{\mathfrak{N}})\psi$. Then (4.11) implies $P^+_{\mathfrak{N}}(M+I)\phi = 0$. Hence, if $\psi \in [\mathrm{Ran}(M^*+I)]^\perp$, then

$$\phi = (P^+_{\mathfrak{N}'_i} + P^+_{\mathfrak{N}})\psi \in \mathrm{Ker}\,\left[P^+_{\mathfrak{N}}(M+I)\phi\right] \quad \text{and} \quad \psi = \phi - P^+_{\mathfrak{N}'_{-i}} M\phi.$$

Let now $\phi \in \mathrm{Ker}\,\left[P^+_{\mathfrak{N}}(M+I)\right]$. Then the vector $\psi = \phi - P^+_{\mathfrak{N}'_{-i}} M\phi$ belongs to $[\mathrm{Ran}(M^*+I)]^\perp$. Indeed,

$$\begin{aligned} -(P^+_{\mathfrak{N}'_{-i}} + P^+_{\mathfrak{N}})\psi &= -P^+_{\mathfrak{N}}\phi + P^+_{\mathfrak{N}'_{-i}} M\phi = P^+_{\mathfrak{N}} M\phi + P^+_{\mathfrak{N}'_{-i}} M\phi \\ &= M\phi = M(P^+_{\mathfrak{N}'_i} + P^+_{\mathfrak{N}})\psi. \end{aligned}$$

Hence,

$$[\mathrm{Ran}(M^*+I)]^\perp = (I - P^+_{\mathfrak{N}'_i} M)\{\mathrm{Ker}\,[P^+_{\mathfrak{N}}(M+I)]\}.$$

It can be shown similarly that

$$[\mathrm{Ran}(M+I)]^\perp = (I - P^+_{\mathfrak{N}'_i} M^*)\{\mathrm{Ker}\,[P^+_{\mathfrak{N}}(M^*+I)]\}. \tag{4.12}$$

The next statement is an application of Theorem 4.1.9 to the case of an m-dissipative extension (cf. Theorem 2.5.7).

Theorem 4.1.10. *Let $T \in \Omega(\dot{A})$ be m-dissipative and regular. Then the operator*

$$PT\restriction(\mathrm{Dom}(T) \cap \mathcal{H}_0)$$

is an m-dissipative extension in $\mathcal{H}_0$ of the operator $P\dot{A}$, where P is the $(\cdot)$-orthogonal projection of $\mathcal{H}$ onto $\mathcal{H}_0 = \overline{\mathrm{Dom}(\dot{A})}$.

Proof. Relations (4.7) and (4.8) are valid for T with a contraction $M \in [\mathfrak{N}'_i \oplus \mathfrak{N}, \mathfrak{N}'_{-i} \oplus \mathfrak{N}]$. Let us show that $\mathrm{Ran}(M^*+I) \cap \mathfrak{N}'_i = \{0\}$. If $\varphi \in \mathfrak{N}'_i$ and $\varphi = (M^*+I)h$, then

$$P^+_{\mathfrak{N}}(M^*+I)h = 0, \quad h \in \mathfrak{N}.$$

It follows that $P_{\mathfrak{N}} M^* h = -h$. Since M^* is a (1)-contraction, we get

$$||M^*h||_1 \le ||h||_1 = ||P^+_{\mathfrak{N}} M^* h||_1 \le ||M^*h||_1.$$

Consequently, $P^+_{\mathfrak{N}_i} M^*h = 0$. This yields the equality $M^*h + h = 0 \iff h = 0$. Hence, $\varphi = 0$.

Now because $\mathrm{Ran}(M^*+I)$ is (1)-closed we get that $\mathrm{Ran}(M^*+I) \neq \mathfrak{M}$. It follows from (4.12) that $\mathcal{N} := \ker(P^+_{\mathfrak{N}}(I+M)) \neq \{0\}$. Let $h \in \mathcal{N}$. Then

$$\begin{aligned} ||P^+_{\mathfrak{N}'_{-i}} Mh||^2 &= \frac{1}{2}||P^+_{\mathfrak{N}'_{-i}} Mh||_1^2 = \frac{1}{2}(||Mh||_1^2 - ||P^+_{\mathfrak{N}} Mh||_1^2) \\ &\le \frac{1}{2}(||h||_1^2 - ||P^+_{\mathfrak{N}} h||_1^2) = \frac{1}{2}(||h||_1^2 - ||P^+_{\mathfrak{N}} h||_1^2) = \frac{1}{2}||P^+_{\mathfrak{N}'_i} h||_1^2 = ||P^+_{\mathfrak{N}'_i} h||^2. \end{aligned}$$

This yields that the operator given by

$$\begin{cases} y = P^+_{\mathfrak{N}'_i} h \\ Wy = P^+_{\mathfrak{N}'_{-i}} Mh \end{cases}, \; h \in \mathcal{N}$$

is well defined and is a $(\cdot)$-contraction with values in $\mathfrak{N}'_{-i}$. Let us show that $\mathrm{Dom}(W) = P^+_{\mathfrak{N}'_i}\mathcal{N} = \mathfrak{N}'_i$. If $\{P^+_{\mathfrak{N}'_i} h_n\}$, where $\{h_n\} \subset \mathcal{N}$, is a (1)-Cauchy sequence, then, since W is a contraction, we get that $\{P^+_{\mathfrak{N}'_{-i}} Mh_n\}$ and hence, $\{(M+I)h_n\}$ are (1)-Cauchy sequences as well. It follows that $\{h_n\}$ (1)-converges. Let $h = \lim_{n\to\infty} h_n = h$. Then $h \in \mathcal{N}$ and $\lim_{n\to\infty} P^+_{\mathfrak{N}'_i} h_n = P^+_{\mathfrak{N}'_i} h \in \mathrm{Dom}(W)$. It follows from the (1)-orthogonal decomposition

$$(M^* + I)\mathfrak{N} \oplus \mathcal{N} = \mathfrak{N}'_i \oplus \mathfrak{N}$$

that if the vector $x \in \mathfrak{N}'_i$ is orthogonal to $\mathrm{Dom}(W)$, then $x = (M^*+I)f$, where $f \in \mathfrak{N}$. Consequently, $P^+_{\mathfrak{N}'_{-i}} Mf = x$ and $P^+_{\mathfrak{N}} M^* f = -f$. The last equality implies (since M^* is a (1)-contraction) that $M^* f = -f$, and, hence, $x = 0$. Thus, $\mathrm{Dom}(W) = \mathfrak{N}'_i$. This yields the relation

$$\mathrm{Dom}(T) \cap \mathcal{H}_0 = \mathrm{Dom}(\dot{A}) \dot{+} (I + W)\mathfrak{N}'_i.$$

Therefore, $PT\restriction(\mathrm{Dom}(T) \cap \mathcal{H}_0)$ is an m-dissipative extension in $\mathcal{H}$ of densely defined in $\mathcal{H}_0$ symmetric operator PA. □

Let T be a closed densely defined operator in a Hilbert space $\mathcal{H}$. The operator $\dot{A}$ defined as

$$\mathrm{Dom}(\dot{A}) = \{f \in \mathrm{Dom}(T) \cap \mathrm{Dom}(T^*) : Tf = T^* f\}, \; \dot{A}f = Tf, \; f \in \mathrm{Dom}(\dot{A})$$

is called the **maximal common symmetric part of T and T^***.

Theorem 4.1.11. *Let $T \in \Omega(\dot{A})$, $-i \in \rho(T)$ and let M be the operator defining T by (4.7). If $\dot{A}$ is a maximal common symmetric part of T and T^*, then*

$$\ker(MM^* - I) = \ker(M^*M - I) = \{0\},$$

and the linear manifold $\mathrm{Dom}(T) + \mathrm{Dom}(T^*)$ *is dense in* $\mathcal{H}_+$.

Proof. Let us assume that $\ker(MM^* - I) \neq \{0\}$. Then there is $\varphi \neq 0$, $\varphi \in \mathfrak{N}'_{-i} \oplus \mathfrak{N}$ such that $MM^*\varphi = \varphi$. Let $\psi = M^*\varphi$ and observe that $M\psi = \varphi$ then. It is easy to see that

$$P^+_{\mathfrak{N}'_i} M^*\varphi = P^+_{\mathfrak{N}'_i}\psi, \; P^+_{\mathfrak{N}} M^*\varphi = P^+_{\mathfrak{N}}\psi, \; P^+_{\mathfrak{N}'_{-i}} M\psi = P^+_{\mathfrak{N}'_{-i}}\varphi, \; P^+_{\mathfrak{N}} M\psi = P^+_{\mathfrak{N}}\varphi,$$

which implies

$$(M + I)\psi = (M^* + I)\varphi, \quad P^+_{\mathfrak{N}}(M^*\varphi - \varphi) = P^+_{\mathfrak{N}}(\psi - M\psi).$$

It follows from (4.7) and (4.8) that for an element

$$f = (M^* + I)\varphi = (M + I)\psi,$$

the equality $Tf = T^*f$ holds true. But according to Theorem 4.1.9, $f \notin \mathrm{Dom}(\dot{A})$ and hence $f = 0$ yielding $\varphi = 0$.

Let $h \in \mathfrak{M}$ be (1)-orthogonal to the linear manifold $\mathrm{Ran}(M + I)$. Then from (4.12) one has

$$M^*(P^+_{\mathfrak{N}'_{-i}} + P^+_{\mathfrak{N}})h = -(P^+_{\mathfrak{N}'_i} + P^+_{\mathfrak{N}})h.$$

Similarly any $g \in \mathfrak{M}$ (1)-orthogonal to $\mathrm{Ran}(M^* + I)$ satisfies the condition

$$M(P^+_{\mathfrak{N}'_i} + P^+_{\mathfrak{N}})g = -(P^+_{\mathfrak{N}'_{-i}} + P^+_{\mathfrak{N}})g.$$

It follows that if $\psi \in \mathfrak{M}$ is (1)-orthogonal to $\mathrm{Ran}(M + I) + \mathrm{Ran}(M^* + I)$, then

$$M^*(P^+_{\mathfrak{N}'_{-i}} + P^+_{\mathfrak{N}})\psi = -(P^+_{\mathfrak{N}'_i} + P^+_{\mathfrak{N}})\psi \quad \text{and} \quad M(P^+_{\mathfrak{N}'_i} + P^+_{\mathfrak{N}})\psi = -(P^+_{\mathfrak{N}'_{-i}} + P^+_{\mathfrak{N}})\psi.$$

Hence, for $f = (P^+_{\mathfrak{N}'_{-i}} + P^+_{\mathfrak{N}})\psi$ we have $MM^*f = f$. Since $\ker(MM^* - I) = \{0\}$, we get $f = 0$ and $\psi = 0$. This yields the density of $\mathrm{Ran}(M + I) + \mathrm{Ran}(M^* + I)$ in $\mathfrak{M}$. Consequently, $\mathrm{Dom}(T) + \mathrm{Dom}(T^*)$ is dense in $\mathcal{H}_+$. □

Observe that the equality $\ker(M^*M - I) = \{0\}$ (respect., $\ker(MM^* - I) = \{0\}$) implies $\ker(MM^* - I) = \{0\}$ (respect., $\ker(M^*M - I) = \{0\}$). Indeed, if $(MM^* - I)f = 0$, then from the equality

$$M^*(MM^* - I)f = (M^*M - I)M^*f$$

follows that $M^*f = 0$. But $||M^*f||_1^2 = ||f||_1^2$. Hence, $f = 0$.

Theorem 4.1.12. *Let $\dot{A}$ be a regular O-operator and let $T \in \Omega(\dot{A})$, $-i \in \rho(T)$. Then $\mathrm{Dom}(T)$ and $\mathrm{Dom}(T^*)$ are dense in $\mathcal{H}_+$. Moreover, if T is regular, then:*

1. *The equalities*

$$\begin{aligned} &\mathrm{Dom}(T) = \mathrm{Dom}(T^*) = \mathcal{H}_+, \\ &T = \dot{A}P^+_{\mathrm{Dom}(\dot{A})} + (\dot{A}^* + (\dot{A}\dot{A}^* + I)^{-1}S)P^+_{\mathfrak{N}}, \\ &T^* = \dot{A}P^+_{\mathrm{Dom}(\dot{A})} + (\dot{A}^* + (\dot{A}\dot{A}^* + I)^{-1}S^*)P^+_{\mathfrak{N}}, \end{aligned} \tag{4.13}$$

   are valid, where $S \in [\mathfrak{N}, \mathfrak{N}]$ and $(-i)$ is a regular point of S;

2. *$\mathrm{Re}\,T = (T + T^*)/2$ is a regular self-adjoint extension of $\dot{A}$;*

3. *the operator $\mathrm{Im}\,T = (T - T^*)/2i$ is $(\cdot,\cdot)$ bounded and $(\cdot)$-essentially self-adjoint in $\mathcal{H}$.*

Moreover, formulas (4.13) establish a one-to-one correspondence between all regular quasi-self-adjoint extensions $T \in \Omega(\dot{A})$, $-i \in \rho(T)$ and all operators $S \in [\mathfrak{N}, \mathfrak{N}]$ such that $-i \in \rho(S)$.

Proof. We have $\mathfrak{N}'_i = \mathfrak{N}'_{-i} = \{0\}$. Suppose $(-i) \in \rho(T)$. Then the operator M corresponding to T by Theorem 4.1.9 belongs to $[\mathfrak{N}, \mathfrak{N}]$ and possesses properties (4.6). It follows that

$$\overline{\mathrm{Ran}(M+I)}^{(+)} = \overline{\mathrm{Ran}(M^*+I)}^{(+)} = \mathfrak{N}.$$

From (4.7) and (4.8) we get $\overline{\mathrm{Dom}(T)}^{(+)} = \overline{\mathrm{Dom}(T^*)}^{(+)} = \mathcal{H}_+$. Suppose that T is regular. Then the linear manifolds $(M+I)\mathfrak{N}$ and $(M^*+I)\mathfrak{N}$ are (1)-closed. Equalities (4.6) now imply that $(M+I)\mathfrak{N} = (M^*+I)\mathfrak{N} = \mathfrak{N}$. It follows that $\mathrm{Dom}(T) = \mathrm{Dom}(T^*) = \mathcal{H}_+$. Define the operator $S \in [\mathfrak{N}, \mathfrak{N}]$ via

$$S = i(I-M)(I+M)^{-1}.$$

Then $(-i) \in \rho(S)$, $S^* = -i(I-M^*)(I+M^*)^{-1}$, and formulas (4.7) and (4.8) become (4.13). Hence,

$$\mathrm{Dom}(\mathrm{Re}\, T) = \mathcal{H}_+, \quad \mathrm{Re}\, T = \dot{A}P^+_{\mathrm{Dom}(\dot{A})} + (\dot{A}^* + (\dot{A}\dot{A}^* + I)^{-1}\mathrm{Re}\, S)P^+_{\mathfrak{N}},$$

where $\mathrm{Re}\, S = (S+S^*)/2 \in [\mathfrak{N}, \mathfrak{N}]$. By Theorem 2.5.8 the operator $\mathrm{Re} T$ is a regular self-adjoint extension of $\dot{A}$. For $\mathrm{Im}\, T$ we have

$$\mathrm{Dom}(\mathrm{Im}\, T) = \mathcal{H}_+, \; \mathrm{Im}\, T = ((\dot{A}\dot{A}^* + I)^{-1}\mathrm{Im}\, S)P^+_{\mathfrak{N}},$$

where $\mathrm{Im}\, S = (S - S^*)/2i$. Consequently, $\mathrm{Im}\, Tg = 0$ for all $g \in \mathrm{Dom}(\dot{A})$. Using decompositions $\mathcal{H}_+ = \mathrm{Dom}(\dot{A}) \oplus \mathfrak{N}$, $\mathcal{H} = \mathcal{H}_0 \dot{+} \mathfrak{N}$ (see Proposition 2.4.2), and symmetry of the operator $\mathrm{Im}\, T$ we obtain that $\mathrm{Im}\, T$ is essentially self-adjoint and $(\cdot,\cdot)$-bounded.

If $S \in [\mathfrak{N}, \mathfrak{N}]$ and $-i \in \rho(S)$, then $M := (iI - S)(S + iI)^{-1} \in [\mathfrak{N}, \mathfrak{N}]$,

$$\mathrm{Ran}(I+M) = \mathfrak{N}, \quad \ker(I+M) = \{0\}, \quad S(I+M) = i(I-M).$$

From Theorem 4.1.9 we get that T given by (4.13) is a regular quasi-self-adjoint extension of $\dot{A}$ of the class $\Omega(\dot{A})$ and $-i \in \rho(T)$. □

It should be noted that if $\dot{A}$ is an O-operator and $T \in \Omega(\dot{A})$ is an arbitrary regular quasi-self-adjoint extension, then $\mathrm{Dom}(T) = \mathrm{Dom}(T^*)$. Indeed, let $\lambda = a + ib \in \rho(T)$, where $a, b \in \mathbb{R}$, and $b \neq 0$. Then the operator

$$\mathrm{Dom}(\tilde{T}) = \mathrm{Dom}(T), \quad \tilde{T} := -b^{-1}(T - aI)$$

is a regular quasi-self-adjoint extension of the regular symmetric operator

$$\mathrm{Dom}(\tilde{\dot{A}}) = \mathrm{Dom}(\dot{A}), \quad \tilde{\dot{A}} = -b^{-1}(\dot{A} - aI),$$

and $-i \in \tilde{T}$. Clearly, $\mathrm{Dom}(\tilde{T}^*) = \mathrm{Dom}(T)$, $\tilde{T}^* = -b^{-1}(T^* - aI)\dot{A}$, the operator $\tilde{\dot{A}}$ is an O-operator, and $\mathrm{Dom}(\tilde{\dot{A}}^*) = \mathrm{Dom}(\dot{A}^*) = \mathcal{H}_+$. It follows from Theorem 4.1.12 that $\mathrm{Dom}(\tilde{T}) = \mathrm{Dom}(\tilde{T}^*) = \mathcal{H}_+$.

Theorem 4.1.13. *If a quasi-self-adjoint extension T of a regular O-operator $\dot{A}$ is $(+,-)$-continuous, then T is $(+,\cdot)$-continuous (and, consequently, a regular extension of $\dot{A}$).*

Proof. If T is $(+,-)$-continuous, then so is T^*. According to Theorem 4.1.12 both $\mathrm{Dom}(T)$ and $\mathrm{Dom}(T^*)$ are $(+)$-dense in $\mathcal{H}_+$. Hence both T and T^* can be extended to $\mathcal{H}_+$ by $(+,-)$-continuity to operators $\overline{T}$ and $\overline{T^*}$ in $[\mathcal{H}_+,\mathcal{H}_-]$, respectively. It is clear that $\overline{T}$ and $\overline{T^*}$ are adjoint to each other operators in $[\mathcal{H}_+,\mathcal{H}_-]$. Thus, $\overline{T}\in\mathcal{E}(\dot{A})$. By Theorem 3.1.6 and Theorem 3.1.8 the operators $\overline{T}$ and $\overline{T^*}$ are $(+,\cdot)$-continuous. This yields $(+,\cdot)$-continuity to operators T and T^*. Therefore, T is a regular extension of $\dot{A}$. □

Suppose that $\dot{A}$ is a regular symmetric operator but is not an O-operator. If the deficiency indices of A are finite, then $\mathrm{codim}\,\overline{\mathrm{Dom}(\dot{A})}<\infty$ and, according to Theorem 4.1.3, all T from the class $\Omega(\dot{A})$ are regular. In this case the equalities $\mathrm{Dom}(T)=\mathcal{H}_+$ and $\mathrm{Dom}(T^*)=\mathcal{H}_+$ can not hold if both semi-deficiency numbers are non-zero. Indeed, let $T\in\Omega(\dot{A})$ with $-i\in\rho(T)$ be determined by the operator $M\in[\mathfrak{N}'_i\oplus\mathfrak{N},\mathfrak{N}'_{-i}\oplus\mathfrak{N}]$ in accordance with Theorem 4.1.9. Since

$$\begin{aligned}\dim\mathrm{Ran}(M+I)&=\dim\mathfrak{N}'_i+\dim\mathfrak{N}=\dim\mathfrak{N}_i,\\ \dim\mathrm{Ran}(M^*+I)&=\dim\mathfrak{N}'_{-i}+\dim\mathfrak{N}=\dim\mathfrak{N}_{-i},\end{aligned}$$

and $\dim\mathfrak{M}=\dim\mathfrak{N}'_i+\dim\mathfrak{N}'_{-i}+\dim\mathfrak{N}$, we get $\mathrm{Dom}(T)\neq\mathcal{H}_+$ and $\mathrm{Dom}(T^*)\neq\mathcal{H}_+$. If $\mathfrak{N}'_i=\{0\}$ (respect., $\mathfrak{N}'_{-i}=\{0\}$, then $\mathrm{Dom}(T^*)=\mathcal{H}_+$ and $\mathrm{Dom}(T)\neq\mathcal{H}_+$ (respect., $\mathrm{Dom}(T)=\mathcal{H}_+$ and $\mathrm{Dom}(T^*)\neq\mathcal{H}_+$).

4.2 Quasi-self-adjoint bi-extension

In the following we will suppose that closed symmetric operator $\dot{A}$ is regular. We begin with its definition.

Definition 4.2.1. An operator $\mathbb{A}\in\mathcal{E}(\dot{A})$ is called a **quasi-self-adjoint bi-extension** (or **q.s.-a. bi-extension**) of a regular $T\in\Omega(\dot{A})$ if

$$\mathbb{A}\supset T\supset\dot{A}\quad\text{and}\quad\mathbb{A}^*\supset T^*\supset\dot{A}.$$

Definition 4.2.2. We say that a quasi-self-adjoint bi-extension $\mathbb{A}$ of an operator $T\in\Omega(\dot{A})$ has the **range property (R)** if

$$\mathrm{Ran}(\mathbb{A}-\lambda_0 I)\supseteq\mathrm{Ran}(\mathbb{A}-\mathbb{A}^*),\quad\mathrm{Ran}(\mathbb{A}^*-\bar{\lambda}_0 I)\supseteq\mathrm{Ran}(\mathbb{A}-\mathbb{A}^*)$$

for some $\lambda_0\in\rho(T)$.

Definition 4.2.3. We say that the operator T of the class $\Omega(\dot{A})$ belongs to the **class** $R(\dot{A})$ if T admits q.s.-a. bi-extensions with the range property (R).

In particular, if T is a regular self-adjoint extension of $\dot{A}$ having a self-adjoint bi-extension $\mathbb{A}$, then $T \in R(\dot{A})$ and $\mathbb{A}$ is a q.s.-a. bi-extension with the range property (R) of T. In Chapter 3 it is established that a regular self-adjoint extension of $\dot{A}$ admits self-adjoint bi-extensions and the description of all such bi-extensions is obtained (see Theorem 3.3.8). In this section we are going to find necessary and sufficient conditions for a regular $T \in \Omega(\dot{A})$ to be in the class $R(\dot{A})$. We will also give a description of all q.s.-a. bi-extensions $\mathbb{A}$ with the range property (R) of $T \in R(\dot{A})$.

Lemma 4.2.4. *If $\mathbb{A}$ is a quasi-self-adjoint bi-extension of T, then T and T^* are quasi-kernels of $\mathbb{A}$ and $\mathbb{A}^*$, respectively.*

Proof. If $f \in \mathrm{Dom}(T)$, then $\mathbb{A}f = Tf \in \mathcal{H}$. Conversely, let $\mathbb{A}f \in \mathcal{H}$. Then for any $g \in \mathrm{Dom}(T^*)$,

$$(T^*g, f) = (\mathbb{A}^*g, f) = (g, \mathbb{A}f),$$

which implies that $f \in \mathrm{Dom}(T)$ and $Tf = \mathbb{A}f$. □

Let $\mathfrak{J} = (P^+_{\mathfrak{N}'_i} - P^+_{\mathfrak{N}'_{-i}})\upharpoonright\mathfrak{M}$ be the operator in $\mathfrak{M}$ (see (3.10)). Clearly $\mathfrak{J}$ is (1)-self-adjoint and $\mathfrak{J}^2 = (P^+_{\mathfrak{N}'_i} + P^+_{\mathfrak{N}'_{-i}})\upharpoonright\mathfrak{M}$. Every bi-extension $\mathbb{A}$ of a regular closed symmetric operator $\dot{A}$ can be represented in the form (3.13)

$$\mathbb{A} = \dot{A}P^+_{\mathrm{Dom}(\dot{A})} + \left(\dot{A}^* + \mathcal{R}^{-1}(S - \frac{i}{2}\mathfrak{J})\right)P^+_{\mathfrak{M}},$$

where S belongs to $[\mathfrak{M}, \mathfrak{M}]$, and the adjoint operator $\mathbb{A}^*$ is given by

$$\mathbb{A}^* = \dot{A}P^+_{\mathrm{Dom}(\dot{A})} + \left(\dot{A}^* + \mathcal{R}^{-1}(S^* - \frac{i}{2}\mathfrak{J})\right)P^+_{\mathfrak{M}},$$

with (1)-adjoint operator S^*. The next statement immediately follows from Theorem 4.1.9, Proposition 3.1.11, and the equality $\mathcal{R}^{-1}\upharpoonright\mathfrak{N} = 2(\dot{A}\dot{A}^* + I)\upharpoonright\mathfrak{N}$ (see (2.26) and (2.7)).

Theorem 4.2.5. *Let $T \in \Omega(\dot{A})$, $-i \in \rho(T)$, be a regular quasi-self-adjoint extension of a regular closed symmetric operator $\dot{A}$. An operator*

$$\mathbb{A} = \dot{A}P^+_{\mathrm{Dom}(\dot{A})} + \left(\dot{A}^* + \mathcal{R}^{-1}(S - \frac{i}{2}\mathfrak{J})\right)P^+_{\mathfrak{M}},$$

is a quasi-self-adjoint bi-extension of T if and only if

$$S(M + I) = \frac{i}{2}(I - M), \quad S^*(M^* + I) = \frac{i}{2}(M^* - I), \tag{4.14}$$

where $M \in [\mathfrak{N}'_i \oplus \mathfrak{N}, \mathfrak{N}'_{-i} \oplus \mathfrak{N}]$ determines T by Theorem 4.1.9, and $M^* \in [\mathfrak{N}'_{-i} \oplus \mathfrak{N}, \mathfrak{N}'_i \oplus \mathfrak{N}]$ is (1)-adjoint to M operator.

Observe that

$$\left(\frac{i}{2}(I-M)f,(I+M^*)h\right)_1=\left((I+M)f,\frac{i}{2}(M^*-I)h\right)_1 \tag{4.15}$$
$$=\frac{i}{2}((f,g)_1-(Mf,M^*g)_1)$$

for all $f\in\mathfrak{N}'_i\oplus\mathfrak{N}$ and all $h\in\mathfrak{N}'_{-i}\oplus\mathfrak{N}$.

In the sequel we will need the following statement.

Lemma 4.2.6. *Let* $T\in\Omega(\dot{A})$. *Then:*

1) (1)*-orthogonal complements to* $\mathrm{Dom}(T)$ *and* $\mathrm{Dom}(T^*)$ *in* $\mathcal{H}_+$ *are given by*

$$\mathcal{H}_+\ominus\mathrm{Dom}(T)=\left\{\varphi\in\mathfrak{M}:\dot{A}^*\varphi\in\mathrm{Dom}(T^*),\ (T^*\dot{A}^*+I)\varphi=-(\dot{A}\dot{A}^*+I)P^+_{\mathfrak{N}}\varphi\right\},$$

$$\mathcal{H}_+\ominus\mathrm{Dom}(T^*)=\left\{\psi\in\mathfrak{M}:\dot{A}^*\psi\in\mathrm{Dom}(T),\ (T\dot{A}^*+I)\psi=-(\dot{A}\dot{A}^*+I)P^+_{\mathfrak{N}}\psi\right\}; \tag{4.16}$$

2) *if* $\mathbb{A}=\dot{A}P^+_{\mathrm{Dom}(\dot{A})}+\left(\dot{A}^*+\mathcal{R}^{-1}(S-\frac{i}{2}\mathfrak{J})\right)P^+_{\mathfrak{M}}$ *is a quasi-self-adjoint bi-extension of* T, *then*

$$P^+_{\mathfrak{N}}S^*\mathfrak{J}\varphi=iP^+_{\mathfrak{N}}\varphi,\quad \varphi\in\mathcal{H}_+\ominus\mathrm{Dom}(T),$$

$$P^+_{\mathfrak{N}}S\mathfrak{J}\psi=iP^+_{\mathfrak{N}}\psi,\quad \psi\in\mathcal{H}_+\ominus\mathrm{Dom}(T^*).$$

Proof. 1) Clearly, $h\in\mathcal{H}_+$ is (+)-orthogonal to $\mathrm{Dom}(T)$ if and only if $(Tf,\dot{A}^*h)=-(f,h)$ for all $f\in\mathrm{Dom}(T)$ that are equivalent to $\dot{A}^*h\in\mathrm{Dom}(T^*)$ and $(T^*\dot{A}^*+I)h=0$. It follows that if $\varphi\in\mathcal{H}_+$ is (1)-orthogonal to $\mathrm{Dom}(T)$, then

$$0=(f,\varphi)_1=(f,(I+P^+_{\mathfrak{N}})\varphi)_+,\quad f\in\mathrm{Dom}(T),$$

which is equivalent to $h=(I+P^+_{\mathfrak{N}})\varphi$ being (+)-orthogonal to $\mathrm{Dom}(T)$. Using that $\dot{A}^*\mathfrak{N}\subset\mathrm{Dom}(\dot{A})$, we get (4.16).

2) Let $\varphi\in\mathcal{H}_+$ be (+)-orthogonal to $\mathrm{Dom}(T)$. Then by (4.16)

$$\dot{A}^*\varphi=i\mathfrak{J}\varphi+\dot{A}^*P^+_{\mathfrak{N}}\varphi\in\mathrm{Dom}(T^*).$$

Consequently, $\mathfrak{J}\varphi\in\mathrm{Dom}(T^*)\cap(\mathfrak{N}'_i\oplus\mathfrak{N}'_{-i})$. Since $\mathbb{A}^*\supset T^*$, we get

$$(P^+_{\mathfrak{N}'_i}+P^+_{\mathfrak{N}'_{-i}})(S^*-\frac{i}{2}\mathfrak{J})\mathfrak{J}\varphi=0.$$

Further

$$T^*\dot{A}^*\varphi=\mathbb{A}^*(i\mathfrak{J}\varphi+\dot{A}^*P^+_{\mathfrak{N}}\varphi)=\dot{A}\dot{A}^*P^+_{\mathfrak{N}}\varphi+i\dot{A}^*\mathfrak{J}\varphi+i\mathcal{R}^{-1}\left(S^*-\frac{i}{2}\mathfrak{J}\right)\mathfrak{J}\varphi$$
$$=-(P^+_{\mathfrak{N}'_i}+P^+_{\mathfrak{N}'_{-i}}+P^+_{\mathfrak{N}})\varphi+(\dot{A}\dot{A}^*+I)(P^+_{\mathfrak{N}}\varphi+2iP^+_{\mathfrak{N}}S^*\mathfrak{J}\varphi).$$

From (4.16) we obtain $P^+_{\mathfrak{N}} S^* \mathfrak{J}\varphi = iP^+_{\mathfrak{N}}\varphi$. □

In order to establish the existence of quasi-self-adjoint bi-extension for a given regular $T \in \Omega(\dot{A})$ we need the following statement.

Proposition 4.2.7. *Let $\mathfrak{K}$ be a Hilbert space and let C_1 and C_2 two bounded linear operators defined on the proper subspaces of $\mathfrak{K}$ such that* $\mathrm{Dom}(C_1) = \mathfrak{K}_1$ *and* $\mathrm{Dom}(C_2) = \mathfrak{K}_2$. *Suppose that*

$$(C_1\phi, \psi)_{\mathfrak{K}} = (\phi, C_2\psi)_{\mathfrak{K}} \quad \textit{for all} \quad \phi \in \mathfrak{K}_1,\ \psi \in \mathfrak{K}_2. \tag{4.17}$$

Let

$$\mathfrak{K} = \mathfrak{K}_1^\perp \oplus \mathfrak{K}_1, \quad \mathfrak{K} = \mathfrak{K}_2^\perp \oplus \mathfrak{K}_2. \tag{4.18}$$

Then all operators $Q \in [\mathfrak{K}, \mathfrak{K}]$ satisfying the condition

$$Q \supset C_1,\ Q^* \supset C_2 \tag{4.19}$$

take, according to decomposition (4.18), *the operator-matrix form*

$$Q = \begin{pmatrix} P_{\mathfrak{K}_2} C_1 & (P_{\mathfrak{K}_1^\perp} C_2)^* \\ P_{\mathfrak{K}_2^\perp} C_1 & H \end{pmatrix}, \tag{4.20}$$

where H is an arbitrary operator in $[\mathfrak{K}_1^\perp, \mathfrak{K}_2^\perp]$. The adjoint Q^* is of the form

$$Q^* = \begin{pmatrix} P_{\mathfrak{K}_1} C_2 & (P_{\mathfrak{K}_2^\perp} C_1)^* \\ P_{\mathfrak{K}_1^\perp} C_2 & H^* \end{pmatrix}, \tag{4.21}$$

where $H^* \in [\mathfrak{K}_2^\perp, \mathfrak{K}_1^\perp]$ is the adjoint to H.

Proof. In view (4.17) the operator given by (4.21) is the adjoint to the operator Q defined by (4.20). Moreover, Q and Q^* satisfy (4.19).

Conversely, if $Q \in [\mathfrak{K}, \mathfrak{K}]$ and $Q \supset C_1$, then Q has an operator-matrix representation of the form

$$Q = \begin{pmatrix} P_{\mathfrak{K}_2} C_1 & X \\ P_{\mathfrak{K}_2^\perp} C_1 & H \end{pmatrix},$$

where $X \in [\mathfrak{K}_1^\perp, \mathfrak{K}_2]$ and $H \in [\mathfrak{K}_1^\perp, \mathfrak{K}_2^\perp]$. Hence,

$$Q^* = \begin{pmatrix} (P_{\mathfrak{K}_2} C_1)^* & (P_{\mathfrak{K}_2^\perp} C_1)^* \\ X^* & H^* \end{pmatrix}.$$

From (4.17) it follows that the adjoint $(P_{\mathfrak{K}_2} C_1)^*$ to $P_{\mathfrak{K}_2} C_1 \in [\mathfrak{K}_1, \mathfrak{K}_2]$ is equal to $P_{\mathfrak{K}_1} C_2 \in [\mathfrak{K}_2, \mathfrak{K}_1]$. Hence, the condition $Q^* \supset C_2$ implies $X^* = P_{\mathfrak{K}_1^\perp} C_2$. So, Q is of the form (4.20). This completes the proof. □

Theorem 4.2.8. *Let $T \in \Omega(\dot{A})$ be a regular quasi-self-adjoint extension of a regular closed symmetric operator $\dot{A}$. Then there exists a bi-extension $\mathbb{A}$ of $\dot{A}$ that is a quasi-self-adjoint bi-extension of T.*

Proof. It is sufficient to prove the statement for the case $-i \in \rho(T)$. Let $M \in [\mathfrak{N}'_i \oplus \mathfrak{N}, \mathfrak{N}'_{-i} \oplus \mathfrak{N}]$ determine T in accordance with Theorem 4.1.9. Let $\mathfrak{K} = \mathfrak{M}$ be equipped with a (1)-inner product,

$$\begin{aligned} &\mathfrak{K}_1 = (I+M)(\mathfrak{N}'_i \oplus \mathfrak{N}), \quad \mathfrak{K}_2 = (I+M^*)(\mathfrak{N}'_{-i} \oplus \mathfrak{N}), \\ &C_1(M+I)f = \frac{i}{2}(I-M)f, \quad f \in \mathfrak{N}'_i \oplus \mathfrak{N}, \\ &C_2(M^*+I)g = \frac{i}{2}(M^*-I)g, \quad g \in \mathfrak{N}'_{-i} \oplus \mathfrak{N}. \end{aligned}$$

Then it follows from (4.15) that C_1 and C_2 satisfy (4.17). Applying Proposition 4.2.7 we obtain the existence of an operator $S \in [\mathfrak{M}, \mathfrak{M}]$ with the property $S \supset C_1$ and $S^* \supset C_2$. Let S be such an operator and let

$$\mathbb{A} = \dot{A}P^+_{\mathrm{Dom}(\dot{A})} + \left(\dot{A}^* + \mathcal{R}^{-1}(S - \frac{i}{2}\mathfrak{J})\right) P^+_{\mathfrak{M}}.$$

Since

$$\mathbb{A}^* = \dot{A}P^+_{\mathrm{Dom}(\dot{A})} + \left(\dot{A}^* + \mathcal{R}^{-1}(S^* - \frac{i}{2}\mathfrak{J})\right) P^+_{\mathfrak{M}},$$

applying Theorem 4.2.5 we get $\mathbb{A} \supset T$ and $\mathbb{A}^* \supset T^*$. □

If $\dot{A}$ is a densely defined closed symmetric operator, then all operators from the class $\Omega(\dot{A})$ with $(-i) \in \rho(T)$ take the form

$$\begin{aligned} \mathrm{Dom}(T) &= \mathrm{Dom}(\dot{A}) \oplus (M+I)\mathfrak{N}_i, \quad T = \dot{A}^* \restriction \mathrm{Dom}(T), \\ \mathrm{Dom}(T^*) &= \mathrm{Dom}(\dot{A}) \oplus (M^*+I)\mathfrak{N}_{-i}, \quad T^* = \dot{A}^* \restriction \mathrm{Dom}(T^*), \end{aligned}$$

where $M \in [\mathfrak{N}_i, \mathfrak{N}_{-i}]$. In this case all quasi-self-adjoint bi-extensions of $T \in \Omega(\dot{A})$ are given by

$$\mathbb{A} = \dot{A}^* + \mathcal{R}^{-1}(S - \frac{i}{2}\mathfrak{J})P^+_{\mathfrak{M}}, \quad \mathbb{A}^* = \dot{A}^* + \mathcal{R}^{-1}(S^* - \frac{i}{2}\mathfrak{J})P^+_{\mathfrak{M}}, \tag{4.22}$$

where $S \in [\mathfrak{M}, \mathfrak{M}]$ satisfies conditions

$$S(M+I) = \frac{i}{2}(I-M), \quad S^*(M^*+I) = \frac{i}{2}(M^*-I).$$

In the densely defined case it is not difficult to derive the general block-matrix form of all such $S : \mathfrak{N}_i \oplus \mathfrak{N}_{-i} \to \mathfrak{N}_i \oplus \mathfrak{N}_{-i}$,

$$S = \begin{pmatrix} \frac{i}{2}I - HM & H \\ -(iI - MH)M & \frac{i}{2}I - MH \end{pmatrix},$$

where $H \in [\mathfrak{N}_{-i}, \mathfrak{N}_i]$ is an arbitrary operator. We are also going to introduce block-operator matrices $S_{\mathbb{A}}$ and $S_{\mathbb{A}^*}$ which will be needed later and are of the form

$$\begin{aligned} S_{\mathbb{A}} &= S - \frac{i}{2}\mathfrak{J} = \begin{pmatrix} -HM & H \\ M(HM - iI) & iI - MH \end{pmatrix}, \\ S_{\mathbb{A}^*} &= S^* - \frac{i}{2}\mathfrak{J} = \begin{pmatrix} -M^*H^* - iI & (M^*H^* - iI)M^* \\ H^* & iI - H^*M^* \end{pmatrix}. \end{aligned} \tag{4.23}$$

As usual we stick to the notations $\operatorname{Re} S = (S + S^*)/2$ and $\operatorname{Im} S = (S - S^*)/2i$ for $S \in [\mathfrak{M}, \mathfrak{M}]$.

Theorem 4.2.9. *Let $\dot{A}$ be a regular closed symmetric operator with non-zero semi-deficiency indices. Let T be a regular extension from the class $\Omega(\dot{A})$. Then $T \in R(\dot{A})$ if and only if there exists a $(\cdot)$-isometric operator U mapping $\mathfrak{N}'_i$ onto $\mathfrak{N}'_{-i}$ such that the equalities*

$$\operatorname{Dom}(T)\dot{+}(I + U)\mathfrak{N}'_i = \operatorname{Dom}(T^*)\dot{+}(I + U)\mathfrak{N}'_i = \mathcal{H}_+ \tag{4.24}$$

hold.

Proof. Necessity. Let $T \in R(\dot{A})$. Let us set

$$\begin{aligned} \mathfrak{M}_T &= \operatorname{Dom}(T) \ominus \operatorname{Dom}(\dot{A}), \quad \mathfrak{M}_T^{\perp} = \mathfrak{M} \ominus \mathfrak{M}_T, \\ \mathfrak{M}_{T^*} &= \operatorname{Dom}(T^*) \ominus \operatorname{Dom}(\dot{A}), \quad \mathfrak{M}_{T^*}^{\perp} = \mathfrak{M} \ominus \mathfrak{M}_{T^*}, \end{aligned}$$

where the orthogonal complements are taken with respect to a(1)-inner product in $\mathcal{H}_+$. Suppose $\mathbb{A} = \dot{A}P^+_{\operatorname{Dom}(\dot{A})} + \left(\dot{A}^* + \mathcal{R}^{-1}(S - \frac{i}{2}\mathfrak{J})\right) P^+_{\mathfrak{M}}$ is a q.s.-a. bi-extension of T with the range property (R). Then for an arbitrary $\lambda \in \rho(T)$ we have

$$\begin{aligned} \mathbb{A} - \lambda I &= (T - \lambda I)P^+_{\operatorname{Dom}(T)} + (\dot{A}^* - \lambda I)P^+_{\mathfrak{M}_T^{\perp}} \\ &\quad + \mathcal{R}^{-1}P^+_{\mathfrak{N}}SP^+_{\mathfrak{M}_T^{\perp}} + \mathcal{R}^{-1}(P^+_{\mathfrak{N}'_i} + P^+_{\mathfrak{N}'_{-i}})(S - \frac{i}{2}\mathfrak{J})P^+_{\mathfrak{M}_T^{\perp}}, \end{aligned} \tag{4.25}$$

and

$$\operatorname{Im} \mathbb{A} = \mathcal{R}^{-1}P^+_{\mathfrak{N}}(\operatorname{Im} S)P^+_{\mathfrak{M}} + \mathcal{R}^{-1}(P^+_{\mathfrak{N}'_i} + P^+_{\mathfrak{N}'_{-i}})(\operatorname{Im} S)P^+_{\mathfrak{M}}.$$

By Definition 4.2.1 the equation

$$(\mathbb{A} - \lambda_0 I)x = (\operatorname{Im} \mathbb{A})f, \ \lambda_0 \in \rho(T),$$

has a solution $x \in \mathcal{H}_+$ for an arbitrary $f \in \mathfrak{M}$. Since

$$\mathcal{R}^{-1}P^+_{\mathfrak{N}}(\operatorname{Im} S)P^+_{\mathfrak{M}}f, \quad \mathcal{R}^{-1}P^+_{\mathfrak{N}}SP^+_{\mathfrak{M}_T^{\perp}}x \in \mathfrak{L} \subset \mathcal{H},$$

it follows from (4.25) that there exists a vector $g \in \mathfrak{M}_T^{\perp}$ such that

$$(P^+_{\mathfrak{N}'_i} + P^+_{\mathfrak{N}'_{-i}})(S - \frac{i}{2}\mathfrak{J})g = (P^+_{\mathfrak{N}'_i} + P^+_{\mathfrak{N}'_{-i}})(\operatorname{Im} S)P^+_{\mathfrak{M}}f.$$

Consequently

$$x = g + (T - \lambda_0 I)^{-1}\left(\mathcal{R}^{-1}P^+_{\mathfrak{N}}(\operatorname{Im} S)f - \mathcal{R}^{-1}P^+_{\mathfrak{N}}Sg - (\dot{A}^* - \lambda_0 I)g\right).$$

Thus, condition $\operatorname{Ran}(\mathbb{A} - \lambda_0 I) \supset \operatorname{Ran}(\operatorname{Im}\mathbb{A})$, yields the inclusion

$$\operatorname{Ran}\left((P^+_{\mathfrak{N}'_i} + P^+_{\mathfrak{N}'_{-i}})(S - \frac{i}{2}\mathfrak{J})\right) \supset \operatorname{Ran}\left((P^+_{\mathfrak{N}'_i} + P^+_{\mathfrak{N}'_{-i}})(\operatorname{Im} S)\right).$$

Therefore,

$$\ker\left(S^*(P^+_{\mathfrak{N}'_i} + P^+_{\mathfrak{N}'_{-i}}) + \frac{i}{2}\mathfrak{J})\right) \subseteq \ker\left((\operatorname{Im} S)(P^+_{\mathfrak{N}'_i} + P^+_{\mathfrak{N}'_{-i}}\right). \tag{4.26}$$

Similarly, the condition $\operatorname{Ran}(\mathbb{A}^* - \bar{\lambda}_0 I) \supset \operatorname{Ran}(\operatorname{Im}\mathbb{A})$ implies

$$\ker\left(S(P^+_{\mathfrak{N}'_i} + P^+_{\mathfrak{N}'_{-i}}) + \frac{i}{2}\mathfrak{J}\right) \subseteq \ker\left((\operatorname{Im} S)(P^+_{\mathfrak{N}'_i} + P^+_{\mathfrak{N}'_{-i}})\right). \tag{4.27}$$

Now from (4.26) and (4.27) we obtain

$$\ker\left(S(P^+_{\mathfrak{N}'_i} + P^+_{\mathfrak{N}'_{-i}}) + \frac{i}{2}\mathfrak{J}\right) = \ker\left(S^*(P^+_{\mathfrak{N}'_i} + P^+_{\mathfrak{N}'_{-i}}) + \frac{i}{2}\mathfrak{J}\right). \tag{4.28}$$

Therefore,

$$\begin{aligned}&\ker\left(S(P^+_{\mathfrak{N}'_i} + P^+_{\mathfrak{N}'_{-i}}) + \tfrac{i}{2}\mathfrak{J}\right) = \ker\left(S^*(P^+_{\mathfrak{N}'_i} + P^+_{\mathfrak{N}'_{-i}}) + \tfrac{i}{2}\mathfrak{J}\right)\\ &\subseteq \ker\left((\operatorname{Re} S)(P^+_{\mathfrak{N}'_i} + P^+_{\mathfrak{N}'_{-i}}) + \tfrac{i}{2}\mathfrak{J}\right).\end{aligned}$$

Since $\operatorname{Re} S$ is a (1)-self-adjoint operator in $\mathfrak{M}$ we get

$$\ker(\operatorname{Re} S + \frac{i}{2}\mathfrak{J}) = \mathfrak{N} \oplus (U + I)\operatorname{Dom}(U), \quad \operatorname{Dom}(U) \subseteq \mathfrak{N}'_i,\ \operatorname{Ran}(U) \subseteq \mathfrak{N}'_{-i},$$

and U is a $(\cdot)$ and (1)-isometry. Let

$$F := \ker\left(S(P^+_{\mathfrak{N}'_i} + P^+_{\mathfrak{N}'_{-i}}) + \frac{i}{2}\mathfrak{J}\right) \ominus \mathfrak{N}. \tag{4.29}$$

Then (4.28) yields $F = (U + I)\hat{\mathfrak{N}}'_i$, where $\hat{\mathfrak{N}}'_i$ is some subspace in $\operatorname{Dom}(U)$.

Let $f \in \mathfrak{M}_T \cap F$. Then, since $f \in \mathfrak{M}_T$, we have

$$((P^+_{\mathfrak{N}'_i} + P^+_{\mathfrak{N}'_{-i}})S - \frac{i}{2}\mathfrak{J})f = 0.$$

On the other hand $f \in F$ and hence,

$$((P^+_{\mathfrak{N}'_i} + P^+_{\mathfrak{N}'_{-i}})S - \frac{i}{2}\mathfrak{J})f = -i\mathfrak{J}f.$$

Therefore, $\mathfrak{M}_T \cap F = \{0\}$. Let $f = \mathfrak{M}_T \dot{+} F$. Then $f = f_1 + f_2$, where $f_1 \in \mathfrak{M}_T$, $f_2 \in F$. Hence,

$$i\mathfrak{J}\left((P^+_{\mathfrak{N}'_i} + P^+_{\mathfrak{N}'_{-i}})S - \frac{i}{2}\mathfrak{J}\right) f = (P^+_{\mathfrak{N}'_i} + P^+_{\mathfrak{N}'_{-i}})f_2 = f_2.$$

It follows that the linear manifold $\mathfrak{M}_T \dot{+} F$ is (1)-closed. Since

$$\ker\left((P^+_{\mathfrak{N}'_i} + P^+_{\mathfrak{N}'_{-i}})(S^* - \frac{i}{2}\mathfrak{J})\right) = \mathfrak{M}_{T^*},$$

taking closure in (+)-metric, we obtain

$$\overline{\operatorname{Ran}\left(S(P^+_{\mathfrak{N}'_i} + P^+_{\mathfrak{N}'_{-i}}) + \frac{i}{2}\mathfrak{J}\right)} = \mathfrak{M}^{\perp}_{T^*}.$$

Let $\psi \in \mathfrak{M}^{\perp}_{T^*}$. Then by Lemma 4.2.6 $\mathfrak{J}\psi \in \mathfrak{M}_T \cap (\mathfrak{N}'_i \oplus \mathfrak{N}'_{-i})$ and $P^+_{\mathfrak{N}} S\mathfrak{J}\psi = iP^+_{\mathfrak{N}}\psi$. Consequently,

$$\left(S(P^+_{\mathfrak{N}'_i} + P^+_{\mathfrak{N}'_{-i}}) + \tfrac{i}{2}\mathfrak{J}\right)\mathfrak{J}\psi = P^+_{\mathfrak{N}} S\mathfrak{J}\psi + i\mathfrak{J}(\mathfrak{J}\psi) + (S - \tfrac{i}{2}\mathfrak{J})\mathfrak{J}\psi = i\psi.$$

Hence

$$\operatorname{Ran}\left(S(P^+_{\mathfrak{N}'_i} + P^+_{\mathfrak{N}'_{-i}}) + \frac{i}{2}\mathfrak{J}\right) = \left(S(P^+_{\mathfrak{N}'_i} + P^+_{\mathfrak{N}'_{-i}}) + \frac{i}{2}\mathfrak{J}\right)\mathfrak{J}\mathfrak{M}^{\perp}_{T^*} = \mathfrak{M}^{\perp}_{T^*}. \tag{4.30}$$

Let $h \in \mathfrak{M} \ominus \left(\mathfrak{M}_T \dot{+} F\right)$. Then (4.30) implies that there exists a vector $\psi \in \mathfrak{M}^{\perp}_{T^*}$ such that

$$\left(S(P^+_{\mathfrak{N}'_i} + P^+_{\mathfrak{N}'_{-i}}) + \frac{i}{2}\mathfrak{J}\right)\mathfrak{J}\psi = \left(S(P^+_{\mathfrak{N}'_i} + P^+_{\mathfrak{N}'_{-i}}) + \frac{i}{2}\mathfrak{J}\right)h.$$

It follows from (4.29) that $h = \mathfrak{J}\psi + f + g$, where $f \in F$, $g \in \mathfrak{N}$. Because $\mathfrak{J}\psi \in \mathfrak{M}_T$ and h is (1)-orthogonal to F and $\mathfrak{M}_T$, we have $h = g \in \mathfrak{N}$. Since $\mathcal{R}^{-1}h \in \mathfrak{L}$, we get that $\mathcal{R}^{-1}h$ is $(\cdot)$-orthogonal to $\operatorname{Dom}(T)$. Therefore, $h = 0$. Hence,

$$\mathfrak{M}_T \dot{+} F = \mathfrak{M}. \tag{4.31}$$

Similarly, one can prove that $\mathfrak{M}_{T^*} \dot{+} F = \mathfrak{M}$. Further, from (4.31) we get that $\operatorname{Ran}\left((P^+_{\mathfrak{N}'_i} + P^+_{\mathfrak{N}'_{-i}})S - \tfrac{i}{2}\mathfrak{J}\right) = \mathfrak{J}F$. Since

$$\ker\left(S^*(P^+_{\mathfrak{N}'_i} + P^+_{\mathfrak{N}'_{-i}}) + \frac{i}{2}\mathfrak{J}\right) \oplus \operatorname{Ran}\left((P^+_{\mathfrak{N}'_i} + P^+_{\mathfrak{N}'_{-i}})S - \frac{i}{2}\mathfrak{J}\right) = \mathfrak{M},$$

then from (4.29) we get $\mathfrak{J}F \oplus F = \mathfrak{N}'_i \oplus \mathfrak{N}'_{-i}$. Now the equality $F = (U + I)\hat{\mathfrak{N}}'_i$, where $\hat{\mathfrak{N}}'_i$ is a subspace in $\operatorname{Dom}(U)$, yields $\hat{\mathfrak{N}}'_i = \mathfrak{N}'_i$ and $\operatorname{Ran}(U) = \mathfrak{N}'_{-i}$. Thus, we have proved the equalities

$$\begin{aligned}\ker\left(S(P^+_{\mathfrak{N}'_i} + P^+_{\mathfrak{N}'_{-i}}) + \tfrac{i}{2}\mathfrak{J}\right) &= \ker\left(S^*(P^+_{\mathfrak{N}'_i} + P^+_{\mathfrak{N}'_{-i}}) + \tfrac{i}{2}\mathfrak{J}\right)\\ &= \ker\left(\operatorname{Re} S(P^+_{\mathfrak{N}'_i} + P^+_{\mathfrak{N}'_{-i}}) + \tfrac{i}{2}\mathfrak{J}\right) = \mathfrak{N} \oplus (I + U)\mathfrak{N}'_i,\end{aligned}$$

and

$$\mathfrak{M}_T \dot{+} (I+U)\mathfrak{N}'_i = \mathfrak{M}_{T^*} \dot{+} (I+U)\mathfrak{N}'_i = \mathfrak{M}, \tag{4.32}$$

where U is a (1)-isometric operator mapping$\mathfrak{N}_i$ onto $\mathfrak{N}'_{-i}$. Clearly, equalities (4.32) are equivalent to (4.24).

Sufficiency. Suppose that (4.32) holds for some $(\cdot)$-isometry $U : \mathfrak{N}'_i \to \mathfrak{N}'_{-i}$, $\operatorname{Ran}(U) = \mathfrak{N}'_{-i}$. For the sake of simplicity let us assume that $-i \in \rho(T)$. Then (4.7) and (4.8) hold, i.e., $\mathfrak{M}_T = (I+M)(\mathfrak{N}'_i \oplus \mathfrak{N})$, and $\mathfrak{M}_{T^*} = (I+M^*)(\mathfrak{N}'_{-i} \oplus \mathfrak{N})$. Let an operator $S \in [\mathfrak{M}, \mathfrak{M}]$ be defined as

$$S((I+M)f + (I+U)\varphi) = \frac{i}{2}((I-M)f - (I-U)\varphi),\ f \in \mathfrak{N}'_i \oplus \mathfrak{N},\ \varphi \in \mathfrak{N}'_i. \tag{4.33}$$

By direct calculations one verifies that the (1)-adjoint $S^* \in [\mathfrak{M}, \mathfrak{M}]$ is given by

$$S^*((I+M^*)h + (I+U)\varphi) = \frac{i}{2}((M^*-I)h - (I-U)\varphi),\ h \in \mathfrak{N}'_{-i} \oplus \mathfrak{N},\ \varphi \in \mathfrak{N}'_i.$$

Then the quasi-kernels of

$$\mathbb{A} = \dot{A}P^+_{\operatorname{Dom}(\dot{A})} + \left(\dot{A}^* + \mathcal{R}^{-1}(S - \frac{i}{2}\mathfrak{J})\right) P^+_{\mathfrak{M}}$$

and

$$\mathbb{A}^* = \dot{A}P^+_{\operatorname{Dom}(\dot{A})} + \left(\dot{A}^* + \mathcal{R}^{-1}(S^* - \frac{i}{2}\mathfrak{J})\right) P^+_{\mathfrak{M}}$$

coincide with T and T^*, respectively. Observe that by construction, we have

$$\begin{aligned} \ker\left(S(P^+_{\mathfrak{N}'_i} + P^+_{\mathfrak{N}'_{-i}}) + \tfrac{i}{2}\mathfrak{J}\right) &= \ker\left(S^*(P^+_{\mathfrak{N}'_i} + P^+_{\mathfrak{N}'_{-i}}) + \tfrac{i}{2}\mathfrak{J}\right) \\ &= \mathfrak{N} \oplus (I+U)\mathfrak{N}'_i \subseteq \ker((\operatorname{Im} S)(P^+_{\mathfrak{N}'_i} + P^+_{\mathfrak{N}'_{-i}})), \end{aligned} \tag{4.34}$$

and

$$\begin{aligned} \operatorname{Ran}\left((P^+_{\mathfrak{N}'_i} + P^+_{\mathfrak{N}'_{-i}})(S - \tfrac{i}{2}\mathfrak{J})\right) &= \operatorname{Ran}\left((P^+_{\mathfrak{N}'_i} + P^+_{\mathfrak{N}'_{-i}})(S^* - \tfrac{i}{2}\mathfrak{J})\right) \\ &= (I-U)\mathfrak{N}'_i \supseteq \overline{\operatorname{Ran}((P^+_{\mathfrak{N}'_i} + P^+_{\mathfrak{N}'_{-i}})\operatorname{Im} S)}. \end{aligned} \tag{4.35}$$

Therefore, $\operatorname{Ran}(\mathbb{A} - \lambda I) \supseteq \operatorname{Ran}(\mathbb{A} - \mathbb{A}^*)$ and $\operatorname{Ran}(\mathbb{A}^* - \bar{\lambda} I) \supseteq \operatorname{Ran}(\mathbb{A} - \mathbb{A}^*)$ for an arbitrary $\lambda \in \rho(T)$. Thus, $\mathbb{A}$ is a q.s.-a. bi-extension of T with the range property (R). Consequently, $T \in R(\dot{A})$. This completes the proof. □

Corollary 4.2.10. *Suppose that an m-dissipative extension T of $\dot{A}$ belongs to the class $R(\dot{A})$. Then each of its q.s.-a. bi-extensions $\mathbb{A}$ with the range property (R) is dissipative* ($\operatorname{Im}(\mathbb{A}h, h) \geq 0$ *for all $h \in \mathcal{H}_+$*).

Proof. Let $\mathbb{A} = \dot{A}P^+_{\mathrm{Dom}(\dot{A})} + \left(\dot{A}^* + \mathcal{R}^{-1}(S - \frac{i}{2}\mathfrak{J})\right) P^+_{\mathfrak{M}}$ be a q.s.-a. bi-extension of T. As proved in Theorem 4.2.9, the operator S is of the form (4.33). Then for $f \in \mathfrak{N}'_i \oplus \mathfrak{N}$ and $\varphi \in \mathfrak{N}'_i$, using that $U : \mathfrak{N}'_i \to \mathfrak{N}'_{-i}$ is a (1)-isometry, we get

$$\mathrm{Im}\,(S((I+M)f + (I+U)\varphi), (I+M)f + (I+U)\varphi)_1 = \frac{1}{2}\left(\|f\|_1^2 - \|Mf\|_1^2\right).$$

Since T is m-dissipative, the operator M is a contraction. So, $\mathrm{Im}\,(\mathbb{A}h, h) \geq 0$ for all $h \in \mathcal{H}_+$. □

4.3 The (*)-extensions and uniqueness theorems

Definition 4.3.1. A q.s.-a. bi-extension $\mathbb{A}$ of an operator $T \in \Omega(\dot{A})$ is called a **(*)-extension** of T if $\mathrm{Re}\,\mathbb{A}$ is a t-self-adjoint bi-extension of $\dot{A}$.

In particular, if $\mathbb{A}$ is a t-self-adjoint bi-extension with quasi-kernel $\hat{A}$, then $\mathbb{A}$ is a (*)-extension of $\hat{A}$.

Theorem 4.3.2. *If T is a regular extension from the class $R(\dot{A})$ and U is a $(\cdot)$-isometric operator, mapping $\mathfrak{N}'_i$ onto $\mathfrak{N}'_{-i}$, such that the equalities (4.24) holds, then the operator S given by (4.33) defines a (*)-extension $\mathbb{A}$ of T. Moreover,*

1. *the operator*

$$\mathrm{Re}\,\mathbb{A} = \dot{A}P^+_{\mathrm{Dom}(\dot{A})} + \left(\dot{A}^* + \mathcal{R}^{-1}(\mathrm{Re}\,S - \frac{i}{2}\mathfrak{J})\right) P^+_{\mathfrak{M}}$$

is a t-self-adjoint bi-extension of $\dot{A}$ and the quasi-kernel $\hat{A}$ of $\mathrm{Re}\,\mathbb{A}$ is given by

$$\mathrm{Dom}(\hat{A}) = \mathrm{Dom}(\dot{A}) \oplus (I + \mathcal{V})(\mathfrak{N}'_i \oplus \mathfrak{N}),$$

where $\mathcal{V} = -(\mathrm{Re}\,S - \frac{i}{2}I)(\mathrm{Re}\,S + \frac{i}{2}I)^{-1} \upharpoonright (\mathfrak{N}'_i \oplus \mathfrak{N})$, the equality

$$U = -(\mathrm{Re}\,S + \frac{i}{2}I)^{-1}(\mathrm{Re}\,S - \frac{i}{2}I) \upharpoonright \mathfrak{N}'_i,$$

holds for the operator U, and

$$\mathrm{Dom}(\hat{A}) \dot{+} (I+U)\mathfrak{N}'_i = \mathcal{H}_+; \tag{4.36}$$

2. *the linear manifolds $\mathrm{Ran}(\mathbb{A} - \lambda I)$ and $\mathrm{Ran}(\mathbb{A}^* - \bar{\lambda} I)$ do not depend on $\lambda \in \rho(T)$ and*

$$\mathrm{Ran}(\mathbb{A} - \lambda I) = \mathrm{Ran}(\mathbb{A}^* - \bar{\lambda} I) = \mathcal{H} \dot{+} \mathcal{R}^{-1}(U - I)\mathfrak{N}'_i, \tag{4.37}$$

$$(\mathbb{A} - \lambda I)\mathfrak{N}_\lambda = (\mathbb{A}^* - \bar{\lambda} I)\mathfrak{N}_{\bar{\lambda}} = \mathfrak{L} \dot{+} \mathcal{R}^{-1}(U - I)\mathfrak{N}'_i, \tag{4.38}$$

for all $\lambda \in \rho(T)$;

3. *the linear manifolds* $\mathrm{Ran}(\mathrm{Re}\,\mathbb{A}-\mu I)$ *do not depend on* $\mu\in\rho(\hat{A})$ *and*

$$\mathrm{Ran}(\mathrm{Re}\,\mathbb{A}-\mu I)=\mathcal{H}\dot{+}\mathcal{R}^{-1}(U-I)\mathfrak{N}'_i, \tag{4.39}$$

$$(\mathrm{Re}\,\mathbb{A}-\mu I)\mathfrak{N}_\mu=\mathfrak{L}\dot{+}\mathcal{R}^{-1}(U-I)\mathfrak{N}'_i \tag{4.40}$$

for all $\mu\in\rho(\hat{A})$;

4. *the inclusion*

$$\mathrm{Ran}(\mathrm{Im}\,\mathbb{A})\subseteq\mathfrak{L}\dot{+}\mathcal{R}^{-1}(U-I)\mathfrak{N}'_i$$

is valid and if $\dot{A}$ *is a maximal common symmetric part of* T *and* T^*, *then the equality*

$$\overline{\mathrm{Ran}(\mathrm{Im}\,\mathbb{A})}^{(-)}=\mathfrak{L}\dot{+}\mathcal{R}^{-1}(U-I)\mathfrak{N}'_i \tag{4.41}$$

holds.

Proof. The statement concerning the operator S is already established in the course of the proof of Theorem 4.2.9. Let us prove statement 1. It follows from (4.34) that

$$\tilde{G}_{\mathrm{Re}\,S}=\ker\left(\mathrm{Re}\,S(P^+_{\mathfrak{N}'_i}+P^+_{\mathfrak{N}'_{-i}})+\frac{i}{2}\mathfrak{J}\right)\supseteq(I+U)\mathfrak{N}'_i\oplus\mathfrak{N}.$$

By Proposition 3.3.4 the operator

$$\left(\dot{A}P^+_{\mathrm{Dom}(\dot{A})}+\dot{A}^*P^+_{\mathfrak{M}}\right)\upharpoonright(\mathrm{Dom}(\dot{A})\oplus\tilde{G}_{\mathrm{Re}\,S})$$

is a closed symmetric extension of $\dot{B}$ defined by (2.19). But

$$\left(\dot{A}P^+_{\mathrm{Dom}(\dot{A})}+\dot{A}^*P^+_{\mathfrak{M}}\right)\upharpoonright(\mathrm{Dom}(\dot{A})\oplus(I+U)\mathfrak{N}'_i\oplus\mathfrak{N})$$

is a self-adjoint extension of $\dot{B}$. Therefore, $\tilde{G}_{\mathrm{Re}\,S}=(I+U)\mathfrak{N}'_i\oplus\mathfrak{N}$. Applying Theorem 3.3.6, we get that

$$\mathrm{Re}\,\mathbb{A}=\dot{A}P^+_{\mathrm{Dom}(\dot{A})}+\left(\dot{A}^*+\mathcal{R}^{-1}(\mathrm{Re}\,S-\frac{i}{2}\mathfrak{J})\right)P^+_{\mathfrak{M}}$$

is a t-self-adjoint bi-extension of $\dot{A}$.

2. Clearly,

$$\mathcal{H}=\mathrm{Ran}(T-\lambda I)\subset\mathrm{Ran}(\mathbb{A}-\lambda I),\quad \mathcal{H}=\mathrm{Ran}(T^*-\bar{\lambda}I)\subset\mathrm{Ran}(\mathbb{A}^*-\bar{\lambda}I)$$

for an arbitrary $\lambda\in\rho(T)$. From (4.25) and (4.35) we get

$$\mathrm{Ran}(\mathbb{A}-\lambda I)\subseteq\mathcal{H}\dot{+}\mathcal{R}^{-1}(U-I)\mathfrak{N}'_i,\quad \mathrm{Ran}(\mathbb{A}^*-\bar{\lambda}I)\subseteq\mathcal{H}\dot{+}\mathcal{R}^{-1}(U-I)\mathfrak{N}'_i.$$

If $f \in \mathfrak{N}_\lambda$, then $\dot{A}^* f = Pf$ and

$$(\dot{A}P^+_{\mathrm{Dom}(\dot{A})} + \dot{A}^* P^+_{\mathfrak{M}})f - \lambda f = P_{\mathfrak{L}}(\dot{A}P^+_{\mathrm{Dom}(\dot{A})} - \lambda I)f.$$

Furthermore, by virtue of (4.35) we obtain

$$\begin{aligned}(\mathbb{A} - \lambda I)f = P_{\mathfrak{L}}(\dot{A}P^+_{\mathrm{Dom}(\dot{A})} &- \lambda I)f + \mathcal{R}^{-1} P^+_{\mathfrak{N}} S P^+_{\mathfrak{M}} f \\ &+ \mathcal{R}^{-1}(P^+_{\mathfrak{N}'_i} + P^+_{\mathfrak{N}'_{-i}})(S - \frac{i}{2}\mathfrak{J})P^+_{\mathfrak{M}} f \in \mathfrak{L} \dot{+} \mathcal{R}^{-1}(U - I)\mathfrak{N}_i.\end{aligned}$$

Similarly, $(\mathbb{A}^* - \bar{\lambda} I)\mathfrak{N}_{\bar{\lambda}} \subseteq \mathfrak{L} \dot{+} \mathcal{R}^{-1}(U - I)\mathfrak{N}_i$. On the other hand the vectors

$$\varphi = (I + U)\phi + (T - \lambda I)^{-1}\left(g - (\dot{A}^* - \lambda I)(I + U)\phi\right)$$

and

$$\psi = (I + U)\phi + (T^* - \bar{\lambda} I)^{-1}\left(g - (\dot{A}^* - \bar{\lambda} I)(I + U)\phi\right)$$

are the solutions of the equations

$$(\mathbb{A} - \lambda I)\varphi = \mathcal{R}^{-1}(U - I)\phi + g \quad \text{and} \quad (\mathbb{A}^* - \bar{\lambda} I)\varphi = \mathcal{R}^{-1}(U - I)\phi + g,$$

where $\phi \in \mathfrak{N}'_i$, $g \in \mathfrak{L}$, respectively. In addition, one can verify that $\varphi \in \mathfrak{N}_\lambda$ and $\psi \in \mathfrak{N}_{\bar{\lambda}}$. Hence, equalities (4.37) and (4.38) are valid.

Statement 3. can be proved similarly with the help of the equality

$$\mathrm{Ran}\left((P^+_{\mathfrak{N}'_i} + P^+_{\mathfrak{N}'_{-i}})\mathrm{Re}\, S - \frac{i}{2}\mathfrak{J}\right) = (I - U)\mathfrak{N}'_i.$$

Observe that the vector

$$\chi = (I + U)\phi + (\hat{A} - \mu I)^{-1}\left(g - (\dot{A}^* - \mu I)(I + U)\phi\right)$$

belongs to $\mathfrak{N}_\mu$ and is the unique solution of the equation $(\mathrm{Re}\,\mathbb{A} - \mu I)\chi = \mathcal{R}^{-1}(U - I)\phi + g$, where $\phi \in \mathfrak{N}'_i$ and $g \in \mathfrak{L}$.

4. Since $\ker(\mathrm{Im}\, S) \supseteq (I + U)\mathfrak{N}'_i$, we get $\overline{\mathrm{Ran}(\mathrm{Im}\, S)}^{(+)} \subseteq \mathfrak{N} \oplus (I - U)\mathfrak{N}'_i$. Consequently, $\mathrm{Ran}(\mathrm{Im}\,\mathbb{A}) \subseteq \mathfrak{L} \dot{+} \mathcal{R}^{-1}(U - I)\mathfrak{N}'_i$. Suppose that $\dot{A}$ is a maximal common symmetric part of T and T^*. Note that $\ker(\mathrm{Im}\,\mathbb{A}) \cap \mathfrak{M}_T = \{0\}$. Indeed, if $h \in \mathfrak{M}_T$ and $(\mathrm{Im}\,\mathbb{A})h = 0$, then, since $\mathbb{A}h = Th$, we get $\mathbb{A}^* h = Th \in \mathcal{H}$. It follows that $h \in \mathrm{Dom}(T^*)$ and $\mathbb{A}^* h = T^* h$. Since $\dot{A}$ is a maximal common symmetric part of T and T^*, we get $h \in \mathrm{Dom}(\dot{A})$. Consequently, $h = 0$. Using decomposition (4.32) and relation (4.47) we get now that

$$\ker(\mathrm{Im}\,\mathbb{A}) = \ker(\mathcal{R}^{-1}(\mathrm{Im}\, S)P^+_{\mathfrak{M}}) = \ker((\mathrm{Im}\, S)(P^+_{\mathfrak{N}'_i} + P^+_{\mathfrak{N}'_{-i}})) = (I + U)\mathfrak{N}'_i.$$

Therefore, taking closure in the $(-)$-metric, we obtain

$$\overline{\mathrm{Ran}(\mathrm{Im}\,\mathbb{A})} = \mathcal{R}^{-1}\overline{\mathrm{Ran}(\mathrm{Im}\, S)} = \mathcal{R}^{-1}(\mathfrak{N} \oplus (U - I)\mathfrak{N}'_i) = \mathfrak{L} \dot{+} \mathcal{R}^{-1}(U - I)\mathfrak{N}'_i. \qquad \square$$

Recall that each bi-extension $\mathbb{A}$ of a regular operator $\dot{A}$ can be written in the form (see (3.12))

$$\mathbb{A} = \dot{B}^* + \mathcal{R}^{-1}\left(S - \frac{i}{2}\mathfrak{J}\right)P^+_{\mathfrak{M}},$$

where $\dot{B}$ is defined by (2.19). Then

$$\mathbb{A} - \dot{B}^* = \mathcal{R}^{-1}\left(S - \frac{i}{2}\mathfrak{J}\right)P^+_{\mathfrak{M}}.$$

Define the linear manifold

$$\mathcal{F}_0 = \left\{\varphi \in \mathcal{H}_- : (\varphi, f) = 0 \quad \text{for all} \quad f \in \operatorname{Dom}(\dot{B})\right\}. \tag{4.42}$$

Since $\operatorname{Dom}(\dot{B}) = \operatorname{Dom}(\dot{A}) \oplus \mathfrak{N}$, $\mathcal{F}_0$ is a subspace in $\mathcal{H}_-$, $\mathcal{F}_0 \cap \mathcal{H} = \{0\}$, and

$$\mathcal{F}_0 = \mathcal{R}^{-1}(\mathfrak{N}'_i \oplus \mathfrak{N}'_{-i}).$$

For a (*)-extension $\mathbb{A}$ of T determined by S of the form (4.33) we have

$$\mathcal{F}_0 \cap \operatorname{Ran}(\mathbb{A} - \dot{B}^*) = \mathcal{R}^{-1}(U - I)\mathfrak{N}'_i.$$

In what follows we will use the set $L_{\mathbb{A}}$,

$$L_{\mathbb{A}} = \mathcal{R}^{-1}(U - I)\mathfrak{N}'_i. \tag{4.43}$$

Thus, we have

$$L_{\mathbb{A}} = \mathcal{F}_0 \cap \operatorname{Ran}(\mathbb{A} - \dot{B}^*). \tag{4.44}$$

In addition, equalities (4.37)–(4.40) can be rewritten as

$$\begin{array}{l} \operatorname{Ran}(\mathbb{A} - \lambda I) = \operatorname{Ran}(\mathbb{A}^* - \bar{\lambda} I) = \mathcal{H} \dot{+} L_{\mathbb{A}}, \\ (\mathbb{A} - \lambda I)\mathfrak{N}_\lambda = (\mathbb{A}^* - \bar{\lambda} I)\mathfrak{N}_{\bar{\lambda}} = \mathfrak{L} \dot{+} L_{\mathbb{A}},\ \lambda \in \rho(T), \\ \operatorname{Ran}(\operatorname{Re}\mathbb{A} - \mu I) = \mathcal{H} \dot{+} L_{\mathbb{A}},\ (\operatorname{Re}\mathbb{A} - \mu I)\mathfrak{N}_\mu = \mathfrak{L} \dot{+} L_{\mathbb{A}},\ \mu \in \rho(\hat{A}). \end{array} \tag{4.45}$$

Theorem 4.3.3. *Let T be a regular extension from the class $\Omega(\dot{A})$ and let $\mathbb{A}$ be a (*)-extension of T. Suppose $\dot{A}$ is a maximal common symmetric part of T and T^*. Then $\mathbb{A}$ is a q.s.-a. bi-extension of T with the range property (R) and, therefore, $T \in R(\dot{A})$.*

Proof. For the sake of simplicity we assume that $-i \in \rho(T)$. Let $M \in [\mathfrak{N}'_i \oplus \mathfrak{N}, \mathfrak{N}'_{-i} \oplus \mathfrak{N}]$ determine T via Theorem 4.1.9. Then equalities (4.14) hold. It follows that

$$\begin{array}{l} (P^+_{\mathfrak{N}'_i} + P^+_{\mathfrak{N}'_{-i}})(S - \frac{i}{2}\mathfrak{J})(M + I)f = 0,\ f \in \mathfrak{N}'_i \oplus \mathfrak{N}, \\ (P^+_{\mathfrak{N}'_i} + P^+_{\mathfrak{N}'_{-i}})(S^* - \frac{i}{2}\mathfrak{J})(M^* + I)h = 0,\ h \in \mathfrak{N}'_{-i} \oplus \mathfrak{N}. \end{array}$$

Since $S = \mathrm{Re}\, S + i\mathrm{Im}\, S$, we get

$$\begin{aligned}(P^+_{\mathfrak{N}'_i} + P^+_{\mathfrak{N}'_{-i}})(\mathrm{Re}\, S - \tfrac{i}{2}\mathfrak{J})(M+I)f &= -i(P^+_{\mathfrak{N}'_i} + P^+_{\mathfrak{N}'_{-i}})(\mathrm{Im}\, S)(M+I)f,\\ (P^+_{\mathfrak{N}'_i} + P^+_{\mathfrak{N}'_{-i}})(\mathrm{Re}\, S - \tfrac{i}{2}\mathfrak{J})(M^*+I)h &= i(P^+_{\mathfrak{N}'_i} + P^+_{\mathfrak{N}'_{-i}})(\mathrm{Im}\, S)(M^*+I)h,\end{aligned}$$

and, therefore,

$$\begin{aligned}&(P^+_{\mathfrak{N}'_i} + P^+_{\mathfrak{N}'_{-i}})(\mathrm{Re}\, S - \tfrac{i}{2}\mathfrak{J})((M+I)f + (M^*+I)h)\\ &= -i(P^+_{\mathfrak{N}'_i} + P^+_{\mathfrak{N}'_{-i}})(\mathrm{Im}\, S)((M+I)f - (M^*+I)h)\end{aligned}$$

for all $f \in \mathfrak{N}'_i \oplus \mathfrak{N}$ and all $h \in \mathfrak{N}'_{-i} \oplus \mathfrak{N}$. By Theorem 4.1.11 the linear manifold $\mathrm{Ran}(M+I) + \mathrm{Ran}(M^*+I)$ is dense in $\mathfrak{M} = \mathfrak{N}'_i \oplus \mathfrak{N}'_{-i} \oplus \mathfrak{N}$. Now, taking closure in the $(+)$-metric, we get the relation

$$\overline{\mathrm{Ran}((P^+_{\mathfrak{N}'_i} + P^+_{\mathfrak{N}'_{-i}})(\mathrm{Re}\, S - \frac{i}{2}\mathfrak{J}))} = \overline{\mathrm{Ran}((P^+_{\mathfrak{N}'_i} + P^+_{\mathfrak{N}'_{-i}})\mathrm{Im}\, S)}.$$

Hence, the equality

$$\ker\left((\mathrm{Re}\, S)(P^+_{\mathfrak{N}'_i} + P^+_{\mathfrak{N}'_{-i}}) + \frac{i}{2}\mathfrak{J}\right) = \ker((\mathrm{Im}\, S)(P^+_{\mathfrak{N}'_i} + P^+_{\mathfrak{N}'_{-i}})) \tag{4.46}$$

holds. Let $\mathbb{A} = \dot{A}P^+_{\mathrm{Dom}(\dot{A})} + \left(\dot{A}^* + \mathcal{R}^{-1}(S - \frac{i}{2}\mathfrak{J})\right)P^+_{\mathfrak{M}}$ be a quasi-self-adjoint bi-extension of T and

$$\mathrm{Re}\,\mathbb{A} = \dot{A}P^+_{\mathrm{Dom}(\dot{A})} + \left(\dot{A}^* + \mathcal{R}^{-1}(\mathrm{Re}\, S - \frac{i}{2}\mathfrak{J})\right)P^+_{\mathfrak{M}}$$

be a t-self-adjoint bi-extension of $\dot{A}$. We set

$$\tilde{G}_{\mathrm{Re}\, S} := \ker\left((\mathrm{Re}\, S)(P^+_{\mathfrak{N}'_i} + P^+_{\mathfrak{N}'_{-i}}) + \frac{i}{2}\mathfrak{J}\right).$$

According to Theorem 3.3.6 the operator

$$\tilde{A} =: \left(\dot{A}P^+_{\mathrm{Dom}(\dot{A})} + \dot{A}^*P^+_{\mathfrak{M}}\right)\restriction(\mathrm{Dom}(\dot{A}) \oplus \tilde{G}_{\mathrm{Re}\, S})$$

is a self-adjoint extension of the operator

$$\mathrm{Dom}(\dot{B}) = \mathrm{Dom}(\dot{A}) \oplus \mathfrak{N}, \quad \dot{B} = \dot{A}P^+_{\mathrm{Dom}(\dot{A})} + \dot{A}^*P^+_{\mathfrak{N}}.$$

It follows that $\dim \mathfrak{N}'_i = \dim \mathfrak{N}'_{-i}$ and there exists a $(\cdot)$-isometric operator U mapping $\mathfrak{N}'_i$ onto $\mathfrak{N}'_{-i}$ such that

$$\mathrm{Dom}(\tilde{A}) = \mathrm{Dom}(\dot{B}) \oplus (I+U)\mathfrak{N}'_i,$$

i.e., $\tilde{G}_{\mathrm{Re}\,S} = (I+U)\mathfrak{N}'_i \oplus \mathfrak{N}$, and

$$(\mathrm{Re}\,S)(I+U)\varphi = -\frac{i}{2}(I-U)\varphi, \quad \varphi \in \mathfrak{N}'_i.$$

Relation (4.46) then yields $(\mathrm{Im}\,S)(I+U)\varphi = 0$, $\varphi \in \mathfrak{N}'_i$. Hence

$$S(I+U)\varphi = S^*(I+U)\varphi = -\frac{i}{2}(I-U)\varphi, \tag{4.47}$$

for all $\varphi \in \mathfrak{N}'_i$. Let the operator $M \in [\mathfrak{N}'_i \oplus \mathfrak{N}, \mathfrak{N}'_{-i} \oplus \mathfrak{N}]$ determine T in accordance with Theorem 4.1.9. Because T and T^* are the quasi-kernels of $\mathbb{A}$ and $\mathbb{A}^*$, respectively, we get

$$S(I+M)f = \frac{i}{2}(I-M)f, \quad S^*(I+M^*)h = \frac{i}{2}(M^*-I)h$$

for all $f \in \mathfrak{N}'_i \oplus \mathfrak{N}$ and all $h \in \mathfrak{N}'_{-i} \oplus \mathfrak{N}$. Let us show that

$$\begin{aligned} (I+M)(\mathfrak{N}'_i \oplus \mathfrak{N}) \dot{+} (I+U)\mathfrak{N}'_i &= \mathfrak{M}, \\ (I+M^*)(\mathfrak{N}'_{-i} \oplus \mathfrak{N}) \dot{+} (I+U)\mathfrak{N}'_i &= \mathfrak{M}. \end{aligned} \tag{4.48}$$

Suppose $\psi \in (I+M)(\mathfrak{N}'_i \oplus \mathfrak{N}) \cap (I+U)\mathfrak{N}'_i$. Then

$$\psi = (I+U)\varphi = (I+M)f,$$

where $\varphi \in \mathfrak{N}'_i$, $f \in \mathfrak{N}'_i \oplus \mathfrak{N}$. Then from (4.47) we obtain

$$(S - \frac{i}{2}\mathfrak{J})(I+U)\varphi = i(U-I)\varphi.$$

On the other hand

$$(S - \frac{i}{2}\mathfrak{J})(I+M)f = P^+_{\mathfrak{N}}(I-M)f.$$

So, $(S - \frac{i}{2}\mathfrak{J})\psi \in \mathfrak{N}'_i \oplus \mathfrak{N}'_{-i}$ and $(S - \frac{i}{2}\mathfrak{J})\psi \in \mathfrak{N}$. It follows that $(I-U)\psi = 0$ if and only if $\psi = 0$. Hence

$$(I+M)(\mathfrak{N}'_i \oplus \mathfrak{N}) \cap (I+U)\mathfrak{N}'_i = \{0\}.$$

Similarly it can be shown that $(I+M^*)(\mathfrak{N}'_{-i} \oplus \mathfrak{N}) \cap (I+U)\mathfrak{N}'_i = \{0\}$. Now suppose that a vector $f \in \mathfrak{M}$ is (1)-orthogonal to $(I+M)(\mathfrak{N}'_i \oplus \mathfrak{N}) \dot{+} (I+U)\mathfrak{N}'_i$. Then from (4.12) we have the equalities

$$f = h - P^+_{\mathfrak{N}'_{-i}} M^* h, \quad P^+_{\mathfrak{N}}(M^*+I)h = 0, \quad h \in \mathfrak{N}'_{-i} \oplus \mathfrak{N},$$

and

$$f = (U-I)\phi + g, \quad \phi \in \mathfrak{N}'_i, \ g \in \mathfrak{N}.$$

It follows that $\phi = -P^+_{\mathfrak{N}'_{-i}} M^* h$, $U\phi = P^+_{\mathfrak{N}'_{-i}} h$. Consequently,

$$(I + U)\phi = (M^* + I)h.$$

Because of $(I + M^*)(\mathfrak{N}'_{-i} \oplus \mathfrak{N}) \cap (I + U)\mathfrak{N}'_i = \{0\}$, we get $\phi = h = 0$. Therefore, the linear manifold $(I + M)(\mathfrak{N}'_i \oplus \mathfrak{N})\dot{+}(I + U)\mathfrak{N}'_i$ is (1)-dense in $\mathfrak{M}$. Similarly, the linear manifold $(I + M^*)(\mathfrak{N}'_{-i} \oplus \mathfrak{N})\dot{+}(I + U)\mathfrak{N}'_i$ is (1)-dense in $\mathfrak{M}$.

Observe that it follows from (4.47) that

$$\left(S + \frac{i}{2}I\right)((M + I)f + (I + U)\varphi) = if + iU\varphi \tag{4.49}$$

for all $f \in \mathfrak{N}'_i \oplus \mathfrak{N}$ and all $\varphi \in \mathfrak{N}'_i$. Consequently, since S is (1)-bounded, we get

$$\begin{aligned}||f||_1^2 + ||\varphi||_1^2 &= \left\|\left(S + \tfrac{i}{2}I\right)((M + I)f + (I + U)\varphi)\right\|_1^2 \\ &\leq c\|(M + I)f + (I + U)\varphi)\|_1^2\end{aligned}$$

for some $c > 0$ and all $f \in \mathfrak{N}'_i \oplus \mathfrak{N}$, $\varphi \in \mathfrak{N}'_i$. This yields that the linear manifold $(I + M)(\mathfrak{N}'_i \oplus \mathfrak{N})\dot{+}(I + U)\mathfrak{N}'_i$ is (1)-closed. Because it is dense in $\mathfrak{M}$, the equality

$$(I + M)(\mathfrak{N}'_i \oplus \mathfrak{N})\dot{+}(I + U)\mathfrak{N}'_i = \mathfrak{M}$$

holds. Similar arguments show that

$$(I + M^*)(\mathfrak{N}'_{-i} \oplus \mathfrak{N})\dot{+}(I + U)\mathfrak{N}'_i = \mathfrak{M}.$$

These equalities are equivalent to (4.24), i.e., $T \in R(\dot{A})$. Moreover,

$$\begin{aligned}\operatorname{Ran}\left((P^+_{\mathfrak{N}'_i} + P^+_{\mathfrak{N}'_{-i}})(S - \frac{i}{2}\mathfrak{J})\right) &= \operatorname{Ran}\left((P^+_{\mathfrak{N}'_i} + P^+_{\mathfrak{N}'_{-i}})(S^* - \frac{i}{2}\mathfrak{J})\right) \\ &= (I - U)\mathfrak{N}'_i = \overline{\operatorname{Ran}((P^+_{\mathfrak{N}'_i} + P^+_{\mathfrak{N}'_{-i}})\operatorname{Im} S)},\end{aligned}$$

where the closure is taken in the (+)-metric. Therefore, $\operatorname{Ran}(\mathbb{A} - \lambda I) \supset \operatorname{Ran}(\operatorname{Im}\mathbb{A})$ and $\operatorname{Ran}(\mathbb{A}^* - \bar{\lambda} I) \supset \operatorname{Ran}(\operatorname{Im}\mathbb{A})$ for each $\lambda \in \rho(T)$. This means that $\mathbb{A}$ is a (∗)-extension of T. □

Remark 4.3.4. Let us make a few remarks.

1) As we have proved above (see (4.49)), if

$$\mathbb{A} = \dot{A}P^+_{\operatorname{Dom}(\dot{A})} + \left(\dot{A}^* + \mathcal{R}^{-1}(S - \frac{i}{2}\mathfrak{J})\right)P^+_{\mathfrak{M}},$$

is a (∗)-extension of $T \in R(\dot{A})$ and $-i \in \rho(T)$, then $-i/2 \in \rho(S)$. Thus $i/2 \in \rho(S^*)$ and if the equality

$$(I + M)(\mathfrak{N}'_i \oplus \mathfrak{N})\dot{+}(I + U)\mathfrak{N}'_i = \mathfrak{M}$$

is valid, then it follows from the equalities $\mathfrak{M} = \mathfrak{N}'_i \oplus \mathfrak{N}'_{-i} \oplus \mathfrak{N}$ and

$$(S^* - \frac{i}{2}I)((I + M^*)g + (I + U)\varphi) = -ig - i\varphi$$

for all $g \in \mathfrak{N}'_{-i} \oplus \mathfrak{N}$ and all $\varphi \in \mathfrak{N}'_i$, that

$$(I + M^*)(\mathfrak{N}'_{-i} \oplus \mathfrak{N}) \cap (I + U)\mathfrak{N}'_i = \{0\}.$$

Moreover,

$$(I + M^*)(\mathfrak{N}'_{-i} \oplus \mathfrak{N}) \dot{+} (I + U)\mathfrak{N}'_i = \mathfrak{M}.$$

Thus, the conditions

$$\mathrm{Dom}(T) \dot{+} (I + U)\mathfrak{N}'_i = \mathcal{H}_+ \quad \text{and} \quad \mathrm{Dom}(T^*) \dot{+} (I + U)\mathfrak{N}'_i = \mathcal{H}_+$$

are equivalent.

2) One can easily derive that equalities (4.48) are equivalent to the following conditions:

(**a**) the operators $UP^+_{\mathfrak{N}'_i} - P^+_{\mathfrak{N}_{-i}} M : \ker(P^+_{\mathfrak{N}}(M + I)) \to \mathfrak{N}'_{-i}$ and $P^+_{\mathfrak{N}}(M + I) : \ker(UP^+_{\mathfrak{N}'_i} - P^+_{\mathfrak{N}'_{-i}} M) \to \mathfrak{N}$ are bijections,

(**b**) the operators $UP^+_{\mathfrak{N}'_i} M^* - P^+_{\mathfrak{N}_{-i}} : \ker(P^+_{\mathfrak{N}}(M^* + I)) \to \mathfrak{N}'_i$ and $P^+_{\mathfrak{N}}(M^* + I) : \ker(UP^+_{\mathfrak{N}'_i} M^* - P^+_{\mathfrak{N}_{-i}}) \to \mathfrak{N}$ are bijections.

When $\dot{A}$ is a closed and densely defined, then the conditions:

(i) the equality
$$(I + M)\mathfrak{N}_i \dot{+} (I + U)\mathfrak{N}_i = \mathfrak{N}_i \oplus \mathfrak{N}_{-i}, \tag{4.50}$$

(ii) the equality
$$(I + M^*)\mathfrak{N}_{-i} \dot{+} (I + U)\mathfrak{N}_i = \mathfrak{N}_i \oplus \mathfrak{N}_{-i}, \tag{4.51}$$

(iii) the operator $M - U : \mathfrak{N}_i \to \mathfrak{N}_{-i}$ is a bijection,

(iv) the operator $I - U^*M : \mathfrak{N}_i \to \mathfrak{N}_i$ is a bijection,

(v) the operator $I - M^*U : \mathfrak{N}_i \to \mathfrak{N}_i$ is a bijection,

are equivalent.

3) By von Neumann's formula (1.15), a $(\cdot)$-unitary mapping U from $\mathfrak{N}'_i$ onto $\mathfrak{N}'_{-i}$ defines a self-adjoint extension $\mathcal{A}_U$ of the symmetric operator $P\dot{A}$ in the space $\mathcal{H}_0$ via

$$\mathrm{Dom}(\mathcal{A}_U) = \mathrm{Dom}(\dot{A}) \dot{+} (I + U)\mathfrak{N}'_i.$$

If U satisfies (4.24), then

$$\mathrm{Dom}(\mathcal{A}_U) \cap \mathrm{Dom}(T) = \mathrm{Dom}(\dot{A}), \; \mathrm{Dom}(\mathcal{A}_U) \cap \mathrm{Dom}(T^*) = \mathrm{Dom}(\dot{A}),$$

and

$$\mathrm{Dom}(\mathcal{A}_U) + \mathrm{Dom}(T) = \mathrm{Dom}(\mathcal{A}_U) + \mathrm{Dom}(T^*) = \mathrm{Dom}(\dot{A}^*) = \mathcal{H}_+. \tag{4.52}$$

4) Denote by $\mathcal{P}_{TU}$ and $\mathcal{P}_{T^*U}$ the projectors onto $\mathrm{Dom}(T)$ and $\mathrm{Dom}(T^*)$, corresponding to the direct decompositions

$$\mathcal{H}_+ = \mathrm{Dom}(T)\dot{+}(I+U)\mathfrak{N}'_i, \tag{4.53}$$

and

$$\mathcal{H}_+ = \mathrm{Dom}(T^*)\dot{+}(I+U)\mathfrak{N}'_i, \tag{4.54}$$

respectively. Then the $(*)$-extension $\mathbb{A}$ of T and its adjoint $\mathbb{A}^*$, corresponding to the choice of the operator U, take the form

$$\begin{aligned}\mathbb{A} &= T\mathcal{P}_{TU} + (\dot{A}^* - \mathcal{R}^{-1}\dot{A}^*)(I - \mathcal{P}_{TU}),\\ \mathbb{A}^* &= T^*\mathcal{P}_{T^*U} + (\dot{A}^* - \mathcal{R}^{-1}\dot{A}^*)(I - \mathcal{P}_{T^*U}).\end{aligned}$$

In addition, if $\dot{A}$ is a maximal common symmetric part of T and T^*, then

$$\ker(\mathrm{Im}\,\mathbb{A}) = \mathrm{Dom}(\mathcal{A}_U).$$

Theorem 4.3.5. *Let the deficiency indices of $\dot{A}$ be equal to $r<\infty$. Suppose that $T \in \Omega(\dot{A})$ and that $\dot{A}$ is a maximal common symmetric part of T and T^*. Then the quasi-self-adjoint bi-extension $\mathbb{A}$ of T has the range property (R) if and only if the conditions*

$$\mathrm{Ran}(\mathrm{Im}\,\mathbb{A}) \supset \mathfrak{L} \quad \textit{and} \quad \dim(\mathrm{Ran}(\mathrm{Im}\,\mathbb{A})) = r \tag{4.55}$$

hold.

Proof. If $\mathbb{A}$ is a $(*)$-extension of T, then the parameter $S \in [\mathfrak{M},\mathfrak{M}]$ is of the form (4.33), where $U \in [\mathfrak{N}'_i, \mathfrak{N}'_{-i}]$ is a (1)-isometry. Then

$$\dim(U-I)\mathfrak{N}'_i = \dim\mathfrak{N}'_i = \dim\mathfrak{N}_i - \dim\mathfrak{N}.$$

Now from (4.41) we get $\dim(\mathrm{Ran}(\mathrm{Im}\,\mathbb{A})) = r$ and $\mathrm{Ran}(\mathrm{Im}\,\mathbb{A}) \supset \mathfrak{L}$.

Conversely, let $\mathbb{A} = \dot{A}P^+_{\mathrm{Dom}(\dot{A})} + \left(\dot{A}^* + \mathcal{R}^{-1}(S - \frac{i}{2}\mathfrak{J})\right)P^+_{\mathfrak{M}}$ be a quasi-self-adjoint extension of T and let conditions (4.55) hold true. Then $\ker(\mathrm{Im}\,S) \subseteq \mathfrak{N}'_i \oplus \mathfrak{N}'_{-i}$. It follows from (4.46) that

$$\ker\left(\mathrm{Re}\,S(P^+_{\mathfrak{N}'_i} + P^+_{\mathfrak{N}'_{-i}}) + \frac{i}{2}\mathfrak{J}\right) = \ker(\mathrm{Im}\,S) \oplus \mathfrak{N}.$$

Since

$$\ker\left(\mathrm{Re}\,S(P^+_{\mathfrak{N}'_i} + P^+_{\mathfrak{N}'_{-i}}) + \frac{i}{2}\mathfrak{J}\right) = (I+W)F_i \oplus \mathfrak{N},$$

where $F_i \subseteq \mathfrak{N}'_i$ and $W : F_i \to \mathfrak{N}'_{-i}$ is a (1)-isometry, we get

$$\dim\ker(\operatorname{Im} S) = \dim F_i \le \dim \mathfrak{N}'_i = r - \dim \mathfrak{N}.$$

On the other hand conditions (4.55) yield

$$\dim\ker(\operatorname{Im} S) = r - \dim \mathfrak{N}.$$

Hence, $F_i = \mathfrak{N}'_i$. From Theorem 3.3.6 we get that $\operatorname{Re}\mathbb{A}$ is a t-self-adjoint bi-extension of $\dot{A}$. Consequently, $\mathbb{A}$ has the range property (R). □

Under the conditions of Theorem 4.3.5 the formula (4.55) implies that $\mathbb{A}$ is a (*)-extension.

Corollary 4.3.6. *Let $\dot{A}$ be a closed densely defined operator with equal finite deficiency indices (r,r). Suppose that $T \in \Omega(\dot{A})$ and that $\dot{A}$ is a maximal common symmetric part of T and T^*. Then the quasi-self-adjoint bi-extension $\mathbb{A}$ of T has the range property (R) if and only if*

$$\dim\operatorname{Ran}(\operatorname{Im}\mathbb{A}) = r.$$

Theorem 4.3.7. *Let the deficiency indices of $\dot{A}$ be equal to $r < \infty$. Suppose that $T \in \Omega(\dot{A})$ and that $\dot{A}$ is a maximal common symmetric part of T and T^*. Then the quasi-self-adjoint bi-extension $\mathbb{A}$ of T is a (*)-extension if and only if conditions* (4.55) *hold.*

Proof. If conditions (4.55) hold, then, as we have already mentioned, Theorem 4.3.5 implies that $\mathbb{A}$ is a (*)-extension. Conversely, let $\mathbb{A}$ be a (*)-extension. Then according to Theorem 4.3.3 it satisfies the range property (R). Thus we can apply Theorem 4.3.5 and obtain conditions (4.55). □

Corollary 4.3.8. *Let $\dot{A}$ be a closed densely defined operator with finite deficiency indices (r,r). Suppose that $T \in \Omega(\dot{A})$ and that $\dot{A}$ is a maximal common symmetric part of T and T^*. Then the quasi-self-adjoint bi-extension $\mathbb{A}$ of T is a (*)-extension of T if and only if*

$$\dim\operatorname{Ran}(\operatorname{Im}\mathbb{A}) = r.$$

The following theorems will be referred at as the **uniqueness theorems**.

Theorem 4.3.9. *Let $T \in R(\dot{A})$ and let $\mathbb{A}'$ and $\mathbb{A}''$ be two (*)-extensions of T. If $\operatorname{Re}\mathbb{A}' = \operatorname{Re}\mathbb{A}''$, then $\mathbb{A}' = \mathbb{A}''$. If $\dot{A}$ is a maximal common symmetric part of T and T^* and $\operatorname{Im}\mathbb{A}' = \operatorname{Im}\mathbb{A}''$, then $\mathbb{A}' = \mathbb{A}''$.*

Proof. Let the operators $\mathbb{A}'$ and $\mathbb{A}''$ be given by

$$\mathbb{A}' = \dot{A}P^+_{\operatorname{Dom}(\dot{A})} + \left(\dot{A}^* + \mathcal{R}^{-1}(S' - \frac{i}{2}\mathfrak{J})\right) P^+_{\mathfrak{M}}$$

and

$$\mathbb{A}'' = \dot{A}P^+_{\operatorname{Dom}(\dot{A})} + \left(\dot{A}^* + \mathcal{R}^{-1}(S'' - \frac{i}{2}\mathfrak{J})\right) P^+_{\mathfrak{M}},$$

respectively, with $S', S'' \in [\mathfrak{M}, \mathfrak{M}]$. Then by Theorem 4.3.2 there are two (1)-isometries $U', U'' \in [\mathfrak{N}'_i, \mathfrak{N}'_{-i}]$, $\mathrm{Ran}(U') = \mathrm{Ran}(U'') = \mathfrak{N}'_{-i}$ such that

$$\begin{aligned} &\ker(S'(P^+_{\mathfrak{N}'_i} + P^+_{\mathfrak{N}'_{-i}}) - \tfrac{i}{2}\mathfrak{J}) = \ker(S'^*(P^+_{\mathfrak{N}'_i} + P^+_{\mathfrak{N}'_{-i}}) - \tfrac{i}{2}\mathfrak{J}) \\ &= \ker((\mathrm{Re}\, S')(P^+_{\mathfrak{N}'_i} + P^+_{\mathfrak{N}'_{-i}}) - \tfrac{i}{2}\mathfrak{J}) = \mathfrak{N} \oplus (I + U')\mathfrak{N}'_i \\ &\subseteq \ker((\mathrm{Im}\, S')(P^+_{\mathfrak{N}'_i} + P^+_{\mathfrak{N}'_{-i}})), \end{aligned}$$

and

$$\begin{aligned} &\ker(S''(P^+_{\mathfrak{N}'_i} + P^+_{\mathfrak{N}'_{-i}}) - \tfrac{i}{2}\mathfrak{J}) = \ker(S''^*(P^+_{\mathfrak{N}'_i} + P^+_{\mathfrak{N}'_{-i}}) - \tfrac{i}{2}\mathfrak{J}) \\ &= \ker((\mathrm{Re}\, S'')(P^+_{\mathfrak{N}'_i} + P^+_{\mathfrak{N}'_{-i}}) - \tfrac{i}{2}\mathfrak{J}) = \mathfrak{N} \oplus (I + U'')\mathfrak{N}'_i \\ &\subseteq \ker((\mathrm{Im}\, S'')(P^+_{\mathfrak{N}'_i} + P^+_{\mathfrak{N}'_{-i}})) \end{aligned}$$

(see equalities (4.34)). If $\mathrm{Re}\,\mathbb{A}' = \mathrm{Re}\,\mathbb{A}''$, then $\mathrm{Re}\, S' = \mathrm{Re}\, S''$. Hence $U' = U''$. By Theorem 4.3.2 $\mathbb{A}' = \mathbb{A}''$.

Suppose $\mathrm{Im}\,\mathbb{A}' = \mathrm{Im}\,\mathbb{A}''$, i.e., $\mathrm{Im}\, S' = \mathrm{Im}\, S''$. Since $\dot{A}$ is a maximal common symmetric part of T and T^*, then by relation (4.46)

$$\ker\left((\mathrm{Re}\, S')(P^+_{\mathfrak{N}'_i} + P^+_{\mathfrak{N}'_{-i}}) + \frac{i}{2}\mathfrak{J}\right) = \ker\left((\mathrm{Re}\, S'')(P^+_{\mathfrak{N}'_i} + P^+_{\mathfrak{N}'_{-i}}) + \frac{i}{2}\mathfrak{J}\right).$$

Since $\mathbb{A}'$ and $\mathbb{A}''$ are $(*)$-extensions, we get the equality $U' = U''$, which implies $\mathbb{A}' = \mathbb{A}''$. □

Theorem 4.3.10. *If a closed symmetric operator $\dot{A}$ has equal and finite deficiency indices, then the classes $\Omega(\dot{A})$ and $R(\dot{A})$ coincide.*

Proof. Clearly, by Theorem 2.4.1 and the remark afterwards the operator $\dot{A}$ is a regular symmetric operator. If $\dot{A}$ is an O-operator, then the statement is already proved (see Theorem 4.1.12). Therefore, we assume that $\mathfrak{N}'_i$ and $\mathfrak{N}'_{-i}$ are nontrivial. Let $\dim \mathfrak{N}_i = \dim \mathfrak{N}_{-i} = r$ and let $\dim \mathfrak{L} = \dim \mathfrak{N} = p < r$. Then $\dim \mathfrak{N}'_i = \dim \mathfrak{N}'_{-i} = r - p$.

If $T \in \Omega(\dot{A})$, then by Theorem 4.1.3 both operators PT and PT^* are closed. Let $M \in [\mathfrak{N}'_i \oplus \mathfrak{N}, \mathfrak{N}'_{-i} \oplus \mathfrak{N}]$ and $M^* \in [\mathfrak{N}'_{-i} \oplus \mathfrak{N}, \mathfrak{N}'_i \oplus \mathfrak{N}]$ define T and T^*, respectively, in accordance with Theorem 4.1.9. Because T and T^* have dense domains in $\mathcal{H}$, the equalities

$$P^+_{\mathfrak{N}}\mathrm{Ran}(M + I) = P^+_{\mathfrak{N}}\mathrm{Ran}(M^* + I) = \mathfrak{N}$$

hold true. We are going to use the following notation:

$$\begin{aligned} &\mathcal{N} := \ker(P^+_{\mathfrak{N}}(M + I)), \quad \mathcal{F} := \ker(P^+_{\mathfrak{N}'_i} \upharpoonright \mathcal{N}), \quad \mathcal{G} = \ker(P^+_{\mathfrak{N}'_{-i}} M \upharpoonright \mathcal{N}), \\ &\mathcal{L} := \mathcal{N} \ominus (\mathcal{F} \dot{+} \mathcal{G}), \quad \mathcal{K} := \mathfrak{N}'_{-i} \ominus (P^+_{\mathfrak{N}'_{-i}} M(\mathcal{F} \oplus \mathcal{L})), \\ &\mathcal{N}_* = \ker(P^+_{\mathfrak{N}}(M^* + I)). \end{aligned}$$

Observe that

1. $\ker(M+I)=\{0\}$ implies $\mathcal{F}\cap\mathcal{G}=\{0\}$,
2. $\dim\mathcal{N}=\dim\mathcal{N}_*=\dim\mathfrak{N}'_{-i}=r-p$,
3. $\dim\mathcal{K}=\dim\mathcal{G}$,
4. $P^+_{\mathfrak{N}_i}\mathcal{G}\cap P^+_{\mathfrak{N}'_i}\mathcal{L}=\{0\}$,
5. $\dim(P^+_{\mathfrak{N}_i}\mathcal{G}\dot{+}P^+_{\mathfrak{N}'_i}\mathcal{L})=\dim(P^+_{\mathfrak{N}'_{-i}}M\mathcal{L}\oplus\mathcal{K})$.

Let U be a (1)-isometry such that $\mathrm{Dom}(U)=\mathfrak{N}'_i$, $\mathrm{Ran}(U)=\mathfrak{N}'_{-i}$, and

$$U(P^+_{\mathfrak{N}_i}\mathcal{G}\dot{+}P^+_{\mathfrak{N}'_i}\mathcal{L})=P^+_{\mathfrak{N}'_{-i}}M\mathcal{L}\oplus\mathcal{K}.$$

Let us consider the following equation with respect to $h\in\mathcal{N}$,

$$UP^+_{\mathfrak{N}'_i}h=\alpha P^+_{\mathfrak{N}'_{-i}}Mh, \tag{4.56}$$

where α is a complex parameter. If h_α is a nontrivial solution of (4.56), then

$$\begin{aligned}g=\alpha P^+_{\mathfrak{N}'_{-i}}Mh_\alpha\in\mathrm{Ran}(P^+_{\mathfrak{N}'_{-i}}M\upharpoonright\mathcal{N})&=P^+_{\mathfrak{N}'_{-i}}M(\mathcal{F}\oplus\mathcal{L})\\&=P^+_{\mathfrak{N}'_{-i}}M\mathcal{F}\dot{+}P^+_{\mathfrak{N}'_{-i}}M\mathcal{L}.\end{aligned}$$

On the other hand

$$g=UP^+_{\mathfrak{N}'_i}h_\alpha\in UP^+_{\mathfrak{N}'_i}\mathcal{N}=UP^+_{\mathfrak{N}'_i}\mathcal{G}\dot{+}UP^+_{\mathfrak{N}'_i}\mathcal{L}.$$

It follows that $g\in\left(P^+_{\mathfrak{N}'_i}M\mathcal{L}\oplus\mathcal{K}\right)\cap\left(P^+_{\mathfrak{N}'_{-i}}M\mathcal{F}\dot{+}P^+_{\mathfrak{N}'_{-i}}M\mathcal{L}\right)$. From the definition of $\mathcal{K}$ we get that $g\in P^+_{\mathfrak{N}'_{-i}}M\mathcal{L}$. Consequently, $h_\alpha\in\mathcal{L}\oplus\mathcal{G}$.

Since the dimensions are finite, it is only possible that either 1) the equation (4.56) has nontrivial solution $h\in\mathcal{N}$ *for all* complex α, or 2) there is *a finite number of α's*, for which (4.56) has nontrivial solutions $h\in\mathcal{N}$. If the case 1) takes place, then choose a sequence $\{\alpha_n\}$ such that $\lim_{n\to\infty}\alpha_n=0$ and let $\{h_n\}$, $\|h_n\|_+=1$, be a corresponding sequence of solutions of (4.56). The sequence $\{h_n\}$ is compact since $\dim\mathfrak{N}_i<\infty$, and hence, we may assume that $\{h_n\}$ converges. Let $\lim_{n\to\infty}h_n=h\in\mathcal{L}\oplus\mathcal{G}$. Then $\|h\|_+=1$. Since

$$\|P^+_{\mathfrak{N}_{-i}}Mh_n\|_1\le\|M\|,$$

we get $\lim_{n\to\infty}\alpha_nP^+_{\mathfrak{N}_{-i}}Mh_n=0$. Hence,

$$\lim_{n\to\infty}UP^+_{\mathfrak{N}_i}h_n=0.$$

Thus, $P^+_{\mathfrak{N}'_i}h=0\Rightarrow h\in\mathcal{F}$ and we arrive at a contradiction. Therefore, case 2) holds, and hence there exists a finite set of complex numbers α's, for which equation (4.56) has a nontrivial solution. Let us denote this set by $\mathfrak{A}$ and assume that $\alpha_0 \notin \mathfrak{A}$ and $|\alpha_0| = 1$. Then the operator

$$\bar{\alpha}_0 U P^+_{\mathfrak{N}_i} - P^+_{\mathfrak{N}_{-i}} M$$

is a one-to-one correspondence between $\mathcal{N}$ and $\mathfrak{N}'_{-i}$. Let

$$\mathcal{U} := \bar{\alpha}_0 U.$$

Suppose also that the equation

$$P_{\mathfrak{N}'_{-i}} y - \mathcal{U} P^+_{\mathfrak{N}'_i} M^* y = 0, \tag{4.57}$$

has a nontrivial solution $f \in \mathcal{N}_*$. Then for all $h \in \mathcal{N}$,

$$\begin{aligned}
0 &= (h, \mathcal{U}^{-1}(P^+_{\mathfrak{N}'_{-i}} f - \mathcal{U} P_{\mathfrak{N}'_i} M^* f))_1 = (\mathcal{U} P^+_{\mathfrak{N}'_i} h, f)_1 - (h, P_{\mathfrak{N}'_i} M^* f)_1 \\
&= (\mathcal{U} P^+_{\mathfrak{N}'_i} h, f)_1 - (h, M^* f - P^+_{\mathfrak{N}} M^* f)_1 \\
&= (\mathcal{U} P^+_{\mathfrak{N}'_i} h, f)_1 - (Mh, f)_1 - (h, P^+_{\mathfrak{N}} f)_1 \\
&= (\mathcal{U} P^+_{\mathfrak{N}'_i} h, f)_1 - (Mh, f)_1 - (P^+_{\mathfrak{N}} h, f)_1 \\
&= (\mathcal{U} P^+_{\mathfrak{N}'_i} h, f)_1 - (Mh, f)_1 + (P^+_{\mathfrak{N}} Mh, f)_1 = ((\mathcal{U} P^+_{\mathfrak{N}'_i} - P^+_{\mathfrak{N}'_{-i}} M)h, f)_1.
\end{aligned}$$

This implies that $f \in \mathfrak{N}$ and, hence, $P^+_{\mathfrak{N}'_i} M^* f = 0$, i.e., $M^* f \in \mathfrak{N}$. Because $f + M^* f = 0$, we get $f = 0$. Therefore, equation (4.57) has only the trivial solution. This is equivalent to the operator

$$P^+_{\mathfrak{N}'_{-i}} - \mathcal{U} P_{\mathfrak{N}'_i} M^*$$

being a one-to-one correspondence between $\mathcal{N}_*$ and $\mathfrak{N}'_{-i}$. Now by the construction we get that

$$\operatorname{Ran}(M + I) \cap \operatorname{Ran}(I + \mathcal{U}) = \operatorname{Ran}(M^* + I) \cap \operatorname{Ran}(I + \mathcal{U}) = \{0\}.$$

Since $\dim \operatorname{Ran}(M + I) = \dim \operatorname{Ran}(M^* + I) = r$, $\dim \operatorname{Ran}(I + \mathcal{U}) = p$, we get

$$\operatorname{Ran}(M + I) \dot{+} \operatorname{Ran}(I + \mathcal{U}) = \operatorname{Ran}(M^* + I) \dot{+} \operatorname{Ran}(I + \mathcal{U}) = \mathfrak{M}.$$

By Theorem 4.2.9 the operator T belongs to the class $R(\dot{A})$. We note that any $T \in R(\dot{A})$ has infinitely many $(*)$-extensions. □

If the semi-deficiency indices are both infinite, the situation is different.

Theorem 4.3.11. *Let $\dot{A}$ be a closed densely defined symmetric operator with equal infinite deficiency indices. Then it admits a quasi-self-adjoint extension $T \in \Omega(\dot{A})$ that does not have a $(*)$-extension.*

Proof. We will construct an example of an operator $T \in \Omega(\dot{A})$ that does not have a (*)-extension. Let $\dot{A}$ be closed, densely defined, symmetric operator in $\mathcal{H}$ with equal infinite deficiency indices. Let $M \in [\mathfrak{N}_i, \mathfrak{N}_i]$, where $\mathfrak{N}_{\pm i}$ are deficiency spaces of $\dot{A}$. By Remark 4.3.4, if $\mathcal{U}$ is an isometry of $\mathfrak{N}_i$ onto $\mathfrak{N}_{-i}$ such that

$$(M+I)\mathfrak{N}_i \dot{+} (I+\mathcal{U})\mathfrak{N}_i = \mathfrak{N}_i \oplus \mathfrak{N}_{-i},$$

then the operator $M - \mathcal{U}$ is bijection from $\mathfrak{N}_i$ onto $\mathfrak{N}_{-i}$.

Also let U be a partial isometry from $\mathfrak{N}_i$ into $\mathfrak{N}_{-i}$ such that

$$\dim(\operatorname{Ran}(U))^{\perp} = \dim\ker(U^*) = \infty.$$

Also let N be a compact and self-adjoint operator in $\mathfrak{N}_i$ such that

$$(N\varphi,\varphi)_+ \geq -\varepsilon(\varphi,\varphi)_+, \quad (0<\varepsilon<1),\ \varphi \in \mathfrak{N}_i,$$

and 0 is not an eigenvalue of N. Considering the operator $M = U(N+I)$, we have $M^*M - I = N(N+2I)$. Since 0 is not an eigenvalue and (-2) is a regular point for N, then $M^*M - I$ is invertible and compact. Moreover,

$$MM^* - I = U(N+I)^2U^* - I.$$

Now let $(MM^*-I)h = 0$. Then $h = U(N+I)^2U^*h$ and hence $N(N+2I)U^*h = 0$. Consequently, $h = 0$. That is why $MM^* - I$ is invertible but not compact, since the operator $MM^* - I$ coincides with $-I$ on the infinite-dimensional subspace $\ker U^*$.

Suppose there exists a (+)-isometric operator $\mathcal{U}$ from $\mathfrak{N}_i$ on $\mathfrak{N}_{-i}$ such that $M - \mathcal{U}$ is an isomorphism from $\mathfrak{N}_i$ on $\mathfrak{N}_{-i}$. Let G be a subspace of $\mathfrak{N}_i$ such that $(M - \mathcal{U})G = \ker U^*$. Then $\dim G = \dim\ker U^* = \infty$. Furthermore, for all $\varphi \in G$ and $\phi \in \mathfrak{N}_i$,

$$((M-\mathcal{U})\varphi, M\phi)_+ = 0.$$

Hence $M^*\mathcal{U}\varphi = M^*M\varphi$. The adjoint operator $M^* - \mathcal{U}^*$ is an bijection from $\mathfrak{N}_{-i}$ on $\mathfrak{N}_i$. Then $M^*\mathcal{U} - I$ is also an bijection from $\mathfrak{N}_i$ on $\mathfrak{N}_i$. Then there exists a $\xi > 0$ such that

$$\|(M^*\mathcal{U} - I)\varphi\|_+ \geq \xi\|\varphi\|_+, \quad (\forall\varphi \in \mathfrak{N}_i).$$

In particular, the last inequality takes place for all $\varphi \in G$. But then

$$\xi\|\varphi\|_+ \leq \|(M^*M - I)\varphi\|_+, \quad (\varphi \in G).$$

Since the operator $M^*M - I$ is compact and $\dim G = \infty$, then the last inequality leads to a contradiction. Thus, there is no (+)-isometric operator $\mathcal{U}$ from $\mathfrak{N}_i$ on $\mathfrak{N}_{-i}$ such that $M - \mathcal{U}$ is an bijection from $\mathfrak{N}_i$ on $\mathfrak{N}_{-i}$. Applying Theorems 4.2.9, 4.3.3, and Remark 4.3.4 we conclude that (4.50) fails to be true and the quasi-self-adjoint extension T of $\dot{A}$ given by

$$\operatorname{Dom}(T) = \operatorname{Dom}(\dot{A}) \oplus (I+M)\mathfrak{N}_i,$$

does not admit a (*)-extension. □

4.4 The $(*)$-extensions in the densely-defined case

Let $\dot{A}$ be a closed densely defined symmetric operator in $\mathcal{H}$. In this case the $(+)$-orthogonal decomposition of $\mathcal{H}_+ = \mathrm{Dom}(\dot{A})$,

$$\mathcal{H}_+ = \mathrm{Dom}(\dot{A}) \oplus \mathfrak{N}_i \oplus \mathfrak{N}_{-i},$$

holds. The operator $\dot{A}^*$ maps $\mathfrak{M} = \mathfrak{N}_i \oplus \mathfrak{N}_{-i}$ onto itself and $\dot{A}^{*2}f = -f$ for all $f \in \mathfrak{M}$. Let T be its quasi-self-adjoint extension, i.e., $T \supset \dot{A}$ and $T^* \supset \dot{A}$. These conditions are equivalent to

$$\dot{A} \subset T \subset \dot{A}^*.$$

Recall that

$$\mathfrak{M}_T = \mathrm{Dom}(T) \ominus \mathrm{Dom}(\dot{A}),$$

and $\mathfrak{M}_T^{\perp} = \mathcal{H}_+ \ominus \mathfrak{M}_T$. Clearly, $\mathfrak{M}_T$ and $\mathfrak{M}_T^{\perp}$ are subspaces of $\mathfrak{M}$ and

$$\mathfrak{M}_T \oplus \mathfrak{M}_T^{\perp} = \mathfrak{M}.$$

The equality

$$\mathfrak{M}_T^{\perp} = \dot{A}^*\mathfrak{M}_{T^*} = T^*\mathfrak{M}_{T^*}$$

holds. Indeed, $f \in \mathfrak{M}_T^{\perp}$ if and only if $(Ty, \dot{A}^*f) = (y, -f)$ for all $y \in \mathrm{Dom}(T)$. The latter is equivalent to $T^*\dot{A}^*f = -f \in \mathrm{Dom}(T^*) \cap \mathfrak{M} = \mathfrak{M}_{T^*} \iff f \in \dot{A}^*\mathfrak{M}_{T^*}$.

Definition 4.4.1. Let $\dot{A}$ be a closed densely defined symmetric operator. Two quasi-self-adjoint extensions T_1 and T_2 of the operator $\dot{A}$ are called **relatively prime (disjoint)** if

$$\mathrm{Dom}(T_1) \cap \mathrm{Dom}(T_2) = \mathrm{Dom}(\dot{A}) \tag{4.58}$$

and are called **mutually transversal** if they are disjoint and

$$\mathrm{Dom}(T_1) + \mathrm{Dom}(T_2) = \mathcal{H}_+. \tag{4.59}$$

Note that in the case of finite deficiency indices the conditions (4.58) and (4.59) are equivalent. Clearly, the disjointness of T and T^* is equivalent to $\dot{A}$ being maximal common symmetric part of T and T^*. It is also not hard to see that if T_1 and T_2 are mutually transversal, then

$$\mathrm{Dom}(T_1) \dot{+} \mathfrak{M}_{T_2} = \mathrm{Dom}(T_2) \dot{+} \mathfrak{M}_{T_1} = \mathcal{H}_+.$$

If $T \in \Omega(\dot{A})$, then relations (4.4) and (4.5) hold, i.e.,

$$\begin{aligned}
&\mathrm{Dom}(T) = \mathrm{Dom}(\dot{A}) + (I - M_\lambda))\mathfrak{N}_{\bar{\lambda}},\\
&T(g + (I - M_\lambda)\psi) = \dot{A}g + \bar{\lambda}\psi - \lambda M_\lambda \psi, \quad g \in \mathrm{Dom}(\dot{A}),\ \psi \in \mathfrak{N}_{\bar{\lambda}},
\end{aligned}$$

where

$$M_\lambda = (T - \bar{\lambda}I)(T - \lambda I)^{-1} \upharpoonright \mathfrak{N}_{\bar{\lambda}} : \mathfrak{N}_{\bar{\lambda}} \to \mathfrak{N}_\lambda, \quad \lambda \in \rho(T),\ \mathrm{Im}\,\lambda \neq 0.$$

Theorem 4.4.2. 1) *Let operators T_1 and T_2 belong to the class $\Omega(\dot{A})$. Let $\rho(T_1)\cap\rho(T_2)\neq\emptyset$. Then*

$$\left((T_1-\lambda I)^{-1}-(T_2-\lambda I)^{-1}\right)\mathcal{H}\subseteq\mathfrak{N}_\lambda,\quad \lambda,\bar\lambda\in\rho(T_1)\cap\rho(T_2),\ \mathrm{Im}\,\lambda\neq 0.$$

2) *If T_1 and T_2 are mutually transversal, then for any $\lambda\in\rho(T_1)\cap\rho(T_2)$, $\mathrm{Im}\,\lambda\neq 0$ the operator*

$$\left[(T_1-\lambda I)^{-1}-(T_2-\lambda I)^{-1}\right]\upharpoonright\mathfrak{N}_{\bar\lambda}$$

is an isomorphism of the spaces $\mathfrak{N}_{\bar\lambda}$ and $\mathfrak{N}_\lambda$. Conversely, if the operator $\left[(T_1-\lambda I)^{-1}-(T_2-\lambda I)^{-1}\right]\upharpoonright\mathfrak{N}_{\bar\lambda}$ is an isomorphism of $\mathfrak{N}_{\bar\lambda}$ and $\mathfrak{N}_\lambda$ for some $\lambda\in\rho(T_1)\cap\rho(T_2)$, $\mathrm{Im}\,\lambda\neq 0$, then T_1 and T_2 are mutually transversal.

3) *Let $T\in\Omega(\dot{A})$. In order for T and T^* to be mutually transversal it is necessary for all, and sufficient for at least one, $\lambda\in\rho(T)$, $\mathrm{Im}\,\lambda\neq 0$ that the operator $I-M_{\bar\lambda}^*M_\lambda$ is an isomorphism of the deficiency subspace $\mathfrak{N}_{\bar\lambda}$.*

Proof. Since both sets $\rho(T_1)$ and $\rho(T_2)$ are open, one can always find a non-real point λ of the set $\rho(T_1)\cap\rho(T_2)$. Since $T_k^*\supset\dot{A}$ $(k=1,2)$, then

$$(T_k^*-\bar\lambda I)^{-1}(\dot{A}-\bar\lambda I)g=g,\qquad (k=1,2,\ g\in\mathrm{Dom}(\dot{A})).$$

Hence for an arbitrary $f\in\mathcal{H}$,

$$\begin{aligned}&\left([(T_1-\lambda I)^{-1}-(T_2-\lambda I)^{-1}]f,\dot{A}g-\bar\lambda g\right)\\&\quad=(f,[(T_1^*-\bar\lambda I)^{-1}-(T_2^*-\bar\lambda I)^{-1}](\dot{A}g-\bar\lambda g))=0.\end{aligned}$$

Then the following inclusion holds:

$$\left((T_1-\lambda I)^{-1}-(T_2-\lambda I)^{-1}\right)\mathcal{H}\subseteq\mathfrak{N}_\lambda.$$

Also

$$\mathrm{Dom}(T_k)=\mathrm{Dom}(\dot{A})\dotplus(I-M_\lambda^{(k)})\mathfrak{N}_{\bar\lambda},\quad k=1,2.$$

If $\mathrm{Dom}(T_1)+\mathrm{Dom}(T_2)=\mathcal{H}_+$ and $\mathrm{Dom}(T_1)\cap\mathrm{Dom}(T_2)=\mathrm{Dom}(\dot{A})$, then the von Neumann formulas imply

$$\mathfrak{N}_\lambda\dotplus\mathfrak{N}_{\bar\lambda}=(I-M_\lambda^{(1)})\mathfrak{N}_\lambda\dotplus(I-M_\lambda^{(2)})\mathfrak{N}_{\bar\lambda}.$$

If $f_\lambda\in\mathfrak{N}_\lambda$, then there are two uniquely defined vectors f_1 and f_2 such that

$$f=(I-M_\lambda^{(1)})f_1+(I-M_\lambda^{(2)})f_2.$$

Therefore, one concludes that the operator $M_\lambda^{(1)}-M_\lambda^{(2)}$ is an isomorphism of spaces $\mathfrak{N}_{\bar\lambda}$ and $\mathfrak{N}_\lambda$. Furthermore,

$$\begin{aligned}M_\lambda^{(1)}-M_\lambda^{(2)}&=\left[(T_1-\bar\lambda I)(T_1-\lambda I)^{-1}-(T_2-\bar\lambda I)(T_2-\lambda I)^{-1}\right]\upharpoonright\mathfrak{N}_{\bar\lambda}\\&=2i\mathrm{Im}\,\lambda\left[(T_1-\lambda I)^{-1}-(T_2-\lambda I)^{-1}\right]\upharpoonright\mathfrak{N}_{\bar\lambda}.\end{aligned}$$

Conversely, if $M_\lambda^{(1)} - M_\lambda^{(2)} = \left[(T_1 - \lambda I)^{-1} - (T_2 - \lambda I)^{-1}\right] \upharpoonright \mathfrak{N}_{\bar\lambda}$ is an isomorphism of spaces $\mathfrak{N}_{\bar\lambda}$ and $\mathfrak{N}_\lambda$, then

$$\mathfrak{N}_\lambda \dotplus \mathfrak{N}_{\bar\lambda} = (I - M_\lambda^{(1)})\mathfrak{N}_{\bar\lambda} \dotplus (I - M_\lambda^{(2)})\mathfrak{N}_{\bar\lambda},$$

which implies the transversality of T_1 and T_2.

The second statement can be proved similarly. □

Using Theorem 4.2.9 and the definition of transversality we obtain the following result.

Theorem 4.4.3. *Let $\dot A$ be a closed densely defined symmetric operator and let $T \in \Omega(\dot A)$. Then $T \in R(\dot A)$ if and only if $\dot A$ has equal deficiency indices and there exists a self-adjoint extension $\tilde A$ of $\dot A$ transversal to T. Moreover, the formulas*

$$\begin{aligned} \mathbb{A} &= \dot A^* - \mathcal{R}^{-1}\dot A^*(I - \mathcal{P}_{T\tilde A}), \\ \mathbb{A}^* &= \dot A^* - \mathcal{R}^{-1}\dot A^*(I - \mathcal{P}_{T^*\tilde A}), \end{aligned} \tag{4.60}$$

set a bijection between the set of all q.s.-a. bi-extensions of $T \in R(\dot A)$ with the range property (R) and their adjoints and all self-adjoint extensions $\tilde A$ of the operator $\dot A$ that are transversal to T. Here $\mathcal{P}_{T\tilde A}$ and $\mathcal{P}_{T^*\tilde A}$ are the projectors in $\mathcal{H}_+$ onto $\operatorname{Dom}(T)$ *and* $\operatorname{Dom}(T^*)$, *corresponding to the direct decompositions*

$$\mathcal{H}_+ = \operatorname{Dom}(T) \dotplus \mathfrak{M}_{\tilde A}, \quad \mathcal{H}_+ = \operatorname{Dom}(T^*) \dotplus \mathfrak{M}_{\tilde A}. \tag{4.61}$$

In addition, if $\dot A$ is maximal common symmetric part of T and T^*, then

$$\ker(\operatorname{Im}\mathbb{A}) = \operatorname{Dom}(\widetilde A).$$

If a quasi-self-adjoint bi-extension $\mathbb{A}$ of T has a form (4.60), we say that $\mathbb{A}$ is **generated** by $\tilde A$.

Theorem 4.4.4. *If $\mathbb{A}$ is a q.s.-a. bi-extension of $T \in \Omega(\dot A)$ with the range property (R) generated by $\tilde A$ via* (4.60), *then* $\operatorname{Re}\mathbb{A}$ *is a t-self-adjoint bi-extension generated by $\tilde A$. Moreover, if*

$$\operatorname{Dom}(\tilde A) = \operatorname{Dom}(\dot A) \oplus (I + U)\mathfrak{N}_i,$$

$-i \in \rho(T)$, *and* $\operatorname{Dom}(T) = \operatorname{Dom}(\dot A) \oplus (I + M)\mathfrak{N}_i$, *then the operator $W \in [\mathfrak{N}_i, \mathfrak{N}_{-i}]$ given by*

$$W = -(I - MU^*)^{-1}(I + MM^* - 2MU^*)(I + MM^* - 2UM^*)^{-1}(U - M) \tag{4.62}$$

$(\cdot)$-unitarily maps $\mathfrak{N}_i$ onto $\mathfrak{N}_{-i}$ and

$$\operatorname{Dom}(\hat A) = \operatorname{Dom}(\dot A) \oplus (I + W)\mathfrak{N}_i,$$

where $\hat A$ is the quasi-kernel of $\operatorname{Re}\mathbb{A}$.

Proof. Let $\mathbb{A} = \dot{A}^* + \mathcal{R}^{-1}(S - \frac{i}{2}\mathfrak{J})P^+_{\mathfrak{M}}$ be a q.s.-a. bi-extension of $T \in \Omega(\dot{A})$ with the range property (R) generated by $\tilde{A}$. Then the first statement follows from Theorem 4.3.2 and equality (4.36).

Let $h \in \mathfrak{M}$. Since $\mathfrak{M} = (I + M^*)\mathfrak{N}_{-i}\dot{+}(I + U)\mathfrak{N}_i$ we have

$$h = (I + M^*)\psi + (I + U)\phi, \ \psi \in \mathfrak{N}_{-i}, \phi \in \mathfrak{N}_i.$$

Due to the decomposition

$$\mathfrak{M} = (I + M)\mathfrak{N}_i\dot{+}(I + U)\mathfrak{N}_i,$$

every vector $(I + M^*)\psi$, $\psi \in \mathfrak{N}_{-i}$, has unique representation

$$(I + M^*)\psi = (I + M)\varphi + (I + U)\chi,$$

where $\varphi, \chi \in \mathfrak{N}_i$. Therefore, $\chi = (U - M)^{-1}(I - MM^*)\psi$. Since

$$(S - \frac{i}{2}\mathfrak{J})(I + M) = 0, \ (S^* - \frac{i}{2}\mathfrak{J})(I + M^*) = 0,$$

and (see (4.47))

$$(\operatorname{Re} S + \frac{i}{2}\mathfrak{J})(I + U) = (S^* + \frac{i}{2}\mathfrak{J})(I + U) = (S + \frac{i}{2}\mathfrak{J})(I + U) = 0,$$

we obtain

$$\begin{aligned}(\operatorname{Re} S - \tfrac{i}{2}\mathfrak{J})h &= (\operatorname{Re} S - \tfrac{i}{2}\mathfrak{J})(I + M^*)\psi + (\operatorname{Re} S - \tfrac{i}{2}\mathfrak{J})(I + U)\phi \\ &= \tfrac{1}{2}(S - \tfrac{i}{2}\mathfrak{J})(I + M^*)\psi - i(I - U)\phi.\end{aligned}$$

Furthermore,

$$(S - \frac{i}{2}\mathfrak{J})(I + M^*)\psi = (S - \frac{i}{2}\mathfrak{J})(I + M)\varphi + (S - \frac{i}{2}\mathfrak{J})(I + U)\chi = -i(I - U)\chi.$$

Therefore,

$$(\operatorname{Re} S - \frac{i}{2}\mathfrak{J})h = -\frac{i}{2}(I - U)\chi - i(I - U)\phi,$$

and thus $(\operatorname{Re} S - \frac{i}{2}\mathfrak{J})h = 0$ if and only if $\phi = -\frac{1}{2}\chi$. Hence, if $h \in \ker(\operatorname{Re} S - \frac{i}{2}\mathfrak{J})$, then

$$h = (I + M^*)\psi - \frac{1}{2}(I + U)(U - M)^{-1}(I - MM^*)\psi. \tag{4.63}$$

Conversely, if $\psi \in \mathfrak{N}_{-i}$, then one can verify that the vector $h \in \mathfrak{M}$ given by (4.63) belongs to $\ker(\operatorname{Re} S - \frac{i}{2}\mathfrak{J})$. The equality (4.63) yields

$$\begin{aligned}P^+_{\mathfrak{N}_i}h &= M^*\psi - \frac{1}{2}(U - M)^{-1}(I - MM^*)\psi \\ &= \frac{1}{2}(U - M)^{-1}(2UM^* - I - MM^*)\psi, \\ P^+_{\mathfrak{N}_{-i}}h &= \psi - \frac{1}{2}U(U - M)^{-1}(I - MM^*)\psi \\ &= \frac{1}{2}(I - MU^*)^{-1}(I + MM^* - 2MU^*)\psi, \qquad \psi \in \mathfrak{N}_{-i}.\end{aligned}$$

Observe that if $B = I - MU^*$, then

$$I + MM^* - 2MU^* = BB^* - 2i\mathrm{Im}\, B,\ I + MM^* - 2UM^* = BB^* + 2i\mathrm{Im}\, B.$$

Because $B^{-1} \in [\mathfrak{N}_{-i}, \mathfrak{N}_{-i}]$, we obtain that

$$(I + MM^* - 2MU^*)^{-1}, (I + MM^* - 2UM^*)^{-1} \in [\mathfrak{N}_{-i}, \mathfrak{N}_{-i}].$$

Hence, the operator

$$W = -(I - MU^*)^{-1}(I + MM^* - 2MU^*)(I + MM^* - 2UM^*)^{-1}(U - M)$$

is well defined and $\mathrm{Dom}(W) = \mathfrak{N}_i$, $\mathrm{Ran}(W) = \mathfrak{N}_{-i}$. By construction we have

$$\ker(\mathrm{Re}\, S - \frac{i}{2}\mathfrak{J}) = (I + W)\mathfrak{N}_i.$$

Since $\mathbb{A} = \dot{A}^* + \mathcal{R}^{-1}(\mathrm{Re}\, S - \frac{i}{2}\mathfrak{J})P^+_{\mathfrak{M}}$ is a t-self-adjoint bi-extension of $\dot{A}$, the operator W is $(\cdot)$ and then $(+)$-unitary mapping $\mathfrak{N}_i$ onto $\mathfrak{N}_{-i}$, and $\mathrm{Dom}(\hat{A}) = \mathrm{Dom}(\dot{A}) \oplus (I + W)\mathfrak{N}_i$ (see Theorems 3.3.6 and 3.4.5). □

Corollary 4.4.5. *If W is given by* (4.62), *then*

$$W - M = -(I - MM^*)(I + MM^* - 2UM^*)^{-1}(U - M).$$

Proof. Due to the equality

$$(U - M)(U^* - M^*) = I + MM^* - UM^* - MU^*,$$

for $W - U$ we get

$$\begin{aligned} W - U &= -U\left((U - M)^{-1}(I + MM^* - 2MU^*)(I + MM^* - 2UM^*)^{-1}\right.\\ &\qquad \left.\times (U - M) + I\right)\\ &= -U(U - M)^{-1}\left((I + MM^* - 2MU^*)(I + MM^* - 2UM^*)^{-1} + I\right)\\ &\qquad \times (U - M)\\ &= -2U(U - M)^{-1}(I + MM^* - MU^* - UM^*)(I + MM^* - 2UM^*)^{-1}\\ &\qquad \times (U - M)\\ &= -2(I - UM^*)(I + MM^* - 2UM^*)^{-1}(U - M). \end{aligned}$$

Furthermore, since $W - M = (W - U) + (U - M)$, we have

$$\begin{aligned} W - M &= -\left(2(I - UM^*)(I + MM^* - 2UM^*)^{-1} - I\right)(U - M)\\ &= -\left(2(I - UM^*) - (I + MM^* - 2UM^*)\right)(I + MM^* - 2UM^*)^{-1}(U - M)\\ &= -(I - MM^*)(I + MM^* - 2UM^*)^{-1}(U - M). \end{aligned}$$

□

The next statement reinforces Theorem 4.3.9.

Theorem 4.4.6. *Let $T \in R(\dot{A})$ and $\dot{A}$ be a maximal common symmetric part of T and T^*. If $\mathbb{A}_1$ and $\mathbb{A}_2$ are $(*)$-extensions of T such that the quasi-kernels of real parts of* $\mathrm{Re}\,\mathbb{A}_1$ *and* $\mathrm{Re}\,\mathbb{A}_2$ *coincide, then $\mathbb{A}_1 = \mathbb{A}_2$.*

Proof. Without loss of generality we may suppose that $-i \in \rho(T)$. Then

$$\mathrm{Dom}(T) = \mathrm{Dom}(\dot{A}) \oplus (I + M)\mathfrak{N}_i, \ \mathrm{Dom}(T^*) = \mathrm{Dom}(\dot{A}) \oplus (I + M^*)\mathfrak{N}_{-i}.$$

Let $\mathbb{A}_1$ and $\mathbb{A}_2$ be generated by $(+)$-unitary mappings $U_1, U_2 : \mathfrak{N}_i \to \mathfrak{N}_{-i}$ ($\mathrm{Ran}(U_1) = \mathrm{Ran}(U_2) = \mathfrak{N}_{-i}$). By assumption the quasi-kernels of $\mathrm{Re}\,\mathbb{A}_1$ and $\mathrm{Re}\,\mathbb{A}_2$ coincide with some self-adjoint extension $\hat{A}$ of $\dot{A}$. Let

$$\mathrm{Dom}(\hat{A}) = \mathrm{Dom}(\dot{A}) \oplus (I + W)\mathfrak{N}_i.$$

Then by Theorem 4.4.4 and Corollary 4.4.5 we have the equality

$$\begin{aligned}(I - MM^*)&(I + MM^* - 2U_1M^*)^{-1}(U_1 - M)\\ &= (I - MM^*)(I + MM^* - 2U_2M^*)^{-1}(U_2 - M).\end{aligned}$$

Since $\dot{A}$ is a maximal common symmetric part of T and T^*, the operator $I - MM^*$ has zero null-space. Hence,

$$(I + MM^* - 2U_1M^*)^{-1}(U_1 - M) = (I + MM^* - 2U_2M^*)^{-1}(U_2 - M)$$

or equivalently

$$(U_1 - M)^{-1}(I + MM^* - 2U_1M^*) = (U_2 - M)^{-1}(I + MM^* - 2U_2M^*).$$

Using the relations

$$\begin{aligned} I + MM^* - 2U_1M^* &= I - MM^* - 2(U_1 - M)M^*,\\ I + MM^* - 2U_2M^* &= I - MM^* - 2(U_2 - M)M^*,\end{aligned}$$

we obtain

$$(U_1 - M)^{-1}(I - MM^*) = (U_2 - M)^{-1}(I - MM^*).$$

This yields $(U_1 - M)^{-1} = (U_2 - M)^{-1}$. Therefore, $U_1 = U_2$ and $\mathbb{A}_1 = \mathbb{A}_2$. □

Theorem 4.4.7. *If $\mathbb{A}$ is a quasi-self-adjoint bi-extension of $T \in \Omega(\dot{A})$ generated by $\tilde{A}$ via (4.60), then for all $\phi = h + f \in \mathcal{H}_+$, $h \in \mathrm{Dom}(T)$, and $f \in \mathrm{Dom}(\tilde{A})$ we have*

$$(\mathbb{A}\phi, \phi) = (Th, h) + (\tilde{A}f, f) + 2\mathrm{Re}\,(Th, f).$$

Proof. Let $\mathbb{A} = \dot{A}^* - \mathcal{R}^{-1}\dot{A}^*(I - \mathcal{P}_{T\tilde{A}})$ according to (4.60). Note that for any $f \in \mathrm{Dom}(\tilde{A})$ we have that $P_{T\tilde{A}}f \in \mathrm{Dom}(\dot{A})$ and hence

$$(\mathcal{P}_{T\tilde{A}}f, Th) = (\dot{A}\mathcal{P}_{T\tilde{A}}f, h),$$

for any $h \in \mathrm{Dom}(T)$. Besides,

$$\left(\mathcal{R}^{-1}\dot{A}^*(I-\mathcal{P}_{T\tilde{A}})f, g\right) = 0, \qquad \forall f, g \in \mathrm{Dom}(\tilde{A}).$$

Indeed, since $\mathrm{Dom}(\tilde{A}) = \mathrm{Dom}(\dot{A}) \oplus (U+I)\mathfrak{N}_i$, where U is a unitary operator from $\mathfrak{N}_i$ onto $\mathfrak{N}_{-i}$, we have $(I-\mathcal{P}_{T\tilde{A}})f = (I+U)\varphi$, for $\varphi \in \mathfrak{N}_i$, and

$$\dot{A}^*(I-\mathcal{P}_{T\tilde{A}})f = i(I-U)\varphi.$$

Moreover, from $(+)$-orthogonality of $(U+I)\mathfrak{N}_i$ and $(U-I)\mathfrak{N}_i$ we obtain the desired equation. Further,

$$\begin{aligned}(\mathbb{A}\phi,\phi) &= (Th+\tilde{A}f-\mathcal{R}^{-1}\tilde{A}(I-\mathcal{P}_{T\tilde{A}})f, h+f)\\ &= (Th,h)+(\tilde{A}f,f)+(Th,f)-(\tilde{A}(I-\mathcal{P}_{T\tilde{A}})f,h)_+ + (\tilde{A}f,h),\end{aligned}$$

and

$$\begin{aligned}(\tilde{A}(I-\mathcal{P}_{T\tilde{A}})f,h)_+ &= (\tilde{A}(I-\mathcal{P}_{T\tilde{A}})f,h) - ((I-\mathcal{P}_{T\tilde{A}})f,Th)\\ &= (\tilde{A}f,h)-(\dot{A}\mathcal{P}_{T\tilde{A}}f,h)-(f,Th)+(\mathcal{P}_{T\tilde{A}}f,Th)\\ &= (\tilde{A}f,h)-(f,Th).\end{aligned}$$

Consequently, $(\mathbb{A}\phi,\phi) = (Th,h)+(\tilde{A}f,f)+2\mathrm{Re}\,(Th,f)$. □

4.5 Resolvents of quasi-self-adjoint extensions

Let C be a linear operator in $\mathcal{H}$. Following Chapter 1, we denote by $\rho(C)$ the resolvent set of C. For every $\lambda \in \rho(C)$, the **resolvent** of C is the operator

$$R_\lambda(C) = (C-\lambda I)^{-1}$$

that is defined on entire $\mathcal{H}$ and $(\cdot)$-continuous. We also recall that we write $\Omega(\dot{A})$ for the class of all quasi-self-adjoint extensions T (of a closed symmetric operator $\dot{A}$) with nonempty resolvent sets.

Theorem 4.5.1. *Let $T \in \Omega(\dot{A})$. Then for every $\lambda \in \rho(T)$ the operator $R_\lambda(T)$ is $(\cdot,+)$-continuous and is a holomorphic function of λ with values in $[\mathcal{H},\mathcal{H}_+]$.*

Proof. Based on Theorems 4.1.2 and 4.1.3 we have $R_\lambda(T)\mathcal{H} = \mathrm{Dom}(T) \subset \mathcal{H}_+$ and

$$\dot{A}^*R_\lambda(T) = PT(T-\lambda I)^{-1} = P[I+\lambda R_\lambda(T)], \tag{4.64}$$

where P is a $(\cdot)$-orthogonal projection operator in $\mathcal{H}$ onto $\mathcal{H}_0 = \overline{\mathrm{Dom}(\dot{A})}$. Hence

$$\begin{aligned}\|R_\lambda(T)f\|_+^2 &= \|R_\lambda(T)f\|^2 + \|\dot{A}^*R_\lambda(T)f\|^2\\ &\le \|R_\lambda(T)f\|^2 + \|f+\lambda R_\lambda(T)f\|^2 \le \gamma\|f\|^2\end{aligned}$$

for some $\gamma > 0$, i.e., $R_\lambda(T) \in [\mathcal{H}, \mathcal{H}_+]$. Furthermore, if $f \in \mathcal{H}$ and $g \in \mathcal{H}_+$, then the function

$$\begin{aligned}(R_\lambda(T)f, g)_+ &= (R_\lambda(T)f, g) + (\dot{A}^* R_\lambda(T)f, \dot{A}^* g)\\ &= (R_\lambda(T)f, g) + (f + \lambda R_\lambda(T)f, \dot{A}^* g)\end{aligned}$$

is holomorphic on $\rho(T)$. Therefore $R_\lambda(T)$ is a holomorphic function with values in $[\mathcal{H}, \mathcal{H}_+]$. □

Proposition 4.5.2. *If T_1 and T_2 belong to $\Omega(\dot{A})$ and $\lambda \in \rho(T_1) \cap \rho(T_2)$, then* $\operatorname{Ran}(R_\lambda(T_1) - R_\lambda(T_2)) \subset \mathfrak{N}_\lambda$.

Proof. Since $T_k^* \supset \dot{A}$ $(k = 1, 2)$, then

$$R_\lambda(T_k^*)(\dot{A} - \bar{\lambda} I)g = g, \qquad (k = 1, 2,\ g \in \operatorname{Dom}(\dot{A})).$$

Hence for an arbitrary $f \in \mathcal{H}$,

$$\Big([R_\lambda(T_1) - R_\lambda(T_2)]f, \dot{A}g - \bar{\lambda}g\Big) = (f, [R_{\bar{\lambda}}(T_1^*) - R_{\bar{\lambda}}(T_2^*)](\dot{A}g - \bar{\lambda}g)) = 0,$$

and this completes the proof. □

For a rigged Hilbert space $\mathcal{H}_+ \subset \mathcal{H} \subset \mathcal{H}_-$ we call an operator C **bi-continuous** if $C \in [\mathcal{H}_-, \mathcal{H}]$ and $C \upharpoonright \mathcal{H} \in [\mathcal{H}, \mathcal{H}_+]$. Now we will show that the resolvents of operators in $\Omega(\dot{A})$ can be extended to $\mathcal{H}_-$ by continuity and study the properties of such resolvents.

Theorem 4.5.3. *Let $T \in \Omega(\dot{A})$ and $\lambda \in \rho(T)$. Then:*

1. *$R_\lambda(T)$ can be extended to $\mathcal{H}_-$ by $(-, \cdot)$-continuity;*
2. *the extended operator $\hat{R}_\lambda(T)$ is bi-continuous and is a holomorphic function of $\lambda \in \rho(T)$ that belongs to $[\mathcal{H}_-, \mathcal{H}]$;*
3. *for all $h \in \mathcal{H}_0$,*
$$\hat{R}_\lambda(T)(\bar{\dot{A}}h - \lambda h) = h, \tag{4.65}$$
where $\bar{\dot{A}}$ is the extension of $\dot{A}$ to $\mathcal{H}_0 = \overline{\operatorname{Dom}(\dot{A})}$ by $(\cdot, -)$-continuity.

Proof. 1. Since $T^* \in \Omega(\dot{A})$ and $\bar{\lambda} \in \rho(T^*)$, then the operator $R_{\bar{\lambda}}(T^*)$ belongs to $[\mathcal{H}, \mathcal{H}_+]$. Its adjoint $[R_{\bar{\lambda}}(T^*)]^*$ then belongs to $[\mathcal{H}_-, \mathcal{H}]$. If $f, g \in \mathcal{H}$, then

$$([R_{\bar{\lambda}}(T^*)]^* f, g) = (f, R_{\bar{\lambda}}(T^*)g) = (R_\lambda(T)f, g),$$

i.e., $[R_{\bar{\lambda}}(T^*)]^* f = R_\lambda(T)f$ for all $f \in \mathcal{H}$. Therefore $\hat{R}_\lambda(T) = [R_{\bar{\lambda}}(T^*)]^*$ is an extension of the operator $R_\lambda(T)$ on $\mathcal{H}_-$ by $(-, \cdot)$-continuity. Note that it follows from the definition of $\hat{R}_\lambda(T)$ that

$$(\hat{R}_\lambda(T)f, g) = (f, R_{\bar{\lambda}}(T^*)g) \qquad (\forall f \in \mathcal{H}_-,\ g \in \mathcal{H}). \tag{4.66}$$

2. The operator $\hat{R}_\lambda(T)$ is clearly bi-continuous. Since $R_{\bar{\lambda}}(T^*)$ is a holomorphic function of $\bar{\lambda}$ that belongs to $[\mathcal{H}, \mathcal{H}_+]$, then $\hat{R}_\lambda(T)$ is a holomorphic on $\rho(T)$ function that belongs to $[\mathcal{H}_-, \mathcal{H}]$.

3. According to (4.64) $(\dot{A}^* - \lambda P)R_{\bar{\lambda}}(T^*) = P$. Passing to the adjoint we get (4.65). □

Theorem 4.5.4. $\operatorname{Ran}(\hat{R}_\lambda(T)) = \mathcal{H}$ *if and only if* T^* *is* $(+,\cdot)$*-continuous.*

Proof. The $(+,\cdot)$-continuity of T^* means that the operator $R_{\bar{\lambda}}(T^*) \in [\mathcal{H}, \mathcal{H}_+]$ has $(+,\cdot)$-continuous inverse. The latter is equivalent to the fact that

$$[R_{\bar{\lambda}}(T^*)]^* = \hat{R}_\lambda(T)$$

maps $\mathcal{H}_-$ onto $\mathcal{H}$. □

Theorem 4.5.5. *Let* T_1 *and* T_2 *belong to* $\Omega(\dot{A})$ *and* $\lambda \in \rho(T_1)$, $\mu \in \rho(T_2)$. *Then* $[\hat{R}_\lambda(T_1) - \hat{R}_\mu(T_2)] \in [\mathcal{H}_-, \mathcal{H}_+]$ *and the* $[\mathcal{H}_-, \mathcal{H}_+]$*-valued function* $[\hat{R}_\lambda(T_1) - \hat{R}_\mu(T_2)]$ *is holomorphic in* λ *and* μ *on* $\rho(T_1) \times \rho(T_2)$.

Proof. Indeed, let $f \in \operatorname{Dom}(\dot{A})$ and $g \in \mathcal{H}_-$. Then $[\hat{R}_\lambda(T_1) - \hat{R}_\mu(T_2)]g \in \mathcal{H}$ and

$$\big(\dot{A}f, [\hat{R}_\lambda(T_1) - \hat{R}_\mu(T_2)]g\big) = \big([R_{\bar{\lambda}}(T_1^*) - R_{\bar{\mu}}(T_2^*)]\dot{A}f, g\big).$$

Since $T_1^* \supset \dot{A}$, then

$$R_{\bar{\lambda}}(T_1^*)\dot{A}f = R_{\bar{\lambda}}(T_1^*)T_1^* f = f + \bar{\lambda}R_{\bar{\lambda}}(T_1^*)f,$$

and similarly

$$R_{\bar{\mu}}(T_2^*)\dot{A}f = R_{\bar{\mu}}(T_1^*)T_1^* f = f + \bar{\mu}R_{\bar{\mu}}(T_2^*)f.$$

Hence

$$\begin{aligned}\big(\dot{A}f, [\hat{R}_{\bar{\lambda}}(T_1^*) - \hat{R}_{\bar{\mu}}(T_2^*)]g\big) &= \big(\bar{\lambda}R_{\bar{\lambda}}(T_1^*)f - \bar{\mu}R_{\bar{\mu}}(T_2^*)f, g\big)\\ &= \big(f, [\lambda R_\lambda(T_1) - \mu\hat{R}_\mu(T_2)]g\big).\end{aligned}$$

Furthermore, $[R_\lambda(T_1) - \hat{R}_\mu(T_2)]g \in \mathcal{H}_+$ and

$$\dot{A}^*\left(\hat{R}_\lambda(T_1) - \hat{R}_\mu(T_2)\right)g = P\left(\lambda\hat{R}_\lambda(T_1) - \mu\hat{R}_\mu(T_2)\right)g. \tag{4.67}$$

Thus, $\hat{R}_\lambda(T_1) - \hat{R}_\mu(T_2) \in [\mathcal{H}_-, \mathcal{H}_+]$. The analyticity of the function $\hat{R}_\lambda(T_1) - \hat{R}_\mu(T_2)$ follows from the fact that for any $g \in \mathcal{H}_-$ both functions $[\hat{R}_\lambda(T_1) - \hat{R}_\mu(T_2)]g$ and $\dot{A}^*[\hat{R}_\lambda(T_1) - \hat{R}_\mu(T_2)]g$ are holomorphic. □

We note that it follows from the above theorem and (4.66) that

$$[\hat{R}_\lambda(T_1) - \hat{R}_\mu(T_2)]^* = \hat{R}_{\bar{\lambda}}(T_1^*) - \hat{R}_{\bar{\mu}}(T_2^*) \tag{4.68}$$

in $[\mathcal{H}_-, \mathcal{H}_+]$.

In the case when $\lambda = \mu$ Theorem 4.5.5 can be refined as follows.

Theorem 4.5.6. *Let T_1 and T_2 belong to $\Omega(\dot{A})$ and $\lambda \in \rho(T_1) \cap \rho(T_2)$. Then $[\hat{R}_\lambda(T_1) - \hat{R}_\lambda(T_2)]\mathcal{H}_- \subseteq \mathfrak{N}_\lambda$.*

Proof. Using (4.66) with $\lambda = \mu$ we have

$$(\dot{A}g - \bar{\lambda}g, [\hat{R}_\lambda(T_1) - \hat{R}_\lambda(T_2)]f) = 0,$$

for all $g \in \mathrm{Dom}(\dot{A})$ and $f \in \mathcal{H}_-$. □

We recall (see (3.1)) that the orthogonal complement of $\mathrm{Dom}(\dot{A})$ ($\subset \mathcal{H}_+$) is denoted by $\mathcal{F}$ ($\subset \mathcal{H}_-$), the linear manifold $\mathcal{B}$ is given by

$$\mathcal{B} = \mathcal{H} + \mathcal{F} = \mathcal{H}_0 \dotplus \mathcal{F}, \tag{4.69}$$

and the projection onto $\mathcal{H}_0$ parallel to $\mathcal{F}$ is denoted by Π.

Lemma 4.5.7. *Let T belong to $\Omega(\dot{A})$ and $\lambda \in \rho(T)$. A vector $f \in \mathcal{H}_-$ belongs to $\mathcal{F}$ if and only if $\hat{R}_\lambda(T)f \in \mathfrak{N}_\lambda$.*

Proof. The proof immediately follows from the relation

$$(\dot{A}g - \bar{\lambda}g, \hat{R}_\lambda(T)f) = (g, f), \qquad (g \in \mathrm{Dom}(\dot{A}),\ f \in \mathcal{H}_-).$$ □

Corollary 4.5.8. *The following statements hold true:*

1. $\hat{R}_\lambda(T)\mathcal{F} \subset \mathfrak{N}_\lambda$;
2. *If* $\mathrm{Ran}(\hat{R}_\lambda(T)) = \mathcal{H}$, *then* $\hat{R}_\lambda(T)\mathcal{F} = \mathfrak{N}_\lambda$;
3. *If $f \in \mathcal{F}$, then the vector function $\hat{R}_\lambda(T)f$ is $(+)$-holomorphic on $\rho(T)$.*

The last condition is true because for every $h \in \mathcal{H}_+$ the scalar function

$$\begin{aligned}(\hat{R}_\lambda(T)f, h)_+ &= (\hat{R}_\lambda(T)f, h) + (\dot{A}^*\hat{R}_\lambda(T)f, \dot{A}^*h)\\ &= (\hat{R}_\lambda(T)f, h) + \lambda(P\hat{R}_\lambda(T)f, \dot{A}^*h),\end{aligned}$$

is holomorphic on $\rho(T)$.

Theorem 4.5.9. *Let $T \in \Omega(\dot{A})$ and $\lambda \in \rho(T)$. Then:*

1. *$\hat{R}_\lambda(T)f$ belongs to $\mathcal{H}_+$ if and only if $f \in \mathcal{B}$;*
2. *the operator $(\dot{A}^* - \lambda P)\hat{R}_\lambda(T)\restriction\mathcal{B}$ coincides with Π.*

Proof. According to Lemma 4.5.7 and Theorem 4.5.1 $\hat{R}_\lambda(T)\mathcal{B} \subset \mathcal{H}_+$. Conversely, let $f \in \mathcal{H}_-$ and $\hat{R}_\lambda(T)\mathcal{B} \subset \mathcal{H}_+$. We set

$$\varphi = (\dot{A}^* - \lambda P)\hat{R}_\lambda(T)f, \quad \psi = f - \varphi.$$

For an arbitrary $g \in \mathrm{Dom}(\dot{A})$ we have

$$\begin{aligned}(\psi, g) &= (f, g) - ((\dot{A}^* - \lambda P)\hat{R}_\lambda(T)f, g)\\ &= (f, g) - (f, R_{\bar{\lambda}}(T^*)(\dot{A}g - \lambda g)) = 0,\end{aligned}$$

since $T^* \supset \dot{A}$. Therefore, $f = \varphi + \psi \in \mathcal{H}_0 + \mathcal{F} = \mathcal{B}$ and

$$\Pi f = \varphi = (\dot{A}^* - \lambda P)\hat{R}_\lambda(T)f. \qquad \square$$

It follows from the proof that $(\dot{A}^* - \lambda P)\hat{R}_\lambda(T)g$ is independent of the choice of $T \in \Omega(\dot{A})$ and $\lambda \in \rho(T)$.

Let F be a $(-)$-subspace of $\mathcal{F}$, $T \in \Omega(\dot{A})$, and $\lambda \in \rho(T)$. By Lemma 4.5.7 the operator $\hat{R}_\lambda(T)$ maps $\mathcal{H}_0 + F$ in $\mathcal{H}_+$. Let

$$Q = \hat{R}_\lambda(T)\restriction(\mathcal{H}_0 + F).$$

Then the following theorem holds.

Theorem 4.5.10. *The operator Q is $(-,+)$-closed.*

Proof. Let $\{h_n\} \subset \mathrm{Dom}(Q)$, $h_n \xrightarrow{(-)} h(\in \mathcal{H})$, and $Qh_n \xrightarrow{(-)} f(\in \mathcal{H}_+)$. It follows from $(+,\cdot)$-continuity of the operator $(\dot{A}^* - \lambda P)\restriction\mathcal{H}_+$ that

$$(\dot{A}^* - \lambda P)Qh_n \to (\dot{A}^* - \lambda P)f \in \mathcal{H}_0.$$

Using the proof of Theorem 4.5.9 we have $(\dot{A}^* - \lambda P)Qh_n = \mathcal{P}h_n$ and hence

$$h_n' = h_n - \mathcal{P}h_n \in F.$$

Since F is $(-)$-closed we have

$$h_n' \xrightarrow{(-)} h - (\dot{A}^* - \lambda P)f \in F.$$

Thus, $h \in \mathcal{H}_0 + F = \mathrm{Dom}(\dot{A})$. Furthermore, the $(-,\cdot)$-continuity of $\hat{R}_\lambda(T)$ yields $\hat{R}_\lambda(T)h_n \to \hat{R}_\lambda(T)h$, i.e., $f = \hat{R}_\lambda(T)h = Qh$. $\square$

Let $T_1, T_2 \in \Omega(\dot{A})$ and $\lambda_k \in \rho(T_k)$ $(k = 1,2)$. It follows from (4.68) that for arbitrary $f, g \in \mathcal{H}_-$,

$$([\hat{R}_{\lambda_1}(T_1) - \hat{R}_{\lambda_2}(T_2)]f, g) = (f, [\hat{R}_{\bar{\lambda}_1}(T_1^*) - \hat{R}_{\bar{\lambda}_2}(T_2^*)]g).$$

If $f, g \in \mathcal{B}$, then the last equation can be re-written as

$$(\hat{R}_{\lambda_1}(T_1)f, g) - (f, \hat{R}_{\bar{\lambda}_1}(T_1^*)g) = (\hat{R}_{\lambda_2}(T_2)f, g) - (f, \hat{R}_{\bar{\lambda}_2}(T_2^*)g).$$

Therefore it is clear that the sesquilinear functional

$$\Omega(f,g) = \frac{1}{2i}\left[(\hat{R}_\lambda(T)f, g) - (f, \hat{R}_{\bar{\lambda}}(T^*)g)\right], \tag{4.70}$$

which is defined on $\mathcal{B} \times \mathcal{B}$, is independent of the choice of $T \in \Omega(\dot{A})$ and $\lambda \in \rho(T)$. This functional is also symmetric. Indeed, taking into account that T^* belongs to

$\Omega(\dot{A})$ together with T, we replace T with T^* and λ with $\bar{\lambda}$ in the right-hand side of (4.70)

$$\Omega(f,g) = \frac{1}{2i}\left[(\hat{R}_{\bar{\lambda}}(T^*)f,g) - (f,\hat{R}_{\lambda}(T)g)\right].$$

Comparing with (4.70) we get $\Omega(f,g) = \overline{\Omega(g,f)}$. It follows from (4.66) that $\Omega(f,g) = 0$ if either f or g belongs to $\mathcal{H}$. Therefore, it is sufficient to consider $\Omega(f,g)$ for $f,g \in \mathcal{F}$.

Theorem 4.5.11. *The functional $\Omega(f,g)\restriction(\mathcal{F}\times\mathcal{F})$ is $(-)$-continuous in both arguments.*

Proof. Let $\{f_n\},\{g_n\} \subset \mathcal{F}$, $f_n \xrightarrow{(-)} 0$, and $g_n \xrightarrow{(-)} 0$. Then $\hat{R}_\lambda(T)f_n \xrightarrow{(\cdot)} 0$. Since, according to Lemma 4.5.7, $\hat{R}_\lambda(T)f_n \in \mathfrak{N}_\lambda$ and $\mathfrak{N}_\lambda$ is poly-closed, then $\hat{R}_\lambda(T)f_n \xrightarrow{(+)} 0$ and hence, $(\hat{R}_\lambda(T)f_n, g_n) \to 0$. Similarly, $(f_n, \hat{R}_{\bar{\lambda}}(T^*)g_n) \to 0$, and thus $\Omega(f_n,g_n) \to 0$ as $n \to \infty$. □

Let us now derive an explicit formula for $\Omega(f,g)\restriction(\mathcal{F}\times\mathcal{F})$ using the $(-)$-orthogonal decomposition (2.18). It follows from Theorem 4.5.11 that $\Omega(f,g) = 0$ if either $f \in \overline{\mathfrak{L}}^{(-)}$ or $g \in \overline{\mathfrak{L}}^{(-)}$. Thus we only need to consider the case when $f,g \in \mathcal{R}^{-1}\mathfrak{N}'_{\pm i}$. First we will show that $\mathfrak{N}'_i$ is $(+)$-orthogonal to $\mathfrak{N}_{-i}$. Indeed, if $\varphi \in \mathfrak{N}'_i$ and $\psi \in \mathfrak{N}_{-i}$, then $\dot{A}^*\varphi = i\varphi$, $\dot{A}^*\psi = -iP\psi$, and

$$(\varphi,\psi)_+ = (\varphi,\psi) + (\dot{A}^*\varphi, \dot{A}^*\psi) = (\varphi,\psi) + (i\varphi, -iP\psi) = 0.$$

Similarly one shows that $\mathfrak{N}'_{-i}$ is $(+)$-orthogonal to $\mathfrak{N}_i$.

Let $\varphi \in \mathfrak{N}'_{\pm i}$. Applying (2.7) we get

$$\mathcal{R}^{-1}\varphi = \varphi + A\dot{A}^*\varphi = \varphi + A(\pm i\varphi) = \pm i(A\varphi \mp i\varphi).$$

Let T be such a quasi-self-adjoint extension of operator $\dot{A}$ that i is its regular point (for example one can take the maximal symmetric extension of $\dot{A}$). Then using (4.65) and (2.1) we have that $\hat{R}_i(T)\mathcal{R}^{-1}\varphi = i\varphi$ for all $\varphi \in \mathfrak{N}'_i$. If $\psi \in \mathcal{H}_+$, then

$$(\hat{R}_i(T)\mathcal{R}^{-1}\varphi, \mathcal{R}^{-1}\psi) = i(\varphi, \mathcal{R}^{-1}\psi) = i(\mathcal{R}^{-1}\varphi, \mathcal{R}^{-1}\psi)_- = i(\varphi,\psi)_+.$$

On the other hand,

$$(\mathcal{R}^{-1}\varphi, \hat{R}_{-i}(T^*)\mathcal{R}^{-1}\psi) = (\varphi, \hat{R}_{-i}(T^*)\mathcal{R}^{-1}\psi)_+.$$

If $\psi \in \mathfrak{M}$, then according to Corollary 4.5.8, $\hat{R}_{-i}(T^*)\mathcal{R}^{-1}\psi \in \mathfrak{N}_i$ and hence

$$(\varphi, \hat{R}_{-i}(T^*)\mathcal{R}^{-1}\psi)_+ = 0.$$

Therefore,

$$\Omega(\mathcal{R}^{-1}\varphi, \mathcal{R}^{-1}\psi) = \frac{1}{2}(\varphi,\psi)_+, \qquad (\forall\varphi \in \mathfrak{N}'_i,\ \psi \in \mathfrak{M}).$$

In particular,

$$\Omega(\mathcal{R}^{-1}\varphi, \mathcal{R}^{-1}\psi) = \begin{cases} \frac{1}{2}(\varphi,\psi)_+, & \forall\varphi,\psi \in \mathfrak{N}'_i; \\ 0, & \forall\varphi \in \mathfrak{N}'_i,\ \psi \in \mathfrak{N}'_{-i}. \end{cases} \tag{4.71}$$

Similarly, if T is a quasi-self-adjoint extension of $\dot{A}$ with a regular point $(-i)$, we have

$$\Omega(\mathcal{R}^{-1}\varphi, \mathcal{R}^{-1}\psi) = \begin{cases} 0, & \forall\varphi \in \mathfrak{N}'_{-i},\ \psi \in \mathfrak{N}'_i; \\ -\frac{1}{2}(\varphi,\psi)_+, & \forall\varphi,\psi \in \mathfrak{N}'_{-i}. \end{cases} \tag{4.72}$$

Formulas (4.71) and (4.72) provide the desired representation for

$$\Omega \upharpoonright \mathcal{R}^{-1}(\mathfrak{N}'_i \oplus \mathfrak{N}'_{-i}).$$

We note that the functional Ω is identically zero if and only if $\dot{A}$ is an O-operator.

Now we establish a connection between $\hat{R}_\lambda(T)$ and $(\mathbb{A}-\lambda I)^{-1}$ for $T \in R(\dot{A})$ and its quasi-self-adjoint extension $\mathbb{A}$ with the range property (R). Recall that relations (4.45) hold, where $L_\mathbb{A} = \mathcal{F}_0 \cap \operatorname{Ran}(\mathbb{A} - \dot{B}^*)$ (see (4.44)).

Theorem 4.5.12. *Let $\dot{A}$ be a regular symmetric operator with equal deficiency indices. If $T \in R(\dot{A})$ and $\mathbb{A}$ is a quasi-self-adjoint bi-extension of T with the range property (R), then the operator $\mathbb{A} - \lambda I$ is invertible for all $\lambda \in \rho(T)$ and $(\mathbb{A}-\lambda I)^{-1}$ can be extended to $\mathcal{H}_-$ by $(-,\cdot)$-continuity. The extended operator coincides with $\hat{R}_\lambda(T)$ and $\hat{R}_\lambda(T)\mathcal{H}_- = \mathcal{H}$. Moreover,*

$$\begin{aligned} &\hat{R}_\lambda(T)\upharpoonright(\mathcal{H}\dot{+}L_\mathbb{A}) = (\mathbb{A}-\lambda I)^{-1}, \quad \hat{R}_{\bar\lambda}(T^*)\upharpoonright(\mathcal{H}\dot{+}L_\mathbb{A}) = (\mathbb{A}^*-\bar\lambda I)^{-1}, \\ &\hat{R}_\mu(\hat{A})\upharpoonright(\mathcal{H}\dot{+}L_\mathbb{A}) = (\operatorname{Re}\mathbb{A} - \mu I)^{-1} \end{aligned}$$

for $\lambda \in \rho(T)$ and $\mu \in \rho(\hat{A})$ ($\hat{A}$ is the quasi-kernel of $\operatorname{Re}\mathbb{A}$*). In addition, the resolvents $(\mathbb{A}-\lambda I)^{-1}$, $(\mathbb{A}^*-\bar\lambda I)^{-1}$, and $(\operatorname{Re}\mathbb{A}-\mu I)^{-1}$ map $(-,+)$-continuously the linear manifold $\mathfrak{L}\dot{+}L_\mathbb{A}$ onto $\mathfrak{N}_\lambda$, $\mathfrak{N}_{\bar\lambda}$, and $\mathfrak{N}_\mu$, correspondingly.*

Proof. Since $\mathbb{A}^* - \bar\lambda I \supset T^* - \bar\lambda I$ and $\bar\lambda \in \rho(T^*)$, then

$$(\mathbb{A}^* - \bar\lambda I)\mathcal{H}_+ \supset (T^* - \bar\lambda I)\operatorname{Dom}(T^*) = \mathcal{H}.$$

It follows that $\ker(\mathbb{A}-\lambda I) = \{0\}$ and by Theorem 2.1.5 the operator $(\mathbb{A}-\lambda I)^{-1}$ admits an extension on $\mathcal{H}_-$ by $(-,\cdot)$-continuity. Since $(\mathbb{A}-\lambda I)^{-1} \supset (T-\lambda I)^{-1}$, then the extension of $(\mathbb{A}-\lambda I)^{-1}$ by $(-,\cdot)$-continuity coincides with the extension of $(T-\lambda I)^{-1}$ by $(-,\cdot)$-continuity, i.e., with $\hat{R}_\lambda(T)$. It follows from Theorems 3.1.8 and 4.1.13 that operator T^* is $(+,\cdot)$-continuous. By Theorem 4.5.4,

$$\hat{R}_\lambda(T)\mathcal{H}_- = \mathcal{H}.$$

The remaining statements follow from Theorem 4.3.2 (see (4.37), (4.39), (4.38), (4.40)) and the equivalence of $(+)$ and $(\cdot)$ norms on the deficiency subspaces. □

Corollary 4.5.13. *$\hat{R}_\lambda(T)(\mathbb{A}-\lambda I)h = h$ for all $h \in \mathcal{H}_+$.*

In the following theorem we establish a connection between extensions of operator $(\mathbb{A}-\lambda I)^{-1}$ by $(-,-)$-continuity and by $(-,\cdot)$-continuity.

Theorem 4.5.14. *If $\mathbb{A} \in \mathcal{E}(\dot{A})$ and the operator $(\mathbb{A}-\lambda I)^{-1}$ is $(-,-)$-continuous, then it is $(-,\cdot)$-continuous.*

Proof. By Theorem 2.1.6 if $C \in [\mathcal{H}_+, \mathcal{H}_-]$, then C^{-1} is $(-,-)$-continuous if and only if $\operatorname{Ran}(C^*) \supset \mathcal{H}_+$. Hence, $\operatorname{Ran}(\mathbb{A}^* - \bar{\lambda} I) \supset \mathcal{H}_+$ and thus $\operatorname{Ran}(\mathbb{A}^* - \bar{\lambda} I) \supset \mathfrak{N}_\lambda$. On the other hand, $\mathbb{A}^* - \bar{\lambda} I \supset \dot{A} - \bar{\lambda} I$ which implies

$$\operatorname{Ran}(\mathbb{A}^* - \bar{\lambda} I) \supset \mathfrak{M}_\lambda.$$

Since $\mathcal{H} = \mathfrak{M}_{\bar{\lambda}} \oplus \mathfrak{N}_\lambda$, then $\operatorname{Ran}(\mathbb{A}^* - \bar{\lambda} I) \supset \mathcal{H}$. Therefore, $(\mathbb{A}-\lambda I)^{-1}$ is $(-,\cdot)$-continuous. □

The following theorem refines Theorem 4.5.14.

Theorem 4.5.15. *Let $\mathbb{A}$ be a self-adjoint bi-extension of symmetric operator $\dot{A}$. The necessary and sufficient condition for operator $(\mathbb{A}-\lambda I)^{-1}$ to be $(-,-)$-continuous is to be $(-,\cdot)$-continuous.*

Proof. Taking into account Theorem 4.5.14 we only need to prove that $(-,\cdot)$-continuity implies $(-,-)$-continuity. Let $(\mathbb{A}-\lambda I)^{-1}$ be $(-,\cdot)$-continuous. Then

$$\|(\mathbb{A}-\lambda I)^{-1}\| \le k\,\|f\|_-, \quad k>0,\ \forall f \in \operatorname{Ran}(\mathbb{A}-\lambda I).$$

Since for all $g \in \mathcal{H}$, $\|g\|_- \le \|g\|$, then it follows that

$$\|(\mathbb{A}-\lambda I)^{-1}\|_- \le k\,\|f\|_-, \quad k>0,\ \forall f \in \operatorname{Ran}(\mathbb{A}-\lambda I),$$

i.e., $(\mathbb{A}-\lambda I)^{-1}$ is $(-,-)$-continuous. □

Corollary 4.5.16. *Let $\mathbb{A} = \mathbb{A}^* \in \mathcal{E}(\dot{A})$ where $\dot{A}$ has finite and equal semi-deficiency indices. If for some λ operator $(\mathbb{A}-\lambda I)^{-1}$ is $(-,-)$-continuous, then $\mathbb{A}$ is a self-adjoint bi-extension of operator $\dot{A}$.*

The proof of the corollary easily follows from Theorem 4.5.15.

Lemma 4.5.17. *If $\mathbb{A} \in \mathcal{E}(\dot{A})$, then for every λ we have $(\mathbb{A}-\lambda I)\mathfrak{N}_\lambda \subset \mathcal{F}$.*

Proof. Let $g \in \mathfrak{N}_\lambda$ and $f \in \operatorname{Dom}(\dot{A})$. Then

$$0 = (\dot{A}f - \bar{\lambda} f, g) = (\mathbb{A} - \bar{\lambda} f, g) = (f, \mathbb{A}^* g - \lambda g),$$

and hence $\mathbb{A}^* g - \lambda g \in \mathcal{F}$. □

Theorem 4.5.18. *If $\mathbb{A} \in \mathcal{E}(\dot{A})$ and the operator $(\mathbb{A}-\lambda I)^{-1}$ can be extended by $(-,-)$-continuity to the operator R_λ, then $R_\lambda \mathcal{F} = \mathfrak{N}_\lambda$.*

Proof. It follows from Lemma 4.5.17 that $R_\lambda \mathcal{F} \supset \mathfrak{N}_\lambda$. In order to prove the inverse inclusion we note that according to Theorem 4.5.14 the operator R_λ is $(-,\cdot)$-continuous. Let $f \in \mathcal{F}$, then there exists such a sequence $\{h_n\} \subset \mathcal{H}_+$ that

$$f = (-)\lim_{n\to\infty} (\mathbb{A} - \lambda I)h_n, \qquad h_n \to R_\lambda f.$$

For an arbitrary $g \in \mathrm{Dom}(\dot{A})$ we have

$$(h_n, (\dot{A} - \bar{\lambda} I)g) = (h_n, (\mathbb{A}^* - \bar{\lambda} I)g) = ((\mathbb{A} - \lambda I)h_n, g).$$

Passing to the limit as $n \to \infty$ we obtain

$$(R_\lambda f, (\dot{A} - \bar{\lambda} I)g) = (f, g) = 0,$$

and hence $R_\lambda f$ is orthogonal to $\mathfrak{M}_\lambda$ and $R_\lambda f \in \mathfrak{N}_\lambda$. □

Chapter 5

The Livšic Canonical Systems with Bounded Operators

In this chapter we present the foundations of the theory of the Livšic canonical open systems with bounded state-space (main) operators. We provide an analysis of such a type of systems in terms of transfer functions and their linear-fractional transformations. We also consider couplings of these systems and present multiplication and factorization theorems of the transfer functions.

5.1 The Livšic canonical system and the Brodskiĭ theorem

Let A be a bounded linear operator in a Hilbert space $\mathcal{H}$, $K \in [E, \mathcal{H}]$, and J be a bounded, self-adjoint, and unitary operator in E, where E is another Hilbert space with $\dim E < \infty$.[1] Let also $\operatorname{Im} A = KJK^*$ and $L^2_{[0,\tau_0]}(E)$ be the Hilbert space of E-valued functions equipped with an inner product

$$(\varphi, \psi)_{L^2_{[0,\tau_0]}(E)} = \int_0^{\tau_0} (\varphi, \psi)_E \, dt, \quad \left(\varphi(t),\ \psi(t) \in L^2_{[0,\tau_0]}(E)\right).$$

Consider the system of equations

$$\begin{cases} i\frac{d\chi}{dt} + A\chi(t) = KJ\psi_-(t), \\ \chi(0) = x \in \mathcal{H}, \\ \psi_+ = \psi_- - 2iK^*\chi(t). \end{cases} \tag{5.1}$$

The following lemma holds.

[1] Here and in the proof of Lemma 5.1.1 we assume that $\dim E < \infty$. However, Lemma 5.1.1 also holds true for the case of a separable infinite-dimensional Hilbert space E.

Lemma 5.1.1. *If for a given continuous in E function $\psi_-(t) \in L^2_{[0,\tau_0]}(E)$ we have that $\chi(t) \in \mathcal{H}$ and $\psi_+(t) \in L^2_{[0,\tau_0]}(E)$ satisfy (5.1), then a system of the form (5.1) satisfies the metric conservation law*

$$2\|\chi(\tau)\|^2 - 2\|\chi(0)\|^2 = \int_0^\tau (J\psi_-, \psi_-)_E\, dt - \int_0^\tau (J\psi_+, \psi_+)_E\, dt, \qquad \tau \in [0, \tau_0]. \quad (5.2)$$

Proof. Using the first equation of system (5.1) we obtain

$$\begin{aligned}
\frac{d}{dt}(\chi(t), \chi(t)) &= (\chi'(t), \chi(t)) + (\chi(t), \chi'(t)) = ((-iKJ\psi_-(t) - A\chi(t)), \chi(t)) \\
&\quad + (\chi(t), (-iKJ\psi_-(t) - A\chi(t))) \\
&= -(iKJ\psi_-(t), \chi(t)) - (\chi(t), iKJ\psi_-(t)) - (iA^*\chi(t), \chi(t)) + (iA\chi(t), \chi(t)) \\
&= -(iKJ\psi_-(t), \chi(t)) - (\chi(t), iKJ\psi_-(t)) - (2KJK^*\chi(t), \chi(t)).
\end{aligned}$$

Now we apply the second equation and continue

$$\begin{aligned}
\frac{d}{dt}(\chi(t), \chi(t)) &= -(iKJ(\psi_+(t) + 2iK^*\chi(t)), \chi(t)) \\
&\quad - (\chi(t), iKJ(\psi_+(t) + 2iK^*\chi(t))) - (2KJK^*\chi(t), \chi(t)) \\
&= -(iKJ\psi_+(t), \chi(t)) - (iKJ\psi_-(t), \chi(t)) \\
&= -(iJ\psi_+(t), K^*\chi(t))_E - (iJ\psi_-(t), K^*\chi(t))_E \\
&= -(iJ\psi_+(t), (\psi_-(t) - \psi_+(t))/2i)_E - (iJ\psi_-(t), (\psi_-(t) - \psi_+(t))/2i)_E \\
&= \frac{1}{2}(J\psi_-(t), \psi_-(t))_E - \frac{1}{2}(J\psi_+(t), \psi_+(t))_E.
\end{aligned}$$

Taking into account that $\psi_\pm(t) \in L^2_{[0,\tau_0]}(E)$ and $\chi(t)$ is continuously differentiable, we integrate both sides from 0 to $\tau \in [0, \tau_0]$, and multiply by 2 to obtain (5.2). □

Given an input vector $\psi_- = \varphi_- e^{izt} \in E$, we seek solutions to the system (5.1) as an output vector $\psi_+ = \varphi_+ e^{izt} \in E$ and a state-space vector $\chi(t) = xe^{izt} \in \mathcal{H}$. Substituting the expressions for $\psi_\pm(t)$ and $\chi(t)$ allows us to cancel exponential terms and convert the system (5.1) to the stationary form

$$\begin{cases} (A - zI)x = KJ\varphi_-, \\ \varphi_+ = \varphi_- - 2iK^*x, \end{cases} \qquad z \in \rho(A), \quad (5.3)$$

which is called **the Livšic canonical system**. Here $\varphi_- \in E$ is an input vector, $\varphi_+ \in E$ is an output vector, and x is a state-space vector in $\mathcal{H}$. The spaces $\mathcal{H}$ and E are called **state** and **input-output**, and the operators A, K, J are **state-space**, **channel**, and **directing**, respectively. The relation

$$KJK^* = \operatorname{Im} A \quad (5.4)$$

implies

$$\operatorname{Ran}(\operatorname{Im} A) \subseteq \operatorname{Ran}(K). \quad (5.5)$$

The subspace $\overline{\operatorname{Ran}(K)}$ is called the **channel subspace**. Briefly the Livšic canonical system (5.3) can be written as an array[2]

$$\Theta = \begin{pmatrix} A & K & J \\ \mathcal{H} & & E \end{pmatrix}, \tag{5.6}$$

which we will sometimes refer to as a canonical system. The following theorem is due to Brodskiĭ [89].

Theorem 5.1.2. *If A is a bounded linear operator acting in a separable Hilbert space $\mathcal{H}$, and $\mathcal{G}$ is any subspace containing $\overline{\operatorname{Ran}(\operatorname{Im} A)}$, then there exists a canonical system of the form (5.6) for which A is a state-space operator and $\mathcal{G}$ is the channel subspace.*

Proof. The operator $\operatorname{Im} A$ maps $\overline{\operatorname{Ran}(\operatorname{Im} A)}$ into itself and annihilates its orthogonal complement. Consider the spectral decomposition

$$\operatorname{Im} A\restriction \overline{\operatorname{Ran}(\operatorname{Im} A)} = \int_a^b t\, d\Sigma(t).$$

Putting

$$K_0 = \int_a^b |t|^{1/2} d\Sigma(t), \quad J_0 = \int_a^b \operatorname{sign} t\, d\Sigma(t),$$

we get

$$J_0 = J_0^*, \quad J_0^2 = I, \quad K_0 J_0 K_0^* = \operatorname{Im} A\restriction \overline{\operatorname{Ran}(\operatorname{Im} A)}.$$

Next, we construct the orthogonal sum

$$E = \overline{\operatorname{Ran}(\operatorname{Im} A)} \oplus \mathcal{H}_1 \oplus \mathcal{H}_2,$$

where $\mathcal{H}_1$ and $\mathcal{H}_2$ are Hilbert spaces whose dimensions coincide with the dimension of the space $\mathcal{H}_0 = \mathcal{G} \ominus \overline{\operatorname{Ran}(\operatorname{Im} A)}$. Suppose that U is some isometric mapping of $\mathcal{H}_2$ onto $\mathcal{H}_1$ and that K_1 is a bounded linear mapping of $\mathcal{H}_1$ into $\mathcal{H}_2$ with dense range in $\mathcal{H}_0$. Then $K_2 = K_1 U \in [\mathcal{H}_2, \mathcal{H}_1]$ and $K_2 K_2^* h = K_1 K_1^* h$, $h \in \mathcal{H}_0$.

Let $K \in [E, \mathcal{H}]$ and $J \in [E, E]$ be defined as follows:

$$Kg = \begin{cases} K_0 g, & g \in \overline{\operatorname{Ran}(\operatorname{Im} A)}, \\ K_1 g, & g \in \mathcal{H}_1, \\ K_2 g, & g \in \mathcal{H}_2, \end{cases} \qquad Jg = \begin{cases} J_0 g, & g \in \overline{\operatorname{Ran}(\operatorname{Im} A)}, \\ g, & g \in \mathcal{H}_1, \\ -g, & g \in \mathcal{H}_2. \end{cases}$$

It is easy to see that $J = J^*$, $J^2 = I_E$, and $\operatorname{Ran}(K)$ is dense in $\mathcal{G}$. Since

$$K^* h = \begin{cases} K_0^* h, & h \in \overline{\operatorname{Ran}(\operatorname{Im} A)}, \\ (K_1^* + K_2^*) h, & h \in \mathcal{H}_0, \\ 0, & h \perp \mathcal{G}, \end{cases}$$

[2]In operator theory the array (5.6) is also referred as an **operator colligation.**

we have KJK^*. Thus

$$\Theta = \begin{pmatrix} A & K & J \\ \mathcal{H} & & E \end{pmatrix}$$

is the desired system. □

A construction of a system based upon a given linear operator A is called an **inclusion of** A into a system. It is clear that this operation is not unique. It is also not hard to see that if $\dim \operatorname{Ran}(\operatorname{Im} A) < \infty$, then A can be included into a system Θ with operator K such that $\operatorname{Ran}(K) = \dim \operatorname{Ran}(\operatorname{Im} A)$ or into a system with an invertible operator K. The systems with these conditions will play an important role in the next chapters.

5.2 Minimal canonical systems

A bounded linear operator A on a Hilbert space $\mathcal{H}$ is called **prime** if $\mathcal{H}$ cannot be represented as an orthogonal sum of two subspaces $\mathcal{G}$ and $\mathcal{G}^\perp (\neq 0)$ with the following properties:

1. $\mathcal{G}$ and $\mathcal{G}^\perp$ are invariant to A;
2. A induces a self-adjoint operator in $\mathcal{G}^\perp$.

Theorem 5.2.1. *The closed linear span $\mathcal{G}$ of the form*

$$\mathcal{G} = \text{c. l. s.}\{A^n(\operatorname{Im} A)h,\ n = 0, 1, \ldots;\ h \in \mathcal{H}\} \tag{5.7}$$

and its orthogonal complement $\mathcal{G}^\perp = \mathcal{H} \ominus \mathcal{G}$ are invariant with respect to A. The operator A induces a prime operator in $\mathcal{G}$, and a self-adjoint operator in $\mathcal{G}^\perp$. Moreover, the operator A is prime if and only if $\mathcal{H} = \mathcal{G}$.

Proof. By definition $A\mathcal{G} \subseteq \mathcal{G}$. Therefore $A^*\mathcal{G}^\perp \subseteq \mathcal{G}^\perp$. Moreover, since $\operatorname{Ran}(\operatorname{Im} A) \subset \mathcal{G}$, we have $\mathcal{G}^\perp \subseteq \ker(\operatorname{Im} A)$. Thus $Ah = A^*h$ for all $h \in \mathcal{G}^\perp$ and $A\mathcal{G}^\perp \subseteq \mathcal{G}^\perp$ and $A\restriction \mathcal{G}^\perp$ is a self-adjoint operator in $\mathcal{G}^\perp$.

Let $A_1 := A\restriction \mathcal{G}$. We suppose that $\mathcal{G} = \mathcal{G}_1 \oplus \mathcal{G}_0$, where: 1) $\mathcal{G}_1$ and $\mathcal{G}_0$ are invariant with respect to A_1; 2) A_1 induces in $\mathcal{G}_0$ a self-adjoint operator. Then A induces in $\mathcal{G}_0 \oplus \mathcal{G}^\perp$ a self-adjoint operator, $\operatorname{Ran}(\operatorname{Im} A) \subset \mathcal{G}_1$, and therefore

$$A^n(\operatorname{Im} A)\, h \in \mathcal{G}_1, \quad (n = 0, 1, \ldots;\ h \in \mathcal{H}).$$

It follows that $\mathcal{G}_0 = \{0\}$. □

Corollary 5.2.2. *The space $\mathcal{H}$ can be represented in one and only one way in the form of an orthogonal sum of subspaces $\mathcal{G}$ and $\mathcal{G}^\perp$ which are invariant with respect to A and in which A induces a prime and a self-adjoint operator, respectively.*

Let us consider a canonical system Θ of the form (5.6) and denote by $\mathcal{F}$ the closed linear span of vectors $A^n Kg$, i.e.,

$$\mathcal{F} = \text{c.l.s.}\{A^n Kg,\ n = 0, 1, \ldots;\ g \in E\}. \tag{5.8}$$

The subspaces $\mathcal{F}$ and $\mathcal{F}^\perp = \mathcal{H} \ominus \mathcal{F}$ are called **principal** and **excess** subspaces, respectively. It easily follows from Theorem 5.2.1 and relation (5.5) that each of the subspaces $\mathcal{F}$ and $\mathcal{F}^\perp$ is invariant with respect to A and A^*, and that $A\upharpoonright \mathcal{F}^\perp$ is a self-adjoint in $\in \mathcal{F}^\perp$.

A canonical system Θ of the form (5.6) is said to be **minimal** if $\mathcal{F} = \mathcal{H}$ and **excess** or **non-minimal** otherwise. For a canonical system to be minimal it is sufficient that its state-space operator is prime. The converse statement is generally speaking not true. Indeed, letting $\mathcal{G} = \mathcal{H}$ in Theorem 5.1.2, we find that every bounded linear operator may be included in a minimal canonical system.

Having a system

$$\Theta = \begin{pmatrix} A & K & J \\ \mathcal{H} & & E \end{pmatrix},$$

we may construct a new canonical system

$$\Theta^* = \begin{pmatrix} A^* & K & -J \\ \mathcal{H} & & E \end{pmatrix},$$

which is called **adjoint** to Θ. The state subspaces of the systems Θ and Θ^* coincide. Indeed, since the subspace $\mathcal{F}$ is invariant with respect to A^* and $\operatorname{Ran} K \subseteq \mathcal{F}$, we have

$$A^{*n} Kg \in \mathcal{F}, \quad (n = 0, 1, \ldots;\ g \in \mathcal{H}),$$

or $\mathcal{F}_* \subseteq \mathcal{F}$, where $\mathcal{F}_*$ is the principle subspace of Θ^* of the form (5.8) where A is replaced with A^*. Similarly one shows that $\mathcal{F} \subseteq \mathcal{F}_*$. Therefore $\mathcal{F} = \mathcal{F}_*$.

Lemma 5.2.3. *Suppose that*

$$\Theta = \begin{pmatrix} A & K & J \\ \mathcal{H} & & E \end{pmatrix},$$

is a canonical system. If the subspace $\mathcal{H}_0 \subset \mathcal{H}$ is invariant with respect to A and orthogonal to $\operatorname{Ran}(K)$, *then it belongs to the excess subspace $\mathcal{F}^\perp$.*

Proof. For any vector $h \in \mathcal{H}_0$ the equation

$$(h, A^{*n} Kg) = (A^n h, Kg) = 0, \quad (n = 0, 1, \ldots;\ g \in \mathcal{H}),$$

holds, and hence $h \perp \mathcal{F}_*$. Since $\mathcal{F} = \mathcal{F}_*$, we have $h \perp \mathcal{F}$, i.e., $h \in \mathcal{F}^\perp$. □

Theorem 5.2.4. *The Livšic canonical system*

$$\Theta = \begin{pmatrix} A & K & J \\ \mathcal{H} & & E \end{pmatrix},$$

is non-minimal if and only if there exists a nontrivial subspace $\mathcal{H}_0 \subset \mathcal{H}$ which is invariant with respect to A and orthogonal to $\operatorname{Ran}(K)$.

Proof. If Θ is non-minimal, then $\mathcal{F}^\perp$ is different from zero, invariant with respect to A, and orthogonal to $\operatorname{Ran}(K)$. The sufficiency follows from Lemma 5.2.3. □

Suppose that the system Θ of the form (5.6) is not minimal. Define $A_\mathcal{F} = A\restriction\mathcal{F}$ and $A_{\mathcal{F}^\perp} = A\restriction\mathcal{F}^\perp$. We obtain the systems

$$\Theta_\mathcal{F} = \begin{pmatrix} A_\mathcal{F} & K & J \\ \mathcal{F} & & E \end{pmatrix} \text{ and } \Theta_{\mathcal{F}^\perp} = \begin{pmatrix} A_{\mathcal{F}^\perp} & 0 & J \\ \mathcal{F}^\perp & & E \end{pmatrix},$$

which are called respectively the **principal** and **excess** parts of the canonical system Θ. It is easy to see that $\Theta_\mathcal{F}$ is a minimal canonical system.

5.3 Couplings of canonical systems

First of all we give the definition of equal canonical systems.

Definition 5.3.1. We say that two canonical systems

$$\Theta = \begin{pmatrix} A & K & J \\ \mathcal{H} & & E \end{pmatrix} \quad \text{and} \quad \Theta' = \begin{pmatrix} A' & K' & J' \\ \mathcal{H}' & & E' \end{pmatrix}$$

are **equal**, and write $\Theta = \Theta'$, if $\mathcal{H} = \mathcal{H}'$, $E = E'$, $A = A'$, $K = K'$, and $J = J'$.

Consider two canonical systems

$$\Theta_1 = \begin{pmatrix} A_1 & K_1 & J \\ \mathcal{H}_1 & & E \end{pmatrix} \quad \text{and} \quad \Theta_2 = \begin{pmatrix} A_2 & K_2 & J \\ \mathcal{H}_2 & & E \end{pmatrix},$$

for which the input-output spaces and directing operators coincide, and denote by P_1 and P_2 the orthoprojections onto $\mathcal{H}_1$ and $\mathcal{H}_2$ acting in the space $\mathcal{H} = \mathfrak{H}_1 \oplus \mathcal{H}_2$. Introduce the operators

$$A = A_1 P_1 + A_2 P_2 + 2iK_1 J K_2^* P_2, \quad K = K_1 + K_2, \tag{5.9}$$

operating respectively in $\mathcal{H}$ and from E into $\mathcal{H}$. Since

$$\begin{aligned} K_1 J K_1^* &= \frac{A_1 - A_1^*}{2i}, & K_2 J K_2^* &= \frac{A_2 - A_2^*}{2i} \\ A^* = A_1^* P_1 + A_2^* P_2 &- 2iK_2 J K_1^* P_1, & K^* &= K_1^* P_1 + K_2^* P_2, \end{aligned}$$

we have

$$\begin{aligned} \frac{A - A^*}{2i} &= K_1 J K_1^* P_1 + K_2 J K_2^* P_2 + K_1 J K_2^* P_2 + K_2 J K_1^* P_1 \\ &= (K_1 + K_2) J (K_1^* P_1 + K_2^* P_2) = KJK^*. \end{aligned}$$

It follows that

$$\Theta = \begin{pmatrix} A & K & J \\ \mathcal{H} & & E \end{pmatrix}$$

is a canonical system. This system is called the **coupling** of the systems Θ_1 and Θ_2. We will write $\Theta = \Theta_1\Theta_2$.

One shows by a direct verification that the following formulas hold:

$$(\Theta_1\Theta_2)\Theta_3 = \Theta_1(\Theta_2\Theta_3), \quad (\Theta_1\Theta_2)^* = \Theta_2^*\Theta_1^*. \tag{5.10}$$

If $\Theta_1\Theta_2 = \Theta_1\Theta_3$ or $\Theta_2\Theta_1 = \Theta_3\Theta_1$, then $\Theta_2 = \Theta_3$.

Now we will introduce a notion of a projection of a canonical system. Let us select in the state space $\mathcal{H}$ of the system

$$\Theta = \begin{pmatrix} A & K & J \\ \mathcal{H} & & E \end{pmatrix}$$

a subspace $\mathcal{H}_0$, and define in it an operator $A_0h = P_0Ah$ $(h \in \mathcal{H}_0)$, where P_0 is the orthoprojection onto $\mathcal{H}_0$. Moreover, we construct a mapping $K_0 = P_0K$ of the space E into $\mathcal{H}_0$. Since for any $h \in \mathcal{H}_0$ we have

$$A_0^*h = P_0A^*h, \quad K_0^*h = K^*h,$$
$$K_0JK_0^*h = P_0KJK^*h = P_0\frac{A - A^*}{2i}h = \frac{A_0 - A_0^*}{2i}h,$$

then the array

$$\Theta_0 = \begin{pmatrix} A_0 & K_0 & J \\ \mathcal{H}_0 & & E \end{pmatrix}$$

is a canonical system of the form (5.6). The system Θ_0 is called the **projection** of the system Θ onto the subspace $\mathcal{H}_0$ and is denoted by $\Theta_0 = \mathrm{pr}_{\mathcal{H}_0}\Theta$. Observe that the relations

$$\mathrm{pr}_{\mathcal{H}_0}\Theta^* = (\mathrm{pr}_{\mathcal{H}_0}\Theta)^*, \quad \mathrm{pr}_{\mathcal{H}_1}\Theta = \mathrm{pr}_{\mathcal{H}_1}(\mathrm{pr}_{\mathcal{H}_2}\Theta), \quad (\mathcal{H}_1 \subseteq \mathcal{H}_2), \tag{5.11}$$

follow directly from the definition of the system projection. If

$$\Theta_1 = \begin{pmatrix} A_1 & K_1 & J \\ \mathcal{H}_1 & & E \end{pmatrix}, \Theta_2 = \begin{pmatrix} A_2 & K_2 & J \\ \mathcal{H}_2 & & E \end{pmatrix}, \Theta = \Theta_1\Theta_2 = \begin{pmatrix} A & K & J \\ \mathcal{H} & & E \end{pmatrix},$$

then, as is shown by formulas (5.9), Θ_1 and Θ_2 are projections of the system Θ onto $\mathcal{H}_1$ and $\mathcal{H}_2$ respectively, and $\mathcal{H}_1$ is invariant with respect to A.

Conversely, every canonical system Θ is the coupling of its projections

$$\Theta_1 = \begin{pmatrix} A_1 & K_1 & J \\ \mathcal{H}_1 & & E \end{pmatrix} \quad \text{and} \quad \Theta_2 = \begin{pmatrix} A_2 & K_2 & J \\ \mathcal{H}_2 & & E \end{pmatrix},$$

onto an arbitrary subspace $\mathcal{H}_1$ invariant with respect to A and its orthogonal complement $\mathcal{H}_2$. Indeed, denoting by P_j the orthoprojection onto $\mathcal{H}_j$ $(j = 1, 2)$, we get

$$P_2AP_1 = 0, \quad P_1A^*P_2 = 0,$$
$$P_1AP_2 = 2iP_1\frac{A - A^*}{2i}P_2 = 2iP_1KJK^*P_2 = 2iK_1JK_2^*P_2,$$
$$A = (P_1 + P_2)A(P_1 + P_2) = A_1P_1 + A_2P_2 + 2iK_1JK_2^*P_2.$$

In particular, every canonical system is equal to the coupling of its principal and excess parts.

A canonical system Θ_0 is called a **left (right) divisor** of a system Θ if it is a projection of Θ onto a subspace $\mathcal{H}_0 \subseteq \mathcal{H}$ which is invariant with respect to A (A^*). Suppose that Θ is a canonical system and that the subspaces

$$0 = \mathcal{H}_0 \subset \mathcal{H}_1 \subset \mathcal{H}_2 \subset \cdots \subset \mathcal{H}_n = \mathcal{H}$$

are invariant with respect to A. Since the subspace $\mathcal{H}_{k-1}$ is invariant with respect to $A_k = A \restriction \mathcal{H}_k$ $(k = 2, 3, \ldots, n-1)$, from formula (5.11) we obtain

$$\Theta = \mathrm{pr}_{\mathcal{H}_1} \Theta \, \mathrm{pr}_{\mathcal{H}_2 \ominus \mathcal{H}_1} \Theta \ldots \mathrm{pr}_{\mathcal{H}_n \ominus \mathcal{H}_{n-1}} \Theta. \tag{5.12}$$

Below we will discuss the coupling of minimal canonical systems.

Lemma 5.3.2. *If a canonical system*

$$\Theta = \begin{pmatrix} A & K & J \\ \mathcal{H} & & E \end{pmatrix},$$

is the coupling of the systems

$$\Theta_1 = \begin{pmatrix} A_1 & K_1 & J \\ \mathcal{H}_1 & & E \end{pmatrix} \quad \text{and} \quad \Theta_2 = \begin{pmatrix} A_2 & K_2 & J \\ \mathcal{H}_2 & & E \end{pmatrix},$$

then

$$\mathcal{F}_j^\perp = \mathcal{F}^\perp \cap \mathcal{H}_j, \quad (j = 1, 2). \tag{5.13}$$

Proof. The subspace $\mathcal{F}_1^\perp$ is invariant with respect to A and orthogonal to $\mathrm{Ran}(K_1)$. It is also orthogonal to $\mathrm{Ran}(K)$, since $\mathrm{Ran}(K_1)$ is the projection of $\mathrm{Ran}(K)$ onto $\mathcal{H}_1$. By Lemma 5.2.3, $\mathcal{F}_1^\perp \subseteq \mathcal{F}^\perp \cap \mathcal{H}_1$. On the other hand, the subspace $\mathcal{F}^\perp \cap \mathcal{H}_1$ is invariant with respect to A_1 and orthogonal to $\mathrm{Ran}(K_1)$. Applying Lemma 5.2.3 to the system Θ_1, we find that $\mathcal{F}^\perp \cap \mathcal{H}_1 \subseteq \mathcal{F}_1^\perp$. Thus equation (5.13) is proved for $j = 1$. It remains to note that $\Theta^* = \Theta_2^* \Theta_1^*$, so that, from what has already been proved, $\mathcal{F}_{*,2}^\perp = \mathcal{F}_*^\perp \cap \mathcal{H}_2$. Inasmuch as $\mathcal{F}_{*,2}^\perp = \mathcal{F}_2^\perp$ and $\mathcal{F}_*^\perp = \mathcal{F}^\perp$, we have $\mathcal{F}_2^\perp = \mathcal{F}^\perp \cap \mathcal{H}_2$. □

Theorem 5.3.3. *If* $\Theta = \Theta_1 \Theta_2 \cdots \Theta_n$, *then*

$$\mathcal{F}_j^\perp = \mathcal{F}^\perp \cap \mathcal{H}_j, \quad (j = 1, 2, \ldots, n),$$

where $\mathcal{H}_j$ *is the state space of the canonical system* Θ_j.

Proof. In view of Lemma 5.3.2, $\mathcal{F}_j^\perp = \mathcal{F}_{1,2,\ldots,j}^\perp \cap \mathcal{H}_j$, and

$$\mathcal{F}_{1,2,\ldots,j}^\perp = \mathcal{F}^\perp \cap (\mathcal{H}_1 \oplus \mathcal{H}_2 \oplus \cdots \oplus \mathcal{H}_j),$$

where $\mathcal{F}_{1,2,\ldots,j}^\perp$ is the excess subspace of the coupling $\Theta_1 \Theta_2 \cdots \Theta_j$. Accordingly,

$$\mathcal{F}_j^\perp = \mathcal{F}^\perp \cap (\mathcal{H}_1 \oplus \mathcal{H}_2 \oplus \cdots \oplus \mathcal{H}_j) \cap \mathcal{H}_j = \mathcal{F}^\perp \cap \mathcal{H}_j. \qquad \square$$

As a consequence of Theorem 5.3.3 we obtain the following statement.

Theorem 5.3.4. *If $\Theta = \Theta_1\Theta_2\cdots\Theta_n$ is a minimal canonical system, then all the systems Θ_j $(j = 1, 2, \ldots, n)$ are minimal.*

One can see that a coupling of minimal canonical systems may turn out to be a non-minimal system. Moreover, the following theorem holds.

Theorem 5.3.5. *Let A_0 be a self-adjoint bounded operator in a separable Hilbert space $\mathcal{H}_0$. Then there exist minimal canonical systems Θ_1 and Θ_2 such that A_0 is a state-space operator of the excess portion of the coupling $\Theta_1\Theta_2$.*

Proof. Consider the orthogonal sum $\tilde{\mathcal{H}} = \mathcal{H}_0 \oplus \mathcal{H}$, where $\mathcal{H}$ is some Hilbert space whose dimension is equal to the dimension of the space $\mathcal{H}_0$. Suppose that we are given an isometric mapping U of $\mathcal{H}_0$ onto $\mathcal{H}$ and that the operator $A = UA_0U^{-1}$ is included in the minimal canonical system

$$\Theta = \begin{pmatrix} A & K & J \\ \mathcal{H} & & E \end{pmatrix}.$$

It is not hard to see that the system

$$\tilde{\Theta} = \Theta\Theta_0 = \begin{pmatrix} \tilde{A} & K & J \\ \tilde{\mathcal{H}} & & E \end{pmatrix}, \text{ where } \Theta_0 = \begin{pmatrix} A_0 & K & J \\ \mathcal{H}_0 & & E \end{pmatrix},$$

has the principal part Θ and the excess part Θ_0. The subspace $\mathcal{H}_1$ of the space $\tilde{\mathcal{H}}$ consisting of vectors of the form $f + Uf$ $(f \in \mathcal{H}_0)$ is invariant with respect to $\tilde{A}$, since

$$\tilde{A}(f + Uf) = A_0 f + AUf = A_0 f + UA_0 f, \quad (f \in \mathcal{H}_0).$$

Accordingly,

$$\tilde{\Theta} = \Theta_1\Theta_2, \quad (\Theta_1 = \mathrm{pr}_{\mathcal{H}_1}\tilde{\Theta},\ \Theta_2 = \mathrm{pr}_{\mathcal{H}_2}\tilde{\Theta},\ \mathcal{H}_2 = \tilde{\mathcal{H}} \ominus \mathcal{H}_1).$$

Let $\tilde{\mathcal{F}}^{\perp}$ be the excess subspace of $\tilde{\Theta}$. Then, since $\mathcal{H}_0 = \tilde{\mathcal{F}}^{\perp}$ and $\mathcal{H}_j \cap \mathcal{H}_0 = 0$ $(j = 1, 2)$, by Lemma 5.3.2 Θ_1 and Θ_2 are minimal canonical systems. □

Now we will define and discuss the spectrum of the coupling of canonical systems.

Theorem 5.3.6. *Suppose that the Livšic canonical system*

$$\Theta = \begin{pmatrix} A & K & J \\ \mathcal{H} & & E \end{pmatrix},$$

is the coupling of the systems

$$\Theta_1 = \begin{pmatrix} A_1 & K_1 & J \\ \mathcal{H}_1 & & E \end{pmatrix} \quad \text{and} \quad \Theta_2 = \begin{pmatrix} A_2 & K_2 & J \\ \mathcal{H}_2 & & E \end{pmatrix}.$$

If λ is a regular point for the operators A_1 and A_2, then it is regular for the operator A as well, while

$$\begin{aligned}(A-\lambda I)^{-1} = (A_1-\lambda I)^{-1}P_1 + (A_2-\lambda I)^{-1}P_2 \\ - 2i(A_1-\lambda I)^{-1}K_1JK_2^*(A_2-\lambda I)^{-1}P_2,\end{aligned} \tag{5.14}$$

where P_1 and P_2 are orthoprojections onto $\mathcal{H}_1$ and $\mathcal{H}_2$.

Proof. From the first of equalities (5.9) we have

$$A-\lambda I = (A_1-\lambda I)P_1 + (A_2-\lambda I)P_2 + 2iK_1JK_2^*P_2. \tag{5.15}$$

Now it is easy to verify that the right side of (5.14) is an operator which is both left and right invertible for the operator (5.15). The theorem is proved. □

We agree that by the **spectrum of a canonical system** we will mean the spectrum of its state-space operator. Theorem 5.3.6 means that the spectrum of the coupling of two canonical systems is contained in the union of the spectra of the factors.

Lemma 5.3.7. *Suppose that $\mathcal{H}_0$ is an invariant subspace of the bounded linear operator A acting in the space $\mathcal{H}$, and $\mathbb{O}$ is a bounded open connected piece of the complex plane $\mathbb{C}$, all of whose points are regular for A. If at least one point $\lambda_0 \in \mathbb{O}$ is regular for the operator A_0 induced by A in $\mathcal{H}_0$, then all the points of $\mathbb{O}$ have the same property.*

Proof. We note in preparation that the point $\lambda \in \mathbb{O}$ is regular for the operator A_0 if and only if $(A-\lambda I)^{-1}\mathcal{H}_0 \subseteq \mathcal{H}_0$.

Let $\mathcal{O}_{\lambda_0}$ be a sufficiently small neighborhood of λ_0 contained in $\mathbb{O}$ and such that the Taylor expansion

$$R_\lambda = R_{\lambda_0} + (\lambda-\lambda_0)R_{\lambda_0}^2 + (\lambda-\lambda_0)^2R_{\lambda_0}^3 + \cdots, \quad \lambda \in \mathcal{O}_{\lambda_0},\ R_\lambda = (A-\lambda I)^{-1}$$

uniformly converges. Then $R_{\lambda_0}\mathcal{H}_0 \subseteq \mathcal{H}_0$ implies $R_\lambda\mathcal{H}_0 \subseteq \mathcal{H}_0$ for all $\lambda \in \mathcal{O}_{\lambda_0}$. Hence $(R_\lambda h, f) = 0$ for all $h \in \mathcal{H}_0$, all $f \in \mathcal{H} \ominus \mathcal{H}_0$, and all $\lambda \in \mathcal{O}_{\lambda_0}$. Because the set $\mathbb{O}$ is connected and open, and the function $(R_\lambda h, f)$ is holomorphic on $\mathbb{O}$, we get

$$(R_\lambda h, f) = 0 \quad \text{for all} \quad \lambda \in \mathbb{O}.$$

So, $R_\lambda\mathcal{H}_0 \subseteq \mathcal{H}_0$ for all $\lambda \in \mathbb{O}$. □

The following is an easy consequence of Theorem 5.3.6, Lemma 5.3.7, and equation (5.10).

Theorem 5.3.8. *Suppose that Θ is the coupling of the canonical systems Θ_1 and Θ_2. If the set of regular points of the operator A is connected, then the spectrum of Θ is equal to the union of the spectra of Θ_1 and Θ_2.*

Now we introduce a notion of unitarily equivalent canonical systems. We recall that an operator A_1 acting in a space $\mathcal{H}_1$ is said to be unitarily equivalent to the operator A_2 in $\mathcal{H}_2$ if there exists an isometric mapping U of $\mathcal{H}_1$ onto $\mathcal{H}_2$ such that $UA_1 = A_2U$. We say that the canonical system

$$\Theta_1 = \begin{pmatrix} A_1 & K_1 & J \\ \mathcal{H}_1 & & E \end{pmatrix},$$

is **unitarily equivalent** to the system

$$\Theta_2 = \begin{pmatrix} A_2 & K_2 & J \\ \mathcal{H}_2 & & E \end{pmatrix},$$

if there exists an isometric mapping U of the space $\mathcal{H}_1$ onto $\mathcal{H}_2$ such that

$$UA_1 = A_2U, \quad UK_1 = K_2. \tag{5.16}$$

Obviously, the relation of unitary equivalence is reflexive, symmetric, and transitive. It is also easy to see that if one of two unitarily equivalent canonical systems is minimal, then so is the other. If

$$\Theta = \begin{pmatrix} A & K & J \\ \mathcal{H} & & E \end{pmatrix}$$

is a minimal canonical system, and if for some unitary operator U in $\mathcal{H}$ the equations

$$UA = AU, \quad UK = K$$

are satisfied, then

$$UA^nKg = A^nKg, \quad (n = 0, 1, \dots; \; g \in \mathcal{H}),$$

which means that $U = I_{\mathcal{H}}$. Using this remark, we arrive at the following conclusion: if

$$\Theta_1 = \begin{pmatrix} A_1 & K_1 & J \\ \mathcal{H}_1 & & E \end{pmatrix} \quad \text{and} \quad \Theta_2 = \begin{pmatrix} A_2 & K_2 & J \\ \mathcal{H}_2 & & E \end{pmatrix}$$

are unitarily equivalent minimal canonical systems, then the isometric mapping satisfying conditions (5.16) is defined uniquely.

5.4 Transfer functions of canonical systems

Consider the Livsič canonical system

$$\Theta = \begin{pmatrix} A & K & J \\ \mathcal{H} & & E \end{pmatrix}.$$

Taking into account (5.3), we call the function of complex variable z,

$$W_\Theta(z) = I - 2iK^*(A - zI)^{-1}KJ, \tag{5.17}$$

the **transfer function** of the canonical system Θ. It is easy to see that if $\varphi_- \in E$ is an input vector and $\varphi_+ \in E$ is an output vector of the system Θ, then $\varphi_+ = W_\Theta(z)\varphi_-$. The function $W_\Theta(z)$ is obviously defined and holomorphic on the set $\rho(A)$ of regular points of the operator A, and its values are bounded linear operators acting in the input-output space E.

Theorem 5.4.1. *Suppose that*

$$\Theta = \begin{pmatrix} A & K & J \\ \mathcal{H} & & E \end{pmatrix}$$

is the coupling of the Livšic canonical systems

$$\Theta_1 = \begin{pmatrix} A_1 & K_1 & J \\ \mathcal{H}_1 & & E \end{pmatrix} \quad \text{and} \quad \Theta_2 = \begin{pmatrix} A_2 & K_2 & J \\ \mathcal{H}_2 & & E \end{pmatrix}.$$

If z is a regular point for the operators A_1 and A_2, then

$$W_{\Theta_1\Theta_2}(z) = W_{\Theta_1}(z)W_{\Theta_2}(z). \tag{5.18}$$

Proof. Denote by P_j $(j = 1, 2)$ the orthoprojection onto $\mathcal{H}_j$ acting in the space $\mathcal{H} = \mathcal{H}_1 \oplus \mathcal{H}_2$. Applying Theorem 5.3.6, we obtain

$$\begin{aligned}
W_{\Theta_1\Theta_2}(z) &= I - 2iK^*(A - zI)^{-1}KJ \\
&= I - 2iK^*P_1(A_1 - zI)^{-1}P_1KJ - 2iK^*P_2(A_2 - zI)^{-1}P_2KJ \\
&\quad + (2i)^2K^*P_1(A_1 - zI)^{-1}K_1JK_2^*(A_2 - zI)^{-1}P_2KJ \\
&= I - 2iK_1^*(A_1 - zI)^{-1}K_1J - 2iK_2^*(A_2 - zI)^{-1}K_2J \\
&\quad + (2i)^2K_1^*(A_1 - zI)^{-1}K_1JK_2^*(A_2 - zI)^{-1}K_2J \\
&= [I - 2iK_1^*(A_1 - zI)^{-1}K_1J][I - 2iK_2^*(A_2 - zI)^{-1}K_2J] \\
&= W_{\Theta_1}(z)W_{\Theta_2}(z).
\end{aligned}$$

□

Corollary 5.4.2. *If the state-space operator of the canonical system*

$$\Theta = \begin{pmatrix} A & K & J \\ \mathcal{H} & & E \end{pmatrix}$$

has invariant subspaces $0 = \mathcal{H}_0 \subset \mathcal{H}_1 \subset \cdots \subset \mathcal{H}_n = \mathcal{H}$ and z is a regular point for the state-space operators of the canonical systems $\Theta_j = \mathrm{pr}_{\mathcal{H}_j \ominus \mathcal{H}_{j-1}} \Theta$ $(j = 1, 2, \ldots, n)$, then

$$W_\Theta(z) = W_{\Theta_1}(z)W_{\Theta_2}(z)\ldots W_{\Theta_n}(z). \tag{5.19}$$

The proof follows from formula (5.12).

We will now state the criteria for unitary equivalence of canonical systems. First we note that if the canonical systems

$$\Theta_1 = \begin{pmatrix} A_1 & K_1 & J \\ \mathcal{H}_1 & & E \end{pmatrix} \quad \text{and} \quad \Theta_2 = \begin{pmatrix} A_2 & K_2 & J \\ \mathcal{H}_2 & & E \end{pmatrix}$$

are unitarily equivalent, then the set $\rho(A_1)$ of regular points of A_1 coincides with the set $\rho(A_2)$ of regular points of A_2, and $W_{\Theta_1}(z) = W_{\Theta_2}(z)$ $(z \in \rho(A_1))$. Indeed, in view of (5.16),

$$\begin{aligned} W_{\Theta_2}(z) &= I - 2iK_2^*(A_2 - zI)^{-1}K_2J \\ &= I - 2iK_1^*U^{-1}[U(A_1 - zI)^{-1}U^{-1}]UK_1J \\ &= I - 2iK_1^*(A_1 - zI)^{-1}K_1J = W_{\Theta_1}(z). \end{aligned}$$

Theorem 5.4.3. *Suppose that*

$$\Theta_1 = \begin{pmatrix} A_1 & K_1 & J \\ \mathcal{H}_1 & & E \end{pmatrix} \quad \textit{and} \quad \Theta_2 = \begin{pmatrix} A_2 & K_2 & J \\ \mathcal{H}_2 & & E \end{pmatrix}$$

are minimal canonical systems. If in some neighborhood G of infinity $W_{\Theta_1}(z) = W_{\Theta_2}(z)$, then Θ_1 and Θ_2 are unitarily equivalent.

Proof. It is given in the statement of the theorem that

$$K_1^*(A_1 - zI)^{-1}K_1 = K_2^*(A_2 - zI)^{-1}K_2, \quad (z \in G).$$

Then for $z, \zeta \in G$ and $j = 1, 2$ we have

$$\begin{aligned} &(A_j - zI)^{-1} - (A_j^* - \bar{\zeta}I)^{-1} = (A_j^* - \bar{\zeta}I)^{-1}[(A_j^* - \bar{\zeta}I) - (A_j - zI)] \times (A_j - zI)^{-1} \\ &= (z - \bar{\zeta})(A_j^* - \bar{\zeta}I)^{-1}(A_j - zI)^{-1} - 2i(A_j^* - \bar{\zeta}I)^{-1}K_jJK_j^*(A_j - zI)^{-1}, \end{aligned}$$

which means that

$$\begin{aligned} (z - \bar{\zeta})K_1^*(A_1^* - \bar{\zeta}I)^{-1}(A_1 - zI)^{-1}K_1 &= K_1^*(A_1 - zI)^{-1}K_1 \\ &\quad - K_1^*(A_1 - \bar{\zeta}I)^{-1}K_1 + 2iK_1^*(A_1 - \bar{\zeta}I)^{-1}K_1JK_1^*(A_1 - zI)^{-1}K_1 \\ &= K_2^*(A_2 - zI)^{-1}K_2 - K_2^*(A_2 - \bar{\zeta}I)^{-1}K_2 \\ &\quad + 2iK_2^*(A_2 - \bar{\zeta}I)^{-1}K_2JK_2^*(A_2 - zI)^{-1}K_2 \\ &= (z - \bar{\zeta})K_2^*(A_2^* - \bar{\zeta}I)^{-1}(A_2 - zI)^{-1}K_2. \end{aligned}$$

This yields

$$K_1^*(A_1^* - \bar{\zeta}I)^{-1}(A_1 - zI)^{-1}K_1 = K_2^*(A_2^* - \bar{\zeta}I)^{-1}(A_2 - zI)^{-1}K_2, \quad (z, \zeta \in G).$$

Using the expansion

$$(A_j - zI)^{-1} = -\frac{I}{z} - \frac{A_j}{z^2} - \frac{A_j^2}{z^3} - \dots, \quad (|z| > \|A_j\|),$$

we arrive at the equation

$$(A_1^m K_1 g, A_1^n K_1 g') = (A_1^m K_1 g, A_2^n K_2 g'), \ (m,n=0,1,\ldots;\ g,\ g' \in E). \tag{5.20}$$

Denote by $\mathcal{G}_j$ $(j=1,2)$ the linear span of vectors of the form $A_j^m K_j g$ $(m = 0,1,\ldots;\ g \in E)$ and consider mapping U of the set $\mathcal{G}_1$ onto $\mathcal{G}_2$ which assigns to each vector of the form $\sum_{n=0}^{l} A_1^n K_1 g_n$ the vector $\sum_{n=0}^{l} A_2^n K_2 g_n$. In view of (5.20) the mapping U is isometric. Since $\mathcal{G}_1$ and $\mathcal{G}_2$ are dense in $\mathcal{H}_1$ and $\mathcal{H}_2$ respectively, U can be extended by continuity to an isometric mapping of the space $\mathcal{H}_1$ onto $\mathcal{H}_2$. Denoting the extended mapping again by U, we obtain the equation

$$UA_1^n K_1 = A_2^n K_2, \qquad (n=0,1,\ldots).$$

Thus $UK_1 = K_2$, and, moreover,

$$UA_1 A_1^n K_1 = A_2 A_2^n K_2 = A_2 U A_1^n K_1 \qquad (n=0,1,\ldots),$$

i.e., $UA_1 = A_2 U$. □

Corollary 5.4.4. *Suppose that*

$$\Theta_1 = \begin{pmatrix} A_1 & K_1 & J \\ \mathcal{H}_1 & & E \end{pmatrix} \quad \text{and} \quad \Theta_2 = \begin{pmatrix} A_2 & K_2 & J \\ \mathcal{H}_2 & & E \end{pmatrix}$$

are minimal canonical systems. If in a neighborhood G of infinity $W_{\Theta_1}(z) = W_{\Theta_2}(z)$, then $\rho(A_1) = \rho(A_2)$ and $W_{\Theta_1}(z) \equiv W_{\Theta_2}(z)$, $(z \in \rho(A_1))$.

Note that in Theorem 5.4.3 we cannot get along without the requirement of minimality of the systems. For example, let us consider a non-minimal system Θ with principal part $\Theta_{\mathcal{F}}$ and excess part $\Theta_{\mathcal{F}^\perp}$. Obviously Θ and $\Theta_{\mathcal{F}}$ cannot be unitarily equivalent. At the same time $W_{\Theta_{\mathcal{F}^\perp}}(z) \equiv I$, and, by Theorem 5.4.1, there exists a neighborhood of infinity in which

$$W_\Theta(z) = W_{\Theta_{\mathcal{F}}}(z) W_{\Theta_{\mathcal{F}^\perp}}(z) = W_{\Theta_{\mathcal{F}}}(z).$$

Now we will discuss analytic properties of a transfer function. Suppose that Θ is some canonical system. Then

$$W_\Theta(z) J W_\Theta^*(\zeta) - J = 2i(\bar{\zeta} - z) K^* (A - zI)^{-1} (A^* - \bar{\zeta})^{-1} K, \ z, \zeta \in \rho(A). \tag{5.21}$$

Indeed, since

$$\begin{aligned}(A - zI)^{-1} - (A^* - \bar{\zeta})^{-1} &= (A - zI)^{-1}[(A^* - \bar{\zeta} I) - (A - zI)](A^* - \bar{\zeta})^{-1} \\ &= (z - \bar{\zeta})(A - zI)^{-1}(A^* - \bar{\zeta})^{-1} - 2i(A - zI)^{-1} K J K^* (A^* - \bar{\zeta})^{-1},\end{aligned}$$

we have

$$\begin{aligned}
&W_\Theta(z)JW_\Theta^*(\zeta) - J \\
&= [I - 2iK^*(A - zI)^{-1}KJ]J[I - 2iJK^*(A^* - \bar{\zeta}I)^{-1}K] - J \\
&= -2iK^*[(A - zI)^{-1} - (A^* - \bar{\zeta}I)^{-1} + 2i(A - zI)^{-1}KJK^*(A^* - \bar{\zeta}I)^{-1}]K \\
&= 2i(\bar{\zeta} - z)K^*(A - zI)^{-1}(A^* - \bar{\zeta})^{-1}K.
\end{aligned}$$

In particular, if z, $\bar{z} \in \rho(A)$, then

$$W_\Theta(z)JW_\Theta^*(\bar{z}) - J = 0. \tag{5.22}$$

Moreover, at each point $z \in \rho(A)$,

$$W_\Theta(z)JW_\Theta^*(z) - J = 4\operatorname{Im} z K^*(A - zI)^{-1}(A^* - \bar{z}I)^{-1}K,$$

which means that

$$W_\Theta(z)JW_\Theta^*(z) - J \geq 0 \quad (\operatorname{Im} z > 0,\ z \in \rho(A)), \tag{5.23}$$
$$W_\Theta(z)JW_\Theta^*(z) - J \leq 0 \quad (\operatorname{Im} z < 0,\ z \in \rho(A)). \tag{5.24}$$

Similarly, using the identity

$$\begin{aligned}
(A - zI)^{-1} - (A^* - \bar{\zeta})^{-1} = (z - \bar{\zeta})&(A - zI)^{-1}(A^* - \bar{\zeta})^{-1} \\
&- 2i(A - zI)^{-1}KJK^*(A^* - \bar{\zeta})^{-1},
\end{aligned}$$

we obtain

$$W_\Theta^*(\zeta)JW_\Theta(z) - J = 2i(\bar{\zeta} - z)JK^*(A^* - \bar{\zeta}I)^{-1}(A - zI)^{-1}KJ, \tag{5.25}$$

which shows that

$$W_\Theta^*(\bar{z})JW_\Theta(z) - J = 0, \qquad z,\ \bar{z} \in \rho(A), \tag{5.26}$$

and

$$\begin{aligned}
W_\Theta^*(z)JW_\Theta(z) - J &\geq 0 \quad (\operatorname{Im} z > 0,\ z \in \rho(A)), \\
W_\Theta^*(z)JW_\Theta(z) - J &\leq 0 \quad (\operatorname{Im} z < 0,\ z \in \rho(A)).
\end{aligned}$$

If z, $\bar{z} \in \rho(A)$, then by (5.22) and (5.26) the operator $W_\Theta(z)$ has a bounded inverse

$$W_\Theta^{-1}(z) = JW_\Theta^*(\bar{z})J. \tag{5.27}$$

Since there exists a neighborhood of the point at infinity in which the resolvent of A decomposes into a series

$$(A - zI)^{-1} = -\frac{I}{z} - \frac{A}{z^2} - \cdots,$$

which converges in norm, it follows that in this same neighborhood

$$W_\Theta(z) = I + \frac{2i}{z} K^* K J + \cdots .$$

Now we will introduce a linear-fractional transformation of the transfer function. We assign to a canonical system Θ the operator-function

$$V_\Theta(z) = K^*(\operatorname{Re} A - zI)^{-1}K, \qquad \operatorname{Re} A = (1/2)(A + A^*). \tag{5.28}$$

The function $V_\Theta(z)$ is holomorphic on the set $\rho(\operatorname{Re} A)$, and its values, as those of the function $W_\Theta(z)$, are operators acting in E. We note that $\rho(\operatorname{Re} A)$ contains all non-real points. From the equality

$$V_\Theta(z) - V_\Theta^*(z) = 2i\operatorname{Im} z K^*(\operatorname{Re} A - \bar{z}I)^{-1}(\operatorname{Re} A - zI)^{-1}K,$$

we obtain that

$$\frac{V_\Theta(z) - V_\Theta^*(z)}{2i} \geq 0 \quad (\operatorname{Im} z > 0), \qquad \frac{V_\Theta(z) - V_\Theta^*(z)}{2i} \leq 0 \quad (\operatorname{Im} z < 0),$$

and

$$V_\Theta(z) = V_\Theta^*(z), \qquad (\operatorname{Im} z = 0,\ z \in \rho(\operatorname{Re} A)). \tag{5.29}$$

At each point of the set $\rho(A) \cap \rho(\operatorname{Re} A)$ there exist the operators $(W_\Theta(z) + I)^{-1}$ and $(I + iV_\Theta(z)J)^{-1}$, while

$$V_\Theta(z) = i(W_\Theta(z) + I)^{-1}(W_\Theta(z) - I)J = i(W_\Theta(z) - I)(W_\Theta(z) + I)^{-1}J, \tag{5.30}$$

$$W_\Theta(z) = (I + iV_\Theta(z)J)^{-1}(I - iV_\Theta(z)J) = (I - iV_\Theta(z)J)(I + iV_\Theta(z)J)^{-1}. \tag{5.31}$$

Indeed, since

$$(\operatorname{Re} A - zI)^{-1} - (A - zI)^{-1} = i(A - zI)^{-1}\operatorname{Im} A(\operatorname{Re} A - zI)^{-1},$$

we have

$$K^*(\operatorname{Re} A - zI)^{-1}K - K^*(A - zI)^{-1}K = iK^*(A - zI)^{-1}KJK^*(\operatorname{Re} A - zI)^{-1}K.$$

Thus

$$V_\Theta(z) + \frac{i}{2}(I - W_\Theta(z))J = \frac{1}{2}(I - W_\Theta(z))V_\Theta(z),$$

so that

$$(W_\Theta(z) + I)(I + iV_\Theta(z)J) = 2I. \tag{5.32}$$

Similarly, starting from the relations

$$(\operatorname{Re} A - zI)^{-1} - (A - zI)^{-1} = i(\operatorname{Re} A - zI)^{-1}\operatorname{Im} A(A - zI)^{-1},$$

we get

$$(I + iV_\Theta(z)J)(W_\Theta(z) + I) = 2I. \tag{5.33}$$

In view of (5.32) and (5.33) each of the operators $W_\Theta(z) + I$ and $I + iV_\Theta(z)J$ has a bounded inverse for $z \in \rho(A) \cap \rho(\operatorname{Re} A)$. Formulas (5.30) and (5.31) follow easily from (5.32).

5.5 Class Ω_J and its realization

In what follows all the matrices will be considered as linear operators in the space $\mathbb{C}^n$ when it deems necessary. The following theorem holds.

Theorem 5.5.1. *Let $V(z)$ be an $(n \times n)$ matrix-valued function in a Hilbert space $\mathbb{C}^n$ that has an integral representation*[3]

$$V(z) = \int_a^b \frac{1}{t-z} d\sigma(t), \tag{5.34}$$

where $\sigma(t)$ is a non-negative, non-decreasing $(n \times n)$ matrix-function in $\mathbb{C}^n$ defined on a finite interval $[a,b]$. Then $V(z)$ can be realized in the form

$$V(z) = i(W_\Theta(z) - I)(W_\Theta(z) + I)^{-1} J, \tag{5.35}$$

where $W_\Theta(z)$ is a transfer function of a minimal canonical system of the form (5.6), z, $(\operatorname{Im} z \neq 0)$ is such that $W_\Theta(z)$ is defined, and $J = J^* = J^{-1}$ is an arbitrary pre-assigned directing operator.

Proof. Let $C_{[a,b]}(\mathbb{C}^n)$ be the the set of all continuous on $[a,b]$, $\mathbb{C}^n$-valued functions and let $\widetilde{L}^2_{[a,b]}(\mathbb{C}^n, d\sigma)$ be the completion of $C_{[a,b]}(\mathbb{C}^n)$ with respect to the semi-inner product

$$(\vec{f}, \vec{g}) = \int_a^b \vec{f}(t) d\sigma(t) \vec{g}^*(t).$$

Then the Hilbert space $L^2_{[a,b]}(\mathbb{C}^n, d\sigma)$ is the quotient space

$$\widetilde{L}^2_{[a,b]}(\mathbb{C}^n, d\sigma) / \ker(p),$$

where

$$\ker(p) = \{\vec{f} \in \widetilde{L}^2_{[a,b]}(\mathbb{C}^n, d\sigma) : p(\vec{f}) = \|\vec{f}\| = 0\}.$$

Consider in $L^2_{[a,b]}(\mathbb{C}^n, d\sigma)$ the operator

$$(\mathcal{A}\vec{f})(t) = t\vec{f}(t) + 2i \int_a^b \vec{f}(t) d\sigma(t) J, \tag{5.36}$$

where $\vec{f}(t) = (f_1(t), \dots, f_n(t))$ is a row-vector function $L^2_{[a,b]}(\mathbb{C}^n, d\sigma)$. Here the directing operator $J = [j_{\alpha\beta}]$ is $((n \times n))$ signature matrix in $\mathbb{C}^n$ with the property that $J = J^* = J^{-1}$. Obviously,

$$(\operatorname{Im} \mathcal{A})\, \vec{f} = \sum_{\alpha,\beta=1}^{n} (\vec{f}, \vec{h}_\alpha) j_{\alpha\beta} \vec{h}_\beta = \int_a^b \vec{f}(t) d\sigma(t) J = KJK^* \vec{f}.$$

[3]Functions of this type belong to the class of Herglotz-Nevanlinna functions that will be studied in detail in Chapter 6 and further on.

Also,

$$Kg = \sum_{\alpha=1}^{n} (g, \vec{h}_\alpha)_{\mathbb{C}^n} \vec{h}_\alpha(t),\ g \in \mathbb{C}^n, \quad K^* \vec{f} = ((\vec{f}, \vec{h}_1), \ldots, (\vec{f}, \vec{h}_1)),$$

$(\cdot,\cdot)_{\mathbb{C}^n}$ is the dot product in $\mathbb{C}^n$, and $\vec{h}_k = \vec{h}_k(t)$ $(k = 1, \ldots, n)$ is a vector whose k-th component is 1 and the rest are zeros.

Consider the system Θ with the state-space operator $\mathcal{A}$ of the form (5.36)

$$\Theta = \begin{pmatrix} \mathcal{A} & K & J \\ L^2_{[a,b]}(\mathbb{C}^n, d\sigma) & & \mathbb{C}^n \end{pmatrix}, \tag{5.37}$$

Since the real part of the operator $\mathcal{A}$ is an operator of multiplication by independent variable $(\operatorname{Re} \mathcal{A} f)(t) = tf(t)$, then

$$V_\Theta(z) = K^*(\operatorname{Re} \mathcal{A} - zI)^{-1} K = \int_a^b \frac{1}{t-z} d\sigma(t) = V(z).$$

Applying (5.30), we get

$$V(z) = \int_a^b \frac{1}{t-z} d\sigma(t) = V_\Theta(z) = i(W_\Theta(z) - I)(W_\Theta(z) + I)^{-1} J,$$

where $W_\Theta(z)$ is a transfer function of a system Θ of the form (5.37). We can also show that this system Θ is minimal because its state-space operator $\mathcal{A}$ is prime. In order to do that we will use Corollary 5.2.1 and show that

$$\text{c.l.s.}\{\mathcal{A}^k \vec{h}_\alpha,\ k = 0, 1, \ldots; \alpha = 1, \ldots, n\} = L^2_{[a,b]}(\mathbb{C}^n, d\sigma).$$

Consider the vector function

$$\vec{h}_\alpha(z,t) = \frac{\vec{h}_\alpha(W_\Theta(z) + I)}{2i(t-z)}, \quad |z| > \|\mathcal{A}\|,\ \operatorname{Im} z \neq 0.$$

Then

$$\int_a^b \vec{h}_\alpha(z,t) d\sigma(t) J = \frac{\vec{h}_\alpha(W_\Theta(z) + I)}{2i} \int_a^b \frac{d\sigma(t)}{t-z} J = \frac{\vec{h}_\alpha(W_\Theta(z) + I)}{2i} V(z) J$$
$$= \frac{1}{2} \vec{h}_\alpha(W_\Theta(z) - I).$$

Hence,

$$\mathcal{A}\vec{h}_\alpha(z,t) = z\vec{h}_\alpha(z,t) + (t-z)\vec{h}_\alpha(z,t) + i\int_a^b \vec{h}_\alpha(z,t) d\sigma(t) J$$
$$= z\vec{h}_\alpha(z,t) + \frac{\vec{h}_\alpha(W_\Theta(z) + I)}{2i} - \frac{\vec{h}_\alpha(W_\Theta(z) - I)}{2i}$$
$$= z\vec{h}_\alpha(z,t) - i\vec{h}_\alpha,$$

or

$$(\mathcal{A}-zI)^{-1}\vec{h}_\alpha = i\vec{h}_\alpha(z,t) = \frac{\vec{h}_\alpha(W_\Theta(z)+I)}{2(t-z)}. \tag{5.38}$$

Let $(\mathcal{A}^k\vec{h}_\alpha, \vec{f}) = 0$, $k = 0,1,\ldots;\alpha = 1,\ldots,n$. Then $((\mathcal{A}-zI)^{-1}\vec{h}_\alpha, \vec{f}) = 0$, for $\alpha = 1,\ldots,n$ and (5.38) implies that

$$\vec{h}_\alpha(W_\Theta(z)+I)\int_a^b \frac{d\sigma(t)\vec{f}^*(t)}{t-z} = 0.$$

Since $\det(W_\Theta(z)+I) \not\equiv 0$, then the integral term in the above equation is zero. We will show that this implies that $\vec{f}(t) = \vec{0}$. First we observe that

$$\frac{1}{t-z} = \left(-\frac{1}{z}\right)\frac{1}{1-t/z} = \left(-\frac{1}{z}\right)\sum_{k=0}^{\infty}\left(\frac{t}{z}\right)^k = -\sum_{k=0}^{\infty}\left(\frac{t^k}{z^{k+1}}\right),$$

for large enough $|z|$. Thus,

$$0 = \int_a^b \frac{d\sigma(t)\vec{f}^*(t)}{t-z} = -\sum_{k=0}^{\infty}\frac{1}{z^{k+1}}\int_a^b t^k d\sigma(t)\vec{f}^*(t),$$

which yields

$$\int_a^b t^k\vec{h}_\alpha d\sigma(t)\vec{f}^*(t) = 0, \quad \forall k = 0,1,\ldots;\ \alpha = 1,\ldots,n. \tag{5.39}$$

According to the Weierstrass approximation theorem, every continuous function on a finite interval $[a,b]$ can be uniformly approximated by a polynomial function. Consequently, the linear span of monomials $\{t^k\}$, $k = 0,1,\ldots$ is dense in $L^2_{[a,b]}(\mathbb{C}, d\sigma)$. Therefore, the linear span of vectors $\{t^k\vec{h}_\alpha,\ k = 0,1,\ldots;\ \alpha = 1,\ldots,n\}$ is dense in $L^2_{[a,b]}(\mathbb{C}^n, d\sigma)$ and hence (5.39) yields $\vec{f}(t) = \vec{0}$. □

Remark 5.5.2. It is not hard to see that Theorem 5.5.1 can also be proved for an operator-valued function $V(z)$ of the form (5.34) whose values are bounded linear operators in an arbitrary finite-dimensional Hilbert space E.

The following theorem is well known [89].

Theorem 5.5.3. *For the function $V(z)$, whose values are bounded linear operators in a finite-dimensional Hilbert space E, to admit the representation*

$$V(z) = \int_a^b \frac{dF(s)}{s-z},$$

outside the finite interval $[a,b]$ of the real axis and $F(s)$ $(a \le x \le b)$ is a non-negative, non-decreasing and bounded operator-function, it is necessary and sufficient that $V(z)$ satisfies the following conditions:

1. *$V(z)$ is holomorphic outside $[a,b]$;*
2. *$V(\infty)=0$;*
3. *in the upper half-plane $V(z)$ has a non-negative imaginary part;*
4. *$V(z)$ takes self-adjoint values on the intervals $(-\infty,a)$ and $(b,+\infty)$ of the real axis.*

Suppose that the linear operator J, acting in a Hilbert space E, satisfies the conditions $J=J^*$ and $J^2=I$. We will say that the function of a complex variable $W(z)$, whose values are bounded linear operators in E, **belongs to the class Ω_J** if it has the following properties:

1. $W(z)$ is holomorphic in some neighborhood G_W of the point at infinity;
2. $\lim_{z\to\infty}\|W(z)-I\|=0$;
3. for all $z\in G_W$ the operator $W(z)+I$ has a bounded inverse, while the operator-function

$$V(z)=i(W(z)+I)^{-1}(W(z)-I)J=i(W(z)-I)(W(z)+I)^{-1}J, \quad (5.40)$$

satisfies the conditions of Theorem 5.5.3.

In view of(5.40) we have

$$(W(z)+I)(I+iV(z)J)=(I+iV(z)J)(W(z)+I)=2I, \quad (z\in G_W).$$

Thus at each $z\in G_W$ the operator $(I+iV(z)J)$ has a bounded inverse, and

$$W(z)=(I+iV(z)J)^{-1}(I-iV(z)J)=(I-iV(z)J)(I+iV(z)J)^{-1}. \quad (5.41)$$

It can be easily shown that a transfer function of any canonical system Θ lies in the class Ω_J.

Let $W(z)$ be a function in a finite-dimensional Hilbert space E. Then the following theorem holds.

Theorem 5.5.4. *If the operator-function $W(z)$ belongs to the class Ω_J, then there exists the Livšic canonical system Θ with directing operator J such that $W_\Theta(z)\equiv W(z)$ in some neighborhood of the point at infinity.*

Proof. It follows from conditions (1)-(3) that the function $V(z)$ satisfies the requirements of Theorem 5.5.3 and hence there exists a non-negative non-decreasing function $F(t)$ $(-\infty<a\le t\le b<+\infty)$ in the finite-dimensional Hilbert space E such that

$$V(z)=\int_a^b\frac{dF(t)}{t-z}, \qquad (z\notin[a,b]).$$

Consequently, we can apply Remark 5.5.2 in conjunction with Theorem 5.5.1 to obtain the system Θ of the form (5.6) such that (5.35) holds. Then, since $V(z)=$

$V_\Theta(z)$, by formulas (5.41) and (5.31) there exists a neighborhood of the point at infinity in which

$$W(z) = (I+iV(z)J)^{-1}(I-iV(z)J) = (I-iV(z)J)(I+iV(z)J)^{-1} = W_\Theta(z). \qquad \square$$

Corollary 5.5.5. *If the function $W(z)$ belongs to the class Ω_J, then in some neighborhood of infinity it satisfies relations* (5.22), (5.23), (5.24), *and* (5.26)–(5.27).

Corollary 5.5.6. *There exists a neighborhood of infinity in which the function $W(z)$ decomposes into a series of the form*

$$W(z) = I + \frac{2i}{z}HJ + \cdots,$$

where $H \geq 0$. If $H = 0$, then $W(z) \equiv I$.

Corollary 5.5.7. *Along with each two operator-functions the class Ω_J contains their coupling. If $W_1(z) \in \Omega_J$, $W_2(z) \in \Omega_J$, and $W_1(z)W_2(z) = I$ $(z \in G_{W_1} \cap G_{W_2})$, then $W_1(z) \equiv W_2(z) \equiv I$.*

The proof of the first statement follows from Theorem 5.4.1, and that of the second from Corollary 5.5.6.

We have already mentioned that the transfer operator-function of a non-minimal canonical system and its principal part coincide in some neighborhood of the point at infinity. This leads to the following result.

Theorem 5.5.8. *If $W(z) \in \Omega_J$, then there exists a minimal canonical system Θ with a direction operator J such that, in some neighborhood of the point at infinity, $W_\Theta(z) \equiv W(z)$.*

5.6 Finite-dimensional state-space case

Suppose the system

$$\Theta = \begin{pmatrix} A & K & J \\ \mathcal{H} & & E \end{pmatrix}$$

is such that *both* state-space $\mathcal{H}$ and input-output space E are finite-dimensional with $\dim \mathcal{H} = n$ and $\dim E = m$. We will refer to this type of systems as the Livšic canonical **finite-dimensional systems**. Clearly, any operator A in a finite-dimensional space $\mathcal{H}$ can be included in a finite-dimensional system system as a state-space operator.

Suppose that J is a linear operator in E, ($\dim E < \infty$) such that $J = J^* = J^{-1}$. We say that the function $W(z)$ with values in E belongs to the **class** Ω_J^m if:

1. $W(z)$ is holomorphic in a region G_W that is a complex plane with a finite number of points (poles of $W(z)$) removed;
2. $\lim_{z\to\infty} \|W(z) - I\| = 0$;

3. $W^*(z)JW(z) - J \geq 0$, $(\operatorname{Im} z > 0,\ z \in G_W)$;
4. $W^*(z)JW(z) - J = 0$, $(\operatorname{Im} z = 0,\ z \in G_W)$.

The class Ω_J^m contains the product of any two if its elements. It is not hard to see that the transfer function of a finite-dimensional system Θ belongs to this class.

Theorem 5.6.1. *Every function $W(z) \in \Omega_J^m$ is a transfer function of the Livšic minimal canonical finite-dimensional system.*

Proof. By Theorem 5.5.4 there exists a minimal system Θ for which $\rho(A) = G_W$, and $W_\Theta(z) = W(z)$, $(z \in G_W)$. It remains to prove that $\mathcal{H}$ is finite-dimensional. It was proved earlier in this chapter that at the points of the set $G_A^{(0)} = \rho(A) \cap \rho(\operatorname{Re} A)$ the operator $W_\Theta(z) + I$ has an inverse, and

$$V_\Theta(z) = K^*(\operatorname{Re} A - zI)^{-1}K = i(W_\Theta(z) + I)^{-1}(W_\Theta(z) - I)J.$$

At the same time, since all scalar products of the form $(W_\Theta(z)f, g)$ are rational functions, the operator $W_\Theta(z)+I$ can fail to have an inverse only at a finite number of points, and only those points may be in the spectrum of the operator $\operatorname{Re} A$. Accordingly, the orthogonal resolution of identity $\Sigma(s)$ corresponding to $\operatorname{Re} A$ is a piece-wise constant function with a finite number of jumps, and therefore the closure $\mathcal{G}_0$ of the linear span of vectors if the form $\Sigma(s)Kg$, $(-\infty < s < \infty,\ g \in E)$ is finite dimensional. On the other hand, $\mathcal{G}_0$ is invariant with respect to A, since

$$A\Sigma(s)Kg = (\operatorname{Re} A + i\operatorname{Im} A)\Sigma(s)Kg = \int_{-\infty}^{s} t\, d\Sigma(t)Kg + iKJK^*\Sigma(s)Kg \in \mathcal{G}_0,$$

which means that $A^nKg \in \mathcal{G}_0$, $(n = 0, 1, \ldots;\ g \in E)$. In view of the minimality of Θ, $\mathcal{G}_0 = \mathcal{H}$. □

Now we consider decomposition of functions of the class Ω_J^m into factors.

Lemma 5.6.2. *Suppose that $\mathcal{H}$ is an n-dimensional space and that A is a linear operator in $\mathcal{H}$. Then there exist subspaces*

$$\begin{gathered} 0 = \mathcal{H}_0 \subset \mathcal{H}_1 \subset \mathcal{H}_2 \subset \cdots \subset \mathcal{H}_{n-1} \subset \mathcal{H}_n = \mathcal{H}, \\ (\dim \mathcal{H}_k = k, \quad k = 1, 2, \ldots, n-1), \end{gathered} \tag{5.42}$$

invariant with respect to A.

Proof. Construct in $\mathcal{H}$ an orthonormal basis $e_1, \ldots, e_n$ such that the vector e_1 is an eigenvector for A, and the vector e_{k+1} $(k = 1, \ldots, n-1)$ is an eigenvector for the operator P_kA, where P_k is the orthoprojection operator onto the orthogonal complement to the linear span $\mathcal{H}_k$ of vectors $e_1, \ldots, e_k$. Then

$$\begin{aligned} Ae_1 &= z_1 e_1, \\ Ae_2 &= \alpha_{21} e_1 + z_2 e_2, \\ &\dots\dots\dots\dots\dots\dots \\ Ae_n &= \alpha_{n1} e_1 + \alpha_{n2} e_2 + \cdots + z_n e_n, \end{aligned}$$

which means that the subspaces $\mathcal{H}_k$ $(k = 1, \dots, n-1)$ are invariant with respect to A. □

Theorem 5.6.3. *Suppose that the function $W(z) \in \Omega_J^m$. Then it may be represented in the form*

$$W(z) = \left(I + \frac{2i\sigma_1}{z - z_1} P_1 J\right)\left(I + \frac{2i\sigma_2}{z - z_2} P_2 J\right) \cdots \left(I + \frac{2i\sigma_n}{z - z_n} P_n J\right), \tag{5.43}$$

where z_j are complex numbers, σ_j are positive numbers, and the P_j are one-dimensional orthogonal projection operators in E satisfying the condition

$$P_j J P_j = \frac{\operatorname{Im} z_j}{\sigma_j} P_j, \quad j = 1, \dots, n. \tag{5.44}$$

Proof. It was proved in Theorem 5.6.1 that the function $W(z)$ is a transfer function for some minimal finite-dimensional system Θ. By Lemma 5.6.2 the space $\mathcal{H}$ contains a subspace (5.42) invariant with respect to A. Using formula (5.19) we have

$$W(z) = W_{\Theta_1}(z) W_{\Theta_2}(z) \cdots W_{\Theta_n}(z), \quad \Theta_j = \operatorname{pr}_{\mathcal{H}_j \ominus \mathcal{H}_{j-1}} \Theta, \ (j = 1, \dots, n).$$

Suppose that A_j and K_j are the state-space and channel operators of the system Θ_j. Since the operator A_j acts in the one-dimensional space $\mathcal{H}_j \ominus \mathcal{H}_{j-1}$, we have

$$W_{\Theta_j}(z) = I - 2iK_j^*(A_j - zI)^{-1} K_j J = I + \frac{2i}{z - z_j} K_j^* K_j J.$$

In view of Theorem 5.3.4 the operator $K_j^* K_j$ is not zero. Accordingly, $K_j^* K_j = \sigma_j P_j$, where $\sigma_j > 0$ and P_j is the orthoprojection onto the one-dimensional subspace. Moreover,

$$P_j J P_j = \frac{1}{\sigma_j^2} K_j^* K_j J K_j^* K_j = \frac{1}{\sigma_j^2} K_j^* \operatorname{Im} A_j K_j = \frac{\operatorname{Im} z_j}{\sigma_j^2} K_j^* K_j = \frac{\operatorname{Im} z_j}{\sigma_j} P_j.$$

The theorem is proved. □

We note that each function of the form

$$W_0(z) = I + \frac{2i\sigma_0}{z - z_0} P_0 J, \quad \left(\sigma_0 > 0, \ P_0 J P_0 = \frac{\operatorname{Im} z_0}{\sigma_0} P_0\right), \tag{5.45}$$

where P_0 is an orthoprojection operator onto a one-dimensional subspace, belongs to the class Ω_J^m. Indeed, from the easily verified equation

$$W_0^*(z) J W_0(z) - J = \frac{4\sigma_0 \operatorname{Im} z}{|z - z_0|^2} J P_0 J,$$

it follows that the function $W_0^*(z)JW_0(z) - J$ is positive in the upper half-plane and equal to zero on the real axis. Thus, the class Ω_J^m coincides with the collection of all possible products of simplest factors of the form (5.45).

It also directly follows from Theorem 5.6.3 and relations (5.43) and (5.44) that if $\dim E = 1$ and $J = 1$, then any function $W(z)$ from the class Ω_1^1 takes the form

$$W(z) = \prod_{k=1}^{n} \frac{z - \bar{z}_k}{z - z_k}, \quad \operatorname{Im} z_k > 0, \quad k = 1, \dots, n, \tag{5.46}$$

that is applied to a vector of a one-dimensional space E.

Let us consider the matrix

$$\vec{A} = \begin{pmatrix} \alpha_1 + \frac{i}{2}\beta_1^2 & i\beta_1\beta_2 & \cdot & \cdot & \cdot & i\beta_1\beta_n \\ 0 & \alpha_2 + \frac{i}{2}\beta_2^2 & \cdot & \cdot & \cdot & i\beta_2\beta_n \\ \cdot & \cdot & \cdot & \cdot & \cdot & \cdot \\ 0 & 0 & \cdot & \cdot & \cdot & \alpha_n + \frac{i}{2}\beta_n^2 \end{pmatrix}, \tag{5.47}$$

where $\{\alpha_k\}_{k=1}^n$ are real numbers and $\{\beta_k\}_{k=1}^n$ are positive numbers. By direct calculations one finds that

$$\operatorname{Im} \vec{A} = \frac{1}{2} \begin{pmatrix} \beta_1^2 & \beta_1\beta_2 & \cdot & \cdot & \cdot & \beta_1\beta_n \\ \beta_1\beta_2 & \beta_2^2 & \cdot & \cdot & \cdot & \beta_2\beta_n \\ \cdot & \cdot & \cdot & \cdot & \cdot & \cdot \\ \beta_1\beta_n & \beta_2\beta_n & \cdot & \cdot & \cdot & \beta_n^2 \end{pmatrix}.$$

We can consider $\vec{A}$ as an operator $\vec{A} : \mathbb{C}^n \to \mathbb{C}^n$ that applies to column-vectors in $\mathbb{C}^n$. Clearly $\vec{A}$ is a prime operator. Then by Theorem 5.1.2, $\vec{A}$ can be included into a system

$$\Theta_{\vec{A}} = \begin{pmatrix} \vec{A} & K & 1 \\ \mathbb{C}^n & & \mathbb{C} \end{pmatrix},$$

where $K\,c = c\,\vec{g}$ and

$$\vec{g} = \frac{1}{\sqrt{2}} \begin{pmatrix} \beta_1 \\ \beta_2 \\ \cdot \\ \cdot \\ \cdot \\ \beta_n \end{pmatrix}.$$

Clearly, $\operatorname{Im} \vec{A}\vec{f} = KK^*\vec{f} = \left(\vec{f}, \vec{g}\right)\vec{g}$, $\vec{f} \in \mathbb{C}^n$. Applying Theorem 5.6.3 and (5.46) we obtain the following formula for the transfer function of $\Theta_{\vec{A}}$:

$$W_{\Theta_{\vec{A}}}(z) = \prod_{k=1}^{n} \frac{z - \alpha_k + i\beta_k^2/2}{z - \alpha_k - i\beta_k^2/2}.$$

The following theorem by Livsič holds.

Theorem 5.6.4. *Let A be a prime dissipative operator in $\mathcal{H}$, $(\dim \mathcal{H} = n)$ with a rank-one imaginary part $\operatorname{Im} A$. Then A is unitarily equivalent to the operator $\vec{A} : \mathbb{C}^n \to \mathbb{C}^n$, where $\vec{A}$ is defined by (5.47).*

Proof. Let $z_k = \alpha_k + i\beta_k^2/2$, where $\{\alpha_k\}_{k=1}^n$ are real numbers, be the eigenvalues of the operator A. Since A is prime then $\{\beta_k\}_{k=1}^n$ can be chosen as positive numbers. According to Theorem 5.1.2, A can be included into a system

$$\Theta = \begin{pmatrix} A & K & 1 \\ \mathcal{H} & & \mathbb{C} \end{pmatrix}.$$

Then we can apply (5.46) and conclude that the transfer function $W_\Theta(z) = W_{\Theta_{\vec{A}}}(z)$. Thus we can apply Theorem 5.3.4 that yields the unitary equivalence of systems $\Theta_{\vec{A}}$ and Θ and hence the operators A and $\vec{A}$. □

5.7 Examples

We conclude this chapter with several simple illustrations.

Example. We consider an operator A in $\mathcal{H} = \mathbb{C}^2$ defined as a matrix

$$A = I + iI = \begin{pmatrix} 1 & 0 \\ 0 & 1 \end{pmatrix} + i \begin{pmatrix} 1 & 0 \\ 0 & 1 \end{pmatrix} = \begin{pmatrix} 1+i & 0 \\ 0 & 1+i \end{pmatrix}.$$

Then its adjoint

$$A^* = I - iI = \begin{pmatrix} 1 & 0 \\ 0 & 1 \end{pmatrix} - i \begin{pmatrix} 1 & 0 \\ 0 & 1 \end{pmatrix} = \begin{pmatrix} 1-i & 0 \\ 0 & 1-i \end{pmatrix},$$

and $\operatorname{Im} A = I$. If $h_1 = \begin{pmatrix} 1 \\ 0 \end{pmatrix}$ and $h_2 = \begin{pmatrix} 0 \\ 1 \end{pmatrix}$, and $f = \begin{pmatrix} f_1 \\ f_2 \end{pmatrix} \in \mathbb{C}^2$, then clearly

$$\operatorname{Im} A\, f = If = f = (f, h_1)h_1 + (f, h_2)h_2.$$

Thus we can introduce an operator $K : E = \mathbb{C}^2 \to \mathcal{H} = \mathbb{C}^2$ such that

$$K\,c = c_1 h_1 + c_2 h_2, \quad K^* f = \begin{pmatrix} (f, h_1) \\ (f, h_2) \end{pmatrix}, \quad c = \begin{pmatrix} c_1 \\ c_2 \end{pmatrix} \in E,\ f = \begin{pmatrix} f_1 \\ f_2 \end{pmatrix} \in \mathcal{H}. \tag{5.48}$$

At this point we are ready to form a system Θ of the form (5.6) where

$$\Theta = \begin{pmatrix} A & K & I \\ \mathbb{C}^2 & & \mathbb{C}^2 \end{pmatrix},$$

with all the components defined above and $J = I$. It is clear that $\operatorname{Im} A = KK^*$. Taking into account that $\operatorname{Re} A = I$ we calculate

$$V_\Theta(z) = K^*(\operatorname{Re} A - zI)^{-1} K = \begin{pmatrix} \dfrac{1}{1-z} & 0 \\ 0 & \dfrac{1}{1-z} \end{pmatrix}, \tag{5.49}$$

and

$$W_\Theta(z) = I - 2iK^*(A - zI)^{-1}K = \begin{pmatrix} \dfrac{1-z-i}{1-z+i} & 0 \\ 0 & \dfrac{1-z-i}{1-z+i} \end{pmatrix}. \tag{5.50}$$

Example. We are going to slightly modify the previous example. Now let A be defined as

$$A = \begin{pmatrix} 1 & 0 \\ 0 & 1 \end{pmatrix} + i \begin{pmatrix} 1 & 0 \\ 0 & -1 \end{pmatrix} = \begin{pmatrix} 1+i & 0 \\ 0 & 1-i \end{pmatrix}.$$

Then its adjoint

$$A^* = \begin{pmatrix} 1 & 0 \\ 0 & 1 \end{pmatrix} - i \begin{pmatrix} 1 & 0 \\ 0 & -1 \end{pmatrix} = \begin{pmatrix} 1-i & 0 \\ 0 & 1+i \end{pmatrix},$$

and $\operatorname{Im} A = \begin{pmatrix} 1 & 0 \\ 0 & -1 \end{pmatrix}$. Let h_1, h_2, and f be the same as in Example 5.7, then

$$\operatorname{Im} A\, f = (f, h_1)h_1 - (f, h_2)h_2.$$

Let $K : E = \mathbb{C}^2 \to \mathcal{H} = \mathbb{C}^2$ be defined by (5.48) and set

$$J = \begin{pmatrix} 1 & 0 \\ 0 & -1 \end{pmatrix}.$$

We are forming a system

$$\Theta = \begin{pmatrix} A & K & J \\ \mathbb{C}^2 & & \mathbb{C}^2 \end{pmatrix},$$

with all the components defined above. It is clear that $\operatorname{Im} A = KJK^*$. Taking into account that both $\operatorname{Re} A = I$ and K are the same as in the previous example we note that $V_\Theta(z)$ is again defined by (5.49) but evaluating $W_\Theta(z)$ yields a different expression,

$$W_\Theta(z) = \begin{pmatrix} \dfrac{1-z-i}{1-z+i} & 0 \\ 0 & \dfrac{1-z+i}{1-z-i} \end{pmatrix}. \tag{5.51}$$

We note that in Example 5.7 the transfer function $W_\Theta(z)$ given in (5.50) belongs to the class Ω_I while $W_\Theta(z)$ from Example 5.7 defined by (5.51) is a member of Ω_J class. Moreover, both functions (5.50) and (5.51) are linear-fractional transformations (5.30) of the same function $V_\Theta(z)$ defined in (5.49).

Example. Let us consider the following operator in the space $L^2_{[0,l]}$:

$$Af = 2i \int_x^l f(t)dt, \quad f \in L^2_{[0,l]}.$$

Its adjoint is

$$A^* f = -2i \int_0^x f(t)\, dt,$$

and

$$\operatorname{Im} Af = \frac{A - A^*}{2i} f = \int_0^l f(t)\, dt.$$

Let $E = \mathbb{C}$, $\mathcal{H} = L^2_{[0,l]}$, and $J = 1$. We introduce an operator $K : E \to \mathcal{H}$ as

$$K\, c = c, \quad K^* f = (f, 1) = \int_0^l f(t)\, dt.$$

We include A in the system

$$\Theta = \begin{pmatrix} A & K & 1 \\ L^2_{[0,l]} & & \mathbb{C} \end{pmatrix}.$$

By direct calculations one finds that

$$(A - zI)^{-1} f(x) = -\frac{1}{z} f(x) - \frac{2i}{z^2} \int_x^l e^{\frac{2i}{z}(t-x)} f(t)\, dt.$$

Using the formula above we obtain

$$W_\Theta(z) = e^{2il/z},$$

and

$$V_\Theta(z) = i \frac{e^{2il/z} - 1}{e^{2il/z} + 1} = i \tanh(il/z).$$

Chapter 6

The Herglotz-Nevanlinna Functions and Rigged Canonical Systems

In this chapter we focus on Herglotz-Nevanlinna functions. After providing all the preliminary results, we introduce the Livšic rigged canonical system (L-system) and its impedance function. We find the necessary and sufficient conditions for a given Herglotz-Nevanlinna function to be realized as the impedance function of an L-system. The properties of the state-space operator of an L-system based on a given impedance function are determined.

6.1 The Herglotz-Nevanlinna functions and their representations

A scalar function $\phi(z)$ that is holomorphic in the upper and lower half-planes is called a **Herglotz-Nevanlinna function** [1] if $\operatorname{Im}\phi(z) \geq 0$ $(\operatorname{Im} z > 0)$ and $\phi(\bar{z}) = \overline{\phi(z)}$. Below we will use some well-known results about the integral representation of a scalar and operator-valued Herglotz-Nevanlinna function. We start off with the following theorem [3].

[1] In addition to the presently used name of Herglotz-Nevanlinna functions one can also find the names Pick, Nevanlinna, Herglotz, Nevanlinna-Pick, and R-functions (sometimes depending on the geographical origin of authors and occasionally whether the open upper half-plane $\mathbb{C}_+$ or the conformaly equivalent open unit disk D is involved).

Theorem 6.1.1. *A finite in the upper half-plane function $\phi(z)$ admits the representation*

$$\phi(z) = a + b\,z + \int_{-\infty}^{\infty} \frac{1+tz}{t-z}\,d\tau(t), \tag{6.1}$$

where $b \geq 0$ and a are two real constants, and $\tau(t)$ is a non-decreasing function with bounded variation, if and only if $\phi(z)$ is holomorphic and has non-negative imaginary part in the upper half-plane $\operatorname{Im} z > 0$. If, in addition, one applies the normalization conditions

$$\tau(t-0) = \tau(t), \quad \tau(-\infty) = 0,$$

then the function $\tau(t)$ is uniquely determined.

As it turns out, the integral representation (6.1) becomes much more convenient if one replaces the function $\tau(t)$ with the function $\sigma(t)$ such that

$$d\sigma(t) = (1+t^2)d\tau(t).$$

Then the integral representation (6.1) takes form

$$\phi(z) = a + bz + \int_{-\infty}^{\infty} \left(\frac{1}{t-z} - \frac{t}{1+t^2}\right) d\sigma(t), \tag{6.2}$$

where $\operatorname{Im} z \neq 0$ and the integral is absolutely convergent while $\sigma(t)$ is such that

$$\int_{-\infty}^{\infty} \frac{d\sigma(t)}{1+t^2} < \infty.$$

The function $\sigma(t)$ can be determined using a given function $\phi(z)$. We have

$$\lim_{y\to 0} \frac{1}{\pi} \int_{t_1}^{t_2} \operatorname{Im} \phi(x+iy)\,dx = \frac{\sigma(t_2-0)+\sigma(t_2+0)}{2} - \frac{\sigma(t_1-0)+\sigma(t_1+0)}{2}. \tag{6.3}$$

The constants a and b can also be uniquely determined. In particular,

$$b = \lim_{y\to\infty} \frac{\phi(iy)}{iy} = \lim_{y\to\infty} \frac{\operatorname{Im} \phi(iy)}{y}.$$

In what follows, the function $\sigma(t)$ will be normalized by the conditions

$$\sigma(t) = \frac{\sigma(t+0)+\sigma(t-0)}{2}, \qquad \sigma(0) = 0. \tag{6.4}$$

After (6.4) the function $\sigma(t)$ in the representation (6.2) is determined uniquely. In particular, the inversion formula (6.3) becomes

$$\sigma(t_1) - \sigma(t_2) = \lim_{y\to 0} \frac{1}{\pi} \int_{t_1}^{t_2} \operatorname{Im}\phi(x+iy)\, dx,$$

and for any real t,

$$\sigma(t) = \lim_{y\to 0} \frac{1}{\pi} \int_{0}^{t} \operatorname{Im}\phi(x+iy)\, dx.$$

It follows from (6.2) that for any real y,

$$y\operatorname{Im}\phi(iy) = b\, y^2 + \int_{-\infty}^{\infty} \frac{y^2}{t^2+y^2}\, d\sigma(t). \tag{6.5}$$

We will need the following proposition [159] that can be obtained from (6.5).

Theorem 6.1.2. *For any Herglotz-Nevanlinna function $\phi(z)$ the following statements are true:*

1. *the function $y\operatorname{Im}\phi(iy)$ is a non-decreasing function on the interval $(0,+\infty)$;*
2. *the equality*

$$\lim_{y\to+\infty} y\operatorname{Im}\phi(iy) = \sup_{y>0} y\operatorname{Im}\phi(iy),$$

   holds. Moreover, both sides of this relation can be finite or infinite;

3. *if $b = 0$ in (6.1), then*

$$\lim_{y\to+\infty} y\operatorname{Im}\phi(iy) = \sup_{y>0} y\operatorname{Im}\phi(iy) = \int_{-\infty}^{\infty} d\sigma(t).$$

Using standard methods for operator theory we can re-write integral representation (6.2) for the case of an operator-valued Herglotz-Nevanlinna function $V(z)$ whose values are bounded linear operators in a Hilbert space E. We have

$$V(z) = Q + zX + \int_{-\infty}^{\infty} \left(\frac{1}{t-z} - \frac{t}{1+t^2}\right) dG(t), \tag{6.6}$$

where $Q = Q^*$ and $X \geq 0$ are linear operators in $[E,E]$, and $G(t)$ is a non-decreasing operator-function on $(-\infty,+\infty)$ for which

$$\int_{-\infty}^{+\infty} \frac{(dG(t)f,f)_E}{1+t^2} < \infty, \quad \forall f\in E.$$

We also note that if $\{z_k\}$ $(k = 1, \dots, n)$ is an arbitrary sequence of non-real complex numbers and h_k is any sequence of vectors in finite-dimensional E, then (6.6) is equivalent to

$$\sum_{k,l=1}^{n} \left(\frac{V(z_k) - V(\bar{z}_l)}{z_k - \bar{z}_l} h_k, h_l \right)_E \geq 0, \tag{6.7}$$

and hence $V(z)$ is a Herglotz-Nevanlinna function if and only if (6.7) holds for an arbitrary choice of $\{z_k\}$ and $\{h_k\}$.

In the sequel, if $V(z)$ $(\operatorname{Im} z \neq 0)$ is a function with values in $[E, E]$, then we say that $V(z)$ **belongs to a certain function class** whenever the function $(V(z)f, f)$ belongs to the corresponding scalar function class for all $f \in E$.

6.2 Extended resolvents and resolution of identity

Let $\dot{A}$ be a closed symmetric operator with equal defect numbers in $\mathcal{H}$ and let A be a self-adjoint extension of $\dot{A}$ in $\mathcal{H}$. The operator function $R_z = (A - zI)^{-1}$ is called a **canonical resolvent** of $\dot{A}$ and

$$R_z = \int_{-\infty}^{\infty} \frac{dE(t)}{t - z}, \quad (\operatorname{Im} z \neq 0), \tag{6.8}$$

where $E(t)$ is the resolution of identity or the spectral function of A. In this case we call $E(t)$ the corresponding **canonical spectral function** of $\dot{A}$.

Let $\mathcal{H}_+ \subset \mathcal{H} \subset \mathcal{H}_-$ be a rigged Hilbert space generated by the operator $\dot{A}$ (see Section 2.2). We denote by $\hat{R}_z$ the $(-, \cdot)$-continuous extension of the operator R_z from $\mathcal{H}_-$ into $\mathcal{H}$ which is adjoint to $R_{\bar{z}}$, i.e.,

$$(\hat{R}_z f, g) = (f, R_{\bar{z}} g), \quad (f \in \mathcal{H}_-, \, g \in \mathcal{H}).$$

Obviously, $\hat{R}_z f = R_z f$ for $f \in \mathcal{H}$, so that $\hat{R}_z$ is an extension of R_z from $\mathcal{H}$ to $\mathcal{H}_-$ with respect to $(-, \cdot)$-continuity. The function $\hat{R}_z$, $(\operatorname{Im} z \neq 0)$ is called the **extended canonical resolvent** of the operator $\dot{A}$. The properties of the resolvents R_z and $\hat{R}_z$ were established in Section 4.5. We note that $\hat{R}_z - \hat{R}_\zeta \in [\mathcal{H}_-, \mathcal{H}_+]$, $\dot{A}^*(\hat{R}_z - \hat{R}_\zeta) = P(z\hat{R}_z - \zeta\hat{R}_\zeta)$ (see (4.67)), and the Hilbert identity

$$\hat{R}_z - \hat{R}_\zeta = (z - \zeta) R_z \hat{R}_\zeta = (z - \zeta) R_\zeta \hat{R}_z,$$

holds.

A complex scalar function $\sigma(t)$ defined on the axis $-\infty < t < +\infty$ is assigned to the **class** $\mathfrak{S}^{(k)}$, $(k = 0, 1, 2)$, if it is of bounded variation on every finite interval and

$$\int_{-\infty}^{+\infty} \frac{|d\sigma(t)|}{1 + |t|^k} < \infty.$$

In particular, $\mathfrak{S}^{(0)}$ is the class of complex functions of bounded variation on the real axis.

With each function $\sigma(t)$ we can associate an interval function $\sigma(\Delta)$ in the usual manner. It was mentioned in Section 6.1 that a scalar Herglotz-Nevanlinna function admits an integral representation of the form (6.2). Suppose $\phi(z)$ is such a scalar Herglotz-Nevanlinna function. Then $\phi(z)$ can be written in the form (6.2) where a is a real number, $b \geq 0$, and $\sigma(t)$ is a non-decreasing function in $\mathfrak{S}^{(2)}$. We will also be interested in a subclass of Herglotz-Nevanlinna functions that has a representation

$$\phi(z) = \int_{-\infty}^{+\infty} \frac{d\sigma(t)}{t-z}, \tag{6.9}$$

where $\sigma(t)$ is a non-decreasing function in $\mathfrak{S}^{(0)}$.

Let R_z be the canonical resolvent of a closed symmetric operator and $\hat{R}_z$ the corresponding extended canonical resolvent. It follows from the integral representation (6.8) that $(R_z f, f)$ is a Herglotz-Nevanlinna function of the form (6.9) for all $f \in \mathcal{H}$. In particular, the operator

$$\frac{1}{2i(\operatorname{Im} z)}(R_z - R_{\bar{z}})$$

is a non-negative operator for all z with $\operatorname{Im} z \neq 0$. This property is passed to the corresponding extended canonical resolvent $\hat{R}_z$ since the operator

$$\frac{1}{2i(\operatorname{Im} z)}(\hat{R}_z - \hat{R}_{\bar{z}})$$

is $(-,+)$-continuous (see Theorem 4.5.5). However, the quadratic functional $(\hat{R}_z f, f)$ is not a Herglotz-Nevanlinna function for all $f \in \mathcal{B}$, where $\mathcal{B}$ is defined by (4.69). The following theorem clarifies the role of the functional Ω defined by equality (4.70) written for $\hat{R}_z = \hat{R}_z(A)$,

$$\Omega(f,g) = \frac{1}{2i}\left[(\hat{R}_z f, g) - (f, \hat{R}_{\bar{z}} g)\right], \quad f, g \in \mathcal{B},$$

where $\mathcal{B}$ is defined by (4.69).

Theorem 6.2.1. *Let $\hat{R}_z$ be an extended canonical resolvent of a closed symmetric operator $\dot{A}$. Then for any $f \in \mathcal{B}$ the function*

$$\phi_f(z) = (\hat{R}_z f, f) - i\Omega(f, f)$$

is a Herglotz-Nevanlinna function.

Proof. The analyticity of the function $\phi_f(z)$ is obvious for all z with $\operatorname{Im} z \neq 0$. We have

$$\phi_f(z) = \frac{1}{2}[(\hat{R}_z f, f) + (f, \hat{R}_{\bar{z}} f)], \qquad (\operatorname{Im} z \neq 0),$$

that implies $\phi_f(\bar{z}) = \overline{\phi_f(z)}$. Furthermore,

$$\begin{aligned}\operatorname{Im}\phi_f(z) &= \operatorname{Im}(\hat{R}_z f, f) - \Omega(f,f)\\ &= \frac{1}{2i}\left(\left[(\hat{R}_z f, f) - \overline{(\hat{R}_z f, f)}\right] - \left[(\hat{R}_z f, f) - (f, \hat{R}_z f\right]\right)\\ &= \frac{1}{2i}\left[(\hat{R}_z f, f) - (\hat{R}_{\bar{z}} f, f)\right] = \frac{1}{2i}((\hat{R}_z - \hat{R}_{\bar{z}})f, f) = \operatorname{Im} z\|\hat{R}_z f\|^2. \quad \square\end{aligned}$$

Thus the number $i\Omega(f,f)$ characterizes the "measure of deviation" of $(\hat{R}_z f, f)$ from the class of Herglotz-Nevanlinna functions. This measure of deviation does not depend on the choice of the extended canonical resolvent $\hat{R}_z$, but only upon the vector $f \in \mathcal{B}$.

Let, as above, $\dot{A}$ be a closed symmetric operator with equal defect numbers in $\mathcal{H}$ and let A be a self-adjoint extension of $\dot{A}$ in $\mathcal{H}$. Let us denote by $\aleph$ the family of all finite intervals Δ of the real axis. We consider the extension of the interval canonical spectral function $E(\Delta)$ of $\dot{A}$ to the space $\mathcal{H}_-$.

Theorem 6.2.2. *If $\Delta \in \aleph$, then $E(\Delta)\mathcal{H} \subset \mathcal{H}_+$ and the operators $E(\Delta)$ and $E(\Delta)\dot{A}$ are $(\cdot,+)$-continuous.*

Proof. Let $h \in \mathcal{H}$ and $f \in \operatorname{Dom}(\dot{A})$. We have

$$(\dot{A}f, E(\Delta)h) = (Af, E(\Delta)h) = (f, AE(\Delta)h) = (f, PAE(\Delta)h).$$

This implies that

$$E(\Delta)h \in \mathcal{H}_+ \quad \text{and} \quad \dot{A}^* E(\Delta)h = PAE(\Delta)h. \tag{6.10}$$

Moreover, for every $g \in \operatorname{Dom}(\dot{A})$,

$$E(\Delta)\dot{A}g = E(\Delta)Ag = AE(\Delta)g, \tag{6.11}$$

and, according to (6.8),

$$\dot{A}^* E(\Delta)\dot{A}g = PAE(\Delta)Ag = PA^2E(\Delta)g. \tag{6.12}$$

For $\Delta \in \aleph$ the operators $AE(\Delta)$ and $A^2E(\Delta)$ are bounded, i.e., for all $h \in \mathcal{H}$,

$$\|AE(\Delta)h\| \le \alpha\|h\|, \quad \|A^2E(\Delta)h\| \le \beta\|h\|, \qquad \alpha, \beta > 0.$$

Thus using (6.10), (6.11), and (6.12) we have

$$\begin{aligned}\|E(\Delta)h\|_+^2 &= \|E(\Delta)h\|^2 + \|\dot{A}^*E(\Delta)h\|^2 \le (1+\alpha^2)\|h\|^2,\\ \|E(\Delta)\dot{A}g\|_+^2 &= \|AE(\Delta)g\|^2 + \|PA^2E(\Delta)g\|^2 \le (\alpha^2+\beta^2)\|g\|^2,\end{aligned} \tag{6.13}$$

which implies the statement of the theorem. $\square$

We denote by $\hat{E}(\varDelta)$ the $(-,\cdot)$-continuous operator from $\mathcal{H}_-$ to $\mathcal{H}$ that is adjoint to $E(\varDelta) \in [\mathcal{H}, \mathcal{H}_+]$. Thus,

$$(\hat{E}(\varDelta)f, g) = (f, E(\varDelta)g) \quad (f \in \mathcal{H}_-,\, g \in \mathcal{H}).$$

One can easily see that $\hat{E}(\varDelta)f = E(\varDelta)f$, $\forall f \in \mathcal{H}$, so that $\hat{E}(\varDelta)$ is the extension of $E(\varDelta)$ by continuity. We note that (6.13) implies that

$$\|\hat{E}(\varDelta)f\|^2 \le (1+\alpha^2)\|f\|_-^2, \quad (f \in \mathcal{H}_-). \tag{6.14}$$

We say that $\hat{E}(\varDelta)$, as a function of $\varDelta \in \aleph$, is the **extended canonical spectral function** of $\dot{A}$ (or the extended resolution of identity) corresponding to the self-adjoint extension A (or to the original spectral function $E(\varDelta)$). Notice that from the equalities $E(\varDelta_1)E(\varDelta_2) = E(\varDelta_1 \cap \varDelta_2)$, $E(\varDelta)R_z = R_zE(\varDelta)$ we get

$$E(\varDelta_1)\hat{E}(\varDelta_2) = \hat{E}(\varDelta_1 \cap \varDelta_2), \quad \varDelta_1, \varDelta_2 \in \aleph \tag{6.15}$$

and

$$E(\varDelta)\hat{R}_z = R_z\hat{E}(\varDelta). \tag{6.16}$$

Theorem 6.2.3. 1. *$\hat{E}(\varDelta) \in [\mathcal{H}_-, \mathcal{H}_+]$ for all $\varDelta \in \aleph$ and all $f \in \mathcal{H}_-$ we have*

$$(\hat{E}(\varDelta)f, f) = ||\hat{E}(\varDelta)f||^2 \ge 0,$$

and

$$\varDelta_1 \subset \varDelta_2 \Rightarrow ||\hat{E}(\varDelta_1)f||^2 \le ||\hat{E}(\varDelta_2)f||^2.$$

2. *Let*

$$\mathcal{F}_A = \{\varphi \in \mathcal{H}_- : (\varphi, g) = 0 \quad \textit{for all} \quad g \in \mathrm{Dom}(A)\},$$

then

$$(\hat{E}(\varDelta)\varphi, \varphi) = 0 \quad \textit{for all} \quad \varDelta \in \aleph \iff \varphi \in \mathcal{F}_A, \tag{6.17}$$

and

$$\sup\left\{(\hat{E}(\varDelta)f, f),\ \varDelta \in \aleph\right\} = +\infty \iff f \notin \mathcal{H}\dot{+}\mathcal{F}_A. \tag{6.18}$$

Proof. 1. Let $\phi \in \mathrm{Dom}(\dot{A})$ and $f \in \mathcal{H}_-$. By Theorem 6.2.2,

$$|(\dot{A}\phi, \hat{E}(\varDelta)f)| = |(E(\varDelta)\dot{A}\phi, f)| \le \|E(\varDelta)\dot{A}\phi\|_+ \cdot \|f\|_- \le \sqrt{\alpha^2+\beta^2}\|\phi\| \cdot \|f\|_-.$$

This yields that $\hat{E}(\varDelta)f \in \mathcal{H}_+$ and $\|\dot{A}^*\hat{E}(\varDelta)f\| \le \sqrt{\alpha^2+\beta^2} \cdot \|f\|_-$. Further, using (6.14) we get

$$\|\hat{E}(\varDelta)f\|_+^2 = \|\hat{E}(\varDelta)f\|^2 + \|\dot{A}^*\hat{E}(\varDelta)f\|^2 \le (1+2\alpha^2+\beta^2)\|f\|_-^2,$$

which implies that $\hat{E}(\varDelta) \in [\mathcal{H}_-, \mathcal{H}_+]$. Since $(E(\varDelta)f, f) \ge 0$ for all $f \in \mathcal{H}$, then $(\hat{E}(\varDelta)f, f) \ge 0$ for all $f \in \mathcal{H}_-$ and thus $\hat{E}(\varDelta)$ is a non-negative operator. Moreover, the equality $E(\varDelta)\hat{E}(\varDelta) = \hat{E}(\varDelta)$ yields

$$(\hat{E}(\varDelta)f, f) = ||\hat{E}(\varDelta)f||^2,$$

for all $\varDelta \in \aleph$ and all $f \in \mathcal{H}_-$. If $\varDelta_1 \subset \varDelta_2$, then the inequality $||E(\varDelta_1)h||^2 \leq ||E(\varDelta_2)h||^2$ for all, $h \in \mathcal{H}$ yields

$$||\hat{E}(\varDelta_1)f||^2 \leq ||\hat{E}(\varDelta_2)f||^2, \ f \in \mathcal{H}_-.$$

2. Since $\hat{E}(\varDelta) \in [\mathcal{H}_+, \mathcal{H}_-]$, $\hat{E}(\varDelta)$ is the extension of $E(\varDelta)$, $E(\varDelta)H \subset \mathrm{Dom}(A)$, and $\mathcal{H}$ is dense in $\mathcal{H}_-$, we get that $\mathrm{Ran}(\hat{E}(\varDelta)) \subset \overline{\mathrm{Dom}(A)}$ for all $\varDelta \in \aleph$, where $\overline{\mathrm{Dom}(A)}$ is the closure of $\mathrm{Dom}(A)$ in $\mathcal{H}_+$. Hence, if $\varphi \in \mathcal{F}_A$, then $(\hat{E}(\varDelta)\varphi, \varphi) = 0$ for all $\varDelta \in \aleph$.

Observe that the linear manifold

$$\mathcal{L}_A = \{E(\varDelta)g, \ g \in \mathrm{Dom}(A), \ \varDelta \in \aleph\},$$

is dense in $\overline{\mathrm{Dom}(A)}$ with respect to the (+)-norm. Indeed, if $h \in \overline{\mathrm{Dom}(A)}$ and $(h, E(\varDelta)g)_+ = 0$ for all $g \in \mathrm{Dom}(A)$ and all $\varDelta \in \aleph$, then

$$(h, E(\varDelta)g) + (\dot{A}^* h, PE(\varDelta)Ag) = 0.$$

Letting $\varDelta \to \mathbb{R}$, one has

$$(h, g) + (\dot{A}^* h, PAg) = (h, g)_+ = 0, \ g \in \mathrm{Dom}(A).$$

Hence $h = 0$. Suppose that $(\hat{E}(\varDelta)\varphi, \varphi) = 0$ for all $\varDelta \in \aleph$. Since $||\hat{E}(\varDelta)\varphi||^2 = (\hat{E}(\varDelta)\varphi, \varphi) = 0$, we get that $\hat{E}(\varDelta)\varphi = 0$. Therefore,

$$(\varphi, E(\varDelta)g) = (\mathcal{R}\varphi, E(\varDelta)g)_+ = 0,$$

for all $\varDelta \in \aleph$ and all $h \in \mathcal{H}$. Taking $g \in \mathrm{Dom}(A)$, we obtain that $\varphi \in \mathcal{F}_A$. Thus, (6.17) holds true.

If $h \in \mathcal{H}$ and $\varphi \in \mathcal{F}_A$, then $\hat{E}(\varDelta)(h + \varphi) = \hat{E}(\varDelta)h$ and

$$\sup_{\varDelta \in \aleph} \left\{ (\hat{E}(\varDelta)(h + \varphi), h + \varphi)) \right\} < \infty.$$

Let $f \in \mathcal{H}_-$. Assume $\sup\limits_{\varDelta \in \aleph} \left\{ (\hat{E}(\varDelta)f, f) \right\} < \infty$. Using the equality $||\hat{E}(\varDelta)\varphi||^2 = (\hat{E}(\varDelta)\varphi, \varphi)$, we get

$$\sup_{\varDelta \in \aleph} \left\{ ||\hat{E}(\varDelta)f|| \right\} < \infty.$$

Therefore, the set $\{\hat{E}(\varDelta)f, \ \varDelta \in \aleph\}$ is weakly compact, i.e., there exists $h \in \mathcal{H}$ such that

$$\lim_{n \to \infty} (\hat{E}(\varDelta_n)f, g) = (h, g),$$

for all $g \in \mathcal{H}$ and for some sequence $\varDelta_1 \subset \varDelta_2 \subset \cdots \subset \varDelta_n \subset \cdots$, $\lim\limits_{n \to \infty} \varDelta_n = \mathbb{R}$. If $\varDelta \in \aleph$, then there exists n_0 such that $\varDelta \subset \varDelta_n$ for all $n > n_0$. Due to the equalities $E(\varDelta)\hat{E}(\varDelta_n) = \hat{E}(\varDelta)$ for $n > n_0$ and

$$(E(\varDelta)\hat{E}(\varDelta_n)f, g) = (\hat{E}(\varDelta_n)f, E(\varDelta)g),$$

we have $(\hat{E}(\Delta)f, g) = (h, E(\Delta)g)$. Thus

$$(f - h, E(\Delta)g) = 0,$$

for all $\Delta \in \aleph$ and all $g \in \mathcal{H}$. Taking into account that the set $\mathcal{L}_A$ is dense in $\overline{\mathrm{Dom}(A)}$ with respect to $(+)$-norm, we obtain that $f - h \in \mathcal{F}_A$. □

Corollary 6.2.4.

$$\int_{-\infty}^{+\infty} d(\hat{E}(t)f, f) = +\infty \quad \textit{if and only if} \quad f \notin \mathcal{H} \dot{+} \mathcal{F}_A.$$

Proof. The statement follows from the equality

$$\int_{\Delta} d(\hat{E}(\tau)f, f) = (\hat{E}(\Delta)f, f)$$

and relation (6.18). □

It is well known that the complex scalar measure $(E(\Delta)f, g)$ is a complex function of bounded variation on the real axis. However, $(\hat{E}(\Delta)f, g)$ may be unbounded for $f, g \in \mathcal{H}_-$.

Theorem 6.2.5. *If $f \in \mathcal{H}_-$ and $g \in \mathcal{H}$, then $(\hat{E}(\Delta)f, g) \in \mathfrak{S}^{(1)}$.*

Proof. Based on (2.1) and (6.10) we have

$$\begin{aligned}(\hat{E}(\Delta)f, g) &= (f, E(\Delta)g) = (\mathcal{R}f, E(\Delta)g)_+ \\ &= (\mathcal{R}f, E(\Delta)g) + (\dot{A}^*\mathcal{R}f, \dot{A}^*E(\Delta)g) \\ &= (\mathcal{R}f, E(\Delta)g) + (\dot{A}^*\mathcal{R}f, AE(\Delta)g).\end{aligned}$$

All that remains is to show that $(AE(\Delta)g, f) \in \mathfrak{S}^{(1)}$ for any $h \in \mathcal{H}$. We have

$$\int_{-\infty}^{+\infty} \frac{|d(AE(\Delta)g, h)|}{1 + |t|} = \int_{-\infty}^{+\infty} \frac{|t|}{1 + |t|} |(dE(\Delta)g, h)| \leq \int_{-\infty}^{+\infty} |(dE(\Delta)g, h)|.$$ □

Note that we have also proved that $(E(\Delta)g, x)_+ \in \mathfrak{S}^{(1)}$ for all $g \in \mathcal{H}$ and $x \in \mathcal{H}_+$. Also applying Theorem 6.2.3 we conclude that $(\hat{E}(\Delta)f, g)$ makes sense for all $f, g \in \mathcal{H}_-$.

Now we will see that equation (6.8) remains true for extended canonical resolvents and extended canonical spectral functions. It follows from (4.64) and

(6.9) that $\dot{A}^*R_z = P(I + zR_z)$, where P is an orthogonal projection of $\mathcal{H}$ onto $\mathrm{Dom}(\dot{A})$. This implies

$$\begin{aligned}\dot{A}^*R_z &= PA(A - zI)^{-1}\restriction\mathcal{H}\\ &= PA\int\limits_{-\infty}^{+\infty}\frac{dE(t)}{t-z} = \int\limits_{-\infty}^{+\infty}\frac{d(PAE(t))}{t-z} = \int\limits_{-\infty}^{+\infty}\frac{d(\dot{A}^*E(t))}{t-z}.\end{aligned}$$

Hence equality (6.8) takes place in $[\mathcal{H},\mathcal{H}_+]$. That is why for any $f\in\mathcal{H}_-$ and $g\in\mathcal{H}$ we have

$$(\hat{R}_z f, g) = (f, \hat{R}_{\bar{z}} g) = (f, \int\limits_{-\infty}^{+\infty}\frac{dE(t)g}{t-\bar{z}}) = \int\limits_{-\infty}^{+\infty}\frac{d(f,E(t)g)}{t-z} = \int\limits_{-\infty}^{+\infty}\frac{d(\hat{E}(t)f,g)}{t-z}.$$

Therefore, the representation

$$\hat{R}_z = \int\limits_{-\infty}^{+\infty}\frac{d\hat{E}(t)}{t-z},$$

takes place in $[\mathcal{H}_-,\mathcal{H}]$. In addition, from (6.15) we get

$$E(\Delta)\hat{R}_z = R_z\hat{E}(\Delta) = \int\limits_{\Delta}\frac{d\hat{E}(t)}{t-z}. \tag{6.19}$$

Proposition 6.2.6. *For each $f\in\mathcal{H}_-$ the equality*

$$\lim_{\Delta\to\mathbb{R}} E(\Delta)(\hat{R}_z - \hat{R}_\zeta)f = (\hat{R}_z - \hat{R}_\zeta)f,$$

holds in $\mathcal{H}_+$.

Proof. Using (4.67) and (6.16) we have

$$\begin{aligned}&\dot{A}^*(E(\Delta)(\hat{R}_z - \hat{R}_\zeta) - (\hat{R}_z - \hat{R}_\zeta)) = \dot{A}^*(R_z - R_\zeta)\hat{E}(\Delta) - \dot{A}^*(\hat{R}_z - \hat{R}_\zeta)\\ &= P(zR_z - \zeta R_\zeta)\hat{E}(\Delta) - P(z\hat{R}_z - \zeta\hat{R}_\zeta) = P(E(\Delta) - I_{\mathcal{H}})(z\hat{R}_z - \zeta\hat{R}_\zeta).\end{aligned}$$

It follows that $\lim\limits_{\Delta\to\mathbb{R}} \dot{A}^*E(\Delta)(\hat{R}_z - \hat{R}_\zeta)f = \dot{A}^*(\hat{R}_z - \hat{R}_\zeta)f$ in $\mathcal{H}$. Since

$$\lim_{\Delta\to\mathbb{R}} E(\Delta)(\hat{R}_z - \hat{R}_\zeta)f = (\hat{R}_z - \hat{R}_\zeta)f \quad \text{in } \mathcal{H},$$

and $||g||_+^2 = ||g||^2 + ||A^*g||^2$, $g\in\mathcal{H}_+$, we obtain

$$\lim_{\Delta\to\mathbb{R}} E(\Delta)(\hat{R}_z - \hat{R}_\zeta)f = (\hat{R}_z - \hat{R}_\zeta)f \quad \text{in } \mathcal{H}_+. \qquad \square$$

From (6.15), (6.19), and Proposition 6.2.6 it follows that for each $f \in \mathcal{H}_-$,

$$(\hat{R}_z - \hat{R}_\zeta)f = \int_{-\infty}^{+\infty} \left(\frac{1}{t-z} - \frac{1}{t-\zeta} \right) d\hat{E}(t)f,$$

where the integral in the right-hand side converges in $\mathcal{H}_+$. In particular,

$$(\hat{R}_z - \hat{R}_\zeta)f, g) = \int_{-\infty}^{+\infty} \left(\frac{1}{t-z} - \frac{1}{t-\zeta} \right) d(\hat{E}(t)f, g). \tag{6.20}$$

for all $f, g \in \mathcal{H}_-$. If $f, g \in \mathcal{H}_-$ then the expression $(\hat{R}_z f, g)$ may not be defined, and the integral

$$\int_{-\infty}^{+\infty} \frac{d(\hat{E}(t)f, g)}{t-z}$$

may diverge. However, a certain regularization both of the extended canonical resolvent and of the integral allows us to change $\hat{R}_z$ to obtain an integral representation in a form similar to (6.8).

Theorem 6.2.7. *There exists a bi-continuous self-adjoint operator H such that for any extended canonical resolvent $\hat{R}_z$ the operator $\hat{R}_{z,H} = \hat{R}_z - H$ is an operator-valued Herglotz-Nevanlinna function with values in $[\mathcal{H}_-, \mathcal{H}_+]$.*

Proof. Let $\hat{R}_z^{(0)}$ be an extended canonical resolvent and z_0 be a non-real number. Set

$$H = \frac{1}{2} \left(\hat{R}_{z_0}^{(0)} + \hat{R}_{\bar{z}_0}^{(0)} \right), \tag{6.21}$$

and observe that H is bi-continuous and self-adjoint. For an arbitrary extended canonical resolvent $\hat{R}_z$ the operator-function

$$\hat{R}_{z,H} = \hat{R}_z - H = (\hat{R}_z - \hat{R}_z^{(0)}) + \frac{1}{2} \left[\left(\hat{R}_z^{(0)} - \hat{R}_{z_0}^{(0)} \right) + \left(\hat{R}_z^{(0)} - \hat{R}_{\bar{z}_0}^{(0)} \right) \right],$$

belongs to the class $[\mathcal{H}_-, \mathcal{H}_+]$ while being holomorphic in the upper and lower half-planes (see Theorem 4.5.5). We also note that $(\hat{R}_{z,H})^* = \hat{R}_{\bar{z},H}$.

For an $f \in \mathcal{H}_-$ we have

$$\frac{\operatorname{Im} (\hat{R}_{z,H} f, f)}{\operatorname{Im} z} = \frac{((\hat{R}_{z,H} - \hat{R}_{\bar{z},H})f, f)}{z - \bar{z}} = \frac{((\hat{R}_z - \hat{R}_{\bar{z}})f, f)}{z - \bar{z}}.$$

Referring to our discussion before Theorem 6.2.1, we conclude that the last expression is non-negative for all $f \in \mathcal{H}_-$ and $\operatorname{Im} z \neq 0$. Thus, $(\hat{R}_{z,H} f, f)$ is a Herglotz-Nevanlinna function for all $f \in \mathcal{H}_-$. □

The operator H in Theorem 6.2.7 is called a **regularizing operator**. As it was shown in the proof of the last theorem we can always choose a regularizing operator to be defined by (6.21) where z_0 is an arbitrary non-real number and $\hat{R}^{(0)}_{z_0}$ is an extended canonical resolvent.

Theorem 6.2.8. *The function $\hat{R}_{z,H}$ admits a weak integral representation*

$$\hat{R}_{z,H} = Q_H + \int\limits_{-\infty}^{+\infty} \left(\frac{1}{t-z} - \frac{t}{1+t^2}\right) d\hat{E}(t), \tag{6.22}$$

where Q_H is some self-adjoint operator in $[\mathcal{H}_-, \mathcal{H}_+]$ (depending on the choice of H) and $(\hat{E}(t)f, g)$ belongs to the class $\mathfrak{S}^{(2)}$.

Proof. According to Theorem 6.2.7, $\hat{R}_{z,H} = \hat{R}_z - H$ is $[\mathcal{H}_-, \mathcal{H}_+]$-valued Herglotz-Nevanlinna. Let

$$H_0 = \frac{1}{2}\left(\hat{R}_i + \hat{R}_{-i}\right),$$

and let $\hat{R}_{z,H_0} = \hat{R}_z - H_0 = R_{z,H} + (H - H_0)$. Using (6.20) we have that

$$\begin{aligned}(\hat{R}_{z,H_0} f, g) &= \left(\left(\hat{R}_z - \frac{1}{2}\left(\hat{R}_i + \hat{R}_{-i}\right)\right) f, g\right) \\ &= \int\limits_{-\infty}^{+\infty} \left(\frac{1}{t-z} - \frac{t}{1+t^2}\right) (d\hat{E}(t) f, g),\end{aligned}$$

for all $f, g \in \mathcal{H}_-$. This yields (6.22) with $Q = H_0 - H$. Since

$$\int\limits_{-\infty}^{+\infty} \frac{d(\hat{E}(t)f, g)}{1+t^2} = \frac{1}{2i}\left(\left(\hat{R}_i - \hat{R}_{-i}\right) f, g\right),$$

for all $f, g \in \mathcal{H}_-$. Hence $d(\hat{E}(t)f, g)$ belongs to the class $\mathfrak{S}^{(2)}$. □

Now we can interpret the preceding results of this section in the following way. Let $\hat{R}_z$ be an extended canonical resolvent of a closed symmetric operator $\dot{A}$ and let $\hat{E}(\Delta)$ be the corresponding extended canonical spectral function. Then for any $f, g \in \mathcal{H}_-$,

$$\int\limits_{-\infty}^{+\infty} \frac{|d(\hat{E}(t)f, g)|}{1+t^2} < \infty,$$

and the following integral representation holds:

$$\hat{R}_z - \frac{\hat{R}_i + \hat{R}_{-i}}{2} = \int\limits_{-\infty}^{+\infty} \left(\frac{1}{t-z} - \frac{t}{1+t^2}\right) d\hat{E}(t). \tag{6.23}$$

We will refine integral representation (6.23) for the case of the resolvent of a t-self-adjoint bi-extension of a regular symmetric operator $\dot{A}$. After that we will establish some additional properties of the corresponding extended canonical spectral function.

Lemma 6.2.9. *Let $\mathbb{A}$ be a t-self-adjoint bi-extension of a regular symmetric operator $\dot{A}$ with the quasi-kernel $\hat{A}$ and let $\hat{E}(\Delta)$ be the extended canonical spectral function of $\hat{A}$. Then for every $f \in \mathcal{H}\dot{+}L_{\mathbb{A}}$, $f \neq 0$, and for every $g \in \mathcal{H}_-$ there is an integral representation*

$$(\mathbb{R}_z f, g) = \int_{-\infty}^{+\infty} \left(\frac{1}{t-z} - \frac{t}{1+t^2}\right) d(\hat{E}(t)f, g) + \frac{1}{2}((\hat{R}_i + \hat{R}_{-i})f, g), \tag{6.24}$$

where $L_{\mathbb{A}}$ is defined by (4.44) and $\mathbb{R}_z = (\mathbb{A} - zI)^{-1}$.

Proof. Due to Theorem 4.5.12 for $\hat{R}_z = \hat{R}_z(\hat{A})$ we have

$$\hat{R}_z \upharpoonright (\mathcal{H} \dot{+} L_{\mathbb{A}}) = (\mathbb{A} - zI)^{-1}.$$

Therefore, $\hat{R}_z(\mathcal{H}\dot{+}L_{\mathbb{A}}) \subseteq \mathcal{H}_+$. Hence for all $g \in \mathcal{H}_-$ and $f \in \mathcal{H} \dot{+} L_{\mathbb{A}}$, $(\mathbb{R}_z f, g)$ is defined. Consequently, using (6.23) we obtain (6.24) where $\hat{E}(t)$ is the extended canonical spectral function of $\hat{A}$. $\square$

Theorem 6.2.10. *Let $\mathbb{A}$ be a t-self-adjoint bi-extension of a regular symmetric operator $\dot{A}$ with the quasi-kernel $\hat{A}$ and let $\hat{E}(\Delta)$ be the canonical spectral function of $\hat{A}$. Then for any $f \in L_{\mathbb{A}} \dot{+} \mathfrak{L}$, $f \neq 0$,*

$$\int_{-\infty}^{+\infty} d(\hat{E}(t)f, f) = \infty, \qquad \textit{if} \quad f \notin \mathfrak{L}, \tag{6.25}$$

and

$$\int_{-\infty}^{+\infty} d(\hat{E}(t)f, f) < \infty, \qquad \textit{if} \quad f \in \mathfrak{L}. \tag{6.26}$$

Moreover, there exist real constants $\beta > 0$ and $\alpha > 0$ such that

$$\alpha \|f\|_-^2 \leq \int_{-\infty}^{+\infty} \frac{d(\hat{E}(t)f, f)}{1+t^2} \leq \beta \|f\|_-^2, \tag{6.27}$$

for all $f \in L_{\mathbb{A}} \dot{+} \mathfrak{L}$, where $L_{\mathbb{A}}$ and $\mathfrak{L}$ are defined by (4.44) and (2.12), respectively.

Proof. Since $\mathbb{A}$ is a t-self-adjoint bi-extension of $\dot{A}$, then according to (3.13) it takes a form $\mathbb{A} = \dot{A}P^+_{\mathrm{Dom}(\dot{A})} + \left(\dot{A}^* + \mathcal{R}^{-1}(S - \frac{i}{2}\mathfrak{J})\right) P^+_{\mathfrak{M}}$. Let us choose a point z

with $\operatorname{Im} z \neq 0$ to be a regular point of the operator $\mathbb{A}$ and consider function $\phi(z)$ defined for all $f \in L_{\mathbb{A}} \dot{+} \mathfrak{L} \subset \operatorname{Ran}(\mathbb{A} - zI)$ (see (4.44) and Theorem 4.3.2) by the formula:

$$\phi(z) = \big((\mathbb{A} - zI)^{-1} f, f\big).$$

It can be seen that $\phi(z) = \overline{\phi(\bar{z})}$ and $\operatorname{Im}\phi(z) = \operatorname{Im} z \,\big\|(\mathbb{A} - \bar{z}I)^{-1} f\big\|^2$, which means that $\phi(z)$ is a Herglotz-Nevanlinna function and according to Lemma 6.2.9 has the integral representation

$$\phi(z) = \int_{-\infty}^{+\infty} \left(\frac{1}{t-z} - \frac{t}{1+t^2}\right) d\big(\hat{E}(t) f, f\big) + \frac{1}{2}\big((\hat{R}_i + \hat{R}_{-i}) f, f\big).$$

This representation implies that

$$\lim_{\eta \to \infty} \frac{\operatorname{Im}\phi(i\eta)}{\eta} = 0.$$

Let $\operatorname{Dom}(\hat{A})$ be given by

$$\operatorname{Dom}(\hat{A}) = \operatorname{Dom}(\dot{A}) \oplus (I + M)(\mathfrak{N}_i' \oplus \mathfrak{N}).$$

According to Theorem 3.3.8, Theorem 4.3.2 and formulas (3.25), (4.33), (4.43) we have

$$(I + M)(\mathfrak{N}_i' \oplus \mathfrak{N}) \dot{+} (I + U)\mathfrak{N}_i' = \mathfrak{M}, \quad S(I + U)\varphi_i = -\frac{i}{2}(I - U)\varphi_i,$$

and $L_{\mathbb{A}} = \mathcal{R}^{-1}(U - I)\mathfrak{N}_i'$. The subspace $(U - I)\mathfrak{N}_i'$ is (1)-orthogonal to $(I + U)\mathfrak{N}_i'$. Let

$$\mathcal{F}_{\hat{A}} = \left\{\varphi \in \mathcal{H}_- : (\varphi, g) = 0 \quad \text{for all} \quad g \in \operatorname{Dom}(\hat{A})\right\}.$$

If $\varphi \in L_{\mathbb{A}} \cap \mathcal{F}_{\hat{A}}$, then the vector $\mathcal{R}\varphi \in (I - U)\mathfrak{N}_i'$ is (1)-orthogonal to $\operatorname{Dom}(\hat{A})$ and therefore, $\mathcal{R}\varphi$ is (1)-orthogonal to $\mathcal{H}_+$. Hence $\varphi = 0$, i.e., $L_{\mathbb{A}} \cap \mathcal{F}_{\hat{A}} = \{0\}$. Due to Theorem 6.2.3 and Corollary 6.2.4 we get (6.25) and (6.25).

Since $(\mathbb{A} + iI)^{-1} f \in \mathfrak{N}_{-i}$ for all $f \in L_{\mathbb{A}} \dot{+} \mathfrak{L}$, the norms $\|\cdot\|$ and $\|\cdot\|_+$ are equivalent on $\mathfrak{N}_{\pm i}$ and so are the norms $\|\cdot\|$ and $\|\cdot\|_-$. Therefore

$$\alpha\|f\|_-^2 \leq \operatorname{Im}\phi(i) \leq \beta\|f\|_-^2, \quad \beta > 0, \alpha > 0 - \text{const}.$$

Combining this with

$$\operatorname{Im}\phi(i) = \frac{1}{2i}\big((\hat{R}_i - \hat{R}_{-i}) f, f\big) = \int_{-\infty}^{+\infty} \frac{d(\hat{E}(t) f, f)}{1 + t^2},$$

we obtain the relation (6.27). □

Corollary 6.2.11. *In the settings of Theorem* 6.2.10 *for all* $f,g \in L_{\mathbb{A}} \dotplus \mathfrak{L}$,

$$\left|\left(\frac{\hat{R}_i + \hat{R}_{-i}}{2} f, g\right)\right| \le k\sqrt{\int_{-\infty}^{+\infty} \frac{d(\hat{E}(t)f,f)}{1+t^2}} \cdot \sqrt{\int_{-\infty}^{+\infty} \frac{d(\hat{E}(t)g,g)}{1+t^2}}, \tag{6.28}$$

where $k > 0$ *is a constant.*

Proof. If $f \in L_{\mathbb{A}} \dotplus \mathfrak{L}$, then

$$(\mathbb{A} - iI)^{-1}f + (\mathbb{A} + iI)^{-1}f = 2(\mathbb{A} - iI)^{-1}\mathbb{A}(\mathbb{A} + iI)^{-1}f.$$

From Theorem 4.5.12 we get

$$((\mathbb{A} - zI)^{-1}f, g) = (f, (\mathbb{A} - \bar{z}I)^{-1}g),$$

and, since $\mathbb{A}(\mathbb{A} + iI)^{-1}f \in \mathcal{H} \dotplus L_{\mathbb{A}}$, we have

$$\begin{aligned}\left|\left(\frac{\hat{R}_i + \hat{R}_{-i}}{2} f, g\right)\right| &= \left|(\mathbb{A}(\mathbb{A} + iI)^{-1}f, (\mathbb{A} + iI)^{-1}g)\right| \\ &\le \|\mathbb{A}(\mathbb{A} + iI)^{-1}f\|_- \cdot \|(\mathbb{A} + iI)^{-1}g\|_+ \le c\|f\|_- \cdot \|g\|_-.\end{aligned}$$

It follows from Theorem 6.2.10 that there is a constant $\alpha > 0$ such that

$$\|f\|_-^2 \le \alpha \int_{-\infty}^{+\infty} \frac{d(\hat{E}(t)f,f)}{1+t^2}.$$

A similar inequality holds for $\|g\|_-^2$. This completes the proof. □

6.3 Definition of an L-system

In this section we introduce the Livšic rigged canonical system or L-system. In order to do that we first define a class of state-space operators.

Definition 6.3.1. An unbounded operator T acting in the Hilbert space $\mathcal{H}$ belongs to the **class** Λ if $\rho(T) \neq \emptyset$ and the maximal common symmetric part of T and T^* has finite and equal deficiency indices.

Definition 6.3.2. Let $\dot{A}$ be a symmetric operator with finite and equal deficiency indices. An operator T of the class $\Omega(\dot{A})$ belongs to the **class** $\Lambda(\dot{A})$ if $\dot{A}$ is the maximal common symmetric part of T and T^*.

Thus, if $T \in \Lambda$, then the operators T and T^* belong to the class $\Lambda(\dot{A})$, where $\dot{A}$ is a maximal common symmetric part of T and T^*. On the other hand for a fixed operator $\dot{A}$ the class $\Lambda(\dot{A})$ is a subclass of the class Λ. In the sequel, given $T \in \Lambda$, by $\dot{A}$ we will denote the maximal common symmetric part of T and T^*. Since $\dot{A}$

has finite deficiency indices, the operator $\dot{A}$ is regular. Therefore the operator T from the class Λ is quasi-self-adjoint and a regular extension of $\dot{A}$ (see Theorem 4.1.3). By Theorem 4.3.10 each operator T from the class Λ admits a $(*)$-extension in rigged Hilbert space $\mathcal{H}_+ \subset \mathcal{H} \subset \mathcal{H}_-$ constructed by the means of the operator $\dot{A}^*$ (see Section 2.2). Any $(*)$-extension $\mathbb{A}$ of $T \in \Lambda$ together with $\mathbb{A}^*$ are bounded linear operators from $\mathcal{H}_+$ into $\mathcal{H}_-$ and hence we can call the operators

$$\operatorname{Re}\mathbb{A} = \frac{1}{2}(\mathbb{A} + \mathbb{A}^*) \text{ and } \operatorname{Im}\mathbb{A} = \frac{1}{2i}(\mathbb{A} - \mathbb{A}^*), \tag{6.29}$$

real and imaginary parts of T as well as real and imaginary parts of $\mathbb{A}$. Since $T \in \Lambda$ can have many different $(*)$-extensions, then T can have many different real and imaginary parts.

Let $T \in \Lambda$, K be a bounded linear operator from a finite-dimensional Hilbert space E into $\mathcal{H}_-$, $K^* \in [\mathcal{H}_+, E]$, and $J = J^* = J^{-1} \in [E, E]$. Consider the following singular system of equations:

$$\begin{cases} i\frac{d\chi}{dt} + T\chi(t) = KJ\psi_-(t), \\ \chi(0) = x \in \operatorname{Dom}(T), \\ \psi_+ = \psi_- - 2iK^*\chi(t). \end{cases} \tag{6.30}$$

Given an input vector $\psi_- = \varphi_- e^{izt} \in E$, we seek solutions to the system (6.30) as an output vector $\psi_+ = \varphi_+ e^{izt} \in E$, and a state-space vector $\chi(t) = xe^{izt} \in \operatorname{Dom}(T)$. Substituting the expressions for $\psi_\pm(t)$ and $\chi(t)$ allows us to cancel exponential terms and convert the system (6.30) to the form

$$\begin{cases} (T - zI)x = KJ\varphi_-, \\ \varphi_+ = \varphi_- - 2iK^*x, \end{cases} \qquad z \in \rho(T). \tag{6.31}$$

The choice of the operator K in the above system is such that $KJ\varphi_- \in \mathcal{B} \subset \mathcal{H}_-$, where $\mathcal{B}$ is defined by (4.69). Therefore the first equation of (6.31) does not, in general, have a regular solution $x \in \operatorname{Dom}(T)$. It has, however, a generalized solution $x \in \mathcal{H}_+$ that can be obtained in the following way. If $z \in \rho(T)$, then we can use the density of $\mathcal{H}$ in $\mathcal{H}_-$ and therefore there is a sequence of vectors $\{\alpha_n\} \in \mathcal{H}$ that approximates $KJ\varphi_-$ in the $(-)$-metric. In this case the state space vector $x = \hat{R}_z(T)KJ\varphi_- \in \mathcal{H}$ is understood as $\lim_{n\to\infty}(T - zI)^{-1}\alpha_n$, where $\hat{R}_z(T)$ is the extended to $\mathcal{H}_-$ by $(-,\cdot)$-continuity resolvent $(T - zI)^{-1}$. But then we can apply Theorem 4.5.9 to conclude that $x \in \mathcal{H}_+$. This explains the expression K^*x in the second line of (6.31). In order to satisfy the condition $\operatorname{Im} T = KJK^*$ we perform the *regularization* of system (6.31) and use $\mathbb{A} \in [\mathcal{H}_+, \mathcal{H}_-]$, a $(*)$-extension of T such that $\operatorname{Im}\mathbb{A} = KJK^*$. This leads to the system

$$\begin{cases} (\mathbb{A} - zI)x = KJ\varphi_-, \\ \varphi_+ = \varphi_- - 2iK^*x, \end{cases} \qquad z \in \rho(T), \tag{6.32}$$

where φ_- is an input vector, φ_+ is an output vector, and x is a state-space vector of the system. System (6.32) is the stationary version of the system

$$\begin{cases} i\frac{d\chi}{dt} + \mathbb{A}\chi(t) = KJ\psi_-(t), \\ \chi(0) = x \in \mathcal{H}_+, \\ \psi_+ = \psi_- - 2iK^*\chi(t). \end{cases} \tag{6.33}$$

Let $L^2_{[0,\tau_0]}(E)$ be the Hilbert space of E-valued functions equipped with an inner product

$$(\varphi, \psi)_{L^2_{[0,\tau_0]}(E)} = \int_0^{\tau_0} (\varphi, \psi)_E \, dt, \quad \left(\varphi(t),\, \psi(t) \in L^2_{[0,\tau_0]}(E)\right).$$

The following lemma proves the metric conservation law for systems of the form (6.33).

Lemma 6.3.3. *If for a given continuous in E function $\psi_-(t) \in L^2_{[0,\tau_0]}(E)$ we have that a $(+)$-continuous and strongly $(\cdot)$-differentiable function $\chi(t) \in \mathcal{H}_+$ and $\psi_+(t) \in L^2_{[0,\tau_0]}(E)$ satisfy* (6.33), *then a system of the form* (6.33) *satisfies the metric conservation law*

$$2\|\chi(\tau)\|^2 - 2\|\chi(0)\|^2 = \int_0^\tau (J\psi_-, \psi_-)_E \, dt - \int_0^\tau (J\psi_+, \psi_+)_E \, dt, \quad \tau \in [0, \tau_0]. \tag{6.34}$$

Proof. Taking into account that $\mathbb{A}, \mathbb{A}^* \in [\mathcal{H}_+, \mathcal{H}_-]$, $\operatorname{Im} \mathbb{A} = KJK^*$, and $K \in [E, \mathcal{H}_-]$, $K^* \in [\mathcal{H}_+, E]$, we rely on the proof of Lemma 5.1.1. It follows from the statement of our Lemma that the function $\frac{d\chi}{dt}$ is both $(-)$- and $(\cdot)$-continuous. We will show that the function $\frac{d}{dt}(\chi(t), \chi(t))$ is continuous. We have

$$\frac{d}{dt}(\chi(t), \chi(t)) = (\chi'(t), \chi(t)) + (\chi(t), \chi'(t)).$$

Consequently, for $t, t_0 \in [0, \tau_0]$ we have

$$\begin{aligned} &|(\chi(t), \chi'(t)) - (\chi(t_0), \chi'(t_0))| \\ &= |(\chi(t), \chi'(t)) - (\chi(t_0), \chi'(t)) + (\chi(t_0), \chi'(t)) - (\chi(t_0), \chi'(t_0))| \\ &= |(\chi(t) - \chi(t_0), \chi'(t)) + (\chi(t_0), \chi'(t) - \chi'(t_0))| \\ &\le \|(\chi(t) - \chi(t_0)\|_+ \|\chi'(t)\|_- + \|\chi(t_0)\|_+ \|\chi'(t) - \chi'(t_0))\|_- \\ &\le C_1 \|(\chi(t) - \chi(t_0)\|_+ + C_2 \|\chi'(t) - \chi'(t_0))\|_-, \end{aligned}$$

where $C_1 > 0$ and $C_2 > 0$ are constants. Applying $(+)$-continuity of $\chi(t)$ and $(-)$-continuity of $\chi'(t)$, we obtain that $(\chi(t), \chi'(t))$ and hence $\frac{d}{dt}(\chi(t), \chi(t))$ is continuous. Now we can apply the algebraic steps described in detail in the proof of Lemma 5.1.1 (replacing A with $\mathbb{A}$) to obtain (6.34). □

We will refer to systems (6.31)-(6.32) as **rigged canonical systems**.

Definition 6.3.4. A rigged canonical system of the form (6.31)-(6.32) with $T \in \Lambda$ is called **the Livšic rigged canonical system** or **L-system** if there exists an imaginary part $\operatorname{Im}\mathbb{A}$ of T with $\operatorname{Im}\mathbb{A} = KJK^*$ and $\operatorname{Ran}(\operatorname{Im}\mathbb{A}) = \operatorname{Ran}(K)$.

According to Theorem 4.3.9, any $(*)$-extension $\mathbb{A}$ of a given operator $T \in \Lambda$ with a fixed imaginary part $\operatorname{Im}\mathbb{A} = KJK^*$ is defined uniquely. Thus, any L-system is well-defined. In the case when T is a bounded operator, its imaginary part is defined naturally and uniquely, and as a result we obtain a canonical system of the Livšic type described in Chapter 5.

It is more convenient to write an L-system in the form of one of two arrays:

$$\Theta = \begin{pmatrix} T & K & J \\ \mathcal{H}_+ \subset \mathcal{H} \subset \mathcal{H}_- & & E \end{pmatrix}, \tag{6.35}$$

or

$$\Theta = \begin{pmatrix} \mathbb{A} & K & J \\ \mathcal{H}_+ \subset \mathcal{H} \subset \mathcal{H}_- & & E \end{pmatrix}, \tag{6.36}$$

where

(1) $\mathbb{A}$ is a $(*)$-extension of an operator T of the class Λ;

(2) $J = J^* = J^{-1} \in [E, E], \quad \dim E < \infty$;

(3) $\operatorname{Im}\mathbb{A} = KJK^*$, where $K \in [E, \mathcal{H}_-]$, $K^* \in [\mathcal{H}_+, E]$, and

$$\operatorname{Ran}(K) = \operatorname{Ran}(\operatorname{Im}\mathbb{A}). \tag{6.37}$$

This is justified by the following statement.

Proposition 6.3.5. *Let $\dot{A}$ be a closed symmetric operator with finite and equal deficiency numbers in the Hilbert space $\mathcal{H}$, $T \in \Lambda(\dot{A})$, and $\mathbb{A}$ be an arbitrary $(*)$-extension of T. Then there exists an L-system of the form (6.36) for which $\mathbb{A}$ is the state-space operator.*

Proof. Let $\mathbf{A} = \mathcal{R}\mathbb{A}$. Then $\mathbf{A} \in [\mathcal{H}_+, \mathcal{H}_+]$ and $\operatorname{Im}\mathbf{A} = \mathcal{R}\operatorname{Im}\mathbb{A}$. Set $\mathcal{G} = \operatorname{Ran}(\operatorname{Im}\mathbf{A})$. By Theorem 5.1.2 there exist a Livsič canonical system

$$\boldsymbol{\Theta} = \begin{pmatrix} \mathbf{A} & \mathbf{K} & J \\ \mathcal{H}_+ & & E \end{pmatrix},$$

for which $\mathbf{A}$ is a state-space operator and $\mathcal{G}$ is the channel subspace, i.e.,

$$\mathbf{K} \in [E, \mathcal{H}_+],\ \mathbf{K}^* \in [\mathcal{H}_+, E],\ \operatorname{Ran}(\mathbf{K}) = \mathcal{G} = \operatorname{Ran}(\operatorname{Im}\mathbf{A}),\ \operatorname{Im}\mathbf{A} = \mathbf{K}J\mathbf{K}^*.$$

Set $K = \mathcal{R}^{-1}\mathbf{K} \in [E, \mathcal{H}_-]$. Then $\operatorname{Ran}(K) = \operatorname{Ran}(\operatorname{Im}\mathbb{A})$ and

$$\Theta = \begin{pmatrix} \mathbb{A} & K & J \\ \mathcal{H}_+ \subset \mathcal{H} \subset \mathcal{H}_- & & E \end{pmatrix}$$

is an L-system. $\square$

As in the case of canonical systems with bounded operators in Chapter 5, the operator K is called a *channel operator* and J is called a *directing operator.* A system Θ of the form (6.35)-(6.36) is called a **scattering L-system** if $J = I$.

Remark 6.3.6. Clearly, for a scattering L-systems or for an L-system with the invertible operator K the condition (6.37) is satisfied automatically.

The following theorem shows that for the case of the densely defined operator $\dot{A}$ there is a set of conditions on operator K that guarantee that a system of the form (6.31) becomes an L-system.

Theorem 6.3.7. *Let Θ be a system of the form* (6.31) *with $T \in \Lambda$, $(-i) \in \rho(T)$, $\mathrm{Dom}(\dot{A}) = \mathcal{H}$, $\ker(K) = \{0\}$, $\mathrm{Ran}(K) \subset \mathcal{F}$, where $\mathcal{F}$ is defined by* (4.69), *and $\mathrm{Dom}(T)$ parameterized by an operator M via Theorem* 4.1.9. *Then Θ is an L-system if and only if K satisfies*

$$2iKJK^* = \mathcal{R}^{-1}\begin{pmatrix} -HM + M^*H^* + iI & H - (M^*H^* + iI)M^* \\ M(HM - iI) - H^* & iI - MH + H^*M^* \end{pmatrix} P^+_{\mathfrak{M}}, \tag{6.38}$$

where

$$H = i(I - M^*M)^{-1}[(I - M^*U)(I - U^*M)^{-1}U^* - M^*], \tag{6.39}$$

$\mathcal{R}$ is a Riesz-Berezansky operator, and $U \in [\mathfrak{N}_i, \mathfrak{N}_{-i}]$ is any isometry such that $U^*M - I$ is a homeomorphism.

Proof. Let Θ be an L-system. Then there is a $(*)$-extension $\mathbb{A}$ of T whose real part contains a quasi-kernel $\hat{A}$ parameterized by von Neumann's formula (1.13) with an isometry U. Consequently there exists an operator $H \in [\mathfrak{N}_i, \mathfrak{N}_{-i}]$ such that $\mathbb{A}$ and $\mathbb{A}^*$ are determined by (4.22), (4.23). Furthermore,

$$\frac{1}{2}(S_{\mathbb{A}} + S_{\mathbb{A}^*}) = \frac{1}{2}\begin{pmatrix} -HM + M^*H^* - iI & H + (M^*H^* + iI)M^* \\ M(HM - iI) + H^* & iI - MH - H^*M^* \end{pmatrix} \tag{6.40}$$

Applying Theorem 3.4.10 and formula (3.37) to the operator matrix (6.40) we obtain the following equations:

$$\begin{aligned} -HM + M^*H^* - iI &= -[H + (M^*H^* + iI)M^*]U, \\ M(HM - iI) + H^* &= -[iI - MH - H^*M^*]U. \end{aligned} \tag{6.41}$$

If we set $\tilde{M} = U^*M$ and $\tilde{H} = HU$, then we obtain a system of operator equations

$$\begin{aligned} \tilde{H}^*(I - \tilde{M}^*) + \tilde{M}\tilde{H}(\tilde{M} - I) &= i(\tilde{M}^* - I), \\ \tilde{M}^*\tilde{H}^*(\tilde{M}^* - I) + \tilde{H}(I - \tilde{M}) &= i(I - \tilde{M}^*), \end{aligned} \tag{6.42}$$

and operator $\tilde{H}$ is its solution. Reversing the argument we conclude that $\mathrm{Re}\,\mathbb{A}$ has a self-adjoint quasi-kernel if and only if a system of operator equations (6.42) has a solution. Using simple algebraic manipulations we solve system (6.42) to obtain the solution of the form

$$Y = i(I - \tilde{M}^*\tilde{M})^{-1}[(I - \tilde{M}^*)(I - \tilde{M})^{-1} - \tilde{M}^*)]. \tag{6.43}$$

Combining definitions of $\tilde{M}$, $\tilde{H}$, (6.43), and (6.41) we obtain (6.39). In order to confirm that $U^*M - I$ is a homeomorphism we apply Theorem 4.4.4, Corollary 4.4.5, and Remark 4.3.4 in that order. Since $\dot{A}$ has finite and equal deficiency indices and is a maximal symmetric part of T and T^*, we can use the fact that $U^*M - I$ is a homeomorphism to show that $I - \tilde{M}^*\tilde{M}$ and $I - \tilde{M}$ have bounded inverses.

Conversely, let K satisfy (6.38)-(6.39) and let $U^*M - I$ be a homeomorphism. Then we have the solvability of the systems (6.42) and (6.42) which implies the existence of a self-adjoint quasi-kernel for $\mathrm{Re}\,\mathbb{A}$, where $\mathbb{A}$ is a $(*)$-extension we can construct using (4.22) and (4.23). Applying the uniqueness Theorem 4.3.9 we see that our operator T has a unique $(*)$-extension $\mathbb{A}$ whose real part has a self-adjoint quasi-kernel $\hat{A}$. Because $\ker(K) = \{0\}$, relation (6.37) is satisfied. Formulas (6.38), (4.22), and (4.23) will provide us with the condition $\mathbb{A} - \mathbb{A}^* = 2iKJK^*$ necessary to complete the Definition 6.3.4 of an L-system. □

We note that the operator U in the statement of Theorem 6.3.7 above is the parameter in the von Neumann formula (1.13) describing the self-adjoint quasi-kernel $\hat{A}$ of the real part of the $(*)$-extension $\mathbb{A}$ that defines the imaginary part of the operator T.

Taking into account Theorem 4.3.2, relations (4.45), (6.37), and following Chapter 5, we associate with an L-system Θ the operator-valued function

$$W_\Theta(z) = I - 2iK^*(\mathbb{A} - zI)^{-1}KJ, \quad z \in \rho(T), \tag{6.44}$$

which is called a **transfer operator-valued function** of the L-system Θ. Following Section 5.4 one can easily show that for the transfer operator-function of the system Θ of the form (6.36) the identities similar to (5.21), (5.25) are valid, i.e., for all $z, \zeta \in \rho(T)$,

$$\begin{aligned} W_\Theta(z)JW_\Theta^*(\zeta) - J &= 2i(\bar{\zeta} - z)K^*(\mathbb{A} - zI)^{-1}(\mathbb{A}^* - \bar{\zeta}I)^{-1}K, \\ W_\Theta^*(\zeta)JW_\Theta(z) - J &= 2i(\bar{\zeta} - z)JK^*(\mathbb{A}^* - \bar{\zeta}I)^{-1}(\mathbb{A} - zI)^{-1}KJ. \end{aligned} \tag{6.45}$$

Therefore,

$$\begin{aligned} W_\Theta^*(z)JW_\Theta(z) - J \geq 0, &\quad (\mathrm{Im}\, z > 0, z \in \rho(T)), \\ W_\Theta^*(z)JW_\Theta(z) - J = 0, &\quad (\mathrm{Im}\, z = 0, z \in \rho(T)), \\ W_\Theta^*(z)JW_\Theta(z) - J \leq 0, &\quad (\mathrm{Im}\, z < 0, z \in \rho(T)). \end{aligned} \tag{6.46}$$

Similar relations take place if we change $W_\Theta(z)$ to $W_\Theta^*(z)$ in (6.46). Thus, the transfer operator-valued function of the system Θ of the form (6.36) is J-contractive in the lower half-plane on the set of regular points of an operator T and J-unitary on real regular points of an operator T. In addition

$$W_\Theta^{-1}(z) = JW_\Theta^*(\bar{z})J, \quad \text{if} \quad z, \bar{z} \in \rho(T).$$

Let Θ be an L-system of the form (6.36). We consider the operator-valued function

$$V_\Theta(z) = K^*(\operatorname{Re}\mathbb{A} - zI)^{-1}K. \tag{6.47}$$

We note that both (6.44) and (6.47) are well defined due to Theorem 4.3.2. The transfer operator-function $W_\Theta(z)$ of the system Θ and an operator-function $V_\Theta(z)$ of the form (6.47) are connected by the relations valid for $\operatorname{Im} z \neq 0$, $z \in \rho(T)$,

$$\begin{aligned} V_\Theta(z) &= i[W_\Theta(z) + I]^{-1}[W_\Theta(z) - I]J, \\ W_\Theta(z) &= [I + iV_\Theta(z)J]^{-1}[I - iV_\Theta(z)J], \end{aligned} \tag{6.48}$$

that can be easily derived following the algebraic steps of Section 5.4. The function $V_\Theta(z)$ defined by (6.47) is called the **impedance function** of an L-system Θ of the form (6.36).

At this point we would like to extend our definition of a scalar Herglotz-Nevanlinna function that we gave in Section 6.1. An operator-function $V(z)$ in a finite-dimensional Hilbert space E is called an operator-valued **Herglotz-Nevanlinna function** if it is holomorphic in the upper and lower half-planes, $\operatorname{Im} V(z) \geq 0$ when $\operatorname{Im} z > 0$, and $V(\bar{z}) = V^*(z)$.

Let $\dot{A}$ be a symmetric operator with equal finite deficiency indices and A be its self-adjoint extension in a Hilbert space $\mathcal{H}$. Consider a system Δ of the form

$$\begin{cases} (A - zI)x = K\psi_-, \\ \psi_+ = K^*x, \end{cases} \tag{6.49}$$

where K is a bounded linear operator from a finite-dimensional Hilbert space E into $\mathcal{H}_-$, $K^* \in [\mathcal{H}_+, E]$, ψ_- is an input vector, ψ_+ is an output vector, and x is a state-space vector of the system. The rigged Hilbert triplet $\mathcal{H}_+ \subset \mathcal{H} \subset \mathcal{H}_-$ in the above system is generated by the symmetric operator $\dot{A}$. The choice of the operator K in the above definition is such that $K\psi_- \in \mathcal{B} \subset \mathcal{H}_-$, where $\mathcal{B}$ is defined by (4.69). If $z \in \rho(A)$, then we can use the density of $\mathcal{H}$ in $\mathcal{H}_-$ and therefore there is a sequence of vectors $\{\beta_n\} \subset \mathcal{H}$ that approximates $K\psi_-$ in the $(-)$-metric. In this case the state-space vector $x = \hat{R}_z(A)K\psi_- \in \mathcal{H}$ is understood as $\lim_{n\to\infty}(A - zI)^{-1}\beta_n$, where $\hat{R}_z(A)$ is the extended to $\mathcal{H}_-$ by $(-,\cdot)$-continuity resolvent $(A - zI)^{-1}$. Apply Theorem 4.5.9 to conclude that $x \in \mathcal{H}_+$. This explains the expression K^*x in the second line of (6.31).

If $z \in \rho(A)$, then we can solve the first equation of the system (6.49) for x and substitute it into the second equation. This leads to the definition of the corresponding transfer operator-valued function associated with the system Δ, that is

$$V_\Delta(z) = K^*\hat{R}_z(A)K, \qquad z \in \rho(A). \tag{6.50}$$

A system Δ of the form (6.49) is called an **impedance system** if its transfer function is Herglotz-Nevanlinna.

If Θ is an L-system of the form (6.31)-(6.36) with the state-space operator $\mathbb{A}$, then the system Δ_Θ of the form

$$\begin{cases} (\hat{A} - zI)x = K\psi_-, \\ \psi_+ = K^*x, \end{cases} \tag{6.51}$$

is called an **impedance system associated with the L-system** Θ if $\hat{A}$ is the quasi-kernel of $\operatorname{Re}\mathbb{A}$ and both operators $\mathbb{A}$ and K are elements of the system Θ. In the next section we will show that Δ_Θ is indeed an impedance system.

To justify the definitions of impedance system we present an example of a system of the form (6.49) that is not impedance.

Example. Let $\mathcal{H} = L^2_{[0,1]}$ and let

$$\dot{A}f = i\frac{df}{dt},$$
$$\operatorname{Dom}(\dot{A}) = \left\{ f(t) \in L^2_{[0,l]} \,\middle|\, f(t) - \text{abs. cont.}, f'(t) \in L^2_{[0,l]}, f(0) = f(1) = 0 \right\},$$

be a symmetric operator in $\mathcal{H}$. Its adjoint

$$\dot{A}^* f = i\frac{df}{dt},$$

is without boundary conditions and $\mathcal{H}_+ = \operatorname{Dom}(\dot{A}^*) = W_2^1$ is a Sobolev space with the scalar product

$$(f, g)_+ = \int_0^1 f(t)\overline{g(t)}\, dt + \int_0^1 f'(t)\overline{g'(t)}\, dt.$$

We construct the rigged Hilbert space $W_2^1 \subset L^2_{[0,1]} \subset (W_2^1)_-$. It is easy to see that $\overline{\operatorname{Dom}(\dot{A})} = \mathcal{H}$ and its orthogonal complement $\mathcal{F}$ is 2-dimensional. We can choose a basis for $\mathcal{F}$ by picking elements $\delta(t)$ and $\delta(t-1)$ in $(W_2^1)_-$ that generate (+)-continuous linear functionals

$$(f(t), \delta(t)) = f(0) \quad \text{and} \quad (f(t), \delta(t-1)) = f(1).$$

Let also

$$Af = i\frac{df}{dt},$$
$$\operatorname{Dom}(A) = \left\{ f(t) \in L^2_{[0,l]} \,\middle|\, f(t) - \text{abs. cont.}, f'(t) \in L^2_{[0,l]}, f(0) = -f(1) \right\},$$

be a self-adjoint extension of $\dot{A}$. Now we set $E = \mathbb{C}$, $Kc = cg(\xi)$, where $g(\xi) = \frac{1}{\sqrt{2}}[e^{-i\xi}\delta(t-1) - \delta(t)]$, and form a system Δ of the form (6.49). Then

$$R_z(A)f = ie^{-izt}\left(\int_0^1 \frac{f(s)e^{izs}ds}{1 + e^{iz}} - \int_0^t f(s)e^{izs}ds \right), \quad f \in \mathcal{H},$$

and for $f \in \mathcal{H}$ we have

$$(\hat{R}_z(A)\delta(t), f) = (\delta(t), R_z(A)f) = \overline{R_{\bar{z}}(A)f\Big|_{t=0}} = \frac{-i}{1+e^{-iz}} \int_0^1 \overline{f(s)} e^{-izs} ds$$
$$= -\frac{ie^{iz}}{1+e^{iz}} \int_0^1 e^{-izs} \overline{f(s)} ds.$$

This implies

$$\hat{R}_z(A)\delta(t) = \frac{-ie^{iz}}{1+e^{iz}} e^{-izt}.$$

Similarly, one finds that

$$\hat{R}_z(A)\delta(t-1) = \frac{ie^{iz}}{1+e^{iz}} e^{-izt}.$$

Moreover,

$$(\hat{R}_z(A)\delta(t), \delta(t)) = \frac{-ie^{iz}}{1+e^{iz}}, \quad (\hat{R}_z(A)\delta(t-1), \delta(t-1)) = \frac{i}{1+e^{iz}},$$
$$(\hat{R}_z(A)\delta(t), \delta(t-1)) = \frac{-i}{1+e^{iz}}, \quad (\hat{R}_z(A)\delta(t-1), \delta(t)) = \frac{ie^{iz}}{1+e^{iz}}.$$

Now let us set $\xi = \pi$, then we have $g(\pi) = \frac{1}{\sqrt{2}}[-\delta(t-1) - \delta(t)]$. By direct calculations using the above relations we obtain

$$V_\Delta(z)c = K^* \hat{R}_z(A)Kc = (\hat{R}_z(A)g(\pi), g(\pi))c = \frac{2i}{1+e^{iz}}c, \quad c \in \mathbb{C},$$

which is not a Herglotz-Nevanlinna function. If, however, we set $\xi = 0$, then $g(0) = \frac{1}{\sqrt{2}}[\delta(t-1) - \delta(t)]$ and the resulting system is an impedance one with

$$V_\Delta(z)c = (\hat{R}_z(A)g(0), g(0))c = -i\tanh(iz/2)c, \quad c \in \mathbb{C},$$

that is a Herglotz-Nevanlinna function.

The following theorem is similar to Theorem 6.3.7 for the case of impedance systems.

Theorem 6.3.8. *Let Δ be a system of the form (6.49) where A is a self-adjoint extension of an operator $\dot{A}$ with $\overline{\mathrm{Dom}(\dot{A})} = \mathcal{H}$ and finite equal deficiency indices, $\mathrm{Ran}(K) \subset \mathcal{F}$, where $\mathcal{F}$ is defined by (4.69). Then Δ is an impedance system associated with a scattering L-system Θ ($J = I$) if and only if*

$$2iKK^* = \mathcal{R}^{-1} \begin{pmatrix} -HM + M^*H^* + iI & H - (M^*H^* + iI)M^* \\ M(HM - iI) - H^* & iI - MH + H^*M^* \end{pmatrix} P^+_{\mathfrak{M}},$$

where M is an arbitrary contraction such that $(I - M^*M)$ is invertible,

$$H = i(I - M^*M)^{-1}[(I - M^*U)(I - U^*M)^{-1}U^* - M^*],$$

$\mathcal{R}$ is a Riesz-Berezansky operator, and $U \in [\mathfrak{N}_i, \mathfrak{N}_{-i}]$ is the parameter in the von Neumann formula (1.13) *describing A such that $U^*M - I$ is a homeomorphism.*

Most of the proof of Theorem 6.3.8 directly follows from Theorem 6.3.7. The rest of the proof becomes obvious when in the next section we will show that the impedance function of an L-system is Herglotz-Nevanlinna.

6.4 Realizable Herglotz-Nevanlinna operator-functions. Class $N(R)$

As we mentioned at the end of Section 6.1, an operator-valued Herglotz-Nevanlinna function, acting on Hilbert space E ($\dim E < \infty$) admits an integral representation

$$V(z) = Q + zX + \int_{-\infty}^{+\infty} \left(\frac{1}{t-z} - \frac{t}{1+t^2}\right) dG(t), \tag{6.52}$$

where $Q = Q^*$, $X \geq 0$ in the Hilbert space E, $G(t)$ is a non-decreasing operator-function on $(-\infty, +\infty)$ for which

$$\int_{-\infty}^{+\infty} \frac{(dG(t)f, f)_E}{1+t^2} < \infty, \quad \forall f \in E.$$

Definition 6.4.1. An operator-valued Herglotz-Nevanlinna function $V(z) \in [E, E]$, ($\dim E < \infty$) belongs to the class $N(R)$ if in the representation (6.52) we have

$$\begin{aligned} &\text{i)} \quad X = 0, \\ &\text{ii)} \quad Qf = \int_{-\infty}^{+\infty} \frac{t}{1+t^2} dG(t)f \end{aligned} \tag{6.53}$$

for all $f \in E$ such that

$$\int_{-\infty}^{+\infty} (dG(t)f, f)_E < \infty. \tag{6.54}$$

Hence the integral representation (6.52) for the class $N(R)$ becomes

$$V(z) = Q + \int_{-\infty}^{+\infty} \left(\frac{1}{t-z} - \frac{t}{1+t^2}\right) dG(t). \tag{6.55}$$

Remark 6.4.2. It can be easily proved that $V(z) \in N(R)$ if and only if

1. $(\operatorname{Im} V(i)f, f)_E \geq 0$ for all $f \in E$,

2. the equality

$$\lim_{\eta\to+\infty}\frac{\operatorname{Im}V(i\eta)}{\eta}=0,$$

holds,

3. if $f\in E\setminus\{0\}$ and $\lim\limits_{\eta\to+\infty}\eta\operatorname{Im}(V(i\eta)f,f)_E<\infty$, then $\lim\limits_{\eta\to+\infty}V(i\eta)f=0$.

Theorem 6.4.3. *Let Θ be an L-system of the form (6.35)–(6.36). Then its impedance function $V_\Theta(z)$ of the form (6.47), (6.48) belongs to the class $N(R)$. Moreover, $V_\Theta(z)$ is also the transfer function of the associated to Θ impedance system Δ_Θ of the form (6.51).*

Proof. First we will show that $V_\Theta(z)$ is a Herglotz-Nevanlinna function. Let G_z be a neighborhood of a point z, $(\operatorname{Im}z\neq 0)$ and $z,\zeta\in G_z$. Then,

$$\begin{aligned}V_\Theta(z)-V_\Theta(\zeta)&=K^*(\operatorname{Re}\mathbb{A}-zI)^{-1}K-K^*(\operatorname{Re}\mathbb{A}-\zeta I)^{-1}K\\&=(z-\zeta)K^*(\operatorname{Re}\mathbb{A}-zI)^{-1}(\operatorname{Re}\mathbb{A}-\zeta I)^{-1}K,\end{aligned}$$

and

$$\frac{V_\Theta(z)-V_\Theta(\zeta)}{z-\zeta}=K^*(\operatorname{Re}\mathbb{A}-zI)^{-1}(\operatorname{Re}\mathbb{A}-\zeta I)^{-1}K,$$

all $z,\zeta\in G_z$. Therefore, letting $z\to\zeta$ we can say that $V_\Theta(z)$ is holomorphic in G_z. Now we can conclude that $V_\Theta(z)$ is holomorphic in any one of the half-planes. It is obvious that $V^*_\Theta(z)=V_\Theta(\bar z)$. Furthermore,

$$\operatorname{Im}V_\Theta(z)=\operatorname{Im}z\,K^*(\operatorname{Re}\mathbb{A}-\bar zI)^{-1}(\operatorname{Re}\mathbb{A}-zI)^{-1}K.$$

Let now $D_z=(\operatorname{Re}\mathbb{A}-zI)^{-1}K$, then it is easy to see that the adjoint operator D_z^* is given by $D_z^*=K^*(\operatorname{Re}\mathbb{A}-\bar zI)^{-1}$. Therefore, we have $\operatorname{Im}V_\Theta(z)=(\operatorname{Im}z)D_z^*D_z$ which implies that $\operatorname{Im}V_\Theta(z)\geq 0$ when $\operatorname{Im}z>0$. Hence, we can conclude that $V_\Theta(z)$ is an operator Herglotz-Nevanlinna function and admits representation (6.52).

Now we observe that by construction the function $V_\Theta(z)$ of the form (6.47), (6.48) is the transfer function of the associated impedance system Δ_Θ of the form (6.51) whose state space operator $A=\hat A$, where $\hat A$ is the quasi-kernel of the operator $\operatorname{Re}\mathbb{A}$ of our system Θ and operator K of the system Δ_Θ is equal to the channel operator K of the system Θ.

At this point all we have to show that the conditions *i)* and *ii)* in the Definition 6.4.1 of the class $N(R)$ hold. Let $\hat E(\delta)$ be the extended canonical spectral function of the self-adjoint operator A of the system Δ_Θ. According to Theorem 3.3.8 there is a t-self-adjoint bi-extension $\mathbb{A}$ of $\dot A$ whose quasi-kernel is A. This allows us to apply Lemma 6.2.9 for all $f,g\in E$ to get

$$(V_\Theta(z)f,g)_E=\int\limits_{-\infty}^{+\infty}\left(\frac{1}{t-z}-\frac{t}{1+t^2}\right)(d\hat G(t)f,g)_E+(\hat Qf,g)_E,\tag{6.56}$$

where

$$\hat{G}(\delta) = K^* \hat{E}(\delta) K,$$

$\delta \in \aleph$ ($\aleph$ is the set of all finite intervals on the real axis), and

$$\hat{Q} = \frac{1}{2} K^* (\hat{R}_i(A) + \hat{R}_{-i}(A)) K = \frac{1}{2} [V_\Theta(-i) + V_\Theta^*(-i)].$$

From Theorem 6.2.10, we have that for all $f \in E$ with $Kf \in \mathfrak{L}$,

$$\int_{-\infty}^{\infty} (d\hat{G}(t)f, f)_E < \infty,$$

and

$$\alpha \|Kf\|_-^2 \le \int_{-\infty}^{+\infty} \frac{(d\hat{G}(t)f, f)_E}{1+t^2} \le \beta \|Kf\|_-^2, \tag{6.57}$$

where $\alpha > 0$ and $\beta > 0$ are constants. Moreover, (6.28) implies that

$$\left| \left(\hat{Q}f, g \right)_E \right| \le C \sqrt{\int_{-\infty}^{+\infty} \frac{(d\hat{G}(t)f, f)_E}{1+t^2}} \cdot \sqrt{\int_{-\infty}^{+\infty} \frac{(d\hat{G}(t)g, g)_E}{1+t^2}}. \tag{6.58}$$

By (6.56) we have for any $f, g \in E$,

$$(V_\Theta(z)f, g)_E = (\hat{Q}f, g)_E + \int_{-\infty}^{+\infty} \left(\frac{1}{t-z} - \frac{t}{1+t^2} \right) (d\hat{G}(t)f, g)_E. \tag{6.59}$$

On the other hand (6.52) implies

$$(V_\Theta(z)f, g)_E = (Qf, g)_E + z(Xf, g)_E + \int_{-\infty}^{+\infty} \left(\frac{1}{t-z} - \frac{t}{1+t^2} \right) (dG(t)f, g)_E. \tag{6.60}$$

Comparing (6.59) and (6.60) we get $(Qf, g)_E = (\hat{Q}f, g)_E$, $(Xf, g)_E = 0$, and $(G(\delta)f, g) = (\hat{G}(\delta)f, g)$ ($\delta \in \aleph$), for all $f, g \in E$. Taking into account the uniqueness of representation (6.56), we find that $X = 0$ and $G(\delta) = \hat{G}(\delta)$ ($\delta \in \aleph$). Thus,

$$V_\Theta(z) = Q + \int_{-\infty}^{+\infty} \left(\frac{1}{t-z} - \frac{t}{1+t^2} \right) dG(t). \tag{6.61}$$

Since $\hat{E}(\delta)$ coincides with $E(\delta)$ on $\mathfrak{L}$, then for any $h \in \mathcal{G}$, we have

$$\int_{-\infty}^{+\infty} (dG(t)h, h)_E < \infty, \; Kh \in \mathfrak{L}.$$

From Theorem 6.2.10 we get

$$\int_{-\infty}^{+\infty} (dG(t)h,h)_E = \infty, \;\; Kh \notin \mathfrak{L}.$$

Further, since

$$Q = \frac{1}{2}\left[V_\Theta(i) + V_\Theta(-i)\right] = \frac{1}{2}\left[K^*(\hat{R}_i(A) + \hat{R}_{-i}(A))K\right],$$

we have $\operatorname{Ran}(Q) \subseteq \operatorname{Ran}(K^*) \subseteq E$. Now formula (6.58) yields

$$|(Qf,g)_E| \leq C\|f\|_E \cdot \|g\|_E, \qquad f,g \in E.$$

On the other hand, if $Kh \in \mathfrak{L}$, then

$$\begin{aligned} Qh &= \frac{1}{2}\left[K^*((A+iI)^{-1} + (A-iI)^{-1})Kh\right] \\ &= K^* \int_{-\infty}^{+\infty} \frac{t}{1+t^2}\, dE(t)Kh = \int_{-\infty}^{+\infty} \frac{t}{1+t^2}\, dG(t)h. \end{aligned}$$

Thus we conclude that $V_\Theta(z)$ belongs to the class $N(R)$. This completes the proof. □

Corollary 6.4.4. *Let Θ be an L-system and $V_\Theta(z)$ be its impedance function. Then the channel operator K of Θ is invertible if and only if*

$$(\operatorname{Im} V_\Theta(i)f,f)_E = \int_{-\infty}^{+\infty} \frac{(dG(t)f,f)_E}{1+t^2} > 0, \quad \forall f \in E \setminus \{0\}, \tag{6.62}$$

where $G(t)$ is the measure from representation (6.61).

Proof. By Theorem 6.4.3 $V_\Theta(z) \in N(R)$ and has the representation (6.61). The proof follows directly from (6.57). □

Remark 6.4.5. Let Θ be an L-system from the statement of Theorem 6.4.3 and $V_\Theta(z)$ be its impedance function. Let

$$\mathcal{G} = \ker(K) = \{f \in E \mid Kf = 0\}, \quad \mathcal{G}^\perp = E \ominus G. \tag{6.63}$$

Suppose $h = h_1 + h_2 \in E$, $h_1 \in \mathcal{G}$, $h_2 \in \mathcal{G}^\perp$, and $K_2 := K \upharpoonright \mathcal{G}^\perp$. Then $Kh = Kh_2 = K_2h_2$ and for $u \in \mathcal{H}_+$ we have

$$\begin{aligned} (K^*u,h)_E &= (u,Kh) = (u,K_2h_2) = (K_2^*u,h_2)_E = (K_2^*u,h_1+h_2)_E \\ &= (K_2^*u,h). \end{aligned}$$

Hence, $K^* = K_2^*$ and thus

$$\begin{aligned} V_\Theta(z)h &= K^*(\operatorname{Re}\mathbb{A} - zI)^{-1}K(h_1 + h_2) = K^*(\operatorname{Re}\mathbb{A} - zI)^{-1}K_2h_2 \\ &= K_2^*(\operatorname{Re}\mathbb{A} - zI)^{-1}K_2h_2, \quad V_\Theta(z)h_1 = 0. \end{aligned}$$

Consequently, since $\mathcal{G}$ is invariant under $V_\Theta(z)$ and $V_\Theta(\bar{z})$, $V_\Theta(z)$ can be written in a diagonal block-matrix form with respect to decomposition (6.63)

$$V_\Theta(z) = \begin{pmatrix} 0 & 0 \\ 0 & V_{\Theta,2}(z) \end{pmatrix}, \qquad V_{\Theta,2}(z) = K_2^*(\operatorname{Re}\mathbb{A} - zI)^{-1}K_2, \tag{6.64}$$

where K_2 is invertible.

Theorem 6.4.6. *Let $\dot{A}$ be a closed symmetric operator with equal and finite deficiency indices, acting on a Hilbert space $\mathcal{H}$. Let $\hat{A}$ be a self-adjoint extension of $\dot{A}$ and $\mathbb{B}$ be a t-self-adjoint bi-extension of $\dot{A}$ with quasi-kernel $\hat{A}$. Suppose E is a Hilbert space, $K \in [E, \mathcal{H}_-]$, $\ker(K) = \{0\}$, and $\operatorname{Ran}(K) = \mathfrak{L}\dot{+}L_\mathbb{B}$. Define*

$$V(z) = K^*(\mathbb{B} - zI)^{-1}K.$$

Assume $J \in [E, E]$, $J = J^* = J^{-1}$, and the operator $I + iV(-i)J$ is invertible. Define the operator

$$\mathbb{A} = \mathbb{B} + iKJK^*.$$

Then $\mathbb{A}$ is a $(*)$-extension of some $T \in \Lambda(\dot{A})$, $-i \in \rho(T)$. Moreover, if we form the L-system,

$$\Theta = \begin{pmatrix} \mathbb{A} & K & J \\ \mathcal{H}_+ \subset \mathcal{H} \subset \mathcal{H}_- & & E \end{pmatrix},$$

then

$$W_\Theta(z) = (I + iV(z)J)^{-1}(I - iV(z)J),$$

for all $z \in \rho(T)$, $\operatorname{Im} z \neq 0$.

Proof. From $\ker(K) = \{0\}$, $\operatorname{Ran}(K) = \mathfrak{L}\dot{+}L_\mathbb{B}$ and (4.43),(4.44), (4.45) we get that the dimension of E is equal to the deficiency number of $\dot{A}$. The equality

$$I + iV(-i)J = J(I + iJV(-i))J,$$

implies that $I + iJV(-i)$ is invertible. Let

$$\Gamma = (I + iJV(-i))^{-1}.$$

The equality $(\mathbb{B} + iI)^{-1}KE = \mathfrak{N}_{-i}$ yields that $\ker(\mathbb{A} + iI) \subseteq \mathfrak{N}_{-i}$. Moreover, since

$$(\mathbb{A} + iI)(\mathbb{B} + iI)^{-1}K\Gamma = K\Gamma + iKJK^*(\mathbb{B} + iI)^{-1}K\Gamma = K(I + iJV(-i))\Gamma = K$$

we obtain $\ker(\mathbb{A} + iI) = \{0\}$ and $(\mathbb{A} + iI)^{-1}K = (\mathbb{B} + iI)^{-1}K\Gamma$. Let

$$\Upsilon = (I - iJV^*(-i))^{-1} = (I - iJV(i))^{-1}.$$

Similarly, it can be shown that $\ker(\mathbb{A}^* - iI) = \{0\}$ and

$$(\mathbb{A}^* - iI)^{-1}K = (\mathbb{B} - iI)^{-1}K\Upsilon.$$

Suppose that $\mathbb{B}$ is of the form (3.13), that is

$$\mathbb{B} = \dot{A}P^+_{\mathrm{Dom}(\dot{A})} + A^*P^+_{\mathfrak{M}} + \mathcal{R}^{-1}\left(S - \frac{i}{2}\mathfrak{J}\right)P^+_{\mathfrak{M}},$$

where S is a (1)-self-adjoint operator in $\mathfrak{M}$. For $\varphi \in \mathfrak{N}_i$ we have $\dot{A}^*\varphi = iP\varphi$ and hence

$$\begin{aligned}(\mathbb{A} + iI)\varphi &= \left(\dot{A}P^+_{\mathrm{Dom}(\dot{A})} + A^*P^+_{\mathfrak{M}} + \mathcal{R}^{-1}\left(S - \tfrac{i}{2}\mathfrak{J}\right)P^+_{\mathfrak{M}}\right)\varphi + i\varphi\\ &= (P\dot{A}P^+_{\mathrm{Dom}(\dot{A})} + A^*P^+_{\mathfrak{M}})\varphi + (I-P)\dot{A}P^+_{\mathrm{Dom}(\dot{A})}\varphi + i\varphi + \mathcal{R}^{-1}\left(S - \tfrac{i}{2}\mathfrak{J}\right)P^+_{\mathfrak{M}}\varphi\\ &= 2i\varphi - i(I-P)\varphi + (I-P)\dot{A}P^+_{\mathrm{Dom}(\dot{A})}\varphi + \mathcal{R}^{-1}\left(S - \tfrac{i}{2}\mathfrak{J}\right)P^+_{\mathfrak{M}}\varphi.\end{aligned}$$

The vector

$$-i(I-P)\varphi + (I-P)\dot{A}P^+_{\mathrm{Dom}(\dot{A})}\varphi + \mathcal{R}^{-1}\left(S - \frac{i}{2}\mathfrak{J}\right)P^+_{\mathfrak{M}}\varphi,$$

belongs to $\mathfrak{L}\dot{+}L_{\mathbb{B}} = \mathrm{Ran}(K)$. Hence, $(\mathbb{A} + iI)\varphi - 2i\varphi \in \mathrm{Ran}(K)$. Therefore, there exists $h \in \mathfrak{N}_{-i}$ such that

$$(\mathbb{A} + iI)\varphi - 2i\varphi = (\mathbb{A} + iI)h,$$

i.e., $(\mathbb{A} + iI)(\varphi - h) = 2i\varphi$. Hence, $(\mathbb{A} + iI)\mathcal{H}_+ \supset \mathfrak{N}_i$. Since

$$(\mathbb{A} + iI)\mathrm{Dom}(\dot{A}) = (\dot{A} + iI)\mathrm{Dom}(\dot{A}),$$

and $(\dot{A} + iI)\mathrm{Dom}(\dot{A}) \oplus \mathfrak{N}_i = \mathcal{H}$, we have $(\mathbb{A} + iI)\mathcal{H}_+ \supset \mathcal{H}$.

Similarly, one can show that $(\mathbb{A}^* - iI)\mathcal{H}_+ \supset \mathcal{H}$. Therefore using Theorem 2.1.5 we conclude that the operators $(\mathbb{A} + iI)^{-1}$ and $(\mathbb{A}^* - iI)^{-1}$ are $(-,\cdot)$-continuous. Let us define

$$\begin{aligned}\mathrm{Dom}(T) &= (\mathbb{A} + iI)^{-1}\mathcal{H}, \quad T = \mathbb{A}\restriction \mathrm{Dom}(T),\\ \mathrm{Dom}(T_1) &= (\mathbb{A}^* - iI)^{-1}\mathcal{H}, \quad T_1 = \mathbb{A}^*\restriction \mathrm{Dom}(T_1).\end{aligned} \tag{6.65}$$

It is easy to see that $\mathrm{Dom}(T)$ and $\mathrm{Dom}(T_1)$ are dense in $\mathcal{H}$ and that the operators $(\mathbb{A} + iI)^{-1}\restriction\mathcal{H}$ and $(\mathbb{A}^* - iI)^{-1}\restriction\mathcal{H}$ are $(\cdot,\cdot)$-continuous.

The points $(-i)$ and (i) are regular points for the operators T and T_1 respectively. This implies that $T_1 = T^*$. The operators T and T^* are quasi-kernels of operators $\mathbb{A}$ and $\mathbb{A}^*$, respectively, and $\mathrm{Re}\,\mathbb{A} = \mathbb{B}$ is a self-adjoint bi-extension of the operator $\dot{A}$. Let $\dim\mathfrak{N}_i = \dim\mathfrak{N}_{-i} = r$, $\dim\mathfrak{L} = d$. Then

$$\dim\mathfrak{N}'_i = \dim\mathfrak{N}'_{-i} = r - d, \quad \dim\mathfrak{N} = d, \quad \dim\mathfrak{M} = 2r - d.$$

Since $\dim \operatorname{Ran}(K) = \dim(\mathfrak{L}\dot{+}L_{\mathbb{B}}) = r$, we find that $\dim(\ker(K^*) \cap \mathfrak{M}) = r - d$. Because $\ker(K^*) \supseteq \mathfrak{J}\mathfrak{R}L_{\mathbb{B}}$ and $\dim(\mathfrak{J}\mathfrak{R}L_{\mathbb{B}}) = r - d$, we get that $\dot{A}$ is a maximal common symmetric part of T and T^*. It follows then that $T \in \Lambda(\dot{A})$.

Finally for the L-system

$$\Theta = \begin{pmatrix} & \mathbb{A} & & K & J \\ \mathcal{H}_+ \subset \mathcal{H} \subset \mathcal{H}_- & & & & E \end{pmatrix},$$

we have $V(z) = V_\Theta(z) = K^*(\operatorname{Re}\mathbb{A} - zI)^{-1}K$. Now equalities (6.48) give $W_\Theta(z) = (I + iV(z)J)^{-1}(I - iV_\Theta(z)J)$ for all non-real regular points z of T. This completes the proof. □

6.5 Realization of the class $N(R)$

In this section we establish the main realization results.

Theorem 6.5.1. *Let an operator-valued function $V(z)$ in a finite-dimensional Hilbert space E belong to the class $N(R)$ such that* (6.62) *holds. Then $V(z)$ is realizable as a transfer function of an impedance system Δ of the form* (6.49).

Proof. We will use several steps to prove this theorem.

Step 1. First we construct a model Hilbert space and a model self-adjoint operator. Let $C_{00}(E,(-\infty,+\infty))$ be the set of continuous compactly supported vector-valued functions $f(t)$ $(-\infty < t < +\infty)$ with values in a finite dimensional Hilbert space E. We introduce an inner product $(\cdot,\cdot)$ defined by

$$(f,g) = \int_{-\infty}^{+\infty} (dG(t)f(t), g(t))_E,$$

for all $f, g \in C_{00}(E,(-\infty,+\infty))$. In order to construct a Hilbert space, we identify with zero all functions $f(t)$ such that $(f,f) = 0$. Then we make the completion and obtain the new Hilbert space $L^2_G(E)$. Let us note that the set $C_{00}(E,(-\infty,+\infty))$ is dense in $L^2_G(E)$. Moreover, if $f(t)$ is continuous and

$$\int_{-\infty}^{+\infty} (dG(t)f(t), f(t))_E < \infty, \tag{6.66}$$

then $f(t)$ belongs to $L^2_G(E)$. The characterization of the Hilbert space $L^2_G(E)$ has been obtained by I.S. Kac [158]. If we set

$$\sigma(t) = \operatorname{trace}(G(t)),$$

then $\sigma(t)$ is also a non-decreasing function and, moreover, the operator measure dG is absolutely continuous with respect to the scalar measure $d\sigma$. By the Radon-Nikodym theorem there exists a $d\sigma$-measurable density $\Psi(t) \in [E,E]$, $\Psi(t) \geq 0$

almost everywhere with respect to $d\sigma$, such that

$$G(\Delta) = \int_{\Delta} \Psi(t)d\sigma(t), \ \Delta \in \aleph.$$

Let $\tilde{L}^2_G(E)$ be the set of all $d\sigma$-measurable E-valued functions f on $\mathbb{R}$ such that

$$\|f\|_0^2 := \int_{\mathbb{R}} (\Psi(t)f(t), f(t))_E d\sigma(t) < \infty. \tag{6.67}$$

If we set

$$\mathcal{N}_0 = \{f : ||f||_0 = 0\},$$

then the equality

$$L^2_G(E) = \widetilde{L}^2_G(E)/\mathcal{N}_0 \tag{6.68}$$

holds. Notice that

$$0 < \int\limits_{\mathbb{R}} \frac{(\Psi(t)h, h)_E \, d\sigma(t)}{1+t^2} < \infty,$$

for all $h \in E \setminus \{0\}$.

Let $\mathfrak{D}_0$ be the set of the continuous vector-valued (with values in E) functions $f(t)$ such that in addition to (6.66), we have

$$\int\limits_{-\infty}^{+\infty} t^2 (dG(t)f(t), f(t))_E < \infty.$$

Since $C_{00} \subset \mathfrak{D}_0$, it follows that $\mathfrak{D}_0$ is dense in $L^2_G(E)$. We introduce an operator A on $\mathfrak{D}_0$ in the following way

$$Af(t) = tf(t). \tag{6.69}$$

Below we denote again by A the closure of the symmetric operator A in (6.69). Now A is a self-adjoint operator in $L^2_G(E)$. Let $\mathfrak{H}_+ = \mathrm{Dom}(A)$ and define the inner product

$$(f, g)_{\mathfrak{H}_+} = (f, g) + (Af, Ag) \tag{6.70}$$

for all $f, g \in \mathfrak{H}_+$. It is clear that $\mathfrak{H}_+$ is a Hilbert space with norm $\|\cdot\|_{\mathfrak{H}_+}$ generated by the inner product (6.70). We equip the space $L^2_G(E)$ with spaces $\mathfrak{H}_+$ and $\mathfrak{H}_-$ and get

$$\mathfrak{H}_+ \subset L^2_G(E) \subset \mathfrak{H}_-.$$

By R we denote the corresponding Riesz-Berezansky operator, $\mathsf{R} \in [\mathfrak{H}_-, \mathfrak{H}_+]$.

Consider the following subspaces of the space E:

$$E_1 = \left\{ h \in E : \int_{-\infty}^{+\infty} (dG(t)h, h)_E < \infty \right\},$$
$$E_2 = \left\{ f \in E : \int_{-\infty}^{+\infty} \frac{(dG(t)f, h)_E}{1+t^2} = 0 \quad \text{for all} \quad h \in E_1 \right\}. \tag{6.71}$$

Clearly, the direct decomposition

$$E = E_1 \dot{+} E_2$$

holds. If $h \in E_1$, then (6.66) implies that the function $h(t) = h$ is an element of the space $L^2_G(E)$. On the other hand, if $h \in E$ and $h \notin E_1$, then $h(t)$ does not belong to $L^2_G(E)$. It can be shown that any function $h(t) = h \in E$ can be identified with an element of $\mathfrak{H}_-$. Indeed, since for all $h \in E$,

$$\int_{-\infty}^{+\infty} \frac{(dG(t)h, h)_E}{1+t^2} < \infty, \tag{6.72}$$

the function

$$\mathbf{h}(t) = \frac{h}{\sqrt{1+t^2}},$$

belongs to the space $L^2_G(E)$. Letting $f(t) \in \mathfrak{D}_0$, we have

$$\int_{-\infty}^{+\infty} (1+t^2)(dG(t)f(t), f(t))_E < \infty.$$

Therefore, the function $\mathbf{f}(t) = \sqrt{1+t^2} f(t)$ belongs to the space $L^2_G(E)$ and hence

$$(\mathbf{f}, \mathbf{h}) = \int_{-\infty}^{+\infty} (dG(t)\mathbf{f}(t), \mathbf{h}(t))_E.$$

Furthermore,

$$|(\mathbf{f}, \mathbf{h})| \leq \|\mathbf{f}\| \cdot \|\mathbf{h}\| = \sqrt{\int_{-\infty}^{+\infty} (1+t^2)(dG(t)f(t), f(t))_E} \cdot \sqrt{\int_{-\infty}^{+\infty} \frac{(dG(t)h), h)}{1+t^2}}$$
$$= \sqrt{\int_{-\infty}^{+\infty} \frac{(dG(t)h, h)}{1+t^2}} \, \|f\|_{\mathfrak{H}_+}.$$

Also,

$$\int_{-\infty}^{+\infty} (dG(t)f(t), h(t))_E = \int_{-\infty}^{+\infty} \left(\sqrt{1+t^2}dG(t)f(t), \frac{h}{\sqrt{1+t^2}}\right)_E$$

$$= \int_{-\infty}^{+\infty} (dG(t)\mathbf{f}(t), \mathbf{h}(t))_E = (\mathbf{f}, \mathbf{h}).$$

Therefore,

$$F_h(f) = \int_{-\infty}^{+\infty} (dG(t)f(t), h(t))_E,$$

is a continuous linear functional on $\mathfrak{H}_+$, for $f \in \mathfrak{D}_0$. Since $\mathfrak{D}_0$ is dense in $\mathfrak{H}_+$, $h(t) = h$ belongs to $\mathfrak{H}_-$.

We calculate the Riesz-Berezansky operator R on the vectors $h(t) = h$, $h \in E$. By the definition of R, for all $f \in \mathfrak{H}_+$ we have $(f, h) = (f, \mathsf{R}h)_{\mathfrak{H}_+}$. Hence, for all $f \in \mathfrak{D}_0$,

$$(f, h) = \int_{-\infty}^{+\infty} (dG(t)f(t), h(t))_E = \int_{-\infty}^{+\infty} (1+t^2)\left(dG(t)f(t), \frac{h(t)}{1+t^2}\right)_E$$

$$= \left(f, \frac{h(t)}{1+t^2}\right)_{\mathfrak{H}_+} = (f, \mathsf{R}h)_{\mathfrak{H}_+}.$$

Thus

$$\mathsf{R}h = \frac{h}{1+t^2}, \qquad h \in E. \tag{6.73}$$

Let us mention some properties of the operator A. It is easy to see that for all $g \in \mathfrak{H}_+$, we have that $\|Ag\| \leq \|g\|_{\mathfrak{H}_+}$. Taking this into account we obtain

$$\|Af\|_{\mathfrak{H}_-} = \sup_{g \in \mathfrak{H}_+} \frac{|(Af, g)|}{\|g\|_{\mathfrak{H}_+}} = \sup_{g \in \mathfrak{H}_+} \frac{|(f, Ag)|}{\|g\|_{\mathfrak{H}_+}} \leq \sup_{g \in \mathfrak{H}_+} \frac{\|f\| \cdot \|Ag\|}{\|g\|_{\mathfrak{H}_+}} \leq \|f\|.$$

Hence, the operator A is $(\cdot, -)$-continuous. Denote by A the extension of the operator A to $\mathcal{H}$ with respect to $(\cdot, -)$-continuity. Now,

$$(A - zI)^{-1}g - (A - \zeta I)^{-1}g = (z - \zeta)(A - zI)^{-1}(A - \zeta I)^{-1}g,$$

holds for all $g \in \mathfrak{H}_-$ and in particular

$$(A - iI)^{-1}g - (A + iI)^{-1}g = 2i(A - iI)^{-1}(A + iI)^{-1}g,$$

and

$$\|(A - iI)^{-1}g\|^2 = \|(A + iI)^{-1}g\|^2,$$

for all g in $\mathfrak{H}_-$. It follows from (6.72) that the element

$$f(t) = \frac{f}{t-z}, \quad f \in E,$$

belongs to the space $L^2_G(E)$. It is easy to show that, for all $h \in E$,

$$(A - zI)^{-1}h = \frac{h}{t-z}, \qquad (\mathrm{Im} z \neq 0). \tag{6.74}$$

Step 2. Now we are going to construct a model symmetric operator and its adjoint. Let $\mathfrak{H}_+$ be the Hilbert space constructed in **Step 1** and let

$$\mathrm{Dom}(\dot{A}) = \{f \in \mathfrak{H}_+ : (f,h) = 0 \quad \text{for all} \quad h \in E\} = \mathfrak{H}_+ \ominus \mathsf{R}E, \tag{6.75}$$

where by $\ominus$ we mean orthogonality in $\mathfrak{H}_+$. We define an operator $\dot{A}$ on $\mathrm{Dom}(\dot{A})$ by the following expression:

$$\dot{A} = A \upharpoonright \mathrm{Dom}(\dot{A}). \tag{6.76}$$

Obviously, $\dot{A}$ is a closed symmetric operator. We note that if $E_1 = 0$, then $\mathrm{Dom}(\dot{A})$ is dense in $L^2_G(E)$. Define $\mathcal{H}_0 = \overline{\mathrm{Dom}(\dot{A})}$ and let P be the orthogonal projection in $\mathcal{H} = L^2_G(E)$ onto $\mathcal{H}_0$. Since $\mathrm{codim}\,\mathrm{Dom}(\dot{A}) = \dim E_1 < \infty$, Theorem 2.4.1 and Theorem 4.1.3 yield that operators $P\dot{A}$ and PA are closed, i.e., $\dot{A}$ is a regular symmetric operator and A is its regular self-adjoint extension. Define

$$A_1 = A \upharpoonright \mathrm{Dom}(A_1), \qquad \mathrm{Dom}(A_1) = \mathfrak{H}_+ \ominus \mathsf{R}E_1.$$

The following obvious inclusions hold: $\dot{A} \subset A_1 \subset A$. It is easy to see that $\mathrm{Dom}(A_1) = \mathrm{Dom}(\dot{A}) \oplus \mathsf{R}E_2$, $\overline{\mathrm{Dom}(A_1)} = \mathcal{H}_0$ and A_1 is a closed symmetric operator. Indeed, if we identify the space E with the space of functions $h(t) = h$, $h \in E$, then we would obtain $L^2_G(E) \ominus \mathcal{H}_0 = E_1$. Since

$$\int\limits_{-\infty}^{+\infty} \frac{(dG(t)h, g)_E}{1+t^2} = 0,$$

and

$$\mathsf{R}g = \frac{g}{1+t^2},$$

for all $h \in E_1$, $g \in E_2$, we find that E_1 is $(\cdot)$-orthogonal to $\mathsf{R}E_2$ and hence $\overline{\mathrm{Dom}(A_1)} = \mathcal{H}_0$. We denote by A_1^* the adjoint of the operator A_1.

Now we are going to find the defect subspaces $\mathfrak{N}_i$ and $\mathfrak{N}_{-i}$ of the operator $\dot{A}$. Since the subspace $E \in \mathfrak{H}_-$ is $(\cdot)$-orthogonal to $\mathrm{Dom}(\dot{A})$, we have that $(A \pm iI)^{-1}E = \mathfrak{N}_{\pm i}$. Moreover, by (6.74) we have

$$(A \pm iI)^{-1}h = \frac{h}{t \pm i}, \qquad h \in E.$$

Therefore

$$\mathfrak{N}_{\pm i} = \left\{ f(t) \in L^2_G(E),\ f(t) = \frac{h}{t \pm i}, \quad h \in E \right\}.$$

Similarly, the defect subspaces of the operator A_1 are

$$\mathfrak{N}^0_{\pm i} = \left\{ f(t) \in L^2_G(E),\ f(t) = \frac{h}{t \pm i},\ h \in E_1 \right\}.$$

Obviously, $\mathfrak{N}^0_z \subset \mathfrak{D}_0$ because

$$\int\limits_{-\infty}^{+\infty} \frac{t^2}{|t-z|^2}(dG(t)h,h)_E \le K(z) \int\limits_{-\infty}^{+\infty} (dG(t)h,h)_E < \infty, \quad h \in E_1.$$

Taking into account that

$$\mathrm{Dom}(A_1^*) = \mathrm{Dom}(\dot{A}) \dotplus \mathfrak{N}^0_i \dotplus \mathfrak{N}^0_{-i},$$

we can conclude that $\mathrm{Dom}(A_1^*) \subseteq \mathrm{Dom}(A)$. At the same time, the inclusion $A_1 \subset A$ implies that $\mathrm{Dom}(A_1^*) \supset \mathrm{Dom}(A)$. Therefore we obtain $\mathrm{Dom}(A_1^*) = \mathrm{Dom}(A)$ and $PA = A_1^*$.

Since A is the self-adjoint extension of operator $\dot{A}$ we find by (1.33) that

$$\mathrm{Dom}(A) = \mathrm{Dom}(\dot{A}) \dotplus (I-U)\mathfrak{N}_i,$$

for some admissible isometric operator U acting from $\mathfrak{N}_i$ into $\mathfrak{N}_{-i}$. It is easy to check that $U(A-iI)^{-1}h = (A+iI)^{-1}h$, for all h in E. Consequently, the operator U has the form

$$U\left(\frac{h}{t-i}\right) = \frac{h}{t+i}, \quad h \in E.$$

Straightforward calculations show that

$$A(I-U)\left(\frac{h}{t-i}\right) = t\frac{h}{t-i} - t\frac{h}{t+i} = \frac{2it}{t^2+1}h.$$

Let $\dot{A}^*$ be the adjoint to $\dot{A}$. The linear manifold $\mathcal{H}_+ = \mathrm{Dom}(\dot{A}^*)$ becomes a Hilbert space with the inner product

$$(x,y)_+ = (x,y) + (\dot{A}^*x, \dot{A}^*y), \quad x,y \in \mathcal{H}_+.$$

Let $\mathcal{H}_+ \subset L^2_G(E) \subset \mathcal{H}_-$ be the rigged triplet and let $\mathcal{R}$ be the corresponding Riesz-Berezansky operator. Since PA is a closed operator, then $\mathfrak{H}_+$ is a subspace of $\mathcal{H}_+$. By Theorem 2.5.7, $\mathcal{H}_+ = \mathrm{Dom}(A) \dotplus (U+I)\hat{\mathfrak{N}}_i$, where

$$\hat{\mathfrak{N}}_i = \{\varphi \in \mathfrak{N}_i,\ (U-I)\varphi \in \mathcal{H}_0\}, \tag{6.77}$$

$\mathcal{H}_0 = \overline{\mathrm{Dom}(\dot{A})}$, and U is an admissible isometric operator related to A. Taking into account that

$$(U-I)\left(\frac{h}{t-i}\right) = \frac{-2ih}{t^2+1}, \quad h \in E,$$

we conclude that

$$\hat{\mathfrak{N}}_i = \left\{\frac{g}{t-i}, \; g \in E_2 = E \ominus E_1\right\}.$$

Therefore,

$$\mathrm{Dom}(\dot{A}^*) = \mathrm{Dom}(A) \dotplus \left\{\frac{tg}{t^2+1}\right\}, \quad g \in E_2.$$

Step 3. In this step we will construct a special self-adjoint bi-extension whose quasi-kernel coincides with the operator A. Applying (2.15), we have

$$\mathcal{H}_+ = \mathrm{Dom}(\dot{A}) \oplus \mathfrak{N}'_i \oplus \mathfrak{N}'_{-i} \oplus \mathfrak{N},$$

where $\mathfrak{N}'_{\pm i}$ are semi-deficiency spaces of the operator $\dot{A}$, $\mathfrak{N} = \mathcal{R}E_1$, and

$$\overline{\mathrm{Dom}(\dot{A})} \oplus E_1 = \mathcal{H} = L^2_G(E).$$

We begin by setting

$$(f,g)_1 = (f,g)_+ + (P^+_{\mathfrak{N}} f, P^+_{\mathfrak{N}} g)_+, \quad \text{for all} \;\; f,g \in \mathcal{H}_+.$$

Here $P^+_{\mathfrak{N}}$ is an orthoprojection of $\mathcal{H}_+$ onto $\mathfrak{N}$. Obviously, the norm $\|\cdot\|_1$ is equivalent to $\|\cdot\|_+$. We denote by $\mathcal{H}_{+1}$ the space $\mathcal{H}_+$ equipped with the norm $\|\cdot\|_1$, so that $\mathcal{H}_{+1} \subset \mathcal{H} \subset \mathcal{H}_{-1}$ is the corresponding rigged space with Riesz-Berezansky operator $\mathfrak{R}$. By Theorem 2.5.5 there exists a (1)-isometric operator $\mathcal{U}$ such that

$$\mathrm{Dom}(A) = \mathrm{Dom}(\dot{A}) \oplus (\mathcal{U}+I)(\mathfrak{N}'_i \oplus \mathfrak{N}),$$

where $\mathrm{Dom}(\mathcal{U}) = \mathfrak{N}'_i \oplus \mathfrak{N}$, $\mathrm{Ran}(\mathcal{U}) = \mathfrak{N}'_{-i} \oplus \mathfrak{N}$ and (-1) is a regular point for the operator $\mathcal{U}$. (For further convenience we have substituted $\mathcal{U}$ in the statement of Theorem 2.5.5 by $-\mathcal{U}$). Moreover,

$$\begin{cases} \varphi = i(I + P^+_{\mathfrak{N}'_i})(\dot{A}^* + iI)^{-1}\chi, \\ \mathcal{U}\varphi = i(I + P^+_{\mathfrak{N}'_{-i}})(\dot{A}^* - iI)^{-1}U\chi, \\ \varphi \in \mathrm{Dom}(\mathcal{U}), \; \chi \in \mathfrak{N}_i. \end{cases}$$

Here U is the isometric operator described in (6.77) of **Step** 2. Consequently we obtain

$$\begin{cases} \chi = -\frac{i}{2}(\dot{A}^* + iI)(I + P^+_{\mathfrak{N}})\varphi, \\ U\chi = -\frac{i}{2}(\dot{A}^* - iI)(I + P^+_{\mathfrak{N}})\mathcal{U}\varphi, \\ \varphi \in \mathrm{Dom}(\mathcal{U}), \; \chi \in \mathfrak{N}_i. \end{cases}$$

It follows that

$$\chi - U\chi = \varphi + \mathcal{U}\varphi + iA^* P^+_{\mathfrak{N}}(\mathcal{U} - I)\varphi,$$
$$\chi + U\chi = \varphi - \mathcal{U}\varphi - iA^* P^+_{\mathfrak{N}}(I + \mathcal{U})\varphi.$$

Since $\chi - U\chi = \varphi + \mathcal{U}\varphi + iA^* P^+_{\mathfrak{N}}(\mathcal{U} - I)\varphi$, we find that $\chi - U\chi \in \mathcal{H}_0$ if and only if $P^+_{\mathfrak{N}}(\mathcal{U} + I)\varphi = 0$. This follows from the fact that $\dot{A}^* P^+_{\mathfrak{N}}(\mathcal{U} - I)\varphi \in \mathrm{Dom}(\dot{A}) \subset \mathcal{H}_0$ and from the direct decomposition $\mathcal{H} = \mathcal{H}_0 \dotplus \mathfrak{N}$ (see Proposition 2.4.2). Let us note that if $P^+_{\mathfrak{N}}(\mathcal{U} + I)\varphi = 0$, then $\chi + U\chi = \varphi - \mathcal{U}\varphi$. Thus,

$$\hat{\mathfrak{N}}_i = \{f = (\dot{A}^* + iI)(I + P^+_{\mathfrak{N}})\varphi,\ \ P^+_{\mathfrak{N}}(\mathcal{U} + I)\varphi = 0\}.$$

Let

$$N = \ker P^+_{\mathfrak{N}}(I + \mathcal{U}).$$

Then we have

$$\mathcal{H}_+ = \mathrm{Dom}(A) \dotplus (I - \mathcal{U})N. \tag{6.78}$$

We denote by $\mathcal{P}$ the projection operator of $\mathcal{H}_+$ onto $\mathrm{Dom}(A)$ with respect to the decomposition (6.78). Let $\mathcal{P} = I - \mathcal{P}$. Since $\mathrm{Dom}(A) = \mathfrak{H}_+$, we have $\mathcal{P} \in [\mathcal{H}_+, \mathfrak{H}_+]$. We will denote by $\mathcal{P}^* \in [\mathfrak{H}_-, \mathcal{H}_-]$ the adjoint operator to $\mathcal{P}$, i.e., $(\mathcal{P}f, g) = (f, \mathcal{P}^* g)$, for all $f \in \mathcal{H}_+$, $g \in \mathcal{H}_-$. If $\phi \in \hat{\mathfrak{N}}_i$, then $\phi + U\phi = (I - \mathcal{U})\varphi$, for $\varphi \in N$, and

$$\begin{aligned}\dot{A}^*(I - \mathcal{U})\varphi &= iP^+_{\mathfrak{N}'_i}\varphi + iP^+_{\mathfrak{N}'_{-i}}\mathcal{U}\varphi + AP^+_{\mathfrak{N}}(I - \mathcal{U})\varphi\\ &= i(\mathcal{U} + I)\varphi + \dot{A}^* P^+_{\mathfrak{N}}(I - \mathcal{U})\varphi.\end{aligned}$$

This implies $\dot{A}^*(I + U)\phi = i(\phi - U\phi)$. Hence

$$\dot{A}^*\left(\frac{tg}{t^2 + 1}\right) = -\frac{g}{t^2 + 1}, \quad g \in E_2. \tag{6.79}$$

Let Q be the operator in the definition of the class $N(R)$. We introduce a new operator Φ acting in the following way:

$$\Phi f = -Q\,\mathsf{R}^{-1}\dot{A}^*\mathcal{P}f, \quad f \in \mathcal{H}_+. \tag{6.80}$$

In order to show that $\Phi \in [\mathcal{H}_+, E]$, we consider the following calculation for $f \in \mathcal{H}_+$:

$$\begin{aligned}\|\Phi f\|_E &= \sup_{g \in E}\frac{|(\Phi f, g)_E|}{\|g\|_E} = \sup_{g \in E}\frac{|(Q\mathsf{R}^{-1}\dot{A}^*\mathcal{P}f, g)_E|}{\|g\|_E}\\ &= \sup_{g \in E}\frac{|(\mathsf{R}^{-1}\dot{A}^*\mathcal{P}f, Qg)_E|}{\|g\|_E} \le \sup_{g \in E}\frac{\|\mathsf{R}^{-1}\dot{A}^*\mathcal{P}f\|_E \cdot \|Qg\|_E}{\|g\|_E}\\ &\le \beta\|\dot{A}^*\mathcal{P}f\|_{\mathfrak{H}_+} \le \alpha\|\dot{A}^*\mathcal{P}f\|_{\mathcal{H}_+}, \quad \alpha, \beta \text{ - positive constants.}\end{aligned}$$

Here we used that $\mathcal{P}f \subset \mathrm{Dom}(A)$, for all $f \in \mathfrak{H}_+$, formulas (6.73) and (6.79), and the equivalence of the norms $\|\cdot\|_{\mathfrak{H}_+}$ and $\|\cdot\|_+ = \|\cdot\|_{\mathcal{H}_+}$. For $f \in \mathcal{H}_+$, we have $\mathcal{P}f = (I-\mathcal{U})\varphi$, $\varphi \in N$ and

$$\dot{A}^*\mathcal{P}f = i(\mathcal{U}+I)\varphi + iA^*P^+_{\mathfrak{N}}(\mathcal{U}-I)\varphi.$$

Hence

$$\begin{aligned}\|\dot{A}^*P^+_{\mathfrak{N}}(\mathcal{U}-I)\varphi\|_+^2 &= \|\dot{A}^*P^+_{\mathfrak{N}}(\mathcal{U}-I)\varphi\|^2 + \|\dot{A}^*\dot{A}^*P^+_{\mathfrak{N}}(\mathcal{U}-I)\varphi\|^2\\ &= \|\dot{A}^*P^+_{\mathfrak{N}}(\mathcal{U}-I)\varphi\|^2 + \|PP^+_{\mathfrak{N}}(\mathcal{U}-I)\varphi\|^2\\ &\le \|\dot{A}^*P^+_{\mathfrak{N}}(\mathcal{U}-I)\varphi\|^2 + \|P^+_{\mathfrak{N}}(\mathcal{U}-I)\varphi\|^2\\ &= \|P^+_{\mathfrak{N}}(\mathcal{U}-I)\varphi\|_+^2,\end{aligned}$$

and

$$\begin{aligned}\|i(\mathcal{U}+I)\varphi + i\dot{A}^*P^+_{\mathfrak{N}}(\mathcal{U}-I)\varphi\|_+^2 &= \|\dot{A}^*P^+_{\mathfrak{N}}(\mathcal{U}-I)\varphi\|_+^2 + \|\varphi+\mathcal{U}\varphi\|_+^2\\ \le \|P^+_{\mathfrak{N}}(\mathcal{U}-I)\varphi\|_+^2 + \|\varphi+\mathcal{U}\varphi\|_+^2 &= \|\varphi-\mathcal{U}\varphi\|_+^2.\end{aligned}$$

This implies that there exists a constant $a>0$ such that

$$\|\dot{A}^*\mathcal{P}f\| \le \|\mathcal{P}f\|_+ \le a\|f\|_+, \quad \forall f \in \mathcal{H}_+.$$

Therefore, for some constant $b>0$ we have $\|\Phi f\|_{\le} b\|f\|_+$, $\forall f \in \mathcal{H}_+$. Thus, $\Phi \in [\mathcal{H}_+, E]$.

Let Φ^* be the adjoint operator to Φ, i.e., $\Phi^* \in [E, \mathcal{H}_-]$ and for all $f \in \mathcal{H}_+$, $g \in E$, $(\Phi f, g)_E = (f, \Phi^* g)$. Since $\Phi(\mathrm{Dom}(A)) = 0$, $\mathrm{Ran}(\Phi^*)$ is $(\cdot)$-orthogonal to $\mathrm{Dom}(A)$. Letting $\mathfrak{M} = \mathfrak{N}'_{-i} \oplus \mathfrak{N}'_i \oplus \mathfrak{N}$ and using (6.78) we obtain

$$\mathfrak{M} = (\mathcal{U}+I)(\mathfrak{N}'_i \oplus \mathfrak{N}) \dot{+} (I-\mathcal{U})N.$$

In the space $\mathfrak{M}$ we define an operator S in the following way:

$$\begin{aligned}S(\varphi+\mathcal{U}\varphi) &= \frac{i}{2}(I-\mathcal{U})\varphi, \quad \varphi \in \mathfrak{N}'_i \oplus \mathfrak{N},\\ S(\psi-\mathcal{U}\psi) &= \left[-\mathcal{R}(\Phi^*+\mathcal{P}^*)\mathrm{R}^{-1}\dot{A}^* + \frac{i}{2}(P^+_{\mathfrak{N}'_i} - P^+_{\mathfrak{N}'_{-i}})\right](\psi-\mathcal{U}\psi),\end{aligned} \tag{6.81}$$

where $\psi \in N$. In order to show that S is a (1)-self-adjoint operator on $\mathfrak{M}$ we must verify the equalities

(**a**) $(S(\varphi+\mathcal{U}\varphi), \varphi+\mathcal{U}\varphi)_1 = (\varphi+\mathcal{U}\varphi, S(\varphi+\mathcal{U}\varphi))_1, \quad \varphi \in \mathfrak{N}'_i \oplus \mathfrak{N}$,
(**b**) $(S(\psi-\mathcal{U}\psi), \psi-\mathcal{U}\psi)_1 = (\psi-\mathcal{U}\psi, S(\psi-\mathcal{U}\psi))_1, \quad \psi \in N$,
(**c**) $(S(\varphi+\mathcal{U}\varphi), \psi-\mathcal{U}\psi)_1 = (\varphi+\mathcal{U}\varphi, S(\psi-\mathcal{U}\psi))_1 \quad \varphi \in \mathfrak{N}'_i \oplus \mathfrak{N},\ \psi \in N$.

The equality (**a**) easily follows from the (1)-unitary property of the operator $\mathcal{U}$. Let us check (**b**). Since $P^+_{\mathfrak{N}}(\psi+\mathcal{U}\psi) = 0$, for $\psi \in N$ we get

$$\left(P^+_{\mathfrak{N}'_i} - P^+_{\mathfrak{N}'_{-i}}\right)(\psi-\mathcal{U}\psi) = \psi+\mathcal{U}\psi, \quad \psi \in N$$

and $(\psi+\mathcal{U}\psi),\psi-\mathcal{U}\psi)_1=0$. Since $\mathcal{P}(I-\mathcal{U})N=0$, we have

$$(\mathcal{R}\mathcal{P}^*\mathsf{R}^{-1}\dot{A}^*(\psi-\mathcal{U}\psi),\psi-\mathcal{U}\psi)=(\mathsf{R}^{-1}\dot{A}^*(\psi-\mathcal{U}\psi),\mathcal{P}(\psi-\mathcal{U}\psi))=0.$$

This allows us to consider only the term of (6.81) that contains Φ^*. Because Q is a self-adjoint operator in E, we have

$$\begin{aligned}
&(S(\psi-\mathcal{U}\psi),\psi-\mathcal{U}\psi)_1=(-\mathcal{R}\Phi^*\mathsf{R}^{-1}\dot{A}^*(\psi-\mathcal{U}\psi),(\psi-\mathcal{U}\psi)_1\\
&=(\mathsf{R}^{-1}\dot{A}^*(\psi-\mathcal{U}\psi),-\Phi(\psi-\mathcal{U}\psi))_E\\
&=(\mathsf{R}^{-1}\dot{A}^*(\psi-\mathcal{U}\psi),Q\mathsf{R}^{-1}\dot{A}^*\mathcal{P}(\psi-\mathcal{U}\psi))_E\\
&=(Q\mathsf{R}^{-1}\dot{A}^*(\psi-\mathcal{U}\psi),\mathsf{R}^{-1}\dot{A}^*(\psi-\mathcal{U}\psi))_E\\
&=((\psi-\mathcal{U}\psi),-\Phi^*\mathcal{R}^{-1}\dot{A}^*(\psi-\mathcal{U}\psi))_E\\
&=((\psi-\mathcal{U}\psi),-\mathcal{R}\Phi^*\mathcal{R}^{-1}\dot{A}^*(\psi-\mathcal{U}\psi))_1\\
&=((\psi-\mathcal{U}\psi),S(\psi-\mathcal{U}\psi))_1.
\end{aligned}$$

Now we will show that (**c**) holds. The relation $P^+_{\mathfrak{N}}(\psi+\mathcal{U}\psi)=0$ implies $P^+_{\mathfrak{N}}\psi=-P^+_{\mathfrak{N}}\mathcal{U}\psi$. Also, $(\varphi,\psi)_1=(\mathcal{U}\varphi,\mathcal{U}\psi)_1$, since $\mathcal{U}$ is a (1)-isometric mapping. Hence,

$$\begin{aligned}
(S(\varphi+\mathcal{U}\varphi),\psi-\mathcal{U}\psi)_1&=\frac{i}{2}((I-\mathcal{U})\varphi,\psi-\mathcal{U}\psi)_1\\
&=i(\varphi,\psi)_1-\frac{i}{2}(\varphi,\mathcal{U}\psi)_1-\frac{i}{2}(\mathcal{U}\varphi,\psi)_1\\
&=i(\varphi,\psi)_1-\frac{i}{2}(\varphi,P^+_{\mathfrak{N}}\mathcal{U}\psi)_1-\frac{i}{2}(\mathcal{U}\varphi,P^+_{\mathfrak{N}}\psi)_1\\
&=i(\varphi,\psi)_1+\frac{i}{2}(P^+_{\mathfrak{N}}(I-\mathcal{U})\varphi,\psi)_1.
\end{aligned}$$

Also, note that

$$\left(\varphi+\mathcal{U}\varphi,\frac{i}{2}(P^+_{\mathfrak{N}'_i}-P^+_{\mathfrak{N}'_{-i}})(\psi-\mathcal{U}\psi)\right)_1=-\frac{i}{2}\left(\varphi+\mathcal{U}\varphi,\psi+\mathcal{U}\psi\right)_1,$$

and

$$\begin{aligned}
(\varphi+\mathcal{U}\varphi,S(\psi-\mathcal{U}\psi)_1&=(\varphi+\mathcal{U}\varphi,-\mathcal{R}(\Phi^*+\mathcal{P}^*)\mathsf{R}^{-1}\dot{A}^*(\psi-\mathcal{U}\psi))_1\\
&\quad-\frac{i}{2}(\varphi+\mathcal{U}\varphi,\psi+\mathcal{U}\psi)_1.
\end{aligned}$$

Next, recall that $\operatorname{Ran}(\Phi^*)$ is $(\cdot)$-orthogonal to $\operatorname{Dom}(A)$ and

$$\varphi+\mathcal{U}\varphi\in\operatorname{Dom}(A)=\operatorname{Dom}(\dot{A})\oplus(\mathcal{U}+I)(\mathfrak{N}'_i\oplus\mathfrak{N}).$$

Moreover,

$$\begin{aligned}
&\dot{A}^*(\psi-U\psi)=iP^+_{\mathfrak{N}'_i}\psi+iP^+_{\mathfrak{N}'_{-i}}\mathcal{U}\psi+\dot{A}^*P^+_{\mathfrak{N}}(\psi-U\psi)\\
&=i(\psi+\mathcal{U}\psi)+\dot{A}^*P^+_{\mathfrak{N}}(\psi-U\psi)\in\operatorname{Dom}(A).
\end{aligned}$$

Thus,

$$(\varphi + \mathcal{U}\varphi, \mathfrak{R}\Phi^*\mathrm{R}^{-1}\dot{A}^*(\psi - \mathcal{U}\psi))_1 = (\varphi + \mathcal{U}\varphi, \Phi^*\mathrm{R}^{-1}\dot{A}^*(\psi - \mathcal{U}\psi)) = 0,$$

$$\begin{aligned}(\varphi + \mathcal{U}\varphi, -\mathfrak{R}\mathcal{P}^*\mathrm{R}^{-1}\dot{A}^*(\psi - \mathcal{U}\psi))_1 &= -(\varphi + \mathcal{U}\varphi, \dot{A}^*(\psi - \mathcal{U}\psi))_{\mathfrak{H}_+}\\ &= -(\varphi + \mathcal{U}\varphi, \dot{A}^*(\psi - \mathcal{U}\psi)) - (A(\varphi + \mathcal{U}\varphi), A\dot{A}^*(\psi - \mathcal{U}\psi)).\end{aligned}$$

Applying Theorem 2.5.5 we obtain

$$A(\varphi + \mathcal{U}\varphi) = \dot{A}^*(\varphi + \mathcal{U}\varphi) + i(\dot{A}\dot{A}^* + I)P^+_{\mathfrak{N}}(I - \mathcal{U})\varphi,$$

$$\dot{A}^*(\psi - \mathcal{U}\psi) = i(I + \mathcal{U})\psi + \dot{A}^*P^+_{\mathfrak{N}}(I - \mathcal{U})\psi,$$

$$\begin{aligned}A\dot{A}^*(\psi - \mathcal{U}\psi) &= \dot{A}\dot{A}^*P^+_{\mathfrak{N}}(I - \mathcal{U})\psi + i\dot{A}^*(\mathcal{U} + I)\psi - (\dot{A}\dot{A}^* + I)P^+_{\mathfrak{N}}(I - \mathcal{U})\psi\\ &= i\dot{A}^*(\mathcal{U} + I)\psi - P^+_{\mathfrak{N}}(I - \mathcal{U})\psi.\end{aligned}$$

The above identities yield

$$(\varphi + \mathcal{U}\varphi, \dot{A}^*(\psi - \mathcal{U}\psi))_{\mathfrak{H}_+} = (\varphi + \mathcal{U}\varphi, i(\psi + \mathcal{U}\psi))_1 - i(P^+_{\mathfrak{N}}(I - \mathcal{U})\varphi, \psi)_1.$$

Hence,

$$\begin{aligned}(\varphi + \mathcal{U}\varphi, -\mathfrak{R}\mathcal{P}^*\mathrm{R}^{-1}\dot{A}^*(\psi - \mathcal{U}\psi)) &= i(\varphi + \mathcal{U}\varphi, \psi + \mathcal{U}\psi)_1 + i(P^+_N(I - \mathcal{U})\varphi, \psi)_1\\ \left(\varphi + \mathcal{U}\varphi, \frac{i}{2}(\psi + \mathcal{U}\psi)\right)_1 &= -\frac{i}{2}(\varphi + \mathcal{U}\varphi, \psi + \mathcal{U}\psi)_1,\end{aligned}$$

and

$$\begin{aligned}&(\varphi + \mathcal{U}\varphi, S(\psi - \mathcal{U}\psi))_1\\ &= i(\varphi + \mathcal{U}\varphi, \psi + \mathcal{U}\psi)_1 + i(P^+_{\mathfrak{N}}(I - \mathcal{U})\varphi, \psi)_1 - \frac{i}{2}(\varphi + \varphi, \psi + \mathcal{U}\psi)_1\\ &= i(\varphi, \psi)_1 + \frac{i}{2}(\mathcal{U}\varphi, \psi)_1 + \frac{i}{2}(\varphi, \mathcal{U}\psi)_1 + i(P^+_{\mathfrak{N}}(I - \mathcal{U})\varphi, \psi)_1\\ &= i(\varphi, \psi)_1 + \frac{i}{2}(P^+_{\mathfrak{N}}(I - \mathcal{U})\varphi, \psi)_1\\ &= (S(\varphi + \mathcal{U}\varphi), \psi - \mathcal{U}\psi)_1.\end{aligned}$$

This shows that S is a (1)-self-adjoint operator in $\mathfrak{M}$. Applying (3.13) we obtain that a self-adjoint bi-extension of the operator $\dot{A}$ is defined by the formula

$$\mathbb{B} = \dot{A}P^+_{\mathrm{Dom}(\dot{A})} + \left(\dot{A}^* + \mathfrak{R}^{-1}(S - \frac{i}{2}\mathfrak{J})\right)P^+_{\mathfrak{M}}, \tag{6.82}$$

where S is given by (6.81). Obviously, if $f = g + (\mathcal{U} + I)\varphi$, $\varphi \in \mathfrak{N}'_i \oplus \mathfrak{N}$, and $g \in \mathrm{Dom}(\dot{A})$, then $\mathbb{B}f = Af$. This means that the quasi-kernel of the operator $\mathbb{B}$ coincides with A and hence $\mathbb{B}$ is t-self-adjoint.

Step 4. Define a linear operator $K \in [E, \mathcal{H}_-]$ by the formula

$$Kh = (\mathcal{P}^* + \Phi^*)P_{E_2}h + \hat{I}P_{E_1}h, \quad h \in E, \tag{6.83}$$

where P_{E_2} and P_{E_1} are projections of the space E onto E_2 and E_1 in (6.71), respectively, and $\hat{I}$ is an embedding of E into $\mathcal{H}_-$. Let z be a regular point of the operator A. By Theorem 4.5.12 we have $\hat{R}_z = \overline{(\mathbb{B} - zI)^{-1}}$. Also, note that

$$(\hat{R}_z f, g) = (f, (A - \bar{z}I)^{-1}g), \quad \forall f \in \mathcal{H}_-, \, g \in \mathcal{H}.$$

As it was shown in **Step 1** (see (6.74))

$$(A - zI)^{-1}h = \frac{h}{t-z}, \quad \forall h \in E,$$

where E is considered as a subspace of $\mathfrak{H}_-$. Clearly,

$$\begin{aligned}(\hat{R}_z\mathcal{P}^*h, g) &= (\mathcal{P}^*h, (A - \bar{z}I)^{-1}g) \\ &= (h, (A - \bar{z}I)^{-1}g) = ((A - zI)^{-1}h, g), \ \forall h \in E, \, g \in \mathcal{H} = L^2_G(E).\end{aligned}$$

It follows that

$$\hat{R}_z\mathcal{P}^*h = \frac{h}{t-z}, \quad \forall h \in E.$$

Since $\Phi(\mathrm{Dom}(A)) = 0$, then $\Phi(A - \bar{z}I)^{-1}g = 0$ for all $g \in \mathcal{H}$, and we have

$$(\hat{R}_z\Phi^*h, g) = \big(\Phi^*h, (A - \bar{z}I)^{-1}g\big) = (h, \Phi(A - \bar{z}I)^{-1}g) = 0.$$

Thus,

$$\hat{R}_z(\Phi^* + \mathcal{P}^*)h = \hat{R}_z\mathcal{P}^*h = \frac{h}{t-z}, \ h \in E.$$

If $h = h_1 + h_2$, where $h_1 \in E_1$ and $h_2 \in E_2$, then

$$\begin{aligned}\hat{R}_zKh_1 &= \hat{R}_z\mathcal{P}^*h_1 = \frac{h_1}{t-z}, \quad h_1 \in E_1, \\ \hat{R}_zKh_2 &= \hat{R}_zh_2 = \frac{h_2}{t-z}, \quad h_2 \in E_2,\end{aligned}$$

This implies that the operator K is invertible. Indeed, if $Kh = 0$, then

$$(\mathcal{P}^* + \Phi^*)h_1 = -\hat{I}h_2,$$

and $\hat{R}_zKh = 0$. Hence, $\hat{R}_z(\mathcal{P}^* + \Phi^*)h_1 = -\hat{R}_zh_2$. That is,

$$\frac{h_1}{t-z} = \frac{h_2}{t-z}, \quad h = h_1 + h_2,$$

which implies that $h = 0$. We should also note that $\hat{R}_zK \in [E, \mathcal{H}_+]$, since $\hat{R}_z$ maps $\mathrm{Ran}(K)$ into $\mathcal{H}_+$ continuously.

Step 5. Let $K^* \in [\mathcal{H}_+, E]$ be the adjoint to the operator K in (6.83), i.e., $(Kf, g) = (f, K^*g)$, $f \in E$, $g \in \mathcal{H}_+$. We consider an operator-valued function $\hat{V}$ defined by

$$\hat{V}(z) = K^*\hat{R}_z K, \quad \mathrm{Im} z \neq 0. \tag{6.84}$$

We will prove that $V(z) = \hat{V}(z)$, $\mathrm{Im}\, z \neq 0$. Obviously, $(\hat{V}(z)g, h)_E = (\hat{R}_z Kg, Kh)$ for $g \in E$, $h \in E$, $g = g_1 + g_2$, $h = h_1 + h_2$, where $g_1, h_1 \in E_1$, $g_2, h_2 \in E_2$. Therefore,

$$\begin{aligned}
(\hat{R}_z Kg, Kh) &= (\hat{R}_z(\mathcal{P}^* + \Phi^*)g_2 + \hat{R}_z g_1, (\mathcal{P}^* + \Phi^*)h_2 + \hat{I}h_1) \\
&= (\hat{R}_z \mathcal{P}^* g_2 + \hat{R}_z g_1, (\mathcal{P}^* + \Phi^*)h_2 + \hat{I}h_1) \\
&= (\hat{R}_z \mathcal{P}^* g_2, \mathcal{P}^* h_2) + (\hat{R}_z \mathcal{P}^* g_2, \Phi^* h_2) + (\hat{R}_z \mathcal{P}^* g_2, h_1) \\
&\quad + (\hat{R}_z g_1, \mathcal{P}^* h_2) + (\hat{R}_z g_1, \Phi^* h_2) + (\hat{R}_z g_1, h_1) \\
&= (\mathcal{P}\hat{R}_z \mathcal{P}^* g_2, h_2) + (\Phi \hat{R}_z \mathcal{P}^* g_2, h_2)_E + (\hat{R}_z \mathcal{P}^* g_2, h_1) \\
&\quad + (\hat{R}_z g_1, h_2) + (\Phi \hat{R}_z g_1, h_2)_E + (\hat{R}_z g_1, h_1).
\end{aligned}$$

We also have

$$\hat{R}_z \mathcal{P}^* g_2 = \frac{g_2}{t - z} \notin \mathfrak{H}_+.$$

Consider an element

$$\frac{g_2}{t - z} - \frac{tg_2}{t^2 + 1} = \frac{g_2}{(t - z)(t^2 + 1)} + \frac{ztg_2}{(t - z)(t^2 + 1)}, \quad g_2 \in E_2.$$

Clearly,

$$\int_{-\infty}^{+\infty} \frac{t^2}{|t - z|^2(t^2 + 1)} \cdot \frac{(dG(t)g_2, g_2)_E}{1 + t^2} < \infty,$$

and

$$\int_{-\infty}^{+\infty} \frac{|z|^2 t^4}{|t - z|^2(t^2 + 1)} \cdot \frac{(dG(t)g_2, g_2)_E}{1 + t^2} < \infty.$$

Hence

$$\frac{g_2}{t - z} - \frac{tg_2}{t^2 + 1} \in \mathrm{Dom}(A).$$

Moreover,

$$\frac{tg_2}{t^2 + 1} \in (I - \mathcal{U})N, \quad g_2 \in E_2.$$

This implies

$$\mathcal{P}\left\{\frac{g_2}{t - z}\right\} = \frac{g_2}{t - z} - \frac{tg_2}{t^2 + 1}, \quad \mathfrak{P}\left\{\frac{g_2}{t - z}\right\} = \frac{tg_2}{t^2 + 1}.$$

Consequently,

$$(\mathcal{P}\hat{R}_z\mathcal{P}^*g_2,h_2)=\int\limits_{-\infty}^{+\infty}\left(\frac{1}{t-z}-\frac{t}{t^2+1}\right)(dG(t)g_2,h_2)_E.$$

We also have that

$$(\Phi\hat{R}_z\mathcal{P}^*g_2,h_2)_E=-(Q\mathsf{R}^{-1}\dot{A}^*\mathcal{P}\hat{R}_z\mathcal{P}^*g_2,h_2)_E=-(\mathsf{R}^{-1}\dot{A}^*\mathcal{P}\hat{R}_z\mathcal{P}g_2,Qh_2)_E.$$

From (6.73) and (6.79)

$$\mathsf{R}^{-1}\dot{A}^*\mathcal{P}\hat{R}_z\mathcal{P}^*g_2=\mathsf{R}^{-1}\left(-\frac{g_2}{t^2+1}\right)=-g_2,$$

implying $(\Phi\hat{R}_z\mathcal{P}^*g_2,h_2)_E=(g_2,Qh_2)_E=(Qg_2,h_2)_E$. Furthermore,

$$\begin{aligned}(\hat{R}_z\mathcal{P}^*g_2,h_1)&=\int\limits_{-\infty}^{+\infty}\left(\frac{1}{t-z}\right)(dG(t)g_2,h_1)_E\\&=\int\limits_{-\infty}^{+\infty}\left(\frac{1}{t-z}\right)(dG(t)g_2,h_1)_E-(g_2,Qh_1)_E+(Qg_2,h_1)_E\\&=\int\limits_{-\infty}^{+\infty}\left(\frac{1}{t-z}\right)(dG(t)g_2,h_1)-\left(g_2,\int\limits_{-\infty}^{+\infty}\frac{t}{t^2+1}dG(t)h_1\right)_E+(Qg_2,h_1)_E\\&=\int\limits_{-\infty}^{+\infty}\left(\frac{1}{t-z}-\frac{t}{t^2+1}\right)(dG(t)g_2,h_1)_E+(Qg_2,h_1)_E.\end{aligned}$$

Since $\Phi\hat{R}_zg_1=0$, we have

$$\begin{aligned}(\hat{R}_zg_1,h_2)&=\int\limits_{-\infty}^{+\infty}\left(\frac{1}{t-z}\right)(dG(t)g_1,h_2)_E-(Qg_1,h_2)_E+(Qg_1,h_2)_E\\&=\int\limits_{-\infty}^{+\infty}\left(\frac{1}{t-z}-\frac{t}{t^2+1}\right)(dG(t)g_1,h_2)_E+(Qg_1,h_2)_E.\end{aligned}$$

Similarly,

$$(\hat{R}_zg_1,h_1)=\int\limits_{-\infty}^{+\infty}\left(\frac{1}{t-z}-\frac{t}{t^2+1}\right)(dG(t)g_1,h_1)_E+(Qg_1,h_1)_E.$$

Finally,

$$(\hat{R}_z g, h) = \int_{-\infty}^{+\infty} \left(\frac{1}{t-z} - \frac{t}{t^2+1}\right)(dG(t)g, h)_E + (Qg, h)_E,$$

and hence,

$$(\hat{V}(z)g, h) = \int_{-\infty}^{+\infty} \left(\frac{1}{t-z} - \frac{t}{t^2+1}\right)(dG(t)g, h)_E + (Qg, h)_E. \tag{6.85}$$

Since the integral representation (6.85) of the function $\hat{V}(z)$ completely matches the one of $V(z) \in N(R)$ in (6.55), we conclude that $\hat{V}(z) \equiv V(z)$. But by (6.84), $\hat{V}(z)$ is the transfer function of the impedance system Δ of the form (6.49) with the self-adjoint state space operator A that is the quasi-kernel of operator $\mathbb{B}$ defined by (6.82) and operator K is defined by (6.83). Therefore, we have constructed an impedance system whose transfer function coincides with $V(z)$. This completes the proof. □

Theorem 6.5.2. *Let an operator-valued function $V(z)$ in a Hilbert space E belong to the class $N(R)$ such that (6.62) holds. Then $V(z)$ can be realized as the impedance function of an L-system Θ of the form (6.31)–(6.36) with invertible channel operator K and preassigned directing operator J for which $I + iV(-i)J$ is invertible.*

Proof. In the proof of this theorem we will heavily rely on the construction developed in the proof of Theorem 6.5.1. Let

$$\mathbb{D} = KJK^*, \tag{6.86}$$

where $J \in [E, E]$ satisfies $J = J^* = J^{-1}$ and K is defined by (6.83). Since $\operatorname{Ran}(K)$ is orthogonal to $\operatorname{Dom}(\dot{A})$, then $\mathbb{D}\phi = 0$ for all $\phi \in \operatorname{Dom}(\dot{A})$. Moreover, $(\mathbb{D}f, g) = (f, \mathbb{D}g)$ for all $f \in \mathcal{H}_+$, $g \in \mathcal{H}_+$. We define an operator $\mathbb{A}$ by

$$\mathbb{A} = \mathbb{B} + i\mathbb{D}, \tag{6.87}$$

where $\mathbb{B}$ is defined by (6.82) and $\mathbb{D}$ by (6.86). We will prove that $\mathbb{A}$ is a $(*)$-extension of some operator T of the class Λ. Let us show that $(\mathbb{B} + iI)\hat{R}_{\pm i}Kg = Kg$, for all $g \in E$, where $\mathbb{B}$ is a t-self-adjoint bi-extension defined by (6.82). By Theorem 4.5.12, the equation $(\mathbb{B} - zI)x = f$ has a unique solution x for any

$$f \in \operatorname{Ran}\left[\mathcal{R}^{-1}\left(S - \frac{i}{2}P^+_{\mathfrak{N}'_i} + \frac{i}{2}P^+_{\mathfrak{N}'_{-i}}\right)\right] + E_1.$$

We are going to show that in fact

$$\operatorname{Ran}(K) = \operatorname{Ran}\left[\mathcal{R}^{-1}\left(S - \frac{i}{2}P^+_{\mathfrak{N}'_i} + \frac{i}{2}P^+_{\mathfrak{N}'_{-i}}\right)\right] + E_1.$$

If $\psi \in N$, then

$$\left(S - \frac{i}{2}P^{+}_{\mathfrak{N}'_i} + \frac{i}{2}P^{+}_{\mathfrak{N}'_{-i}}\right)(\psi - \mathcal{U}\psi) = \mathfrak{R}(\Phi^* + \mathcal{P}^*)\mathsf{R}^{-1}\dot{A}^*(\psi - \mathcal{U}\psi).$$

Using (6.79) we can conclude that $\mathsf{R}^{-1}(I - \mathcal{U})N = E_2$, and hence

$$\operatorname{Ran}\left[\mathfrak{R}^{-1}\left(S - \frac{i}{2}P^{+}_{\mathfrak{N}'_i} + \frac{i}{2}P^{+}_{\mathfrak{N}'_{-i}}\right)\right](I - \mathcal{U})N = (\Phi^* + \mathcal{P}^*)E_2.$$

Letting $P^+ = P^{+}_{\mathfrak{N}'_i} + P^{+}_{\mathfrak{N}'_{-i}}$, we have

$$P^+\left(S - \frac{i}{2}P^{+}_{\mathfrak{N}'_i} + \frac{i}{2}P^{+}_{\mathfrak{N}'_{-i}}\right)(I + \mathcal{U})\varphi = 0, \ \varphi \in \mathfrak{M}.$$

Therefore, $E_1 + \operatorname{Ran}\left[\mathsf{R}^{-1}\left(S - \frac{i}{2}P^{+}_{\mathfrak{N}'_i} + \frac{i}{2}P^{+}_{\mathfrak{N}'_{-i}}\right)\right] = \operatorname{Ran}(K)$. Since $\hat{R}_z = \overline{(\mathbb{B} - zI)^{-1}}$, the above calculations imply

$$(\mathbb{B} - zI)^{-1}Kg = \hat{R}_z Kg,$$

for all $g \in E$. For $\operatorname{Im} z \neq 0$ we have that $\hat{R}_z KE = \mathfrak{N}_z$ is the defect space of the operator $\dot{A}$. Therefore $(\mathbb{B} + iI)\hat{R}_{\pm i}Kg = Kg$ and $\hat{R}_{\pm i}KE = \mathfrak{N}_{\pm i}$. It follows that

$$V(z) = K^*(\mathbb{B} - zI)^{-1}K, \ \operatorname{Im} z \neq 0.$$

Now we apply Theorem 6.4.6. □

Remark 6.5.3. In the proof of Theorem 6.5.2 above we have shown that $(-i)$ is a regular point for the constructed operator T. Hence $V_\Theta(z)$ is a linear-fractional transformation of the form (6.48) of the transfer function $W_\Theta(z)$ in some neighborhood of the point $(-i)$.

Now we will show that condition (6.62) that was used in the statements of Theorems 6.5.1 and 6.5.2 can be released.

Theorem 6.5.4. *Let an operator-valued function $V(z)$ in a Hilbert space E belong to the class $N(R)$. Then $V(z)$ can be realized as:*

1. *transfer function of an impedance system Δ of the form* (6.49),
2. *impedance function of an L-system Θ of the form* (6.35)–(6.36) *with a preassigned directing operator J for which $I + iV(-i)J$ is invertible.*

Proof. Let $V(z) \in N(R)$ and have a representation (6.55). Define the subspaces $\mathcal{G}$ and $\mathcal{G}^\perp$ of E by the formula

$$\mathcal{G} = \left\{ f \in E \mid \int_{-\infty}^{+\infty} \frac{(dG(t)f, f)_E}{1 + t^2} = 0 \right\}, \quad \mathcal{G}^\perp = E \ominus \mathcal{G}. \tag{6.88}$$

The definition of $\mathcal{G}$ in (6.88) implies that, for any interval $\delta \in \aleph$,

$$\int_\delta \frac{(dG(t)f, f)_E}{1+t^2} = 0.$$

Taking into account that $1/(1+t^2)$ increases on the left real semi-axis and decreases on the right semi-axis, we have

$$\int_\delta (G(t)f, f)_E = (G(\delta)f, f)_E = 0,$$

for all $f \in \mathcal{G}$. Thus $G(\delta)f = 0$, $\forall f \in \mathcal{G}$, $\delta \in \aleph$. Considering this and integral representation (6.53) of Q we have that both operator Q and the measure $G(\delta)$ in the representation (6.55) have block-matrix form with respect to decomposition (6.88),

$$Q = \begin{pmatrix} 0 & 0 \\ 0 & Q_2 \end{pmatrix}, \qquad G(\delta) = \begin{pmatrix} 0 & 0 \\ 0 & G_2(\delta) \end{pmatrix}, \qquad \delta \in \aleph. \tag{6.89}$$

Consequently, our function $V(z)$ can also be written as

$$V(z) = \begin{pmatrix} 0 & 0 \\ 0 & V_2(z) \end{pmatrix}, \quad V_2(z) = Q_2 + \int_{-\infty}^{+\infty} \left(\frac{1}{t-z} - \frac{t}{1+t^2} \right) dG_2(t), \tag{6.90}$$

where Q_2 and $G_2(t)$ are described by (6.89). But by construction $V_2(z)$ is such that the inequality (6.62) holds. Therefore, by Theorem 6.5.1, $V_2(z)$ can be realized as a transfer function of an impedance system Δ of the form (6.49), and by Theorem 6.5.2 as an impedance function of an L-system Θ of the form (6.36)-(6.36) with a preassigned directing operator J for which $I + iV(-i)J$ is invertible. In order to realize the function $V(z)$ by the same type of systems that realize $V_2(z)$, we only need to replace the input-output space $\mathcal{G}$ of the realizing systems by E and channel operator $K_2 \in [\mathcal{G}, \mathcal{H}_-]$ by $K \in [E, \mathcal{H}_-]$, where $K = \begin{pmatrix} 0 \\ K_2 \end{pmatrix}$. Clearly, $\operatorname{Ran}(K) = \operatorname{Ran}(K_2) = \operatorname{Ran}(\operatorname{Im} \mathbb{A})$. □

Theorem 6.5.5. *Let an operator-valued function $V(z)$ in a Hilbert space E belong to the class $N(R)$. Then $V(z)$ can be realized as an impedance function of a scattering ($J = I$) L-system Θ of the form* (6.35)–(6.36).

Proof. It can be seen that when $J = I$ the invertibility condition for the operator $I + iV(-i)J$ is satisfied automatically. Indeed, it follows from representation (6.55) that

$$V(-i) = -i \int_{-\infty}^{+\infty} \frac{dG(t)}{1+t^2} + Q = -iB + Q,$$

where $B \geq 0$ is the integral in the above formula. Suppose $(I + iV(-i))f = 0$ for some $f \in E$. Then

$$(I + iV(-i))f = f + Bf + iQf = 0, \quad f \in E,$$

and $((I + iV(-i))f, f) = (f, f) + (Bf, f) + i(Qf, f) = 0$ for $f \in E$. Consequently, $(f, f) + (Bf, f) = 0$ and, since B is a non-negative operator, we have $f = 0$. Thus, $I + iV(-i)$ is invertible. The rest follows from Theorem 6.5.2. □

The following theorem deals with the realization of two realizable operator-valued Herglotz-Nevanlinna functions that differ from each other only by constant terms in representation (6.55).

Theorem 6.5.6. *Let the operator-valued functions*

$$V_1(z) = Q_1 + \int_{-\infty}^{+\infty} \left(\frac{1}{t - z} - \frac{t}{1 + t^2} \right) dG(t), \tag{6.91}$$

and

$$V_2(z) = Q_2 + \int_{-\infty}^{+\infty} \left(\frac{1}{t - z} - \frac{t}{1 + t^2} \right) dG(t),$$

belong to the class $N(R)$. Then they can be realized as impedance functions of L-systems

$$\Theta_1 = \begin{pmatrix} \mathbb{A}_1 & K_1 & I \\ \mathcal{H}_+ \subset \mathcal{H} \subset \mathcal{H}_- & & E \end{pmatrix}, \quad (\mathbb{A}_1 \supset T_1), \tag{6.92}$$

and

$$\Theta_2 = \begin{pmatrix} \mathbb{A}_2 & K_2 & I \\ \mathcal{H}_+ \subset \mathcal{H} \subset \mathcal{H}_- & & E \end{pmatrix}, \quad (\mathbb{A}_2 \supset T_2),$$

respectively, so that the operators T_1 and T_2 acting on the Hilbert space $\mathcal{H}$ are both extensions of the symmetric operator $\dot{A}$ defined in this Hilbert space.

Proof. Applying Theorem 6.5.2 to the function $V_1(z)$, we obtain a system Θ_1 of the type (6.92). The corresponding symmetric operator A_1 constructed in Steps 1 and 2 of the proof of Theorem 6.5.2 satisfies the formulas (6.75) and (6.76). The construction of A_1 doesn't involve the operator Q_1 from (6.91). It is easy to see that the corresponding rigged Hilbert space $\mathcal{H}_+^{(1)} \subset \mathcal{H}^{(1)} \subset \mathcal{H}_-^{(1)}$ was built without the use of the operator Q_1 too.

Similarly, if we apply Theorem 6.5.2 to the function $V_2(z)$ we get the corresponding symmetric operator $A_2 = A_1$ and the same rigged Hilbert space. This happens because the operator-functions $V_1(z)$ and $V_2(z)$ differ from each other only by the constant terms Q_1 and Q_2. Setting $\dot{A} = A_1 = A_2$, we can conclude that T_1 and T_2 are both extensions of the symmetric operator $\dot{A}$. □

6.6 Minimal realization and the theorem on bi-unitary equivalence

Following Chapter 5, we call a closed linear operator in a Hilbert space $\mathcal{H}$ a **prime operator** if there exists no non-trivial reducing invariant subspace of $\mathcal{H}$ on which it induces a self-adjoint operator. A rigged canonical system Θ of the form (6.31)-(6.36) is said to be a **minimal L-system** if its symmetric operator $\dot{A}$ is a prime operator.

Theorem 6.6.1. *Let the operator-valued function $V(z)$ belong to the class $N(R)$. Then it can be realized as an impedance function of a minimal L-system Θ of the form* (6.31)–(6.36) *with a preassigned directing operator J for which $I+iV(-i)J$ is invertible.*

Proof. Theorem 6.5.2 provides us with a possibility of realization for a given operator-valued function $V(z)$ from the class $N(R)$. Let us assume that its symmetric operator $\dot{A}$ has a non-trivial reducing invariant subspace $\mathcal{H}^1 \subset \mathcal{H}$ on which it generates the self-adjoint operator A_1. Then we can write the $(\cdot)$-orthogonal decomposition

$$\mathcal{H} = \mathcal{H}^0 \oplus \mathcal{H}^1, \qquad \dot{A} = \dot{A}_0 \oplus A_1,$$

where $\dot{A}_0$ is an operator induced by $\dot{A}$ on $\mathcal{H}^0$. Now let us consider an operator $T \supset \dot{A}$ as in the definition of the system Θ. We have

$$T = T_0 \oplus A_1,$$

where $T_0 \supset \dot{A}_0$. Indeed, since A_1 is a self-adjoint operator it can not be extended any further. Clearly, $\overline{\mathrm{Dom}(A_1)} = \mathcal{H}^1$. Similarly, $T^* = T_0^* \oplus A_1$, where $T_0^* \supset \dot{A}_0$. Furthermore,

$$\mathcal{H}_+ = \mathcal{H}_+^0 \oplus \mathcal{H}_+^1 = \mathrm{Dom}(\dot{A}_0^*) \oplus \mathrm{Dom}(A_1).$$

We now show that the same holds in the $(+)$-orthogonality sense. Indeed, if $f_0 \in \mathcal{H}_+^0$, $f_1 \in \mathcal{H}_+^1 = \mathrm{Dom}(A_1)$, then

$$(f_0, f_1)_+ = (f_0, f_1) + (\dot{A}^* f_0, \dot{A}^* f_1) = (f_0, f_1) + (\dot{A}_0^* f_0, A_1 f_1) = 0 + 0 = 0.$$

Consequently, we have

$$\mathcal{H}_+ \subset \mathcal{H} \subset \mathcal{H}_- = \mathcal{H}_+^0 \oplus \mathcal{H}_+^1 \subset \mathcal{H}^0 \oplus \mathcal{H}^1 \subset \mathcal{H}_-^0 \oplus \mathcal{H}_-^1.$$

Similarly, we obtain $\mathbb{A} = \mathbb{A}_0 \oplus A_1$ and $\mathbb{A}^* = \dot{A}_0 \oplus A_1$. Therefore,

$$\frac{\mathbb{A} - \mathbb{A}^*}{2i} = \frac{(\mathbb{A}_0 \oplus A_1) - (\mathbb{A}_0^* \oplus A_1)}{2i} = \frac{\mathbb{A}_0 - \mathbb{A}_0^*}{2i} \oplus \frac{A_1 - A_1}{2i} = \frac{\mathbb{A}_0 - \mathbb{A}_0^*}{2i} \oplus O,$$

where O is the zero operator in $[\mathcal{H}_+^1, \mathcal{H}_-^1]$. This implies that

$$KJK^* = K_0 J K_0^* \oplus O.$$

Let P_+^0 be an orthoprojection operator of $\mathcal{H}_+$ onto $\mathcal{H}_+^0$ and set $K = K_0$. Now $K^* = K_0^* P_+^0$, since for all $f \in E$, $g \in \mathcal{H}_+$, then we have

$$
\begin{aligned}
(Kf, g) &= (K_0 f, g) = (K_0 f, g_0 + g_1) = (K_0 f, g_0) + (K_0 f, g_1) \\
&= (K_0 f, g_0) = (f, K_0^* g_0) = (f, K_0^* P_+^0 g).
\end{aligned}
$$

Next, consider $h \in E$ and $\phi = \phi_0 + \phi_1$ in $\mathcal{H}_+$ such that $(\mathbb{A} - zI)P_+^0\phi = Kh$. Then

$$
\begin{aligned}
(\mathbb{A}_0 \oplus A_1 - zI)P_+^0\phi &= K_0 h, \\
\mathbb{A}_0\phi_0 - z\phi_0 &= K_0 h, \\
(\mathbb{A} - zI)\phi_0 &= K_0 h, \\
\phi_0 &= (\mathbb{A}_0 - zI)^{-1} K_0 h.
\end{aligned}
$$

On the other hand, $\phi_0 = (\mathbb{A} - zI)^{-1} Kh$. Therefore

$$
(\mathbb{A} - zI)^{-1} Kh = (\mathbb{A}_0 - zI)^{-1} K_0 h,
$$

and

$$
K^*(\mathbb{A} - zI)^{-1} Kh = K_0^*(\mathbb{A}_0 - zI)^{-1} K_0 h.
$$

This means that the transfer operator-functions of our system Θ and of the system

$$
\Theta_0 = \begin{pmatrix} \mathbb{A}_0 & K_0 & J \\ \mathcal{H}_+ \subset \mathcal{H} \subset \mathcal{H}_- & & E \end{pmatrix} \tag{6.93}
$$

coincide. This proves the statement of the theorem. □

Corollary 6.6.2. *If an operator-valued function $V(z)$ belongs to the class $N(R)$, then $V(z)$ can be realized as an impedance function of a minimal scattering ($J = I$) L-system Θ of the form* (6.36).

The proof of the corollary immediately follows from Theorems 6.5.5 and 6.6.1.

Remark 6.6.3. Let Θ be an L-system of the form (6.31)-(6.36). Then the minimal L-system Θ_0 of the form (6.93) constructed in the proof of Theorem 6.6.1 is called the **principal part** of the L-system Θ. It follows directly from the proof of Theorem 6.6.1 that both the transfer and impedance functions of any L-system Θ coincide with the transfer and impedance functions of its principal part Θ_0, respectively.

The following lemma gives a criterion of primeness for a closed symmetric operator.

Lemma 6.6.4. *A closed symmetric operator $\dot{A}$ is prime if and only if*

$$
\mathcal{N} = \underset{z \neq \bar{z}}{c.l.s.}\, \mathfrak{N}_z = \mathcal{H}, \tag{6.94}
$$

where $\mathfrak{N}_z$ is the deficiency subspace of $\dot{A}$.

Proof. Let $\mathcal{M} = \bigcap\limits_{z\neq\bar{z}} \mathfrak{M}_z$, where $\mathfrak{M}_z = (\dot{A} - zI)\mathrm{Dom}(\dot{A})$. It is easy to see that $\mathcal{H} = \mathcal{N} \oplus \mathcal{M}$. We show that

$$(\dot{A} - zI)^{-1}\mathcal{M} = \mathcal{M} \cap \mathrm{Dom}(\dot{A}) \quad \text{for all} \quad z,\ \mathrm{Im}\, z \neq 0. \tag{6.95}$$

Let $\mathrm{Im}\, z_0 \neq 0$ and $\phi \in \mathcal{M}$. Then $h_{z_0} = (\dot{A} - z_0 I)^{-1}\phi \in \mathrm{Dom}(\dot{A})$. If $\mathrm{Im}\, z \neq 0$ and $z \neq z_0$, then

$$(\dot{A} - zI)h_{z_0} = (\dot{A} - z_0 I)h_{z_0} + (z_0 - z)h_{z_0} = \phi + (z_0 - z)h_{z_0}.$$

Since $(\dot{A} - zI)h_{z_0} \in \mathfrak{M}_z$ and $\phi \in \mathfrak{M}_z$, we get $h_{z_0} \in \mathfrak{M}_z$. Thus, $h_{z_0} \in \mathrm{Dom}(\dot{A}) \cap \mathfrak{M}_z$ for all $z \neq z_0$, $\mathrm{Im}\, z \neq 0$.

Let T_0 be a quasi-self-adjoint extension of $\dot{A}$ having a regular point $\bar{z}_0$. For example,

$$\mathrm{Dom}(T_0) = \mathrm{Dom}(\dot{A}) \dot{+} \mathfrak{N}_{z_0},\ T_0(h + \psi_{z_0}) = \dot{A}h + \bar{z}_0\psi_{z_0},\ h \in \mathrm{Dom}(\dot{A}),\ \psi_{z_0} \in \mathfrak{N}_{z_0}.$$

One can easily verify that the operator

$$U_{zz_0} = (T_0 - \bar{z}_0 I)(T_0 - \bar{z}I)^{-1} = I + (\bar{z}_0 - \bar{z})(T_0 - \bar{z}I)^{-1}$$

maps $\mathfrak{N}_{\bar{z}_0}$ into $\mathfrak{N}_{\bar{z}}$ for all $\bar{z} \in \rho(T_0)$. Clearly,

$$\lim_{z\to z_0} U_{zz_0} f = f \quad \text{for all} \quad f \in \mathcal{H}.$$

Further, for all $f_{\bar{z}_0} \in \mathfrak{N}_{\bar{z}_0}$, using that $h_{z_0} = (\dot{A} - z_0 I)^{-1}\phi \in \mathfrak{M}_z$, $z \neq z_0$, we have $(h_{z_0}, U_{zz_0} f_{\bar{z}_0}) = 0$. It follows that

$$(h_{z_0}, f_{\bar{z}_0}) = \lim_{z\to \bar{z}_0} ((\dot{A} - z_0 I)^{-1}\phi, U_{zz_0} f_{\bar{z}_0}) = 0.$$

Therefore, $h_{z_0} \in \mathfrak{M}_z$ for all z, $\mathrm{Im}\, z \neq 0$. Since $h_{z_0} \in \mathrm{Dom}(\dot{A})$, we get

$$(\dot{A} - z_0 I)^{-1}\mathcal{M} \subseteq \mathcal{M} \cap \mathrm{Dom}(\dot{A}).$$

Conversely, if $f \in \mathcal{M} \cap \mathrm{Dom}(\dot{A})$, then since

$$\phi = (\dot{A} - z_0 I)f = (\dot{A} - zI)f + (z - z_0)f,$$

we have $(\dot{A} - z_0 I)(\mathcal{M} \cap \mathrm{Dom}(\dot{A})) \subseteq \mathcal{M}$ and thus, (6.95) holds. This means that the symmetric operator $\dot{A}\restriction(\mathcal{M} \cap \mathrm{Dom}(\dot{A}))$, as acting on $\mathcal{M}$, has regular points in the open upper and lower half plains. It follows that $\dot{A}\restriction(\mathcal{M} \cap \mathrm{Dom}(\dot{A}))$ is a self-adjoint operator in $\mathcal{M}$. □

Remark 6.6.5. One can see from the proof of Lemma 6.6.4 that if G is an open subset of $\mathbb{C}$ such that $\mathbb{C}_+ \cap G \neq \emptyset$, and $\mathbb{C}_- \cap G \neq \emptyset$, then $\underset{z\in G}{c.l.s.}\, \mathfrak{N}_z = \mathcal{H}$ implies that $\dot{A}$ is prime.

Lemma 6.6.6. *Let $\dot{A}$ be a closed, symmetric operator with equal deficiency indices in the Hilbert space $\mathcal{H}$. Let A be an arbitrary self-adjoint extension of $\dot{A}$ and let $E(t)$ and $R_\lambda = R_\lambda(A)$ be the resolution of identity and the resolvent of A, respectively. Then for $z \in \mathbb{C}$, $\operatorname{Im} z \neq 0$, the following statements are equivalent:*

(i) *$\dot{A}$ is prime;*

(ii) *$c.l.s.\{R_\lambda \mathfrak{N}_{z_0}, \operatorname{Im}\lambda \neq 0\} = \mathcal{H}$;*

(iii) *$c.l.s\{E(\Delta)\mathfrak{N}_{z_0}, \Delta \in \aleph\} = \mathcal{H}$, where $\mathfrak{N}_{z_0}$ is the defect subspace of $\dot{A}$.*

Proof. Let $\mathcal{L}$ be a subspace of $\mathcal{H}$. Define

$$\mathcal{H}_0 = c.l.s\{E(\Delta)\mathcal{L}, \Delta \in \aleph\}, \quad \mathcal{H}_0' = c.l.s.\{R_\lambda\mathcal{L}, \operatorname{Im}\lambda \neq 0\}.$$

Since

$$R_\lambda = \int_{\mathbb{R}} \frac{dE(t)}{t-\lambda},$$

we get $R_\lambda\mathcal{L} \subset \mathcal{H}_0'$, and hence, $\mathcal{H}_0' \subseteq \mathcal{H}_0$. On the other hand, clearly, the subspace $\mathcal{H}_0'$ reduces R_λ for all non-real λ, and hence $\mathcal{H}_0'$ and $\mathcal{H} \ominus \mathcal{H}_0'$ are invariant with respect to A. In addition, in view of the equality

$$\lim_{\eta\to\infty} i\eta R_{i\eta} f = -f, \quad f \in \mathcal{H},$$

we get $\mathcal{L} \subset \mathcal{H}_0'$. It follows that $E(\Delta)\mathcal{L} \subset \mathcal{H}_0'$ for all $\Delta \in \aleph$. Hence, $\mathcal{H}_0 \subseteq \mathcal{H}_0'$. Thus, $\mathcal{H}_0 = \mathcal{H}_0'$. Taking $\mathcal{L} = \mathfrak{N}_{z_0}$, we get (ii) $\iff$ (iii). Let us prove. (i) $\iff$ (ii). The latter follows from

$$(I - (\lambda - z_0)R_\lambda)\,\mathfrak{N}_{z_0} = \mathfrak{N}_\lambda,$$

which leads to the equality

$$R_{\lambda_1}\mathfrak{N}_{z_0} + R_{\lambda_2}\mathfrak{N}_{z_0} + \cdots + R_{\lambda_n}\mathfrak{N}_{z_0} = \mathfrak{N}_{z_0} + \mathfrak{N}_{\lambda_1} + \cdots + \mathfrak{N}_{\lambda_n}. \qquad \square$$

With the help of Lemma 6.6.4 we can offer an alternative proof of Theorem 6.6.1 and provide a valuable property of the model symmetric operator $\dot{A}$ constructed in the proof of Theorem 6.5.1.

Theorem 6.6.7. *The symmetric operator $\dot{A}$ of the form (6.75)–(6.76) is prime. Consequently, any operator-valued function $V(z) \in N(R)$ can be realized as impedance function of a minimal L-system Θ of the form (6.31)–(6.36) with the prime symmetric operator $\dot{A}$ of the form (6.75)–(6.76) and a preassigned directing operator J for which $I + iV(-i)J$ is invertible.*

Proof. Let $G(t)$ be the function from the integral representation of $V(z)$, $\sigma(t) = \operatorname{trace}(G(t))$. Recall that in the proof of Theorem 6.5.1 by $\Psi(t)$ we used the Radon-Nikodym derivative

$$\Psi(t) = \frac{dG(t)}{d\sigma(t)}.$$

Let $L^2_G(E)$ be the Hilbert space constructed in the proof of Theorem 6.5.1. Let A be a self-adjoint operator in $L^2_G(E)$ defined by (6.69), i.e.,

$$\mathrm{Dom}(A) = \left\{f \in L^2_G(E) : tf(t) \in L^2_G(E)\right\}, \quad Af(t) = tf(t),$$

and $\dot{A}$ be a symmetric operator of the form (6.75)–(6.76). The defect subspaces of $\dot{A}$ are described in the second step of the proof of Theorem 6.5.1 and are given by

$$\mathfrak{N}_z = \left\{\frac{e}{t-z},\ e \in E\right\}, \quad \mathrm{Im}\, z \neq 0.$$

Suppose that there is a function $h \in L^2_G(E)$ such that

$$(\mathfrak{N}_z, h)_{L^2_G(E)} = 0,\ \mathrm{Im}\, z \neq 0.$$

Then using (6.67) and (6.68), we get

$$\int_{\mathbb{R}} \frac{(\Psi(t)e, h(t))_E}{t-z}\, d\sigma(t) = 0,\ e \in E,\ \mathrm{Im}\, z \neq 0.$$

Let

$$\gamma_e(t) = (\Psi(t)e, h(t))_E = (e, \Psi(t)h(t))_E,\ t \in \mathbb{R}.$$

Then we have

$$\int_{\mathbb{R}} \frac{\gamma_e(t)d\sigma(t)}{t-z} = 0 \quad \text{for all} \quad z \text{ such that } \mathrm{Im}\, z \neq 0.$$

It follows that

$$\int_{\mathbb{R}} \left(\frac{1}{t-z} - \frac{1}{t-\xi}\right) \gamma_e(t)d\sigma(t) = 0,\ \mathrm{Im}\, z \neq 0,\ \mathrm{Im}\, \xi \neq 0.$$

Then

$$\int_{\mathbb{R}} \frac{1}{(t-z)(t-\xi)} \gamma_e(t)d\sigma(t) = 0.$$

Fix $\xi = i$ and set

$$f_e(t) := \frac{\gamma_e(t)}{t-i}.$$

Hence

$$\int\limits_{\mathbb{R}} \frac{f_e(t)\, d\sigma(t)}{t-z} = 0,\ \mathrm{Im}\, z \neq 0. \tag{6.96}$$

Since

$$\begin{aligned} |f_e(t)| &= \left(\frac{\Psi(t)e}{t-i}, h(t)\right)_E \leq \sqrt{\frac{(\Psi(t)e, e)_E}{|t-i|^2}\,(\Psi(t)h(t), h(t))_E} \\ &\leq \frac{1}{2}\left(\frac{(\Psi(t)e, e)_E}{1+t^2} + (\Psi(t)h(t), h(t))_E\right), \end{aligned}$$

and

$$\int_{\mathbb{R}} \frac{(\Psi(t)e,e)_E}{1+t^2} d\sigma(t) < \infty, \quad \int_{\mathbb{R}} (\Psi(t)h(t),h(t))_E d\sigma(t) < \infty,$$

the function $f_e(t)$ belongs to the Banach space $L^1(\mathbb{R}, d\sigma)$. Relations (6.96) yield

$$\int_{\mathbb{R}} \left(\frac{1}{t-z} - \frac{1}{t-\bar{z}} \right) f_e(t) d\sigma(t) = 0, \ \operatorname{Im} z \neq 0.$$

Set

$$\mu_e(t) := \int_{-\infty}^{t} f_e(\tau) d\sigma(\tau).$$

Then we have

$$\int_{\mathbb{R}} \frac{y}{(x-t)^2+y^2} d\mu_e(t) = 0, \ y \neq 0.$$

Because the complex measure $d\mu_e$ has finite variation, the latter equality yields (see [166])

$$\int_{\mathbb{R}} g(t)\, d\mu_e(t) = 0$$

for all continuous and bounded on $\mathbb{R}$ functions g. In particular,

$$\int_{\mathbb{R}} \frac{g(t)}{t-z} d\mu_e(t) = 0$$

for all non-real z and all bounded and continuous on $\mathbb{R}$ functions g. Take a continuous function g with a compact support and define a new measure

$$\nu_e(t) := \int_{-\infty}^{t} g(t) d\mu_e(t).$$

Then $d\nu_e$ has a compact support and bounded variation. Since

$$\int_{\mathbb{R}} \frac{d\nu_e(t)}{t-z} = 0 \quad \text{for all nonreal} \quad z,$$

we obtain $d\nu_e = 0$ [132]. It follows that $\gamma_e(t) = 0$ almost everywhere with respect to $d\sigma$. Since $e \in E$ is arbitrary, we have $h(t) \in \ker(\Psi(t))$ almost everywhere with respect to $d\sigma$. Due to (6.67) and (6.68) we get $h = 0$ in $L^2_G(E)$. Thus,

$$c.l.s.\{\mathfrak{N}_z, \ \operatorname{Im} z \neq 0\} = L^2_G(E),$$

and $\dot{A}$ is prime. □

The lemma below establishes the link between the primeness of operators T and $\dot{A}$.

Lemma 6.6.8. *If $\dot{A}$ is a maximal common symmetric part of T and T^*, then operator T is prime if and only if $\dot{A}$ is prime.*

Proof. Let $\mathcal{H}^1$ be a non-trivial reducing invariant subspace for T. Then $\mathcal{H} = \mathcal{H}^1 \oplus \mathcal{H}^2$ and $T_1 = T\upharpoonright\mathcal{H}^1$ is a self-adjoint in $\mathcal{H}^1$ operator. Let P_k $(k = 1, 2)$ be the orthogonal projection operator on subspaces $\mathcal{H}^k$, $(k = 1, 2)$, respectively. Furthermore, for all $g \in \operatorname{Dom}(T)$, $P_k g \in \operatorname{Dom}(T)$ and $TP_k g \in \mathcal{H}^k$, $(k = 1, 2)$. If $f \in \mathcal{H}^1 \cap \operatorname{Dom}(T)$, then for all $g \in \operatorname{Dom}(T)$ we have

$$(Tf, g) = (Tf, P_1 g) = (f, TP_1 g) = (f, TP_1 g) + (f, TP_2 g) = (f, Tg).$$

This implies that, $f \in \operatorname{Dom}(T^*)$ and $T^* f = Tf$. Since

$$\operatorname{Dom}(\dot{A}) = \operatorname{Dom}(T) \cap \operatorname{Dom}(T^*),$$

then we have $\operatorname{Dom}(T) \cap \mathcal{H}^1 \subset \operatorname{Dom}(\dot{A})$ and consequently $\dot{A}$ induces a self-adjoint operator on $\mathcal{H}^1$. Clearly, if $\dot{A}$ is not prime, then T is not prime either. □

The next statement immediately follows from Remark 6.6.5 and Lemma 6.6.8.

Lemma 6.6.9. *Let T be a quasi-self-adjoint extension of $\dot{A}$ such that $\rho(T) \cap \mathbb{C}_+$ and $\rho(T) \cap \mathbb{C}_-$ are non-empty. Let also G be an open subset of $\rho(T)$ such that $G \cap \mathbb{C}_- \neq \emptyset$ and $G \cap \mathbb{C}_+ \neq \emptyset$. Then the operator T is prime if and only if*

$$\underset{z \in G}{c.l.s.}\, \mathfrak{N}_z = \mathcal{H},$$

where $\mathfrak{N}_z$ is the deficiency subspace of $\dot{A}$ that is the maximal common symmetric part of T and T^*.

Theorem 6.6.10. *Let*

$$\Theta_1 = \begin{pmatrix} \mathbb{A}_1 & K_1 & J \\ \mathcal{H}_{+1} \subset \mathcal{H}_1 \subset \mathcal{H}_{-1} & & E \end{pmatrix} \text{ and } \Theta_2 = \begin{pmatrix} \mathbb{A}_2 & K_2 & J \\ \mathcal{H}_{+2} \subset \mathcal{H}_2 \subset \mathcal{H}_{-2} & & E \end{pmatrix}, \tag{6.97}$$

be two minimal L-systems with

$$\mathbb{A}_1 \supset T_1 \supset \dot{A}_1, \quad \mathbb{A}_1^* \supset T_1^* \supset \dot{A}_1,$$
$$\mathbb{A}_2 \supset T_2 \supset \dot{A}_2, \quad \mathbb{A}_2^* \supset T_2^* \supset \dot{A}_2.$$

Let also operators T_1 and T_2 be such that $(\rho(T_1) \cap \rho(T_2)) \cap \mathbb{C}_\pm \neq \emptyset$. If the transfer functions $W_{\Theta_1}(z)$ and $W_{\Theta_2}(z)$ satisfy the condition

$$W_{\Theta_1}(z) = W_{\Theta_2}(z), \quad z \in \rho(T_1) \cap \rho(T_2), \tag{6.98}$$

then there exists an isometric operator U from $\mathcal{H}_1$ onto $\mathcal{H}_2$ such that $U_+ = U|_{\mathcal{H}_{+1}}$ is an isometry from $\mathcal{H}_{+1}$ onto $\mathcal{H}_{+2}$, $U_- = (U_+^*)^{-1}$ is an isometry from $\mathcal{H}_{-1}$ onto $\mathcal{H}_{-2}$, and

$$UT_1 = T_2 U, \quad U_- \mathbb{A}_1 = \mathbb{A}_2 U_+, \quad U_- K_1 = K_2. \tag{6.99}$$

Proof. It follows from (6.44) and (6.98) that

$$K_1^*(\mathbb{A}_1 - zI)^{-1}K_1 = K_2^*(\mathbb{A}_2 - zI)^{-1}K_2. \tag{6.100}$$

Since for $j = 1, 2$ we have

$$\begin{aligned}(\mathbb{A}_j - zI)^{-1} - (\mathbb{A}_j^* - \bar{\zeta}I)^{-1} &= (\mathbb{A}_j^* - \bar{\zeta}I)^{-1}[(\mathbb{A}_j^* - \bar{\zeta}I) - (\mathbb{A}_j - zI)](\mathbb{A}_j - zI)^{-1} \\ &= (z - \bar{\zeta})(\mathbb{A}_j^* - \bar{\zeta}I)^{-1}(\mathbb{A}_j - zI)^{-1} - 2i(\mathbb{A}_j^* - \bar{\zeta}I)^{-1}K_jJK_j^*(\mathbb{A}_j - zI)^{-1},\end{aligned}$$

then

$$\begin{aligned}(z - \bar{\zeta})K_1^*(\mathbb{A}_1^* - \bar{\zeta}I)^{-1}(\mathbb{A}_1 - zI)^{-1}K_1 &= K_1^*(\mathbb{A}_1 - zI)^{-1}K_1 \\ &\quad - K_1^*(\mathbb{A}_1^* - \bar{\zeta}I)^{-1}K_1 + 2iK_1^*(\mathbb{A}_1^* - \bar{\zeta}I)^{-1}K_1JK_1^*(\mathbb{A}_1 - zI)^{-1}K_1 \\ &= K_2^*(\mathbb{A}_2 - zI)^{-1}K_2 - K_2^*(\mathbb{A}_2^* - \bar{\zeta}I)^{-1}K_2 \\ &\quad + 2iK_2^*(\mathbb{A}_2^* - \bar{\zeta}I)^{-1}K_2JK_2^*(\mathbb{A}_2 - zI)^{-1}K_2 \\ &= (z - \bar{\zeta})K_2^*(\mathbb{A}_2^* - \bar{\zeta}I)^{-1}(\mathbb{A}_2 - zI)^{-1}K_2.\end{aligned}$$

Therefore,

$$K_1^*(\mathbb{A}_1^* - \bar{\zeta}I)^{-1}(\mathbb{A}_1 - zI)^{-1}K_1 = K_2^*(\mathbb{A}_2^* - \bar{\zeta}I)^{-1}(\mathbb{A}_2 - zI)^{-1}K_2. \tag{6.101}$$

Now we can apply (4.45) and get

$$(\mathbb{A}_1 - zI)^{-1}K_1E = \mathfrak{N}_z^{(1)}, \quad (\mathbb{A}_2 - zI)^{-1}K_2E = \mathfrak{N}_z^{(2)}. \tag{6.102}$$

Because the systems Θ_1 and Θ_2 are minimal, then by Lemma 6.6.8 the operators T_1 and T_2 are prime. Thus, using $(\rho(T_1) \cap \rho(T_2)) \cap \mathbb{C}_\pm \neq \emptyset$ yields

$$\underset{z\in\rho(T_1)\cap\rho(T_2)}{c.l.s.} \mathfrak{N}_z^{(1)} = \mathcal{H}_1, \qquad \underset{z\in\rho(T_1)\cap\rho(T_2)}{c.l.s.} \mathfrak{N}_z^{(2)} = \mathcal{H}_2.$$

Define linear manifolds

$$\mathcal{D}_j = l.s.\{(\mathbb{A}_j - zI)^{-1}K_jE,\ z \in \rho(T_1) \cap \rho(T_2)\},\ j = 1, 2.$$

Then $\mathcal{D}_1$ and $\mathcal{D}_2$ are dense in $\mathcal{H}_1$ and $\mathcal{H}_2$, respectively. Now define a linear operator $U : \mathcal{D}_1 \to \mathcal{D}_2$:

$$U\left(\sum(\mathbb{A}_1 - z_jI)^{-1}K_1f_j\right) = \left(\sum(\mathbb{A}_2 - z_jI)^{-1}K_2f_j\right),$$

where $\{z_j\} \subset \rho(T_1) \cap \rho(T_2)$, $\{f_j\} \subset E$. Then equality (6.101) yields that the operator U is isometric and maps $\mathcal{D}_1$ onto $\mathcal{D}_2$. It follows that U admits a unitary continuation mapping from $\mathcal{H}_1$ onto $\mathcal{H}_2$. We preserve the notation U for this continuation. Taking into account Hilbert's identity for resolvents

$$(\mathbb{A}_j - zI)^{-1} - (\mathbb{A}_j - \xi I)^{-1} = (z - \xi)(\mathbb{A}_j - \xi I)^{-1}(\mathbb{A}_j - zI)^{-1}, \quad (j = 1, 2)$$

we get that

$$U(T_1 - zI)^{-1}\left(\sum(\mathbb{A}_1 - z_jI)^{-1}K_1f_j\right) = (T_2 - zI)^{-1}\left(\sum(\mathbb{A}_2 - z_jI)^{-1}K_2f_j\right)$$
$$= (T_2 - zI)^{-1}U\left(\sum(\mathbb{A}_1 - z_jI)^{-1}K_1f_j\right),$$

for each $z \in \rho(T_1) \cap \rho(T_2)$. Hence,

$$U(T_1 - zI)^{-1} = (T_2 - zI)^{-1}U, \tag{6.103}$$

and

$$UT_1 = T_2U. \tag{6.104}$$

Since $U((\mathbb{A}_1 - zI)^{-1}K_1 = (\mathbb{A}_2 - zI)^{-1}K_2$, the operator U maps $\mathfrak{N}_z^{(1)}$ onto $\mathfrak{N}_z^{(2)}$. Therefore U maps $(\dot{A}_1 - \bar{z}I)\mathrm{Dom}(\dot{A}_1)$ onto $(\dot{A}_2 - \bar{z}I)\mathrm{Dom}(\dot{A}_2)$. Relations (6.103) and (6.104) imply

$$U\mathrm{Dom}(\dot{A}_1) = \mathrm{Dom}(\dot{A}_2), \quad U\dot{A}_1 = \dot{A}_2U,$$

and $U\mathrm{Dom}(\dot{A}_1^*) = \mathrm{Dom}(\dot{A}_2^*)$, $\quad U\dot{A}_1^* = \dot{A}_2^*U$. Because

$$(f,g)_{\mathcal{H}_{+1}} = (f,g)_{\mathcal{H}_1} + (\dot{A}_1^*f, \dot{A}_1^*g)_{\mathcal{H}_1}, \quad (f,g)_{\mathcal{H}_{+2}} = (f,g)_{\mathcal{H}_2} + (\dot{A}_2^*f, \dot{A}_2^*g)_{\mathcal{H}_2},$$

the operator U is an isometry of $\mathcal{H}_{+1}$ onto $\mathcal{H}_{+2}$. Since $\mathrm{Re}\,\mathbb{A}_j$, $(j = 1,2)$ is a self-adjoint bi-extension of the symmetric operator A_j, the operators

$$\widehat{A}_jf = \mathrm{Re}\,\mathbb{A}_jf, \qquad D(\widehat{A}_j) = \{f \in \mathcal{H}_{+j} : \mathrm{Re}\,\mathbb{A}_jf \in \mathcal{H}_j\}, \tag{6.105}$$

are self-adjoint in $\mathcal{H}_j$, $(j = 1,2)$. These operators are self-adjoint extensions of $\dot{A}_j$ $(j = 1,2)$. Let

$$U_+ = U\restriction\mathcal{H}_{+1} \in [\mathcal{H}_{+1}, \mathcal{H}_{+2}].$$

Then $U_+^* \in [\mathcal{H}_{-2}, \mathcal{H}_{-1}]$ and U_+^* isometrically maps $\mathcal{H}_{-2}$ onto $\mathcal{H}_{-1}$. Put $U_- = (U_+^*)^{-1}$. Clearly $(U_+^*)^{-1}$ is an extension of U onto $\mathcal{H}_{-1}$. Thus, we have obtained a triplet of operators (U_+, U, U_-) that maps isometrically the triplet $(\mathcal{H}_{+1}, \mathcal{H}_1, \mathcal{H}_{-1})$ onto the triplet $(\mathcal{H}_{+2}, \mathcal{H}_2, \mathcal{H}_{-2})$. Equation (6.103) implies that $U_+(\mathbb{A}_1 - zI)^{-1} = (\mathbb{A}_2 - zI)^{-1}U_-$. Taking into account that $U_-|_{\mathcal{H}_{+1}} = U_+$, we have $(\mathbb{A}_2 - zI)U_+ = U_-(\mathbb{A}_1 - zI)$. Finally, $\mathbb{A}_2 = U_-\mathbb{A}_1U_+^{-1}$ and

$$\mathrm{Re}\,\mathbb{A}_2 = U_-\mathrm{Re}\,\mathbb{A}_1U_+^{-1}, \quad \mathrm{Im}\,\mathbb{A}_2 = U_-\mathrm{Im}\,\mathbb{A}_1U_+^{-1}.$$

From (6.105) and the description of self-adjoint bi-extensions of the symmetric operators one concludes $\widehat{A}_2 = U\widehat{A}_1U^{-1}$. This means that the operators $\widehat{A}_2$ and $\widehat{A}_1$ are unitary equivalent. □

Corollary 6.6.11. *Let Θ_1 and Θ_2 be the two L-systems from the statement of Theorem* 6.6.10. *Then the mapping U described in the conclusion of the theorem is unique.*

Proof. First let us make an observation that if $\Theta = \begin{pmatrix} \mathbb{A} & K & J \\ \mathcal{H}_+\subset\mathcal{H}\subset\mathcal{H}_- & & E \end{pmatrix}$ is a minimal L-system such that $U_-\mathbb{A} = \mathbb{A}U_+$ and $U_-K = K$, where U is an isometry mapping described in theorem 6.6.10, then $U = I$. Indeed, we know (see (6.102)) that

$$(\mathrm{Re}\,\mathbb{A} - \lambda I)^{-1}KE = \mathfrak{N}_\lambda. \tag{6.106}$$

We have

$$\begin{aligned} U(\mathrm{Re}\,\mathbb{A} - \lambda I)^{-1}Ke &= U_+(\mathrm{Re}\,\mathbb{A} - \lambda I)^{-1}Ke = (\mathrm{Re}\,\mathbb{A} - \lambda I)^{-1}U_-Ke \\ &= (\mathrm{Re}\,\mathbb{A} - \lambda I)^{-1}Ke, \qquad \forall e \in E,\ \lambda \neq \bar{\lambda}. \end{aligned}$$

Combining the above equation with (6.94) and (6.106) we obtain $U = I$.

Now let Θ_1 and Θ_2 be the two minimal L-systems from the statement of Theorem 6.6.10. Suppose there are two isometric mappings U_1 and U_2 guaranteed by Theorem 6.6.10. Then the relations

$$\mathbb{A}_2 = U_{-,1}\mathbb{A}_1U_{+,1}^{-1}, \quad U_{-,1}K_1 = K_2, \quad \mathbb{A}_2 = U_{-,2}\mathbb{A}_1U_{+,2}^{-1}, \quad U_{-,2}K_1 = K_2,$$

lead to

$$\mathbb{A}_1U_{+,1}^{-1}U_{+,2} = U_{-,1}^{-1}U_{-,2}\mathbb{A}_1, \quad U_{-,1}^{-1}U_{-,2}K = K.$$

Since Θ_1 is minimal then $U_1^{-1}U_2 = I$ and hence $U_1 = U_2$. This proves the uniqueness of U. □

Two L-systems of the form (6.97) are called **bi-unitarily equivalent** if there exists a triplet of operators (U_+, U, U_-) that isometrically maps the triplet $(\mathcal{H}_{+1}, \mathcal{H}_1, \mathcal{H}_{-1})$ onto the triplet $(\mathcal{H}_{+2}, \mathcal{H}_2, \mathcal{H}_{-2})$ in a way that (6.99) holds and $\mathbb{A}_2 = U_-\mathbb{A}_1U_+^{-1}$. For the remainder of the text we will refer to Theorem 6.6.10 as the theorem on bi-unitary equivalence.

Corollary 6.6.12. *If two L-systems Θ_1 and Θ_2 satisfying the conditions of Theorem* 6.6.10 *are bi-unitary equivalent, then their transfer functions $W_{\Theta_1}(z)$ and $W_{\Theta_2}(z)$ coincide on $z \in \rho(T_1) \cap \rho(T_2)$, i.e.,* (6.98) *holds.*

The corollary is proved by reversing the argument in the proof of Theorem 6.6.10.

Chapter 7

Classes of realizable Herglotz-Nevanlinna functions

In this chapter we are going to introduce three distinct subclasses $N_0(R)$, $N_1(R)$, and $N_{01}(R)$ of the class of functions $N(R)$ realizable as impedance functions of L-systems, that was studied in Chapter 6. We give complete proofs of direct and inverse realization theorems in each subclass. We show that each subclass is characterized by a different property of the state-space operator in the corresponding realizing L-system. Based on this partition of the class $N(R)$, we introduce the corresponding structure of subclasses $\Omega_0(R, J)$, $\Omega_1(R, J)$, and $\Omega_{01}(R, J)$ on the class $\Omega(R, J)$ consisting of functions realizable as transfer functions of L-systems. Multiplication and coupling theorems are then proved for each subclass of $\Omega(R, J)$.

We also establish the connection between the impedance functions of L-system and boundary triplets. In addition to that we consider the Krein-Langer Q-functions of a densely defined symmetric operator and show that they belong to the class $N_0(R)$ and thus can be realized as impedance functions of L-systems.

7.1 Sub-classes of the class $N(R)$ and their realizations

Let E be a finite-dimensional Hilbert space. We start the section by introducing the following subclasses of the class $N(R)$ from Section 6.4. We recall that an operator-valued Herglotz-Nevanlinna function $V(z)$ in E belongs to the class $N(R)$ if in formula (6.52), that is,

$$V(z) = Q + zX + \int_{-\infty}^{+\infty} \left(\frac{1}{t-z} - \frac{t}{1+t^2} \right) dG(t),$$

we have $X = 0$ and

$$Qh = \int_{-\infty}^{+\infty} \frac{t}{1+t^2} dG(t)h,$$

for all $h \in E$ such that $\int_{-\infty}^{+\infty} (dG(t)h, h)_E < \infty$.

Definition 7.1.1. An operator-valued Herglotz-Nevanlinna function $V(z)$ in E of the class $N(R)$ is said to be a member of the **subclass** $N_0(R)$ if in the representation (6.52)

$$\int_{-\infty}^{+\infty} (dG(t)h, h)_E = \infty, \qquad (h \in E,\, h \neq 0). \tag{7.1}$$

Obviously, any function $V(z) \in N_0(R)$ has the representation

$$V(z) = Q + \int_{-\infty}^{+\infty} \left(\frac{1}{t-z} - \frac{t}{1+t^2}\right) dG(t), \qquad (Q = Q^*), \tag{7.2}$$

where $G(t)$ satisfies (7.1) and Q is an arbitrary self-adjoint operator in the Hilbert space E.

Definition 7.1.2. An operator-valued Herglotz-Nevanlinna function $V(z)$ in E of the class $N(R)$ is said to be a member of the **subclass** $N_1(R)$ if in the representation (6.52)

$$\int_{-\infty}^{+\infty} (dG(t)h, h)_E < \infty, \qquad (h \in E).$$

It follows from the definition of the class $N(R)$ that the operator-valued function $V(z)$ of the class $N_1(R)$ has a representation

$$V(z) = \int_{-\infty}^{+\infty} \frac{1}{t-z} dG(t). \tag{7.3}$$

Definition 7.1.3. An operator-valued Herglotz-Nevanlinna function $V(z)$ in E of the class $N(R)$ is said to be a member of the **subclass** $N_{01}(R)$ if the subspace

$$E_1 = \left\{ h \in E : \int_{-\infty}^{+\infty} (dG(t)h, h)_E < \infty \right\} \tag{7.4}$$

possesses a property: $E_1 \neq \{0\}$, $\quad E_1 \neq E$. [1]

[1] The definition of the class $N_{01}(R)$ implies that it does not exist if $\dim E = 1$.

One may notice that $N(R)$ is a union of three distinct subclasses $N_0(R)$, $N_1(R)$ and $N_{01}(R)$. The following theorem is an analogue of Theorem 6.5.2 for the class $N_0(R)$.

Theorem 7.1.4. *Let Θ be an L-system of the form (6.36) with an invertible channel operator K and a densely-defined symmetric operator $\dot{A}$. Then its impedance function $V_\Theta(z)$ of the form (6.47) belongs to the class $N_0(R)$.*

Proof. Relying on Theorem 6.4.3 we conclude that an operator-valued function $V_\Theta(z)$ of the system Θ mentioned in the statement belongs to the class $N(R)$. Since $N_0(R)$ is a subclass of $N(R)$, it is sufficient to show that

$$\int_{-\infty}^{+\infty} (dG(t)h, h)_E = \infty, \qquad (h \in E, h \neq 0).$$

According to Theorem 6.2.10, if for some vector $h \in E$, $h \neq 0$ we have that $Kh \notin \mathfrak{L}$ where $\mathfrak{L} = \mathcal{H} \ominus \overline{\mathrm{Dom}(\dot{A})}$, then

$$\int_{-\infty}^{+\infty} (dG(t)h, h)_E = \infty, \quad \text{where} \quad G(t) = K^* \hat{E}(t) K, \tag{7.5}$$

$\hat{E}(t)$ is an extended canonical spectral function of the operator $\hat{A}$. Here $\hat{A}$ is the quasi-kernel of the operator $\mathrm{Re}\,\mathbb{A}$.

The fact that $\dot{A}$ is a closed symmetric operator with dense domain ($\overline{\mathrm{Dom}(\dot{A})} = \mathcal{H}$) implies that $\mathfrak{L} = \{0\}$. Thus, for any $h \in E$ such that $h \neq 0$ we have $Kh \notin \mathfrak{L}$, and (7.5) holds. Therefore, $V_\Theta(z) \in N_0(R)$. □

It directly follows from Theorem 7.1.4 above and Remark 6.4.5 that if the operator K in the statement of Theorem 7.1.4 is not invertible and $\ker(K) = \mathcal{G}$, then $V_\Theta(z)$ can be written as a diagonal block-matrix (6.64) with respect to decomposition (6.63), i.e.,

$$V_\Theta(z) = \begin{pmatrix} 0 & 0 \\ 0 & V_{\Theta,2}(z) \end{pmatrix},$$

where $V_{\Theta,2}(z) \in N_0(R)$.

Theorem 7.1.5 below is a version of Theorem 6.5.2 for the class $N_0(R)$.

Theorem 7.1.5. *Let an operator-valued function $V(z)$ in a Hilbert space E belong to the class $N_0(R)$. Then it can be realized as an impedance function of an L-system Θ with an invertible channel operator K, a preassigned directing operator J for which $I + iV(-i)J$ is invertible, a densely defined symmetric operator $\dot{A}$, and* $\mathrm{Dom}(T) \neq \mathrm{Dom}(T^*)$.

Proof. Since $N_0(R)$ is a subclass of $N(R)$, then all conditions of Theorem 6.5.2 are satisfied and the operator-valued function $V(z) \in N_0(R)$ can be realized as an impedance function of the L-system Θ constructed in the proof of Theorem 6.5.2. Thus, all we have to show is that $\overline{\mathrm{Dom}(\dot{A})} = \mathcal{H}$ and $\mathrm{Dom}(T) \neq \mathrm{Dom}(T^*)$.

Since $V(z)$ is a member of the class $N_0(R)$, then (7.1) holds for all non-zero $h \in E$ and hence $E_1 = \{0\}$, where E_1 is defined by (7.4). As it was shown in Step 2 of the proof of Theorem 6.5.1, any element $h \in E$ can be considered as an element $h = h(t) \in \mathfrak{H}_-$ and the following formula holds:

$$L^2_G(E) \ominus \overline{\mathrm{Dom}(\dot{A})} = E_1. \tag{7.6}$$

The right-hand side of (7.6) is trivial in our case and we can conclude that

$$\overline{\mathrm{Dom}(\dot{A})} = L^2_G(E) = \mathcal{H}.$$

Following the proof of Theorem 6.5.2 we construct the realizing L-system Θ with the operator T defined by (6.65). Let us show that $\mathrm{Dom}(T) \neq \mathrm{Dom}(T^*)$. Indeed, since $T \in \Lambda(\dot{A})$, then the densely-defined operator $\dot{A}$ is the maximal common symmetric part of T and T^*. Therefore, we can apply Lemmas 4.1.4, 4.2, and Theorem 4.1.11 to get the condition $\mathrm{Dom}(T) \neq \mathrm{Dom}(T^*)$. The proof of the theorem is complete. □

Corollary 7.1.6. *If $V(z)$ belongs to the class $N_0(R)$, then it can be realized as an impedance function of a scattering ($J = I$) L-system with an invertible channel operator K, a densely-defined closed symmetric operator $\dot{A}$, and* $\mathrm{Dom}(T) \neq \mathrm{Dom}(T^*)$.

Proof. The proof immediately follows from Theorems 6.5.5 and 7.1.5. □

Similar results for the class $N_1(R)$ can be obtained in the next two theorems.

Theorem 7.1.7. *Let Θ be an L-system of the form* (6.36), *where $\dot{A}$ is a symmetric O-operator and* $\mathrm{Dom}(T) = \mathrm{Dom}(T^*)$. *Then its impedance function $V_\Theta(z)$ of the form* (6.47) *belongs to the class $N_1(R)$.*

Proof. As in Theorem 7.1.4 we already know that the operator-valued function $V_\Theta(z)$ belongs to the class $N(R)$. Therefore it is enough to show that

$$\int_{-\infty}^{+\infty} (dG(t)h, h)_E < \infty,$$

for all $h \in E$ and (7.3) holds. Since it is given that $\dot{A}$ is a closed symmetric O-operator, we can use Theorem 4.1.12 saying that, for the system Θ,

$$\mathrm{Dom}(T) = \mathrm{Dom}(T^*) = \mathcal{H}_+ = \mathrm{Dom}(\dot{A}^*).$$

This fact implies that the $(*)$-extension $\mathbb{A}$ coincides with operator T. Consequently, $\mathbb{A}^* = T^*$ and our system Θ has a form

$$\Theta = \begin{pmatrix} T & K & J \\ \mathcal{H}_+ \subset \mathcal{H} \subset \mathcal{H}_- & & E \end{pmatrix},$$

where

$$\operatorname{Im} T = \frac{T - T^*}{2i} = KJK^*.$$

Taking into account that $\dim E < \infty$ and $K : E \to \mathcal{H}_-$ we conclude that $\dim(\operatorname{Ran}(\operatorname{Im} T)) < \infty$. Let

$$T = \operatorname{Re} T + i \operatorname{Im} T, \qquad T^* = \operatorname{Re} T - i \operatorname{Im} T,$$

where $\operatorname{Re} T = (1/2)(T + T^*)$. In our case the operator K acts from the space E into the space $\mathcal{H}$. Therefore $Kh = g \in \mathcal{H}$ for all $h \in E$. For the operator-valued function $V_\Theta(z)$ we can derive an integral representation for all $f \in E$,

$$\begin{aligned} \left(V_\Theta(z)f, f\right)_E &= \left(K^*(\operatorname{Re} T - zI)^{-1}Kf, f\right)_E = \left(K^* \int_{-\infty}^{+\infty} \frac{dE(t)}{t - z} Kf, f\right)_E \\ &= \int_{-\infty}^{+\infty} \frac{d\left(K^*E(t)Kf, f\right)_E}{t - z}, \end{aligned} \tag{7.7}$$

where $E(t)$ is the resolution of identity of the operator $\operatorname{Re} T$. Let

$$G(t) = K^*E(t)K.$$

Then

$$\begin{aligned} \int_{-\infty}^{+\infty} d(G(t)h, h) &= \int_{-\infty}^{+\infty} d(K^*E(t)Kh, h) = \int_{-\infty}^{+\infty} d(E(t)Kh, Kh) \\ &= \int_{-\infty}^{+\infty} d(E(t)g, g) = (g, g) \int_{-\infty}^{+\infty} dE(t) = (g, g) \\ &= (Kh, Kh) = (K^*Kh, h) = (\operatorname{Im} Th, h) < \infty, \end{aligned}$$

for all $h \in E$. Using the last relation and (7.7) we obtain the representation (7.3). □

Theorem 7.1.8. *Let an operator-valued function $V(z)$ in a Hilbert space E belong to the class $N_1(R)$. Then it can be realized as an impedance function of an L-system Θ with a preassigned directing operator J for which $I + iV(-i)J$ is invertible, a symmetric O-operator $\dot{A}$ with non-dense domain, and* $\operatorname{Dom}(T) = \operatorname{Dom}(T^*)$.

Proof. Similarly to Theorem 7.1.5 we can say that since $N_1(R)$ is a subclass of $N(R)$, then it is sufficient to show that operator $\dot{A}$ is a closed symmetric O-operator with a non-dense domain and $\mathrm{Dom}(T) = \mathrm{Dom}(T^*)$.

Once again we rely on the proof of Theorem 6.5.1 and use formula (7.6), together with the fact that $E_1 = E$, (where E_1 is defined by (6.71)), to obtain $\overline{\mathrm{Dom}(\dot{A})} \neq \mathcal{H} = L^2_G(E)$. Let $\mathfrak{N}'_{\pm i}$ be the semi-defect subspaces of operator $\dot{A}$ and $\mathfrak{N}^0_{\pm i}$ be the defect subspaces of operator A_1, described in Step 2 of the proof of Theorem 6.5.1. It was also shown there that

$$\mathfrak{N}_{\pm i} = \left\{ f(t) \in L^2_G(E),\ f(t) = \frac{h}{t \pm i},\quad h \in E \right\},$$

$$\mathfrak{N}^0_{\pm i} = \left\{ f(t) \in L^2_G(E),\ f(t) = \frac{h}{t \pm i},\ h \in E_1 \right\},$$

and $\mathfrak{N}'_{\pm i} = \mathfrak{N}_{\pm i} \ominus \mathfrak{N}^0_{\pm i}$, where $\mathfrak{N}{\pm i}$ are the defect spaces of the operator $\dot{A}$. In our case, however, $E = E_1$ and therefore the above formulas yield that $\mathfrak{N}_{\pm i} = \mathfrak{N}^0_{\pm i}$. Consequently, $\mathfrak{N}'_{\pm i} = \{0\}$ and hence $\dot{A}$ is an O-operator.

Note that $\dot{A}$ is also a regular symmetric operator. Thus, Theorem 4.1.12 is applicable and gives

$$\mathrm{Dom}(T) = \mathrm{Dom}(T^*).$$

This completes the proof of the theorem. □

Corollary 7.1.9. *If a function $V(z)$ belongs to the class $N_1(R)$, then it can be realized as an impedance function of a scattering $(J = I)$ L-system with a symmetric non-densely defined O-operator $\dot{A}$ and* $\mathrm{Dom}(T) = \mathrm{Dom}(T^*)$.

Proof. The proof of the corollary follows from Theorems 6.5.5 and 7.1.8. □

The following two theorems will complete our framework by establishing direct and inverse realization results for the remaining subclass of realizable operator-valued Herglotz-Nevanlinna functions $N_{01}(R)$.

Theorem 7.1.10. *Let Θ be an L-system of the form* (6.36) *with a symmetric non-densely defined operator $\dot{A}$ and* $\mathrm{Dom}(T) \neq \mathrm{Dom}(T^*)$. *Then its impedance function $V_\Theta(z)$ of the form* (6.47) *belongs to the class $N_{01}(R)$.*

Proof. We know that $V_\Theta(z)$ belongs to the class $N(R)$. To prove the statement of the theorem we only have to show that in the direct decomposition $E = E_1 \dotplus E_2$ (see Step 1 of the proof of Theorem 6.5.1) both components, defined in (6.71), are non-zero. In other words we have to show the existence of such vectors $h \in E$ that

$$\int_{-\infty}^{+\infty} d(G(t)h, h) = \infty, \tag{7.8}$$

and vectors $f \in E$, $f \neq 0$ that

$$\int_{-\infty}^{+\infty} d(G(t)f,f) < \infty. \tag{7.9}$$

Let $\mathcal{H}_0 = \overline{\operatorname{Dom}(\dot{A})}$ and $\mathfrak{L} = \mathcal{H} \ominus \mathcal{H}_0$. Since $\overline{\operatorname{Dom}(\dot{A})} = \mathcal{H}_0 \neq \mathcal{H}$, $\mathfrak{L}$ is non-empty. $K^{-1}\mathfrak{L}$ is obviously a subset of E. Moreover, according to Theorem 6.2.10 for all $f \in K^{-1}\mathfrak{L}$, (7.9) holds. Thus, $K^{-1}\mathfrak{L}$ is a non-zero subset of E_1.

Now we have to show that the vectors satisfying (7.8) make a non-zero subset of E as well. Indeed, the definition of L-system and condition $\operatorname{Dom}(T) \neq \operatorname{Dom}(T^*)$ implies that $\operatorname{Ran}(K) = \operatorname{Ran}(\operatorname{Im}\mathbb{A}) \subseteq L_{\mathbb{A}} \dot{+} \mathfrak{L}$, where L_A was defined by (4.44). Therefore, there exist $g \in \mathcal{H}_-$, $g \notin \mathfrak{L}$, $f \in E$ such that $Kf = g \notin \mathfrak{L}$. Then, according to Theorem 6.2.10, for this $f \in E$, (7.8) holds. The proof of the theorem is complete. □

Theorem 7.1.11. *Let an operator-valued function $V(z)$ in a Hilbert space E belong to the class $N_{01}(R)$. Then it can be realized as an impedance function of an L-system Θ with a preassigned directing operator J for which $I + iV(-i)J$ is invertible, a symmetric non-densely defined operator $\dot{A}$, and* $\operatorname{Dom}(T) \neq \operatorname{Dom}(T^*)$.

Proof. Once again all we have to show is that $\overline{\operatorname{Dom}(\dot{A})} \neq \mathcal{H}$. We have already mentioned (7.6) that $L^2_G(E) \ominus \overline{\operatorname{Dom}(\dot{A})} = E_1$. This implies that $\operatorname{Dom}(\dot{A})$ is dense in $\mathcal{H}$ if and only if $E_1 = 0$. Since the class $N_{01}(R)$ assumes the existence of non-zero vectors $f \in E$ such that (7.9) is true, we can conclude that $E_1 \neq 0$ and therefore $\overline{\operatorname{Dom}(\dot{A})} \neq \mathcal{H}$.

In the proofs of Theorems 7.1.5 and 7.1.8 we have shown that $\operatorname{Dom}(T) = \operatorname{Dom}(T^*)$ in the case when $E_2 = 0$. If $E_2 \neq 0$, then $\operatorname{Dom}(T) \neq \operatorname{Dom}(T^*)$. The definition of the class $N_{01}(R)$ implies that $E_2 \neq 0$. Thus we have $\operatorname{Dom}(T) \neq \operatorname{Dom}(T^*)$. The proof is complete. □

Corollary 7.1.12. *If a function $V(z)$ belongs to the class $N_{01}(R)$, then it can be realized as an impedance function of a scattering ($J = I$) L-system Θ with a symmetric non-densely defined operator $\dot{A}$ and* $\operatorname{Dom}(T) \neq \operatorname{Dom}(T^*)$.

Proof. The proof of the corollary follows from Theorems 6.5.5 and 7.1.11. □

7.2 Class $\Omega(R,J)$. The Potapov-Ginzburg Transformation

In this section we introduce the class $\Omega(R,J)$ of J-contractive in a half-plane operator-valued functions that are the transfer functions of L-systems.

Definition 7.2.1. An operator-valued function $W(z)$ in a finite-dimensional Hilbert space E holomorphic in the domain G_W in the lower half-plane is said to be a member of the **class** $\Omega(R,J)$, where $J \in [E,E]$, $J = J^* = J^{-1}$, if

1. the operator $I - W(z_0)$ is invertible for some $z_0 \in G_W$,
2. the operator-valued function

$$V(z) = i(W(z)+I)^{-1}(W(z)-I)J, \quad z \in G_W, \tag{7.10}$$

admits a holomorphic continuation to the Herglorz-Nevanlinna function from the class $N(R)$ (in the lower half-plane).

It follows from condition 2 that, for $z \in G_W$,

$$\begin{aligned}\operatorname{Im} V(z) &= (W(z)+I)^{-1}(W(z)JW^*(z)-J)(W^*(z)+I)^{-1}\\ &= J(W^*(z)+I)^{-1}(W^*(z)JW(z)-J)(W(z)+I)^{-1}J.\end{aligned}$$

Therefore,

$$W(z)JW^*(z) - J \leq 0, \quad W^*(z)JW(z) - J \leq 0, \quad z \in G_W,$$

i.e., the function $W(z)$ is J-contractive.

Theorem 7.2.2. *Let $W(z) \in \Omega(R,J)$. Then there exists an L-system*

$$\Theta = \begin{pmatrix} \mathbb{A} & K & J \\ \mathcal{H}_+ \subset \mathcal{H} \subset \mathcal{H}_- & & E \end{pmatrix},$$

such that $\ker(K) = \{0\}$, $W(z) = W_\Theta(z)$ *for* $z \in G_W$.

Proof. It follows from Definition 7.2.1 that the operator $V(z_0)$ is invertible. Since $V(z)$ admits a holomorphic continuation to a function from the class $N(R)$, the subspace $\ker(V(z))$ does not depend on the choice z from the lower half-plane (see (6.64), (6.90)). Applying Theorem 6.5.2 we obtain the desired statement. □

Definition 7.2.3. An operator-valued function $W(z)$ of the class $\Omega(R,J)$ belongs to the **class** $\Omega_0(R,J)$ **(resp.** $\Omega_1(R,J)$, $\Omega_{01}(R,J)$**)** if the operator-valued function $V(z)$ defined by (7.10) belongs to the class $N_0(R)$ (resp. $N_1(R,J)$, $N_{01}(R,J)$). [2]

The theorem below is a version of **Potapov-Ginzburg transformation**. Together with Corollary 7.2.5 it establishes the relation between contractive and J-contractive in the half-plane operator-valued functions from the classes $\Omega(R,J)$, $\Omega_0(R,J)$, $\Omega_1(R,J)$, and $\Omega_{01}(R,J)$.

[2]The definition of the class $\Omega_{01}(R)$ implies that it does not exist if $\dim E = 1$.

Theorem 7.2.4. *Let operator-valued function $W(z)$ belong to the class $\Omega(R,J)$. Let also P^+ and P^- be a pair of orthogonal projections of the form*

$$P^+ = \frac{1}{2}(I+J) \qquad \text{and} \qquad P^- = \frac{1}{2}(I-J).$$

Then there exists an operator-function $\Sigma(z)$ of the class $\Omega(R,I)$ such that

$$W(z) = (P^+\Sigma(z) - P^-)(P^+ - P^-\Sigma(z))^{-1}.$$

Proof. Let

$$V(z) = i[W(z)+I]^{-1}[W(z)-I]J.$$

Since $W(z)$ belongs to $\Omega(R,J)$ we have that $V(z)$ belongs to the class $N(R)$ and thus, by Theorem 6.5.2, can be realized as an impedance function of a scattering L-system

$$\Theta' = \begin{pmatrix} \mathbb{A}' & K' & I \\ \mathcal{H}'_+ \subset \mathcal{H}' \subset \mathcal{H}'_- & & E \end{pmatrix}.$$

The latter implies that

$$V(z) = V_{\Theta'}(z) = K'^*(\mathrm{Re}\,\mathbb{A}' - zI)^{-1}K' = i[W_{\Theta'}(z)+I]^{-1}[W_{\Theta'}(z)-I],$$

for $z \in \rho(T')$ where

$$W_{\Theta'}(z) = I - 2iK'^*(\mathbb{A}' - zI)^{-1}K'.$$

It is clear that $W_{\Theta'}(z)$ belongs to the $\Omega(R,I)$ class. Therefore

$$i[W(z)+I]^{-1}[W(z)-I]J = i[W_{\Theta'}(z)+I]^{-1}[W_{\Theta'}(z)-I], \quad z \in \rho(T'),$$

where $W(z) \in \Omega(R,J)$ and $W_{\Theta'}(z) \in \Omega(R,I)$. Now let $\Sigma(z) \equiv W_{\Theta'}(z)$. Then

$$\begin{aligned}(W(z)+I)^{-1}(W(z)-I)J &= [\Sigma(z)+I]^{-1}[\Sigma(z)-I] \\ &= [\Sigma(z)-I][\Sigma(z)+I]^{-1}.\end{aligned}$$

It follows that

$$[W(z)-I]J[\Sigma(z)+I] = [W(z)+I][\Sigma(z)-I].$$

Taking into account that $P^+ - P^- = J$ and $P^+ + P^- = I$ we obtain

$$[W(z)-I](P^+ - P^-)[\Sigma(z)+I] = [W(z)+I]J[\Sigma(z)-I],$$

or

$$\begin{aligned}W(z)P^+\Sigma(z) - W(z)P^-\Sigma(z) - P^+\Sigma(z) + P^-\Sigma(z)& \\ + W(z)P^+ - W(z)P^- - P^+ + P^-& \\ = W(z)\Sigma(z) - W(z) + \Sigma(z) - I&,\end{aligned}$$

$$W(z)[P^+\Sigma(z) - P^-\Sigma(z) + 2P^+ - \Sigma(z)] = [P^+\Sigma(z) - P^-\Sigma(z) + \Sigma(z) - 2P^-],$$

or

$$W(z)[2P^+ - 2P^-\Sigma(z)] = [2P^+\Sigma(z) - 2P^-].$$

Cancellation yields

$$W(z)[P^+ - P^-\Sigma(z)] = [P^+\Sigma(z) - P^-]. \tag{7.11}$$

Let us show that operator $[P^+ - P^-\Sigma(z)]$ is invertible. We choose $\phi \in E$ such that

$$[P^+ - P^-\Sigma(z)]\phi = 0. \tag{7.12}$$

Then (7.11) implies

$$[P^+\Sigma(z) - P^-]\phi = 0. \tag{7.13}$$

We apply $P+$ to both sides of (7.12) and obtain

$$P^+[P^+ - P^-\Sigma(z)]\phi = 0,$$

or $P^+\phi = 0$. Similarly, we apply P^- to both sides of (7.13) and get that $P^-\phi = 0$. Thus $\phi = 0$ and operator $[P^+ - P^-\Sigma(z)]$ is invertible. Using this we obtain

$$W(z) = [P^+\Sigma(z) - P^-][P^+ - P^-\Sigma(z)]^{-1},$$

that proves the theorem. □

Corollary 7.2.5. *Let the operator-valued function $W(z)$ belong to the class $\Omega_0(R,J)$ (resp. $\Omega_1(R,J)$, $\Omega_{01}(R,J)$), $P^+ = 1/2(I+J)$, $P^- = 1/2(I-J)$. Then there exists an operator-valued function $\Sigma(z)$ from $\Omega_0(R,J)$ (resp. $\Omega_1(R,J)$, $\Omega_{01}(R,J)$) class such that*

$$W(z) = [P^+\Sigma(z) - P^-][P^+ - P^-\Sigma(z)]^{-1}.$$

Corollary 7.2.5 can be proved in exactly the same way as Theorem 7.2.4.

7.3 Multiplication Theorems for $\Omega(R,J)$ classes

In this section we state and prove multiplication theorems for the operator-valued functions of $\Omega(R,J)$ class.

Let

$$\Theta_1 = \begin{pmatrix} \mathbb{A}_1 & K_1 & J \\ \mathcal{H}_{+1} \subset \mathcal{H}_1 \subset \mathcal{H}_{-1} & & E \end{pmatrix} \text{ and } \Theta_2 = \begin{pmatrix} \mathbb{A}_2 & K_2 & J \\ \mathcal{H}_{+2} \subset \mathcal{H}_2 \subset \mathcal{H}_{-2} & & E \end{pmatrix}$$

be two L-systems with $\ker(K_1) = \ker(K_2) = \{0\}$. Let

$$\mathcal{H}_+ = \mathcal{H}_{+1} \oplus \mathcal{H}_{+2}, \quad \mathcal{H} = \mathcal{H}_1 \oplus \mathcal{H}_2, \quad \mathcal{H}_- = \mathcal{H}_{-1} \oplus \mathcal{H}_{-2},$$

and $P_k : \mathcal{H} \to \mathcal{H}_k$, $P_k^+ : \mathcal{H}_+ \to \mathcal{H}_{+k}$, and $P_k^- : \mathcal{H}_- \to \mathcal{H}_{-k}$ $(k = 1,2)$ denote the set of orthoprojections. In the space $\mathcal{H}$ we introduce an operator

$$\dot{\mathcal{A}} = \dot{A}_1 \oplus \dot{A}_2, \tag{7.14}$$

where $\dot{A}_1 \subset T_1 \subset \mathbb{A}_1$, $\dot{A}_2 \subset T_2 \subset \mathbb{A}_2$ are correspondent elements of Θ_1 and Θ_2, respectively. Moreover, $\mathcal{H}_{+1} = \mathrm{Dom}(\dot{A}_1^*)$ and $\mathcal{H}_{+2} = \mathrm{Dom}(\dot{A}_2^*)$. Consequently,

$$\dot{\mathcal{A}}^* = \dot{A}_1^* \oplus \dot{A}_2^*,$$

and $\mathcal{H}_+ = \mathrm{Dom}(\mathcal{A}^*) = \mathrm{Dom}(\dot{A}_1^*) \oplus \mathrm{Dom}(\dot{A}_2^*)$. Now define $\mathbf{A} : \mathcal{H}_+ \to \mathcal{H}_-$

$$\mathbf{A} = \mathbb{A}_1 P_1^+ + \mathbb{A}_2 P_2^+ + 2iK_1JK_2^*P_2^+. \tag{7.15}$$

Then the adjoint $\mathbf{A}^* : \mathcal{H}_+ \to \mathcal{H}_-$ is given by

$$\mathbf{A}^* = \mathbb{A}_1^* P_1^+ + \mathbb{A}_2^* P_2^+ - 2iK_2JK_1^*P_1^+. \tag{7.16}$$

Both operators $\mathbf{A}$ and $\mathbf{A}^*$ are extensions of the operator $\dot{\mathcal{A}}$. Put

$$\mathbf{K} = K_1 + K_2. \tag{7.17}$$

Clearly, $\mathbf{K} \in [E, \mathcal{H}_-]$, $\ker(\mathbf{K}) = \{0\}$, and $\mathbf{K}^* \in [\mathcal{H}_+, E]$ is of the form

$$\mathbf{K}^* = K_1^* P_1^+ + K_2^* P_2^+. \tag{7.18}$$

In addition $\mathbf{A} - \mathbf{A}^* = 2i\mathbf{K}J\mathbf{K}^*$. From (7.15)–(7.18) it follows that the resolvents $(\mathbf{A} - zI)^{-1}$ and $(\mathbf{A}^* - \bar{z}I)^{-1}$ are defined for $z \in \rho(T_1) \cap \rho(T_2)$ on the linear manifold $\mathcal{H} + \mathrm{Ran}(\mathbf{K})$ and take the form

$$\begin{aligned}(\mathbf{A} - zI)^{-1} =& (\mathbb{A}_1 - zI)^{-1} P_1^- + (\mathbb{A}_2 - zI)^{-1} P_2^- \\ &- 2i(\mathbb{A}_1 - zI)^{-1} K_1JK_2^*(\mathbb{A}_2 - zI)^{-1} P_2^-,\end{aligned} \tag{7.19}$$

$$\begin{aligned}(\mathbf{A}^* - \bar{z}I)^{-1} =& (\mathbb{A}_1^* - \bar{z}I)^{-1} P_1^- + (\mathbb{A}_2^* - \bar{z}I)^{-1} P_2^- \\ &+ 2i(\mathbb{A}_2^* - zI)^{-1} K_2JK_1^*(\mathbb{A}_1 - zI)^{-1} P_1^-.\end{aligned} \tag{7.20}$$

In particular, $(\mathbf{A} - zI)^{-1}\mathcal{H} \subset \mathcal{H}$, $(\mathbf{A}^* - \bar{z}I)^{-1}\mathcal{H} \subset \mathcal{H}$. Let T and T_* be quasi-kernels of $\mathbf{A}$ and $\mathbf{A}^*$, respectively, i.e.,

$$\begin{aligned}\mathrm{Dom}(T) &= \{f \in \mathcal{H}_+ : \mathbf{A}f \in \mathcal{H}\}, \quad T = \mathbf{A} \upharpoonright \mathrm{Dom}(T), \\ \mathrm{Dom}(T_*) &= \{f \in \mathcal{H}_+ : \mathbf{A}^*f \in \mathcal{H}\}, \quad T_* = \mathbf{A}^* \upharpoonright \mathrm{Dom}(T_*).\end{aligned}$$

Then for $z \in \rho(T_1) \cap \rho(T_2)$ operators T and T_* have the resolvent

$$(T - zI)^{-1} = (\mathbf{A} - zI)^{-1} \upharpoonright \mathcal{H}, \ (T_* - \bar{z}I)^{-1} = (\mathbf{A}^* - \bar{z}I)^{-1} \upharpoonright \mathcal{H}.$$

Thus $T_* = T^*$. Obviously, $T \supset \dot{\mathcal{A}}$ and $T^* \supset \dot{\mathcal{A}}$. So, the operator $\mathbf{A}$ is a quasi-self-adjoint bi-extension of T with the range property (R), therefore T belongs to the class $R(\dot{\mathcal{A}})$ (see Definitions 4.2.2 and 4.2.3). It follows from Theorem 4.2.9 and Theorem 4.3.2 that $\mathrm{Re}\,\mathbf{A}$ is a t-self-adjoint bi-extension of $\dot{\mathcal{A}}$ in (7.14), i.e., the quasi-kernel B of $\mathrm{Re}\,\mathbf{A}$ is a self-adjoint extension of $\dot{\mathcal{A}}$. Let us set

$$\mathrm{Dom}(\dot{A}) = \{g \in \mathcal{H}_+ : Tg = T^*g\},$$

and define an operator $\dot{A}$ on $\mathrm{Dom}(\dot{A})$ via

$$\dot{A}g = \mathbf{A}g, \qquad g \in \mathrm{Dom}(\dot{A}). \tag{7.21}$$

Clearly, $\dot{A}$ is a maximal symmetric part of T and T^* and by constructions we have $\dot{\mathcal{A}} \subset \dot{A} \subset T$ and $\dot{\mathcal{A}} \subset \dot{A} \subset T^*$. We are going to show that

$$\begin{aligned}\mathrm{Dom}(\dot{A}) = \{g \in \mathcal{H}_+ : g = g_1 + g_2,\ g_1 \in \mathrm{Dom}(T_1^*),\\ g_2 \in \mathrm{Dom}(T_2) \text{ and } K_1^*g_1 + K_2^*g_2 = 0\}.\end{aligned} \tag{7.22}$$

Let us pick an element $g = g_1 + g_2 \in \mathrm{Dom}(\dot{A})$, where $g_1 \in \mathcal{H}_{+1}$, $g_2 \in \mathcal{H}_{+2}$. We are going to show that g fits the description in (7.22). It follows from (7.15) and (7.16) that

$$\mathbb{A}_1 g_1 + \mathbb{A}_2 g_2 + 2iK_1JK_2^*g_2 = \mathbb{A}_1^*g_1 + \mathbb{A}_2^*g_2 - 2iK_2JK_1^*g_1 \in \mathcal{H}. \tag{7.23}$$

Moreover,

$$\begin{aligned}\mathbb{A}_1 g_1 + 2iK_1JK_2^*g_2 &= \mathbb{A}_1^*g_1 \in \mathcal{H}_{-1},\\ \mathbb{A}_2 g_2 &= \mathbb{A}_2^*g_2 - 2iK_2JK_1^*g_1 \in \mathcal{H}_{-2}.\end{aligned} \tag{7.24}$$

From (7.24) we get

$$\begin{aligned}&\mathbb{A}_1 g_1 - \mathbb{A}_1^* g_1 = -2iK_1JK_2^*g_2, \quad \frac{1}{2i}(\mathbb{A}_1 - \mathbb{A}_1^*)g_1 = -K_1JK_2^*g_2,\\ &K_1JK_1^*g_1 = -K_1JK_2^*g_2, \quad K_1JK_1^*g_1 + K_1JK_2^*g_2 = 0,\\ &K_1J(K_1^*g_1 + K_2^*g_2) = 0.\end{aligned}$$

Since the operator K_1 is invertible, we obtain

$$K_1^*g_1 + K_2^*g_2 = 0,$$

or $K_1^*g_1 = -K_2^*g_2$. Also, (7.23) and (7.24) imply $\mathbb{A}_2 g_2 - \mathbb{A}_1^* g_1 \in \mathcal{H}$ that yields $g_1 \in \mathrm{Dom}(T_1^*)$ and $g_2 \in \mathrm{Dom}(T_2)$.

Let us show now that if $g_1 \in \mathrm{Dom}(T_1^*)$, $g_2 \in \mathrm{Dom}(T_2)$ and $K_1^*g_1 + K_2^*g_2 = 0$, then $g = g_1 + g_2$ belongs to $\mathrm{Dom}(\dot{A})$. We have

$$\begin{aligned}\mathbf{A}g &= \mathbb{A}_1 g_1 + \mathbb{A}_2 g_2 + 2iK_1JK_2^*g_2 = \mathbb{A}_1 g_1 + T_2 g_2 - 2iK_1JK_1^*g_1\\ &= \mathbb{A}_1 g_1 + T_2 g_2 - \mathbb{A}_1 g_1 + \mathbb{A}_1^* g_1 = T_2 g_2 + T^* g_1 \in \mathcal{H}.\end{aligned}$$

Therefore, $\mathbf{A}g = \mathbf{A}^*g$ belongs to $\mathcal{H}$ or $Tg = T^*g$, implying that $g \in \mathrm{Dom}(\dot{A})$. Let $\dot{A}^*$ be an adjoint to opewrator $\dot{A}$. Then $\mathrm{Dom}(\dot{A}^*) \subset \mathrm{Dom}(\dot{\mathcal{A}}^*) = \mathcal{H}_+$. Let us set

$$H_+ = \mathrm{Dom}(\dot{A}^*),$$

and construct a new rigged Hilbert space

$$H_+ \subset \mathcal{H} \subset H_-. \tag{7.25}$$

It is easy to see that the following inclusions take place:

$$\begin{array}{c} H_+ \hookrightarrow \mathcal{H}_+ \subset \mathcal{H} \subset \mathcal{H}_- \hookrightarrow H_- \\ \cap \\ H_- \end{array}$$

Let us denote by γ an embedding operator acting from H_+ into $\mathcal{H}_+$:

$$\gamma : H_+ \hookrightarrow \mathcal{H}_+, \quad \gamma f = f, \ \forall f \in H_+. \tag{7.26}$$

Let us define an adjoint operator γ^* as $\gamma^* : \mathcal{H}_- \hookrightarrow H_-$ We have $\gamma^* h = h$ for all $h \in \mathcal{H}$. Indeed, for all $f \in H_+$, $h \in \mathcal{H}$,

$$(f, h) = (\gamma f, h) = (f, \gamma^* h).$$

Let the operator $\mathbb{A} \in [H_+, H_-]$ be defined as

$$\mathbb{A} = \gamma^* \mathbf{A} \upharpoonright H_+. \tag{7.27}$$

Since for all $f, g \in H_+$,

$$\begin{aligned} (\mathbb{A} f, g) &= (\gamma^* \mathbf{A} f, g) = (\mathbf{A} f, \gamma g) \\ &= (\mathbf{A} f, g) = (f, \mathbf{A}^* g) = (f, \gamma^* \mathbf{A}^* g) = (f, \mathbb{A}^* g), \end{aligned}$$

we get

$$\mathbb{A}^* = \gamma^* \mathbf{A}^* \upharpoonright H_+.$$

Let $f \in \mathrm{Dom}(T)$. Then $\mathbf{A} f \in \mathcal{H}$ and $\mathbb{A} f = \gamma^* \mathbf{A} f = \gamma^* T f = T f \in \mathcal{H}$. Thus, $\mathbb{A} \supset T$. Similarly, $\mathbb{A}^* \supset T^*$. Let us show now that operator $\mathbb{A}$ defined by (7.27) can be included as a state-space operator into an L-system Θ. Define $K \in [E, H_-]$ as

$$K = \gamma^* \mathbf{K}, \tag{7.28}$$

where $\mathbf{K}$ is given by (7.17). Then, clearly, $K^* = \mathbf{K}^* \upharpoonright H_+$. Furthermore,

$$\frac{1}{2i}(\mathbb{A} - \mathbb{A}^*) = \frac{1}{2i}(\gamma^* \mathbf{A} - \gamma^* \mathbf{A}^*) = \frac{1}{2i}\gamma^*(\mathbf{A} - \mathbf{A}^*) = \gamma^* \mathbf{K} J \mathbf{K}^* = K J K^*.$$

Thus, $\mathbb{A} \supset T \supset \dot{A}$, $\mathbb{A}^* \supset T^* \supset \dot{A}$, $\mathrm{Im}\mathbb{A} = KJK^*$, where K is defined by (7.28). Moreover, since $\mathrm{Re}\,\mathbf{A}$ is a t-self-adjoint bi-extension of $\dot{\mathcal{A}}$ with the quasi-kernel $B = B^*$, the operator $\mathrm{Re}\,\mathbb{A} = \gamma^* \mathrm{Re}\,\mathbf{A}$ is a t-self-adjoint bi-extension of $\dot{A}$ with the same quasi-kernel B, i.e., $\mathbb{A}$ is a $(*)$-extension of T and hence $T \in \Lambda(\dot{A})$. Let us show additionally that

$$(\mathbb{A} - zI)^{-1} K = (\mathbf{A} - zI)^{-1} \mathbf{K}, \tag{7.29}$$

and
$$(\mathbb{A}-zI)^{-1}KE=\mathfrak{N}_z(\dot{A}) \tag{7.30}$$
for $z\in\rho(T)$, where $\mathfrak{N}_z(\dot{A})$ is the deficiency subspace of $\dot{A}$. Equality (7.29) follows from $(\mathbb{A}-zI)(\mathbf{A}-z)^{-1}\mathbf{K}=\gamma^*(\mathbf{A}-zI)(\mathbf{A}-zI)^{-1}\mathbf{K}=\gamma^*\mathbf{K}=K$. In order to prove (7.30) suppose $f\in\mathcal{H}$ and $(f,(\dot{A}-zI)^{-1}Kh)$ for all $h\in E$. Then there exists $g\in\operatorname{Dom}(T^*)$ such that $f=(T^*-\bar{z}I)g$. Then by (7.20) we have
$$f=(\mathbb{A}_1^*-\bar{z}I)P_1^+g+(\mathbb{A}_2^*-\bar{z}I)P_2^+g-2iK_2JK_1^*P_1^+g.$$
Further using (7.19) and (7.29) we get
$$\begin{aligned}0&=((T^*-\bar{z}I)g,(\mathbb{A}-zI)^{-1}Kh)=((T^*-\bar{z}I)g,(\mathbf{A}-zI)^{-1}\mathbf{K}h)\\&=((\mathbb{A}_1^*-\bar{z}I)P_1^+g,(\mathbb{A}_1-zI)^{-1}P_1^-\mathbf{K}h)\\&\quad+2i((\mathbb{A}_1^*-\bar{z}I)P_1^+g,(\mathbb{A}_1-zI)^{-1}K_1JK_2^*(\mathbb{A}_2-zI)^{-1}P_2^-\mathbf{K}h)\\&\quad+((\mathbb{A}_2^*-\bar{z}I)P_2^+g-2iK_2JK_1^*P_1^+g,(\mathbb{A}_2-zI)^{-1}P_2^-\mathbf{K}h)\\&=(P_1^+g,P_1^-\mathbf{K}h-2iK_1JK_2^*(\mathbb{A}-zI)^{-1}P_2^-\mathbf{K}h)\\&\quad+(P_2^+g,P_2^-\mathbf{K}h)-2(iK_2JK_1^*P_1^+g,(\mathbb{A}_2-zI)^{-1}P_1^-\mathbf{K}h)\\&=(g,\mathbf{K}h)=(\mathbf{K}^*g,h)_E.\end{aligned}$$

Hence $\mathbf{K}^*g=0$ and therefore, $\operatorname{Im}\mathbf{A}g=\mathbf{K}J\mathbf{K}^*g=0$. Since $g\in\operatorname{Dom}(T^*)$, we get $Tg-T^*g=0$, i.e., $g\in\operatorname{Dom}(\dot{A})$ and $f=(\dot{A}-\bar{z}I)g$. Thus $(\mathbb{A}-zI)^{-1}KE\supseteq\mathfrak{N}_z(\dot{A})$. On the other hand if $g\in\operatorname{Dom}(\dot{A})$, then $\operatorname{Im}\mathbb{A}g=\operatorname{Im}\mathbf{A}g=0$. It follows that $K_1J\mathbf{K}^*g=K_2J\mathbf{K}^*g=0$. Since $\ker(K_1)=\{0\}$ and $\ker(K_2)=\{0\}$, we get $\mathbf{K}^*g=0$. Hence
$$((\dot{A}-\bar{z}I)g,(\mathbb{A}-zI)^{-1}Kh)=(\mathbf{A}-zI)^{-1}\mathbf{K}h)=(g,\mathbf{K}h)=(\mathbf{K}^*g,h)_E=0,$$
for all $h\in E$. This means that $(\mathbb{A}-zI)^{-1}KE\subseteq\mathfrak{N}_z(\dot{A})$ and thus (7.30) holds. From Theorem 4.3.2, the equality $\operatorname{Im}\mathbb{A}=KJK^*$, and equality (7.30) we get that $\operatorname{Ran}(K)=\operatorname{Ran}(\operatorname{Im}\mathbb{A})$ and the system
$$\Theta=\begin{pmatrix}\mathbb{A} & K & J\\ H_+\subset\mathcal{H}\subset H_- & & E\end{pmatrix}, \tag{7.31}$$
is an L-system. Let now
$$\begin{aligned}W_{\Theta_1}(z)&=I-2iK_1^*(\mathbb{A}_1-zI)^{-1}K_1J,\\W_{\Theta_2}(z)&=I-2iK_2^*(\mathbb{A}_2-zI)^{-1}K_2J,\end{aligned}$$
be transfer operator-valued functions of the L-systems Θ_1 and Θ_2, respectively. We introduce a new auxiliary system
$$\boldsymbol{\theta}=\begin{pmatrix}\mathbf{A} & \mathbf{K} & J\\ \mathcal{H}_+\subset\mathcal{H}\subset\mathcal{H}_- & & E\end{pmatrix},$$

where all its components are described above. Let us note that $\boldsymbol{\theta}$ does not exactly satisfy the Definition 6.3.4 of the L-system because operator $\dot{\mathcal{A}}$ is not the maximal symmetric part of T and T^* and thus $T \notin \Lambda(\dot{\mathcal{A}})$. System $\boldsymbol{\theta}$ will, however, suffice for our purposes. Let

$$W_{\boldsymbol{\theta}}(z) = I - 2i\mathbf{K}^*(\mathbf{A} - zI)^{-1}\mathbf{K}J, \quad z \in \rho(T_1) \cap \rho(T_2),$$

be the transfer operator-valued function of the system $\boldsymbol{\theta}$. This implies

$$\begin{aligned}
W_{\boldsymbol{\theta}}(z) &= I - 2i\mathbf{K}^*(\mathbf{A} - zI)^{-1}\mathbf{K}J \\
&= I - 2i\mathbf{K}^* P_1^-(\mathbb{A}_1 - zI)^{-1} P_1^- \mathbf{K}J - 2i\mathbf{K}^* P_2^-(\mathbb{A}_2 - zI)^{-1} P_2^- \mathbf{K}J \\
&\quad + (2i)^2 \mathbf{K}^* P_1^-(\mathbb{A}_1 - zI)^{-1} K_1 J K_2^*(\mathbb{A}_2 - zI)^{-1} P_2^- \mathbf{K}J \\
&= I + 2iK_1^*(\mathbb{A}_1 - zI)^{-1} K_1 J - 2iK_2^*(\mathbb{A}_2 - zI)^{-1} K_2 J \\
&\quad + (2i)^2 K_1^*(\mathbb{A}_1 - zI)^{-1} K_1 J K_2^*(\mathbb{A}_2 - zI)^{-1} K_2 J \\
&= (I - 2iK_1^*(\mathbb{A}_1 - zI)^{-1} K_1 J)(I - 2iK_2^*(\mathbb{A}_2 - zI)^{-1} K_2 J) \\
&= W_{\Theta_1}(z) \cdot W_{\Theta_2}(z).
\end{aligned}$$

We have just shown that

$$W_{\boldsymbol{\theta}}(z) = W_{\Theta_1}(z) \cdot W_{\Theta_2}(z).$$

Let now

$$W_\Theta(z) = I - 2iK^*(\mathbb{A} - zI)^{-1}KJ, \quad z \in \rho(T_1) \cap \rho(T_2),$$

be the transfer function of the L-system Θ defined by (7.31). We will show now that

$$W_\Theta(z) = W_{\boldsymbol{\theta}}(z) = W_{\Theta_1}(z) \cdot W_{\Theta_2}(z).$$

Equality (7.29) yields

$$K^*(\mathbb{A} - zI)^{-1}K = \mathbf{K}^*(\mathbf{A} - zI)^{-1}\mathbf{K}.$$

Hence $W_{\boldsymbol{\theta}}(z) = W_\Theta(z)$. Now we have $W_\Theta(z) = W_{\Theta_1}(z) \cdot W_{\Theta_2}(z)$.

Definition 7.3.1. An L-system Θ of the form (7.31) is called a **coupling of two L-systems** ($\Theta = \Theta_1 \cdot \Theta_2$)

$$\Theta_1 = \begin{pmatrix} \mathbb{A}_1 & K_1 & J \\ \mathcal{H}_{+1} \subset \mathcal{H}_1 \subset \mathcal{H}_{-1} & & E \end{pmatrix} \text{ and } \Theta_2 = \begin{pmatrix} \mathbb{A}_2 & K_2 & J \\ \mathcal{H}_{+2} \subset \mathcal{H}_2 \subset \mathcal{H}_{-2} & & E \end{pmatrix} \tag{7.32}$$

with $\ker(K_1) = \ker(K_2) = \{0\}$, if operators $\mathbb{A}$, K and rigged space $H_+ \subset \mathcal{H} \subset H_-$ are defined by the formulas (7.27), (7.28) and (7.25), respectively.

Theorem 7.3.2. *Let an L-system Θ be the coupling of two L-systems Θ_1 and Θ_2 of the form (7.32). Then if $z \in \rho(T_1) \cap \rho(T_2)$,*

$$W_\Theta(z) = W_{\Theta_1}(z) \cdot W_{\Theta_2}(z).$$

The proof of Theorem 7.3.2 was constructively obtained above.

We recall that if a function $W(z)$ belongs to the class $\Omega(R,J)$, then according to Definition 7.2.1, by G_W we denote the domain in the lower half-plane where $W(z)$ is holomorphic.

Theorem 7.3.3. *Let operator-valued functions $W_1(z)$ and $W_2(z)$ in the Hilbert space E belong to the class $\Omega(R,J)$ and $G_{W_1}\cap G_{W_2}\neq\emptyset$. Then the operator-valued function*

$$W(z)=W_1(z)\cdot W_2(z)$$

is defined on $G_{W_1}\cap G_{W_2}$ and belongs to the class $\Omega(R,J)$.

Proof. Since the operator-valued functions $W_1(z)$ and $W_2(z)$ belong to the class $\Omega(R,J)$, then there exist two L-system Θ_1 and Θ_2 such that $W_1(z)=W_{\Theta_1}(z)$ and $W_2(z)=W_{\Theta_2}(z)$ for $z\in\rho(T_1)\cap\rho(T_2)$. We note that, in view of Remark 6.5.3, the set $\rho(T_1)\cap\rho(T_2)$ is a non-empty set in $\mathbb{C}$ containing some neighborhood of the point $(-i)$. Thus, we can say that $W(z)$ is defined on some domain of the complex plane. Let system Θ be the coupling of Θ_1 and Θ_2. Then according to Theorem 7.3.2,

$$W_\Theta(z)=W_{\Theta_1}(z)\cdot W_{\Theta_2}(z)=W_1(z)\cdot W_2(z)=W(z).$$

Thus, for the function $W(z)$ there exists an L-system Θ with $W(z)=W_\Theta(z)$ for $z\in\rho(T_1)\cap\rho(T_2)$. Therefore, $W(z)=W_1(z)\cdot W_2(z)\in\Omega(R,J)$. □

The next theorem establishes a similar result for the class $\Omega_0(R,J)$.

Theorem 7.3.4. *Let operator-valued functions $W_1(z)$ and $W_2(z)$ in the Hilbert space E belong to the class $\Omega_0(R,J)$ and $G_{W_1}\cap G_{W_2}\neq\emptyset$. Then the operator-valued function*

$$W(z)=W_1(z)\cdot W_2(z)$$

is defined on $G_{W_1}\cap G_{W_2}$ and belongs to the class $\Omega_0(R,J)$.

Proof. According to Theorem 7.3.3 we have $W(z)$ defined on some domain of the complex plane and belonging to $\Omega(R,J)$. Therefore, there exists an L-system

$$\Theta=\begin{pmatrix}\mathbb{A} & K & J\\ H_+\subset\mathcal{H}\subset H_- & & E\end{pmatrix}\tag{7.33}$$

such that $W(z)=W_\Theta(z)$ for $z\in\rho(T)$. Hence it would be enough to show that if $\mathbb{A}\supset T\supset\dot{A}$, $\mathbb{A}^*\supset T^*\supset\dot{A}$ are correspondent elements of Θ then $\overline{\mathrm{Dom}(\dot{A})}=\mathcal{H}$ and $\mathrm{Dom}(T)\neq\mathrm{Dom}(T^*)$.

Since $W_1(z)$ and $W_2(z)$ both belong to the class $\Omega_0(R,J)$, then corresponding systems Θ_1 and Θ_2 have a property that $\overline{\mathrm{Dom}(\dot{A}_1)}=\mathcal{H}_1$ and $\overline{\mathrm{Dom}(\dot{A}_2)}=\mathcal{H}_2$. Let operator $\mathcal{A}$ be defined by (7.2) and

$$\mathrm{Dom}(\mathcal{A})=\mathrm{Dom}(\dot{A}_1)\oplus\mathrm{Dom}(\dot{A}_2).\tag{7.34}$$

Considering the closure of the equality (7.34) yields $\overline{\mathrm{Dom}(\mathcal{A})} = \mathcal{H}_1 \oplus \mathcal{H}_2 = \mathcal{H}$. As it was shown above, $\mathcal{A} \subset \dot{A}$ or $\mathrm{Dom}(\mathcal{A}) \subset \mathrm{Dom}(\dot{A}) \subset \mathcal{H}$. Hence, $\overline{\mathrm{Dom}(\dot{A})} = \mathcal{H}$. It was shown in the proof of Theorem 7.1.5 that $\overline{\mathrm{Dom}(\dot{A})} = \mathcal{H}$ already implies that $\mathrm{Dom}(T) \neq \mathrm{Dom}(T^*)$. Thus, $W_\Theta(z) = W_{\Theta_1}(z) \cdot W_{\Theta_2}(z)$ belongs to the class $\Omega_0(R,J)$. □

Theorem 7.3.5. *Let operator-valued functions $W_1(z)$ and $W_2(z)$ in the Hilbert space E belong to the class $\Omega_1(R,J)$ and $G_{W_1} \cap G_{W_2} \neq \emptyset$. Then their product*

$$W(z) = W_1(z) \cdot W_2(z)$$

is defined on $G_{W_1} \cap G_{W_2}$ and belongs to the class $\Omega_1(R,J)$.

Proof. Using the same argument as in Theorem 7.3.3 we conclude that $W(z)$ is defined on some domain of the complex plane, belongs to $\Omega(R,J)$ class, and is realizable by an L-system Θ of the type (7.33) such that $W(z) = W_\Theta(z)$ for $z \in \rho(T)$. It only remains to show that the L-system Θ has the property $\overline{\mathrm{Dom}(\dot{A})} \neq \mathcal{H}$ and $\mathrm{Dom}(T) = \mathrm{Dom}(T^*)$.

Since both $W_1(z)$ and $W_2(z)$ belong to the class $\Omega_1(R,J)$ then correspondent L-systems Θ_1 and Θ_2 possesses properties $\overline{\mathrm{Dom}(\dot{A}_1)} \neq \mathcal{H}$, $\mathrm{Dom}(T_1) = \mathrm{Dom}(T_1^*)$ and $\overline{\mathrm{Dom}(\dot{A}_2)} \neq \mathcal{H}$, $\mathrm{Dom}(T_2) = \mathrm{Dom}(T_2^*)$. Due to Theorem 7.3.3 $\Theta = \Theta_1 \cdot \Theta_2$. In the proof of Theorem 7.1.8 we have shown that L-systems with the above condition have reduced form. Namely,

$$\Theta_1 = \begin{pmatrix} T_1 & K_1 & J \\ \mathcal{H}_{+1} \subset \mathcal{H}_1 \subset \mathcal{H}_{-1} & & E \end{pmatrix} \text{ and } \Theta_2 = \begin{pmatrix} T_2 & K_2 & J \\ \mathcal{H}_{+2} \subset \mathcal{H}_2 \subset \mathcal{H}_{-2} & & E \end{pmatrix}$$

where $\mathrm{Im}\, T_j = K_j J K_j^*$, $(j = 1, 2)$. Consequently, the domain of the state-space operator of the system $\Theta = \Theta_1 \cdot \Theta_2$ is determined by the formula

$$\mathrm{Dom}(T) = \mathrm{Dom}(T_1) \oplus \mathrm{Dom}(T_2).$$

Using the fact that $\mathrm{Dom}(T_1) = \mathrm{Dom}(T_1^*)$ and $\mathrm{Dom}(T_2) = \mathrm{Dom}(T_2^*)$ we conclude that $\mathrm{Dom}(T) = \mathrm{Dom}(T^*)$. This implies (see the argument in the proof of Theorem 7.1.5) that $\overline{\mathrm{Dom}(\dot{A})} \neq \mathcal{H}$. Thus, $W(z) \in \Omega_1(R,J)$. □

The next result for the class $\Omega_{01}(R,J)$ is not that straightforward and an additional condition is required.

Theorem 7.3.6. *Let operator-valued functions $W_1(z)$ and $W_2(z)$ in the Hilbert space E belong to the class $\Omega_{01}(R,J)$ and $G_{W_1} \cap G_{W_2} \neq \emptyset$. Then the product $W(z) = W_1(z) \cdot W_2(z)$ is defined on $G_{W_1} \cap G_{W_2}$ and belongs to the class $\Omega_{01}(R,J)$ if and only if the set*

$$\begin{aligned} \mathfrak{D} = \big\{ g = g_1 + g_2 \in \mathcal{H}_1 \oplus \mathcal{H}_2 \big|\, & g_1 \in \mathrm{Dom}(T_1^*),\ g_2 \in \mathrm{Dom}(T_2), \\ & K_1^* g_1 + K_2^* g_2 = 0 \big\}, \end{aligned} \tag{7.35}$$

is not dense in $\mathcal{H} = \mathcal{H}_1 \oplus \mathcal{H}_2$. *Here* T_1, T_2, K_1, K_2, $\mathcal{H}_1$ *and* $\mathcal{H}_2$ *are correspondent elements of the L-systems* Θ_1 *and* Θ_2 *related to the functions* $W_1(z)$ *and* $W_2(z)$.

Proof. Since $\Omega(R, J)$ is a union of three distinct classes $\Omega_0(R, J)$, $\Omega_1(R, J)$, and $\Omega_{01}(R, J)$, Theorem 7.3.3 guarantees that $W(z)$ is defined on some domain in $\mathbb{C}$ and belongs to one of the indicated subclasses. The set $\mathfrak{D}$ defined by (7.35) actually coincides with the domain of operator $\dot{A}$ defined in (7.21). Therefore, since $\mathfrak{D}$ is not dense in $\mathcal{H} = \mathcal{H}_1 \oplus \mathcal{H}_2$, then $\overline{\mathrm{Dom}(\dot{A})} \neq \mathcal{H}$ and $W(z)$ is certainly not in the $\Omega_0(R, J)$ class.

Let us assume that $W(z)$ belongs to $\Omega_1(R, J)$. Then the L-system $\Theta = \Theta_1 \cdot \Theta_2$ has a property $\mathcal{H}_+ = \mathrm{Dom}(T) = \mathrm{Dom}(T^*)$. The operator T here is actually the quasi-kernel of the operator $\mathbb{A}$ of the L-system Θ. Hence, for all $g \in \mathcal{H}_+$, $g = g_1 + g_2$, $g_1 \in \mathcal{H}_{+1}$, $g_2 \in \mathcal{H}_{+2}$,

$$\mathbb{A}_1 g_1 + \mathbb{A}_2 g_2 + 2iK_1 J K_2^* g_2 \in \mathcal{H}, \quad \mathbb{A}_2^* g_2 + \mathbb{A}_2^* g_2 - 2iK_2 J K_1^* g_2 \in \mathcal{H},$$

where $\mathcal{H} = \mathcal{H}_1 \oplus \mathcal{H}_2$ and all operators belong to the correspondent systems Θ_1 and Θ_2. Since g_2 is an arbitrary element of $\mathcal{H}_{+2}$, then we can choose it equal to 0. Thus the first relation yields $g_1 \in \mathrm{Dom}(T_1)$ for all $g_1 \in \mathcal{H}_{+1}$. Because g_1 is arbitrary we have that

$$\mathrm{Dom}(T_1) = \mathcal{H}_{+1} = \mathrm{Dom}(\dot{A}_1^*).$$

Taking into account that $W_1(z) \in \Omega_{01}(R, J)$ we get a contradiction. Hence, the product of $W_1(z)$ and $W_2(z)$ under the assumption of the theorem belongs to the class $\Omega_{01}(R, J)$. □

Remark 7.3.7. It is not hard to show that if the set $\mathfrak{D}$ in the statement of the Theorem 7.3.6 is dense in $\mathcal{H}_1 \oplus \mathcal{H}_2$, then $W(z) = W_1(z) \cdot W_2(z)$ belongs to the class $\Omega_0(R, J)$.

The theorem below describes properties of the mixed products of two operator-valued functions of $\Omega(R, J)$ class.

Theorem 7.3.8. *Let operator-valued functions* $W_1(z)$ *and* $W_2(z)$ *in the Hilbert space* E $(\dim E > 1)$ *belong to the classes* $\Omega_0(R, J)$ *and* $\Omega_1(R, J)$, *respectively, and* $G_{W_1} \cap G_{W_2} \neq \emptyset$. *Then their product* $W(z) = W_1(z) \cdot W_2(z)$ *is defined on* $G_{W_1} \cap G_{W_2}$ *and belongs to the class* $\Omega_{01}(R, J)$ *if and only if the set* $\mathfrak{D}$ *in* (7.35) *is not dense in* $\mathcal{H} = \mathcal{H}_1 \oplus \mathcal{H}_2$.

We omit the proof of this theorem because it is similar to the one of the Theorem 7.3.6. As before we should note that if the set $\mathfrak{D}$ in the statement of Theorem 7.3.8 is not dense in $\mathcal{H}_1 \oplus \mathcal{H}_2$, then $W(z) = W_1(z) \cdot W_2(z)$ belongs to the class $\Omega_0(R, J)$. Furthermore, Theorem 7.3.8 holds even if we consider product $W(z) = W_2(z) \cdot W_1(z)$.

Theorem 7.3.9. *Let operator-valued functions $W_1(z)$ and $W_2(z)$ in the Hilbert space E ($\dim E > 1$) belong to the classes $\Omega_0(R,J)$ and $\Omega_1(R,J)$, respectively, and $G_{W_1} \cap G_{W_2} \neq \emptyset$. Let also*

$$W_1(z) \cdot W_2(z) = W_2(z) \cdot W_1(z) = W(z). \tag{7.36}$$

Then operator-valued function $W(z)$ is defined on $G_{W_1} \cap G_{W_2}$ and belongs to the class $\Omega_{01}(R,J)$.

Proof. Condition (7.36) implies that

$$\mathbf{A} = \mathbb{A}_1 P_1^+ + \mathbb{A}_2 P_2^+ + 2iK_1 J K_2^* P_2^+ = \mathbb{A}_2 P_2^+ + \mathbb{A}_1 P_1^+ + 2iK_2 J K_1^* P_1^+.$$

The obvious cancellation yields $K_1 J K_2^* P_2^+ = K_2 J K_1^* P_1^+$. Left- and right-hand sides of this equality belong to H_{-1} and H_{-2}, respectively. Hence the equality may hold only if $K_1 J K_2^* P_2^+ = K_2 J K_1^* P_1^+ = 0$. Thus,

$$\mathbf{A} = \mathbb{A}_1 P_1^+ + \mathbb{A}_2 P_2^+,$$

and we are actually dealing with operator $\mathbf{A}$ of a block-diagonal structure. Now let $f = f_1 + f_2$ be an element of $\mathrm{Dom}(T)$, then

$$Tf = \mathbf{A}f = \mathbb{A}_1 f_1 + \mathbb{A}_2 f_2 \in \mathcal{H},$$

but $\mathbb{A}_2 f_2 \in \mathcal{H}$, and therefore, $\mathbb{A}_1 f_1 \in \mathcal{H}$, or

$$Tf = T_1 f_1 + T_2 f_2,$$

Similarly,

$$T^* g = T_1^* g_1 + T_2^* g_2,$$

where $g = g_1 + g_2$ is an element of $\mathrm{Dom}(T^*)$. In other words we have just shown that $f \in \mathrm{Dom}(T)$ implies $f_1 \in \mathrm{Dom}(T_1)$ and $f_2 \in \mathrm{Dom}(T_2)$, where $f = f_1 + f_2$. Conversely, if $f_1 \in \mathrm{Dom}(T_1)$ and $f_2 \in \mathrm{Dom}(T_2)$, then $f = f_1 + f_2 \in \mathrm{Dom}(T)$, i.e.,

$$\mathrm{Dom}(T) = \mathrm{Dom}(T_1) \oplus \mathrm{Dom}(T_2).$$

Similarly shown,

$$\mathrm{Dom}(T^*) = \mathrm{Dom}(T_1^*) \oplus \mathrm{Dom}(T_2^*).$$

But since $\mathrm{Dom}(T_1) \neq \mathrm{Dom}(T_1^*)$ and $\mathrm{Dom}(T_2) = \mathrm{Dom}(T_2^*)$, then $\mathrm{Dom}(T) \neq \mathrm{Dom}(T^*)$. It is also not hard to see that, under these circumstances,

$$\mathrm{Dom}(\dot{A}) = \mathrm{Dom}(\dot{A}_1) \oplus \mathrm{Dom}(\dot{A}_2).$$

Hence, if $\overline{\mathrm{Dom}(\dot{A}_2)} \neq \mathcal{H}_2$, then $\overline{\mathrm{Dom}(\dot{A})} \neq \mathcal{H}$. It follows then that $W(z)$ belongs to the class $\Omega_{01}(R,J)$. □

7.4 Boundary triplets and self-adjoint bi-extensions

Let $\dot{A}$ be a closed densely defined symmetric operator in $\mathcal{H}$ with equal deficiency numbers.

Definition 7.4.1. The triplet $\Pi = \{\mathcal{N}, \Gamma_1, \Gamma_0\}$ is called a **boundary triplet** for $\dot{A}^*$ if $\mathcal{N}$ is a Hilbert space and Γ_0, Γ_1 are bounded linear operators from the Hilbert space $\mathcal{H}_+ = \mathrm{Dom}(\dot{A}^*)$ (with the inner product (2.6)) into $\mathcal{N}$ such that the mapping

$$\Gamma := \langle \Gamma_0, \Gamma_1 \rangle : \mathcal{H}_+ \to \mathcal{N} \oplus \mathcal{N},$$

is surjective and the abstract Green's identity

$$\left(\dot{A}^* f, g\right) - \left(f, \dot{A}^* g\right) = (\Gamma_1 f, \Gamma_0 g)_{\mathcal{N}} - (\Gamma_0 f, \Gamma_1 g)_{\mathcal{N}}, \tag{7.37}$$

holds for all f, $g \in \mathcal{H}_+$.

A boundary triplet for $\dot{A}^*$ exists since the deficiency numbers are assumed to be equal [164], [142], [143].

It follows from Definition 7.4.1 (see [101], [103]) that the operators

$$\mathrm{Dom}(A_k) := \mathrm{Ker}\,\Gamma_k, \quad A_k := \dot{A}^* \upharpoonright \mathrm{Dom}(A_k), \quad (k = 0, 1), \tag{7.38}$$

are self-adjoint extensions of $\dot{A}$. Moreover, they are transversal, i.e.,

$$\mathrm{Dom}(\dot{A}^*) = \mathrm{Dom}(A_0) + \mathrm{Dom}(A_1).$$

Notice that if $\Pi = \{\mathcal{N}, \Gamma_1, \Gamma_0\}$ is a boundary triplet for $\dot{A}^*$, then $\Pi' = \{\mathcal{N}, -\Gamma_0, \Gamma_1\}$ is the boundary triplet for $\dot{A}^*$ too.

Let $\mathcal{N}$ be a Hilbert space whose dimension is equal to the deficiency number of $\dot{A}$. We choose a self-adjoint extension A of $\dot{A}$.

Definition 7.4.2. The function $\Gamma(z) \in [\mathcal{N}, \mathcal{H}]$ is called the **γ-field**, corresponding to A if

1. the operator $\Gamma(z)$ isomorphically maps $\mathcal{N}$ onto $\mathfrak{N}_z$ for all $z \in \rho(A)$, where $\mathfrak{N}_z$ are the defect subspaces of $\dot{A}$,
2. for every z, $\zeta \in \rho(A)$ the identity
$$\Gamma(z) = \Gamma(\zeta) + (z - \zeta)(A - zI)^{-1}\Gamma(\zeta) \tag{7.39}$$
holds.

Due to (7.39) the γ-field $\Gamma(z)$ is a holomorphic operator-valued function of z in $\rho(A)$. Two γ-fields $\Gamma(z)$ and $\hat{\Gamma}(z)$ corresponding to the same self-adjoint extension A are connected by the relation $\Gamma(z) = \hat{\Gamma}(z)\hat{X}$, where $\hat{X}$ is an isomorphism of Hilbert spaces $\mathcal{N}$ and $\hat{\mathcal{N}}$ [175].

The γ-field corresponding to A can be constructed as follows: fix $\zeta_0 \in \rho(A)$ and let $\Gamma^{(0)} \in [\mathcal{N}, \mathcal{H}]$ be a bijection of $\mathcal{N}$ and $\mathfrak{N}_{\zeta_0}$. Then clearly, the function

$$\Gamma(z) = (A - \zeta_0 I)(A - zI)^{-1}\Gamma^{(0)} = \Gamma^{(0)} + (z - \zeta_0)(A - zI)^{-1}\Gamma^{(0)},\ z \in \rho(A)$$

is a γ-field corresponding to A.

Let $\Pi = \{\mathcal{N}, \Gamma_1, \Gamma_0\}$ be a boundary triplet for $\dot{A}^*$. Define an operator-valued function

$$\gamma_0(z) = \left(\Gamma_0 \restriction \mathfrak{N}_z\right)^{-1},\ z \in \rho(A_0). \tag{7.40}$$

It can be derived from (7.37) and (7.40) (see [101], [103]) that

$$\begin{aligned} \gamma_0(z) &= (A_0 - \zeta I)(A_0 - zI)^{-1}\gamma_0(\zeta), \\ \gamma_0^*(\bar{z}) &= \Gamma_1(A_0 - zI)^{-1},\ z, \zeta \in \rho(A_0). \end{aligned} \tag{7.41}$$

Thus, the function γ_0 forms the γ-field, corresponding to A_0.

Definition 7.4.3. Let $\Pi = \{\mathcal{N}, \Gamma_1, \Gamma_0\}$ be a boundary triplet for $\dot{A}^*$. An operator-valued function $M_0(z) \in [\mathcal{N}, \mathcal{N}]$ defined by the relation

$$M_0(z) := \Gamma_1\gamma_0(z), \quad z \in \rho(A_0), \tag{7.42}$$

is called the **Weyl-Titchmarsh function corresponding to the boundary triplet** Π.

Notice that (7.42) implies the relations

$$M_0(z)\Gamma_0 f_z = \Gamma_1 f_z,\ f_z \in \mathfrak{N}_z,\ z \in \rho(A_0).$$

From (7.41) one can also get

$$M_0(\bar{z}) = M_0^*(z), \quad M_0(z) - M_0^*(\zeta) = (z - \bar{\zeta})\gamma_0^*(\zeta)\gamma_0(z), \quad z, \zeta \in \rho(A_0).$$

Similarly, one defines another Weyl-Titchmarsh function related to the boundary triplet $\Pi = \{\mathcal{N}, \Gamma_1, \Gamma_0\}$ or with $\Pi = \{\mathcal{N}, -\Gamma_0, \Gamma_1\}$. Set

$$\gamma_1(z) := \left(\Gamma_1 \restriction \mathfrak{N}_z\right)^{-1},\ z \in \rho(A_1)$$

and

$$M_1(z) = -\Gamma_0\gamma_1(z). \tag{7.43}$$

The function M_1 is connected with the function M_0 via

$$M_1(z) = -M_0^{-1}(z),\ \operatorname{Im} z \neq 0. \tag{7.44}$$

Now we are going to establish connections between self-adjoint bi-extensions and boundary triplets. The proposition below immediately follows from Definition 7.4.1.

Proposition 7.4.4. *Let $\dot{A}$ be a closed densely defined symmetric operator with equal deficiency indices in the Hilbert space $\mathcal{H}$. Suppose $\mathcal{N}$ is a Hilbert space, $\Gamma_0, \Gamma_1 \in [\mathcal{H}_+, \mathcal{N}]$, and the operator $\langle \Gamma_0, \Gamma_1 \rangle \in [\mathcal{H}_+, \mathcal{N} \oplus \mathcal{N}]$ is surjective. Then the following statements are equivalent:*

(i) $\Pi = \{\mathcal{N}, \Gamma_1, \Gamma_0\}$ *is the boundary triplet for* $\dot{A}^*$*;*

(ii) *the sesquilinear form*

$$w_0(f, g) := (\dot{A}^* f, g) + (\Gamma_0 f, \Gamma_1 g)_{\mathcal{N}}, \quad f, g \in \mathcal{H}_+ = \mathrm{Dom}(\dot{A}^*)$$

is Hermitian, i.e., $w_0(f, g) = \overline{w_0(g, f)}$*;*

(iii) *the sesquilinear form*

$$w_1(f, g) := (\dot{A}^* f, g) - (\Gamma_1 f, \Gamma_0 g)_{\mathcal{N}}, \quad f, g \in \mathcal{H}_+ = \mathrm{Dom}(\dot{A}^*)$$

is Hermitian.

The following theorem sets up the connection between boundary triplets and t-self-adjoint bi-extensions.

Theorem 7.4.5. *Let $\dot{A}$ be a closed densely-defined symmetric operator with equal deficiency numbers in the Hilbert space $\mathcal{H}$. Consider the rigged Hilbert space $\mathcal{H}_+ \subset \mathcal{H} \subset \mathcal{H}_-$ generated by $\dot{A}$.*

1. *Let $\Pi = \{\mathcal{N}, \Gamma_1, \Gamma_0\}$ for $\dot{A}^*$ be a boundary triplet for $\dot{A}^*$. Define operators $\mathbb{A}_0$ and $\mathbb{A}_1$,*

$$\mathbb{A}_0 := \dot{A}^* + \Gamma_1^\times \Gamma_0, \quad \mathbb{A}_1 := \dot{A}^* - \Gamma_0^\times \Gamma_1, \tag{7.45}$$

where $\Gamma_0^\times$ and $\Gamma_1^\times \in [\mathcal{N}, \mathcal{H}_-]$ are the adjoint operators to Γ_0 and Γ_1, respectively. Then:

(i) *$\mathbb{A}_0$ and $\mathbb{A}_1$ belong to $[\mathcal{H}_+, \mathcal{H}_-]$ and are t-self-adjoint bi-extensions of $\dot{A}$. Moreover,*

$$\mathbb{A}_0 \supset A_0, \ \mathbb{A}_1 \supset A_1,$$

(ii) *the Weyl-Titchmarsh function defined by (7.42) is given by*

$$M_0(z) = \Gamma_1 (\mathbb{A}_0 - zI)^{-1} \Gamma_1^\times, \tag{7.46}$$

(iii) *the Weyl-Titchmarsh function defined by (7.43) is given by*

$$M_1(z) = \Gamma_0 (\mathbb{A}_1 - zI)^{-1} \Gamma_0^\times, \quad z \in \rho(A_1). \tag{7.47}$$

2. *If $\mathbb{A}_0$ is a t-self-adjoint bi-extension of a self-adjoint extension A_0 of $\dot{A}$, then there exists a boundary triplet $\Pi = \{\mathcal{N}, \Gamma_1, \Gamma_0\}$ for $\dot{A}^*$ such that $\dot{A}^* \upharpoonright \ker \Gamma_0 = A_0$ and $\mathbb{A}_0 = \dot{A}^* + \Gamma_1^\times \Gamma_0$.*

Proof. 1. The statement (i) follows from Proposition 7.4.4 and relations (7.38). To prove (ii) we let $f_z \in \mathfrak{N}_z$, $\operatorname{Im} z \neq 0$. Then

$$(\mathbb{A}_0 - zI)f_z = (\dot{A}^* - zI)f_z + \Gamma_1^\times \Gamma_0 f_z = \Gamma_1^\times \Gamma_0 f_z.$$

Let $h = \Gamma_0 f_z$. Then by (7.40) we have $f_z = \gamma_0(z)h$ and

$$(\mathbb{A}_0 - zI)\gamma_0(z)h = \Gamma_1^\times h.$$

Consequently, $\gamma_0(z) = (\mathbb{A}_0 - zI)^{-1}\Gamma_1^\times$. Now from (7.42) we get (7.46). Let us prove (iii). From (7.45) we obtain

$$(\mathbb{A}_1 - zI)f_z = -\Gamma_0^\times \Gamma_1 f_z, \quad f_z \in \mathfrak{N}_z, \quad z \in \rho(A_1).$$

It follows that $\operatorname{Ran}(\Gamma_0^\times) \subset \operatorname{Ran}(\mathbb{A}_1 - zI)$ for all $z \in \rho(A_1)$. Since

$$(\mathbb{A}_0 - zI)^{-1}\Gamma_1^\times \Gamma_0 f_z = f_z,$$

for all $f_z \in \mathfrak{N}_z$, $z \in \rho(A_0)$, we get $\Gamma_0(\mathbb{A}_0 - zI)^{-1}\Gamma_1^\times \Gamma_0 f_z = \Gamma_0 f_z$. The equalities $\mathcal{H}_+ = \operatorname{Dom}(A_0)\dot{+}\mathfrak{N}_z$, $\ker(\Gamma_0) = \operatorname{Dom}(A_0)$, and $\operatorname{Ran}(\Gamma_0) = \mathcal{N}$ imply

$$\Gamma_0(\mathbb{A}_0 - z)^{-1}\Gamma_1^\times = I, \quad z \in \rho(A_0).$$

Similarly one obtains $(\mathbb{A}_1 - zI)^{-1}\Gamma_0^\times \Gamma_1 f_z = -f_z$ for all $f_z \in \mathfrak{N}_z$, $z \in \rho(A_1)$, and

$$\Gamma_1(\mathbb{A}_1 - z)^{-1}\Gamma_0^\times = -I, \quad z \in \rho(A_1).$$

Notice that for $z \in \rho(A_0) \cap \rho(A_1)$ we have

$$M_1(z)M_0(z) = \Gamma_0(\mathbb{A}_1 - z)^{-1}\Gamma_0^\times \Gamma_1(\mathbb{A}_0 - z)^{-1}\Gamma_1^\times = -\Gamma_0(\mathbb{A}_0 - z)^{-1}\Gamma_1^\times = -I.$$

Similarly, $M_0(z)M_1(z) = -I$.

2. Since $\mathbb{A}_0$ is a t-self-adjoint bi-extension of $\dot{A}$ containing A_0 as a quasi-kernel, the operator $\mathbb{A}_0$ is defined by means of a self-adjoint extension of A_1 transversal to A_0 via formula (see Theorem 3.4.9)

$$\mathbb{A}_0 = \dot{A}^* - \mathcal{R}^{-1}\dot{A}^*\mathcal{P}_{A_0A_1} = \dot{A}^* - \mathcal{R}^{-1}A_1\mathcal{P}_{A_0A_1}, \tag{7.48}$$

where $\mathcal{P}_{A_0A_1}$ is a projector in $\mathcal{H}_+$ onto $(I + \mathcal{U}_1)\mathfrak{N}_i$ corresponding to the decompositions of the form (3.35) and (3.36) with $\mathcal{U} = -U$ or

$$\mathcal{H}_+ = \operatorname{Dom}(A_0)\dot{+}(I + \mathcal{U}_1)\mathfrak{N}_i,$$

and $\operatorname{Dom}(A_1) = \operatorname{Dom}(\dot{A}) \oplus (I + \mathcal{U}_1)\mathfrak{N}_i$. Let $\mathcal{P}_{A_1A_0}$ be a projector in $\mathcal{H}_+$ onto $(I + \mathcal{U}_0)\mathfrak{N}_i$ corresponding to the decomposition

$$\mathcal{H}_+ = \operatorname{Dom}(A_1)\dot{+}(I + \mathcal{U}_0)\mathfrak{N}_i,$$

and $\mathrm{Dom}(A_0) = \mathrm{Dom}(\dot{A}) \oplus (I + \mathcal{U}_0)\mathfrak{N}_i$. Define

$$\mathbb{A}_1 = \dot{A}^* - \mathcal{R}^{-1}\dot{A}^*\mathcal{P}_{A_1A_0} = \dot{A}^* - \mathcal{R}^{-1}A_0\mathcal{P}_{A_1A_0}. \tag{7.49}$$

Then $\mathbb{A}_1$ is a t-self-adjoint bi-extension of $\dot{A}$ containing A_1 as a quasi-kernel. The connection with $\mathbb{A}_0$ is given by the relation (3.31), i.e.,

$$(\mathbb{A}_1 f, g) = (\dot{A}^* f, g) + (f, \dot{A}^* g) - (\mathbb{A}_0 f, g), \quad f, g \in \mathcal{H}_+,$$

which is equivalent to the equality

$$(\mathbb{A}_0 - \dot{A}^*)^* = \dot{A}^* - \mathbb{A}_1, \tag{7.50}$$

where $(\mathbb{A}_0 - \dot{A}^*)^* \in [\mathcal{H}_+, \mathcal{H}_-]$. Let subspaces $\Phi_0 \subset \mathcal{H}_-$, $\Phi_1 \subset \mathcal{H}_-$ be defined as

$$\Phi_0 := \mathrm{Ran}(\mathbb{A}_0 - \dot{A}^*), \quad \Phi_1 := \mathrm{Ran}(\mathbb{A}_1 - \dot{A}^*).$$

From (3.34) we get

$$\Phi_0 = \mathcal{R}^{-1}(I - \mathcal{U}_1)\mathfrak{N}_i, \quad \Phi_1 = \mathcal{R}^{-1}(I - \mathcal{U}_0)\mathfrak{N}_i.$$

Then $\dim \Phi_0 = \dim \mathfrak{N}_i$. Choose a Hilbert space $\mathcal{N}$ with $\dim \mathcal{N} = \dim \Phi_0$. Let $K_1 \in [\mathcal{N}, \mathcal{H}_-]$ be a bijection onto Φ_0. It follows from (7.50)that the relation

$$K_0 K_1^\times f = (\dot{A}^* - \mathbb{A}_1) f, \quad f \in \mathcal{H}_+, \tag{7.51}$$

defines $K_0 \in [\mathcal{N}, \mathcal{H}_-]$, which is a bijection onto Φ_1. Finally let

$$\Gamma_0 := K_0^\times, \quad \Gamma_1 = K_1^\times.$$

Then $\Gamma_0, \Gamma_1 \in [\mathcal{H}_+, \mathcal{N}]$ and

$$\mathbb{A}_0 = A^* + \Gamma_1^\times \Gamma_0, \quad \mathbb{A}_1 = A^* - \Gamma_0^\times \Gamma_1. \tag{7.52}$$

Because A_0 and A_1 are mutually transversal, the operator

$$\langle \Gamma_0, \Gamma_1 \rangle : \mathcal{H}_+ \to \mathcal{N} \oplus \mathcal{N}$$

is a surjection. Thus, $\Pi = \{\mathcal{N}, \Gamma_1, \Gamma_0\}$ is a boundary triplet for $\dot{A}^*$ and $\dot{A}^* \upharpoonright \ker \Gamma_0 = A_0$. □

7.5 The Krein-Langer Q-functions and their realizations

Let $\dot{A}$ be a densely-defined closed symmetric operator in a Hilbert space $\mathcal{H}$ with finite equal deficiency indices and let again $\mathcal{N}$ be a Hilbert space whose dimension is equal to the deficiency number of $\dot{A}$. Let us also choose a self-adjoint extension A of $\dot{A}$.

Definition 7.5.1. Let $\Gamma(z) \in [\mathcal{N}, \mathcal{H}]$ be a γ-field corresponding to A. An operator-valued function $Q(z) \in [\mathcal{N}, \mathcal{N}])$ with the property

$$Q(z) - Q^*(\zeta) = (z - \bar{\zeta})\Gamma^*(\zeta)\Gamma(z), \quad z, \zeta \in \rho(A) \tag{7.53}$$

is called the **Kreĭn-Langer Q-function** of $\dot{A}$ corresponding to the γ-field $\Gamma(z)$.

It follows from (7.53) that

$$Q(z) = C - i\mathrm{Im}\,\zeta_0\Gamma^*_{\zeta_0}\Gamma_{\zeta_0} + (z - \bar{\zeta}_0)\Gamma^*_{\zeta_0}\Gamma_z,$$

where $C = \mathrm{Re}\,Q(\zeta_0) \in [\mathcal{N}, \mathcal{N}]$ is a self-adjoint operator. Thus, the Q-function is defined up to the bounded self-adjoint term in $\mathcal{N}$ and is a Herglotz-Nevanlinna function. Moreover, for every z, $\mathrm{Im}\,z \neq 0$ the operator $-i\,\mathrm{Im}\,z\,(Q(z) - Q^*(z))$ is positive definite. Hence, $-Q^{-1}(z)$, ($\mathrm{Im}\,z \neq 0$) is a Herglotz-Nevanlinna function too. From (7.39) we get

$$Q(z) = C - i\mathrm{Im}\,\zeta_0\Gamma^*_{\zeta_0}\Gamma_{\zeta_0} + (z - \bar{\zeta}_0)\Gamma^*_{\zeta_0}\left(\Gamma_{\zeta_0} + (z - \zeta_0)(A - zI)^{-1}\Gamma_{\zeta_0}\right). \tag{7.54}$$

The γ-fields and corresponding Q-functions can be defined by means of the resolvents of t-self-adjoint bi-extensions of $\dot{A}$. Indeed, consider the rigged Hilbert space $\mathcal{H}_+ \subset \mathcal{H} \subset \mathcal{H}_-$ generated by the operator $\dot{A}$. Let A be a self-adjoint extension of $\dot{A}$ and let $\mathbb{A}$ be an arbitrary t-self-adjoint bi-extension of $\dot{A}$. Choose a Hilbert space $\mathcal{N}$ with $\dim \mathcal{N} = \dim \mathfrak{N}_i$ and an arbitrary operator $K \in [\mathcal{N}, \mathcal{H}_-]$ such that Kmaps $\mathcal{N}$ isomorphically onto the subspace $\mathrm{Ran}(\mathbb{A} - \dot{A}^*)$. Put

$$\Gamma(z) := (\mathbb{A} - zI)^{-1}K, \quad z \in \rho(A). \tag{7.55}$$

Then clearly $\Gamma(z)$ maps $\mathcal{N}$ isomorphically onto $\mathfrak{N}_z$ for each $z \in \rho(A)$, satisfies (7.39), and $\Gamma^*(z) = K^\times(A - \bar{z}I)^{-1}$, where $K^\times \in [\mathcal{H}_+, \mathcal{N}]$ is the adjoint operator. Define a $[\mathcal{N}, \mathcal{N}]$-valued function

$$Q(z) := K^\times\Gamma(z) = K^\times(\mathbb{A} - zI)^{-1}K, \quad z \in \rho(A). \tag{7.56}$$

Then

$$\begin{aligned} Q(z) - Q^*(\zeta) &= K^\times\left((\mathbb{A} - I)^{-1} - (\mathbb{A} - \bar{\zeta}I)^{-1}\right)K \\ &= (z - \bar{\zeta})K^\times(A - \bar{\zeta}I)^{-1}(\mathbb{A} - zI)^{-1}K = (z - \bar{\zeta})\Gamma^*(\zeta)\Gamma(z). \end{aligned}$$

Thus, the functions defined by (7.55) and (7.56) form a γ-field, corresponding to A and a Q-function. Clearly, the function

$$C + K^\times(\mathbb{A} - z)^{-1}K$$

with an arbitrary self-adjoint operator $C \in [\mathcal{N}, \mathcal{N}]$ is also a Q-function, corresponding to the γ-field (7.55).

Theorem 7.5.2. *Let A_0 be a self-adjoint extension of $\dot{A}$, $\Gamma(z) \in [\mathcal{N}, \mathcal{H}]$ be the γ-field corresponding to A_0, and let $Q(z)$ be the Q-function of $\dot{A}$ corresponding to the γ-field $\Gamma(z)$. Then there exists a boundary triplet $\Pi = \{\mathcal{N}, \Gamma_1, \Gamma_0\}$ of $\dot{A}^*$ whose Weyl-Titchmarsh function M_0 coincides with Q.*

Proof. Let

$$\mathrm{Dom}(A_0) = \mathrm{Dom}(\dot{A}) \oplus (I + \mathcal{U}_0)\mathfrak{N}_i,$$

where $\mathcal{U}_0$ isometrically maps $\mathfrak{N}_i$ onto $\mathfrak{N}_{-i}$. The operator $\Gamma(i)$ isomorphically maps $\mathcal{N}$ onto $\mathfrak{N}_i$. Hence one may consider $\Gamma(i)$ as an element of $[\mathcal{N}, \mathfrak{N}_i]$. Let $\Gamma^*(i) \in [\mathfrak{N}_i, \mathcal{N}]$ be the adjoint of $\Gamma(i)$. Define a unitary operator $U_0 \in [\mathfrak{N}_i, \mathfrak{N}_i]$ by

$$U_0 := -\left(iI + \Gamma^{*-1}(i)\mathrm{Re}\, Q(i)\Gamma^{-1}(i)\right)\left(iI - \Gamma^{*-1}(i)\mathrm{Re}\, Q(i)\Gamma^{-1}(i)\right)^{-1}.$$

Set $\mathcal{U}_1 = \mathcal{U}_0 U_0$ and define a self-adjoint extension A_1 of $\dot{A}$ via

$$\mathrm{Dom}(A_1) = \mathrm{Dom}(\dot{A}) \oplus (I + \mathcal{U}_1)\mathfrak{N}_i, \quad A_1 := \dot{A}^* | \mathrm{Dom}(A_1).$$

We have

$$\mathcal{U}_0 - \mathcal{U}_1 = 2i\mathcal{U}_0\left(iI - \Gamma^{*-1}(i)\mathrm{Re}\, Q(i)\Gamma^{-1}(i)\right)^{-1}. \tag{7.57}$$

It follows that the operator $\mathcal{U}_1 - \mathcal{U}_0$ is an isomorphism of $\mathfrak{N}_i$ and $\mathfrak{N}_{-i}$. Therefore, the self-adjoint extensions A_0 and A_1 are transversal. Let t-self-adjoint bi-extensions $\mathbb{A}_0$ and $\mathbb{A}_1$ be defined by (7.48) and (7.49). Then $\mathbb{A}_0 \supset A_0$ and $\mathbb{A}_1 \supset A_1$. Set

$$K_1 = (\mathbb{A}_0 - iI)\Gamma(i), \quad \Gamma_1 = K_1^{\times} = \Gamma^{\times}(i)(\mathbb{A}_0 + iI) \in [\mathcal{H}_+, \mathcal{N}].$$

Here $\Gamma^{\times}(i) \in [\mathcal{H}_-, \mathcal{N}]$ is the adjoint operator for $\Gamma(i) \in [\mathcal{N}, \mathcal{H}_+]$. For $z \in \rho(A_0)$ one has

$$(\mathbb{A}_0 - zI)^{-1}K_1 = (\mathbb{A}_0 - z)^{-1}(\mathbb{A}_0 - iI)\Gamma(i) = \Gamma(i) + (z - i)(A_0 - zI)^{-1}\Gamma(i).$$

It follows that

$$\Gamma(z) = (\mathbb{A}_0 - zI)^{-1}K_1. \tag{7.58}$$

Define the operator $K_0 \in [\mathcal{N}, \mathcal{H}_+]$ by (7.51) and put $\Gamma_0 = K_0^{\times}$. As it has been proved above, the collection $\Pi = \{\mathcal{N}, \Gamma_1, \Gamma_0\}$ is a boundary triplet for $\dot{A}^*$ and

$$M_0(z) = \Gamma_1(\mathbb{A}_0 - zI)^{-1}\Gamma_1^{\times} = K_1^{\times}(\mathbb{A}_0 - zI)^{-1}K_1,$$

is the Weyl-Titchmarsh function corresponding to Π. Let us show that $M_0(z) = Q(z)$ for $z \in \rho(A_0)$. Notice that from (7.53) and (7.58) we get

$$\begin{aligned}
Q(z) &= Q^*(i) + (z + i)\Gamma^*(i)\Gamma(z) \\
&= Q^*(i) + (z + i)K_1^{\times}(\mathbb{A}_0 + iI)^{-1}(\mathbb{A}_0 - zI)^{-1}K_1 \\
&= Q^*(i) + K_1^{\times}(\mathbb{A}_0 - zI)^{-1}K_1 - K_1^{\times}(\mathbb{A}_0 + iI)^{-1}K_1.
\end{aligned}$$

If $z=i$, then $2i\mathrm{Im}\, Q(i)=K_1^\times(\mathbb{A}_0-iI)^{-1}K_1-K_1^\times(\mathbb{A}+iI)^{-1}K_1$. Therefore

$$Q(z)=M_0(z)+\mathrm{Re}\, Q(i)-\mathrm{Re}\, K_1^\times(\mathbb{A}_0-iI)^{-1}K_1. \tag{7.59}$$

Furthermore,

$$\begin{aligned}\left(\mathrm{Re}\, K_1^\times(\mathbb{A}_0-iI)^{-1}K_1h,g\right)_{\mathcal{N}}&=(\mathbb{A}_0(\mathbb{A}-iI)^{-1}K_1h,(\mathbb{A}-iI)^{-1}K_1g)\\ &=(\mathbb{A}_0\Gamma(i)h,\Gamma(i)g),\quad h,g\in\mathcal{N}.\end{aligned}$$

By simple calculations we get

$$\mathcal{P}_{A_0A_1}\varphi=(I+\mathcal{U}_1)(\mathcal{U}_0-\mathcal{U}_1)^{-1}\mathcal{U}_0\varphi,\quad \varphi\in\mathfrak{N}_i.$$

Hence for $\varphi,\psi\in\mathfrak{N}_i$,

$$\begin{aligned}(\mathbb{A}_0\varphi,\psi)&=(i\varphi-\mathcal{R}^{-1}A_1\mathcal{P}_{A_0A_1}\varphi,\psi)=i(\varphi,\psi)-(A_1\mathcal{P}_{A_0A_1}\varphi,\psi)_+\\ &=i(\varphi,\psi)-i((I-\mathcal{U}_1)(\mathcal{U}_0-\mathcal{U}_1)^{-1}\mathcal{U}_0\varphi,\psi)_+\\ &=i(\varphi,\psi)-i((\mathcal{U}_0-\mathcal{U}_1)^{-1}\mathcal{U}_0\varphi,\psi)_+\\ &=i(\varphi,\psi)-2i((\mathcal{U}_0-\mathcal{U}_1)^{-1}\mathcal{U}_0\varphi,\psi)\\ &=(if-2i(\mathcal{U}_0-\mathcal{U}_1)^{-1}\mathcal{U}_0\varphi,\psi).\end{aligned}$$

It follows that for $h,g\in\mathcal{N}$,

$$(\mathbb{A}_0\Gamma(i)h,\Gamma(i)g)=(i\Gamma^*(i)\Gamma(i)h-2i\Gamma^*(i))(\mathcal{U}_0-\mathcal{U}_1)^{-1}\mathcal{U}_0\Gamma(i)h,g)_{\mathcal{N}},$$

and (7.57) yields

$$\mathrm{Re}\, Q(i)=i\Gamma^*(i)\Gamma(i)f-2i\Gamma^*(i))(\mathcal{U}_0-\mathcal{U}_1)^{-1}\mathcal{U}_0\Gamma(i).$$

Taking into account (7.59), we obtain $Q(z)=M_0(z)$, $z\in\rho(A_0)$. □

The next statement immediately follows from Theorems 7.4.5, 7.5.2, and relations (7.43), (7.47), and (7.44).

Theorem 7.5.3. *Let $\Gamma(z)\in[\mathcal{N},\mathcal{H}]$ be a γ-field corresponding to a self-adjoint extension A of $\dot{A}$ and let $Q(z)$ be a Q-function corresponding to $\Gamma(z)$. Then there exist a t-self-adjoint bi-extension $\mathbb{A}$ of $\dot{A}$ and an operator $K\in[\mathcal{N},\mathcal{H}_-]$ which isomorphically maps $\mathcal{N}$ onto* $\mathrm{Ran}(\mathbb{A}-\dot{A}^*)$ *such that*

$$\Gamma(z)=(\mathbb{A}-zI)^{-1}K,\quad Q(z)=K^\times(\mathbb{A}-zI)^{-1}K,\quad z\in\rho(A).$$

In addition, the function $-Q^{-1}(z)$ takes the form

$$-Q^{-1}(z)=K^{'\times}(\mathbb{A}'-zI)^{-1}K',\quad \mathrm{Im}\, z\neq 0,$$

where:

1. *$\mathbb{A}'$ is a self-adjoint bi-extension of $\dot{A}$ defined as*

$$\mathbb{A}' = \dot{A}^* + (\dot{A}^* - \mathbb{A})^*,$$

2. *the quasi-kernel of $\mathbb{A}'$ is a self-adjoint extension of $\dot{A}$ transversal to A,*

3. *the operator K' is defined by the relation $K'K^\times = \dot{A}^* - \mathbb{A}'$.*

Remark 7.5.4. Theorem 7.5.3 above implies that a Krein-Langer Q-function corresponding to a self-adjoint extension A of $\dot{A}$ can be realized as a transfer function of an impedance system Δ of the form (6.49). This system Δ contains the state space operator A, $E = \mathcal{N}$, and operator K is the one described in the statement of Theorem 7.5.3. It also follows from Theorem 7.1.4 that any Krein-Langer Q-function $Q(z)$ as well as $-Q^{-1}(z)$ belongs to the class $N_0(R)$. Consequently, by Theorem 7.1.5, both $Q(z)$ and $-Q^{-1}(z)$ can be realized as impedance functions of L-systems. We can also conclude that if an invertible $V(z) \in N_0(R)$, then $-V^{-1}(z) \in N_0(R)$.

Let $\Pi = \{\mathcal{N}, \Gamma_1, \Gamma_0\}$ be a boundary triplet for $\dot{A}^*$ and let $C = C^* \in [\mathcal{N}, \mathcal{N}]$. Then $\Pi_C = \{\mathcal{N}, \Gamma_1 + C\Gamma_0, \Gamma_0\}$ is also a boundary triplet for $\dot{A}^*$. With Π_C are associated two self-adjoint bi-extensions

$$\mathbb{A}_{0C} = \dot{A}^* + (\Gamma_1^\times + \Gamma_0^\times C)\Gamma_0 = \mathbb{A}_0 + \Gamma_0^\times C\Gamma_0,$$

and

$$\mathbb{A}_{1C} = \dot{A}^* - \Gamma_0^\times(\Gamma_1 + C\Gamma_0) = \mathbb{A}_1 - \Gamma_0^\times C\Gamma_0,$$

where $\mathbb{A}_0$ and $\mathbb{A}_1$ are given by (7.52). Note that $\mathbb{A}_{0C} \supset A_0$ and $\mathbb{A}_{1C} \supset A_{1C}$, where

$$\operatorname{Dom}(A_{1C}) = \{f \in \operatorname{Dom}(\dot{A}^*) : \Gamma_1 f + C\Gamma_0 f = 0\}, \quad A_{1C} = \dot{A}^* \upharpoonright \operatorname{Dom}(A_{1C}).$$

The Weyl-Titchmarsh function $M_C(z) = (\Gamma_1 + C\Gamma_0)\gamma_0(z)$ corresponding to Π_C is of the form

$$M_C(z) = M_0(z) + C.$$

On the other hand by Theorem 7.4.5 (see (7.46)) we have

$$\begin{aligned} M_C(z) &= (\Gamma_1 + C\Gamma_0)(\mathbb{A}_{0C} - zI)^{-1}(\Gamma_1^\times + \Gamma_0^\times C) \\ &= (\Gamma_1 + C\Gamma_0)(\mathbb{A}_0 + \Gamma_0^\times C\Gamma_0 - zI)^{-1}(\Gamma_1^\times + \Gamma_0^\times C). \end{aligned}$$

Thus, for all $z \in \rho(A_0)$ we obtain

$$\Gamma_1(\mathbb{A}_0 - zI)^{-1}\Gamma_1^\times + C = (\Gamma_1 + C\Gamma_0)(\mathbb{A}_0 + \Gamma_0^\times C\Gamma_0 - zI)^{-1}(\Gamma_1^\times + \Gamma_0^\times C).$$

7.6 Examples

Example. This example is to illustrate the realization in $N_0(R)$ class. Let

$$Tx = \frac{1}{i}\frac{dx}{dt}, \tag{7.60}$$

with $\mathrm{Dom}(T) = \left\{x(t) \,\middle|\, x(t) - \text{ abs. continuous}, x'(t) \in L^2_{[0,l]}, x(0) = 0\right\}$ be the differential operator in $\mathcal{H} = L^2_{[0,l]}$ $(l > 0)$. Obviously,

$$T^*x = \frac{1}{i}\frac{dx}{dt},$$

with $\mathrm{Dom}(T^*) = \left\{x(t) \,\middle|\, x(t) - \text{ abs. continuous}, x'(t) \in L^2_{[0,l]}, x(l) = 0\right\}$ is its adjoint. Consider a symmetric operator $\dot{A}$,

$$\begin{aligned} \dot{A}x &= \frac{1}{i}\frac{dx}{dt}, \\ \mathrm{Dom}(\dot{A}) &= \left\{x(t) \,\middle|\, x(t) - \text{ abs. continuous}, x'(t) \in L^2_{[0,l]}, x(0) = x(l) = 0\right\}, \end{aligned} \tag{7.61}$$

and its adjoint $\dot{A}^*$,

$$\begin{aligned} \dot{A}^*x &= \frac{1}{i}\frac{dx}{dt}, \\ \mathrm{Dom}(\dot{A}^*) &= \left\{x(t) \,\middle|\, x(t) - \text{ abs. continuous}, x'(t) \in L^2_{[0,l]}\right\}. \end{aligned} \tag{7.62}$$

Then $\mathcal{H}_+ = \mathrm{Dom}(\dot{A}^*) = W^1_2$ is a Sobolev space with scalar product

$$(x,y)_+ = \int_0^l x(t)\overline{y(t)}\,dt + \int_0^l x'(t)\overline{y'(t)}\,dt.$$

Construct rigged Hilbert space $W^1_2 \subset L^2_{[0,l]} \subset (W^1_2)_-$ and consider operators

$$\mathbb{A}x = \frac{1}{i}\frac{dx}{dt} + ix(0)\left[\delta(t-l) - \delta(t)\right], \quad \mathbb{A}^*x = \frac{1}{i}\frac{dx}{dt} + ix(l)\left[\delta(t-l) - \delta(t)\right], \tag{7.63}$$

where $x(t) \in W^1_2$, $\delta(t)$, $\delta(t-l)$ are delta-functions and elements of $(W^1_2)_-$ that generate functionals by the formulas $(x, \delta(t)) = x(0)$ and $(x, \delta(t-l)) = x(l)$. It is easy to see that $\mathbb{A} \supset T \supset \dot{A}$, $\mathbb{A}^* \supset T^* \supset \dot{A}$, and

$$\Theta_1 = \begin{pmatrix} \frac{1}{i}\frac{dx}{dt} + ix(0)[\delta(t-l) - \delta(t)] & K & -1 \\ W^1_2 \subset L^2_{[0,l]} \subset (W^1_2)_- & & \mathbb{C} \end{pmatrix} \quad (J = -1),$$

is an L-system where

$$\begin{aligned} Kc &= c \cdot \frac{1}{\sqrt{2}}[\delta(t-l)-\delta(t)], \quad (c \in \mathbb{C}), \\ K^*x &= \left(x, \frac{1}{\sqrt{2}}[\delta(t-l)-\delta(t)]\right) = \frac{1}{\sqrt{2}}[x(l)-x(0)], \end{aligned} \tag{7.64}$$

and $x(t) \in W_2^1$. Also

$$\operatorname{Im}\mathbb{A} = -\left(\cdot, \frac{1}{\sqrt{2}}[\delta(t-l)-\delta(t)]\right)\frac{1}{\sqrt{2}}[\delta(t-l)-\delta(t)].$$

The transfer function of this system can be found as

$$W_{\Theta_1}(z) = I - 2iK^*(\mathbb{A}-zI)^{-1}KJ = e^{izl}.$$

Consider the following Herglotz-Nevanlinna function (hyperbolic tangent)

$$V(z) = -i\tanh\left(\frac{i}{2}zl\right).$$

Obviously this function can be realized as

$$\begin{aligned} V(z) &= -i\tanh\left(\frac{i}{2}zl\right) = -i\frac{e^{\frac{i}{2}zl}-e^{-\frac{i}{2}zl}}{e^{\frac{i}{2}zl}+e^{-\frac{i}{2}zl}} = -i\frac{e^{izl}-1}{e^{izl}+1} \\ &= i\left[W_{\Theta_1}(z)+I\right]^{-1}\left[W_{\Theta_1}(z)-I\right]J \qquad (J=-1). \end{aligned}$$

Now let us consider another L-system

$$\Theta_2 = \begin{pmatrix} \frac{1}{i}\frac{dx}{dt} + ix(l)[\delta(t-l)-\delta(t)] & K & 1 \\ W_2^1 \subset L^2_{[0,l]} \subset (W_2^1)_- & & \mathbb{C} \end{pmatrix} \quad (J=1), \tag{7.65}$$

whose state-space operator is $\mathbb{A}^*$ and $J=1$. Similar reasoning confirms that

$$W_{\Theta_2}(z) = e^{-izl}, \tag{7.66}$$

but

$$V(z) = -i\tanh\left(\frac{i}{2}zl\right) = i\left[W_{\Theta_2}(z)+I\right]^{-1}\left[W_{\Theta_2}(z)-I\right]J.$$

This shows that $V(z) \in N_0(R)$ is the impedance function of the L-system Θ_2 in (7.65) with the transfer function $W_{\Theta_2}(z)$ of the form (7.66).

Example. Consider a bounded linear operator in $\mathbb{C}^2$ defined by

$$T = \begin{pmatrix} i & i \\ -i & 1 \end{pmatrix}. \tag{7.67}$$

Let x and φ be the elements of $\mathbb{C}^2$ such that

$$x = \begin{pmatrix} x_1 \\ x_2 \end{pmatrix} \quad \text{and} \quad \varphi = \begin{pmatrix} 1 \\ 0 \end{pmatrix}.$$

Obviously,

$$T^* = \begin{pmatrix} -i & i \\ -i & 1 \end{pmatrix}.$$

It is clear that $\mathrm{Dom}(T) = \mathrm{Dom}(T^*) = \mathbb{C}^2$. Let $J = 1$. Now we can find

$$\operatorname{Im} T = \frac{T - T^*}{2i} = \begin{pmatrix} 1 & 0 \\ 0 & 0 \end{pmatrix}.$$

and show that φ above is the only channel vector such that

$$\operatorname{Im} T\, x = \begin{pmatrix} 1 & 0 \\ 0 & 0 \end{pmatrix} \begin{pmatrix} x_1 \\ x_2 \end{pmatrix} = (x, \varphi)\varphi.$$

Consider a symmetric operator $\dot{A}$ of the form

$$\dot{A}x = \begin{pmatrix} 0 & i \\ -i & 1 \end{pmatrix} \begin{pmatrix} x_1 \\ x_2 \end{pmatrix}, \quad \mathrm{Dom}(\dot{A}) = \left\{ x \in \mathbb{C}^2 \mid (x, \varphi) = 0 \right\}. \tag{7.68}$$

Obviously, $\dot{A}$ is the maximal common symmetric part of T and T^*. Thus, operator T can be included in the system

$$\Theta = \begin{pmatrix} T & K & 1 \\ \mathbb{C}^2 & & \mathbb{C} \end{pmatrix}, \tag{7.69}$$

with

$$Kc = c\varphi = \begin{pmatrix} c \\ 0 \end{pmatrix}, \quad K^*x = (x, \varphi) = x_1, \quad c \in \mathbb{C}, \quad x = \begin{pmatrix} x_1 \\ x_2 \end{pmatrix} \in \mathbb{C}^2,$$

Then $W_\Theta(z)$ is represented by the formula

$$W_\Theta(z) = \frac{z^2 + (1-i)z - 1 - i}{z^2 - (1+i)z - 1 + i}. \tag{7.70}$$

The impedance function $V_\Theta(z)$ is a Herglotz-Nevanlinna function

$$V_\Theta(z) = \frac{1 - z}{z^2 - z - 1}$$

and hence belongs to the class $N_1(R)$ according to Theorem 7.1.7.

Example. In order to illustrate the realization in $N_{01}(R)$ class we will use Examples 7.6 and 7.6. Consider an L-system

$$\Theta = \begin{pmatrix} \mathbb{A} & K & J \\ W_2^1 \oplus \mathbb{C}^2 \subset L^2_{[0,l]} \oplus \mathbb{C}^2 \subset (W_2^1)_- \oplus \mathbb{C}^2 & & \mathbb{C}^2 \end{pmatrix}$$

where $\mathbb{A}$ is a diagonal block-matrix

$$\mathbb{A} = \begin{pmatrix} \mathbb{A}_1 & 0 \\ 0 & T \end{pmatrix},$$

with

$$\mathbb{A}_1 = \frac{1}{i}\frac{dx}{dt} + ix(0)\left[\delta(t-l) - \delta(t)\right],$$

of the form (7.63) from Example 7.6, and T of the form (7.67) from Example 7.6. The corresponding symmetric operator $\dot{A}$ in this case takes the form

$$\dot{A} = \begin{pmatrix} \dot{A}_1 & 0 \\ 0 & \dot{A}_2 \end{pmatrix},$$

where $\dot{A}_1$ and $\dot{A}_2$ are defined by (7.62) and (7.68), respectively. Clearly, $\overline{\mathrm{Dom}(\dot{A})} \neq L^2_{[0,l]} \oplus \mathbb{C}^2$. Operator $K \in [\mathbb{C}^2, (W_2^1)_- \oplus \mathbb{C}^2]$ here is defined as a diagonal operator block-matrix

$$K = \begin{pmatrix} K_1 & 0 \\ 0 & K_2 \end{pmatrix},$$

where operators K_1 and K_2 are from Examples 7.6 and 7.6, respectively. Moreover, $K^* \in [W_2^1 \oplus \mathbb{C}^2, \mathbb{C}^2]$ and J are defined by

$$K^* = \begin{pmatrix} K_1^* & 0 \\ 0 & K_2^* \end{pmatrix}, \qquad J = \begin{pmatrix} -1 & 0 \\ 0 & 1 \end{pmatrix}.$$

We find that

$$\mathrm{Im}\,\mathbb{A}\,f = \begin{pmatrix} \mathrm{Im}\,\mathbb{A}_1 & 0 \\ 0 & \mathrm{Im}\,T \end{pmatrix} \begin{pmatrix} f_1 \\ f_2 \end{pmatrix} = \begin{pmatrix} -(f_1, g)\,g \\ (f_2, \varphi)\,\varphi \end{pmatrix} = KJK^*, \quad f \in W_2^1 \oplus \mathbb{C}^2,$$

where $g = \frac{1}{\sqrt{2}}[\delta(t-l) - \delta(t)]$ and $\varphi = \begin{pmatrix} 1 \\ 0 \end{pmatrix}$ are the channel vectors from Examples 7.6 and 7.6, respectively. It can be easily shown that

$$W_\Theta(z) = \begin{pmatrix} e^{izl} & 0 \\ 0 & \frac{z^2+(1-i)z-1-1}{z^2-(1+i)z-1+i} \end{pmatrix}$$

and

$$V_\Theta(z) = \begin{pmatrix} -i\tanh\left(\frac{i}{2}zl\right) & 0 \\ 0 & \frac{1-z}{z^2-z-1} \end{pmatrix}.$$

According to Theorem 7.1.10, the function $V_\Theta(z)$ above belongs to the class $N_{01}(R)$.

Example. This example illustrates multiplication Theorem 7.3.8. Let the L-system Θ_1 be given by (7.65) with all the components constructed in Example 7.6. Its transfer function $W_{\Theta_1}(z)$ is presented in (7.66). Clearly, by construction $W_{\Theta_1}(z) \in \Omega_0(R,1)$. Similarly, let Θ_2 be the L-system of the form (7.69) with the transfer function $W_{\Theta_2}(z)$ given by (7.70). It follows then from derivations in Example 7.6 that $W_{\Theta_2}(z) \in \Omega_1(R,1)$. Consider

$$W(z) = W_{\Theta_1}(z) \cdot W_{\Theta_2}(z) = (e^{-izl}) \cdot \frac{z^2+(1-i)z-1-i}{z^2-(1+i)z-1+i}.$$

As we have already mentioned in Section 7.2, the class $\Omega_{01}(R,J)$ does not exist for the case of scalar functions when $\dim E = 1$. Thus the function $W(z)$ above can only belong to either $\Omega_0(R,1)$ or $\Omega_1(R,1)$. The latter, however, is impossible because of Theorem 7.3.6 and Remark 7.3.7. Hence, $W(z) \in \Omega_0(R,1)$.

Chapter 8

Normalized L-Systems

In this chapter we consider special types of L-systems and study the properties of their transfer functions. In the first two sections we will use the notion of an auxiliary canonical system to prove a theorem about the constant J-unitary factor. The theorem states that if an operator-valued function $W(z)$ belongs to the class $\Omega_0(R,J)$ described in Section 7.2, then for an arbitrary J-unitary operator B the functions $W(z)B$ and $BW(z)$ again belong to the same class $\Omega_0(R,J)$. Consequently, they can be realized as transfer functions of the same type of L-system. We will construct this new realizing system and show that it contains the same unbounded operator T but different channel operators K.

In the remainder of the chapter we deal with specially constructed L-systems, so called *normalized L-systems*. It will be shown that under certain conditions on the state-space operator of such an L-system Θ, one can always find another L-system Θ' so that the two transfer functions are related with a property $W_\Theta(\lambda) = W_{\Theta'}\left(\frac{1}{\lambda}\right)$.

8.1 Auxiliary canonical system

Let $\mathcal{H}$ be a separable Hilbert space and let $\dot{A}$ be a closed symmetric operator whose domain is dense in $\mathcal{H}$, i.e., $\overline{\mathrm{Dom}(\dot{A})} = \mathcal{H}$. We also assume that $\dot{A}$ has finite and equal deficiency indices.

Now let T be an operator of the class $\Lambda(\dot{A})$ defined on $\mathrm{Dom}(T)$ with the resolvent set $\rho(T)$. Consider $\mathcal{H}_T = \mathrm{Dom}(T)$ as a Hilbert space $\mathcal{H}_T$ with the inner product

$$(u,v)_T := (u,v) + (Tu,Tv), \quad u,v \in \mathcal{H}_T.$$

Let $\mathcal{H}_T \subset \mathcal{H} \subset \mathcal{H}'_T$ be the rigged Hilbert space. Since T is a bounded operator from $\mathcal{H}_T$ into $\mathcal{H}$, the adjoint operator $T^\times$ is bounded from $\mathcal{H}$ into $\mathcal{H}'_T$ and satisfies the condition $(Tu,f) = (u,T^\times f)$ for all $u \in \mathcal{H}_T$ and all $f \in \mathcal{H}$. Let T^* be the adjoint of T in $\mathcal{H}$. Then obviously $T^\times \supset T^*$. The resolvent $\mathrm{R}_z(T) = (T - zI)^{-1}$ of

T is a bounded operator from $\mathcal{H}$ onto $\operatorname{Dom}(T) = \mathcal{H}_T$. It follows that the resolvent $\mathrm{R}_{\bar z}(T^*) = (T^* - \bar z I)^{-1}$ of T^* has the continuation $\overline{\mathrm{R}_{\bar z}(T^*)}$ on $\mathcal{H}'_T$ and maps $\mathcal{H}'_T$ onto $\mathcal{H}$. Moreover, $\overline{\mathrm{R}_{\bar z}(T^*)} = (T^\times - \bar z I)^{-1}$ (see Section 6.2). We know that according to the definition of the class $\Lambda(\dot A)$, $\operatorname{Dom}(\dot A) = \ker(T - T^\times)$. Define

$$\Phi := \overline{\operatorname{Ran}(T - T^\times)} \quad \text{(the closure in } \mathcal{H}'_T). \tag{8.1}$$

Thus $\Phi \cap \mathcal{H} = \{0\}$. Then

$$\operatorname{Dom}(\dot A) = \ker(T - T^\times) = \{g \in \mathcal{H}_T : (g, \varphi) = 0 \quad \text{for all} \quad \varphi \in \Phi\}.$$

Since $T \in \Lambda(\dot A)$, then $T \supset \dot A$, $T^* \supset \dot A$. Let as above $\mathfrak{N}_z = \ker(\dot A^* - zI)$ be the defect subspaces of $\dot A$. Then the following lemma takes place.

Lemma 8.1.1. *If z is a regular point of the operator T^*, then*

$$(T^\times - zI)^{-1}\Phi = \mathfrak{N}_z. \tag{8.2}$$

Proof. For an arbitrary $g \in \operatorname{Dom}(\dot A)$ we have

$$((\dot A - \bar z I)g, (T^\times - zI)^{-1} f) = (g, f).$$

Hence for $f = (1/2i)(T - T^\times)h$, $h \in \operatorname{Dom}(T)$ and the fact that $Tg = T^*g$, we have

$$\left((\dot A - \bar z I)g, (T^\times - zI)^{-1} f\right) = \left(g, \frac{(T - T^\times)}{2i} h\right) = 0. \tag{8.3}$$

If $\phi_z \in \mathfrak{N}_z$, then for a $g \in \operatorname{Dom}(\dot A)$,

$$\left((T^\times - zI)\phi_z, g\right) = (\phi_z, (T - \bar z I)g) = (\phi_z, (\dot A - \bar z I)g) = 0.$$

Relation (8.3) and the inequality

$$c(z)\|f\| \le \|(T^* - zI)f\|_{\mathcal{H}'_T} \le d(z)\|f\|, \quad c(z) > 0, d(z) > 0$$

completes the proof. □

According to Remark 4.3.4 our assumptions on operator $\dot A$ imply that there exists a self-adjoint extension A of $\dot A$ in $\mathcal{H}$ such that (see (4.52))

$$\operatorname{Dom}(T) + \operatorname{Dom}(A) = \operatorname{Dom}(\dot A^*). \tag{8.4}$$

Following the procedure from Section 2.2 we construct an operator-generated rigging using the operator $\dot A$. As usual we denote by $\mathcal{H}_+$ the Hilbert space $\operatorname{Dom}(\dot A^*)$ equipped by the inner product

$$(u, v)_+ = (u, v) + (\dot A^* u, \dot A^* v).$$

Then, according to (2.9), a (+)-orthogonal decomposition holds:

$$\mathcal{H}_+ = \mathrm{Dom}(\dot{A}) \oplus \mathfrak{N}_i \oplus \mathfrak{N}_{-i}.$$

By the von Neumann formula (1.15)

$$\mathrm{Dom}(A) = \mathrm{Dom}(\dot{A}) \oplus (I + \mathcal{U})\mathfrak{N}_i, \;\; A = \dot{A}^* \upharpoonright \mathrm{Dom}(A),$$

where $\mathcal{U}$ is an isometry from $\mathfrak{N}_i$ onto $\mathfrak{N}_{-i}$. Since A satisfies condition (8.4), the direct decomposition

$$\mathcal{H}_+ = \mathcal{H}_T \dot{+} (I + \mathcal{U})\mathfrak{N}_i \tag{8.5}$$

holds. Let $\mathcal{H}_{T^*} = \mathrm{Dom}(T^*)$, $(u, v)_{T^*} = (u, v) + (T^*u, T^*)$, $u, v \in \mathrm{Dom}(T^*)$ and let $\mathcal{H}_{T^*} \subset \mathcal{H} \subset \mathcal{H}'_{T^*}$ be the corresponding rigging. The operator $\mathcal{T} : \mathcal{H} \to \mathcal{H}'_{T^*}$ is defined as the adjoint to T^*:

$$(T^*v, f) = (v, (T^*)^\times f) = (v, \mathcal{T} f).$$

It is easy to see that $T \subset \mathcal{T}$. Note that the subspace in $\mathcal{H}'_{T^*}$,

$$\Phi_* := \overline{\mathrm{Ran}(\mathcal{T} - T^*)} \quad (\text{ the closure is taken in } \mathcal{H}'_{T^*}), \tag{8.6}$$

satisfies the condition $\Phi_* \cap \mathcal{H} = \{0\}$.

The Hilbert spaces $\mathcal{H}_T$ and $\mathcal{H}_{T^*}$ are subspaces of the Hilbert space $\mathcal{H}_+$ and one can easily show that if (8.4) holds, then $\mathrm{Dom}(T^*) + \mathrm{Dom}(A) = \mathrm{Dom}(\dot{A}^*)$ and by (4.54)

$$\mathcal{H}_+ = \mathcal{H}_{T^*} \dot{+} (I + \mathcal{U})\mathfrak{N}_i. \tag{8.7}$$

We denote by $\mathcal{P}$ and $\mathcal{P}_*$ the skew projections in $\mathcal{H}_+$ onto $\mathcal{H}_T$ and $\mathcal{H}_{T^*}$ with respect to the decompositions (8.5) and (8.7), respectively. Let

$$\mathcal{H}_+ \subset \mathcal{H} \subset \mathcal{H}_-$$

be the rigged Hilbert triplet and let $\mathcal{P}^\times : \mathcal{H}'_T \to \mathcal{H}_-$ and $\mathcal{P}_*^\times : \mathcal{H}'_{T^*} \to \mathcal{H}_-$ be the adjoint operators. These operators are defined by the relations

$$(\mathcal{P}u, \varphi) = (u, \mathcal{P}^\times \varphi), \;\; u \in \mathcal{H}_+, \varphi \in \mathcal{H}'_T \tag{8.8}$$

and

$$(\mathcal{P}_* v, \psi) = (v, \mathcal{P}_*^\times \psi), \;\; u \in \mathcal{H}_+, \psi \in \mathcal{H}'_{T^*}. \tag{8.9}$$

Observe that in view of (8.5) and (8.7) the operators $\mathcal{P}$ and $\mathcal{P}_*$ are isomorphisms from $\mathcal{H}_{T^*}$ onto $\mathcal{H}_T$ and $\mathcal{H}_T$ onto $\mathcal{H}_{T^*}$, respectively. Moreover the relations

$$\begin{aligned} \mathcal{P}\mathcal{P}_* f &= f, \;\; f \in \mathcal{H}_T, \\ \mathcal{P}_*\mathcal{P} g &= g, \;\; g \in \mathcal{H}_{T^*} \end{aligned} \tag{8.10}$$

hold, and

$$\mathcal{P}^\times \Phi = \mathcal{P}_*^\times \Phi_* = \Psi,$$

where

$$\Psi := \{\varphi \in \mathcal{H}_- : \ (\varphi, h) = 0 \quad \text{for all} \quad h \in \operatorname{Dom}(A)\}.$$

It is easy to see that if $\dot{A}$ has finite deficiency indices (n, n), then Φ is an n-dimensional subspace. In this case, by direct check one obtains that $\ker(\mathcal{P}^\times \upharpoonright \Phi) = \{0\}$, $\ker(\mathcal{P}_*^\times \upharpoonright \Phi_*) = \{0\}$, and $\Psi = \mathcal{P}^\times \Phi$ is n-dimensional as well.

Theorem 8.1.2. *Let the operator $\mathbb{A}_\mathcal{P} : \mathcal{H}_+ \to \mathcal{H}_-$ be defined by the formula*

$$\mathbb{A}_\mathcal{P} = \dot{A}^* + \mathcal{P}^\times (T^\times - \dot{A}^*)(I - \mathcal{P}). \tag{8.11}$$

Then its adjoint $\mathbb{A}_\mathcal{P}^* : \mathcal{H}_+ \to \mathcal{H}_-$ acts by the rule

$$\mathbb{A}_\mathcal{P}^* = \dot{A}^* + \mathcal{P}_*^\times (\mathcal{T} - \dot{A}^*)(I - \mathcal{P}_*). \tag{8.12}$$

Moreover, $T \subset \mathbb{A}_\mathcal{P}$, $T^* \subset \mathbb{A}_\mathcal{P}^*$ and the following relations hold:

$$\begin{aligned}
&\mathbb{A}_\mathcal{P} = \dot{A}^* + \mathcal{P}_*^\times (\mathcal{T} - \dot{A}^*), \quad \mathbb{A}_\mathcal{P}^* = \dot{A}^* + \mathcal{P}^\times (T^\times - \dot{A}^*),\\
&\mathbb{A}_\mathcal{P} - \mathbb{A}_\mathcal{P}^* = \mathcal{P}^\times (T - T^\times)\mathcal{P}, \quad \mathbb{A}_\mathcal{P} - \mathbb{A}_\mathcal{P}^* = \mathcal{P}_*^\times (\mathcal{T} - T^*)\mathcal{P}_*,\\
&(\mathbb{A}_\mathcal{P} - zI)^{-1}\mathcal{P}_*^\times \varphi = (\mathcal{T} - zI)^{-1}\varphi, \quad \varphi \in \Phi_*,\\
&(\mathbb{A}_\mathcal{P}^* - \bar{z}I)^{-1}\mathcal{P}^\times \varphi = (T^\times - \bar{z}I)^{-1}\varphi, \quad \varphi \in \Phi,\\
&\big((\mathbb{A}_\mathcal{P} - zI)^{-1}\psi, \psi\big) = \big(\psi, (\mathbb{A}_\mathcal{P}^* - \bar{z}I)^{-1}\psi\big), \quad \psi \in \Psi, \quad z \in \rho(T).
\end{aligned} \tag{8.13}$$

Proof. Let $\mathbb{A}_\mathcal{P}$ be an operator defined by (8.11). Then for $f \in \operatorname{Dom}(T) = \mathcal{H}_T$ we get $\mathbb{A}_\mathcal{P} f = Tf$, i.e., $T \subset \mathbb{A}_\mathcal{P}$. Since $\dot{A}^* \upharpoonright (I + \mathcal{U})\mathfrak{N}_i$ is a symmetric operator, we have

$$(\dot{A}^*(I - \mathcal{P})u, (I - \mathcal{P})v) = ((I - \mathcal{P})u, \dot{A}^*(I - \mathcal{P})v), \ u, v \in \mathcal{H}_+.$$

Therefore for all $u, v \in \mathcal{H}_+$,

$$\begin{aligned}
(\mathbb{A}_\mathcal{P} u, v) - (u, \mathbb{A}_\mathcal{P} v) &= (T\mathcal{P}u + \dot{A}^*(I - \mathcal{P})u, \mathcal{P}v + (I - \mathcal{P})v)\\
&\quad - (\mathcal{P}u + (I - \mathcal{P})u, T\mathcal{P}v + \dot{A}^*(I - \mathcal{P})v) + ((I - \mathcal{P})u, T\mathcal{P}v)\\
&\quad - (T\mathcal{P}u, (I - \mathcal{P})v) - (\dot{A}^*(I - \mathcal{P})u, \mathcal{P}v) + (\mathcal{P}u, \dot{A}^*(I - \mathcal{P})v)\\
&= ((T - T^\times)\mathcal{P}u, \mathcal{P}v) = (\mathcal{P}^\times (T - T^\times)\mathcal{P}u, v).
\end{aligned}$$

Thus, $\mathbb{A}_\mathcal{P} - \mathbb{A}_\mathcal{P}^* = \mathcal{P}^\times (T - T^\times)\mathcal{P}$. It follows that

$$\begin{aligned}
\mathbb{A}_\mathcal{P}^* &= \mathbb{A}_\mathcal{P} - \mathcal{P}^\times (T - T^\times)\mathcal{P} = \dot{A}^* + \mathcal{P}^\times (T^\times - \dot{A}^*)(I - \mathcal{P}) - \mathcal{P}^\times (T - T^\times)\mathcal{P}\\
&= \dot{A}^* + \mathcal{P}^\times (T^\times - \dot{A}^*).
\end{aligned}$$

Further we show that

$$\mathcal{P}^\times (T^\times - \dot{A}^*)(I - \mathcal{P}) = \mathcal{P}_*^\times (\mathcal{T} - \dot{A}^*), \tag{8.14}$$

holds for $f \in \mathcal{H}_T$. Let $f \in (I + \mathcal{U})\mathfrak{N}_i$ and $g \in \mathcal{H}_+$, then

$$\mathcal{P}g - \mathcal{P}_* g = -(I - \mathcal{P})g + (I - \mathcal{P}_*)g \in (I + \mathcal{U})\mathfrak{N}_i.$$

Therefore,

$$\begin{aligned}(\dot{A}^* f, \mathcal{P}g - \mathcal{P}_* g) &= (Af, \mathcal{P}g - \mathcal{P}_* g) = (f, A(\mathcal{P}g - \mathcal{P}_* g)) \\ &= (f, \dot{A}^*(\mathcal{P}g - \mathcal{P}_* g)) = (f, T\mathcal{P}g - T^*\mathcal{P}_* g).\end{aligned}$$

Now we get $((T^\times - \dot{A}^*)f, \mathcal{P}g) = ((\mathcal{T} - \dot{A}^*)f, \mathcal{P}_* g)$. Consequently

$$\mathcal{P}^\times (T^\times - \dot{A}^*)f = \mathcal{P}_*^\times (\mathcal{T} - \dot{A}^*)f, \ f \in (I + \mathcal{U})\mathfrak{N}_i.$$

From (8.11) and (8.14) we obtain (8.12). Let $z \in \rho(T^*)$ and $\varphi_z \in \mathfrak{N}_z$. Then

$$(\mathbb{A}_{\mathcal{P}}^* - zI)\varphi_z = \mathcal{P}^\times (T^\times - \dot{A}^*)\varphi_z = \mathcal{P}^\times (T^\times - zI)\varphi_z.$$

Now from (8.2) we obtain the relation $(\mathbb{A}_{\mathcal{P}}^* - zI)^{-1}\mathcal{P}^\times \varphi = (T^\times - zI)^{-1}\varphi$ for $\varphi \in \Phi$. □

Theorem 8.1.3. *The resolvents $(\mathbb{A}_{\mathcal{P}} - zI)^{-1}$, $(\mathbb{A}_{\mathcal{P}}^* - \bar{z}I)^{-1}$ map $\mathcal{H}\dot{+}\Psi$ onto $\mathcal{H}_+$ for all $z \in \rho(T)$. The real part $\mathrm{Re}\,\mathbb{A}_{\mathcal{P}} = (\mathbb{A}_{\mathcal{P}} + \mathbb{A}_{\mathcal{P}}^\times)/2$ satisfies the condition $\mathrm{Re}\,\mathbb{A}_{\mathcal{P}} \supset \widehat{A}$, where $\widehat{A}$ is a self-adjoint extension of $\dot{A}$ transversal to A. The resolvent $(\mathrm{Re}\,\mathbb{A}_{\mathcal{P}} - zI)^{-1}$ maps $\mathcal{H}\dot{+}\Psi$ onto $\mathcal{H}_+$ for all z, $\mathrm{Im}\, z \neq 0$.*

Proof. If we let $z \in \rho(T^*)$, then $\mathcal{H}_{T^*}\dot{+}\mathfrak{N}_z = \mathcal{H}_+$. This decomposition and the relations $(T^\times - zI)\mathfrak{N}_z = \Phi$, $\mathcal{P}_*\mathfrak{N}_z = (I + \mathcal{U})\mathfrak{N}_i$ imply

$$(T^\times - \dot{A}^*)\mathcal{H}_+ = (T^\times - \dot{A}^*)(I + \mathcal{U})\mathfrak{N}_i = \Phi.$$

Let $(T^\times - \dot{A}^*)^{-1} = \left((T^\times - \dot{A}^*)\restriction (I + \mathcal{U})\mathfrak{N}_i\right)^{-1}$. From (8.11) and (8.13) we get

$$\mathrm{Re}\,\mathbb{A}_{\mathcal{P}} = \dot{A}^* + \mathcal{P}^\times \left((T^\times - \dot{A}^*)(I - \mathcal{P}) + \frac{1}{2}(T^\times - T)\mathcal{P}\right).$$

Let

$$\mathrm{Dom}(\widehat{A}) = \{u \in \mathcal{H}_+ : \ \mathrm{Re}\,\mathbb{A}_{\mathcal{P}} u \in \mathcal{H}\}, \quad \widehat{A}u = \mathrm{Re}\,\mathbb{A}_{\mathcal{P}} u, \ u \in \mathrm{Dom}(\widehat{A}).$$

Then

$$\mathrm{Dom}(\widehat{A}) = \left(I - \frac{1}{2}(T^\times - \dot{A}^*)^{-1}(T^\times - T)\right)\mathcal{H}_T.$$

It follows that $\mathrm{Dom}(\widehat{A})\dot{+}(I + \mathcal{U})\mathfrak{N}_i = \mathcal{H}_+$. Since $\widehat{A}$ is a symmetric extension of $\dot{A}$ the last equality implies the self-adjointness of $\widehat{A}$. □

Remark 8.1.4. The projections $\mathcal{P}$ and $\mathcal{P}_*$ coincide with projections $\mathcal{P}_{TA}$ and $\mathcal{P}_{T^*A}$ defined in Theorem 4.4.3. Moreover, it follows from Theorems 8.1.2 and 8.1.3 that the $(*)$-extension $\mathbb{A}_\mathcal{P}$ in (8.11) and its adjoint $\mathbb{A}^*_\mathcal{P}$ in (8.12) coincide with operators $\mathbb{A}$ and its adjoint $\mathbb{A}^*$ given by (4.60).

Definition 8.1.5. The collection

$$\Theta_0 = \begin{pmatrix} T & K_0 & J \\ \mathcal{H}_T \subset \mathcal{H} \subset \mathcal{H}'_T & & E \end{pmatrix} \tag{8.15}$$

is called the **auxiliary rigged canonical system** if E is a finite-dimensional Hilbert space, J is an operator in E such that $J = J^* = J^{-1}$, K_0 is a bounded linear operator from E into $\mathcal{H}'_T$ such that $\operatorname{Ker} K_0 = \{0\}$, and the identity holds: $T - T^\times = K_0 J K_0^\times$. Here $K_0^\times : \mathcal{H}_T \to E$ is the adjoint to K_0 operator.

Assume that operator $T \in \Lambda(\dot{A})$ and let Θ_0 be an auxiliary rigged canonical system. Suppose that A is a self-adjoint extension of $\dot{A}$ satisfying the condition (8.4). If $\mathcal{P}$ is a corresponding projection determined by (8.5), then

$$\Theta_\mathcal{P} = \begin{pmatrix} \mathbb{A}_\mathcal{P} & \mathcal{P}^\times K_0 & J \\ \mathcal{H}_+ \subset \mathcal{H} \subset \mathcal{H}_- & & E \end{pmatrix} \tag{8.16}$$

is the L-system with the state-space operator $\mathbb{A}_\mathcal{P}$, since by Theorem 8.1.2 we have $\mathbb{A}_\mathcal{P} - \mathbb{A}^*_\mathcal{P} = \mathcal{P}^\times(T - T^\times)\mathcal{P} = \mathcal{P}^\times K_0 J K_0^\times \mathcal{P}$. The operator-valued function

$$W_{\Theta_\mathcal{P}}(z) = I - 2iK_0^\times \mathcal{P}(\mathbb{A}_\mathcal{P} - zI)^{-1}\mathcal{P}^\times K_0 J,$$

is a transfer function of the system $\Theta_\mathcal{P}$. Observe that

$$W^*_{\Theta_\mathcal{P}}(z) = I + 2iJK_0^\times \mathcal{P}(\mathbb{A}^*_\mathcal{P} - \bar{z}I)^{-1}\mathcal{P}^\times K_0.$$

It follows from Theorem 8.1.2 that

$$W^*_{\Theta_\mathcal{P}}(z) = I + 2iJK_0^\times \mathcal{P}(T^\times - \bar{z}I)^{-1}K_0. \tag{8.17}$$

Remark 8.1.6. In a similar manner one can introduce an auxiliary system based upon operator T^* as

$$\Theta_{*,0} = \begin{pmatrix} T^* & K_{*,0} & -J \\ \mathcal{H}_{T^*} \subset \mathcal{H} \subset \mathcal{H}'_{T^*} & & E \end{pmatrix}, \tag{8.18}$$

where $K_{*,0}$ is a bounded linear operator from E into $\mathcal{H}'_{T^*}$ with $\ker(K_{*,0}) = \{0\}$ and such that $T^* - T = K_{*,0} J K_{*,0}^\times$. Here $K_{*,0}^\times : \mathcal{H}_{T^*} \to E$ is adjoint to the $K_{*,0}$ operator. If $\mathcal{P}_*$ is the corresponding projection determined by (8.7), then

$$\Theta_{\mathcal{P}_*} = \begin{pmatrix} \mathbb{A}_{\mathcal{P}_*} & \mathcal{P}_*^\times K_{*,0} & -J \\ \mathcal{H}_+ \subset \mathcal{H} \subset \mathcal{H}_- & & E \end{pmatrix}$$

is an L-system similar to $\Theta_\mathcal{P}$ in (8.16).

8.2 Constant J-unitary factor

In this section we study the change of a given L-system when its transfer function is multiplied by a constant J-unitary factor.

Theorem 8.2.1. *Let Θ_0 be an auxiliary rigged canonical system with the state space operator T and let A_1 and A_2 be two self-adjoint extensions of $\dot{A}$ satisfying the condition of transversality (8.4). If $\mathcal{P}_1$ and $\mathcal{P}_2$ are two corresponding to A_1 and A_2 skew projections defined by decomposition (8.5) and $\Theta_{\mathcal{P}_1}$, $\Theta_{\mathcal{P}_2}$ are two L-systems of the form (8.16) with the state-space operators $\mathbb{A}_{\mathcal{P}_1}$ and $\mathbb{A}_{\mathcal{P}_2}$, then for all $z \in \rho(T)$ the identity*

$$W_{\Theta_{\mathcal{P}_2}}(z) = W_{\Theta_{\mathcal{P}_1}}(z)B$$

holds, where B is a J-unitary operator acting in the Hilbert space E.

Proof. Let $\mathfrak{M}_T = \mathcal{H}_T \ominus \mathrm{Dom}(\dot{A})$ be the orthogonal complement of $\mathrm{Dom}(\dot{A})$ in the Hilbert space $\mathcal{H}_T$. Then $K_0^\times f \neq 0$ for all $f \in \mathfrak{M}_T \setminus \{0\}$. Let us write

$$K_0^{\times[-1]} = \left(K_0^\times \upharpoonright \mathfrak{M}_T\right)^{-1}. \tag{8.19}$$

For all $f \in \mathfrak{M}_T$ we have from (8.17)

$$\begin{aligned} W^*_{\Theta_{\mathcal{P}_m}}(z)JK_0^\times f &= JK_0^\times\left(f + \mathcal{P}_m(T^\times - \bar{z}I)^{-1}(T - T^\times)f\right) \\ &= JK_0^\times \mathcal{P}_m(T^* - \bar{z}I)^{-1}(T - \bar{z}I)f, \quad (m = 1, 2). \end{aligned}$$

From (8.10) we obtain

$$(T^* - \bar{z}I)^{-1}(T - \bar{z}I)f = \mathcal{P}_{*1}K_0^{\times[-1]}JW^*_{\Theta_{\mathcal{P}_1}}(z)JK_0^\times f,$$

and

$$W^*_{\Theta_{\mathcal{P}_2}}(z)JK_0^\times f = JK_0^\times\mathcal{P}_2\mathcal{P}_{*1}K_0^{\times[-1]}JW^*_{\Theta_{\mathcal{P}_1}}(z)JK_0^\times f.$$

Set

$$\mathcal{B} = JK_0^\times\mathcal{P}_2\mathcal{P}_{*1}K_0^{\times[-1]}J. \tag{8.20}$$

The operator $\mathcal{B}$ is obviously a bounded operator in E ($\dim E < \infty$). Thus,

$$W^*_{\Theta_{\mathcal{P}_2}}(z) = \mathcal{B}W^*_{\Theta_{\mathcal{P}_1}}(z).$$

If $h \in E$, then using Theorem 8.1.2 we get

$$\begin{aligned} (J\mathcal{B}h, \mathcal{B}h)_E &= (K_0^\times\mathcal{P}_2\mathcal{P}_{*1}K_0^{\times[-1]}Jh, JK_0^\times\mathcal{P}_2\mathcal{P}_{*1}K_0^{\times[-1]}Jh)_E \\ &= \left((T - T^\times)\mathcal{P}_2\mathcal{P}_{*1}K_0^{\times[-1]}Jh, \mathcal{P}_2\mathcal{P}_{*1}K_0^{\times[-1]}Jh\right) \\ &= \left((\mathbb{A}_{\mathcal{P}_2} - \mathbb{A}^*_{\mathcal{P}_2})\mathcal{P}_{*1}K_0^{\times[-1]}Jh, \mathcal{P}_{*1}K_0^{\times[-1]}Jh\right) \\ &= (\mathcal{P}_{*1}K_0^{\times[-1]}Jh, T^*\mathcal{P}_{*1}K_0^{\times[-1]}Jh) - (T^*\mathcal{P}_{*1}K_0^{\times[-1]}Jh, \mathcal{P}_{*1}K_0^{\times[-1]}Jh) \\ &= \left((\mathcal{T} - T^*)\mathcal{P}_{*1}K_0^{\times[-1]}Jh, \mathcal{P}_{*1}K_0^{\times[-1]}Jh\right) \\ &= \left((\mathbb{A}_{\mathcal{P}_1} - \mathbb{A}^*_{\mathcal{P}_1})K_0^{\times[-1]}Jh, K_0^{\times[-1]}Jh\right) = (K_0JK_0^\times K_0^{\times[-1]}Jh, K_0^{\times[-1]}Jh) \\ &= (Jh, h)_E. \end{aligned}$$

This implies the identity $\mathcal{B}^*J\mathcal{B} = J$. Since $\mathrm{Ran}(\mathcal{B}) = E$, the operator $\mathcal{B}$ is a J-unitary operator in E. The operator $B = \mathcal{B}^*$ is J-unitary as well and $W_{\Theta_{\mathcal{P}_2}}(z) = W_{\Theta_{\mathcal{P}_1}}(z)B$ for all $z \in \rho(T)$. □

Notice that it follows from (6.45) and (8.13) that

$$W_{\Theta_{\mathcal{P}}}(z)JW^*_{\Theta_{\mathcal{P}}}(\zeta) - J = 2i(\bar{\zeta} - z)K_0^*(T - zI)^{-1}(T^\times - \bar{\zeta})^{-1}K_0,\ z, \zeta \in \rho(T).$$

Hence,

$$W_{\Theta_{\mathcal{P}_1}}(z)JW^*_{\Theta_{\mathcal{P}_1}}(\zeta) = W_{\Theta_{\mathcal{P}_2}}(z)JW^*_{\Theta_{\mathcal{P}_2}}(\zeta).$$

If $\rho(T)$ contains points ζ_0 and $\bar{\zeta}_0$, then applying relation (6.46) yields

$$\begin{aligned} W_{\Theta_{\mathcal{P}_2}}(z) &= W_{\Theta_{\mathcal{P}_1}}(z)JW^*_{\Theta_{\mathcal{P}_1}}(\zeta_0)W^{*-1}_{\Theta_{\mathcal{P}_2}}(\zeta_0)J \\ &= W_{\Theta_{\mathcal{P}_1}}(z)JW^*_{\Theta_{\mathcal{P}_1}}(\zeta_0)JW_{\Theta_{\mathcal{P}_2}}(\bar{\zeta}_0). \end{aligned}$$

Put

$$B = JW^*_{\Theta_{\mathcal{P}_1}}(\zeta_0)JW_{\Theta_{\mathcal{P}_2}}(\bar{\zeta}_0).$$

Then $W_{\Theta_{\mathcal{P}_2}}(z) = W_{\Theta_{\mathcal{P}_1}}(z)B$ for all $z \in \rho(T)$. Since

$$W_{\Theta_{\mathcal{P}_1}}(\zeta_0)BJB^*W^*_{\Theta_{\mathcal{P}_1}}(\zeta_0) = W_{\Theta_{\mathcal{P}_1}}(\zeta_0)JW^*_{\Theta_{\mathcal{P}_1}}(\zeta_0)$$

and $W_{\Theta_{\mathcal{P}_1}}(\zeta_0)$ has bounded inverse, we get that B is J-unitary in E.

Now we will establish an important inverse result.

Theorem 8.2.2. *Let E be a finite-dimensional Hilbert space and let J be a self-adjoint and unitary operator in E. If the operator-valued function $W(z)$ in E belongs to the class $\Omega_0(R, J)$ and B is a J-unitary operator in E, then the function*

$$W_B(z) = W(z)B$$

also belongs to the class $\Omega_0(R, J)$.

Proof. According to the definition of the class $\Omega_0(R, J)$ (see Section 7.2) and Theorem 7.2.2 there exists an L-system

$$\Theta = \begin{pmatrix} \mathbb{A} & K & J \\ \mathcal{H}_+ \subset \mathcal{H} \subset \mathcal{H}_- & & E \end{pmatrix} \tag{8.21}$$

whose transfer operator-valued function coincides with $W(z)$. This system can be used to generate the auxiliary system

$$\Theta_0 = \begin{pmatrix} T & K_0 & J \\ \mathcal{H}_T \subset \mathcal{H} \subset \mathcal{H}'_T & & E \end{pmatrix} \tag{8.22}$$

in the following way. Let us set $\mathcal{H}_T = \mathrm{Dom}(T)$ where T is taken from the system Θ in (8.21). Let also $\mathcal{P}$ be a skew projector determined by the decomposition (8.5).

In this case the subspace Φ defined in (8.1) is finite-dimensional and hence, as we mentioned it in the previous section, the projection operator $\mathcal{P}^\times\upharpoonright\Phi$ is invertible, i.e., $(\mathcal{P}^\times\upharpoonright\Phi)^{-1}$ exists.Thus we can define operator $K_0\in[E,\mathcal{H}'_T]$ by the formula

$$K_0=(\mathcal{P}^\times\upharpoonright\Phi)^{-1}K, \tag{8.23}$$

where K is an operator from the system Θ in (8.21). Consequently, $K=\mathcal{P}^\times K_0$, $\mathbb{A}=\mathbb{A}_\mathcal{P}$ of the form (8.11), and our L-system Θ in (8.21) and auxiliary system Θ_0 in (8.22) can be related as in (8.16). One can also show that if $K_0^\times\in[\mathcal{H}_T,E]$, then $K_0^\times=K^*\upharpoonright\mathcal{H}_T$. Observe that

$$\mathcal{H}_T\dot{+}\left(\operatorname{Ker}(\mathbb{A}-\mathbb{A}^*)\ominus\operatorname{Dom}(\dot{A})\right)=\mathcal{H}_+$$

and $\mathcal{P}$ is the projector on $\mathcal{H}_T$ with respect to this decomposition. Let B be a J-unitary operator in E and let

$$\operatorname{Dom}(A_1)=\left(I-K_0^{\times[-1]}B^{*-1}K_0^\times\mathcal{P}\right)\operatorname{Dom}(T^*),\quad A_1=\dot{A}^*\upharpoonright\operatorname{Dom}(A_1), \tag{8.24}$$

where $K_0^{\times[-1]}$ is defined by (8.19) and $\dot{A}$ is from the system Θ in (8.21). Let us show that A_1 is symmetric. For $u=g-K_0^{\times[-1]}B^{*-1}K_0^\times\mathcal{P}g$, $g\in\operatorname{Dom}(T^*)$ we have

$$\begin{aligned}(\dot{A}^*u,u)-(u,\dot{A}^*u)&=(T^*g-TK_0^{\times[-1]}B^{*-1}K_0^\times\mathcal{P}g,g-K_0^{\times[-1]}B^{*-1}K_0^\times\mathcal{P}g)\\&\quad-(g-K_0^{\times[-1]}B^{*-1}K_0^\times\mathcal{P}g,T^*g-TK_0^{\times[-1]}B^{*-1}K_0^\times\mathcal{P}g)=((\mathbb{A}^*-\mathbb{A})g,g)\\&\quad+\left((\mathbb{A}-\mathbb{A}^*)K_0^{\times[-1]}B^{*-1}K_0^\times\mathcal{P}g,K_0^{\times[-1]}B^{*-1}K_0^\times\mathcal{P}g\right)\\&=-2i(\mathcal{P}^\times K_0JK_0^\times\mathcal{P}g,g)\\&\quad+2i(\mathcal{P}^\times K_0JK_0^\times K_0^{\times[-1]}B^{*-1}K_0^\times\mathcal{P}g,K_0^{\times[-1]}B^{*-1}K_0^\times\mathcal{P}g)\\&=-2i(\mathcal{P}^\times K_0JK_0^\times\mathcal{P}g,g)+2i(\mathcal{P}^\times K_0B^{-1}JB^{*-1}K_0^\times\mathcal{P}g,g)=0.\end{aligned}$$

Thus, A_1 is a symmetric operator. Let $\dim E=n$, then $\dim\operatorname{Ran}(K)=\dim\operatorname{Ran}(K_0)=\dim\operatorname{Ran}(\mathbb{A}-\mathbb{A}^*)=n$. Also the defect numbers of $\dot{A}$ are (n,n) and $\dim\left(\operatorname{Dom}(A_1)/\operatorname{Dom}(\dot{A})\right)=n$. Therefore, A_1 is a self-adjoint extension of $\dot{A}$. Moreover, $\operatorname{Dom}(A_1)\cap\operatorname{Dom}(T^*)=\operatorname{Dom}(\dot{A})$. Thus

$$\operatorname{Dom}(T)+\operatorname{Dom}(A_1)=\operatorname{Dom}(T^*)+\operatorname{Dom}(A_1)=\operatorname{Dom}(\dot{A}^*)=\mathcal{H}_+.$$

This means that A_1 is transversal to T. Let $\mathcal{P}_1$ and $\mathcal{P}_{*1}$ be the corresponding skew projections onto $\operatorname{Dom}(T)$ and $\operatorname{Dom}(T^*)$ according to (8.5) and (8.7) written for A_1, respectively. From (8.24) we obtain the equality $K_0^\times\mathcal{P}_1g=B^{*-1}K_0^\times\mathcal{P}g$ for all $g\in\operatorname{Dom}(T^*)$. Therefore, using (8.10) we get

$$K_0^\times\mathcal{P}_1\mathcal{P}_*f=B^{*-1}K_0^\times f\quad\text{for all}\quad f\in\operatorname{Dom}(T).$$

We define operator $\mathbb{A}_{\mathcal{P}_1}$ by means of $\mathcal{P}_1$ and via formula (8.11). If $K_1 = \mathcal{P}_1^\times K_0$, then the L-system

$$\Theta_1 = \begin{pmatrix} \mathbb{A}_{\mathcal{P}_1} & K_1 & J \\ \mathcal{H}_+ \subset \mathcal{H} \subset \mathcal{H}_- & & E \end{pmatrix}$$

has the transfer function $W_{\Theta_1}(z) = I - 2iK_1^\times(\mathbb{A}_{\mathcal{P}_1} - zI)^{-1}K_1 J$ and

$$W_{\Theta_1}^*(z) = I + 2iJK_0^\times \mathcal{P}_1(T^\times - zI)^{-1}K_0.$$

Using the proof of Theorem 8.2.1 we get the equality $W_{\Theta_1}(z) = W(z)B$. □

The same result as in Theorem 8.2.2 holds for $BW(z)$. The next theorem follows from Theorem 8.2.2.

Theorem 8.2.3. *Let Θ be an L-system of the form* (6.31) *defined by*

$$\begin{cases} (T - zI)x = KJ\varphi_-, \\ \varphi_+ = \varphi_- - 2iK^*x, \end{cases} \tag{8.25}$$

with the transfer function $W_\Theta(z)$. If B is a J-unitary operator in the input-output space E of Θ, then the function $W_{\Theta_B}(z) = W_\Theta(z)B$ is the transfer function of the L-system Θ_B of the form

$$\begin{cases} (T - zI)x = K_B J\varphi_-, \\ \varphi_+ = \varphi_- - 2iK_B^*x, \end{cases} \tag{8.26}$$

where

$$\begin{aligned} K_B &= \mathcal{P}_1^\times(\mathcal{P}^\times \upharpoonright \Phi)^{-1}K \\ &= ((K^* \upharpoonright \mathfrak{M}_T)^{-1}JBJ(K^* \upharpoonright \mathfrak{M}_T)^{-1}\mathcal{P}_*^{-1})^\times(\mathcal{P}^\times \upharpoonright \Phi)^{-1}K. \end{aligned} \tag{8.27}$$

Here the operators $\mathcal{P}$, $\mathcal{P}_*$, $\mathcal{P}^\times$, $\mathcal{P}_1$, and the set Φ are defined in the proof of Theorem 8.2.2 *by relations* (8.5), (8.7), (8.8), (8.9), *and* (8.1), *respectively.*

Proof. In order to prove the theorem we follow the main steps of the proof of Theorem 8.2.2. First we build an auxiliary system Θ_0 based on the L-system Θ in (8.25) and then construct the projection $\mathcal{P}_1$. Using the same operator T as in (8.25) we form a new L-system (8.26) as prescribed in the proof of Theorem 8.2.2. This L-system has a new channel operator K_B. In order to obtain representation (8.27) of K_B we utilize the formula (8.20). □

Theorem 8.2.3 above shows that if one multiplies the transfer function of a given L-system by a J-unitary constant factor, then the result is a transfer function of another L-system with the same operator T and a new operator K_B determined by (8.27).

Now let us consider an L-system

$$\Theta = \begin{pmatrix} \mathbb{A} & K & J \\ \mathcal{H}_+ \subset \mathcal{H} \subset \mathcal{H}_- & & E \end{pmatrix},$$

of the form (8.16) with operator T of the class $\Lambda(\dot{A})$, invertible operator K, and $\overline{\mathrm{Dom}(\dot{A})} = \mathcal{H}$. From Section 8.1 we recall that for a system of the form (8.16) there is a skew projection operator $\mathcal{P}$ determined by a decomposition

$$\mathcal{H}_+ = \mathrm{Dom}(T)\dot{+}(I+\mathcal{U})\mathfrak{N}_i$$

that maps $\mathcal{H}_+$ onto $\mathrm{Dom}(T)$. Therefore we can follow the argument of the proof of Theorem 8.2.2 to obtain the auxiliary system (8.22) with the invertible channel operator K_0 defined by (8.23) such that $K = \mathcal{P}^\times K_0$. After that we form the system $\Theta_{\mathcal{P}}$ of the form (8.16) with operator $\mathbb{A}_{\mathcal{P}}$ defined by (8.11). Since $\mathrm{Im}\,\mathbb{A}_{\mathcal{P}} = \mathcal{P}^\times K_0 J K_0^\times \mathcal{P} = KJK^* = \mathrm{Im}\,\mathbb{A}$, we can apply the uniqueness Theorem 4.3.9 to conclude that $\mathbb{A} = \mathbb{A}_{\mathcal{P}}$. Similarly, we can use the reasoning in Remark 8.1.6 and represent the elements of Θ in terms of the skew projector $\mathcal{P}_*$ defined by (8.7). This yields $\mathbb{A}^* = \mathbb{A}_{\mathcal{P}_*}$, $K = \mathcal{P}_*^\times K_{*,0}$, where $K_{*,0}$ is a bounded invertible linear operator from E into $\mathcal{H}'_{T^*}$ defined in (8.18).

Theorem 8.2.4. *Let $T \in \Lambda$, $(-1) \in \rho(T)$, $\mathbb{A}$ be a $(*)$-extension of T. Let S be the fractional-linear transformation of T given by*

$$S = (I-T)(I+T)^{-1}. \tag{8.28}$$

If

$$\Theta = \begin{pmatrix} & \mathbb{A} & K & J \\ \mathcal{H}_+ \subset & \mathcal{H} & \subset \mathcal{H}_- & E \end{pmatrix}$$

is an L-system with $\ker(K) = \{0\}$, *then*

$$\Theta' = \begin{pmatrix} S & \sqrt{2}(I+\mathbb{A})^{-1}K & -J \\ \mathcal{H} & & E \end{pmatrix} \tag{8.29}$$

is an L-system as well and the transfer functions W_Θ and $W_{\Theta'}$ are connected by the relation

$$W_\Theta(z) = W_\Theta(-1)W_{\Theta'}\left(\frac{1-z}{1+z}\right), \quad z \in \rho(T),\ z \neq -1. \tag{8.30}$$

Proof. It is easy to see that $2(I+T)^{-1} - I = (I-T)(I+T)^{-1}$. Taking into account (8.28) we get

$$S = 2(I+T)^{-1} - I,$$

and

$$(S-\lambda I)^{-1} = -\frac{1}{\lambda}(I+T)\left(T - \frac{1-\lambda}{1+\lambda}I\right)^{-1},$$

implying that

$$\begin{aligned}\mathrm{Im}\,S &= 2\mathrm{Im}\,(I+T)^{-1} = 2\mathrm{Im}\,(I+\mathbb{A})^{-1} = -2(I+\mathbb{A})^{-1}\mathrm{Im}\,\mathbb{A}(I+\mathbb{A}^*)^{-1} \\ &= -2(I+\mathbb{A})^{-1}KJK^*(I+\mathbb{A}^*)^{-1}.\end{aligned}$$

Hence, $\operatorname{Im} S = \tilde{K}(-J)\tilde{K}^*$, where

$$\tilde{K} = \sqrt{2}(I + \mathbb{A})^{-1}K, \qquad \tilde{K} \in [E, \mathcal{H}].$$

The transfer function $W_{\Theta'}$ is of the form

$$\begin{aligned} W_{\Theta'}(\lambda) &= I + 4iK^*(I + \mathbb{A}^*)^{-1}(S - \lambda I)^{-1}(I + \mathbb{A})^{-1}KJ \\ &= W_{\Theta'}(\lambda) = I - 2i(z+1)K^*(I + \mathbb{A}^*)^{-1}(\mathbb{A} - zI)^{-1}KJ, \quad \lambda \in \rho(S). \end{aligned}$$

Let

$$z = \frac{1-\lambda}{1+\lambda}.$$

Then

$$\lambda = \frac{1-z}{1+z}, \ \frac{1}{1+\lambda} = z + 1.$$

For $z \in \rho(T)$ one has

$$\begin{aligned} (z+1)(I + \mathbb{A}^*)^{-1}(\mathbb{A} - zI)^{-1} = 2i((I + \mathbb{A}^*)^{-1}KJK^*(\mathbb{A} - zI)^{-1} \\ + (\mathbb{A} - zI)^{-1} - (I + \mathbb{A}^*)^{-1}. \end{aligned}$$

Further

$$\begin{aligned} &JW_\Theta^*(-1)JW_\Theta(z) = (I + 2iK^*(I + \mathbb{A}^*)^{-1}KJ)(I - 2iK^*(\mathbb{A} - zI)^{-1}KJ) \\ &= I - 2iK^*\left((\mathbb{A} - zI)^{-1} - (I + \mathbb{A}^*)^{-1} + 2i(I + \mathbb{A}^*)^{-1}KJK^*(\mathbb{A} - zI)^{-1}\right)KJ \\ &= I - 2i(z+1)K^*(I + \mathbb{A}^*)^{-1}(\mathbb{A} - zI)^{-1}KJ = W_{\Theta'}\left(\frac{1-z}{1+z}\right). \end{aligned}$$

Since $W_\Theta^{-1}(-1) = JW_\Theta^*(-1)J$, we get (8.37). □

Note that since $(-1) \in \rho(T)$ then by properties (6.46) the operator $W_{\Theta_T}(-1)$ is J-unitary.

8.3 The Donoghue transform and impedance functions of scattering L-systems

Let operator-function $V(z)$ act on a finite-dimensional Hilbert space E. If $\alpha \in [E, E]$ is a self-adjoint operator, then the formula

$$[V(z)]_\alpha = e^{-i\alpha}[\cos\alpha + (\sin\alpha)V(z)][\sin\alpha - (\cos\alpha)V(z)]^{-1}e^{i\alpha}, \tag{8.31}$$

defines the **Donoghue transform** of the operator-function $V(z)$.

Theorem 8.3.1. *Let the scattering L-systems* Θ *and* Θ_α *with transfer functions* $W_\Theta(z)$ *and* $W_{\Theta_\alpha}(z)$ *be such that*

$$W_{\Theta_\alpha}(z) = W_\Theta(z)(-e^{2i\alpha}), \tag{8.32}$$

where $\alpha \in [E, E]$ *is a self-adjoint operator. Then the impedance functions* $V_{\Theta_\alpha}(z)$ *and* $V_\Theta(z)$ *are connected by the Donoghue transform* (8.31).

Proof. Formula (8.32) implies

$$-W_{\Theta_\alpha}(z)e^{-2i\alpha} = W_\Theta(z) = [I - i\,V_\Theta(z)][I + i\,V_\Theta(z)]^{-1},$$

that is equivalent to

$$(e^{i\alpha}W_{\Theta_\alpha}(z)e^{-i\alpha}) \cdot e^{-i\alpha}[-I - i\,V_\Theta(z)] = e^{i\alpha}[I - i\,V_\Theta(z)].$$

Dividing both sides by i yields

$$(e^{i\alpha}W_{\Theta_\alpha}(z)e^{-i\alpha})[ie^{-i\alpha} - e^{-i\alpha}\,V_\Theta(z)] = (-i)e^{i\alpha} - e^{i\alpha}\,V_\Theta(z).$$

Let $\mathcal{W} = e^{i\alpha}W_{\Theta_\alpha}(z)e^{-i\alpha}$. Then we apply Euler's formula and get

$$\begin{aligned}\mathcal{W}[i(\cos\alpha - i\sin\alpha) - (\cos\alpha - i\sin\alpha)\,V_\Theta(z)]& \\ = (-i)(\cos\alpha + i\sin\alpha) - (\cos\alpha + i\sin\alpha)\,V_\Theta(z).&\end{aligned}$$

Pulling out i and regrouping yields

$$\begin{aligned}\mathcal{W}[(\sin\alpha + i\cos\alpha) - (\cos\alpha - i\sin\alpha)\,V_\Theta(z)]& \\ = (\sin\alpha - i\cos\alpha) - (\cos\alpha + i\sin\alpha)\,V_\Theta(z).&\end{aligned}$$

Now we multiply both sides from the right by $(\sin\alpha - \cos\alpha\,V_\Theta(z))^{-1}$ and get

$$\begin{aligned}\mathcal{W} + i\mathcal{W}(\cos\alpha + \sin\alpha\,V_\Theta(z))(\sin\alpha - \cos\alpha\,V_\Theta(z))^{-1}& \\ = I - i(\cos\alpha + \sin\alpha\,V_\Theta(z))(\sin\alpha - \cos\alpha\,V_\Theta(z))^{-1}.&\end{aligned}$$

Let $[V_\Theta(z)]_\alpha$ be the Donoghue transform (8.31) of the function $V_\Theta(z)$. Then

$$\mathcal{W} + i\mathcal{W}[V_\Theta(z)]_\alpha = I - i[V_\Theta(z)]_\alpha.$$

The latter gives us

$$\mathcal{W} = e^{i\alpha}W_{\Theta_\alpha}(z)e^{-i\alpha} = [I - i[V_\Theta(z)]_\alpha][I + i[V_\Theta(z)]_\alpha]^{-1},$$

or

$$e^{i\alpha}W_{\Theta_\alpha}(z)e^{-i\alpha}[I + i[V_\Theta(z)]_\alpha] = I - i[V_\Theta(z)]_\alpha.$$

Distributing and multiplying both sides from the left by $e^{-i\alpha}$ and from the right by $e^{i\alpha}$ we obtain

$$W_{\Theta_\alpha}(z) + iW_{\Theta_\alpha}(z)e^{-i\alpha}[V_\Theta(z)]_\alpha e^{i\alpha} = I - ie^{-i\alpha}[V_\Theta(z)]_\alpha\,e^{i\alpha}.$$

Since $W_{\Theta_\alpha}(z) = [I - iV_{\Theta_\alpha}(z)][I + iV_{\Theta_\alpha}(z)]^{-1}$ is equivalent to

$$W_{\Theta_\alpha}(z) + iW_{\Theta_\alpha}(z)V_{\Theta_\alpha}(z) = I - iV_{\Theta_\alpha}(z),$$

the last statement implies that

$$[V_\Theta(z)]_\alpha = (\cos\alpha + \sin\alpha\, V_\Theta(z))(\sin\alpha - \cos\alpha\, V_\Theta(z))^{-1} = V_{\Theta_\alpha}(z),$$

which completes the proof. □

Theorem 8.3.2. *If a function $V(z)$ belongs to the class $N_0(R)$, then its Donoghue transform $[V(z)]_\alpha$ also belongs to $N_0(R)$. Moreover, both $V(z)$ and $[V(z)]_\alpha$ can be realized as impedance functions of minimal scattering systems Θ and Θ_α with the same operator T.*

Proof. Since $V(z) \in N_0(R)$, then it can be realized as the impedance function of a minimal scattering system Θ with the transfer function $W_\Theta(z)$ and properties described in Theorem 7.1.5. Consider the function

$$W_\alpha(z) = [I - i[V(z)]_\alpha][I + i[V(z)]_\alpha]^{-1}.$$

Reversing the argument of the proof of Theorem 8.3.1 we obtain that

$$W_\alpha(z) = W_\Theta(z)(-e^{2i\alpha}).$$

Taking into account that for a self-adjoint operator $\alpha \in [E, E]$, the operator $B = (-e^{2i\alpha})$ is unitary, we apply Theorem 8.2.3 utilizing the last formula. This theorem gives us the existence and description of the L-system $\Theta_\alpha = \Theta_B$ with the desired properties. Since both L-systems Θ and Θ_α share not only operator T but also densely-defined symmetric operator $\dot{A}$, then by Theorem 7.1.4 we have that $[V(z)]_\alpha = V_{\Theta_\alpha}$ belongs to the class $N_0(R)$. □

8.4 Normalized (∗)-extensions and normalized L-systems

Theorem 8.4.1. *Let $T \in \Lambda(\dot{A})$, where $\dot{A}$ is a densely-defined closed symmetric operator. Suppose $0 \in \rho(T)$. Then there is a unique (∗)-extension $\mathbb{A}_0$ of T such that* $\mathrm{Re}\,\mathbb{A}_0$ *is a t-self-adjoint bi-extension of $\dot{A}$ whose quasi-kernel is the operator $\hat{A}_0 = \left(\mathrm{Re}\,(T^{-1})\right)^{-1}$.*

Proof. The operator $\mathrm{Re}\,(T^{-1})$ is invertible since the equality $\mathrm{Re}\,(T^{-1})f = 0$ yields

$$\dot{A}^*\left(\mathrm{Re}\,(T^{-1})f\right) = \frac{1}{2}TT^{-1}f + \frac{1}{2}T^*(T^{-1})^*f = f = 0.$$

Since the operator $\mathrm{Re}\,(T^{-1})$ is a bounded self-adjoint extension of non-densely-defined bounded symmetric operator $\dot{A}^{-1}$, the operator

$$\hat{A}_0 = \left(\mathrm{Re}\,(T^{-1})\right)^{-1}$$

is a self-adjoint extension of the operator $\dot{A}$. Hence, $0 \in \rho(\hat{A}_0)$. Because $0 \in \rho(T) \cap \rho(T^*) \cap \rho(\hat{A}_0)$, the direct decompositions

$$\mathcal{H}_+ = \mathrm{Dom}(T) \dot{+} \mathfrak{N}_0, \quad \mathcal{H}_+ = \mathrm{Dom}(T^*) \dot{+} \mathfrak{N}_0, \quad \mathcal{H}_+ = \mathrm{Dom}(\hat{A}_0) \dot{+} \mathfrak{N}_0$$

hold, where $\mathfrak{N}_0 = \ker(\dot{A}^*)$ and $\mathcal{H}_+ = \mathrm{Dom}(\dot{A}^*)$. Let us define one more self-adjoint extension $\tilde{A}_0$ of $\dot{A}$ via

$$\mathrm{Dom}(\tilde{A}_0) = \mathrm{Dom}(\dot{A}) \dot{+} \mathfrak{N}_0, \quad \tilde{A}_0 = \dot{A}^* \upharpoonright \mathrm{Dom}(\tilde{A}_0).$$

Then, clearly,

$$\mathrm{Dom}(\dot{A}^*) = \mathrm{Dom}(\hat{A}_0) + \mathrm{Dom}(\tilde{A}_0).$$

The domain $\mathrm{Dom}(\tilde{A})$ admits the von Neumann representation

$$\mathrm{Dom}(\tilde{A}_0) = \mathrm{Dom}(\dot{A}) \oplus (I - U_0)\mathfrak{N}_i,$$

where U_0 is an isometry from $\mathfrak{N}_i$ onto $\mathfrak{N}_{-i}$. Thus,

$$\mathcal{H}_+ = \mathrm{Dom}(T) \dot{+} (I - U_0)\mathfrak{N}_i, \tag{8.33}$$

$$\mathcal{H}_+ = \mathrm{Dom}(T^*) \dot{+} (I - U_0)\mathfrak{N}_i, \tag{8.34}$$

$$\mathcal{H}_+ = \mathrm{Dom}(\hat{A}_0) \dot{+} (I - U_0)\mathfrak{N}_i. \tag{8.35}$$

Let $\mathcal{P}_{T\tilde{A}_0}$, $\mathcal{P}_{T^*\tilde{A}_0}$, and $\mathcal{P}_{\hat{A}_0\tilde{A}_0}$ be skew projections in $\mathcal{H}_+$ onto $\mathrm{Dom}(T)$, $\mathrm{Dom}(T^*)$, and $\mathrm{Dom}(\hat{A}_0)$, corresponding to decompositions (8.33), (8.34), (8.35), respectively. Now define a (*)-extension of T generated by $\tilde{A}_0$ (see Theorem 4.4.3 and relations (4.60)) by the formula

$$\mathbb{A}_0 = \dot{A}^* - \mathcal{R}^{-1}\dot{A}^*(I - \mathcal{P}_{T\tilde{A}_0}).$$

Then $\mathbb{A}_0$ is a (*)-extension of T. The adjoint $\mathbb{A}_0^*$ is given by

$$\mathbb{A}_0^* = \dot{A}^* - \mathcal{R}^{-1}\dot{A}^*(I - \mathcal{P}_{T^*\tilde{A}_0}).$$

Let us show that the quasi-kernel of $\mathrm{Re}\,\mathbb{A}_0$ coincides with the operator $\hat{A}_0$. Observe that $\ker(\mathrm{Im}\,\mathbb{A}_0) = \mathrm{Dom}(\tilde{A}_0)$ (see Theorem 4.4.3). Clearly,

$$\ker(\mathrm{Im}\,(T^{-1})) = \mathrm{Ran}(\dot{A}).$$

Hence, $\mathrm{Ran}(\mathrm{Im}\,(T^{-1})) = \mathfrak{N}_0$. It follows that for all $f \in \mathcal{H}$ one has

$$(\mathrm{Im}\,\mathbb{A}_0)(T^{-1} - T^{*-1})f = 0.$$

Further, for $f \in \mathcal{H}$ we have

$$\begin{aligned}(\mathbb{A}_0^* T_0^{-1} + \mathbb{A}_0 T^{*-1})f &= (\mathbb{A}_0 - (\mathbb{A}_0 - \mathbb{A}_0^*))\mathbb{A}_0^{-1} f + (\mathbb{A}_0^* + (\mathbb{A}_0 - \mathbb{A}_0^*)\mathbb{A}_0^{*-1} f \\ &= 2f + 2i(\mathrm{Im}\,\mathbb{A}_0)(T^{*-1} - T^{-1})f = 2f,\end{aligned}$$

$$\begin{aligned}(\mathrm{Re}\,\mathbb{A}_0)(\mathrm{Re}\,(T^{-1}))f &= (\mathrm{Re}\,\mathbb{A}_0)\mathrm{Re}\,(\mathbb{A}_0^{-1})f = \frac{1}{4}\left(2I + (\mathbb{A}_0^* T_0^{-1} + \mathbb{A}_0 T^{*-1}\right) f \\ &= f.\end{aligned}$$

Thus, $(\mathrm{Re}\,\mathbb{A}_0)^{-1}f = (\mathrm{Re}\,(T^{-1}))f = \hat{A}_0^{-1}f$, i.e., the quasi-kernel of $\mathrm{Re}\,\mathbb{A}_0$ is the operator $\hat{A}_0$. Observe that by Theorems 3.4.9 and 4.4.4 the operator $\mathrm{Re}\,\mathbb{A}_0$ is given by

$$\mathrm{Re}\,\mathbb{A}_0 = \dot{A}^* - \mathcal{R}^{-1}\dot{A}^*(I - \mathcal{P}_{\hat{A}_0\tilde{A}_0}).$$

Uniqueness of $(*)$-extension $\mathbb{A}_0$ of T with quasi-kernel $\hat{A}_0$ follows from Theorem 4.4.6. □

A $(*)$-extension $\mathbb{A}_0$ of an operator $T \in \Lambda(\dot{A})$ $(0 \in \rho(T))$ with quasi-kernel $\big(\mathrm{Re}\,(T^{-1})\big)^{-1}$ is called **normalized at point zero**.

Similarly one can construct a $(*)$-extension $\mathbb{A}_{\lambda_0}$ **normalized at a real point** λ_0 of an operator $T \in \Lambda$ having a real regular point $\lambda_0 \neq 0$. Namely, such a $(*)$-extension is generated via Theorem 4.4.3 and formulas (4.60) by a self-adjoint extension $\tilde{A}_{\lambda_0}$ of $\dot{A}$ of the form

$$\mathrm{Dom}(\tilde{A}_{\lambda_0}) = \mathrm{Dom}(\dot{A})\dot{+}\mathfrak{N}_{\lambda_0}.$$

The quasi-kernel of $\mathrm{Re}\,\mathbb{A}_{\lambda_0}$ coincides with $\hat{A}_{\lambda_0} = \big(\mathrm{Re}\,((T-\lambda_0 I)^{-1})\big)^{-1}$.

Proposition 8.4.2. *Let $T \in \Lambda$, $0 \in \rho(T)$, and let $\mathbb{A}$ be a $(*)$-extension of T. Let also*

$$\Theta = \begin{pmatrix} \mathbb{A} & K & J \\ \mathcal{H}_+ \subset \mathcal{H} \subset \mathcal{H}_- & & E \end{pmatrix}$$

be an L-system. Then $W_\Theta(0) = I$ if and only if $\mathbb{A}$ is normalized at point zero.

Proof. We have $W_\Theta(0) = I - 2iK^*\mathbb{A}^{-1}KJ$. Then

$$W_\Theta(0) = I \iff K^*\mathbb{A}^{-1}KJ = 0.$$

Since $\mathbb{A}^{-1}KE = \mathfrak{N}_0$, we get

$$W_\Theta(0) = I \iff K^*\mathfrak{N}_0 = 0.$$

Now the relation $\ker(K^*) = \ker(\mathrm{Im}\,\mathbb{A}) = \mathrm{Dom}(\tilde{A})$, where a self-adjoint extension $\tilde{A}$ generates $\mathbb{A}$, yields

$$K^*\mathfrak{N}_0 = 0 \iff \mathrm{Dom}(\tilde{A}) = \mathrm{Dom}(\dot{A})\dot{+}\mathfrak{N}_0.$$

Therefore, $W_\Theta(0) = I$ if and only if $\mathbb{A} = \mathbb{A}_0$, where $\mathbb{A}_0$ is normalized at point zero. □

Definition 8.4.3. Let $T \in \Lambda(\dot{A})$, $0 \in \rho(T)$, $\dot{A}$ be a densely-defined closed symmetric operator with deficiency numbers (n,n), and let $\mathbb{A}_0$ be a normalized at point zero $(*)$-extension of T. An L-system

$$\Theta_0 = \begin{pmatrix} \mathbb{A}_0 & K & J \\ \mathcal{H}_+ \subset \mathcal{H} \subset \mathcal{H}_- & & \mathbb{C}^n \end{pmatrix},$$

with $\ker(K) = \{0\}$ is called a **normalized at point zero L-system** .

Proposition 8.4.2 shows that an L-system Θ is normalized at point zero if and only if $W_\Theta(0)=I$. Since the quasi-kernel $\hat{A}_0$ of the real part $\mathrm{Re}\,\mathbb{A}_0$ has bounded inverse, the impedance function $V_{\Theta_0}(z)=K^*(\mathrm{Re}\,\mathbb{A}_0-zI)^{-1}K$ of a normalized at zero L-system is holomorphic at zero. Moreover, $V_{\Theta_0}(0)=K^*(\mathrm{Re}\,\mathbb{A}_0)^{-1}K=0$.

Similarly, a normalized at real point λ_0 L-system is characterized by the condition $W_\Theta(\lambda_0)=I$ and is given by

$$\Theta_{\lambda_0}=\begin{pmatrix}\mathbb{A}_{\lambda_0} & K & J\\ \mathcal{H}_+\subset\mathcal{H}\subset\mathcal{H}_- & & \mathbb{C}^n\end{pmatrix},$$

where $\mathbb{A}_{\lambda_0}$ is a (*)-extension normalized at λ_0. In addition $V_{\Theta_{\lambda_0}}$ is holomorphic at λ_0 and $V_{\Theta_{\lambda_0}}(\lambda_0)=0$.

Theorem 8.4.4. *Let $T\in\Lambda(\dot{A})$, $0\in\rho(T)$ and $\dot{A}$ be a densely-defined closed symmetric operator. Let $\mathbb{A}$ be a (*)-extension of T. If*

$$\Theta=\begin{pmatrix}\mathbb{A} & K & J\\ \mathcal{H}_+\subset\mathcal{H}\subset\mathcal{H}_- & & E\end{pmatrix},$$

is an L-system with $\ker(K)=\{0\}$, then

$$\Theta'=\begin{pmatrix}T^{-1} & \mathbb{A}^{*-1}K & -J\\ \mathcal{H} & & E\end{pmatrix}, \tag{8.36}$$

is an L-system as well and the transfer functions W_Θ and $W_{\Theta'}$ are connected by the relation

$$W_\Theta(z)=W_{\Theta'}\left(\frac{1}{z}\right)W_\Theta(0),\quad z\in\rho(T)\setminus\{0\}. \tag{8.37}$$

Thus, if Θ_0 is normalized at zero, then $W_{\Theta_0}(z)=W_{\Theta_0'}\left(\frac{1}{z}\right)$, $z\in\rho(T)\setminus\{0\}$.

Proof. The relation $\mathbb{A}-\mathbb{A}^*=2iKJK^*$ implies

$$T^{-1}-T^{*-1}=\mathbb{A}^{-1}-\mathbb{A}^{*-1}=-2i\mathbb{A}^{*-1}KJK^*\mathbb{A}^{-1}.$$

Hence Θ' of the form (8.36) is an L-system. The transfer function $W_{\Theta'}(z)$ is of the form

$$\begin{aligned}W_{\Theta'}(z)&=I+2iK^*\mathbb{A}^{-1}(T^{-1}-zI)^{-1}\mathbb{A}^{*-1}KJ\\ &=I-\frac{2i}{z}K^*\left(\mathbb{A}-\frac{1}{z}I\right)^{-1}\mathbb{A}^{*-1}KJ.\end{aligned}$$

For $z\in\rho(T)$ one has

$$z(\mathbb{A}-zI)^{-1}\mathbb{A}^{*-1}=(\mathbb{A}-zI)^{-1}-\mathbb{A}^{*-1}+2i(\mathbb{A}-zI)^{-1}KJK^*\mathbb{A}^{*-1}.$$

Further

$$\begin{aligned}W_\Theta(z)JW_\Theta^*(0)J&=(I-2iK^*(\mathbb{A}-zI)^{-1}KJ)(I+2iK^*\mathbb{A}^{*-1}KJ)\\ &=I-2iK^*\left((\mathbb{A}-zI)^{-1}-\mathbb{A}^{*-1}+2i(\mathbb{A}-zI)^{-1}KJK^*\mathbb{A}^{*-1}\right)KJ\\ &=I-2izK^*(\mathbb{A}-zI)^{-1}\mathbb{A}^{*-1}KJ=W_{\Theta'}\left(\frac{1}{z}\right).\end{aligned}$$

Since $W_\Theta^{-1}(0) = JW_\Theta^*(0)J$, we get (8.37). If Θ_0 is a normalized at zero L-system, then by Proposition 8.4.2 we get for the transfer function of the corresponding L-system Θ'_0 the relation $W_{\Theta'_0}\left(\frac{1}{z}\right) = W_{\Theta_0}(z)$, $z \in \rho(T)$, $z \neq 0$. □

One can easily see that if

$$\Theta' = \begin{pmatrix} T^{-1} & K' & J' \\ \mathcal{H} & & E \end{pmatrix},$$

is an L-system with $\ker(K) = \{0\}$, then

$$\Theta = \begin{pmatrix} \mathbb{A} & \mathbb{A}^{*-1}K' & -J' \\ \mathcal{H}_+ \subset \mathcal{H} \subset \mathcal{H}_- & & E \end{pmatrix},$$

is an L-system too.

Similarly, if $\mathbb{A}_{\lambda_0}$ is $(*)$-extension normalized at real point λ_0 and

$$\Theta_{\lambda_0} = \begin{pmatrix} \mathbb{A}_{\lambda_0} & K & J \\ \mathcal{H}_+ \subset \mathcal{H} \subset \mathcal{H}_- & & E \end{pmatrix},$$

is an L-system, then transfer function $W_{\Theta'_{\lambda_0}}$ of the system

$$\Theta'_{\lambda_0} = \begin{pmatrix} (T-\lambda_0 I)^{-1} & (\mathbb{A}^* - \lambda_0 I)^{-1}K & -J \\ \mathcal{H} & & E \end{pmatrix},$$

is connected with W_{Θ_0} by the relation

$$W_{\Theta'_{\lambda_0}}\left(\frac{1}{z-\lambda_0}\right) = W_{\Theta_{\lambda_0}}(z), \quad z \in \rho(T) \setminus \{\lambda_0\}.$$

8.5 Realizations of e^{izl} and $e^{il/z}$ as transfer functions of L-systems

Example. We consider the construction that we have used in Example 7.6. Let

$$Tx = \frac{1}{i}\frac{dx}{dt}, \quad \mathrm{Dom}(T) = \left\{ x(t) \,\middle|\, x(t) - \text{ abs. cont.}, x'(t) \in L^2_{[0,l]},\, x(0) = 0 \right\},$$

and operators $\dot{A}$, $\mathbb{A}$, and K be given by formulas (7.61), (7.63), and (7.64), respectively. Let also $W_2^1 \subset L^2_{[0,l]} \subset (W_2^1)_-$, be the rigged Hilbert space constructed in Example 7.6. With these elements we form a normalized at point zero L-system with $J = -1$,

$$\Theta = \begin{pmatrix} \mathbb{A} & K & -1 \\ W_2^1 \subset L^2_{[0,l]} \subset (W_2^1)_- & & \mathbb{C} \end{pmatrix}, \tag{8.38}$$

As it was shown in Example 7.6 the transfer function of system (8.38) is

$$W_\Theta(z) = I - 2iK^*(\mathbb{A} - zI)^{-1}KJ = e^{izl}. \tag{8.39}$$

Now let us construct L-system Θ' of the form (8.36). It is easy to see that

$$T^{-1}g = i\int_0^t g(u)\,du, \qquad T^{-1*}g = -i\int_t^l g(u)\,du,$$

and hence

$$\frac{T^{-1} - T^{-1*}}{2i}\,g = \int_0^l g(u)\,du = (g,h)h,$$

where $h \equiv -i/\sqrt{2}$. Defining K' via the formula

$$K^{*\prime}x = (x,h),$$

we form

$$\Theta' = \begin{pmatrix} T^{-1} & K' & 1 \\ L^2_{[0,l]} & & \mathbb{C} \end{pmatrix}.$$

Using direct computations we obtain

$$(T^{-1} - zI)^{-1}h = \left(\frac{i}{z}\right) e^{\frac{i}{\sqrt{2}z}t}, \qquad \left((T^{-1} - zI)^{-1}h, h\right) = 2i[e^{\frac{i}{z}l} - 1],$$

and

$$W_{\Theta'}(z) = I - 2iK^{*\prime}(T^{-1} - zI)^{-1}K'J' = 1 - 2i\left((T^{-1} - zI)^{-1}h, h\right) = e^{\frac{i}{z}l}.$$

Example. In this example we will show a simple illustration of a normalized at point ξ L-system. Based on an operator T from the previous example we introduce a new operator

$$T^{(\xi)}x = \frac{1}{i}\frac{dx}{dt} - \xi x,$$

with $\mathrm{Dom}(T^{(\xi)}) = \mathrm{Dom}(T)$ and any real number ξ. Following the steps of Example 8.5 we obtain

$$\begin{aligned} &T^{(\xi)*}x - \xi x = \frac{1}{i}\frac{dx}{dt}, \\ &\mathrm{Dom}(T^{(\xi)*}) = \left\{x(t)\,\middle|\, x(t) - \text{ abs. cont.}, x'(t) \in L^2_{[0,l]}, x(l) = 0\right\}. \end{aligned}$$

Also a symmetric operator $\dot{A}^{(\xi)}$ is given by

$$\begin{aligned} &\dot{A}^{(\xi)}x = \frac{1}{i}\frac{dx}{dt} - \xi x, \\ &\mathrm{Dom}(\dot{A}^{(\xi)}) = \left\{x(t)\,\middle|\, x(t) - \text{ abs. cont.}, \ x'(t) \in L^2_{[0,l]}, x(0) = x(l) = 0\right\}, \end{aligned}$$

and

$$\dot{A}^{(\xi)*}x = \frac{1}{i}\frac{dx}{dt} - \xi x,\ \mathrm{Dom}(\dot{A}^{(\xi)*}) = \left\{x(t)\,\middle|\, x(t) - \text{ abs. cont., } x'(t) \in L^2_{[0,l]}\right\}.$$

The operator $T^{(\xi)}$ has inverse

$$T^{(\xi)^{-1}}g = ie^{i\xi t}\int_0^t e^{-i\xi u}g(u)\,du.$$

Evaluating the imaginary component of $T^{(\xi)^{-1}}$,

$$\frac{T^{(\xi)^{-1}} - T^{(\xi)^{-1*}}}{2i}g = e^{i\xi t}\int_0^l e^{-i\xi u}g(u)\,du = (g,h)h,$$

we obtain $h = ie^{i\xi t}/\sqrt{2}$. Using the rigged Hilbert space triplet

$$W_2^1 \subset L^2_{[0,l]} \subset (W_2^1)_-,$$

we define operators

$$\begin{aligned}\mathbb{A}x &= \frac{1}{i}\frac{dx}{dt} + ix(0)\left[e^{i\xi l}\delta(t-l) - \delta(t)\right],\\ \mathbb{A}^*x &= \frac{1}{i}\frac{dx}{dt} + ie^{-i\xi l}x(l)\left[e^{i\xi l}\delta(t-l) - \delta(t)\right],\end{aligned}$$

and identify $\mathbb{A}$ as a $(*)$-extension normalized at point ξ. Then we construct a normalized at point ξ L-system

$$\Theta^{(\xi)} = \begin{pmatrix} \mathbb{A} & K^{(\xi)} & -1\\ W_1^2 \subset L^2_{[0,l]} \subset (W_2^1)_- & & \mathbb{C}\end{pmatrix},\quad (J=-1), \tag{8.40}$$

where

$$\begin{aligned}K^{(\xi)}c &= c\cdot\frac{1}{\sqrt{2}}[e^{i\xi l}\delta(t-l) - \delta(t)], \quad (c\in\mathbb{C}),\\ K^{(\xi)^*}x &= \left(x, \frac{1}{\sqrt{2}}[e^{i\xi l}\delta(t-l) - \delta(t)]\right) = \frac{1}{\sqrt{2}}[e^{-i\xi l}x(l) - x(0)],\end{aligned}$$

and $x(t) \in W_2^1$. The channel vector $\hat{h}$ of $\mathbb{A}$ is $\hat{h} = (e^{i\xi l}\delta(t-l) - \delta(t))/\sqrt{2}$. Using standard steps one finds

$$\begin{aligned}W_{\Theta^{(\xi)}}(z) &= -2iK^{(\xi)^*}(\mathbb{A} - zI)^{-1}K^{(\xi)}J\\ &= 1 - 2i\left((\mathbb{A} - zI)^{-1}\hat{h}, \hat{h}\right) = e^{i(z-\xi)l}.\end{aligned} \tag{8.41}$$

Now let us construct L-system $\Theta^{(\xi)'}$. Following the lead of Example 8.5 we have

$$\Theta^{(\xi)'} = \begin{pmatrix} {T^{(\xi)}}^{-1} & K^{(\xi)'} & 1 \\ L^2_{[0,l]} & & \mathbb{C} \end{pmatrix},$$

where

$$K^{(\xi)'^*} x = (x, h), \quad x \in L^2_{[0,l]}.$$

Using direct computations one confirms that

$$W_{\Theta^{(\xi)'}}(z) = 1 - i\left(\left({T^{(\xi)}}^{-1} - zI\right)^{-1} h, h\right) = e^{\frac{il}{z-\xi}} = W_{\Theta^{(\xi)}}\left(\frac{1}{z-\xi}\right)$$

Examples 8.5-8.5 can also be used to illustrate Theorems 8.2.1-8.2.3 about constant J-unitary factor. Indeed, according to formula (8.41), the transfer function $W_{\Theta^{(\xi)}}(z)$ of L-system $\Theta^{(\xi)}$ in (8.40) is represented by

$$W_{\Theta^{(\xi)}}(z) = e^{i(z-\xi)l} = e^{izl} \cdot e^{-i\xi l}.$$

By (8.39), $W_\Theta(z) = e^{izl}$ is the transfer function of L-system Θ in (8.38) while $B = e^{-i\xi l}$ (with real ξ) plays the role of the constant J-unitary factor.

Chapter 9

Canonical L-systems with Contractive and Accretive Operators

In this chapter the Kreĭn classical theorem is extended to the case of quasi-self-adjoint contractive extensions of symmetric contractions, and their complete parametrization is given. On its basis we present the solution of the Phillips-Kato restricted extension problem on existence and description of all proper maximal accretive and sectorial extensions of a densely-defined non-negative symmetric operator. The criterion in terms of the impedance function of an L-system for the state-space operator to be a contraction (or so-called θ-co-sectorial contraction) is obtained. We establish the conditions for a given Stieltjes and inverse Stieltjes function to be realized as an impedance function of some L-system. We also verify when the state-space operator of an L-system is maximal accretive or sectorial. The connections between the Friedrichs and Kreĭn-von Neumann extensions and impedance function of L-systems are provided.

9.1 Contractive extensions and their block-matrix forms

We will keep the following notations. By $P_{\mathfrak{H}}$ we always denote the orthogonal projection in a Hilbert space $\mathcal{H}$ onto its subspace $\mathfrak{H}$. By $I_{\mathfrak{H}}$ we denote the restriction of the identity operator I in $\mathcal{H}$ onto a proper subspace $\mathfrak{H}$ of $\mathcal{H}$. For a given contraction $T \in [\mathcal{H}_1, \mathcal{H}_2]$ the operator

$$D_T = (I - T^*T)^{1/2} \in [\mathcal{H}_1, \mathcal{H}_1] \tag{9.1}$$

is called the **defect operator** of T. W write

$$\mathfrak{D}_T = \overline{\operatorname{Ran}(D_T)}. \tag{9.2}$$

It is well known [243] that the defect operators satisfy the following commutation relation: $TD_T = D_{T^*}T$.

If B and C are two bounded self-adjoint operators acting on $\mathcal{H}$, then the notation $B \geq C$ means that the operator $B - C$ is non-negative. The statement below is well known (see [13], [177]).

Lemma 9.1.1. *Let B be a non-negative self-adjoint operator in a Hilbert space $\mathcal{H}$ and let $B^{[-1]}$ be its Moore-Penrose inverse. Then*

$$\begin{aligned} \operatorname{Ran}(B^{1/2}) &= \left\{g \in \mathcal{H} : \sup_{f \in \operatorname{Dom}(B)} \frac{|(f,g)|^2}{(Bf,f)} < \infty\right\}, \\ \|B^{[-1/2]}g\|^2 &= \sup_{f \in \operatorname{Dom}(B)} \frac{|(f,g)|^2}{(Bf,f)}, \quad g \in \operatorname{Ran}(B^{1/2}), \end{aligned} \tag{9.3}$$

and

$$\lim_{\lambda \uparrow 0} \left((B - \lambda I)^{-1}g, g\right) = \begin{cases} \|B^{[-1/2]}g\|^2, & g \in \operatorname{Ran}(B^{1/2}), \\ +\infty, & g \in \mathcal{H} \setminus \operatorname{Ran}(B^{1/2}). \end{cases} \tag{9.4}$$

Proof. Since $\mathcal{H} = \ker B \oplus \overline{\operatorname{Ran}(B)}$ and the **Moore-Penrose inverse** $B^{[-1]}$ is the inverse to the operator $B\restriction \operatorname{Ran}(B)$, we may assume that $\ker B = \{0\}$. Let $u \in \operatorname{Dom}(B^{1/2})$. Since $\operatorname{Dom}(B)$ is dense in $\operatorname{Dom}(B^{1/2})$ (with respect to the graph norm of the operator $B^{1/2}$) and $\operatorname{Dom}(B^{1/2})$ is dense in $\mathcal{H}$, we get

$$\sup_{f \in \operatorname{Dom}(B)} \frac{\left|(f, B^{1/2}u)\right|^2}{(Bf,f)} = \sup_{f \in \operatorname{Dom}(B)} \frac{\left|(B^{1/2}f, u)\right|^2}{\|B^{1/2}f\|} = \|u\|^2.$$

Conversely, let

$$\sup_{f \in \operatorname{Dom}(B)} \frac{|(f,g)|^2}{(Bf,f)} \leq C(g),$$

where $C(g) < \infty$. Then

$$|(\varphi, g)|^2 \leq C(g)\|B^{1/2}\varphi\|^2 \quad \text{for all} \quad \varphi \in \operatorname{Dom}(B^{1/2}).$$

Consequently, $|(B^{-1/2}\psi, g)|^2 \leq C(g)\|\psi\|^2$ for all $\psi \in \operatorname{Dom}(B^{-1/2})$. By the Riesz theorem, we have $g \in \operatorname{Dom}(B^{-1/2}) = \operatorname{Ran}(B^{1/2})$. Suppose that $\|(B - \lambda I)^{-1/2}g\|^2 \leq C(g)$ for $\lambda < 0$. Then

$$\frac{|(f,g)|^2}{(Bf,f)} = \frac{\left|((B-\lambda I)^{1/2}f, (B-\lambda I)^{-1/2}g)\right|^2}{(Bf,f)} \leq \frac{(Bf,f) - \lambda\|f\|^2}{(Bf,f)}C(g).$$

Letting $\lambda \uparrow 0$ we get

$$\frac{|(f,g)|^2}{(Bf,f)} \le C(g) \quad \text{for all} \quad f \in \mathrm{Dom}(B).$$

It follows from (9.3) that $g \in \mathrm{Ran}(B^{1/2})$.

Let $E(t), t \ge 0$ be the resolution of the identity for B. Then for each $h \in \mathrm{Ran}(B)$ and all $\lambda < 0$ we have

$$\left\|B^{1/2}(B-\lambda I)^{-1/2}h - h\right\|^2 = \int_0^\infty \left(\frac{t^{1/2}}{(t-\lambda)^{1/2}} - 1\right)^2 d(E(t)h,h)$$
$$= \lambda^2 \int_0^\infty \frac{d(E(t)h,h)}{(t-\lambda)((t^{1/2}+(t-\lambda)^{1/2})^2} \le \lambda^2 \int_0^\infty \frac{d(E(t)h,h)}{t^2} = \lambda^2 \|B^{-1}h\|^2.$$

Hence,

$$\lim_{\lambda\uparrow 0} B^{1/2}(B-\lambda I)^{-1/2}h = h,$$

for all $h \in \mathrm{Ran}(B)$. Since the operator $B^{1/2}(B-\lambda I)^{-1/2}$ is a contraction for $\lambda < 0$ and the linear manifold $\mathrm{Ran}(B)$ is dense in $\mathcal{H}$, we get

$$\lim_{\lambda\uparrow 0} B^{1/2}(B-\lambda I)^{-1/2}u = u,$$

for all $u \in \mathcal{H}$. In particular, for all $\varphi \in \mathrm{Dom}(B^{1/2})$,

$$\lim_{\lambda\uparrow 0}(B-\lambda I)^{-1/2}B^{1/2}\varphi = \lim_{\lambda\uparrow 0} B^{1/2}(B-\lambda I)^{-1/2}\varphi = \varphi.$$

Thus, relations (9.4) are proved. □

The following lemma is useful for the parametrization of contractions in a block form.

Lemma 9.1.2. *Let $\mathcal{H}$, $\mathfrak{H}$, $\mathfrak{M}$, and $\mathfrak{N}$ be Hilbert spaces, and operators $F \in [\mathcal{H}, \mathfrak{H}]$, $M \in [\mathfrak{M}, \mathfrak{D}_{F^*}]$, $K \in [\mathfrak{D}_F, \mathfrak{N}]$, and $X \in [\mathfrak{D}_M, \mathfrak{D}_{K^*}]$ be contractions. Then the operator G defined by*

$$G = KFM + D_{K^*}XD_M \in [\mathfrak{M}, \mathfrak{N}], \tag{9.5}$$

satisfies the identity

$$\|h\|^2 - \|Gh\|^2 = \left\|D_F M h\right\|^2 + \|D_X D_M h\|^2 + \|(D_K FM - K^* X D_M)\, h\|^2, \tag{9.6}$$

for all $h \in \mathfrak{M}$. In particular, G is a contraction.

Proof. From the definition of G in (9.5) one obtains

$$\begin{aligned} &\|h\|^2 - \|Gh\|^2 = \|h\|^2 - \|(KFM + D_{K^*}XD_M)\,h\|^2 \\ &= \|h\|^2 - \|KFMh\|^2 - \|D_{K^*}XD_Mh\|^2 - 2\mathrm{Re}\,(KFMh, D_{K^*}XD_Mh)\,. \end{aligned} \tag{9.7}$$

The relation $K^*D_{K^*} = D_KK^*$ gives

$$(KFMh, D_{K^*}XD_Mh) = (D_KFMh, K^*XD_Mh)\,.$$

The definition of D_{K^*} shows that

$$-\|KFMh\|^2 = \|D_KFMh\|^2 - \|FMh\|^2,$$

and, likewise,

$$-\|D_{K^*}XD_Mh\|^2 = -\|XD_Mh\|^2 + \|K^*XD_Mh\|^2.$$

Now the right-hand side of (9.7) becomes

$$\begin{aligned} &\|h\|^2 - \|FMh\|^2 - \|XD_Mh\|^2 \\ &\qquad + \|D_KFMh\|^2 + \|K^*XD_Mh\|^2 - 2\mathrm{Re}\,(D_KFMh, K^*XD_Mh) \\ &= \|h\|^2 - \|FMh\|^2 - \|XD_Mh\|^2 + \|(D_KFM - K^*XD_M)\,h\|^2\,. \end{aligned}$$

Finally, observe that

$$\|D_FMh\|^2 = \|Mh\|^2 - \|FMh\|^2,\ \|D_XD_Mh\|^2 = \|h\|^2 - \|Mh\|^2 - \|XD_Mh\|^2.$$

Hence the proof of (9.6) is complete. □

M. Kreĭn has introduced and studied in [172] the following operator transformation. Let H be a bounded non-negative self-adjoint operator in $\mathcal{H}$ and let $\mathfrak{L}$ be a subspace in $\mathcal{H}$. He has proved that the set of all bounded self-adjoint operators C in $\mathcal{H}$ such that

$$0 \le C \le H, \quad \mathrm{Ran}(C) \subset \mathfrak{L},$$

has a maximal element

$$H_{\mathfrak{L}} = H^{\frac{1}{2}} P_\Omega H^{\frac{1}{2}}, \tag{9.8}$$

where P_Ω is the orthogonal projection in $\mathcal{H}$ onto the subspace

$$\Omega = \{f \in \mathcal{H} : H^{\frac{1}{2}} f \in \mathfrak{L}\}.$$

In addition

$$(H_{\mathfrak{L}}f, f) = \inf_{\varphi \in \mathcal{H}\ominus\mathfrak{L}} ((H(f+\varphi), f+\varphi))\,, \tag{9.9}$$

for all $f \in \mathcal{H}$. It follows from (9.8) that $\mathrm{Ran}((H_{\mathfrak{L}})^{\frac{1}{2}}) = \mathrm{Ran}(H^{\frac{1}{2}}) \cap \mathfrak{L}$ and hence

$$H_{\mathfrak{L}} = 0 \iff \mathrm{Ran}(H^{1/2}) \cap \mathfrak{L} = \{0\}. \tag{9.10}$$

The operator $H_{\mathfrak{L}}$ is called the **shorted operator**.

Let $\mathcal{H}$ and $\mathcal{H}'$ be two Hilbert spaces. Suppose that $\mathcal{G}$ is a subspace of $\mathcal{H}$ and $\dot{A} : \mathcal{G} \to \mathcal{H}'$ is a contraction. The operator A defined on $\mathcal{H}$ is called a **contractive extension** of $\dot{A}$ if $A \supset \dot{A}$ and $\|A\| \le 1$. Consider $\dot{A}$ as an operator from $[\mathcal{G}, \mathcal{H}']$. Then $\dot{A}$ has the adjoint $\dot{A}^* \in [\mathcal{H}', \mathcal{G}]$. Let

$$\mathfrak{N} = \mathcal{H} \ominus \mathcal{G}.$$

Theorem 9.1.3. *The formula*

$$A = \dot{A}P_{\mathcal{G}} + D_{\dot{A}^*}KP_{\mathfrak{N}}, \tag{9.11}$$

establishes a one-to-one correspondence between all contractive extensions $A \in [\mathcal{H}, \mathcal{H}']$ of $\dot{A}$ and all contractions $K \in [\mathfrak{N}, \mathfrak{D}_{\dot{A}^*}]$.

Proof. Let operator A be of the form (9.11), where $K \in [\mathfrak{N}, \mathfrak{D}_{\dot{A}^*}]$ is a contraction. Then

$$A^* = \dot{A}^* + K^* D_{\dot{A}^*}.$$

It follows that for all $f \in \mathcal{H}'$,

$$\|A^* f\|^2 = \|\dot{A}^* f\|^2 + \|K^* D_{\dot{A}^*} f\|^2 \le \|\dot{A}^* f\|^2 + \|D_{\dot{A}^*} f\|^2 = \|f\|^2.$$

The norms above and below are understood in the sense of the corresponding Hilbert space. Thus A^* is a contraction. Hence A is a contraction as well. Moreover $A{\upharpoonright}\mathcal{G} = \dot{A}$.

Conversely, if A is a contractive extension of $\dot{A}$, then its adjoint $A^* : \mathcal{H}' \to \mathcal{H}$ is a contraction. Because $\dot{A} \subset A$, we get $P_{\mathcal{G}}A^* = \dot{A}^*$. Therefore the operator A^* takes the form

$$A^* = \dot{A}^* + L,$$

where the range of the operator L is contained in $\mathfrak{N}$. It follows that $\|A^* f\|^2 = \|\dot{A}^* f\|^2 + \|Lf\|^2$ for all $f \in \mathcal{H}'$. Since A^* is a contraction, we obtain

$$\|Lf\|^2 \le \|f\|^2 - \|\dot{A}^* f\|^2, \; f \in \mathcal{H}'.$$

By Theorem 2.1.2 we get $L^* = D_{\dot{A}^*}K$, where $K : \mathfrak{N} \to \mathfrak{D}_{\dot{A}^*}$ is a contraction. □

As a consequence for $A = \dot{A}P_{\mathcal{G}} + D_{\dot{A}^*}KP_{\mathfrak{N}}$ with a contraction $K \in [\mathfrak{N}, \mathfrak{D}_{\dot{A}^*}]$ one has from (9.6) the following relations:

$$\begin{aligned} \|D_A f\|^2 &= \|(D_{\dot{A}}P_{\mathcal{G}} - \dot{A}^* K P_{\mathfrak{N}})f\|^2 + \|D_K P_{\mathfrak{N}} f\|^2, \; f \in \mathcal{H}, \\ \|D_{A^*} g\|^2 &= \|D_{K^*} D_{\dot{A}^*} g\|^2, \; g \in \mathcal{H}'. \end{aligned} \tag{9.12}$$

Because $\dot{A}^* \mathfrak{D}_{\dot{A}^*} \subset \mathfrak{D}_{\dot{A}}$, the first relation in (9.12) gives

$$\inf_{\varphi \in \mathcal{G}} \|D_A(f + \varphi)\|^2 = \|D_K P_{\mathfrak{N}} f\|^2 \quad \text{for all} \quad f \in \mathcal{H}.$$

This means that the shorted operator $(D_A^2)_{\mathfrak{N}}$ of the form (9.8)-(9.9) has a property

$$(D_A^2)_{\mathfrak{N}} = D_K^2 P_{\mathfrak{N}}, \quad \operatorname{Ran}(D_A) \cap \mathfrak{N} = \operatorname{Ran}(D_K). \tag{9.13}$$

The second equality in (9.12) yields

$$\operatorname{Ran}(D_{A^*}) = D_{\dot{A}^*} \operatorname{Ran}(D_{K^*}). \tag{9.14}$$

Suppose now that the Hilbert space $\mathcal{H}'$ is decomposed as $\mathcal{H}' = \mathcal{G}' \oplus \mathfrak{M}$. Then $\dot{A} = \mathcal{A} + C$, where $\mathcal{A} = P_{\mathcal{G}'}\dot{A} \in [\mathcal{G}, \mathcal{G}']$ and $C = P_{\mathfrak{M}}\dot{A} \in [\mathcal{G}, \mathfrak{M}]$. We can rewrite $\dot{A}$ in the block-matrix form

$$\dot{A} = \begin{pmatrix} \mathcal{A} \\ C \end{pmatrix}. \tag{9.15}$$

Since $\dot{A}$ is a contraction, we have $\|\mathcal{A}g\|^2 + \|Cg\|^2 \le \|g\|^2$ for all $g \in \mathcal{G}$. It follows that

$$C = \mathcal{K} D_{\mathcal{A}}, \tag{9.16}$$

where $\mathcal{K} \in [\mathfrak{D}_{\mathcal{A}}, \mathfrak{M}]$ is a contraction. Because $\dot{A}^* = \mathcal{A}^* P_{\mathcal{G}'} + D_{\mathcal{A}} \mathcal{K} P_{\mathfrak{M}}$, we get from (9.6) the relation

$$\|D_{\dot{A}^*} f\|^2 = \|(D_{\mathcal{A}^*} P_{\mathcal{G}'} - \mathcal{A}\mathcal{K}^* P_{\mathfrak{M}}) f\|^2 + \|D_{\mathcal{K}^*} P_{\mathfrak{M}} f\|^2, \ f \in \mathcal{H}'.$$

This yields the relations

$$(D_{\dot{A}^*}^2)_{\mathfrak{M}} = D_{\mathcal{K}^*}^2 P_{\mathfrak{M}}, \tag{9.17}$$

$$\operatorname{Ran}(D_{\dot{A}^*}) \cap \mathfrak{M} = \operatorname{Ran}(D_{\mathcal{K}^*}). \tag{9.18}$$

Here and below $(D_{\dot{A}^*}^2)_{\mathfrak{M}}$ is the shorted operator of the form (9.8)–(9.9). All bounded extensions A of $\dot{A}$ also have the block-matrix form

$$A = \begin{pmatrix} \mathcal{A} & B \\ C & D \end{pmatrix} : \begin{pmatrix} \mathcal{G} \\ \mathfrak{N} \end{pmatrix} \to \begin{pmatrix} \mathcal{G}' \\ \mathfrak{M} \end{pmatrix}.$$

Theorem 9.1.4. *The formula*

$$A = \begin{pmatrix} \mathcal{A} & D_{\mathcal{A}^*} N \\ \mathcal{K} D_{\mathcal{A}} & -\mathcal{K}\mathcal{A}^* N + D_{\mathcal{K}^*} X D_N \end{pmatrix} : \begin{pmatrix} \mathcal{G} \\ \mathfrak{N} \end{pmatrix} \to \begin{pmatrix} \mathcal{G}' \\ \mathfrak{M} \end{pmatrix}, \tag{9.19}$$

establishes a bijective correspondence between all contractive extensions A of the contraction $\dot{A} = \mathcal{A} + \mathcal{K} D_{\mathcal{A}}$ and all pairs

$$\langle N \in [\mathfrak{N}, \mathfrak{D}_{\mathcal{A}^*}], \ X \in [\mathfrak{D}_N, \mathfrak{D}_{\mathcal{K}^*}] \rangle$$

of contractive operators.

Proof. It follows from (9.16) that $\dot{A}^* = \mathcal{A}^* P_{\mathcal{G}'} + D_{\mathcal{A}} \mathcal{K}^* P_{\mathfrak{M}}$. Therefore using (9.6) for all $f \in \mathcal{H}'$ we get

$$\|D_{\dot{A}^*} f\|^2 = \|D_{\mathcal{A}^*} P_{\mathcal{G}'} f - \mathcal{A}\mathcal{K}^* P_{\mathfrak{M}} f\|^2 + \|D_{\mathcal{K}^*} P_{\mathfrak{M}} f\|^2.$$

Thus,

$$\|D_{\dot{A}^*} f\|^2 = \|D_{\mathcal{A}^*} P_{\mathcal{G}'} f - \mathcal{A}\mathcal{K}^* P_{\mathfrak{M}} f\|^2 + \|D_{\mathcal{K}^*} P_{\mathfrak{M}} f\|^2, \ f \in \mathcal{H}'. \tag{9.20}$$

In view of the equality $\mathcal{A} D_{\mathcal{A}} = D_{\mathcal{A}^*} \mathcal{A}$ we get that $\mathcal{A} \mathfrak{D}_{\mathcal{A}} \subset \mathfrak{D}_{\mathcal{A}^*}$ and since $\operatorname{Ran}(\mathcal{K}^*) \subset \mathfrak{D}_{\mathcal{A}}$, (9.20) yields

$$\inf\left\{\|D_{\dot{A}^*}(f - \varphi)\|^2, \ \varphi \in \mathcal{G}\right\} = \|D_{\mathcal{K}^*} P_{\mathfrak{M}} f\|^2, \ f \in \mathcal{H}'. \tag{9.21}$$

Let $\mathfrak{K} = \overline{D_{\dot{A}^*} \mathcal{G}'}$ and $\mathfrak{L} = \mathfrak{D}_{\dot{A}^*} \ominus \mathfrak{K}$. Observe that (see [172])

$$\mathfrak{L} = \left\{ f \in \mathfrak{D}_{\dot{A}^*} : D_{\dot{A}^*} f \in \mathfrak{M} \right\}.$$

From (9.20) and (9.21) we get the equalities

$$\begin{aligned} \|P_{\mathfrak{K}} D_{\dot{A}^*} f\|^2 &= \|D_{\mathcal{A}^*} P_{\mathcal{G}'} f - \mathcal{A}\mathcal{K}^* P_{\mathfrak{M}} f\|^2, \\ \|P_{\mathfrak{L}} D_{\dot{A}^*} f\|^2 &= \|D_{\mathcal{K}^*} P_{\mathfrak{M}} f\|^2, \ f \in \mathcal{H}'. \end{aligned} \tag{9.22}$$

In particular,

$$\|P_{\mathfrak{K}} D_{\dot{A}^*} \varphi\|^2 = \|D_{\mathcal{A}^*} \varphi\|^2, \ \varphi \in \mathcal{G}'.$$

It follows from (9.22) that there are unitary operators $\mathcal{U} \in [\mathfrak{K}, \mathfrak{D}_{\mathcal{A}^*}]$ and $\mathcal{Z} \in [\mathfrak{L}, \mathfrak{D}_{\mathcal{K}^*}]$ such that

$$\begin{aligned} \mathcal{U} P_{\mathfrak{K}} D_{\dot{A}^*} f &= D_{\mathcal{A}^*} P_{\mathcal{G}'} f - \mathcal{A}\mathcal{K}^* P_{\mathfrak{M}} f, \\ \mathcal{Z} P_{\mathfrak{L}} D_{\dot{A}^*} f &= D_{\mathcal{K}^*} P_{\mathfrak{M}} f, \ f \in \mathcal{H}'. \end{aligned} \tag{9.23}$$

Obviously, $\mathcal{U}^* \in [\mathfrak{D}_{\mathcal{A}^*}, \mathfrak{K}]$, $\mathcal{Z}^* \in [\mathfrak{L}, \mathfrak{D}_{\mathcal{K}^*}]$ and $\mathcal{U}^* = \mathcal{U}^{-1}$, $\mathcal{Z}^* = \mathcal{Z}^{-1}$. Then from (9.23) we have

$$D_{\dot{A}^*} = \mathcal{U}^* \left(D_{\mathcal{A}^*} P_{\mathcal{G}'} - \mathcal{A}\mathcal{K}^* P_{\mathfrak{M}} \right) + \mathcal{Z}^* D_{\mathcal{K}^*} P_{\mathfrak{M}},$$

and

$$D_{\dot{A}^*} = (D_{\mathcal{A}^*} - \mathcal{K}\mathcal{A}^*) \mathcal{U} P_{\mathfrak{K}} + D_{\mathcal{K}^*} \mathcal{Z} P_{\mathfrak{L}}. \tag{9.24}$$

Let K be an arbitrary bounded operator from $\mathfrak{N}$ into $\mathfrak{D}_{\dot{A}^*}$. Then $K = P_{\mathfrak{K}} K + P_{\mathfrak{L}} K$. Let $N = \mathcal{U} P_{\mathfrak{K}} K$ and $Y = \mathcal{Z} P_{\mathfrak{L}} K$. Clearly, K is a contraction if and only if the operator $\widetilde{K} = N + Y \in [\mathfrak{M}, \mathfrak{D}_{\mathcal{A}^*} \oplus \mathfrak{D}_{\mathcal{K}^*}]$ is a contraction and $\widetilde{K}$ is a contraction if and only if $Y = X D_N$, where $X \in [\mathfrak{D}_N, \mathfrak{D}_{\mathcal{K}^*}]$ is a contraction. Further, for a contraction $K \in [\mathfrak{N}, \mathfrak{D}_{\dot{A}^*}]$ from (9.24) and for all $h \in \mathfrak{N}$ we get

$$D_{\dot{A}^*} K h = (D_{\mathcal{A}^*} - \mathcal{K}\mathcal{A}^*) N h + D_{\mathcal{K}^*} X D_N h. \tag{9.25}$$

Let $A = \dot{A} P_{\mathcal{G}} + D_{\dot{A}^*} K P_{\mathfrak{N}}$. Then (9.11) and (9.25) yield (9.19). □

Let the unitary operators $\mathcal{U}$ and $\mathcal{Z}$ be defined by (9.23). Define the unitary operator $U \in [\mathfrak{D}_{\dot{A}^*}, \mathfrak{D}_{\mathcal{A}^*} \oplus \mathfrak{D}_{\mathcal{K}^*}]$ by the operator matrix

$$U = \begin{pmatrix} \mathcal{U} & 0 \\ 0 & \mathcal{Z} \end{pmatrix} : \begin{pmatrix} \mathfrak{K} \\ \mathfrak{L} \end{pmatrix} \to \begin{pmatrix} \mathfrak{D}_{\mathcal{A}^*} \\ \mathfrak{D}_{\mathcal{K}^*} \end{pmatrix}.$$

Then the parameters $K \in [\mathfrak{N}, \mathfrak{D}_{\dot{A}^*}]$ in (9.11) and

$$\begin{pmatrix} N \\ XD_N \end{pmatrix} : \mathfrak{N} \to \begin{pmatrix} \mathfrak{D}_{\mathcal{A}^*} \\ \mathfrak{D}_{\mathcal{K}^*} \end{pmatrix}$$

in (9.19) are connected by the relation

$$K = U^* \begin{pmatrix} N \\ XD_N \end{pmatrix} = \begin{pmatrix} \mathcal{U}^* N \\ \mathcal{Z}^* XD_N \end{pmatrix} : \mathfrak{N} \to \begin{pmatrix} \mathfrak{K} \\ \mathfrak{L} \end{pmatrix}. \tag{9.26}$$

Suppose that $\mathcal{G} \subset \mathcal{H}$, $\mathcal{G}' \subset \mathcal{H}'$, $\dot{A} : \mathcal{G} \to \mathcal{H}'$, $B : \mathcal{G}' \to \mathcal{H}$, and the operators $\dot{A}$ and B form a **dual pair**, i.e.,

$$(\dot{A}f, h)_{\mathcal{H}'} = (f, Bh)_{\mathcal{H}} \quad \text{for all} \quad f \in \mathcal{G},\ g \in \mathcal{G}'. \tag{9.27}$$

The operator $A \in [\mathcal{H}, \mathcal{H}']$ is called an **extension of the dual pair** $\langle \dot{A}, B \rangle$ if $A \supset \dot{A}$ and $A^* \supset B$. Theorem 9.1.4 enables us to give the block-matrix form of all contractive extensions of the dual pair of contractions $\langle \dot{A}, B \rangle$. Put

$$\mathcal{B} = P_{\mathcal{G}} B, \quad \mathcal{C}' = P_{\mathfrak{N}} B.$$

Then $\mathcal{B} \in [\mathcal{G}', \mathcal{G}]$, $\mathcal{C}' \in [\mathcal{G}', \mathfrak{N}]$,

$$B = \begin{pmatrix} \mathcal{B} \\ \mathcal{C}' \end{pmatrix},$$

$\mathcal{C}' = \mathcal{M} D_{\mathcal{B}}$, and $\mathcal{M} \in [\mathfrak{D}_{\mathcal{B}}, \mathfrak{N}]$ is a contraction. In addition, in view of (9.27) one has $\mathcal{B}^* = \mathcal{A}$. Thus $D_{\mathcal{B}} = D_{\mathcal{A}^*}$.

Theorem 9.1.5. *The formula*

$$A = \begin{pmatrix} \mathcal{A} & D_{\mathcal{A}^*} \mathcal{M}^* \\ \mathcal{K} D_{\mathcal{A}} & -\mathcal{K} \mathcal{A}^* \mathcal{M}^* + D_{\mathcal{K}^*} X D_{\mathcal{M}^*} \end{pmatrix} : \begin{pmatrix} \mathcal{G} \\ \mathfrak{N} \end{pmatrix} \to \begin{pmatrix} \mathcal{G}' \\ \mathfrak{M} \end{pmatrix}, \tag{9.28}$$

establishes a bijective correspondence between all contractive extensions A of the dual pair of contractions $\langle \dot{A}, B \rangle$ and all contractions $X \in [\mathfrak{D}_{\mathcal{M}^*}, \mathfrak{D}_{\mathcal{K}^*}]$.

Proof. Let A be given by (9.28), then by Theorem 9.1.4 the operator A is a contractive extension of $\dot{A}$. Since $A^* f = (\mathcal{A}^* + \mathcal{M} D_{\mathcal{A}^*}) f = Bf$ for all $f \in \mathcal{G}'$, the operator A^* is a contractive extension of B. Conversely, let A be a contractive extension of a dual pair $\langle \dot{A}, B \rangle$. Since A is a contractive extension of $\dot{A}$, it is of the form (9.19) and since $A^* \upharpoonright \mathcal{G}' = B$, we get $N^* D_{\mathcal{A}^*} = P_{\mathfrak{N}} B = \mathcal{C}' = \mathcal{M} D_{\mathcal{A}^*}$. So, $N^* = \mathcal{M}$ and hence $N = \mathcal{M}^*$. This completes the proof. □

9.2 Quasi-self-adjoint contractive extensions of symmetric contractions

Definition 9.2.1. Let $\alpha \in [0, \pi/2)$ and let

$$\mathcal{S}(\alpha) := \{z \in \mathbb{C} : |\arg z| \le \alpha\}$$

be a sector on the complex plane $\mathbb{C}$ with the vertex at the origin and the semi-angle α. A linear operator T in a Hilbert space $\mathcal{H}$ is called **sectorial with vertex at $z = 0$ and the semi-angle α (α-sectorial)** if

$$|\mathrm{Im}\,(Tf, f)| \le (\tan\alpha)\,\mathrm{Re}\,(Tf, f), \tag{9.29}$$

for all $f \in \mathrm{Dom}(T)$.

Definition 9.2.2. Let $\alpha \in (0, \pi/2)$. We say that a bounded operator $T \in [\mathcal{H}, \mathcal{H}]$ belongs to the **class** $C_{\mathcal{H}}(\alpha)$ if

$$\|T \sin\alpha \pm iI\cos\alpha\| \le 1. \tag{9.30}$$

It is clear that $C_{\mathcal{H}}(\pi/2)$ is the set of all linear contractions in $\mathcal{H}$, the class $C_{\mathcal{H}}(\alpha)$ is a convex and closed (with respect to the strong operator topology) set, which is the intersection of two closed operator balls corresponding to $\pm$. Moreover, by virtue of (9.30) one immediately concludes that

$$T \in C_{\mathcal{H}}(\alpha) \iff -T \in C_{\mathcal{H}}(\alpha) \iff T^* \in C_{\mathcal{H}}(\alpha),$$

and that condition (9.30) is equivalent to

$$(\tan\alpha)\,(\|f\|^2 - \|Tf\|^2) \ge 2|\mathrm{Im}\,(Tf, f)|, \quad f \in \mathcal{H}. \tag{9.31}$$

Together with Definition 9.2.1, it also proves that $T \in C_{\mathcal{H}}(\alpha)$ is equivalent to the statement that $(I - T^*)(I + T)$ is a bounded α-sectorial operator. According to (9.31) it is natural to identify $C_{\mathcal{H}}(0)$ with the set of all self-adjoint contractions. If $\alpha \in (0, \pi/2)$, then the members of the class $C_{\mathcal{H}}(\alpha)$ are also referred to as **α-co-sectorial contractions**.

Let $\dot{A}$ be a non-densely-defined symmetric contraction in a Hilbert space $\mathcal{H}$. An operator $T \in [\mathcal{H}, \mathcal{H}]$ is called a **quasi-self-adjoint contractive extension** of $\dot{A}$ or **qsc-extension** if $T \supset \dot{A}$, $T^* \supset \dot{A}$, and $\|T\| \le 1$. If $T = T^*$, then T is called a **self-adjoint contractive extension** of $\dot{A}$ or **sc-extension**. Below we provide a description of all qsc-extensions of operator $\dot{A}$.

Theorem 9.2.3. *Let $\dot{A}$ be a closed symmetric contraction in $\mathcal{H} = \mathcal{H}_0 \oplus \mathfrak{N}$ with $\mathrm{Dom}(\dot{A}) = \mathcal{H}_0$ and let $\dot{A}$ be decomposed as in* (9.15). *Then:*

(i) *the formula*

$$T = \begin{pmatrix} \mathcal{A} & D_{\mathcal{A}}\mathcal{K}^* \\ \mathcal{K}D_{\mathcal{A}} & -\mathcal{K}\mathcal{A}\mathcal{K}^* + D_{\mathcal{K}^*}XD_{\mathcal{K}^*} \end{pmatrix} : \begin{pmatrix} \mathcal{H}_0 \\ \mathfrak{N} \end{pmatrix} \to \begin{pmatrix} \mathcal{H}_0 \\ \mathfrak{N} \end{pmatrix}, \tag{9.32}$$

sets a one-to-one correspondence between all qsc-extensions T of the symmetric contraction $\dot{A} = \mathcal{A} + \mathcal{K}D_{\mathcal{A}}$ and all contractions X in the subspace $\mathfrak{D}_{\mathcal{K}^*}$;

(ii) *T is a self-adjoint contractive extension of $\dot{A}$ if and only if X in (9.32) is a self-adjoint contraction in $\mathfrak{D}_{\mathcal{K}^*}$;*

(iii) *T in (9.32) belongs to the class $C_{\mathcal{H}}(\alpha)$ if and only if X belongs to the class $C_{\mathfrak{D}_{\mathcal{K}^*}}(\alpha)$, $\alpha \in (0, \pi/2)$.*

Proof. Because $\dot{A}$ is a symmetric contraction, we apply Theorem 9.1.4 to the dual pair $\langle \dot{A}, \dot{A} \rangle$. In that case $\mathcal{M} = \mathcal{K}$ and hence (9.28) takes the form (9.32). Clearly, T is an sc-extension if and only if $X \in [\mathfrak{D}_{\mathcal{K}^*}, \mathfrak{D}_{\mathcal{K}^*}]$ is a self-adjoint contraction. It follows from (9.32) that

$$I - T^*T = \begin{pmatrix} D_{\mathcal{A}}^2 - D_{\mathcal{A}}\mathcal{K}^*\mathcal{K}D_{\mathcal{A}} & -\mathcal{A}D_{\mathcal{A}}\mathcal{K}^* - D_{\mathcal{A}}\mathcal{K}^*D \\ -\mathcal{K}D_{\mathcal{A}}\mathcal{A} - D^*\mathcal{K}D_{\mathcal{A}} & D_{\mathcal{K}^*}^2 - \mathcal{K}\mathcal{A}^2\mathcal{K}^* - D^*D \end{pmatrix},$$

i.e.,

$$\begin{aligned} &\left((I - T^*T)\begin{pmatrix} f \\ h \end{pmatrix}, \begin{pmatrix} f \\ h \end{pmatrix}\right) \\ &\quad = \|D_{\mathcal{K}}(D_{\mathcal{A}}f - \mathcal{A}\mathcal{K}^*h) - \mathcal{K}^*XD_{\mathcal{K}^*}h\|^2 + \|D_XD_{\mathcal{K}^*}h\|^2 \geq 0, \end{aligned} \tag{9.33}$$

for all $f \in \mathcal{H}_0 = \operatorname{Dom}(\dot{A})$, $h \in \mathfrak{N}$. It follows from (9.33) that $T \in C_{\mathcal{H}}(\alpha)$ is equivalent to

$$\begin{aligned} &|(\operatorname{Im} XD_{\mathcal{K}^*}h, D_{\mathcal{K}^*}h)| \\ &\leq \frac{\tan\alpha}{2}\left(\|D_{\mathcal{K}}(D_{\mathcal{A}}f - \mathcal{A}\mathcal{K}^*h) - \mathcal{K}^*XD_{\mathcal{K}^*}h\|^2 + \|D_XD_{\mathcal{K}^*}h\|^2\right) \end{aligned} \tag{9.34}$$

which holds for all $f \in \mathcal{H}_0$, $h \in \mathfrak{N}$. Due to the inclusions

$$\operatorname{Ran}(X) \subset \mathfrak{D}_{\mathcal{K}^*}, \quad \mathcal{K}^*\mathfrak{D}_{\mathcal{K}^*} \subset \mathfrak{D}_{\mathcal{K}} \subseteq \mathfrak{D}_{\mathcal{A}}, \quad \mathcal{A}\mathfrak{D}_{\mathcal{A}} \subset \mathfrak{D}_{\mathcal{A}},$$

one can choose a sequence $\{f_n\}_{n=1}^{\infty} \subset \mathfrak{D}_{\mathcal{A}}$ such that for a given $h \in \mathfrak{N}$ the equality

$$\lim_{n\to\infty} D_{\mathcal{K}}D_{\mathcal{A}}f_n = D_{\mathcal{K}}\mathcal{A}\mathcal{K}^*h + \mathcal{K}^*XD_{\mathcal{K}^*}h, \tag{9.35}$$

holds for all $f \in \mathcal{H}_0$, $h \in \mathfrak{N}$. In view of (9.35) the condition (9.34) is equivalent to

$$|(\operatorname{Im} Xh, h)| \leq \frac{\tan\alpha}{2}\,\|D_Xh\|^2 \iff X \in C_{\mathfrak{D}_{\mathcal{K}^*}}(\alpha). \tag{9.36}$$

for all $h \in \mathfrak{D}_{\mathcal{K}^*}$. □

Define two operators A_μ and A_M, corresponding to $X = -I_{\mathfrak{D}_{\mathcal{K}^*}}$ and $X = I_{\mathfrak{D}_{\mathcal{K}^*}}$ in (9.32), respectively:

$$A_\mu = \begin{pmatrix} \mathcal{A} & D_{\mathcal{A}}\mathcal{K}^* \\ \mathcal{K}D_{\mathcal{A}} & -\mathcal{K}\mathcal{A}\mathcal{K}^* - D_{\mathcal{K}^*}^2 \end{pmatrix}, \tag{9.37}$$

$$A_M = \begin{pmatrix} \mathcal{A} & D_{\mathcal{A}}\mathcal{K}^* \\ \mathcal{K}D_{\mathcal{A}} & -\mathcal{K}\mathcal{A}\mathcal{K}^* + D^2_{\mathcal{K}^*} \end{pmatrix}. \tag{9.38}$$

Then as a consequence of Theorem 9.2.3 we get the following result.

Theorem 9.2.4. *The class of all sc-extensions of $\dot{A}$ forms an operator interval $[A_\mu, A_M]$ with endpoints given by (9.37) and (9.38).*

Proposition 9.2.5. *The operators A_μ and A_M possess the following properties for all $f \in \mathcal{H}$:*

$$\inf_{\varphi \in \mathrm{Dom}(\dot{A})} ((I + A_\mu)(f + \varphi), f + \varphi) = \inf_{\varphi \in \mathrm{Dom}(\dot{A})} ((I - A_M)(f + \varphi), f + \varphi) = 0.$$

Proof. From (9.37) for $f = x + h$, $x \in \mathcal{H}_0$, $h \in \mathfrak{N}$ we have

$$\begin{aligned}
&((I + A_\mu)(x + h), x + h) \\
&= ((I_{\mathcal{H}_0} + \mathcal{A})x, x) + 2\mathrm{Re}\,(D_{\mathcal{A}}\mathcal{K}^* h, x) - (\mathcal{A}\mathcal{K}^* h, \mathcal{K}^* h) + \|\mathcal{K}^* h\|^2 \\
&= \|(I_{\mathcal{H}_0} + \mathcal{A})^{1/2} x\|^2 + 2\mathrm{Re}\,((I_{\mathcal{H}_0} + \mathcal{A})^{1/2} x, (I_{\mathcal{H}_0} - \mathcal{A})^{1/2}\mathcal{K}^* h) \\
&\quad + \|(I_{\mathcal{H}_0} - \mathcal{A})^{1/2}\mathcal{K}^* h\|^2 \\
&= \|(I_{\mathcal{H}_0} + \mathcal{A})^{1/2} x + (I_{\mathcal{H}_0} - \mathcal{A})^{1/2}\mathcal{K}^* h\|^2.
\end{aligned}$$

Hence, if $\varphi \in \mathcal{H}_0$, then

$$((I + A_\mu)(f + \varphi), f + \varphi) = \|(I_{\mathcal{H}_0} + \mathcal{A})^{1/2}(x + \varphi) + (I_{\mathcal{H}_0} - \mathcal{A})^{1/2}\mathcal{K}^* h\|^2.$$

Since $\mathrm{Ran}(\mathcal{K}^*) \subseteq \mathfrak{D}_{\mathcal{A}}$, one can find a sequence $\{\varphi_n\} \subset \mathcal{H}_0$ such that

$$\lim_{n \to \infty} (I_{\mathcal{H}_0} + \mathcal{A})^{1/2}\varphi_n = -(I_{\mathcal{H}_0} + \mathcal{A})^{1/2} x - (I_{\mathcal{H}_0} - \mathcal{A})^{1/2}\mathcal{K}^* h.$$

It follows that $\inf_{\varphi \in \mathrm{Dom}(\dot{A})} ((I + A_\mu)(f + \varphi), f + \varphi) = 0$ for any $f \in \mathcal{H}$. Similarly $\inf_{\varphi \in \mathrm{Dom}(\dot{A})} ((I - A_M)(f + \varphi), f + \varphi) = 0$, $f \in \mathcal{H}$. □

Thus the *sc*-extensions A_μ and A_M coincide with extreme contractive self-adjoint extensions of a symmetric contraction $\dot{A}$ discovered by M. Kreĭn. Following Kreĭn's notations we call the *sc*-extensions A_μ and A_M of a symmetric contraction $\dot{A}$ the **rigid** and the **soft** extensions, respectively. Hence for the extreme *sc*-extensions A_μ and A_M of $\dot{A}$ from Proposition 9.2.5 one has

$$(I + A_\mu)_{\mathfrak{N}} = (I - A_M)_{\mathfrak{N}} = 0,$$

where $(I + A_\mu)_{\mathfrak{N}}$ and $(I - A_M)_{\mathfrak{N}}$ are the shorted operators of the form (9.8)-(9.9). These equalities together with (9.10) yield the next statement.

Proposition 9.2.6. *Let A be an sc-extension of $\dot{A}$ and let $E(x)$ be the resolution of identity for A. Then:*

1. $A = A_\mu$ *if and only if*

$$\int_{-1}^{1} \frac{d(E(x)\varphi,\varphi)}{1+x} = +\infty, \qquad \textit{for all } \varphi \in \mathfrak{N}\setminus\{0\}; \tag{9.39}$$

2. $A = A_M$ *if and only if*

$$\int_{-1}^{1} \frac{d(E(x)\varphi,\varphi)}{1-x} = +\infty, \qquad \textit{for all } \varphi \in \mathfrak{N}\setminus\{0\}. \tag{9.40}$$

It follows from (9.37) and (9.38) that

$$\frac{A_\mu + A_M}{2} = \begin{pmatrix} \mathcal{A} & D_{\mathcal{A}}\mathcal{K}^* \\ \mathcal{K}D_{\mathcal{A}} & -\mathcal{K}\mathcal{A}\mathcal{K}^* \end{pmatrix}, \quad \frac{A_M - A_\mu}{2} = \begin{pmatrix} 0 & 0 \\ 0 & D^2_{\mathcal{K}^*} \end{pmatrix} = D^2_{\mathcal{K}^*} P_{\mathfrak{N}}. \tag{9.41}$$

Applying (9.32) and (9.41) and performing straightforward calculations we obtain the following result.

Theorem 9.2.7. *The formula*

$$T = \frac{1}{2}(A_M + A_\mu) + \frac{1}{2}(A_M - A_\mu)^{1/2} X (A_M - A_\mu)^{1/2}, \tag{9.42}$$

establishes a one-to-one correspondence between the set of all qsc-extensions T of $\dot{A}$ and the set of contractions X on the subspace $\mathfrak{N}_0 = \overline{\operatorname{Ran}(A_M - A_\mu)}$. Moreover,

(i) *the operator T is an sc-extension of $\dot{A}$ if and only if X is a self-adjoint contraction in $\mathfrak{N}_0$,*

(ii) *$T \in C_{\mathcal{H}}(\alpha)$ if and only if $X \in C_{\mathfrak{N}_0}(\alpha)$.*

Thus, the set of all *qsc*-extensions of symmetric contraction $\dot{A}$ forms the operator ball

$$\mathbf{B}\left(\frac{A_M + A_\mu}{2}, \frac{A_M - A_\mu}{2}\right) \tag{9.43}$$

with the center $(A_M + A_\mu)/2$ and the left and right radii $(A_M - A_\mu)/2$.

Definition 9.2.8. Let T be a *qsc*-extension of a symmetric contraction $\dot{A}$. The operator T is said to be **extremal** ***qsc*-extension** if T is an extreme point (see [118]) of the operator ball $\mathbf{B}$ in (9.43).

It is easy to see that an operator T is an extremal *qsc*-extension if and only if the corresponding to T contraction $X \in [\mathfrak{D}_{\mathcal{K}^*}, \mathfrak{D}_{\mathcal{K}^*}]$ in representation (9.32) or in representation (9.42) is an isometry or co-isometry. From (9.33), arguing as in the proof of Theorem 9.2.3, we immediately obtain the relations for the shorted operators $(D^2_T)_{\mathfrak{N}}$ and $(D^2_{T^*})_{\mathfrak{N}}$,

$$\begin{aligned}(D^2_T)_{\mathfrak{N}} &= D_{\mathcal{K}^*}D^2_X D_{\mathcal{K}^*}P_{\mathfrak{N}} = \tfrac{1}{2}(A_M - A_\mu)^{1/2}D^2_X(A_M - A_\mu)^{1/2},\\ (D^2_{T^*})_{\mathfrak{N}} &= D_{\mathcal{K}^*}D^2_{X^*}D_{\mathcal{K}^*}P_{\mathfrak{N}} = \tfrac{1}{2}(A_M - A_\mu)^{1/2}D^2_{X^*}(A_M - A_\mu)^{1/2}.\end{aligned} \tag{9.44}$$

Consequently,

$$\begin{aligned}(D^2_T)_{\mathfrak{N}} = 0 &\iff X \quad \text{is an isometry},\\ (D^2_{T^*})_{\mathfrak{N}} = 0 &\iff X^* \quad \text{is an isometry},\end{aligned}$$

and we arrive at the following result.

Proposition 9.2.9. (1) *The following statements are equivalent:*

(i) *T is an extremal qsc-extension of symmetric contraction $\dot{A}$;*

(ii) *either $(D^2_T)_{\mathfrak{N}} = 0$ or $(D^2_{T^*})_{\mathfrak{N}} = 0$;*

(iii) *either $\operatorname{Ran}(D_T) \cap \mathfrak{N} = \{0\}$ or $\operatorname{Ran}(D_{T^*}) \cap \mathfrak{N} = \{0\}$.*

(2) *The sc-extension T is extremal if and only if the corresponding self-adjoint contraction X in $[\mathfrak{D}_{\mathcal{K}^*}, \mathfrak{D}_{\mathcal{K}^*}]$ is unitary, i.e., $X = X^* = X^{-1}$.*

(3) *If T is an extremal qsc-extension and $T \in C_{\mathcal{H}}(\alpha)$ for some $\alpha < \pi/2$, then T is an extremal sc-extension.*

Remark 9.2.10. If $\dim \mathfrak{N} < \infty$, then $(D^2_T)_{\mathfrak{N}} = 0$ if and only if $(D^2_{T^*})_{\mathfrak{N}} = 0$.

In view of the relations

$$(I \pm X^*)(I \pm X) + D^2_X = 2(I \pm \operatorname{Re}X), \tag{9.45}$$

for any contraction X, we get from (9.42) that there are no non-selfadjoint *qsc*-extensions T with the real parts $\operatorname{Re}T$ equal to A_μ or A_M.

Notice also, that if *sc*-extension A is defined via (9.42) by means of a self-adjoint contraction $X \in [\mathfrak{N}_0, \mathfrak{N}_0]$, then from Proposition 9.2.5 we get

$$\begin{aligned}(I - A)_{\mathfrak{N}} &= \frac{1}{2}(A_M - A_\mu)^{1/2}(I_{\mathfrak{N}_0} - X)(A_M - A_\mu)^{1/2} = A_M - A,\\ (I + A)_{\mathfrak{N}} &= \frac{1}{2}(A_M - A_\mu)^{1/2}(I_{\mathfrak{N}_0} + X)(A_M - A_\mu)^{1/2} = A - A_\mu.\end{aligned} \tag{9.46}$$

From (9.46) and (9.8) we obtain

$$\begin{aligned}A - A_\mu &= (I + A)^{1/2}P_{\Omega_+}(I + A)^{1/2},\\ A_M - A &= (I - A)^{1/2}P_{\Omega_-}(I - A)^{1/2},\end{aligned} \tag{9.47}$$

where $\Omega_\pm := \{f \in \mathcal{H} : (I \pm A)^{1/2}f \in \mathfrak{N}\}$.

Proposition 9.2.11. *Let A be an sc-extension of $\dot{A}$. Then*

$$\begin{aligned}\operatorname{Ran}((I+A_\mu)^{1/2}) &\subseteq \operatorname{Ran}((I+A)^{1/2}),\\ \operatorname{Ran}((I-A_M)^{1/2}) &\subseteq \operatorname{Ran}((I-A)^{1/2}),\end{aligned} \tag{9.48}$$

and the equalities

$$\begin{aligned}\|(I+A)^{-1/2}(I+A_\mu)^{1/2}h\|^2 = \|h\|^2,\ h \in \overline{\operatorname{Ran}((I+A_\mu)^{1/2})},\\ \|(I-A)^{-1/2}(I-A_M)^{1/2}g\|^2 = \|g\|^2,\ g \in \overline{\operatorname{Ran}((I-A_M)^{1/2})},\end{aligned} \tag{9.49}$$

hold.

Proof. Inclusions (9.48) follow from the inequalities $I+A_\mu \le I+A$, $I-A_M \le I-A$ and Theorem 2.1.2. In order to prove (9.49) we use (9.47) which produces the relations

$$\begin{aligned}I+A_\mu &= (I+A)^{1/2}(I-P_{\Omega_+})(I+A)^{1/2},\\ I-A_M &= (I-A)^{1/2}(I-P_{\Omega_-})(I-A)^{1/2}.\end{aligned}$$

Hence

$$\begin{aligned}(I+A_\mu)^{1/2} &= V_+(I-P_{\Omega_+})(I+A)^{1/2},\\ (I-A_M)^{1/2} &= V_-(I-P_{\Omega_-})(I-A)^{1/2},\end{aligned}$$

where V_+ and V_- are isometries, which map $\operatorname{Ran}(I-P_{\Omega_+})$ and $\operatorname{Ran}(I-P_{\Omega_-})$ onto

$$\overline{\operatorname{Ran}((I+A_\mu)^{1/2})} \quad \text{and} \quad \overline{\operatorname{Ran}((I-A_M)^{1/2})},$$

respectively. It follows that

$$(I+A_\mu)^{1/2} = (I+A)^{1/2}V_+^*, \quad (I-A_M)^{1/2} = (I-A)^{1/2}V_-^*,$$

These relations yield (9.49). □

Let us give other formulas for the operators A_μ, A_M, $(A_\mu + A_M)/2$, and $(A_M - A_\mu)/2$. We introduce the following subspaces of $\mathfrak{D}_{\dot{A}^*}$:

$$L_{\dot{A}} = \overline{D_{\dot{A}^*}\operatorname{Dom}(\dot{A})}, \quad L_0 = \mathfrak{D}_{\dot{A}^*} \ominus L_{\dot{A}},$$

where $\mathcal{H}_0 = \operatorname{Dom}(\dot{A})$. Clearly,

$$L_0 = D_{\dot{A}^*}^{-1}(\mathfrak{N}) = \{f \in \mathfrak{D}_{\dot{A}^*} : D_{\dot{A}^*}f \in \mathfrak{N}\},$$

and by (9.14) one has

$$(D_{\dot{A}^*}^2)_{\mathfrak{N}} = D_{\dot{A}^*}P_{L_0}D_{\dot{A}^*}. \tag{9.50}$$

Define the operator $K_{\dot{A}}$ via relation

$$K_{\dot{A}}D_{\dot{A}^*}f_A = P_{\mathfrak{N}}\dot{A}f_{\dot{A}}, \quad f_{\dot{A}} \in \operatorname{Dom}(\dot{A}). \tag{9.51}$$

By (9.15) we have $\dot{A}^*\restriction\mathcal{H}_0 = P_{\mathcal{H}_0}\dot{A} = \mathcal{A}$ and the operator $K_{\dot{A}}$ in (9.51) is a contraction in $[L_{\dot{A}}, \mathfrak{N}]$. Let $K_{\dot{A}}^* \in [\mathfrak{N}, L_{\dot{A}}]$ be the adjoint to $K_{\dot{A}}$ operator.

Proposition 9.2.12. *Let $T_0 = (A_\mu + A_M)/2$. Then the following relations hold:*

$$T_0 = AP_{\mathcal{H}_0} + D_{\dot{A}^*}K^*_{\dot{A}}P_{\mathfrak{N}} = \dot{A}^* + K_A P_{L_{\dot{A}}} D_{\dot{A}^*}, \tag{9.52}$$

$$\frac{A_M - A_\mu}{2} = (D^2_{\dot{A}^*})_{\mathfrak{N}} = (D^2_{T_0})_{\mathfrak{N}} = D^2_{\mathcal{K}^*}P_{\mathfrak{N}} = D^2_{K^*_{\dot{A}}}P_{\mathfrak{N}}, \tag{9.53}$$

$$\begin{aligned} A_\mu &= AP_{\mathcal{H}_0} + D_{\dot{A}^*}(K^*_{\dot{A}}P_{\mathfrak{N}} - P_{L_0}D_{\dot{A}^*}), \\ A_M &= AP_{\mathcal{H}_0} + D_{\dot{A}^*}(K^*_{\dot{A}}P_{\mathfrak{N}} + P_{L_0}D_{\dot{A}^*}). \end{aligned} \tag{9.54}$$

Proof. By (9.11)

$$T_0 = AP_{\mathcal{H}_0} + D_{\dot{A}^*}K^*_{T_0}P_{\mathfrak{N}},$$

with some contraction $K_{T_0} \in [\mathfrak{N}, \mathfrak{D}_{\dot{A}^*}]$. On the other hand the operator

$$\begin{pmatrix} \mathcal{K}^* \\ 0 \end{pmatrix} : \mathfrak{N} \to \begin{pmatrix} \mathfrak{D}_{A^*} \\ \mathfrak{D}_{\mathcal{K}^*} \end{pmatrix}$$

is related to T_0 via formula (9.32). The connection is given by (9.26). Therefore

$$K_{T_0} = U_0^* \mathcal{K}^* \in [\mathfrak{N}, L_{\dot{A}}].$$

Since T_0 is self-adjoint, we get $T_0 = \dot{A}^* + K^*_{T_0}D_{\dot{A}^*}$, $D^2_{K_{T_0}} = D^2_{\mathcal{K}^*}$, and $T_0 f = Af$ for all $f \in \mathcal{H}_0$. Hence $K^*_{T_0}D_{\dot{A}^*}f = P_{\mathfrak{N}}Af$. It follows $K^*_{T_0} = K_{\dot{A}}$, $D^2_{K^*_{\dot{A}}} = D^2_{K_{T_0}}$. Applying (9.17) yields $(D^2_{\dot{A}^*})_{\mathfrak{N}} = D^2_{\mathcal{K}^*}P_{\mathfrak{N}}$ while (9.13) implies that $(D^2_{T_0})_{\mathfrak{N}} = D^2_{K^*_{\dot{A}}}P_{\mathfrak{N}}$. Relations (9.54) follow from (9.52), (9.53), and (9.50). This completes the proof. □

Theorem 9.2.13. *Let $\dot{A}$ be a symmetric contraction with* $\mathrm{Dom}(\dot{A}) = \mathcal{H}_0$ *in a Hilbert space $\mathcal{H}$. Then the following conditions are equivalent:*

(i) *the operator $\dot{A}$ admits a unique qsc-extension;*

(ii) $A_\mu = A_M$*;*

(iii) $\mathrm{Ran}(D_{\dot{A}^*}) \cap \mathfrak{N} = \{0\}$ *;*

(iv) *the operator $\mathcal{K}^*$ is isometric;*

(v)

$$\sup_{\varphi \in \mathcal{H}_0} \frac{\left|(\dot{A}\varphi, h)\right|^2}{\|D_{\dot{A}}\varphi\|^2} = \infty \quad \textit{for all} \quad h \in \mathfrak{N} \setminus \{0\}. \tag{9.55}$$

Proof. By Theorem 9.2.3 (i) $\iff$ (ii) $\iff$ (iv). The equivalence of conditions (iii) and (iv) follows from (9.10) and (9.17). According to Lemma 9.1.1 condition (9.55) is equivalent to

$$\dot{A}^*\mathfrak{N} \cap \mathrm{Ran}(D_{\dot{A}}) = \{0\}.$$

Since (see (9.15)-(9.16)) $P_{\mathfrak{N}}\dot{A} = \mathcal{K}D_{\mathcal{A}}$, we have $\|D_{\mathcal{A}}f\|^2 = \|D_{\mathcal{K}}D_{\mathcal{A}}f\|^2$, $f \in \mathrm{Dom}(\dot{A}) = \mathcal{H}_0$, $\dot{A}^*h = D_{\mathcal{A}}\mathcal{K}^*h$, $h \in \mathfrak{N}$. By the Douglas Theorem 2.1.2 we get

$$\mathrm{Ran}(D_{\dot{A}}) = \mathrm{Ran}(D_{\mathcal{A}}D_{\mathcal{K}}). \tag{9.56}$$

Therefore $\dot{A}^*\mathfrak{N} \cap \mathrm{Ran}(D_{\dot{A}}) = \{0\}$ if and only if $\mathcal{R}(\mathcal{K}^*) \cap \mathrm{Ran}(D_{\mathcal{K}}) = \{0\}$. Since $\mathcal{K}^*D_{\mathcal{K}^*} = D_{\mathcal{K}}\mathcal{K}^*$, we have that $\mathrm{Ran}(\mathcal{K}^*) \cap \mathrm{Ran}(D_{\mathcal{K}}) = \{0\}$ if and only if the operator $\mathcal{K}^*$ is isometric. □

Corollary 9.2.14. *If $A_M = A_\mu$, then a symmetric contraction $\dot{A}$ does not admit non-self-adjoint qsc-extensions.*

The proof of the corollary follows from Theorem 9.2.7 and Theorem 9.2.13.

Corollary 9.2.15. *If $A_M \neq A_\mu$, then for any $\alpha \in (0, \pi/2)$ there is a non-self-adjoint α-co-sectorial qsc-extension of the symmetric contraction $\dot{A}$.*

Proof. If in the formula (9.42) one sets

$$X = i\tan\frac{\alpha}{2}\, I\upharpoonright\mathfrak{N}_0,$$

then by direct computations it is confirmed that

$$\|X\sin\alpha \pm i\cos\alpha\, I\| = 1.$$

This equality implies the α-co-sectorialilty of X. Then by Theorem 9.2.7 the corresponding T belongs to $C_{\mathcal{H}}(\alpha)$. □

Theorem 9.2.16. *Let $\dot{A} = \mathcal{A} + \mathcal{K}D_{\mathcal{A}}$ be a symmetric contraction of the form* (9.15) *written for the decomposition $\mathcal{H} = \mathcal{H}_0 \oplus \mathfrak{N}$, where $\mathcal{H}_0 = \mathrm{Dom}(\dot{A})$. Then the following conditions are equivalent:*

(i) $\mathrm{Ran}(A_M - A_\mu) = \mathfrak{N}$;

(ii) $\|\mathcal{K}\| < 1$;

(iii) $\mathrm{Ran}(D_{\dot{A}^*}) \supset \mathfrak{N}$;

(iv) $\mathrm{Ran}(D_{\dot{A}}) = \mathrm{Ran}(D_{\mathcal{A}})$;

(v) *for all $h \in \mathfrak{N}$ holds*

$$\sup_{\varphi\in\mathcal{H}_0}\frac{\left|(\dot{A}\varphi, h)\right|^2}{\|D_{\dot{A}}\varphi\|^2} < \infty. \tag{9.57}$$

Proof. By (9.17), (9.18), and (9.56) conditions (i), (ii), and (iii) are equivalent for every non-densely-defined symmetric contraction $\dot{A}$. It follows from (9.3) that condition (9.57) is equivalent to $\dot{A}^*\mathfrak{N} \subset \mathrm{Ran}(D_{\dot{A}})$. By the Douglas Theorem 2.1.2 the latter is equivalent to

$$\|P_{\mathfrak{N}}\dot{A}\varphi\| \leq \gamma\|D_{\dot{A}}\varphi\| \quad \text{for all} \quad \varphi \in \mathcal{H}_0,$$

where $\gamma > 0$. Since $P_{\mathfrak{N}}\dot{A} = \mathcal{K}D_{\mathcal{A}}$ the last inequality is equivalent to

$$\|\mathcal{K}\varphi\|^2 \le \frac{\gamma^2}{1+\gamma^2}\|\varphi\|^2,\ \varphi \in \mathfrak{D}_{\mathcal{A}} \iff \|\mathcal{K}\| < 1. \qquad \square$$

Let b be a real number, $-1 < b < 1$. Consider the function

$$z \mapsto w_b = \frac{z-b}{1-bz}$$

and corresponding operator fractional-linear transformation

$$\dot{A} \mapsto w_b(\dot{A}) = (\dot{A} - bI)(I - b\dot{A})^{-1}. \tag{9.58}$$

Proposition 9.2.17. *Let $\dot{A}$ be symmetric contraction. Then $w_b(\dot{A})$ is also symmetric contraction and*

$$(w_b(\dot{A}))_\mu = w_b(A_\mu), \quad (w_b(\dot{A}))_M = w_b(A_M),$$

i.e., the operators $w_b(A_\mu)$ and $w_b(A_\mu)$ are the rigid and the soft sc-extensions of $w_b(\dot{A})$, respectively. Moreover, the function

$$A \mapsto w_b(A) = (A - bI)(I - bA)^{-1}$$

bijectively maps the operator interval $[A_\mu, A_M]$ onto the operator interval $[w_b(A_\mu), w_b(A_M)]$.

Proof. Clearly, $\mathrm{Dom}(w_b(\dot{A})) = (I-b\dot{A})\mathrm{Dom}(\dot{A})$, $\mathcal{H}\ominus\mathrm{Dom}(w_b(\dot{A})) = (I-b\dot{A}^*)^{-1}\mathfrak{N}$, and $w_b(\dot{A})$ is symmetric contraction. Let us find

$$\inf_{\psi\in\mathrm{Dom}(w_b(\dot{A}))} ((I + w_b(A_\mu)(f+\psi), f+\psi)$$

for $f \in \mathcal{H}$. Since

$$\begin{aligned} I + w_b(A_\mu) &= (1-b)(I + A_\mu)(I - bA_\mu)^{-1} \\ &= (1-b)(I - bA_\mu)^{-1}(I - bA_\mu)(I + A_\mu)(I - bA_\mu)^{-1}, \end{aligned}$$

we obtain for each $h \in \mathcal{H}$ that

$$((I + w_b(A_\mu)h, h) = (1-b)\|(I - bA_\mu)^{1/2}(I + A_\mu)^{1/2}(I - bA_\mu)^{-1}h\|^2.$$

Therefore,

$$\begin{aligned} &((I + w_b(A_\mu))f + \psi, f + \psi) \\ &= (1-b)\|(I - bA_\mu)^{1/2}(I + A_\mu)^{1/2}((I - bA_\mu)^{-1}f + (I - bA_\mu)^{-1}\psi)\|^2. \end{aligned}$$

Because $(I - bA_\mu)^{-1}\mathrm{Dom}(w_b(\dot{A})) = (I - b\dot{A})^{-1}\mathrm{Dom}(w_b(\dot{A})) = \mathrm{Dom}(\dot{A})$ and

$$\inf_{\varphi\in\mathrm{Dom}(\dot{A})} ((I + A_\mu)(h+\varphi), h+\varphi) = 0,$$

for all $h \in \mathcal{H}$, we get

$$\inf_{\psi\in \mathrm{Dom}(w_b(\dot{A}))} ((I+w_b(A_\mu))(f+\psi), f+\psi) = \inf_{\psi\in \mathrm{Dom}(w_b(\dot{A}))} (1-b)$$
$$\times \|(I-bA_\mu)^{1/2}(I+A_\mu)^{1/2}((I-bA_\mu)^{-1}f + (I-bA_\mu)^{-1}\psi)\|^2 = 0.$$

Hence $(w_b(A))_\mu = w_b(A_\mu)$. Similarly $(w_b(A))_M = w_b(A_M)$.

Let A_1 and A_2 be two self-adjoint contractions. If $A_1 \geq A_2$, then $bI - b^2A_1 \leq bI - b^2A_2$. Then it follows that

$$(bI - b^2A_1)^{-1} \geq (bI - b^2A_2)^{-1}.$$

This inequality yields $w_b(A_1) \geq w_b(A_2)$. Therefore,

$$A \in [A_\mu, A_M] \Rightarrow w_b(A) \in [w_b(A_\mu), w_b(A_M)].$$

Since the inverse mapping is given by

$$C \mapsto (C+bI)(I+bC)^{-1},$$

we get that $A \in [A_\mu, A_M]$ if and only if $w_b(A) \in [w_b(A_\mu), w_b(A_M)]$. □

Let the Hilbert space $\mathcal{H}$ be decomposed as $\mathcal{H} = \mathcal{H}_1 \oplus \mathcal{H}_2$ and $T \in [\mathcal{H}, \mathcal{H}]$ be decomposed accordingly:

$$T = \begin{pmatrix} T_{11} & T_{12} \\ T_{21} & T_{22} \end{pmatrix}, \quad T_{ij} \in [\mathcal{H}_i, \mathcal{H}_j].$$

Define the operator-valued functions

$$V_T(z) = T_{21}(T_{11} - zI)^{-1}T_{12} - T_{22}, \quad W_T(z) = -zI - V_T(z), \quad z \in \rho(T_{11}). \tag{9.59}$$

By the Schur-Frobenius formula (see [157]) the resolvent $(T-z)^{-1}$ of T can be rewritten in the block form

$$\begin{pmatrix} (T_{11}-zI)^{-1}\left(I+T_{12}W_T(z)^{-1}T_{21}(T_{11}-zI)^{-1}\right) & -(T_{11}-zI)^{-1}T_{12}W_T^{-1}(z) \\ -W_T^{-1}(z)T_{21}(T_{11}-zI)^{-1} & W_T^{-1}(z) \end{pmatrix}, \tag{9.60}$$

for $z \in \rho(T) \cap \rho(T_{11})$. In particular,

$$P_{\mathcal{H}_2}(T-zI)^{-1}\restriction \mathcal{H}_2 = -\left(V_T(z) + zI\right)^{-1}, \quad z \in \rho(T) \cap \rho(T_{11}). \tag{9.61}$$

Following the definition from Section 6.5, we call a non-densely-defined symmetric contraction $\dot{A}$ by a **prime contraction** if there is no non-trivial subspace in $\mathrm{Dom}(\dot{A})$ which is invariant under $\dot{A}$. Since $\dot{A}$ is also symmetric, its primeness is equivalent to $\dot{A}$ being completely non-self-adjoint, i.e., to $\dot{A}$ having no self-adjoint parts.

Lemma 9.2.18. *Let the symmetric contraction $\dot{A} = \mathcal{A} + \mathcal{K}D_{\mathcal{A}}$ in $\mathcal{H} = \mathcal{H}_0 \oplus \mathfrak{N}$, $\mathcal{H}_0 = \operatorname{Dom}(\dot{A})$, be decomposed as in (9.15) with $\mathcal{K} : \mathfrak{D}_{\mathcal{A}} \to \mathfrak{N}$. Then $\dot{A}$ is prime if and only if the subspace*

$$\mathcal{H}_0^s := c.l.s.\left\{ (\mathcal{A} - zI)^{-1}\mathcal{K}^*\mathfrak{N},\ z \in \rho(\mathcal{A}) \right\} = c.l.s.\left\{ \mathcal{A}^n\mathcal{K}^*\mathfrak{N},\ n = 0, 1, \dots \right\} \quad (9.62)$$

coincides with $\mathcal{H}_0$. In this case, $\mathfrak{D}_{\mathcal{A}} = \mathcal{H}_0$, $\mathcal{K} : \mathcal{H}_0 \to \mathfrak{N}$, and $\|\mathcal{A}f\| < \|f\|$ for all $f \in \mathcal{H}_0 \setminus \{0\}$.

Proof. Suppose that $\dot{A}$ is prime. Then clearly $\ker D_{\mathcal{A}} = \{0\}$ or equivalently $\|\mathcal{A}f\| < \|f\|$ for all $f \in \mathcal{H}_0 \setminus \{0\}$, so that $\mathfrak{D}_{\mathcal{A}} = \mathcal{H}_0$ and $\mathcal{K} : \mathcal{H}_0 \to \mathfrak{N}$. Observe, that the subspace $\mathcal{H}_0^s$ in (9.62) and hence also $\mathcal{H}_0 \ominus \mathcal{H}_0^s$, is invariant under $\mathcal{A} = \mathcal{A}^*$. Then the subspace $\mathcal{H}_0 \ominus \mathcal{H}_0^s$ is also invariant under $D_{\mathcal{A}}$. Moreover,

$$\mathcal{H}_0 \ominus \mathcal{H}_0^s = \{ f \in \mathcal{H}_0 : \mathcal{K}\mathcal{A}^n f = 0,\ n = 0, 1, \dots \}. \quad (9.63)$$

It follows that $\mathcal{K}D_{\mathcal{A}}f = 0$ for all $f \in \mathcal{H}_0 \ominus \mathcal{H}_0^s$. Hence, in view of (9.15) $Af = \mathcal{A}f$ for all $f \in \mathcal{H}_0 \ominus \mathcal{H}_0^s$. This means that the subspace $\mathcal{H}_0 \ominus \mathcal{H}_0^s$ is invariant under $\dot{A}$. Since $\dot{A}$ is a prime, one concludes that $\mathcal{H}_0^s = \mathcal{H}_0$.

Conversely, assume that $\mathcal{H}_0^s = \mathcal{H}_0$. Since $\operatorname{Ran}(\mathcal{K}^*) \subset \mathfrak{D}_{\mathcal{A}}$ and $\mathfrak{D}_{\mathcal{A}}$ is invariant under $\mathcal{A}$, the definition of $\mathcal{H}_0^s$ in (9.62) shows that $\mathcal{H}_0^s \subset \mathfrak{D}_{\mathcal{A}}$. Hence, the assumption implies that $\mathcal{H}_0 = \mathfrak{D}_{\mathcal{A}} = \overline{\operatorname{Ran}(D_{\mathcal{A}})}$, so that $\ker D_{\mathcal{A}} = \{0\}$. Now, suppose that $\widetilde{\mathcal{H}}_0 \subset \mathcal{H}_0$ is a subspace which is invariant under $\dot{A}$. Then for every $f \in \widetilde{\mathcal{H}}_0$ one has $\dot{A}f = \mathcal{A}f + \mathcal{K}D_{\mathcal{A}}f \in \widetilde{\mathcal{H}}_0$, so that $\mathcal{K}D_{\mathcal{A}}f = 0$ for all $f \in \widetilde{\mathcal{H}}_0$ and $\dot{A}\restriction\widetilde{\mathcal{H}}_0 = \mathcal{A}\restriction\widetilde{\mathcal{H}}_0$. Hence, $\widetilde{\mathcal{H}}_0$ is invariant under $\mathcal{A}$ and $D_{\mathcal{A}}$. Moreover, since $\ker D_{\mathcal{A}} = \{0\}$ the image $D_{\mathcal{A}}\widetilde{\mathcal{H}}_0$ is dense in $\widetilde{\mathcal{H}}_0$. This implies that $\mathcal{K}\widetilde{\mathcal{H}}_0 = \{0\}$ and since $\mathcal{A}^n\widetilde{\mathcal{H}}_0 \subset \widetilde{\mathcal{H}}_0$ one has $\mathcal{K}\mathcal{A}^n\widetilde{\mathcal{H}}_0 = \{0\}$ for all $n = 0, 1, \dots$, i.e.,

$$\widetilde{\mathcal{H}}_0 \subset \{ f \in \mathcal{H}_0 : \mathcal{K}\mathcal{A}^n f = 0,\ n = 0, 1, \dots \} = \mathcal{H}_0 \ominus \mathcal{H}_0^s = \{0\},$$

cf. (9.63). Therefore, $\dot{A}$ is prime. □

Let T be a *qsc*-extension of $\dot{A}$ in the Hilbert space $\mathcal{H} = \mathcal{H}_0 \oplus \mathfrak{N}$ with $\mathcal{H}_0 = \operatorname{Dom}(\dot{A})$. It is evident that the subspace

$$\mathcal{H}'_T := c.l.s.\left\{ (T - zI)^{-1}\mathfrak{N} : |z| > 1 \right\} = c.l.s.\{ T^n\mathfrak{N} : n = 0, 1, 2, \dots \}, \quad (9.64)$$

is invariant under T, and that the subspace

$$\mathcal{H}''_T := \mathcal{H} \ominus \mathcal{H}'_T,$$

is invariant under T^*. Since $\mathfrak{N} \subset \mathcal{H}'_T$, one obtains

$$\mathcal{H}''_T \subset \mathfrak{N}^\perp = \operatorname{Dom}(\dot{A}) \subset \ker(T - T^*).$$

Therefore the restriction of T^* onto $\mathcal{H}''_T$ is a self-adjoint operator in $\mathcal{H}''_T$. The restriction $T\restriction\mathcal{H}'_T$ $(= P_{\mathcal{H}'_T}T\restriction\mathcal{H}'_T)$ is called the **$\mathfrak{N}$-minimal part of** T. Moreover, T

is said to be $\mathfrak{N}$**-minimal** if the equality $\mathcal{H} = \mathcal{H}'_T$ holds. If T be a *qsc*-extension of $\dot{A}$, then its adjoint T^* is also a *qsc*-extension of $\dot{A}$ and one can associate with it the subspace $\mathcal{H}'_{T^*}$ and the corresponding $\mathfrak{N}$-minimal part of T^*. The next result shows that the $\mathfrak{N}$-minimal parts of T and T^* are *qsc*-extensions of the prime part $\dot{A}\restriction \mathcal{H}_0^s$ of $\dot{A}$ in the same subspace $\mathcal{H}'_T = \mathcal{H}'_{T^*}$.

Proposition 9.2.19. *Let $\dot{A}$ be a symmetric contraction in $\mathcal{H} = \mathcal{H}_0 \oplus \mathfrak{N}$ with $\mathcal{H}_0 =$ $\mathrm{Dom}(\dot{A})$, T be a qsc-extension of $\dot{A}$ in $\mathcal{H}$, and T^* be its adjoint. Then the subspaces $\mathcal{H}'_T$, $\mathcal{H}'_{T^*}$, and $\mathcal{H}_0^s$ of $\mathcal{H} = \mathcal{H}_0 \oplus \mathfrak{N}$ as defined in (9.64) and (9.62) are connected by*

$$(\mathcal{H}' :=)\ \mathcal{H}'_T = \mathcal{H}'_{T^*} = \mathcal{H}_0^s \oplus \mathfrak{N}. \tag{9.65}$$

In particular, the symmetric contraction $\dot{A}$ is prime if and only if the qsc-extension T (or equivalently T^*) of $\dot{A}$ is $\mathfrak{N}$-minimal.

Proof. It follows from the Schur-Frobenius formula (9.60) that

$$(T - z)^{-1}\mathfrak{N} = \begin{pmatrix} -(\mathcal{A} - z)^{-1} D_{\mathcal{A}} \mathcal{K}^* \mathfrak{N} \\ \mathfrak{N} \end{pmatrix}, \quad |z| > 1,$$

which implies that

$$\begin{aligned} c.l.s.&\left\{ (T - zI)^{-1}\mathfrak{N} : |z| > 1 \right\} \\ &= c.l.s.\left\{ (\mathcal{A} - zI)^{-1} D_{\mathcal{A}} \mathcal{K}^* \mathfrak{N} : z \in \rho(\mathcal{A}) \right\} \oplus \mathfrak{N} \\ &= c.l.s.\left\{ D_{\mathcal{A}} (\mathcal{A} - zI)^{-1} \mathcal{K}^* \mathfrak{N} : z \in \rho(\mathcal{A}) \right\} \oplus \mathfrak{N}. \end{aligned}$$

This shows that

$$\mathcal{H}'_T = \overline{D_{\mathcal{A}} \mathcal{H}_0^s} \oplus \mathfrak{N}. \tag{9.66}$$

Since $\mathrm{Ran}(\mathcal{K}^*) \subset \mathfrak{D}_{\mathcal{A}}$ and $\mathfrak{D}_{\mathcal{A}}$ is invariant under $\mathcal{A}$ one has $\mathcal{H}_0^s \subset \mathfrak{D}_{\mathcal{A}}$. In particular, $\mathcal{H}_0^s \cap \ker D_{\mathcal{A}} = \{0\}$, which together with $D_{\mathcal{A}} \mathcal{H}_0^s \subset \mathcal{H}_0^s$ implies that $\overline{D_{\mathcal{A}} \mathcal{H}_0^s} = \mathcal{H}_0^s$. Hence, (9.66) implies the equality $\mathcal{H}'_T = \mathcal{H}_0^s \oplus \mathfrak{N}$. It follows from

$$(T^* - zI)^{-1} - (T - zI)^{-1} = (T - zI)^{-1}[T - T^*](T^* - zI)^{-1}, \quad |z| > 1,$$

and the inclusion $\mathrm{Ran}(T - T^*) \subset \mathfrak{N}$ that

$$(T^* - zI)^{-1}\mathfrak{N} \subset (T - zI)^{-1}\mathfrak{N} \subset \mathcal{H}'_T, \quad |z| > 1.$$

Therefore, $\mathcal{H}'_{T^*} \subset \mathcal{H}'_T$ and the reverse inclusion follows via symmetry. This completes the proof of (9.65).

The last statement is clear from (9.65). □

9.3 The Weyl-Titchmarsh functions of quasi-self-adjoint contractive extensions

Let $\mathcal{G}$ be a Hilbert space. The subclass of Herglotz-Nevanlinna $[\mathcal{G},\mathcal{G}]$-valued functions $V(z)$ holomorphic on the domain $\mathrm{Ext}[-1,1] = \mathbb{C}\setminus[-1,1]$ is denoted by $\mathbf{N}(\mathcal{G},[-1,1])$. It is well known (see [89]) that every function $V(z)$ in $\mathbf{N}(\mathcal{G},[-1,1])$ has an integral representation of the form

$$V(z) = Q + \int_{-1}^{1} \frac{dG(t)}{t-z},$$

where Q is a bounded self-adjoint operator on $\mathcal{G}$ and the $[\mathcal{G},\mathcal{G}]$-valued function $G(t)$ is non-decreasing, non-negative, normalized by $G(-1-0)=0$, and has finite total variation concentrated on $[-1,1]$. Clearly, $V(\infty) = s-\lim\limits_{z\to\infty} V(z) = Q$. The next result is also well known, cf. [89].

Theorem 9.3.1. *Let $\mathcal{G}$ be a Hilbert space and let $V(z)\in\mathbf{N}(\mathcal{G},[-1,1])$. Then there exist a Hilbert space $\mathcal{H}$, a self-adjoint contraction B on $\mathcal{H}$, and $F\in[\mathcal{G},\mathcal{H}]$, such that*

$$V(z) = V(\infty) + F^*(B-zI)^{-1}F, \quad z\in\mathrm{Ext}[-1,1]. \tag{9.67}$$

In what follows the subclass of functions $V(z)$ in $\mathbf{N}(\mathcal{G},[-1,1])$ which have the limit values $V(\pm1)$ in $[\mathcal{G},\mathcal{G}]$ will play a central role. In this case Theorem 9.3.1 can be completed as follows.

Theorem 9.3.2. *Let $\mathcal{G}$ be a Hilbert space and let $V(z)\in\mathbf{N}(\mathcal{G},[-1,1])$. If for all $f\in\mathcal{G}$ the limit values*

$$\lim_{x\uparrow -1}(V(x)f,f), \quad \lim_{x\downarrow 1}(V(x)f,f) \tag{9.68}$$

are finite, then there exist a Hilbert space $\mathcal{H}$, a self-adjoint contraction B in $\mathcal{H}$, and an operator $G\in[\mathcal{G},\mathfrak{D}_B]$, such that

$$V(z) = V(\infty) + G^*D_B^2(B-zI)^{-1}G, \quad z\in\mathrm{Ext}[-1,1], \tag{9.69}$$

where D_B and $\mathfrak{D}_B$ are defined by (9.1) and (9.2), respectively. Conversely, for every function $V(z)$ of the form (9.69) the limit values (9.68) exist for all $f\in\mathcal{G}$ and are finite.

Proof. By Theorem 9.3.1, $V(z)$ has the representation (9.67), where B is a self-adjoint contraction in a Hilbert space $\mathcal{H}$ and $F\in[\mathcal{G},\mathcal{H}]$. Since the limits in (9.68) exist for all $f\in\mathcal{G}$, one concludes that

$$\mathrm{Ran}(F)\subset\mathrm{Ran}(I-B)^{1/2}\cap\mathrm{Ran}(I+B)^{1/2}.$$

Consequently, $\operatorname{Ran}(F) \subset \operatorname{Ran}(D_B)$ and this implies (see Theorem 2.1.2) that $F = D_B G$ for some operator $G \in [\mathcal{G}, \mathfrak{D}_B]$.Taking into account (9.67) and the fact that D_B commutes with B we get (9.69).

Conversely, if $V(z)$ is of the form (9.69), then the inclusions $\operatorname{Ran}(D_B) \subset \operatorname{Ran}((I \pm B)^{1/2})$ imply the existence of the limit values (9.68) for all $f \in \mathcal{G}$ (see (9.4)). $\square$

It follows from Theorem 9.3.2 that

$$\begin{aligned} V(-1) &:= s - \lim_{x \uparrow -1} V(x) = V(\infty) + G^*(I - B)G \in [\mathcal{G}, \mathcal{G}], \\ V(1) &:= s - \lim_{x \downarrow 1} V(x) = V(\infty) - G^*(I + B)G \in [\mathcal{G}, \mathcal{G}], \end{aligned}$$

so that

$$V(-1) + V(1) = 2V(\infty) - 2G^*BG, \quad V(-1) - V(1) = 2G^*G.$$

The operator $T \in [\mathcal{H}, \mathcal{H}]$ is called a **quasi-self-adjoint contraction** (a *qsc*-**operator**) if

$$\|T\| \leq 1, \text{ and } \ker(T - T^*) \neq \{0\}.$$

Let T be a *qsc*-operator and let $\mathcal{G}$ be a subspace of $\mathcal{H}$ such that $\mathcal{G} \supseteq \operatorname{Ran}(T - T^*)$. Then T is a *qsc*-extension of a non-densely-defined contraction $\dot{A}$ defined as

$$\operatorname{Dom}(\dot{A}) = \mathcal{H} \ominus \mathcal{G}, \quad \dot{A} = T \upharpoonright \operatorname{Dom}(\dot{A}).$$

Clearly, $\operatorname{Dom}(\dot{A}) \subset \ker(T - T^*)$.

Let T be a *qsc*-operator in a separable Hilbert space $\mathcal{H}$ and let $\mathcal{G}$ be a subspace of $\mathcal{H}$ such that $\mathcal{G} \supset \operatorname{Ran}(T - T^*)$. The operator-valued function

$$Q_T(z) = P_{\mathcal{G}}(T - zI)^{-1} \upharpoonright \mathcal{G}, \quad |z| > 1, \tag{9.70}$$

where $P_{\mathcal{G}}$ is the orthogonal projection in $\mathcal{H}$ onto $\mathcal{G}$, is said to be the **Weyl-Titchmarsh function associated with T and the subspace** $\mathcal{G}$. Clearly, it has the limit value $Q_T(\infty) = 0$ and the Weyl-Titchmarsh functions of T and T^* in $\mathcal{G}$ are connected by

$$Q_{T^*}(z) = Q_T(\bar{z})^*, \quad |z| > 1.$$

If T is a self-adjoint contraction, then the Weyl-Titchmarsh function $Q_T(z)$ in (9.70) is a Herglotz-Nevanlinna function of the class $\mathbf{N}(\mathcal{G}, [-1, 1])$. The next result contains some basic properties for the Weyl-Titchmarsh function Q_T of a *qsc*-operator T as defined in (9.70).

Proposition 9.3.3. *Let Q_T be Weyl-Titchmarsh function of a qsc-operator T as defined in* (9.70). *Then:*

(i) *Q_T has the following asymptotic expansion*

$$Q_T(z) = -\frac{1}{z}I_{\mathcal{G}} + \frac{1}{z^2}F + o\left(\frac{1}{z^2}\right), \quad z \to \infty, \tag{9.71}$$

where $F = -P_{\mathcal{G}}T\upharpoonright\mathcal{G}$;

(ii) *$Q_T^{-1}(z) \in [\mathcal{G},\mathcal{G}]$ for all $|z| > 1$;*

(iii) *$Q_T^{-1}(z)$ has strong limit values $Q_T^{-1}(\pm 1)$ that are*

$$Q_T^{-1}(-1) = \lim_{z\to -1} Q_T^{-1}(z), \quad Q_T^{-1}(1) = \lim_{z\to 1} Q_T^{-1}(z);$$

(iv) *for all $f, g \in \mathcal{G}$ the following inequality holds:*

$$\begin{aligned} &\left|\left(\left(Q_T^{-1}(-1) + Q_T^{-1}(1)\right) f, g\right)\right|^2 \\ &\quad\leq \left(\left(Q_T^{-1}(-1) - Q_T^{-1}(1)\right) f, f\right) \left(\left(Q_T^{-1}(-1) - Q_T^{-1}(1)\right) g, g\right); \end{aligned} \tag{9.72}$$

(v) *the function $-Q_T^{-1}(z) - F - zI_{\mathcal{G}}$ is an operator-valued Herglotz-Nevanlinna function;*

(vi) *the following conditions are equivalent:*

a) $Q_T \in \mathbf{N}(\mathcal{G}, [-1,1])$,

b) $F = F^*$,

c) $Q_T^{-1}(-1) \geq 0$ *and* $Q_T^{-1}(1) \leq 0$.

Moreover, if T is decomposed as in (9.32) with $\mathcal{H}_0 = \mathcal{H} \ominus \mathcal{G}$ and $\dot{A} = T\upharpoonright\mathcal{H}_0$, then

$$F = \mathcal{K}\mathcal{A}\mathcal{K}^* - D_{\mathcal{K}^*}XD_{\mathcal{K}^*}, \tag{9.73}$$

$$Q_T^{-1}(-1) = D_{\mathcal{K}^*}(X + I_{\mathfrak{D}_{\mathcal{K}^*}})D_{\mathcal{K}^*}, \quad Q_T^{-1}(1) = D_{\mathcal{K}^*}(X - I_{\mathfrak{D}_{\mathcal{K}^*}})D_{\mathcal{K}^*}, \tag{9.74}$$

$$-Q_T^{-1}(z) - F - zI_{\mathcal{G}} = \mathcal{K}(I_{\mathcal{H}_0} - \mathcal{A}^2)(\mathcal{A} - zI_{\mathcal{H}_0})^{-1}\mathcal{K}^*. \tag{9.75}$$

Proof. (i) Clearly, $\lim_{z\to\infty} zQ_T(z)h = \lim_{z\to\infty} zP_{\mathcal{G}}(T-zI)^{-1}h = -h$ for all $h \in \mathcal{G}$. Moreover, for all $h \in \mathcal{G}$,

$$\lim_{z\to\infty} z(I_{\mathcal{G}} + zQ_T(z))h = \lim_{z\to\infty} zP_{\mathcal{G}}T(T - zI)^{-1}h = -P_{\mathcal{G}}Th. \tag{9.76}$$

Hence, Q_T admits the asymptotic expansion (9.71).

(ii) Let $|z| > 1$, $f \in \mathcal{G}$, and $\varphi = (T - zI)^{-1}f$. Then $\|f\| \leq (1 + |z|)\|\varphi\|$ and

$$\begin{aligned} |(Q_T(z)f, f)| &= \left|\left((T - zI)^{-1}f, f\right)\right| = |(\varphi, (T - zI)\varphi)| \\ &= \left|(\varphi, T\varphi) - \bar{z}\|\varphi\|^2\right| \geq \frac{|z| - 1}{(|z| + 1)^2}\|f\|^2. \end{aligned}$$

Since $|(Q_T(z)f,f)| = |(Q_T(z)^*f,f)|$, this implies that

$$\|Q_T(z)f\| \geq \frac{|z|-1}{(|z|+1)^2}\|f\|, \quad \|Q_T(z)^*f\| \geq \frac{|z|-1}{(|z|+1)^2}\|f\|.$$

Therefore, $Q_T^{-1} \in [\mathcal{G},\mathcal{G}]$ for all $|z|>1$.

(iii) Decompose $\mathcal{H} = \mathcal{H}_0 \oplus \mathcal{G}$ and write T in block form as in (9.32), where $\mathcal{A} = P_{\mathcal{H}_0}\dot{A}$ is a self-adjoint contraction in $\mathcal{H}_0$, $D_{\mathcal{A}} = \left(I_{\mathcal{H}_0} - \mathcal{A}^2\right)^{1/2}$, $\mathcal{K} \in [\mathfrak{D}_{\mathcal{A}},\mathcal{G}]$ is a contraction, and X is a contraction in the subspace $\mathfrak{D}_{\mathcal{K}^*} \subset \mathcal{G}$. The formula (9.73) for F is immediate from (9.32). Write $Q_T^{-1}(z)$ as in (9.61),

$$Q_T^{-1}(z) = -V_T(z) - zI_{\mathcal{G}}, \quad |z|>1,$$

where

$$V_T(z) = \mathcal{K}\left[\mathcal{A} + (\mathcal{A} - zI_{\mathcal{H}_0})^{-1}(I_{\mathcal{H}_0} - \mathcal{A}^2)\right]\mathcal{K}^* - D_{\mathcal{K}^*}XD_{\mathcal{K}^*}. \tag{9.77}$$

This shows that the limit values $Q_T^{-1}(\pm 1)$ exist and that they are given by (9.74).

(iv) It follows from (9.74) that

$$\frac{Q_T^{-1}(-1) + Q_T^{-1}(1)}{2} = D_{\mathcal{K}^*}XD_{\mathcal{K}^*}, \quad \frac{Q_T^{-1}(-1) - Q_T^{-1}(1)}{2} = D_{\mathcal{K}^*}^2. \tag{9.78}$$

Since X is a contraction in $\mathfrak{D}_{\mathcal{K}^*}$ we get (9.72).

(v) It follows from (9.73) and (9.77) that (9.75) holds. Clearly, the function in (9.75) is a Herglotz-Nevanlinna function.

The statement (vi) follows from (9.73) and (9.74). □

Corollary 9.3.4. *If* $\mathcal{G} = \mathrm{Ran}(\mathrm{Im}\, T)$, *then the operator* $Q_T^{-1}(-1) - Q_T^{-1}(1)$ *is positive definite. If, also,* $\dim \mathcal{G} < \infty$, *then there exist the limits* $Q_T(\pm 1)$.

Proof. Let T be decomposed as in (9.32). If $\mathcal{G} = \mathrm{Ran}(\mathrm{Im}\, T)$, then the operator $\mathrm{Im}\, T \upharpoonright \mathcal{G}$ is boundedly invertible in $\mathcal{G}$. Since $\mathrm{Im}\, T\upharpoonright \mathcal{G} = D_{\mathcal{K}^*}\mathrm{Im}\, X D_{\mathcal{K}^*}$, it follows that $\mathrm{Ran}(D_{\mathcal{K}^*}) = \mathcal{G}$ and $\mathrm{Im}\, X$ is boundedly invertible in $\mathcal{G}$. Hence, from (9.78) we get that the operator $Q_T^{-1}(-1) - Q_T^{-1}(1)$ is positive definite and

$$(Q_T^{-1}(-1) - Q_T^{-1}(1))^{-1} = \frac{1}{2}D_{\mathcal{K}^*}^{-2}.$$

Suppose $\dim \mathcal{G} < \infty$. Since $\mathrm{Im}\, X$ is invertible, the relations (9.45) yield that the operators $I_{\mathcal{G}} \pm X$ and $I_{\mathcal{G}} \pm \mathrm{Re}\, X$ are invertible. Now from (9.74) we get that $Q_T(\pm 1)$ exist and

$$Q_T(-1) = D_{\mathcal{K}^*}^{-1}(X + I_{\mathcal{G}})^{-1}D_{\mathcal{K}^*}^{-1}, \; Q_T(1) = D_{\mathcal{K}^*}^{-1}(X - I_{\mathcal{G}})^{-1}D_{\mathcal{K}^*}^{-1}. \tag{9.79}$$

□

Corollary 9.3.5. *Let* $\dim \mathcal{G} < \infty$ *and let* $V(z) \in \mathbf{N}(\mathcal{G},[-1,1])$, $V(\infty) = 0$. *If* $\ker(\operatorname{Im} V(z)) = \{0\}$ *for some non-real* z, *then* $V(z)$ *has the representation*

$$V(z) = K^* Q_B(z) K,$$

where B *is a self-adjoint contractive extension of a non-densely-defined symmetric contraction in some Hilbert space* $\mathcal{H}$, *and* $K \in [\mathcal{G},\mathcal{H}]$ *is an invertible operator. Hence, the limit values* $V^{-1}(\pm 1)$ *exist.*

Proof. By Theorem 9.3.1 the function $V(z)$ has the representation

$$V(z) = K^*(B - zI)^{-1} K,$$

where B is a self-adjoint contraction in some Hilbert space $\mathcal{H}$ and $K \in [\mathcal{G},\mathcal{H}]$. From the condition $\ker(\operatorname{Im} V(z)) = \{0\}$ for some non-real z it follows that $\ker(K) = \{0\}$. Let $\mathcal{G} := \operatorname{Ran}(K)$ and let $\mathcal{H}_0 := \mathcal{H} \ominus \mathcal{G}$. Define the symmetric contraction $\dot{A} := B{\upharpoonright}\mathcal{H}_0$. Then B is an sc-extension of $\dot{A}$ and

$$V(z) = K^* Q_B(z) K.$$

Now the statements of the corollary follow from Proposition 9.3.3. $\square$

Let A_μ and A_M be the M. Kreĭn rigid and soft sc-extensions of a symmetric contraction $\dot{A}$ defined by (9.37) and (9.37), respectively and $\mathcal{G} = \mathcal{H} \ominus \operatorname{Dom}(\dot{A})$. Consider the Weyl-Titchmarsh functions of the operators A_μ and A_M,

$$Q_{A_\mu}(z) = P_\mathcal{G}(A_\mu - zI)^{-1}{\upharpoonright}\mathcal{G}, \quad Q_{A_M}(z) = P_\mathcal{G}(A_M - zI)^{-1}{\upharpoonright}\mathcal{G},$$

where $z \in \operatorname{Ext}[-1,1]$. Since

$$\operatorname{Ran}((I_\mathcal{G} + A_\mu)^{1/2}) \cap \mathcal{G} = \operatorname{Ran}((I_\mathcal{G} - A_M)^{1/2}) \cap \mathcal{G} = \{0\},$$

using (9.3) we get the relations

$$\lim_{z\uparrow -1} (Q_{A_\mu}(z)f,f) = +\infty, \quad \lim_{z\downarrow 1}(Q_{A_M}(z)f,f) = -\infty, \ f \in \mathcal{G}\setminus\{0\}.$$

Proposition 9.3.6. *Let* $Q_T(z)$ *be the Weyl-Titchmarsh function of some qsc-operator* T. *Suppose that*

$$\liminf_{\lambda\uparrow -1} |(Q_T(\lambda)f,f)| = \infty, \quad \text{for all } f \in \mathcal{G}\setminus\{0\}, \tag{9.80}$$

or

$$\liminf_{\lambda\downarrow 1} |(Q_T(\lambda)f,f)| = \infty, \quad \text{for all } f \in \mathcal{G}\setminus\{0\}. \tag{9.81}$$

Then Q *is a Herglotz-Nevanlinna function from the class* $\mathbf{N}(\mathcal{G},[-1,1])$ *and* $T = A_\mu$ *or* $T = A_M$.

Proof. We have,

$$(T-\lambda I)^{-1}=(\operatorname{Re}T-\lambda I)^{-1/2}\,(I+iB)^{-1}\,(\operatorname{Re}T-\lambda I)^{-1/2},\ \lambda<-1,$$

where $B=(\operatorname{Re}T-\lambda I)^{-1/2}\operatorname{Im}T(\operatorname{Re}T-\lambda I)^{-1/2}$, and

$$(T-\lambda I)^{-1}=-(\lambda I-\operatorname{Re}T)^{-1/2}\,(I-iB)^{-1}\,(\lambda I-\operatorname{Re}T)^{-1/2},\ \lambda>1,$$

where $B=(\lambda I-\operatorname{Re}T)^{-1/2}\operatorname{Im}T(\lambda I-\operatorname{Re}T)^{-1/2}$. This shows that for all $f\in\mathcal{G}$ and $\lambda<-1$,

$$(Q_T(\lambda)f,f)=\left((I+iB)^{-1}(\operatorname{Re}T-\lambda I)^{-1/2}f,(\operatorname{Re}T-\lambda I)^{-1/2}f\right).$$

Since $\|(I+iB)^{-1}\|\le 1$, one obtains

$$|(Q_T(\lambda)f,f)|\le\left\|(\operatorname{Re}T-\lambda I)^{-1/2}f\right\|^2.$$

Now the assumption (9.80) implies that

$$\liminf_{\lambda\uparrow -1}\left\|(\operatorname{Re}T-\lambda I)^{-1/2}f\right\|^2=\infty,\quad \text{for all } f\in\mathcal{G}\setminus\{0\}.$$

Thus $\operatorname{Ran}((I+\operatorname{Re}T)^{1/2})\cap\mathcal{G}=\{0\}$ (see (9.3)). Since $\operatorname{Re}T$ is a *sc*-extension of $\dot{A}$, one concludes that $\operatorname{Re}T=A_\mu$. Now one has $\operatorname{Im}T=0$ and $T=A_\mu$. The proof of the other statement is similar. □

9.4 Canonical L-systems with contractive state-space operators

Let us consider a canonical L-system Θ of the type (5.6) with a bounded operator T acting on a Hilbert space $\mathcal{H}$,

$$\Theta=\begin{pmatrix} T & K & J\\ \mathcal{H} & & E\end{pmatrix}. \tag{9.82}$$

Here $\dim E=r<\infty$, and we suppose that $\ker K=\{0\}$. Then we have $\operatorname{Im}T=KJK^*$ and the transfer function of Θ is given by

$$W_\Theta(z)=I-2iK^*(T-zI)^{-1}KJ,$$

while the impedance function of Θ is

$$V_\Theta(z)=K^*(\operatorname{Re}T-zI)^{-1}K=i(W_\Theta(z)+I)^{-1}(W_\Theta(z)-I)J. \tag{9.83}$$

Clearly, if T is a contraction, then the function $V_\Theta(z)$ is holomorphic in $\operatorname{Ext}[-1,1]$ and $W_\Theta(z)$ is holomorphic outside of the closed unit disk.

Theorem 9.4.1. *Let the state-space operator T of the L-system Θ of the form (9.82) be a contraction. Then the transfer function $W_\Theta(z)$ and the impedance function $V_\Theta(z)$ satisfy the following conditions:*

1. *the limit values $V_\Theta(\pm 1)$, $W_\Theta(\pm 1)$ exist;*
2. *the operators $W_\Theta(\pm 1)$ are J-unitary;*
3. *the operators $V_\Theta(\pm 1)$ are invertible;*
4. *the operator $V_\Theta^{-1}(-1) - V_\Theta^{-1}(1)$ is invertible;*
5. *the operator*

$$\begin{aligned} \mathcal{K}_J = [V_\Theta^{-1}(-1) - V_\Theta^{-1}(1)]^{-1/2} \left(2iJ + [V_\Theta^{-1}(-1) + V_\Theta^{-1}(1)]\right) \\ \times [V_\Theta^{-1}(-1) - V_\Theta^{-1}(1)]^{-1/2} \end{aligned} \tag{9.84}$$

   is a contraction;
6. Re $(i(W_\Theta(-1)JW_\Theta^*(1) - J)) \geq 0$.

Proof. Let T be a contraction in $\mathcal{H}$. We set

$$\mathcal{G} = \mathrm{Ran}(\mathrm{Im}\, T), \quad \mathcal{H}_0 = \ker(\mathrm{Im}\, T),$$

and introduce an operator $\dot{A}x = Tx$ for $x \in \mathrm{Dom}(\dot{A}) = \mathcal{H}_0$. Clearly, the operator T is a qsc-extension of the non-densely-defined symmetric contraction $\dot{A}$.

Using the definition of the Weyl-Titchmarsh function associated with T and $\mathrm{Re}\, T$ and the subspace $\mathcal{G}$, we apply formula (9.70) and rewrite the functions $W_\Theta(z)$ and $V_\Theta(z)$ as follows:

$$W_\Theta(z) = I - 2iK^* Q_T(z) KJ, \quad V_\Theta(z) = K^* Q_{\mathrm{Re}\, T}(z) K.$$

From Proposition 9.3.3 and Corollary 9.3.4 we get that the limit values $W_\Theta(\pm 1)$, $V_\Theta(\pm 1)$ exist. Since T is a qsc-extension of $\dot{A}$, it takes the form (9.32)

$$T = \begin{pmatrix} \mathcal{A} & D_{\mathcal{A}} \mathcal{K}^* \\ \mathcal{K} D_{\mathcal{A}} & -\mathcal{K}\mathcal{A}\mathcal{K}^* + D_{\mathcal{K}^*} X D_{\mathcal{K}^*} \end{pmatrix} : \begin{pmatrix} \mathcal{H}_0 \\ \mathcal{G} \end{pmatrix} \to \begin{pmatrix} \mathcal{H}_0 \\ \mathcal{G} \end{pmatrix}.$$

Then it follows from (9.79) that

$$\begin{aligned} W_\Theta(-1) &= I - 2iK^* D_{\mathcal{K}^*}^{-1} (X + I_{\mathcal{G}})^{-1} D_{\mathcal{K}^*}^{-1} KJ, \\ W_\Theta(1) &= I - 2iK^* D_{\mathcal{K}^*}^{-1} (X - I_{\mathcal{G}})^{-1} D_{\mathcal{K}^*}^{-1} KJ, \\ V_\Theta(-1) &= K^* D_{\mathcal{K}^*}^{-1} (I_{\mathcal{G}} + \mathrm{Re}\, X)^{-1} D_{\mathcal{K}^*}^{-1} K, \\ V_\Theta(1) &= K^* D_{\mathcal{K}^*}^{-1} (\mathrm{Re}\, X - I_{\mathcal{G}})^{-1} D_{\mathcal{K}^*}^{-1} K, \end{aligned}$$

and

$$\begin{aligned} V_\Theta^{-1}(-1) &= K^{-1} D_{\mathcal{K}^*} (I_{\mathcal{G}} + \mathrm{Re}\, X) D_{\mathcal{K}^*} K^{*[-1]}, \\ V_\Theta^{-1}(1) &= -K^{-1} D_{\mathcal{K}^*} (I_{\mathcal{G}} - \mathrm{Re}\, X) D_{\mathcal{K}^*} K^{*[-1]}, \end{aligned}$$

where $K^{*[-1]} \in [E, \mathrm{Ran}(\mathrm{Im}\,(T))]$ is the Moore-Penrose inverse of K^*. Therefore

$$K^{-1}D^2_{\mathcal{K}^*}K^{*[-1]} = \frac{1}{2}\left(V_\Theta^{-1}(-1) - V_\Theta^{-1}(1)\right),$$
$$\left(\frac{1}{2}\left(V_\Theta^{-1}(-1) - V_\Theta^{-1}(1)\right)\right)^{-1} = K^*D^{-2}_{\mathcal{K}^*}K,$$
$$K^{-1}D_{\mathcal{K}^*}\mathrm{Re}\,X D_{\mathcal{K}^*}K^{*[-1]} = \frac{1}{2}\left(V_\Theta^{-1}(-1) + V_\Theta^{-1}(1)\right).$$

Since $KJK^* = D_{\mathcal{K}^*}\mathrm{Im}\,X D_{\mathcal{K}^*}$, we have $K^{-1}D_{\mathcal{K}^*}\mathrm{Im}\,X D_{\mathcal{K}^*}K^{*[-1]} = J$. Hence,

$$X = \mathrm{Re}\,X + i\mathrm{Im}\,X = D^{-1}_{\mathcal{K}^*}K\left(\frac{1}{2}\left(V_\Theta^{-1}(-1) + V_\Theta^{-1}(1)\right) + iJ\right)K^*D^{-1}_{\mathcal{K}^*}.$$

Furthermore, from

$$\|D_{\mathcal{K}^*}K^{*[-1]}f\|^2_{\mathcal{G}} = \|\frac{1}{\sqrt{2}}\left(V_\Theta^{-1}(-1) - V_\Theta^{-1}(1)\right)^{1/2}f\|^2_E, \quad f \in E$$

we get

$$D_{\mathcal{K}^*}K^{*[-1]} = \frac{1}{\sqrt{2}}U\left(V_\Theta^{-1}(-1) - V_\Theta^{-1}(1)\right)^{1/2}, \tag{9.85}$$

where $U \in [E, \mathcal{G}]$ and unitarily maps E onto $\mathcal{G}$. Now

$$X = 2U\left(V_\Theta^{-1}(-1) - V_\Theta^{-1}(1)\right)^{-1/2}\left(\frac{1}{2}\left(V_\Theta^{-1}(-1) + V_\Theta^{-1}(1)\right) + iJ\right)$$
$$\times\left(V_\Theta^{-1}(-1) - V_\Theta^{-1}(1)\right)^{-1/2}U^* = U\mathcal{K}_J U^*.$$

Since $\|X\| \leq 1$, we get (9.84). From the relations (5.22) written for all real $\lambda \in \mathrm{Ext}[-1, 1]$ we obtain

$$W_\Theta(\lambda)JW^*_\Theta(\lambda) = J, \quad W^*_\Theta(\lambda)JW_\Theta(\lambda) = J$$

and hence the operators $W_\Theta(\pm 1)$ are J-unitary. From the expressions for $W_\Theta(\pm 1)$ one can easily derive that

$$i(W_\Theta(-1)JW^*_\Theta(1) - J) = 4K^*D^{-1}_{\mathcal{K}^*}(I_{\mathcal{G}} + X)^{-1}(I_{\mathcal{G}} - X^*)^{-1}D^{-1}_{\mathcal{K}^*}K.$$

Since $\mathrm{Re}\,((I_{\mathcal{G}} - X^*)(I_{\mathcal{G}} + X)) = D^2_X$, we have

$$\mathrm{Re}\,(i(W_\Theta(-1)JW^*_\Theta(1) - J)) \geq 0. \qquad \square$$

The next theorem is the converse statement.

Theorem 9.4.2. *Let E be a finite-dimensional Hilbert space, $J \in [E,E]$, $J = J^* = J^{-1}$ and let an $[E,E]$-valued function $W(z)$ belong to the class Ω_J. Suppose that the function*

$$V(z) = i(W(z)+I)^{-1}(W(z)-I)J,$$

has a holomorphic continuation to a function from the class $\mathbf{N}(E,[-1,1])$ and $\ker \operatorname{Im} V(z) = \{0\}$ *for some non-real z. If the operator*

$$\begin{aligned}\mathcal{K}_J = [V^{-1}(-1) - V^{-1}(1)]^{-1/2} \left(2iJ + [V^{-1}(-1) - V^{-1}(1)]\right)\\ \times [V^{-1}(-1) - V^{-1}(1)]^{-1/2}\end{aligned}$$

is a contraction, then there exists a minimal L-system Θ of the form (9.82) with contractive state-space operator T such that $W_\Theta(z) = W(z)$. Moreover, T belongs to the class $C_{\mathcal{H}}(\alpha)$ for some $\alpha \in (0,\pi/2)$ if and only if the operator $\mathcal{K}_J \in C_E(\alpha)$.

Proof. By Theorem 5.5.4 there exists a prime system of the form (9.82) whose transfer function $W_\Theta(z)$ is equal to $W(z)$ and $V(z) = K^*(\operatorname{Re} T - zI)^{-1}K$ for $z \in \operatorname{Ext}[-1,1]$. The operator $\operatorname{Re} T$ is a self-adjoint contraction. Set $\mathcal{H}_0 = \ker(K^*)$, $\mathfrak{N} = \mathcal{H} \ominus \mathcal{H}_0$, and $\dot{A} := T \upharpoonright \mathcal{H}_0$. Then $\operatorname{Re} T$ is an sc-extension of $\dot{A}$ and by Corollary 9.3.5, $V(z) = K^* Q_{\operatorname{Re} T}(z) K$ and there exist the limit values $V(\pm 1)$ of the function V. It follows from Corollary 9.3.4 that the operator $V^{-1}(-1) - V^{-1}(1)$ is invertible. Represent $\operatorname{Re} T$ in the form (9.32)

$$\operatorname{Re} T = \begin{pmatrix} \mathcal{A} & D_{\mathcal{A}}\mathcal{K}^* \\ \mathcal{K}D_{\mathcal{A}} & -\mathcal{K}\mathcal{A}\mathcal{K}^* + D_{\mathcal{K}^*} Y D_{\mathcal{K}^*} \end{pmatrix} : \begin{pmatrix} \mathcal{H}_0 \\ \mathfrak{N} \end{pmatrix} \to \begin{pmatrix} \mathcal{H}_0 \\ \mathfrak{N} \end{pmatrix},$$

where $Y \in [\mathfrak{N},\mathfrak{N}]$ is a self-adjoint contraction. Then the formulas

$$\begin{aligned} V^{-1}(-1) &= K^{-1} D_{\mathcal{K}^*}(I_{\mathfrak{N}} + Y) D_{\mathcal{K}^*} K^{*-1}, \\ V^{-1}(1) &= -K^{-1} D_{\mathcal{K}^*}(I_{\mathfrak{N}} - Y) D_{\mathcal{K}^*} K^{*-1} \end{aligned}$$

hold. Therefore, (9.85) holds with some unitary operator $U \in [E, \mathfrak{N}]$. Put

$$Z = D_{\mathcal{K}^*}^{-1} \operatorname{Im} T D_{\mathcal{K}^*}^{-1} = D_{\mathcal{K}^*}^{-1} K J K^* D_{\mathcal{K}^*}^{-1}$$

and let $X = Y + iZ$. Arguing in the reverse direction in the proof of Theorem 9.4.2 we get that the operator X is a contraction in $\mathfrak{N}$. It follows from Theorem 9.2.3 that T is a qsc-extension of $\dot{A}$. Moreover,

$$\mathcal{K} \in C_E(\alpha) \iff X \in C_{\mathfrak{N}}(\alpha) \iff T \in C_{\mathcal{H}}(\alpha). \qquad \square$$

Notice that the condition $\|\mathcal{K}_J\| \le 1$ is equivalent to the condition

$$\operatorname{Re}\,(i(W_\Theta(-1) J W_\Theta^*(1) - J)) \ge 0,$$

and the condition $\mathcal{K}_J \in C_E(\alpha)$ is equivalent to the condition of the operator $i(W_\Theta(-1) J W_\Theta^*(1) - J)$ being α-sectorial.

Example. Let us consider an example that illustrates Theorem 9.4.2. Let $\alpha(x)$ be a non-decreasing function defined on a closed interval $[0, l]$. We define an operator T_α by the formula

$$T_\alpha f(x) = \alpha(x) f(x) + i \int_x^l f(t)\, dt,$$

where $f(x) \in L^2_{[0,l]}$, $l > 0$. It is easy to see that

$$\begin{aligned} \operatorname{Im} T_\alpha f &= \frac{1}{2} \int_0^l f(t)\, dt = (f, g)g, \quad \left(g(x) = \frac{1}{\sqrt{2}} \right), \\ \operatorname{Re} T_\alpha f &= \alpha(x) f(x) + \frac{i}{2} \int_x^l f(t)\, dt - \frac{i}{2} \int_0^x f(t)\, dt. \end{aligned} \tag{9.86}$$

Consider an L-system

$$\Theta_\alpha = \begin{pmatrix} T_\alpha & K_\alpha & I \\ L^2_{[0,l]} & & \mathbb{C} \end{pmatrix},$$

where $K_\alpha c = cg$, $c \in \mathbb{C}$. Let us set $\alpha(x) \equiv 0$ and let

$$T_0 f(x) = i \int_x^l f(t)\, dt, \quad T_0^* f(x) = -i \int_0^x f(t)\, dt. \tag{9.87}$$

From (9.86), (9.87), Theorem 5.2.1, and the fact that a linear span of polynomials is dense in $L^2_{[0,l]}$ we conclude that T_0 is a prime operator. Thus, since $J = 1$, we can apply Theorem 9.4.2 and find those values of l such that the number $\mathcal{K}_J$ in (9.84) does not exceed 1 in absolute value. Clearly (see [89],

$$W_{\Theta_0}(z) = \exp \frac{il}{z}, \quad V_{\Theta_0}(z) = i \frac{W_{\Theta_0}(z) - 1}{W_{\Theta_0}(z) + 1} = -\tan \frac{l}{2z}.$$

Note that the numbers

$$\frac{l}{(2n+1)\pi}, \quad n \in \mathbb{Z}$$

are the eigenvalues of the compact self-adjoint operator $\operatorname{Re} T_0$ [89]. Hence, $\|\operatorname{Re} T_0\| = l/\pi$ and $\|\operatorname{Re} T_0\| \le 1$ if and only if $l \le \pi$. The operator $\operatorname{Re} T_0$ is a self-adjoint extension of a non-densely-defined symmetric operator

$$\operatorname{Dom}(\dot{A}) = \left\{ f(x) \in L^2_{[0,l]} : \int_0^l f(x)\, dx = 0 \right\}, \quad \dot{A} f = i \int_x^l f(t)\, dt. \tag{9.88}$$

Let $l \le \pi$. Then the operator $\dot{A}$ is a contraction and $\mathrm{Re}\, T_0$ is its sc-extension. If $l = \pi$, then

$$\lim_{z\uparrow -1} |V_{\Theta_0}(z)| = \lim_{z\downarrow 1} |V_{\Theta_0}(z)| = \infty.$$

According to Proposition 9.3.6 we get $A_\mu = A_M = \mathrm{Re}\, T_0$. Thus in the case $l = \pi$ the operator $\dot{A}$ has a unique sc-extension. In particular, it follows that in this case $\|T_0\| > 1$. Let $l < \pi$. Then

$$V_{\Theta_0}(1) = -\tan\left(\frac{l}{2}\right), \quad V_{\Theta_0}(-1) = \tan\left(\frac{l}{2}\right).$$

We have

$$V_0^{-1}(-1) + V_0^{-1}(1) = 0, \quad V_0^{-1}(-1) - V_0^{-1}(1) = \frac{2}{\tan\left(\frac{l}{2}\right)}.$$

It is worth mentioning that the Weyl-Titchmarsh function of $\mathrm{Re}\, T_0$ is

$$Q_{\mathrm{Re}\, T_0}(z) = -\frac{2}{l} \tan\left(\frac{l}{2z}\right).$$

Let A_μ and A_M be extremal extensions of the operator $\dot{A}$ given by (9.88). From relations (9.74) we get that

$$\mathrm{Re}\, T_0 = \frac{A_\mu + A_M}{2}$$

and

$$\begin{aligned} A_\mu f &= i \int_x^l f(t)\, dt - \frac{1}{2}\left(i + \frac{1}{\tan\left(\frac{l}{2}\right)}\right) \int_0^l f(t)\, dt, \\ A_M f &= i \int_x^l f(t)\, dt - \frac{1}{2}\left(i - \frac{1}{\tan\left(\frac{l}{2}\right)}\right) \int_0^l f(t)\, dt. \end{aligned}$$

For the operator K_J given by (9.84) we have

$$\begin{aligned} \mathcal{K}_J = {} & [V_{\Theta_0}^{-1}(-1) - V_{\Theta_0}^{-1}(1)]^{-1/2} \left(2iJ + [V_{\Theta_0}^{-1}(-1) + V_{\Theta_0}^{-1}(1)]\right) \\ & \times [V_{\Theta_0}^{-1}(-1) - V_{\Theta_0}^{-1}(1)]^{-1/2} = i \tan\frac{l}{2}. \end{aligned}$$

Then

$$\|\mathcal{K}_J\| \le 1 \iff l \le \frac{\pi}{2}.$$

Thus, the operator T_0 is a contraction if and only if $l \le \pi/2$. By virtue of Theorem 9.4.2 the operator T_0 on $L^2_{[0,l]}$ is an α-co-sectorial if and only if $0 < l < \frac{\pi}{2}$ and in this case $\alpha = l$.

9.5 The restricted Phillips-Kato extension problem

Let us remind readers that a symmetric operator $\dot{B}$ is called **non-negative** if

$$(\dot{B}f, f) \geq 0, \quad \forall f \in \mathrm{Dom}(\dot{B}).$$

Let $\dot{B}$ be a closed densely-defined non-negative operator in a Hilbert space $\mathcal{H}$ and let $\dot{B}^*$ be its adjoint. Consider the sesquilinear form $\tau_{\dot{B}}[f, g] = (\dot{B}f, g)$, $f, g \in \mathrm{Dom}(\dot{B})$. A sequence $\{f_n\} \subset \mathrm{Dom}(\dot{B})$ is called **$\tau_{\dot{B}}$-converging** to the vector $f \in \mathcal{H}$ if

$$\lim_{n\to\infty} f_n = f \quad \text{and} \quad \lim_{n,m\to\infty} \tau_{\dot{B}}[f_n - f_m] = 0.$$

The form $\tau_{\dot{B}}$ is **closable** [163], i.e., there exists a minimal closed extension (the closure) of $\tau_{\dot{B}}$. Following the M. Kreĭn notations we denote by $\dot{B}[\cdot,\cdot]$ the closure of $\tau_{\dot{B}}$ and by $\mathcal{D}[\dot{B}]$ its domain. By definition $\dot{B}[u] = \dot{B}[u, u]$ for all $u \in \mathcal{D}[\dot{B}]$. Because $\dot{B}[u, v]$ is closed, it possesses the property: if

$$\lim_{n\to\infty} u_n = u \quad \text{and} \quad \lim_{n,m\to\infty} \dot{B}[u_n - u_m] = 0,$$

then $\lim_{n\to\infty} \dot{B}[u - u_n] = 0$.

The **Friedrichs extension** B_F of $\dot{B}$ is defined as a non-negative self-adjoint operator associated with the form $\dot{B}[\cdot,\cdot]$ by the First Representation Theorem [163]:

$$(B_F u, v) = \dot{B}[u, v] \quad \text{for all} \quad u \in \mathrm{Dom}(B_F) \quad \text{and for all} \quad v \in \mathcal{D}[\dot{B}].$$

It follows that

$$\mathrm{Dom}(B_F) = \mathcal{D}[\dot{B}] \cap \mathrm{Dom}(\dot{B}^*), \; B_F = \dot{B}^* \upharpoonright \mathrm{Dom}(B_F).$$

The Friedrichs extension B_F is a unique non-negative self-adjoint extension having its domain in $\mathcal{D}[\dot{B}]$. Notice that by the Second Representation Theorem [163] one has

$$\mathcal{D}[\dot{B}] = \mathcal{D}[B_F] = \mathrm{Dom}(B_F^{1/2}), \; \dot{B}[u, v] = (B_F^{1/2}u, B_F^{1/2}v), \; u, v \in \mathcal{D}[\dot{B}].$$

Let $\dot{B}$ be a non-negative closed densely-defined symmetric operator. Consider the family of symmetric contractions

$$\dot{A}^{(a)} = (aI - \dot{B})(aI + \dot{B})^{-1}, \; a > 0,$$

defined on $\mathrm{Dom}(\dot{A}^{(a)}) = (aI + \dot{B})\mathrm{Dom}(\dot{B})$. Notice that the orthogonal complement $\mathfrak{N}^{(a)} = \mathcal{H} \ominus \mathrm{Dom}(\dot{A}^{(a)})$ coincides with the defect subspace $\mathfrak{N}_{-a}$ of the operator $\dot{B}$. Let $\dot{A} = \dot{A}^{(1)}$ and let $b = (1 - a)(a + 1)^{-1}$. Then $b \in (-1, 1)$ and (see (9.58))

$$\dot{A}^{(a)} = (\dot{A} - bI_{\mathcal{H}})(I - b\dot{A})^{-1} = w_b(\dot{A}).$$

Clearly, there is a one-one correspondence given by the Cayley transform

$$B = a(I - A^{(a)})(I + A^{(a)})^{-1}, \quad A^{(a)} = (aI - B)(aI + B)^{-1},$$

between all non-negative self-adjoint extensions B of the operator $\dot{B}$ and all *sc*-extensions $A^{(a)}$ of $\dot{A}^{(a)}$. The next result describe the sesquilinear form $B[u, v]$ by means of the fractional-linear transformation $A = (I - B)(I + B)^{-1}$.

Proposition 9.5.1. (1) *Let B be a non-negative self-adjoint operator and let $A = (I - B)(I + B)^{-1}$ be its Cayley transform. Then*

$$\begin{aligned} \mathcal{D}[B] &= \mathrm{Ran}((I + A)^{1/2}), \\ B[u, v] &= -(u, v) + 2\left((I + A)^{-1/2}u, (I + A)^{-1/2}v\right), \quad u, v \in \mathcal{D}[B]. \end{aligned} \tag{9.89}$$

(2) *Let $\dot{B}$ be a closed densely-defined non-negative symmetric operator and let B be its non-negative self-adjoint extension. If $\dot{A} = (I - \dot{B})(I + \dot{B})^{-1}$, $A = (I - B)(I + B)^{-1}$, then*

$$\mathcal{D}[B] = \mathrm{Ran}(I + A_\mu)^{1/2} \dotplus \mathrm{Ran}(A - A_\mu)^{1/2}. \tag{9.90}$$

Proof. (1). Since $B = (I - A)(I + A)^{-1}$, one obtains with $f = (I + A)h$,

$$\begin{aligned} B[f] &= ((I - A)h, (I + A)h) = -\|(I + A)h\|^2 + 2\|(I + A)^{1/2}h\|^2 \\ &= -\|f\|^2 + 2\|(I + A)^{-1/2}f\|^2. \end{aligned}$$

Now the closure procedure leads to (9.89).

(2) Since A is an *sc*-extension of $\dot{A}$, we get $A_\mu \leq A \leq A_M$. Hence $I + A = I + A_\mu + (A - A_\mu)$. Because $I + A_\mu$ and $A - A_\mu$ are non-negative self-adjoint operators, we get the equality [123]:

$$\mathrm{Ran}((I + A)^{1/2}) = \mathrm{Ran}((I + A_\mu)^{1/2}) + \mathrm{Ran}((A - A_\mu)^{1/2}).$$

Since $\mathrm{Ran}((I+A_\mu)^{1/2}) \cap \mathfrak{N} = \{0\}$, where $\mathfrak{N} = \mathcal{H} \ominus \mathrm{Dom}(\dot{A})$, and $\mathrm{Ran}(A - A_\mu) \subseteq \mathfrak{N}$, we get $\mathrm{Ran}((I + A_\mu)^{1/2}) \cap \mathrm{Ran}((A - A_\mu)^{1/2}) = \{0\}$. Then we arrive at (9.90). □

We note that $\mathrm{Ran}(B^{1/2}) = \mathrm{Ran}((I - A)^{1/2})$. Now let A_μ and A_M be the rigid and the soft extensions of $\dot{A}$. Then the operators

$$B_F = (I - A_\mu)(I + A_\mu)^{-1}, \tag{9.91}$$

and

$$B_K = (I - A_M)(I + A_M)^{-1}, \tag{9.92}$$

are non-negative self-adjoint extensions of $\dot{B}$. It also follows from Proposition 9.2.17 that

$$B_F = a(I - A_\mu^{(a)})(I + A_\mu^{(a)})^{-1}, \quad B_K = a(I - A_M^{(a)})(I + A_M^{(a)})^{-1}.$$

Since, the operators $A_\mu^{(a)}$ and $A_M^{(a)}$ possess the properties

$$\operatorname{Ran}((I + A_\mu^{(a)})^{1/2}) \cap \mathfrak{N}_{-a} = \operatorname{Ran}((I - A_M^{(a)})^{1/2}) \cap \mathfrak{N}_{-a} = \{0\},$$

we get the following result.

Proposition 9.5.2. *Let B be a non-negative self-adjoint extension of $\dot{B}$ and let $E(\lambda)$ be its resolution of identity. Then*

1. *$B = B_F$ if and only if at least for one $a > 0$ (then for all $a > 0$) the relation*

$$\int_0^\infty \lambda (dE(\lambda)\varphi, \varphi) = +\infty, \tag{9.93}$$

holds for each $\varphi \in \mathfrak{N}_{-a} \setminus \{0\}$;

2. *$B = B_K$ if and only if at least for one $a > 0$ (then for all $a > 0$) the relation*

$$\int_0^\infty \frac{(dE(\lambda)\varphi, \varphi)}{\lambda} = +\infty, \tag{9.94}$$

holds for each $\varphi \in \mathfrak{N}_{-a} \setminus \{0\}$.

Theorem 9.5.3. *For $a > 0$,*

$$\begin{aligned}
&\inf_{\psi \in \operatorname{Dom}(\dot{B})} \left\{(B_F(h-\psi), h-\psi) + a\|h-\psi\|^2\right\} = 0, \ \forall h \in \operatorname{Dom}(B_F),\\
&\inf_{\psi \in \operatorname{Dom}(\dot{B})} \left\{\|B_K(g-\psi)\|^2 + a(B_K(g-\psi), g-\psi)\right\} = 0, \ \forall g \in \operatorname{Dom}(B_K).
\end{aligned}$$

Proof. By (9.91) we have

$$\begin{aligned}
&\operatorname{Dom}(B_F) \ni h = (I + A_\mu^{(a)})f, \ B_F h = a(I - A_\mu^{(a)})f,\\
&\operatorname{Dom}(B_K) \ni g = (I + A_M^{(a)})f, \ B_K g = a(I - A_M^{(a)})f, \ f \in \mathcal{H}.
\end{aligned}$$

Then

$$\begin{aligned}
&(B_F(h-\psi), h-\psi) + a\|h-\psi\|^2 = 2a((I + A_\mu)(f-\varphi), f-\varphi),\\
&\|B_K(g-\psi)\|^2 + a(B_K(g-\psi), g-\psi) = 2a^2((I - A_M)(f-\varphi), f-\varphi),
\end{aligned}$$

where $\psi = (I + \dot{A})\varphi$, $\varphi \in \operatorname{Dom}(\dot{A}^{(a)})$. The statement follows from Propositions 9.2.5 and 9.2.17. □

Thus, the self-adjoint extension B_F given by (9.91) coincides with the *Friedrichs extension* of $\dot{B}$. In the sequel we will call the operator B_K defined in (9.92) the **Kreĭn-von Neumann extension** of $\dot{B}$. As we already know (Theorem 9.2.4), the set of all *sc*-extensions of a symmetric contraction $\dot{A}$ forms the operator interval $[A_\mu, A_M]$. This result yields the following theorem established by M. Kreĭn [172].

Theorem 9.5.4. *The following conditions are equivalent:*

(i) *the non-negative self-adjoint operator B is the extension of $\dot{B}$,*

(ii)
$$(B_F + aI)^{-1} \le (B + aI)^{-1} \le (B_K + aI)^{-1},$$
for some (then for all) positive numbers a,

(iii) *$B_F \le B \le B_K$ in the sense of quadratic forms, i.e.,*
$$\begin{aligned} &\mathcal{D}[\dot{B}] \subseteq \mathcal{D}[B] \subseteq \mathcal{D}[B_K], \quad B[u] \ge B_K[u] \quad \text{for all} \quad u \in \mathcal{D}[B], \\ &B[v] = \dot{B}[v] \quad \text{for all} \quad v \in \mathcal{D}[\dot{B}]. \end{aligned}$$

The operator $\dot{B}$ admits a unique non-negative self-adjoint extension if and only if
$$\inf_{v \in \mathrm{Dom}(\dot{B})} \frac{(\dot{B}v, v)}{|(v, \varphi_{-a})|^2} = 0, \tag{9.95}$$
for all non-zero vectors φ_{-a} from the defect subspace $\mathfrak{N}_{-a}$ of $\dot{B}$, where $a > 0$.

Proof. Let B be a non-negative self-adjoint extension of $\dot{B}$ and let $a > 0$. Then $A^{(a)} = (aI - B)(aI + B)^{-1} = -I + 2a(aI + B)^{-1}$ is an *sc*-extension of symmetric contraction $\dot{A}^{(a)} = (aI - \dot{B})(aI + \dot{B})^{-1}$. It follows that
$$A_\mu^{(a)} \le A^{(a)} \le A_M^{(a)} \iff (B_F + aI)^{-1} \le (B + aI)^{-1} \le (B_K + aI)^{-1}.$$
Suppose B is a non-negative self-adjoint operator in $\mathcal{H}$ such that
$$(B_F + aI)^{-1} \le (B + aI)^{-1} \le (B_K + aI)^{-1},$$
for some $a > 0$. Since
$$(B_F + aI)^{-1} f = (B_K + aI)^{-1} f = (\dot{B} + aI)^{-1} f,$$
for all $f \in (\dot{B} + aI)\mathrm{Dom}(\dot{B})$, the latter inequalities yield that B is an extension of $\dot{B}$. By Theorem 9.2.13 the operator $\dot{A}^{(a)}$ admits a unique *sc*-extension if and only if
$$\sup_{\varphi \in \mathrm{Dom}(\dot{A}^{(a)})} \frac{\left|(\dot{A}^{(a)}\varphi, h)\right|^2}{\|D_{\dot{A}^{(a)}}\varphi\|^2} = \infty \quad \text{for all} \quad h \in \mathcal{H} \ominus \mathrm{Dom}(\dot{A}^{(a)}),\ h \ne 0.$$
Because $\mathcal{H} \ominus \mathrm{Dom}(\dot{A}^{(a)}) = \mathfrak{N}_{-a}$ (the defect subspace of $\dot{B}$),
$$\|D_{\dot{A}^{(a)}}\varphi\|^2 = (\dot{B}v, v), \quad \varphi = (aI + \dot{B})v,$$
and $(\dot{A}^{(a)}\varphi, \varphi_{-a}) = ((aI - \dot{B})v, \varphi_{-a}) = 2a(v, \varphi_{-a})$ for all $\varphi_{-a} \in \mathfrak{N}_{-a}$, we get that $\dot{B}$ admits a unique non-negative self-adjoint extension if and only if condition (9.95) is satisfied.

Let B be a non-negative self-adjoint extension of $\dot{B}$ and let

$$\begin{aligned}&A_\mu = (I - B_F)(I + B_F)^{-1}, \quad A_M = (I - B_K)(I + B_K)^{-1},\\ &A = (I - B)(I + B)^{-1}.\end{aligned}$$

Then $I + A_\mu \leq I + A \leq I + A_M$. It follows from Theorem 2.1.2 that

$$\operatorname{Ran}((I + A_\mu)^{1/2}) \subseteq \operatorname{Ran}((I + A)^{1/2}) \subseteq \operatorname{Ran}((I + A_M)^{1/2}),$$

and

$$\begin{aligned}&\|(I + A_M)^{-1/2} f\|^2 \leq \|(I + A)^{-1/2} f\|^2, \ f \in \operatorname{Ran}((I + A)^{1/2}),\\ &\|(I + A)^{-1/2} g\|^2 \leq \|(I + A_\mu)^{-1/2} g\|^2, \ g \in \operatorname{Ran}((I + A_\mu)^{1/2}).\end{aligned}$$

By Proposition 9.5.1 we get $\mathcal{D}[\dot{B}] = \mathcal{D}[B_F] \subseteq \mathcal{D}[B] \subseteq \mathcal{D}[B_K]$, and

$$\begin{aligned}&B[u] \geq B_K[u] \quad \text{for all} \quad u \in \mathcal{D}[B],\\ &B[v] \leq \dot{B}[v] \quad \text{for all} \quad v \in \mathcal{D}[\dot{B}].\end{aligned}$$

In addition, using the first equality from (9.49), we obtain $B[v] = \dot{B}[v]$ for all $v \in \mathcal{D}[\dot{B}]$. □

Remark 9.5.5. Let C_1 and C_2 be two non-negative self-adjoint operators. It is well known (see [163]) that the following conditions are equivalent

(i) $C_1 \leq C_2$ in the sense of quadratic forms;

(ii) $(C_1 + aI)^{-1} \geq (C_2 + aI)^{-1}$ for some (then for all) positive number a.

Remark 9.5.6. It follows from (9.3) and the definition of the Friedrichs extension that

$$\inf_{v \in \operatorname{Dom}(\dot{B})} \frac{(\dot{B}v, v)}{|(v, \varphi_{-a})|^2} = 0 \iff \varphi_{-a} \notin \operatorname{Ran}(B_F^{1/2}).$$

Therefore, $\dot{B}$ admits a unique non-negative self-adjoint extension if and only if $\operatorname{Ran}(B_F^{1/2}) \cap \mathfrak{N}_{-a} = \{0\}$ at least for one (then for all) $a > 0$.

From Proposition 9.5.1 and relations (9.89), (9.90) we obtain

$$\begin{aligned}&\mathcal{D}[B_F] = \operatorname{Ran}((I + A_\mu^{(a)})^{1/2}),\\ &\mathcal{D}[B_K] = \operatorname{Ran}((I + A_\mu^{(a)})^{1/2}) \dot{+} \operatorname{Ran}((A_M^{(a)} - A_\mu^{(a)})^{1/2}), \ a > 0.\end{aligned} \tag{9.96}$$

Moreover, since $\operatorname{Ran}((A^{(a)} - A_\mu^{(a)})^{1/2}) \subseteq \mathfrak{N}_{-a}$, the direct decomposition

$$\mathcal{D}[B] = \mathcal{D}[B_F] \dot{+} (\mathfrak{N}_{-a} \cap \mathcal{D}[B]), \tag{9.97}$$

holds for an arbitrary non-negative self-adjoint extension B of $\dot{B}$.

Using Proposition 9.5.1 and Theorem 9.5.3 we obtain the following theorem by T. Ando and K. Nishio [16].

Theorem 9.5.7. *The following relations describing $\mathcal{D}[B_K]$ and $B_K[u]$ hold:*

$$\begin{aligned} \mathcal{D}[B_K] &= \left\{ u \in \mathcal{H} : \sup_{f \in \mathrm{Dom}(\dot{B})} \frac{|(\dot{B}f, u)|^2}{(\dot{B}f, f)} < \infty \right\}, \\ B_K[u] &= \sup_{f \in \mathrm{Dom}(\dot{B})} \frac{|(\dot{B}f, u)|^2}{(\dot{B}f, f)}, \quad u \in \mathcal{D}[B_K]. \end{aligned} \tag{9.98}$$

Proof. Let $u \in \mathcal{D}[B_K] = \mathrm{Dom}(B_K^{1/2})$. Then for all $f \in \mathrm{Dom}(\dot{B})$ one has

$$\frac{|(\dot{B}f, u)|^2}{(\dot{B}f, f)} = \frac{|(B_K f, u)|^2}{\|B_K^{1/2} f\|^2} = \frac{|(B_K^{1/2} f, B_K^{1/2} u)|^2}{\|B_K^{1/2} f\|^2}.$$

From Theorem 9.5.3 we obtain that $B_K^{1/2}\mathrm{Dom}(\dot{B})$ is a dense set in $\mathrm{Ran}(B_K^{1/2})$. Hence,

$$\sup_{f \in \mathrm{Dom}(\dot{B})} \frac{|(B_K^{1/2} f, B_K^{1/2} u)|^2}{\|B_K^{1/2} f\|^2} = \|B_K^{1/2} u\|^2 = B_K[u].$$

We have proved that $\mathcal{D}[B_K] \subset B_K[u]$. In order to prove the inverse inclusion we use the Cayley transform

$$\dot{A} = (I - \dot{B})(I + \dot{B})^{-1}, \quad A_M = (I - B_K)(I + B_K)^{-1}.$$

Let $u \in \mathcal{H}$ and let

$$\sup_{f \in \mathrm{Dom}(\dot{B})} \frac{|(\dot{B}f, u)|^2}{(\dot{B}f, f)} < \infty.$$

This is equivalent to

$$\sup_{\varphi \in \mathrm{Dom}(\dot{A})} \frac{|((I - \dot{A})\varphi, u)|^2}{\|D_{\dot{A}}\varphi\|^2} < \infty,$$

i.e., $|((I - \dot{A})\varphi, u)|^2 \le c\|D_{\dot{A}}\varphi\|^2$ for all $\varphi \in \mathrm{Dom}(\dot{A})$, where $c > 0$. Since

$$\begin{aligned} &(I - \dot{A})\varphi = (I - A_M)^{1/2}(I - A_M)^{1/2}\varphi, \\ &\|D_{\dot{A}}\varphi\|^2 = \|D_{A_M}\varphi\|^2 = \|(I + A_M)^{1/2}(I - A_M)^{1/2}\varphi\|^2, \end{aligned}$$

and $(I - A_M)^{1/2}\mathrm{Dom}(\dot{A})$ is dense in $\mathrm{Ran}((I - A_M)^{1/2})$ (see Proposition 9.2.5), we have

$$|((I - A_M)h, u)|^2 \le c\|D_{A_M}h\|^2,$$

for all $h \in \mathcal{H}$. It follows that for $h \in \mathcal{H}$,

$$\begin{aligned} &|((I - A_M)^{1/2}h, (I - A_M)^{1/2}u)|^2 \le c\|(I + A_M)^{1/2}(I - A_M)^{1/2}h\|^2 \\ &\iff |(g, (I - A_M)^{1/2}u)|^2 \le c\|(I + A_M)^{1/2}g\|^2, g \in \mathcal{H}. \end{aligned}$$

Now from (9.3) we obtain $(I - A_M)^{1/2}u \in \mathrm{Ran}((I + A_M)^{1/2})$. Since

$$\mathrm{Ran}((I - A_M)^{1/2} \cap \mathrm{Ran}((I + A_M)^{1/2}) = \mathrm{Ran}(D_{A_M}),$$

we get $u \in \mathrm{Ran}((I + A_M)^{1/2})$. By Proposition 9.5.1,

$$\mathcal{D}[B_K] = \mathrm{Ran}((I + A_M)^{1/2}).$$

This completes the proof. □

Remark 9.5.8. The equality $B[u, v] = \dot{B}[u, v]$, $u, v \in \mathcal{D}[\dot{B}]$ and the Second Representation Theorem yield

$$||B^{1/2}u|| = ||B_F^{1/2}u||, \quad u \in \mathrm{Dom}(B_F^{1/2}) = \mathcal{D}[\dot{B}].$$

Let $\dot{u} \in \mathcal{D}[\dot{B}]$ and let $f \in \mathcal{D}[B] \cap \mathrm{Dom}(\dot{B}^*)$. Then

$$(B^{1/2}u, B^{1/2}f) = B[\dot{u}, f] = (\dot{B}\dot{u}, f) = (\dot{u}, \dot{B}^* f).$$

On the other hand for each $u \in \mathcal{D}[\dot{B}]$ there exists a sequence $\{\dot{u}_n\}$ such that

$$\lim_{n\to\infty} \dot{u}_n = u, \ \lim_{n\to\infty} B_F^{1/2}\dot{u}_n = B_F^{1/2}u.$$

Since $B^{1/2}$ is closed and $\mathrm{Dom}(B^{1/2}) \supset \mathrm{Dom}(B_F^{1/2})$, we obtain

$$\lim_{n\to\infty} B^{1/2}\dot{u}_n = B^{1/2}u.$$

Therefore for any non-negative self-adjoint extension B of $\dot{B}$ one has

$$B[u, f] = (u, \dot{B}^* f), \ u \in \mathcal{D}[\dot{B}], \ f \in \mathcal{D}[B] \cap \mathrm{Dom}(\dot{B}^*). \tag{9.99}$$

Example. Let $\mathcal{H} = L_2(\mathbb{R}^2)$. Consider a self-adjoint and non-negative operator

$$(\hat{A}\hat{f})(p) = |p|^2 \hat{f}(p), \quad \mathrm{Dom}(\hat{A}) = \left\{\hat{f}(p) \in L_2(\mathbb{R}^2) : |p|^2 \hat{f}(p) \in L_2(\mathbb{R}^2)\right\},$$

Here and below $p = (p_1, p_1)$, $dp = dp_1 dp_2$, $|p| = \sqrt{p_1^2 + p_2^2}$. Clearly,

$$(\hat{A}^{1/2}\hat{f})(p) = |p|\hat{f}(p), \ \hat{f} \in \mathrm{Dom}(\hat{A}^{1/2}) = \left\{\hat{f}(p) \in L_2(\mathbb{R}^2) : |p|\hat{f}(p) \in L_2(\mathbb{R}^2)\right\}.$$

Let a symmetric operator $\hat{A}_0$ be given by

$$(\hat{A}_0\hat{f})(p) = |p|^2 \hat{f}(p), \quad \mathrm{Dom}(\hat{A}_0) = \{\hat{f} \in \mathrm{Dom}(\hat{A}) : \int_{\mathbb{R}^2} \hat{f}(p)dp = 0\}.$$

The deficiency indices of $\hat{A}_0$ are $(1, 1)$ and

$$\mathfrak{N}_\lambda = \left\{\frac{c}{|p|^2 - \lambda}, \ c \in \mathbb{C}\right\}, \quad \lambda \in \mathbb{C} \setminus \mathbb{R}_+$$

is a defect subspace of $\hat{A}_0$. Obviously, $\mathfrak{N}_{-1} \cap \mathrm{Dom}(\hat{A}^{1/2}) = \{0\}$. Hence, from (9.97) it follows that $\hat{A}$ is the Friedrichs extension of $\hat{A}_0$. Moreover, since

$$\frac{1}{|p|(|p|^2+1)} \notin L_2(\mathbb{R}^2),$$

we get $\mathfrak{N}_{-1} \cap \mathrm{Ran}(\hat{A}^{1/2}) = \{0\}$. Remark 9.5.6 yields that $\hat{A}$ is the unique nonnegative self-adjoint extension of $\hat{A}_0$. The application of the Fourier transform

$$\boldsymbol{\mathcal{F}} f(x) := \hat{f}(p) = \frac{1}{2\pi}\int_{\mathbb{R}^2} f(x)\exp(-ip\cdot x)\,dx,$$

to $\hat{A}_0$ and $\hat{A}$ leads to the operators A_0 and A in $L_2(\mathbb{R}^2, dx)$, $x = (x_1, x_2)$:

$$A_0 = \boldsymbol{\mathcal{F}}^{-1}\hat{A}_0\boldsymbol{\mathcal{F}}, \quad A = \boldsymbol{\mathcal{F}}^{-1}\hat{A}\boldsymbol{\mathcal{F}},$$

$$\mathrm{Dom}(A_0) = \{f \in W_2^2(\mathbb{R}^2), f(0) = 0\},\ A_0 f = -\Delta f,\ f \in \mathrm{Dom}(A_0), \tag{9.100}$$

and

$$\mathrm{Dom}(A) = W_2^2(\mathbb{R}^2), \quad Af = -\Delta f, \quad f \in \mathrm{Dom}(A),$$

where $W_2^2(\mathbb{R}^2)$ is the Sobolev space and Δ is the Laplacian. Since $\boldsymbol{\mathcal{F}}$ is a unitary operator, the operator A_0 has a unique non-negative self-adjoint extension A.

Theorem 9.5.9. *The Friedrichs and Kreĭn-von Neumann extensions of $\dot{B}$ are transversal if and only if*

$$\mathrm{Dom}(\dot{B}^*) \subseteq \mathcal{D}[B_K]. \tag{9.101}$$

Proof. Let $\mathfrak{N}_\lambda$ be the defect subspace of $\dot{B}$. Assume condition (9.101). Then $\mathfrak{N}_\lambda \subset \mathcal{D}[B_K]$ for all $\lambda \in \mathbb{C} \setminus [0, +\infty)$. In particular $\mathfrak{N}_{-1} \subset \mathcal{D}[B_K]$. It follows from (9.96) that

$$\mathrm{Ran}((A_M - A_\mu)^{1/2}) = \mathfrak{N}_{-1}.$$

Hence, also $\mathrm{Ran}(A_M - A_\mu) = \mathfrak{N}_{-1}$. Because

$$2(B_F + I)^{-1} = I + A_\mu,\ 2(B_K + I)^{-1} = I + A_M,$$

we have

$$(B_K + I)^{-1} - (B_F + I)^{-1} = \frac{1}{2}(A_M - A_\mu).$$

Therefore $\mathrm{Ran}\left((B_K + I)^{-1} - (B_F + I)^{-1}\right) = \mathfrak{N}_{-1}$. Now Proposition 3.4.8 yields that B_F and B_K are transversal.

Conversely, suppose that B_F and B_K are transversal, i.e., that $\mathrm{Dom}(B_F) + \mathrm{Dom}(B_K) = \mathrm{Dom}(\dot{B}^*)$. Since

$$\mathrm{Dom}(B_F) \subset \mathcal{D}[B_F] = \mathcal{D}[\dot{B}] \subseteq \mathcal{D}[B_K]$$

and $\mathrm{Dom}(B_K) \subset \mathcal{D}[B_K]$, we obtain $\mathrm{Dom}(\dot{B}^*) \subseteq \mathcal{D}[B_K]$. □

Now we consider the case of bounded non-densely-defined non-negative symmetric operator $\dot{B}$.

Theorem 9.5.10. *Let $\dot{B}$ be a bounded, non-densely-defined, non-negative, and symmetric operator in a Hilbert space $\mathcal{H}$, $\mathrm{Dom}(\dot{B}) = \mathcal{H}_0$. Let $\dot{B}^* \in [\mathcal{H}, \mathcal{H}_0]$ be the adjoint of $\dot{B}$. Put $\dot{B}_0 = P_{\mathcal{H}_0}\dot{B}$, $\mathfrak{L} = \mathcal{H} \ominus \mathcal{H}_0$, where $P_{\mathcal{H}_0}$ is an orthogonal projection in $\mathcal{H}$ onto $\mathcal{H}_0$. Then the following statements are equivalent:*

(i) *$\dot{B}$ admits bounded non-negative self-adjoint extensions in $\mathcal{H}$;*

(ii) $\sup\limits_{f\in\mathcal{H}_0} \dfrac{||\dot{B}f||^2}{(\dot{B}f,f)} < \infty$;

(iii) $\dot{B}^*\mathfrak{L} \subseteq \mathrm{Ran}(\dot{B}_0^{1/2})$.

Proof. Since $(\dot{B}f,f) = ||\dot{B}_0^{1/2}f||^2$, $f \in \mathcal{H}_0$, and

$$\dot{B}^* = \dot{B}_0 P_{\mathcal{H}_0} + \dot{B}^* P_{\mathfrak{L}},$$

conditions (i) and (ii) are equivalent due to Theorem 2.1.2. Suppose $\dot{B}$ admits a bounded non-negative self-adjoint extension B. Then for $f \in \mathcal{H}_0$ one has

$$\begin{aligned}||\dot{B}f||^2 = ||Bf||^2 = ||B^{1/2}B^{1/2}f||^2 &\le ||B^{1/2}||^2||B^{1/2}f||^2 \\ &= ||B^{1/2}||^2(Bf,f) = ||B^{1/2}||^2(\dot{B}f,f) = ||B^{1/2}||^2||\dot{B}_0^{1/2}f||^2.\end{aligned}$$

It follows that statement (ii) holds true.

Now suppose that (iii) is fulfilled. Then the operator $L_0 := \dot{B}_0^{[-1/2]}\dot{B}^*\upharpoonright\mathfrak{L}$ is bounded, where $\dot{B}_0^{[-1/2]}$ is the Moore-Penrose inverse to $\dot{B}_0^{1/2}$. Let $L_0^* \in [\mathcal{H}_0, \mathfrak{L}]$ be the adjoint to L_0. Set

$$\mathcal{B}_0 = \dot{B}P_{\mathcal{H}_0} + (\dot{B}^* + L_0^*L_0)P_{\mathfrak{L}}. \tag{9.102}$$

Then $\mathcal{B}_0$ is a bounded extension of $\dot{B}$ in $\mathcal{H}$. Let $P_{\mathfrak{L}}$ be the orthogonal projection operator in $\mathcal{H}$ onto $\mathfrak{L}$. For $h \in \mathcal{H}$ we have

$$\begin{aligned}&(\mathcal{B}_0h,h) = (\dot{B}P_{\mathcal{H}_0}h + (\dot{B}^* + L_0^*L_0)P_{\mathfrak{L}}h, P_{\mathcal{H}_0}h + P_{\mathfrak{L}}h) \\ &= ||\dot{B}_0^{1/2}P_{\mathcal{H}_0}h||^2 + ||L_0P_{\mathfrak{L}}h||^2 + 2\mathrm{Re}\,(P_{\mathcal{H}_0}h, \dot{B}^*P_{\mathfrak{L}}h) \\ &= ||\dot{B}_0^{1/2}P_{\mathcal{H}_0}h||^2 + ||L_0P_{\mathfrak{L}}h||^2 + 2\mathrm{Re}\,(\dot{B}_0^{1/2}P_{\mathcal{H}_0}h, \dot{B}_0^{[-1/2]}\dot{B}^*P_{\mathfrak{L}}h) \\ &= ||\dot{B}_0^{1/2}P_{\mathcal{H}_0}h + L_0P_{\mathfrak{L}}h||^2.\end{aligned}$$

Thus, $\mathcal{B}_0$ is a non-negative bounded self-adjoint extension of $\dot{B}$. Therefore (i) is equivalent to (iii). □

Remark 9.5.11. 1) It is easy to see that the conditions

1. $\sup\limits_{f\in H_0} \dfrac{||\dot{B}f||^2}{(\dot{B}f,f)} < \infty$,

2. there exists $c > 0$ such that

$$|(\dot{B}f, g)|^2 \leq c(\dot{B}f, f)||g||^2, \ f \in \mathcal{H}_0, g \in \mathcal{H},$$

3. there exists $c > 0$ such that

$$|(\dot{B}f, g)|^2 \leq c(\dot{B}f, f)||g||^2, \ f \in \mathcal{H}_0, g \in \mathfrak{L}$$

are equivalent.

2) If $\mathcal{B}_0$ is given by (9.102) and $W \in [\mathfrak{L}, \mathfrak{L}]$, $W \geq 0$, then the operator

$$B = \mathcal{B}_0 + WP_{\mathfrak{L}},$$

is also a non-negative bounded self-adjoint extension of $\dot{B}$.

A linear operator T on a Hilbert space $\mathcal{H}$ is called **accretive** if

$$\operatorname{Re}(Tf, f) \geq 0, \qquad f \in \operatorname{Dom}(T).$$

An accretive operator is called **maximal accretive** or **m-accretive** if it does not admit accretive extensions in $\mathcal{H}$. The following conditions for accretive operator T are equivalent:

- the operator T is m-accretive;
- the resolvent set $\rho(T)$ contains a point from the open left half-plane;
- the operator T is densely defined and closed, and T^* is an accretive operator.

The resolvent set $\rho(T)$ of m-accretive operators contains the open left half-plane and

$$\|(T - \lambda I)^{-1}\| \leq \frac{1}{|\operatorname{Re}\lambda|}, \qquad \operatorname{Re}\lambda < 0.$$

It is well known [163] that if T is an m-accretive operator, then the one-parameter semigroup

$$U(t) = \exp(-tT), \quad t \geq 0,$$

is contractive. Conversely, if the family $\{U(t)\}_{t \geq 0}$ is a strongly continuous semigroup of bounded operators in a Hilbert space $\mathcal{H}$, with $U(0) = I$ (C_0-semigroup) and $U(t)$ is a contraction for each t, then the generator T of $U(t)$:

$$Tu := \lim_{t \to +0} \frac{(I - U(t))u}{t}, \quad u \in \operatorname{Dom}(T),$$

where domain $\operatorname{Dom}(T)$ is defined by condition:

$$\operatorname{Dom}(T) = \left\{ u \in \mathcal{H} : \lim_{t \to +0} \frac{(I - U(t))u}{t} \text{ exists} \right\},$$

is an m-accretive operator in $\mathcal{H}$. Then the approximation by the Euler formula

$$U(t) = s - \lim_{n\to\infty} \left(I + \frac{t}{n}T\right)^{-n}, \quad t \geq 0,$$

(see [163]) holds in the strong operator topology.

The formulas

$$S = (I - T)(I + T)^{-1}, \quad T = (I - S)(I + S)^{-1},$$

establish a bijective correspondence between all m-accretive operators T and all contractions S such that $\ker(I + S) = \{0\}$. Indeed, if $f = (I + T)^{-1}\phi$, then $S\phi = f - Tf, \quad \phi = f + Tf$, and

$$||\phi||^2 - ||S\phi||^2 = 4\text{Re}\,(Tf, f). \tag{9.103}$$

We call an accretive operator T **α-sectorial** if there exists a value of $\alpha \in (0, \pi/2)$ such that

$$|\text{Im}\,(Tf, f)| \leq \tan\alpha\,\text{Re}\,(Tf, f), \qquad f \in \text{Dom}(T). \tag{9.104}$$

It is known [163] that if T is an m-accretive and α-sectorial operator, then the operator $(-T)$ is a generator of a semi-group of contractions holomorphic in the sector $|\arg\zeta| < \pi/2 - \alpha$, i.e., the Cauchy problem

$$\begin{cases} \dfrac{dx}{dt} + Tx = 0, & \\ x(0) = 0, & x_0 \in \text{Dom}(T), \end{cases} \tag{9.105}$$

generates a contractive semi-group $U(t) = \exp(-tT), t \geq 0$ that can be analytically extended as a semi-group of contractions holomorphic in the sector $|\arg\zeta| < \pi/2 - \alpha$.

For the remainder of this chapter we will mostly consider m-accretive and both m-accretive and α-sectorial operators (**m-α-sectorial**).

Lemma 9.5.12. *Let T be a densely-defined m-accretive operator on a Hilbert space $\mathcal{H}$. Then the operator*

$$S = (I - T)(I + T)^{-1},$$

belongs to the class $C_{\mathcal{H}}(\alpha)$ (is an α-co-sectorial contraction) if and only if T is an α-sectorial operator.

Proof. Let $f = (I + T)^{-1}\phi$, then

$$S\phi = f - Tf, \quad \phi = f + Tf.$$

Hence,

$$f = \frac{1}{2}(I + S)\phi, \quad Tf = \frac{1}{2}(I - S)\phi.$$

Moreover,

$$(Tf,f) = \left(\frac{1}{2}(I-S)\phi, \frac{1}{2}(I+S)\phi\right) = \left(\frac{1}{4}(I+S^*)(I-S)\phi, \phi\right) = (F\phi,\phi),$$

where $F = \frac{1}{4}(I+S^*)(I-S)$. Let us find $\mathrm{Re}\, F$ and $\mathrm{Im}\, F$. Using the definition of operator S we have

$$\mathrm{Re}\, F = \frac{1}{2}\left[\frac{1}{4}(I+S^*)(I-S) + \frac{1}{4}(I-S^*)(I+S)\right] = \frac{1}{4}(I-S^*S),$$

and

$$\mathrm{Im}\, F = \frac{1}{2i}\left[\frac{1}{4}(I+S^*)(I-S) - \frac{1}{4}(I-S^*)(I+S)\right] = \frac{1}{2}\,\frac{S-S^*}{2i}.$$

Now let

$$\cot\alpha\, |\mathrm{Im}\,(Tf,f)| \le \mathrm{Re}\,(Tf,f), \qquad f \in \mathrm{Dom}(T),$$

which would imply that

$$\cot\alpha\, |2(\mathrm{Im}\,(S\phi,\phi)| \le ((I-S^*S)\phi,\phi),$$

which proves the lemma. □

Let $\dot{B}$ be a closed, densely-defined non-negative symmetric operator in Hilbert space $\mathcal{H}$. The **Phillips-Kato extension problem in the restricted sense** consists of existence and description of m-accretive and α-sectorial extensions T of $\dot{B}$ such that

$$\dot{B} \subset T \subset \dot{B}^*. \tag{9.106}$$

Obviously, the operators T satisfying (9.106) are quasi-self-adjoint extensions of symmetric operator $\dot{B}$. One can easily show that the fractional-linear transformation

$$B = (I-A)(I+A)^{-1}, \quad A = (I-B)(I+B)^{-1},$$

establishes a one-to-one correspondence between all quasi-self-adjoint m-accretive extensions B of $\dot{B}$ and all qsc-extensions of $\dot{A} = (I-\dot{B})(I+\dot{B})^{-1}$.

Theorem 9.5.13. *Let $\dot{B}$ be a densely-defined, closed, and non-negative operator. The following conditions are equivalent:*

(i) $B_F \ne B_K$,

(ii) *$\dot{B}$ admits a non-self-adjoint quasi-self-adjoint m-accretive extension,*

(iii) *admits a non-self-adjoint quasi-self-adjoint m-α-sectorial extension for any $\alpha \in (0, \pi/2)$.*

Proof. It follows from (9.91) and (9.92) that $B_F = B_K$, and hence $A_F = A_K$. The proof of the theorem then follows from Corollaries 9.2.14 and 9.2.15. □

The next theorem gives characterizations of quasi-self-adjoint m-accretive extensions.

Theorem 9.5.14. *Let $\dot{B}$ be a non-negative densely-defined symmetric operator and let B be a maximal accretive extension of $\dot{B}$. The following conditions are equivalent:*

(1) $B \subset \dot{B}^*$;

(2) $\mathrm{Dom}(B) \subset \mathcal{D}[B_K]$ *and* $\mathrm{Re}\,(Bf, f) \geq B_K[f]$ *for all* $f \in \mathrm{Dom}(B)$;

(3) $|(\dot{B}g, f)|^2 \leq (\dot{B}g, g)\,\mathrm{Re}\,(Bf, f)$ *for all* $f \in \mathrm{Dom}(B)$, $g \in \mathrm{Dom}(\dot{B})$.

The extension B is quasi-self-adjoint and m-α-sectorial if and only if

$$\mathrm{Dom}(B) \subset \mathrm{Dom}(B_K^{1/2})$$

and the sesquilinear form

$$\omega[f, h] := (Bf, h) - B_K[f, h], \ f, h \in \mathrm{Dom}(B), \tag{9.107}$$

is sectorial with the same semi-angle α and the vertex at the origin.

Proof. If B is an accretive extension of $\dot{B}$, then for all $g \in \mathrm{Dom}(\dot{B})$, for all $f \in \mathrm{Dom}(B)$, and for all $t \in \mathbb{R}$, it follows that

$$\begin{aligned} 0 &\leq \mathrm{Re}\,(B(tg + f), tf + g) \\ &= t^2(\dot{B}g, g) + t\left(\mathrm{Re}\,(\dot{B}g, f) + \mathrm{Re}\,(Bf, g)\right) + \mathrm{Re}\,(Bf, f). \end{aligned}$$

If in addition $B \subset \dot{B}^*$, then $\mathrm{Dom}(B) \subset \mathrm{Dom}(\dot{B}^*)$ and $(\dot{B}g, f) = (g, Bf)$. Hence

$$t^2(\dot{B}g, g) + 2t\mathrm{Re}\,(\dot{B}g, f) + \mathrm{Re}\,(Bf, f) \geq 0,$$

for all $t \in \mathbb{R}$. Now we get $|\mathrm{Re}\,(\dot{B}g, f)|^2 \leq (\dot{B}g, g)\mathrm{Re}\,(Bf, f)$, and therefore

$$|(\dot{B}g, f)|^2 \leq (\dot{B}g, g)\mathrm{Re}\,(Bf, f),$$

for all $g \in \mathrm{Dom}(\dot{B})$ and all $f \in \mathrm{Dom}(B)$, i.e., (1) $\Rightarrow$ (3). The equivalence (3) $\Longleftrightarrow$ (2) follows from (9.98).

Let us show that (3) implies (1). Let $\dot{A} = (I - \dot{B})(I + \dot{B})^{-1}$ and $A = (I - B)(I + B)^{-1}$. Then $\dot{A}$ is a symmetric contraction defined on $\mathrm{Dom}(\dot{A}) = (I + \dot{B})\mathrm{Dom}(\dot{B})$ and A is a contractive extension of $\dot{A}$ defined on $\mathcal{H}$. Then the inequality in (3) can be rewritten as

$$|((I - \dot{A})\varphi, (I + A)h)|^2 \leq ((I - \dot{A})\varphi, (I + \dot{A})\varphi)\mathrm{Re}\,((I - A)h, (I + A)h),$$

for all $\varphi \in \mathrm{Dom}(\dot{A})$ and all $h \in \mathcal{H}$. Using the relation $\dot{A}\varphi = A\varphi$ and simplifying we obtain

$$|(D_A^2\varphi - 2i\mathrm{Im}\,A\varphi, h)|^2 \leq \|D_A\varphi\|^2\|D_Ah\|^2 \quad \text{for all} \quad \varphi \in \mathrm{Dom}(\dot{A}),$$

and all $h \in \mathcal{H}$. From (9.3) we obtain that $D_A^2\varphi - 2i\mathrm{Im}\, A\varphi \in \mathrm{Ran}(D_A)$ for all $\varphi \in \mathrm{Dom}(\dot{A})$ and

$$\|D_A^{[-1]}(D_A^2\varphi - 2i\mathrm{Im}\, A\varphi)\|^2 \leq \|D_A\varphi\|^2,\ \varphi \in \mathrm{Dom}(\dot{A}),$$

where $D_A^{[-1]}$ is the the Moore-Penrose inverse of D_A described in Lemma 9.1.1. Since $D_A^2\varphi \in \mathrm{Ran}(D_A)$, we get $\mathrm{Im}\, A\varphi \in \mathcal{R}(A)$ and

$$\|D_A\varphi - 2iD_A^{[-1]}\mathrm{Im}\, A\varphi\|^2 \leq \|D_A\varphi\|^2, \quad \varphi \in \mathrm{Dom}(\dot{A}).$$

Hence

$$\|D_A\varphi\|^2 + 4\|D_A^{[-1]}\mathrm{Im}\, A\varphi\|^2 \leq \|D_A\varphi\|^2.$$

It follows that $\mathrm{Im}\, A\varphi = 0$ for all $\varphi \in \mathrm{Dom}(\dot{A})$. This means that $A^* \supset \dot{A}$, i.e., A is a *qsc*-extension of $\dot{A}$. Therefore, B is a quasi-self-adjoint extension of $\dot{B}$.

Suppose that the sesquilinear form ω given by (9.107) is α-sectorial with vertex at the origin. Then $\mathrm{Re}\,(Bf, f) \geq B_K[f]$ for all $f \in \mathrm{Dom}(B)$. Therefore B is a maximal accretive and quasi-self-adjoint extension of $\dot{B}$. On the other hand for all $f \in \mathrm{Dom}(B)$ we have

$$\tan\alpha \mathrm{Re}\,(Bf, f) \pm \mathrm{Im}\,(Bf, f) = \tan\alpha \mathrm{Re}\,\omega[f] \pm \mathrm{Im}\,\omega[f] + B_K[f] \geq 0.$$

Hence B is an m-α-sectorial extension of $\dot{B}$.

Conversely, let B be quasi-self-adjoint and an m-α-sectorial extension of $\dot{B}$. Hence $\mathrm{Dom}(B) \subset \mathcal{D}[B_K]$. Since for each $\varphi \in \mathrm{Dom}(\dot{B}) \subset \mathrm{Dom}(B_K)$ and all $f \in \mathcal{D}[B_K]$ one has $B_K[f, \varphi] = (f, \dot{B}\varphi)$ and $B_K[\varphi, f] = (\dot{B}\varphi, f)$, we get

$$\omega[f - \varphi] = \omega[f],$$

for all $f \in \mathrm{Dom}(B)$ and all $\varphi \in \mathrm{Dom}(\dot{B})$. Because

$$\inf_{\varphi \in \mathrm{Dom}(\dot{B})} B_K[f - \varphi] = 0,$$

for all $f \in \mathcal{D}[B_K]$, for given $f \in \mathcal{D}[B_K]$ and for every $\varepsilon > 0$ one can find $\varphi_0 \in \mathrm{Dom}(\dot{B})$ such that

$$B_K[f - \varphi_0] < \varepsilon.$$

It follows that

$$\begin{aligned}&(\tan\alpha)\mathrm{Re}\ \omega[f] \pm \mathrm{Im}\,\omega[f] = (\tan\alpha)\mathrm{Re}\ \omega[f - \varphi_0] \pm \mathrm{Im}\,\omega[f - \varphi_0]\\&= (\tan\alpha)\mathrm{Re}\,(B(f - \varphi_0, f - \varphi_0) \pm \mathrm{Im}\,(B(f - \varphi_0, f - \varphi_0) - B_K[f - \varphi_0]\\&> -\varepsilon.\end{aligned}$$

Since ε is an arbitrary positive number, the form ω is with the semi-angle α and vertex at the origin. □

Definition 9.5.15. Let $\dot{B}$ be a closed densely-defined non-negative operator. A quasi-self-adjoint m-accretive extension B of $\dot{B}$ is called **extremal** if

$$\inf_{\varphi\in \mathrm{Dom}(\dot{B})} \mathrm{Re}\,(B(f-\varphi), f-\varphi) = 0 \quad \text{for all} \quad f\in \mathrm{Dom}(B).$$

From Proposition 9.2.9, Remark 9.2.10, and relations (9.44), (9.103) we immediately obtain the following statement.

Proposition 9.5.16. *A quasi-self-adjoint m-accretive extension B of a densely-defined symmetric operator $\dot{B}$ is extremal if and only if its Cayley transform $T=(I-B)(I+B)^{-1}$ has the property $(D_T^2)_{\mathfrak{N}}=0$, where $\mathfrak{N}=\mathfrak{N}_{-1}$ is the defect subspace of $\dot{B}$. Thus, if B is an extremal extension of $\dot{B}$, then T is an extremal qsc-extension of $\dot{A}=(I-\dot{B})(I+\dot{B})^{-1}$ (see Definition 9.2.8).*

If the deficiency number of $\dot{B}$ is finite, then the following statements are equivalent:

(i) *B is a quasi-self-adjoint m-accretive extremal extension of $\dot{B}$,*

(ii) *B^* is a quasi-self-adjoint m-accretive extremal extension of $\dot{B}$,*

(iii) *T is an extremal qsc-extension of $\dot{A}$,*

(iv) *T^* is an extremal qsc-extension of $\dot{A}$.*

Remark 9.5.17. If T is a quasi-self-adjoint and extremal m-accretive extension such that $T\neq T^*$, then T is not sectorial.

Now we establish limiting properties of the Kreĭn-Langer Q-functions corresponding to the Friedrichs and Kreĭn non-negative self-adjoint extensions B_F and B_K of a non-negative densely-defined symmetric operator $\dot{B}$. We will use Proposition 9.3.6. Let B be a non-negative self-adjoint extension of $\dot{B}$ and let A be its fractional-linear transformation

$$A=(I-B)(I+B)^{-1}.$$

Then the resolvents of B and A are given by

$$(B-\lambda I)^{-1} = -\frac{1}{1+\lambda}(I+A)\left(A-\frac{1-\lambda}{1+\lambda}I\right)^{-1}, \quad \lambda\in\rho(B).$$

It follows that

$$I+(\lambda+1)(B-\lambda I)^{-1} = -\frac{2}{1+\lambda}\left(A-\frac{1-\lambda}{1+\lambda}I\right)^{-1}.$$

Observe that A is an sc-extension of a non-densely-defined symmetric contraction $\dot{A}=(I-\dot{B})(I+\dot{B})^{-1}$.

Furthermore, suppose $\Gamma(\lambda) \in [\mathcal{N}, \mathcal{H}]$ is the γ-field corresponding to B and $Q(\lambda)$ is the Kreĭn-Langer Q-function of $\dot{B}$ corresponding to the γ-field (see Definition 7.4.2 and Definition 7.5.1). Let $z_0 = -1$. Then from (7.54) we obtain

$$Q(\lambda) = C - 2X_0^*\left(A - \frac{1-\lambda}{1+\lambda}I\right)^{-1} X_0,$$

where $X_0 = \Gamma(-1) \in [\mathcal{N}, \mathfrak{N}_{-1}]$, $\mathfrak{N}_{-1}$ is the deficiency subspace of $\dot{B}$. Note that $\mathfrak{N}_{-1} = \mathcal{H} \ominus \operatorname{Dom}(\dot{A})$. Hence

$$Q(\lambda) = C - 2X_0^* Q_A\left(\frac{1-\lambda}{1+\lambda}\right) X_0, \tag{9.108}$$

where Q_A is defined by (9.70).

Theorem 9.5.18. *Let B be a non-negative self-adjoint extension of a non-negative densely-defined symmetric operator $\dot{B}$. Then:*

1) *The operator B coincides with the Friedrichs extension of $\dot{B}$ if and only if*

$$\lim_{x\downarrow-\infty} (Q(x)f, f)_{\mathcal{N}} = -\infty, \quad \forall f \in \mathcal{N}. \tag{9.109}$$

2) *The operator B coincides with the Kreĭn extension of $\dot{B}$ if and only if*

$$\lim_{x\uparrow-0} (Q(x)f, f)_{\mathcal{N}} = +\infty, \quad \forall f \in \mathcal{N}. \tag{9.110}$$

Proof. Let $B = B_F$. Then $A = A_\mu$. Since $\operatorname{Ran}((I + A_\mu)^{1/2}) \cap \mathfrak{N}_{-1} = \{0\}$, from Proposition 9.2.5, equalities (9.9), (9.10), and (9.4), we get

$$\lim_{z\uparrow-1-0} ((A_\mu - zI)^{-1}h, h) = +\infty$$

for all $h \in \mathfrak{N}_{-1}$. Clearly

$$x \to -\infty \iff z = \frac{1-x}{1+x} \to -1-0.$$

It follows from (9.108) that

$$\lim_{x\downarrow-\infty} (Q(x)f, f)_{\mathcal{N}} = -\infty, \quad \forall f \in \mathcal{N}.$$

If $B = B_K$, then $A = A_M$. As before using $\operatorname{Ran}((I - A_\mu)^{1/2}) \cap \mathfrak{N}_{-1} = \{0\}$ one has

$$\lim_{z\uparrow+1+0} ((A_M - zI)^{-1}h, h) = -\infty,$$

for all $h \in \mathfrak{N}_{-1}$. Since

$$x \to -0 \iff z = \frac{1-x}{1+x} \to +1+0,$$

we get $\lim_{x\uparrow-0} (Q(x)f, f)_{\mathcal{N}} = +\infty$ for all $f \in \mathcal{N}$. The converse statements follow from Proposition 9.3.6. $\square$

9.6 Bi-extensions of non-negative symmetric operators

We consider semi-bounded (in particular non-negative) symmetric densely-defined operators $\dot{A}$,

$$(\dot{A}x, x) \geq m(x, x), \qquad x \in \operatorname{Dom}(\dot{A}).$$

According to the classical von Neumann theorem there exists a self-adjoint extension A of $\dot{A}$ with an arbitrary close to m lower bound. It was shown later by Friedreichs that operator $\dot{A}$ actually admits a self-adjoint extension with the same lower bound. In this section we are going to show that for the case of a self-adjoint bi-extension of $\dot{A}$ the analogue of von Neumann's theorem is true while the analogue of the Friedrichs theorem, generally speaking, does not hold.

Theorem 9.6.1. *Let $\dot{A}$ be a semi-bounded operator with a lower bound m and $\hat{A}$ be its symmetric extension with the same lower bound. Then $\dot{A}$ admits a self-adjoint bi-extension $\mathbb{A}$ with the same lower bound and containing $\hat{A}$ ($\mathbb{A} \supset \hat{A}$) if and only if there exists a number $k > 0$ such that*

$$\left|((\hat{A} - mI)f, h)\right|^2 \leq k((\hat{A} - mI)f, f)\, \|h\|_+^2, \tag{9.111}$$

for all $f \in \operatorname{Dom}(\hat{A})$, $h \in \mathcal{H}_+$.

Proof. Let $\mathcal{H}_+ \subseteq \mathcal{H} \subseteq \mathcal{H}_-$ be the rigged triplet generated by $\dot{A}$ and $\mathcal{R}$ be a Riesz-Berezansky operator corresponding to this triplet. In the Hilbert space $\mathcal{H}_+$ consider the operator

$$\dot{B} := \mathcal{R}(\hat{A} - mI), \quad \operatorname{Dom}(\dot{B}) = \operatorname{Dom}(\hat{A}).$$

Then $(\dot{B}f, f)_+ = ((\hat{A}f - mI)f, f) \geq 0$ for all $f \in \operatorname{Dom}(\dot{B})$. Observe that $\mathbb{A}$ is a self-adjoint bi-extension of $\dot{A}$ containing $\hat{A}$ if and only if the operator $B := \mathcal{R}\mathbb{A}$ is a $(+)$-bounded and $(+)$-self-adjoint extension of the operator $\dot{B}$ in $\mathcal{H}_+$. It follows from Theorem 9.5.10 and Remark 9.5.11 that the operator $\dot{B}$ admits $(+)$-non-negative bounded self-adjoint extension in $\mathcal{H}_+$ if and only if there exists $k > 0$ such that

$$|(\dot{B}f, h)_+|^2 \leq k(\dot{B}f, f)_+ ||h||_+^2, \; f \in \operatorname{Dom}(\dot{B}), h \in \mathcal{H}_+.$$

This is equivalent to (9.111) □

Remark 9.5.11 yields that if $\dot{A}$ has at least one self-adjoint bi-extension $\mathbb{A}$ containing $\hat{A}$ with the same lower bound, then it has infinitely many such bi-extensions.

Corollary 9.6.2. *Inequalities* (9.111) *hold if and only if there exists a constant $C > 0$ such that*

$$|((\hat{A} - mI)f, \varphi_a)|^2 \leq C((\hat{A} - mI)f, f)\, \|\varphi_a\|_+^2, \tag{9.112}$$

for all $f \in \operatorname{Dom}(\hat{A})$ and all φ_a such that $(\dot{A}^* - (m - a)I)\varphi_a = 0$, $(a > 0)$.

Proof. Suppose (9.111). Then for $h = \varphi_a \in \ker(\dot{A}^* - (m-a)I))$ we have (9.112). Now let us show that (9.111) follows from (9.112). It is known that there exists a self-adjoint extension A of $\hat{A}$ (for instance, the Friedrichs extension of $\hat{A}$) with the lower bound m. If λ is a regular point for A, then

$$\mathcal{H}_+ = \operatorname{Dom}(A) \dot{+} \mathfrak{N}_\lambda. \tag{9.113}$$

Indeed, if $f \in \mathcal{H}_+ = \operatorname{Dom}(\dot{A}^*)$, then there exists an element $g \in \operatorname{Dom}(A)$ such that $(\dot{A}^* - \lambda I)f = (A - \lambda I)g$. This implies $(\dot{A}^* - \lambda I)(f-g) = 0$ and hence $(f-g) \in \mathfrak{N}_\lambda$ for any $f \in \mathcal{H}_+$ and $g \in \operatorname{Dom}(A)$, which confirms (9.113). Further, applying the Cauchy-Schwartz inequality we obtain

$$\begin{aligned} |((\hat{A} - mI)f, g)|^2 &\le ((\hat{A} - mI)f, f)((A - mI)g, g) \\ &\le C((\hat{A} - mI)f, f)\|g\|_+^2, \quad C > 0, \end{aligned} \tag{9.114}$$

for $f \in \operatorname{Dom}(\hat{A})$ and $g \in \operatorname{Dom}(A)$. Clearly, all the points of the form $(m-a)$, $(a > 0)$ are regular points for A and the points of a regular type for $\dot{A}$. Thus (9.113) implies

$$\mathcal{H}_+ = \operatorname{Dom}(A) \dot{+} \mathfrak{N}_{m-a}. \tag{9.115}$$

Let $h \in \mathcal{H}_+$ be an arbitrary vector. Applying (9.115) we get $h = g + \psi_a$, where $g \in \operatorname{Dom}(A)$ and $\psi_a \in \mathfrak{N}_{m-a}$. Adding up inequalities (9.112) and (9.114) and taking into account that the norms $\|\cdot\|$ and $\|\cdot\|_+$ are equivalent on $\mathfrak{N}_{m-a}$ we get (9.111). □

The following theorem is the analogue of the classical von Neumann's result.

Theorem 9.6.3. *Let ε be an arbitrary small positive number and $\dot{A}$ be a semi-bounded operator with the lower bound m. Then there exist infinitely many semi-bounded self-adjoint bi-extensions with the lower bound $(m-\varepsilon)$.*

Proof. First we show that the inequality

$$|((\dot{A} - (m-\varepsilon)I)f, g)|^2 \le k((\dot{A} - (m-\varepsilon)I)f, f)\|g\|_+^2,$$

holds for all $f \in \operatorname{Dom}(\dot{A})$, $g \in \mathfrak{M}$, and $k > 0$. Indeed,

$$\begin{aligned} &|((\dot{A} - (m-\varepsilon)I)f, g)| = |(f, (\dot{A}^* - (m - \varepsilon I)g)| \\ &\le |(f, \dot{A}^* g)| + |m-\varepsilon| \cdot |(f,g)| \le \|f\| \cdot \|A^* g\| + |m-\varepsilon| \cdot \|f\| \cdot \|g\| \\ &\le \frac{1}{\sqrt{\varepsilon}}((\dot{A} - (m-\varepsilon)I)f, f)^{1/2}\|g\|_+ + \frac{|m-\varepsilon|}{\sqrt{\varepsilon}}((\dot{A} - (m-\varepsilon)I)f, f)^{1/2}\|g\|_+ \\ &= \frac{1 + |m-\varepsilon|}{\sqrt{\varepsilon}}((\dot{A} - (m-\varepsilon)I)f, f)\|g\|_+. \end{aligned}$$

The statement of the theorem follows from Theorem 9.6.1 and Remark 9.5.11. □

Theorem 9.6.4. *A non-negative densely-defined operator $\dot{A}$ admits a non-negative self-adjoint bi-extension if and only if the Friedrichs and Kreĭn-von Neumann extensions of $\dot{A}$ are transversal.*

Proof. Suppose that the Friedrichs extension A_F and the Kreĭn-von Neumann extension A_K of the operator $\dot{A}$ are transversal. Then due to Theorem 9.5.9 the inclusion $\operatorname{Dom}(\dot{A}^*) \subset \mathcal{D}[A_K]$ holds. This means that $\mathcal{H}_+ \subseteq \operatorname{Dom}(A_K^{1/2})$. Since $||h||_+ \geq ||h||$ for all $h \in \mathcal{H}_+$, and $A_K^{1/2}$ is closed in $\mathcal{H}$, the closed graph theorem yields now that $A_K^{1/2} \in [\mathcal{H}_+, \mathcal{H}]$, i.e., there exists a number $c > 0$ such that

$$||A_K^{1/2} u||^2 = A_K[u] \leq c||u||_+^2.$$

It follows that the sesquilinear form $A_K[u,v] = (A_K^{1/2}u, A_K^{1/2}v)$, $u, v \in \mathcal{H}_+$ is bounded on $\mathcal{H}_+$. Therefore, by the Riesz theorem, there exists an operator $\mathbb{A}_K \in [\mathcal{H}_+, \mathcal{H}_-]$ such that

$$(\mathbb{A}_K u, v) = A_K[u,v], \quad u, v \in \mathcal{H}_+,\ u \in \mathcal{H}_+.$$

Due to $A_K[u] \geq 0$ for all $u \in \mathcal{D}[A_K]$, the operator $\mathbb{A}_K$ is non-negative. Since $(A_K u, v) = A_K[u,v]$ for all $u \in \operatorname{Dom}(A_K)$ and all $v \in \mathcal{D}[A_K]$, we get

$$(\mathbb{A}_K u, v) = (A_K u, v), \quad u \in \operatorname{Dom}(A_K),\ v \in \mathcal{H}_+.$$

Hence $\mathbb{A}_K \supset A_K$, i.e., $\mathbb{A}_K$ is t-self-adjoint bi-extension of $\dot{A}$ with quasi-kernel A_K.

Conversely, let $\dot{A}$ admit a non-negative self-adjoint bi-extension. Then from Theorem 9.6.1 we get the equality

$$|(\dot{A}f, h)|^2 \leq k(\dot{A}f, f)||h||_+^2,$$

for all $f \in \operatorname{Dom}(\dot{A})$ and all $h \in \mathcal{H}_+ = \operatorname{Dom}(\dot{A}^*)$, and some $k > 0$. Applying Theorem 9.5.7 we get that $\mathcal{H}_+ \subseteq \mathcal{D}[A_K]$. Now Theorem 9.5.9 yields that A_F and A_K are transversal. □

Corollary 9.6.5. *If a non-negative densely-defined symmetric operator $\dot{A}$ admits a non-negative self-adjoint bi-extension, then it also admits a non-negative t-self-adjoint bi-extension $\mathbb{A}$ containing A_K as a quasi-kernel.*

It follows from Theorem 9.6.4 that if $A_K = A_F$, then the operator $\dot{A}$ does not admit non-negative self-adjoint bi-extensions. Consequently, in this case the analogue of the Friedrichs theorem is not true. The following theorem provides a criterion on when the analogue of the Friedrichs theorem does hold.

Theorem 9.6.6. *A non-negative densely-defined symmetric operator $\dot{A}$ admits a non-negative self-adjoint bi-extensions if and only if*

$$\int_0^\infty t\, d(E(t)h, h) < \infty \quad \textit{for all} \quad h \in \mathfrak{N}_{-a},\ a > 0, \tag{9.116}$$

where $E(t)$ is a spectral function of the Kreĭn-von Neumann extension A_K of $\dot{A}$.

Proof. The inequality (9.116) is equivalent to the inclusion

$$\mathfrak{N}_{-a} \subset \mathrm{Dom}(A_K^{1/2}) = \mathcal{D}[A_K].$$

Since $-a$ is a regular point of A_K, the direct decomposition

$$\mathrm{Dom}(\dot{A}^*) = \mathrm{Dom}(A_K) \dot{+} \mathfrak{N}_{-a},$$

holds. So, from Theorem 9.5.9 we get that (9.116) is equivalent to transversality of A_F and A_K. The latter is equivalent to existence of a non-negative self-adjoint bi-extension of $\dot{A}$ (see Theorem 9.6.4). □

Observe, that since $\mathfrak{N}_{-a}$ is a subspace in $\mathcal{H}$, A_K is closed in $\mathcal{H}$, condition (9.116) is equivalent to the following: there exists a positive number $k > 0$, depending on a, such that

$$\int_0^\infty t\, d(E(t)h, h) < k||h||^2, \quad \forall h \in \mathfrak{N}_{-a},\ a > 0. \tag{9.117}$$

On the other hand, (9.116) is equivalent (see proof of Theorem 9.6.4) to the existence of $k > 0$ such that

$$\int_0^\infty t\, d(E(t)f, f) < k||f||_+^2, \quad \forall f \in \mathrm{Dom}(\dot{A}^*).$$

9.7 Accretive bi-extensions

We recall (see page 301) that a closed, densely-defined, linear operator T in a Hilbert space $\mathcal{H}$ is maximal accretive if it is accretive and T has a regular point in the left half-plane. A bi-extension $\mathbb{A} \in [\mathcal{H}_+, \mathcal{H}_-]$ is called an **accretive bi-extension** if $\mathrm{Re}\,(\mathbb{A}f, f) \geq 0$ for all $f \in \mathcal{H}_+$. In this section we will study the existence of accretive $(*)$-extensions of a given accretive operator $T \in \Omega(\dot{A})$.

Lemma 9.7.1. *Let $\mathbb{A}$ be a $(*)$-extension of an accretive operator $T \in \Omega(\dot{A})$ generated via (4.60) by a self-adjoint operator A. Then $\mathbb{A}$ is accretive if and only if the form*

$$\mathrm{Re}\,(Th, h) + (Ag, g) + 2\mathrm{Re}\,(Th, g), \tag{9.118}$$

is non-negative for all $h \in \mathrm{Dom}(T)$ and $g \in \mathrm{Dom}(A)$.

Proof. Let $\mathbb{A}$ be an accretive $(*)$-extension of an accretive operator $T \in \Omega(\dot{A})$. Then for $h \in \mathrm{Dom}(T)$ and $g \in \mathrm{Dom}(A)$, $\mathrm{Re}\,(\mathbb{A}f, f) \geq 0$, $f = g + h$. It follows from Theorem 4.4.7 that the form (9.118) is non-negative.

On the other hand, if the form (9.118) is non-negative, then for all $h \in \mathrm{Dom}(T)$ and $g \in \mathrm{Dom}(A)$ the transversality of $\mathrm{Dom}(T)$ and $\mathrm{Dom}(A)$ and Theorem 4.4.7 imply that $\mathbb{A}$ is accretive. □

Let A_F and A_K be Friedrich's and Kreĭn-von Neumann's extensions of operator $\dot{A}$ of the form (9.91) and (9.92). By **class** $\Xi(\dot{A})$ we denote the set of all maximal accretive extensions of operator $\dot{A}$. In particular, both A_F and A_K belong to $\Xi(\dot{A})$. It follows from Lemma 9.7.1 that if $T \in \Xi(\dot{A})$ and if $(*)$-extension $\mathbb{A}$ of T generated by A is accretive, then $A \in \Xi(\dot{A})$. On the class $\Xi(\dot{A})$ we define Cayley transform described in the Section 1.4 and given by the formula

$$K(T) = (I - T)(I + T)^{-1}, \qquad T \in \Xi(\dot{A}). \tag{9.119}$$

This Cayley transform sets a one-to-one correspondence between the class $\Xi(\dot{A})$ and a set of contractions $Q \in [\mathcal{H}, \mathcal{H}]$ such that both Q and Q^* are extensions of a symmetric contraction

$$\dot{S} = (I - \dot{A})(I + \dot{A})^{-1},$$

defined on a subspace $\mathrm{Dom}(\dot{S}) = (I + \dot{A})\mathrm{Dom}(\dot{A})$. Clearly, Q and Q^* are both qsc-extensions of $\dot{S}$. Put

$$\mathfrak{N} = \mathcal{H} \ominus \mathrm{Dom}(\dot{S}). \tag{9.120}$$

Notice that $\mathfrak{N} = \mathfrak{N}_{-1} = \ker(\dot{A}^* + I)$ (the deficiency subspace of $\dot{A}$).

Let $S_\mu = K(A_F)$ and $S_M = K(A_K)$. Using Theorem 9.2.7 and formula (9.42) we get that $Q \in [\mathcal{H}, \mathcal{H}]$ is a qsc-extension of a symmetric contraction $\dot{S}$ if and only if it can be represented in the form

$$Q = \frac{1}{2}(S_M + S_\mu) + \frac{1}{2}(S_M - S_\mu)^{1/2} Z (S_M - S_\mu)^{1/2}, \tag{9.121}$$

where Z is a contraction in the subspace $\overline{\mathrm{Ran}(S_M - S_\mu)} \subseteq \mathfrak{N}$.

Clearly, if Z is a self-adjoint contraction, then (9.121) provides a description of all sc-extensions of a symmetric contraction S.

Lemma 9.7.2. 1) *The class $\Xi(\dot{A})$ contains mutually transversal operators if and only if A_F and A_K are mutually transversal.*

2) *Let T_1 and T_2 belong to $\Xi(\dot{A})$. Then T_1 and T_2 are mutually transversal if and only if*

$$(K(T_1) - K(T_2))\mathfrak{N} = \mathfrak{N}.$$

Proof. It follows from (9.121) that

$$K(T_1) - K(T_2) = \frac{1}{2}(S_M - S_\mu)^{1/2}(Z_1 - Z_2)(S_M - S_\mu)^{1/2},$$

where Z_l, $(l = 1, 2)$ are the corresponding to T_l contractions in $\overline{\mathrm{Ran}(S_M - S_\mu)}$. Relation (9.119) yields

$$K(T_1) - K(T_2) = 2\left((I + T_1)^{-1} - (I + T_2)^{-1}\right).$$

Thus

$$(I + T_1)^{-1} - (I + T_2)^{-1} = \frac{1}{4}(S_M - S_\mu)^{1/2}(Z_1 - Z_2)(S_M - S_\mu)^{1/2}.$$

Furthermore, using Theorem 4.4.2 we get that

$$\begin{aligned} &\left((I+T_1)^{-1}-(I+T_2)^{-1}\right)\mathfrak{N}_{-1}=\mathfrak{N}_{-1}\\ \iff &\left\{\begin{array}{l}\overline{\operatorname{Ran}(S_M-S_\mu)}=\operatorname{Ran}(S_M-S_\mu)=\mathfrak{N}=\mathfrak{N}_{-1},\\ \operatorname{Ran}(Z_1-Z_2)\mathfrak{N}=\mathfrak{N}.\end{array}\right.\end{aligned}$$

□

In what follows we assume that A_K and A_F are mutually transversal. Let A_1 and A_2 be two mutually transversal operators from $\Xi(\dot{A})$. Consider a form defined on $\operatorname{Dom}(A_1)\times\operatorname{Dom}(A_2)$ as

$$B(f_1,f_2)=(A_1f_1,f_1)+(A_2f_2,f_2)+2\operatorname{Re}(A_1f_1,f_2), \tag{9.122}$$

where $f_l\in\operatorname{Dom}(A_l)$, $(l=1,2)$. Let

$$\phi_l=\frac{1}{2}(I+A_l)f_l,\qquad S_l\phi_l=\frac{1}{2}(I-A_l)f_l,$$

be the Cayley transform of A_l for $l=1,2$. Then

$$f_l=(I+S_l)\phi_l,\qquad A_lf_l=(I-S_l)\phi_l,\quad (l=1,2). \tag{9.123}$$

Substituting (9.123) into (9.122) we obtain a form defined on $\mathcal{H}\times\mathcal{H}$,

$$\tilde{B}(\phi_1,\phi_2)=\|\phi_1+\phi_2\|^2-\|S_1\phi_1+S_2\phi_2\|^2-2\operatorname{Re}\left((S_1-S_2)\phi_1,\phi_2\right).$$

Let us set

$$F=\frac{1}{2}(S_1-S_2),\quad G=\frac{1}{2}(S_1+S_2),\quad u=\frac{1}{2}(\phi_1+\phi_2),\quad v=\frac{1}{2}(\phi_1-\phi_2). \tag{9.124}$$

Then $\tilde{B}(\phi_1,\phi_2)=4H(u,v)$ where

$$H(u,v)=\|u\|^2+(Fv,v)-(Fu,u)-\|Fv+Gu\|^2. \tag{9.125}$$

Moreover, $F\pm G$ are contractive operators. From the above reasoning we conclude that non-negativity of the form $B(f_1,f_2)$ on $\operatorname{Dom}(A_1)\times\operatorname{Dom}(A_2)$ is equivalent to non-negativity of the form $H(u,v)$ on $\mathcal{H}\times\mathcal{H}$.

Lemma 9.7.3. *The form $H(u,v)$ in (9.125) is non-negative for all $u,v\in\mathcal{H}$ if and only if operator F defined in (9.124) is non-negative.*

Proof. If $H(u,v)\geq 0$ for all $u,v\in\mathcal{H}$ then $H(0,v)\geq 0$ for all $v\in\mathcal{H}$. Hence $(Fv,v)\geq\|Fv\|^2\geq 0$, i.e., $F\geq 0$.

Conversely, let $F\geq 0$. Since both operators $F\pm G$ are self-adjoint contractions, then $-I\leq F+G\leq I$ and $-I\leq F-G\leq I$. This implies $-(I-F)\leq G\leq I-F$, and thus

$$G=(I-F)^{1/2}Z(I-F)^{1/2}, \tag{9.126}$$

where Z is a self-adjoint contraction. Then (9.126) yields that for all $u, v \in \mathcal{H}$,

$$
\begin{aligned}
\|Fv + Gu\| &= \|Fv\|^2 + \|Gu\|^2 + 2\mathrm{Re}\,(Fv, Gu) \\
&\le \|Fv\|^2 + \Big((I-F)Z(I-F)^{1/2}u, Z(I-F)^{1/2}u\Big) \\
&\quad + 2\Big|\Big(F(I-F)^{1/2}Z(I-F)^{1/2}u, v\Big)\Big| \\
&= \|Fv\|^2 + \|Z(I-F)^{1/2}u\|^2 - (FZ(I-F)^{1/2}u, Z(I-F)^{1/2}u) \\
&\quad + 2\Big|\Big(FZ(I-F)^{1/2}u, (I-F)^{1/2}v\Big)\Big| \\
&\le \|Fv\|^2 + \|Z(I-F)^{1/2}u\|^2 - (FZ(I-F)^{1/2}u, Z(I-F)^{1/2}u) \\
&+ (FZ(I-F)^{1/2}u, Z(I-F)^{1/2}u) + (F(I-F)^{1/2}v, (I-F)^{1/2}v) \\
&= \|Fv\|^2 + \|Z(I-F)^{1/2}u\|^2 + (Fv, v) - \|Fv\|^2 \\
&\le (Fv, v) + \|u\|^2 - (Fu, u).
\end{aligned}
$$

Therefore, for all $u, v \in \mathcal{H}$

$$
H(u, v) = \|u\|^2 - (Fu, u) + (Fv, v) - \|Fv + Gu\|^2 \ge 0.
$$

The lemma is proved. □

In what follows we will use Theorem 3.4.9, Theorem 4.4.3, and formulas (6.13).

Theorem 9.7.4. *Let $\mathbb{A} = \dot{A}^* - \mathcal{R}^{-1}\dot{A}^*(I - \mathcal{P}_{\hat{A}A})$ be a self-adjoint $(*)$-extension of a non-negative symmetric operator $\dot{A}$, with a self-adjoint quasi-kernel $\hat{A} \in \Xi(\dot{A})$, and generated (via (4.60)) by a self-adjoint extension A. Then the following statements are equivalent:*

(i) *$\mathbb{A}$ is non-negative,*

(ii) *$\Big(K(\hat{A}) - K(A)\Big) \restriction \mathfrak{N}$, (see (9.119)–(9.120)) is positively defined,*

(iii) *$(\hat{A} + I)^{-1} \ge (A + I)^{-1}$, and $\hat{A}$ is transversal to A,*

(iv) *$\hat{A} \le A$ and $\hat{A}$ is transversal to A.*

Proof. Let $(\mathbb{A}f, f) \ge 0$ for all $f \in \mathcal{H}_+$. Then by Theorem 4.4.7 we have that the form

$$
B(g, h) = (\hat{A}g, h) + (Ah, g) + 2\mathrm{Re}\,(\hat{A}g, h), \quad (g \in \mathrm{Dom}(\hat{A}),\ h \in \mathrm{Dom}(A)),
$$

is non-negative on $\mathrm{Dom}(\hat{A}) \times \mathrm{Dom}(A)$. Consequently, the form $H(u, v)$ given by (9.125) is non-negative for all $u, v \in \mathcal{H}$ where

$$
F = \frac{1}{2}\Big(K(\hat{A}) - K(A)\Big) \text{ and } G = \frac{1}{2}\Big(K(\hat{A}) + K(A)\Big).
$$

Using Lemma 9.7.3 we conclude that $F\upharpoonright\mathfrak{N}\geq 0$ and applying Lemma 9.7.2 yields $F\mathfrak{N}=\mathfrak{N}$. This proves that (i) $\Rightarrow$ (ii). The implication (ii) $\Rightarrow$ (i) can be shown by reversing the argument. Since

$$K(\hat{A})=-I+2(\hat{A}+I)^{-1},\ K(A)=-I+2(A+I)^{-1},$$

we get that (ii) $\iff$ (iii). Remark 9.5.5 yields (iii) $\iff$ (iv). $\square$

Theorem 9.7.5. *A self-adjoint operator $\hat{A}\in\Xi(\dot{A})$ admits non-negative $(*)$-extensions if and only if $\hat{A}$ is transversal to A_F.*

Proof. If $\hat{A}$ is transversal to A_F, then $\Big(K(\hat{A})-K(A_F)\Big)\upharpoonright\mathfrak{N}$ is positively defined. Applying Theorem 9.7.4 we obtain that

$$\mathbb{A}=\dot{A}^*-\mathcal{R}^{-1}\dot{A}^*(I-\mathcal{P}_{\hat{A}A_F}),$$

is a non-negative $(*)$-extension.

Conversely, if $\mathbb{A}=\dot{A}^*-\mathcal{R}^{-1}\dot{A}^*(I-\mathcal{P}_{\hat{A}A})$ is a $(*)$-extension of $\hat{A}$, then via Theorem 9.7.4 we get that $\Big(K(\hat{A})-K(A)\Big)\upharpoonright\mathfrak{N}$ is positively defined. But then due to the chain of inequalities

$$K(\hat{A})\geq K(A)\geq K(A_F),$$

the operator $\Big(K(\hat{A})-K(A_F)\Big)\upharpoonright\mathfrak{N}$ is positively defined as well. According to Lemma 9.7.2, $\hat{A}$ is transversal A_F. $\square$

We note that if $\hat{A}$ is a self-adjoint extension of $\dot{A}$, then all self-adjoint $(*)$-extensions of $\hat{A}$ coincide with t-self-adjoint bi-extensions of $\dot{A}$ with the quasi-kernel $\hat{A}$. Consequently, Theorem 9.7.5 gives the criterion of the existence of a non-negative t-self-adjoint bi-extension of $\dot{A}$ and hence provides the conditions when Friedreichs theorem for t-self-adjoint bi-extensions is true.

Now we focus on the study of non-self-adjoint accretive $(*)$-extensions of operator $T\in\Xi(\dot{A})$.

Lemma 9.7.6. *Let $\mathbb{A}$ be a $(*)$-extension of operator $T\in\Xi(\dot{A})$ generated by an operator $A\in\Xi(\dot{A})$. Then the quasi-kernel $\hat{A}$ of the operator* $\mathrm{Re}\,\mathbb{A}$ *is defined by the formula*

$$\begin{aligned} f&=(Q+I)g+\frac{1}{2}(S+I)(Q^*-S)^{-1}(Q-Q^*)g,\\ \hat{A}f&=(I-Q)g+\frac{1}{2}(I-S)(Q^*-S)^{-1}(Q-Q^*)g, \end{aligned}\tag{9.127}$$

where $g\in\mathcal{H}$, $Q=K(T)$, $Q^*=K(T^*)$, and $S=K(A)$.

Proof. Let $\mathbb{A}=\dot{A}^*-\mathcal{R}^{-1}\dot{A}^*(I-\mathcal{P}_{TA})$ (of the form (4.60)) be a $(*)$-extensions of operator T generated by a self-adjoint extension A. Let

$$\mathrm{Dom}(A)=\mathrm{Dom}(\dot{A})\oplus(\mathcal{U}+I)\mathfrak{N}_i,$$

where $\mathcal{U} \in [\mathfrak{N}_i, \mathfrak{N}_{-i}]$ is a unitary mapping. Suppose $f \in \mathrm{Dom}(\hat{A})$, where $\hat{A}$ is a quasi-kernel of $\mathrm{Re}\,\mathbb{A}$. Due to the transversality of T^* and A and T and A we have

$$f = u + (\mathcal{U} + I)\varphi, \qquad f = v + (\mathcal{U} + I)\psi,$$

where $u \in \mathrm{Dom}(T)$, $v \in \mathrm{Dom}(T^*)$, and $\varphi, \psi \in \mathfrak{N}_i$. Also

$$\begin{aligned} \mathbb{A}f &= Tu + \dot{A}^*(\mathcal{U} + I)\varphi - i\mathcal{R}^{-1}(I - \mathcal{U})\varphi, \\ \mathbb{A}^*f &= T^*v + \dot{A}^*(\mathcal{U} + I)\psi - i\mathcal{R}^{-1}(I - \mathcal{U})\psi, \end{aligned}$$

and

$$\begin{aligned} \hat{A}f &= \frac{1}{2}(\mathbb{A}f + \mathbb{A}^*f) \\ &= \frac{1}{2}\left(Tu + T^*v + \dot{A}^*(\mathcal{U} + I)\varphi + \dot{A}^*(\mathcal{U} + I)\psi - i\mathcal{R}^{-1}(I - \mathcal{U})(\varphi + \psi)\right). \end{aligned}$$

Since $\hat{A}f \in \mathcal{H}$, then $\varphi = -\psi$ and hence any vector $f \in \mathrm{Dom}(\hat{A})$ is uniquely represented in the form

$$f = u + \phi, \quad u \in \mathrm{Dom}(T),\ \phi \in (\mathcal{U} + I)\mathfrak{N}_i,$$

or in the form $f = v - \phi$, $v \in \mathrm{Dom}(T^*)$. By Corollary 4.4.4 $\hat{A}$ is transversal to A and

$$\mathrm{Re}\,\mathbb{A} = \dot{A}^* - \mathcal{R}^{-1}(I - P_{\hat{A}A}).$$

Thus $\mathfrak{M}_{\hat{A}} \dotplus (\mathcal{U} + I)\mathfrak{N}_i = \mathfrak{M}$, where $\mathfrak{M}_{\hat{A}} = \mathrm{Dom}(\hat{A}) \ominus \mathrm{Dom}(\dot{A})$. It follows from

$$\mathfrak{M}_T \dotplus (\mathcal{U} + I)\mathfrak{N}_i = \mathfrak{M}, \quad \text{where} \quad \mathfrak{M}_T = \mathrm{Dom}(T) \ominus \mathrm{Dom}(\dot{A}),$$

that $\mathcal{P}_{\hat{A}A}\mathfrak{M}_T = \mathfrak{M}_{\hat{A}}$ and hence for any $u \in \mathrm{Dom}(T)$ there exists a $\phi \in (\mathcal{U} + I)\mathfrak{N}_i$ and $f \in \mathrm{Dom}(\hat{A})$ such that $f = u + \phi$. Similarly, for any $v \in \mathrm{Dom}(T)$ there exists a $\phi \in (\mathcal{U} + I)\mathfrak{N}_i$ and $f \in \mathrm{Dom}(\hat{A})$ such that $f = u - \phi$. Since

$$\mathrm{Dom}(T) = (I + Q)\mathcal{H},\ \mathrm{Dom}(T^*) = (I + Q^*)\mathcal{H},\ \mathrm{Dom}(A) = (I + S)\mathcal{H},$$

and

$$Q{\upharpoonright}\mathrm{Dom}(\dot{S}) = Q^*{\upharpoonright}\mathrm{Dom}(\dot{S}) = S{\upharpoonright}\mathrm{Dom}(\dot{S}),$$

we conclude that for any $f \in \mathrm{Dom}(\hat{A})$ there are uniquely defined $g, g_* \in \mathcal{H}$ and $h \in \mathfrak{N}$ such that

$$f = (Q + I)g + (S + I)h, \quad f = (Q^* + I)g_* - (S + I)h. \tag{9.128}$$

Conversely, for every $g \in \mathcal{H}$ (respectively, $g_* \in \mathcal{H}$) there are $g_* \in \mathcal{H}$ (respectively, $g \in \mathcal{H}$) and $h \in \mathfrak{N}$, such that (9.128) holds with $f \in \mathrm{Dom}(\hat{A})$. Since $\dot{A}^*(Q + I)g = (I - Q)g$, $\dot{A}^*(I + Q^*)g_* = (I - Q^*)g_*$, and $\dot{A}^*(I + S)h = (I - S)h$, then

$$\hat{A}f = (I - Q)g + (I - S)h, \quad \hat{A}f = (I - Q^*)g_* - (I - S)h. \tag{9.129}$$

From (9.128) and (9.129) we have $2h = g_* - g$ and $2Sh = Q^* g_* - g$, which implies

$$2(Q^* - S)h = (Q - Q^*)g. \tag{9.130}$$

Since T^* and A are mutually transversal, according to Lemma 9.7.2 $(Q^* - S)\restriction\mathfrak{N}$ is an isomorphism on $\mathfrak{N}$. Then (9.130) implies

$$h = \frac{1}{2}(Q^* - S)^{-1}(Q - Q^*)g. \tag{9.131}$$

Substituting (9.131) into (9.128) and (9.129) we obtain (9.127). □

Lemma 9.7.7. *Let $T \in \Xi(\dot{A})$ and $A \in \Xi(\dot{A})$ be a transversal to T self-adjoint operator. If the operator*

$$[K(T) + K(T^*) - 2K(A)]\restriction\mathfrak{N},$$

is an isomorphism of the space $\mathfrak{N}$ (defined in (9.120)), then the quasi-kernel $\hat{A}$ of the real part of the operator $\mathbb{A} = \dot{A}^* - \mathcal{R}^{-1}\dot{A}^*(I - \mathcal{P}_{TA})$ is a Cayley transform of the operator

$$\hat{S} = S + (Q - S)(\mathrm{Re}\, Q - S)^{-1}(Q^* - S),$$

where $S = K(A)$ and $\mathrm{Re}\, Q = (1/2)[K(T) + K(T^*)]$.

Proof. Let $\mathbb{A} = \dot{A}^* - \mathcal{R}^{-1}\dot{A}^*(I - \mathcal{P}_{TA})$. Then by virtue of Lemma 9.7.6, formula (9.127) defines the quasi-kernel $\hat{A}$ of the operator $\mathrm{Re}\,\mathbb{A}$. It also follows from (9.127) that

$$f + \hat{A}f = 2g + (Q^* - S)^{-1}(Q - Q^*)g.$$

Let $P_{\mathfrak{N}}$ and $P_{\dot{S}}$ denote the orthoprojection operators in $\mathcal{H}$ according to (9.120) onto $\mathfrak{N}$ and $\mathrm{Dom}(\dot{S})$, respectively. Then

$$\begin{aligned} 2g + (Q^* - S)^{-1}(Q - Q^*)g &= 2P_{\dot{S}}g + (Q^* - S)^{-1}(2Q^* - 2S + Q - Q^*)P_{\mathfrak{N}}g \\ &= 2P_{\dot{S}}g + 2(Q^* - S)^{-1}(\mathrm{Re}\, Q - S)P_{\mathfrak{N}}g, \end{aligned}$$

and

$$(I + \hat{A})f = 2P_{\dot{S}}g + 2(Q^* - S)^{-1}(\mathrm{Re}\, Q - S)P_{\mathfrak{N}}g. \tag{9.132}$$

From the statement of the lemma we have that $(\mathrm{Re}\, Q - S)\restriction\mathfrak{N}$ is an isomorphism of the space $\mathfrak{N}$. Hence, (9.132) $\mathrm{Ran}(I + \hat{A}) = \mathcal{H}$ and the Cayley transform is well defined for $\hat{A}$. Let

$$\hat{S} = (I - \hat{A})(I + \hat{A})^{-1}.$$

It follows from (9.127) that

$$\begin{aligned} (\hat{S} + I)\phi &= (Q + I)g + \frac{1}{2}(S + I)(Q^* - S)^{-1}(Q - Q^*)g, \\ (I - \hat{S})\phi &= (I - Q)g + \frac{1}{2}(I - S)(Q^* - S)^{-1}(Q - Q^*)g. \end{aligned}$$

Therefore,

$$\phi = g + \frac{1}{2}(Q^* - S)^{-1}(Q - Q^*)g,$$
$$\hat{S}\phi = Qg + \frac{1}{2}S(Q^* - S)^{-1}(Q - Q^*)g$$

and hence

$$\begin{aligned} \hat{S}\phi &= S\phi + (Q - S)P_{\mathfrak{N}}g, \\ \phi &= P_{\dot{S}}g + (Q^* - S)^{-1}(\operatorname{Re} Q - S)P_{\mathfrak{N}}g. \end{aligned} \tag{9.133}$$

Using the second half of (9.133) we have

$$P_{\mathfrak{N}}g = (\operatorname{Re} Q - S)^{-1}(Q^* - S)P_{\mathfrak{N}}\phi. \tag{9.134}$$

Substituting, (9.134) into the first part of (9.133) we obtain

$$\hat{S}\phi = S\phi + (Q - S)(\operatorname{Re} Q - S)^{-1}(Q^* - S)\phi,$$

which proves the lemma. □

Let $T \in \Xi(\dot{A})$. By the **class** Ξ_{AT} we denote the set of all non-negative self-adjoint operators $A \supset \dot{A}$ satisfying the following conditions:

1. $[K(T) + K(T^*) - 2K(A)]\upharpoonright\mathfrak{N}$ is a non-negative operator in $\mathfrak{N}$, where $\mathfrak{N}$ is defined in (9.120);
2. $K(A)+2[K(T)-K(A)][K(T)+K(T^*)-2K(A)]^{-1}[K(T^*)-K(A)] \leq K(A_K)$.[1]

Theorem 9.7.8. *A* $(*)$*-extension of operator* $T \in \Xi(\dot{A})$,

$$\mathbb{A} = \dot{A}^* - \mathcal{R}^{-1}\dot{A}^*(I - \mathcal{P}_{TA}),$$

generated by a self-adjoint operator $A \supset \dot{A}$, *is accretive if and only if* $A \in \Xi_{AT}$.

Proof. We prove the necessity part first. Let $\mathbb{A} = \dot{A}^* - \mathcal{R}^{-1}\dot{A}^*(I - \mathcal{P}_{TA})$ be an accretive $(*)$-extension, then $\operatorname{Re}\mathbb{A}$ is a non-negative $(*)$-extension of the quasi-kernel $\hat{A} \in \Xi(\dot{A})$. But according to Lemma 9.7.6 $\hat{A}$ is defined by formulas (9.127). Since (-1) is a regular point of operator $\hat{A}$, then (9.132) implies that the operator

$$(\operatorname{Re} Q - S)\upharpoonright\mathfrak{N} = \frac{1}{2}[K(T) + K(T^*) - 2K(A)]\upharpoonright\mathfrak{N},$$

is an isomorphism of the space $\mathfrak{N}$. According to Lemma 9.7.7 we have

$$K(\hat{A}) = K(A) + 2[K(T) - K(A)][K(T) + K(T^*) - 2K(A)]^{-1}[K(T^*) - K(A)].$$

Since $K(\hat{A})$ is a self-adjoint contractive extension of $\dot{S}$, then $K(\hat{A}) \leq K(A_K)$. Also, since $\operatorname{Re}\mathbb{A}$ is generated by A and $\operatorname{Re}\mathbb{A} \geq 0$, then by Theorem 9.7.4 the operator

[1]When we write $[K(T) + K(T^*) - 2K(A)]^{-1}$ we mean the operator inverse to $[K(T) + K(T^*) - 2K(A)]\upharpoonright\mathfrak{N}$.

$[K(\hat{A}) - K(A)]\upharpoonright\mathfrak{N}$ is non-negative. Consequently, the operator $[K(T) + K(T^*) - 2K(A)\upharpoonright\mathfrak{N}$ is non-negative as well and we conclude that $A \in \Xi_{AT}$.

Now we prove sufficiency. Let $A \in \Xi_{AT}$, then by Lemma 9.7.7, $\hat{A}$ is a Cayley transform of a self-adjoint extension $\hat{S}$ of the operator $\dot{S}$. Since

$$[\hat{S} - S]\upharpoonright\mathfrak{N} = [K(\hat{A}) - K(A)]\upharpoonright\mathfrak{N},$$

is a non-negative operator, then due to Theorem 9.7.4 the operator $\mathrm{Re}\,\mathbb{A}$ is a non-negative $(*)$-extension of $\hat{A}$. That is why $\mathbb{A}$ is an accretive $(*)$-extension of operator T. □

Theorem 9.7.9. *An operator $T \in \Xi(\dot{A})$ admits accretive $(*)$-extensions if and only if T is transversal to A_F.*

Proof. If T admits accretive $(*)$-extensions, then the class Ξ_{AT} is non-empty, i.e., there exists a self-adjoint operator $A \in \Xi(\dot{A})$ such that the operator

$$[(K(T) + K(T^*) - 2K(A)]\upharpoonright\mathfrak{N},$$

is non-negative. But then $[(K(T) + K(T^*) - 2K(A_F)]\upharpoonright\mathfrak{N}$ is non-negative as well. This yields that $[K(T) - K(A_F)]\upharpoonright\mathfrak{N}$ is an isomorphism of $\mathfrak{N}$. Then by Lemma 9.7.2 T and A_F are mutually transversal. This proves the necessity.

Now let us assume that T and A_F are mutually transversal. We will show that in this case $A_F \in \Xi_{AT}$. By Lemma 9.7.2, $[K(T) - K(A_F)]\upharpoonright\mathfrak{N}$ is an isomorphism of the space $\mathfrak{N}$. Then using formula (9.121) we have

$$K(T) = Q = \frac{1}{2}(S_M + S_\mu) + \frac{1}{2}(S_M - S_\mu)^{1/2} Z (S_M - S_\mu)^{1/2},$$

where $Z \in [\mathfrak{N}, \mathfrak{N}]$ is a contraction. Furthermore,

$$(Q - S_\mu)\upharpoonright\mathfrak{N} = \frac{1}{2}(S_M - S_\mu)^{1/2}(Z + I)(S_M - S_\mu)^{1/2}\upharpoonright\mathfrak{N}.$$

Thus, $Z + I$ is an isomorphism of the space $\mathfrak{N}$. Moreover, $\mathrm{Re}\,Z + I \geq 0$ and for every $f \in \mathfrak{N}$,

$$((\mathrm{Re}\,Z + I)f, f) = \frac{1}{2}\left(\|f\|^2 - \|Zf\|^2 + \|(Z + I)f\|^2\right). \tag{9.135}$$

But since $\|(Z + I)f\|^2 \geq a\|f\|^2$, where $a > 0$, $f \in \mathfrak{N}$, we have

$$((\mathrm{Re}\,Z + I)f, f) \geq b\|f\|^2, \qquad (b > 0).$$

Hence, $\mathrm{Re}\,Z + I$ is a non-negative operator implying that

$$[(1/2)(K(T) + K(T^*) - K(A)]\upharpoonright\mathfrak{N} = \frac{1}{2}(S_M - S_\mu)^{1/2}(\mathrm{Re}\,Z + I)(S_M - S_\mu)^{1/2}\upharpoonright\mathfrak{N},$$

is non-negative too. Also (9.135) implies

$$\mathrm{Re}\, Z + I \geq \frac{1}{2}(Z^* + I)(Z + I).$$

It is easy to see then that $(\mathrm{Re}\, Z + I)^{-1} \leq 2(Z + I)^{-1}(Z^* + I)^{-1}$. Therefore,

$$\frac{1}{2}(Z + I)(\mathrm{Re}\, Z + I)^{-1}(Z^* + I) \leq I. \tag{9.136}$$

Now, since

$$\begin{aligned} K(A_F) &+ 2[K(T) - K(A)][K(T) + K(T^*) - 2K(A)]^{-1}[K(T^*) - K(A)] \\ &= S_\mu + \frac{1}{2}(S_M - S_\mu)^{1/2}(Z + I)(\mathrm{Re}\, Z + I)^{-1}(Z^* + I)(S_M - S_\mu)^{1/2}, \end{aligned}$$

then applying (9.136) we obtain

$$K(A_F) + 2[K(T) - K(A)][K(T) + K(T^*) - 2K(A)]^{-1}[K(T^*) - K(A)] \leq K(A_K).$$

Thus, A_F belongs to the class Ξ_{AT} and applying theorem (9.7.8) we conclude that $\mathbb{A} = \dot{A}^* - \mathcal{R}^{-1}\dot{A}^*(I - \mathcal{P}_{AT})$ is an accretive $(*)$-extension of T. □

Theorem 9.7.10. *Let $T \in \Xi(\dot{A})$ be transversal to A_F. Then an accretive $(*)$-extension $\mathbb{A}$ of T generated by A_F has a property that $\mathrm{Re}\,\mathbb{A} \supset A_K$ if and only if T and T^* are extremal extensions of $\dot{A}$ (see Definition 9.5.15).*

Proof. Suppose $\mathrm{Re}\,\mathbb{A} \supset A_K$ and $\mathrm{Re}\,\mathbb{A} = \dot{A}^* - \mathcal{R}^{-1}\dot{A}^*(I - \mathcal{P}_{TA_F})$. Then by Lemma 9.7.7 we have

$$S_M = S_\mu + (Q - S_\mu)(\mathrm{Re}\, Q - S_\mu)^{-1}(Q^* - S_\mu).$$

Thus,

$$(Z + I)(\mathrm{Re}\, Z + I)^{-1}(Z^* + I) = 2I, \tag{9.137}$$

where

$$Q = K(T) = \frac{1}{2}(S_M + S_\mu) + \frac{1}{2}(S_M - S_\mu)^{1/2} Z (S_M - S_\mu)^{1/2}.$$

It is easy to see that

$$(Z^* + I)(\mathrm{Re}\, Z + I)^{-1}(Z + I) = (Z + I)(\mathrm{Re}\, Z + I)^{-1}(Z^* + I). \tag{9.138}$$

Then it follows from (9.137) and (9.138) that

$$Z^* Z = Z Z^* = I,$$

i.e., Z is a unitary operator in $\mathfrak{N}$. From Propositions 9.5.16 and 9.2.9 we get that both operators T and T^* are extremal m-accretive extensions of $\dot{A}$.

The second part of the theorem is proved by reversing the argument. □

Let $\alpha \in [0, \pi/2)$. A bi-extension $\mathbb{A}$ of a non-negative symmetric operator $\dot{A}$ is called **α-sectorial** if

$$|(\operatorname{Im} \mathbb{A} f, f)| \leq \tan \alpha (\operatorname{Re} \mathbb{A} f, f), \quad f \in \mathcal{H}_+.$$

Clearly, non-negative self-adjoint bi-extensions are 0-sectorial, and formally accretive bi-extensions one can consider as $\pi/2$-sectorial.

Theorem 9.7.11. *Let T be a quasi-self-adjoint and maximal α-sectorial extension of $\dot{A}$. If T is transversal to A_F, then T admits α-sectorial $(*)$-extensions.*

Proof. Let $\mathbb{A}$ be generated via (4.60) by A_F and is given by

$$\mathbb{A} = \dot{A}^* - \mathcal{R}^{-1} \dot{A}^* (I - \mathcal{P}_{T A_F}).$$

By Theorem 9.7.9 the operator $\mathbb{A}$ is an accretive $(*)$-extension with quasi-kernel T. If $f \in \mathcal{H}_+$ is decomposed as $f = \phi + h$, where $\phi \in \operatorname{Dom}(A_F)$, $h \in \operatorname{Dom}(T)$, then by Theorem 4.4.7 we have

$$(\mathbb{A} f, f) = (A_F \phi, \phi) + (Th, h) + 2\operatorname{Re}(Th, \phi).$$

Let $A_K[u, v]$ be the closed form associated with the Kreĭn-von Neumann extension A_K of $\dot{A}$ and let

$$w[h] = (Th, h) - A_K[h], \quad h \in \operatorname{Dom}(T).$$

According to (9.99) we have $(Th, \phi) = A_K[h, \phi]$. Hence, using the equality $(A_F \phi, \phi) = A_K[\phi]$ (see Theorem 9.5.4), one obtains

$$\begin{aligned}(\mathbb{A} f, f) &= A_F[\phi, \phi] + A_K[h] + A_K[h, \phi] + A_K[\phi, h] + w[h] \\ &= A_K[\phi + h] + w[h].\end{aligned}$$

From Theorem 9.5.14 we have, for all $h \in \operatorname{Dom}(T)$,

$$|\operatorname{Im} w[h]| \leq (\tan \alpha) \operatorname{Re} w[h].$$

This yields

$$\begin{aligned}|(\operatorname{Im} \mathbb{A} f, f)| &= |\operatorname{Im} w[h]| \leq \tan \alpha \operatorname{Re} w[h] \\ &\leq \tan \alpha (A_K[\phi + h] + \operatorname{Re} w[h]) = \tan \alpha (\operatorname{Re} \mathbb{A} f, f).\end{aligned}$$

So $\mathbb{A}$ is α-sectorial $(*)$-bi-extension of T. □

9.8 Realization of Stieltjes functions

The scalar versions of the following definition can be found in [159].

Definition 9.8.1. An operator-valued Herglotz-Nevanlinna function $V(z)$ in a finite-dimensional Hilbert space E is called a **Stieltjes function** if $V(z)$ is holomorphic in $\operatorname{Ext}[0,+\infty)$ and

$$\frac{\operatorname{Im}[zV(z)]}{\operatorname{Im} z} \geq 0. \tag{9.139}$$

Consequently, an operator-valued Herglotz-Nevanlinna function $V(z)$ is Stieltjes if $zV(z)$ is also a Herglotz-Nevanlinna function. Applying (6.7) for this case we get that

$$\sum_{k,l=1}^{n} \left(\frac{z_k V(z_k) - \bar{z}_l V(\bar{z}_l)}{z_k - \bar{z}_l} h_k, h_l \right)_E \geq 0, \tag{9.140}$$

for an arbitrary sequence $\{z_k\}$ $(k = 1, \dots, n)$ of $(\operatorname{Im} z_k > 0)$ complex numbers and a sequence of vectors $\{h_k\}$ in E.

Similar to (6.52), the formula holds true for the case of a Stieltjes function. Indeed, if $V(z)$ is a Stieltjes operator-valued function, then

$$V(z) = \gamma + \int_0^\infty \frac{dG(t)}{t - z}, \tag{9.141}$$

where $\gamma \geq 0$ and $G(t)$ is a non-decreasing on $[0,+\infty)$ operator-valued function such that

$$\int_0^\infty \frac{(dG(t)h,h)_E}{1+t} < \infty, \qquad h \in E. \tag{9.142}$$

Theorem 9.8.2. *Let Θ be an L-system of the form (6.31)–(6.36) with a densely-defined non-negative symmetric operator $\dot{A}$. Then the impedance function $V_\Theta(z)$ defined by (6.47) is a Stieltjes function if and only if the operator $\mathbb{A}$ of the L-system Θ is accretive.*

Proof. Let us assume first that $\mathbb{A}$ is an accretive operator, i.e., $(\operatorname{Re}\mathbb{A}f, f) \geq 0$, for all $f \in \mathcal{H}_+$. Let $\{z_k\}$ $(k = 1, \dots, n)$ be a sequence of $(\operatorname{Im} z_k > 0)$ complex numbers and h_k be a sequence of vectors in E. Let us write

$$Kh_k = \delta_k, \quad g_k = (\operatorname{Re}\mathbb{A} - z_k I)^{-1}\delta_k, \quad g = \sum_{k=1}^{n} g_k. \tag{9.143}$$

Since $(\operatorname{Re}\mathbb{A}g, g) \geq 0$, we have

$$\sum_{k,l=1}^{n} (\operatorname{Re}\mathbb{A}g_k, g_l) \geq 0. \tag{9.144}$$

By formal calculations one can have $(\operatorname{Re}\mathbb{A})g_k = \delta_k + z_k(\operatorname{Re}\mathbb{A} - z_k I)^{-1}\delta_k$, and

$$\begin{aligned}\sum_{k,l=1}^{n} (\operatorname{Re}\mathbb{A}g_k, g_l) = \sum_{k,l=1}^{n} \Big[&(\delta_k, (\operatorname{Re}\mathbb{A} - z_l I)^{-1}\delta_l) \\ &+ (z_k(\operatorname{Re}\mathbb{A} - z_k I)^{-1}\delta_k, (\operatorname{Re}\mathbb{A} - z_k I)^{-1}\delta_l) \Big].\end{aligned}$$

Using obvious equalities

$$\big((\mathrm{Re}\,\mathbb{A}-z_kI)^{-1}Kh_k,Kh_l\big)=\big(V_\Theta(z_k)h_k,h_l\big)_E,$$

and

$$\big((\mathrm{Re}\,\mathbb{A}-\bar z_lI)^{-1}(\mathrm{Re}\,\mathbb{A}-z_kI)^{-1}Kh_k,Kh_l\big)=\left(\frac{V_\Theta(z_k)-V_\Theta(\bar z_l)}{z_k-\bar z_l}h_k,h_l\right)_E,$$

we obtain

$$\sum_{k,l=1}^{n}((\mathrm{Re}\,\mathbb{A})g_k,g_l)=\sum_{k,l=1}^{n}\left(\frac{z_kV_\Theta(z_k)-\bar z_lV_\Theta(\bar z_l)}{z_k-\bar z_l}h_k,h_l\right)_E\geq 0,\tag{9.145}$$

which implies that $V_\Theta(z)$ is a Stieltjes function.

Now we prove necessity. First we assume that $\dot A$ is a prime operator. Then the equivalence of (9.145) and (9.144) implies that $(\mathrm{Re}\,\mathbb{A}g,g)\geq 0$ for any g from c.l.s.$\{\mathfrak{N}_z\}$. According to Lemma 6.6.4 and (6.94), a symmetric operator $\dot A$ with the equal deficiency indices is prime if and only if

$$\underset{z\neq\bar z}{c.l.s.}\,\mathfrak{N}_z=\mathcal{H}.$$

Thus $(\mathrm{Re}\,\mathbb{A}g,g)\geq 0$ for any $g\in\mathcal{H}_+$ and therefore $\mathbb{A}$ is an accretive operator.

Now let us assume that $\dot A$ is not a prime operator. Then there exists a subspace $\mathcal{H}^1\subset\mathcal{H}$ on which $\dot A$ generates a self-adjoint operator A_1. Let us denote by $\dot A_0$ an operator induced by $\dot A$ on $\mathcal{H}^0=\mathcal{H}\ominus\mathcal{H}^1$. As we have shown in the proof of Theorem 6.6.1 the decomposition

$$\mathcal{H}_+=\mathcal{H}^0_+\oplus\mathcal{H}^1_+,\quad \mathcal{H}^0_+=\mathrm{Dom}(\dot A_0^*),\ \mathcal{H}^1_+=\mathrm{Dom}(A_1),\tag{9.146}$$

is $(+)$-orthogonal. Since $\dot A$ is a non-negative operator, then

$$(\mathrm{Re}\,\mathbb{A}g,g)=(A_1g,g)=(\dot Ag,g)\geq 0,\qquad \forall g\in\mathcal{H}^1_+=\mathrm{Dom}(A_1).$$

On the other hand, operator $\dot A_0$ is prime in $\mathcal{H}^0$ and hence (by Lemma 6.6.4) $\underset{z\neq\bar z}{c.l.s.}\,\mathfrak{N}^0_z=\mathcal{H}^0$, where $\mathfrak{N}^0_z$ are the deficiency subspaces of the symmetric operator $\dot A_0$ in $\mathcal{H}^0$. Then the equivalence of (9.145) and (9.144) again implies that $(\mathrm{Re}\,\mathbb{A}g,g)\geq 0$ for any $g\in\mathcal{H}^0_+$. Taking into account decomposition (9.146) we conclude that $\mathrm{Re}\,(\mathbb{A}g,g)\geq 0$ holds for all $g\in\mathcal{H}_+$ and hence $\mathbb{A}$ is accretive. □

Let $\alpha\in(0,\frac{\pi}{2})$. Now we introduce **sectorial subclasses** S^α of operator-valued Stieltjes functions. An operator-valued Stieltjes function $V(z)$ belongs to S^α if

$$K_\alpha=\sum_{k,l=1}^{n}\left(\left[\frac{z_kV(z_k)-\bar z_lV(\bar z_l)}{z_k-\bar z_l}-(\cot\alpha)\ V^*(z_l)V(z_k)\right]h_k,h_l\right)_E\geq 0,\tag{9.147}$$

for an arbitrary sequence $\{z_k\}$ $(k = 1, \dots, n)$ of $(\operatorname{Im} z_k > 0)$ complex numbers and a sequence of vectors $\{h_k\}$ in E. For $0 < \alpha_1 < \alpha_2 < \frac{\pi}{2}$, we have

$$S^{\alpha_1} \subset S^{\alpha_2} \subset S,$$

where S denotes the class of all Stieltjes functions (which corresponds to the case $\alpha = \frac{\pi}{2}$), as follows from the inequality

$$K_{\alpha_1} \le K_{\alpha_2} \le K_{\frac{\pi}{2}}.$$

The following theorem refines the result of Theorem 9.8.2 as applied to the class S^α.

Theorem 9.8.3. *Let Θ be a scattering L-system of the form (6.31)–(6.36) with a densely-defined non-negative symmetric operator $\dot{A}$. Then the impedance function $V_\Theta(z)$ defined by (6.47) belongs to the class S^α if and only if the operator $\mathbb{A}$ of the L-system Θ is α-sectorial.*

Proof. The outline of the proof can be replicated from the proof of Theorem 9.8.2. Then all we need is to show that (9.147) is equivalent to relation (9.29) of Definition 9.2.1 of sectoriality. Suppose that $\mathbb{A}$ is α-sectorial, then (9.29) holds for all $g \in \mathcal{H}_+$ and hence

$$\cot\alpha \cdot |(\operatorname{Im}\mathbb{A}\, g, g)| \le (\operatorname{Re}\mathbb{A}\, g, g), \quad g \in \mathcal{H}_+. \tag{9.148}$$

Consequently, it follows from (9.143) that

$$\sum_{k,\ell=1}^{n} (\operatorname{Re}\mathbb{A}\, g_k, g_\ell) \ge (\cot\alpha) \sum_{k,\ell=1}^{n} (\operatorname{Im}\mathbb{A}\, g_k, g_\ell).$$

Taking into account that $\operatorname{Im}\mathbb{A} = KK^*$, we get

$$\begin{aligned} &\sum_{k,\ell=1}^{n} (\operatorname{Re}\mathbb{A}(\operatorname{Re}\mathbb{A} - z_k I)^{-1} K h_k, (\operatorname{Re}\mathbb{A} - z_\ell I)^{-1} K h_\ell) \\ &\qquad \ge \cot\alpha \cdot \sum_{k,\ell=1}^{n} (KK^*(\operatorname{Re}\mathbb{A} - z_k I)^{-1} K h_k, (\operatorname{Re}\mathbb{A} - z_\ell I)^{-1} K h_\ell), \end{aligned}$$

which leads to

$$\begin{aligned} &\sum_{k,\ell=1}^{n} (K^*(\operatorname{Re}\mathbb{A} - \bar{z}_\ell I)^{-1} \operatorname{Re}\mathbb{A}(\operatorname{Re}\mathbb{A} - z_k I)^{-1} K h_k, h_\ell)_E \\ &\ge \quad \cot\alpha \cdot \sum_{k,\ell=1}^{n} (K^*(\operatorname{Re}\mathbb{A} - \bar{z}_\ell I)^{-1} KK^*(\operatorname{Re}\mathbb{A} - z_k I)^{-1} K h_k, h_\ell)_E. \end{aligned}$$

Using resolvent identity

$$\begin{aligned} K^*(\mathrm{Re}\,\mathbb{A}-\bar{z}_\ell I)^{-1}\mathrm{Re}\,\mathbb{A}(\mathrm{Re}\,\mathbb{A}-z_k I)^{-1}K \\ = K^*\,\frac{\bar{z}_\ell(\mathrm{Re}\,\mathbb{A}-\bar{z}_\ell I)^{-1}-z_k(\mathrm{Re}\,\mathbb{A}-z_k I)^{-1}}{\bar{z}_\ell-z_k}K, \end{aligned} \tag{9.149}$$

we obtain (9.147). The converse statement is proved by reversing the argument. Thus we have shown that (9.147) is equivalent to (9.148) and this completes the proof. □

At this point we would like to introduce a special subclass of scalar Stieltjes functions. Let

$$0 \le \alpha_1 \le \alpha_2 \le \frac{\pi}{2}.$$

We say that a scalar Stieltjes function $V(z)$ belongs to the **class** S^{α_1,α_2} if

$$\tan\alpha_1 = \lim_{x\to-\infty} V(x), \qquad \tan\alpha_2 = \lim_{x\to-0} V(x). \tag{9.150}$$

Theorem 9.8.4. *Let Θ be a scattering L-system of the form*

$$\Theta = \begin{pmatrix} \mathbb{A} & K & 1 \\ \mathcal{H}_+ \subset \mathcal{H} \subset \mathcal{H}_- & & \mathbb{C} \end{pmatrix}, \tag{9.151}$$

with a densely-defined non-negative symmetric operator $\dot{A}$. Let also $\mathbb{A}$ be an α-sectorial $(*)$-extension of $T \in \Lambda(\dot{A})$. Then the impedance function $V_\Theta(z)$ defined by (6.47) belongs to the class S^{α_1,α_2}, $\tan\alpha_2 \le \tan\alpha$, and T is $(\alpha_2-\alpha_1)$-sectorial with the exact angle of sectoriality $(\alpha_2-\alpha_1)$.[2]

Proof. Since $\mathbb{A}$ is an α-sectorial $(*)$-extension of T, then (9.148) holds and we can apply Theorem 9.8.3. Then using (9.149) for $z_k = z_\ell = z$ we obtain

$$\mathrm{Im}\,(zV(z)) = (\mathrm{Im}\,z)K^*(\mathrm{Re}\,\mathbb{A}-zI)^{-1}\mathrm{Re}\,\mathbb{A}(\mathrm{Re}\,\mathbb{A}-\bar{z}I)^{-1}K. \tag{9.152}$$

Applying Theorem 9.8.3 again we have

$$\frac{\mathrm{Im}\,(zV_\Theta(z))}{\mathrm{Im}\,z} \ge (\cot\alpha)V_\Theta^*(z)V_\Theta(z). \tag{9.153}$$

It follows from (9.152) that

$$\lim_{z\to x}\frac{\mathrm{Im}\,(zV_\Theta(z))}{\mathrm{Im}\,z} = K^*(\mathrm{Re}\,\mathbb{A}-xI)^{-1}\mathrm{Re}\,\mathbb{A}(\mathrm{Re}\,\mathbb{A}-xI)^{-1}K. \tag{9.154}$$

[2] We say that the angle of sectoriality α is **exact** for an α-sectorial operator T if $\tan\alpha = \sup_{f\in\mathrm{Dom}(T)}\frac{|\mathrm{Im}\,(Tf,f)|}{|\mathrm{Re}\,(Tf,f)|}$.

Here we used the fact that if $\mathbb{A}$ is α-sectorial, then $\mathbb{A}$ is accretive and $x < 0$ is a regular point for the quasi-kernel of $\mathrm{Re}\,\mathbb{A}$. Thus, (9.154) yields

$$\begin{aligned}
V_\Theta(x) - \lim_{z\to x}\frac{\mathrm{Im}\,(zV_\Theta(z))}{\mathrm{Im}\,z} &= K^*(\mathrm{Re}\,\mathbb{A} - xI)^{-1}K \\
&\quad - K^*(\mathrm{Re}\,\mathbb{A} - xI)^{-1}\mathrm{Re}\,\mathbb{A}(\mathrm{Re}\,\mathbb{A} - xI)^{-1}K \\
&= K^*(\mathrm{Re}\,\mathbb{A} - xI)^{-1}[I - \mathrm{Re}\,\mathbb{A}(\mathrm{Re}\,\mathbb{A} - xI)^{-1}]K \\
&= K^*(\mathrm{Re}\,\mathbb{A} - xI)^{-1}[\mathrm{Re}\,\mathbb{A} - xI - \mathrm{Re}\,\mathbb{A}](\mathrm{Re}\,\mathbb{A} - xI)^{-1}K \\
&= -xK^*(\mathrm{Re}\,\mathbb{A} - xI)^{-1}(\mathrm{Re}\,\mathbb{A} - xI)^{-1}K \geq 0.
\end{aligned}$$

Therefore,

$$V_\Theta(x) \geq \lim_{z\to x}\frac{\mathrm{Im}\,(zV_\Theta(z))}{\mathrm{Im}\,z}, \qquad (x < 0). \tag{9.155}$$

Since $V_\Theta(x) > 0$ for $x < 0$, then applying (9.153) and (9.155) yields

$$V_\Theta(x) \geq (\cot\alpha)V_\Theta^2(x),$$

and therefore

$$V_\Theta(x) \leq \tan\alpha, \qquad (x < 0). \tag{9.156}$$

It follows from Theorem 9.8.2 that the impedance function $V_\Theta(z)$ of an L-system with accretive operator $\mathbb{A}$ has the integral representation (9.141), i.e.,

$$V_\Theta(z) = \gamma + \int_0^\infty \frac{dG(t)}{t - z}. \tag{9.157}$$

Then (9.156) and (9.157) yield

$$V_\Theta(x) = \gamma + \int_0^\infty \frac{dG(t)}{t - x} \leq \tan\alpha. \qquad (x < 0).$$

and thus

$$\int_0^\infty \frac{dG(t)}{t} < \infty \quad \text{and} \quad \gamma + \int_0^\infty \frac{dG(t)}{t} \leq \tan\alpha.$$

Let us write

$$\tan\alpha_1 = \gamma, \qquad \tan\alpha_2 = \gamma + \int_0^\infty \frac{dG(t)}{t}. \tag{9.158}$$

Using (9.158) we obtain that $V_\Theta(z) \in S^{\alpha_1,\alpha_2}$ and $\tan\alpha_2 \leq \tan\alpha$.

According to Theorem 8.2.4 for the system Θ of the form (9.151) there is a system Θ' of the form (8.29) with the main operator $S = (I - T)(I + T)^{-1}$ and such that (8.30) holds, i.e.,

$$W_\Theta(z) = W_\Theta(-1)W_{\Theta'}\left(\frac{1 - z}{1 + z}\right), \; z \in \rho(T), \; z \neq -1. \tag{9.159}$$

We also know from Lemma 9.5.12 that T is α-sectorial if and only if S is α-co-sectorial contraction. It follows from Theorems 9.4.1 and 9.4.2 that the exact angle of co-sectoriality of S can be calculated by

$$\cot\beta = \frac{1 + V_{\Theta'}(1)V_{\Theta'}(-1)}{|V_{\Theta'}(-1) - V_{\Theta'}(1)|}. \tag{9.160}$$

Let us compute $V_{\Theta'}(1)$ and $V_{\Theta'}(-1)$ using (9.159) and (6.48). We get

$$V_{\Theta'}(1) = -i(I + W_\Theta^{-1}(-1)W_\Theta(0))^{-1}(W_\Theta^{-1}(-1)W_\Theta(0) - I),$$
$$V_{\Theta'}(-1) = -i(I + W_\Theta^{-1}(-1)W_\Theta(-\infty))^{-1}(W_\Theta^{-1}(-1)W_\Theta(-\infty) - I),$$

and

$$W_\Theta(0) = \frac{1 - iV_\Theta(0)}{1 + iV_\Theta(0)},\ W_\Theta(-\infty) = \frac{1 - iV_\Theta(-\infty)}{1 + iV_\Theta(-\infty)},\ W_\Theta^{-1}(-1) = \frac{1 + iV_\Theta(-1)}{1 - iV_\Theta(-1)}.$$

This yields

$$V_{\Theta'}(1) = \frac{V_\Theta(-1) - V_\Theta(-\infty)}{1 + V_\Theta(-1)V_\Theta(-\infty)}, \quad V_{\Theta'}(-1) = \frac{V_\Theta(-1) - V_\Theta(0)}{1 + V_\Theta(-1)V_\Theta(0)}.$$

Taking into account (9.160) we get

$$\cot\beta = \frac{1 + V_\Theta(0)V_\Theta(-\infty)}{V_\Theta(0) - V_\Theta(-\infty)} = \frac{1 + \tan\alpha_2 \cdot \tan\alpha_1}{\tan\alpha_2 - \tan\alpha_1} = \cot(\alpha_2 - \alpha_1). \qquad \square$$

Corollary 9.8.5. *Let Θ of the form (9.151) be an L-system as in the statement of Theorem 9.8.4 and let α be the exact angle of sectoriality of the operator T of the system Θ. Then $V_\Theta(z) \in S^{0,\alpha}$.*

Proof. According to Theorem 9.8.4 the exact angle of sectoriality is given by $\alpha_2 - \alpha_1$, where

$$\tan\alpha_1 = \lim_{x \to -\infty} V_\Theta(x), \quad \tan\alpha_2 = \lim_{x \to -0} V_\Theta(x).$$

It was also shown that $\tan\alpha \geq \tan\alpha_2$. On the other hand, since in the statement of the current corollary α is the exact angle of sectoriality of T, then $\alpha = \alpha_2 - \alpha_1$ and hence $\tan(\alpha_2 - \alpha_1) \geq \tan\alpha_2$. Therefore, $\alpha_1 = 0$. $\square$

Remark 9.8.6. It follows that under assumptions of Corollary 9.8.5, the impedance function $V_\Theta(z)$ has the form

$$V_\Theta(z) = \int_0^\infty \frac{dG(t)}{t - z}.$$

Theorem 9.8.7. *Let Θ be an L-system of the form (9.151), where $\mathbb{A}$ is a $(*)$-extension of $T \in \Lambda(\dot{A})$ and $\dot{A}$ is a closed densely-defined non-negative symmetric operator with deficiency numbers $(1,1)$. If the impedance function $V_\Theta(z)$ belongs to the class S^{α_1,α_2}, then $\mathbb{A}$ is α-sectorial, where*

$$\tan\alpha = \tan\alpha_2 + 2\sqrt{\tan\alpha_1(\tan\alpha_2 - \tan\alpha_1)}.$$

Proof. Since $V_\Theta(z) \in S^{\alpha_1,\alpha_2}$, we have

$$V_\Theta(z) = \tan\alpha_1 + \int_0^\infty \frac{dG(t)}{t-z} \quad \text{and} \quad \tan\alpha_2 = \tan\alpha_1 + \int_0^\infty \frac{dG(t)}{t}. \tag{9.161}$$

Let $\{z_k\}$, $k = 1,\dots,n$ be arbitrary numbers in $\mathbb{C}_+$ and $\{\xi_k\}$, $k = 1,\dots,n$ be arbitrary complex numbers. By direct substitution one gets

$$\sum_{k,l=1}^n \frac{z_k V_\Theta(z_k) - \bar{z}_l V_\Theta(\bar{z}_l)}{z_k - \bar{z}_l}\bar{\xi}_k\xi_l = (\tan\alpha_1)\left|\sum_{l=1}^n \xi_l\right|^2 + \int_0^\infty \left|\sum_{l=1}^n \frac{\xi_l\sqrt{t}}{t-z_l}\right|^2 dG(t). \tag{9.162}$$

Furthermore,

$$\begin{aligned}\sum_{k,l=1}^n V_\Theta(z_l)V_\Theta(\bar{z}_k)\xi_l\bar{\xi}_k &= \sum_{k,l=1}^n \left(\tan\alpha_1 + \int_0^\infty \frac{dG(t)}{t-z_l}\right)\\ \times\left(\tan\alpha_1 + \int_0^\infty \frac{dG(t)}{t-\bar{z}_k}\right)\xi_l\bar{\xi}_k &= \left|\sum_{l=1}^n (\tan\alpha_1)\xi_l + \int_0^\infty \sum_{l=1}^n \frac{\xi_l\, dG(t)}{t-z_l}\right|^2.\end{aligned} \tag{9.163}$$

It follows from (9.163) that

$$\begin{aligned}&\left|\sum_{l=1}^n \tan\alpha_1\xi_l + \int_0^\infty \sum_{l=1}^n \frac{\xi_l\, dG(t)}{t-z_k}\right| \le \tan\alpha_1\left|\sum_{l=1}^n \xi_l\right| + \left|\int_0^\infty \sum_{l=1}^n \frac{\xi_l dG(t)}{t-z_l}\right|\\
&= \tan\alpha_1\left|\sum_{l=1}^n \xi_l\right| + \left|\int_0^\infty \frac{1}{\sqrt{t}}\sum_{l=1}^n \frac{\xi_l\sqrt{t}dG(t)}{t-z_l}\right|\\
&\le \tan\alpha_1\left|\sum_{l=1}^n \xi_l\right| + \left(\int_0^\infty \frac{dG(t)}{t}\right)^{1/2}\left(\int_0^\infty \left|\sum_{l=1}^n \frac{\xi_l\sqrt{t}}{t-z_l}\right|^2 dG(t)\right)^{1/2}\\
&\le (\tan\alpha_1)^{1/2}\left[\tan\alpha_1\left|\sum_{l=1}^n \xi_l\right|^2 + \int_0^\infty \left|\sum_{l=1}^n \frac{\xi_l\sqrt{t}}{t-z_l}\right|^2 dG(t)\right]^{1/2}\\
&\quad + \left(\int_0^\infty \frac{dG(t)}{t}\right)^{1/2}\left[\tan\alpha_1\left|\sum_{l=1}^n \xi_l\right|^2 + \int_0^\infty \left|\sum_{l=1}^n \frac{\xi_l\sqrt{t}}{t-z_l}\right|^2 dG(t)\right]^{1/2}\\
&= \left[\tan^{1/2}\alpha_1 + \left(\int_0^\infty \frac{dG(t)}{t}\right)^{1/2}\right]\left(\sum_{k,l=1}^n \frac{z_k V_\Theta(z_k) - \bar{z}_l V_\Theta(\bar{z}_l)}{z_k - \bar{z}_l}\bar{\xi}_k\xi_l\right)^{1/2}.\end{aligned}$$

Using (9.162), (9.163) we obtain

$$\sum_{k,l=1}^{n} V_\Theta(z_l)V_\Theta(\bar{z}_k)\xi_l\bar{\xi}_k \leq \left[\tan^{1/2}\alpha_1 + \left(\int_0^\infty \frac{dG(t)}{t}\right)^{1/2}\right]^2 \times \sum_{k,l=1}^{n} \frac{z_k V_\Theta(z_k) - \bar{z}_l V_\Theta(\bar{z}_l)}{z_k - \bar{z}_l}\bar{\xi}_k\xi_l.$$

Applying Theorem 9.8.3 we get that $\mathbb{A}$ is α-sectorial and using (9.161) yields

$$\begin{aligned}\tan\alpha &= \left[\tan^{1/2}\alpha_1 + \left(\int_0^\infty \frac{dG(t)}{t}\right)^{1/2}\right]^2\\ &= \tan\alpha_1 + (\tan\alpha_2 - \tan\alpha_1) + 2\sqrt{\tan\alpha_1}\cdot\sqrt{\tan\alpha_2 - \tan\alpha_1}\\ &= \tan\alpha_2 + 2\sqrt{\tan\alpha_1}\cdot\sqrt{\tan\alpha_2 - \tan\alpha_1}.\end{aligned}$$

□

The next statement immediately follows from Theorems 9.8.4 and 9.8.7.

Theorem 9.8.8. *Let Θ be an L-system of the form (9.151) with a densely-defined non-negative symmetric operator $\dot{A}$. Then $\mathbb{A}$ is α-sectorial $(*)$-extension of an α-sectorial operator $T \in \Lambda(\dot{A})$ with the exact angle $\alpha \in (0, \pi/2)$ if and only if*

$$V_\Theta(z) = \int_0^\infty \frac{dG(t)}{t - z} \in S^{0,\alpha}.$$

Moreover, the angle α can be found via the formula

$$\tan\alpha = \int_0^\infty \frac{dG(t)}{t}. \tag{9.164}$$

Now we define a class of realizable Stieltjes functions. At this point we need to note that since Stieltjes functions form a subset of Herglotz-Nevanlinna functions, then according to Definition 6.4.1 and Theorems 6.4.3 and 6.5.2, we have that the class of all realizable Stieltjes functions is a subclass of $N(R)$. To see the specifications of this class we recall that aside of integral representation (9.141), any Stieltjes function admits a representation (6.52). It was shown in Chapter 7 that a Herglotz-Nevanlinna operator-function can be realized as the impedance function of an L-system if and only if in the representation (6.52) $X = 0$ and

$$Qh = \int_{-\infty}^{+\infty} \frac{t}{1+t^2}\, dG(t)h, \tag{9.165}$$

for all $h \in E$ such that

$$\int_{-\infty}^{\infty} (dG(t)h, h)_E < \infty. \tag{9.166}$$

holds. Considering this we obtain

$$Q = \frac{1}{2}\left[V(-i) + V^*(-i)\right] = \gamma + \int_0^{+\infty} \frac{t}{1+t^2}\,dG(t). \tag{9.167}$$

Combining (9.165) and (9.167) we conclude that $\gamma h = 0$ for all $h \in E$ such that (9.166) holds.

Definition 9.8.9. An operator-valued Stieltjes function $V(z)$ in a finite-dimensional Hilbert space E belongs to the **class** $S(R)$ if in the representation (9.141)

$$\gamma h = 0$$

for all $h \in E$ such that

$$\int_0^{\infty} (dG(t)h, h)_E < \infty. \tag{9.168}$$

We are going to focus though on the subclass $S_0(R)$ of $S(R)$ whose definition is the following.

Definition 9.8.10. An operator-valued Stieltjes function $V(z) \in S(R)$ belongs to the **class** $S_0(R)$ if in the representation (9.141) we have

$$\int_0^{\infty} (dG(t)h, h)_E = \infty, \tag{9.169}$$

for all non-zero $h \in E$.

An L-system Θ of the form (6.31)-(6.36) is called an **accretive L-system** if its operator $\mathbb{A}$ is accretive. The following theorem gives the analogue of the Theorem 7.1.4 for the functions of the class $S_0(R)$.

Theorem 9.8.11. *Let Θ be an accretive L-system of the form (6.31)–(6.36) with an invertible channel operator K and a densely-defined symmetric operator $\dot{A}$. Then its impedance function $V_\Theta(z)$ of the form (6.47) belongs to the class $S_0(R)$.*

Proof. Since our L-system Θ is accretive, then by Theorem 9.8.2, $V_\Theta(z)$ is a Stieltjes function. Now let us show that $V_\Theta(z)$ belongs to $S_0(R)$. It follows from the Theorem 6.2.10 that $E_1 = K^{-1}\mathfrak{L}$, where $\mathfrak{L} = \mathcal{H} \ominus \overline{\mathrm{Dom}(\dot{A})}$ and

$$E_1 = \left\{ h \in E : \int_0^{+\infty} (dG(t)h, h)_E < \infty \right\}.$$

But $\overline{\mathrm{Dom}(\dot{A})} = \mathcal{H}$ and consequently $\mathfrak{L} = \{0\}$. Next, $E_1 = \{0\}$,

$$\int_0^{\infty} (dG(t)h, h)_E = \infty,$$

for all non-zero $h \in E$, and therefore $V_\Theta(z) \in S_0(R)$. □

Inverse realization theorem analogous to the Theorem 7.1.5 can be stated and proved for the classes $S_0(R)$ as well.

Theorem 9.8.12. *Let an operator-valued function $V(z)$ belong to the class $S_0(R)$. Then $V(z)$ can be realized as an impedance function of a minimal accretive L-system Θ of the form* (6.31)–(6.36) *with an invertible channel operator K, a densely-defined non-negative symmetric operator $\dot{A}$,* $\mathrm{Dom}(T) \neq \mathrm{Dom}(T^*)$, *and a preassigned direction operator J for which $I + iV(-i)J$ is invertible.*[3]

Proof. We have already noted that the class of Stieltjes function lies inside the wider class of all Herglotz-Nevanlinna functions. Thus all we actually have to show is that $S_0(R) \subset N_0(R)$ and that the realizing L-system in the proof of the Theorem 7.1.5 appears to be an accretive L-system. The former is rather obvious and follows directly from the definition of the class $S_0(R)$. To see that the realizing L-system is accretive we need to recall that the model L-system Θ was constructed in the proof of the Theorem 7.1.5 and then it was shown that $V_\Theta(z) = V(z)$. But $V(z)$ is a Stieltjes function and hence so is $V_\Theta(z)$. Applying Theorem 9.8.2 yields the desired result. □

Now it directly follows from Theorems 9.8.8 and 9.8.12 that if a function $V(z) \in S_0(R)$ is also a member of the class S^{α_1,α_2}, then it can be realized by an accretive L-system Θ whose operator $\mathbb{A}$ is an α-sectorial $(*)$-extension of an α-sectorial operator T with the exact angle $\alpha \in (0, \pi/2)$ found by (9.164).

Let us define a subclass of the class $S_0(R)$.

Definition 9.8.13. An operator-valued Stieltjes function $V(z)$ of the class $S_0(R)$ is said to be a member of the **class** $S_0^K(R)$ if

$$\int_0^\infty \frac{(dG(t)h,h)_E}{t} = \infty, \tag{9.170}$$

for all non-zero $h \in E$.

Below we state and prove a direct and inverse realization theorem for this subclass.

Theorem 9.8.14. *Let Θ be an accretive L-system of the form* (6.31)–(6.36) *with an invertible channel operator K and a densely-defined symmetric operator $\dot{A}$. If the Kreĭn-von Neumann extension A_K is a quasi-kernel for* $\mathrm{Re}\,\mathbb{A}$, *then the impedance function $V_\Theta(z)$ of the form* (6.47) *belongs to the class $S_0^K(R)$.*

Conversely, if $V(z) \in S_0^K(R)$, then it can be realized as the impedance function of an accretive L-system Θ of the form (6.31)-(6.36) *with* $\mathrm{Re}\,\mathbb{A}$ *containing A_K as a quasi-kernel and a preassigned direction operator J for which $I+iV(-i)J$ is invertible.*

---

[3]We have already mentioned that if $J = I$ this invertibility condition is satisfied automatically.

Proof. We begin with the proof of the second part. First we use realization Theorems 6.5.1, 7.1.5, and 9.8.12 to construct a minimal model L-system Θ whose impedance function is $V(z)$. Then we will show that (9.170) is equivalent to the fact that the self-adjoint operator A defined in (6.69), that we constructed in these theorems to be a quasi-kernel for $\operatorname{Re}\mathbb{A}$, coincides with A_K, that is the Kreĭn-von Neumann extension of the symmetric operator $\dot{A}$ of the form (6.76). Let $L^2_G(E)$ be a model space constructed in the proof or theorem (6.5.1). Let also $E(s)$ be the orthoprojection operator in $L^2_G(E)$ defined by

$$E(s)f(t) = \begin{cases} f(t), & 0 \le t \le s, \\ 0, & t > s, \end{cases} \tag{9.171}$$

where $f(t) \in C_{00}(E, [0, +\infty))$. Here $C_{00}(E, [0, +\infty))$ is the set of continuous compactly supported functions $f(t)$, $([0 < t < +\infty))$ with values in E. Then for the operator A, that is the operator of multiplication by independent variable defined in (6.69) in the proof of Theorem 6.5.1, we have

$$A = \int_0^\infty s\, dE(s), \tag{9.172}$$

and $E(s)$ is the resolution of identity of the operator A. By construction provided in the proof of Theorem 6.5.1, the operator A is the quasi-kernel of $\operatorname{Re}\mathbb{A}$, where $\mathbb{A}$ is an accretive $(*)$-extension of the model system. Let us calculate $(E(s)f(t), f(t))$ and $(Af(t), f(t))$ (here we use $L^2_G(E)$ scalar product).

$$(E(s)f(t), f(t)) = \int_0^\infty (dG(t)E(s)f(t), f(t))_E = \int_0^s (dG(t)f(t), f(t))_E, \tag{9.173}$$

$$(Af(t), f(t)) = \int_0^\infty s\, d\left\{\int_0^s (dG(t)f(t), f(t))_E\right\} = \int_0^\infty s\, d(G(s)x(s), x(s))_E. \tag{9.174}$$

The equality $A = A_K$ holds (see Proposition 9.5.2) if for all $\varphi \in \mathfrak{N}_{-a}$, $\varphi \ne 0$

$$\int_0^\infty \frac{(dE(t)\varphi, \varphi)}{t} = \infty, \tag{9.175}$$

where $\mathfrak{N}_{-a}$ is the deficiency subspace of the operator $\dot{A}$ corresponding to the point $(-a)$, $(a > 0)$. But according to Theorem 6.5.1 (see also (6.74)) we have

$$\mathfrak{N}_z = \left\{\frac{h}{t - z} \in L^2_G(E) \mid h \in E\right\},$$

and hence

$$\mathfrak{N}_{-a} = \left\{\frac{h}{t + a} \in L^2_G(E) \mid h \in E\right\}. \tag{9.176}$$

Taking into account (9.170) we have for all $h \in E$,

$$\int_0^\infty \frac{(dE(s)\varphi,\varphi)_{L^2_G(E)}}{s} = \int_0^\infty \frac{(dE(s)\frac{h}{t+a},\frac{h}{t+a})_{L^2_G(E)}}{s} = \int_0^\infty \frac{(dG(s)h,h)_E}{s(s+a)^2}.$$

Hence the operator $A = A_K$ iff

$$\int_0^\infty \frac{(dG(t)h,h)_E}{t(t+a)^2} = \infty, \quad \forall h \in E,\ h \neq 0. \tag{9.177}$$

Let us transform (9.170)

$$\begin{aligned} \int_0^\infty \frac{(dG(t)h,h)_E}{t} &= \int_0^\infty \frac{(t+a)^2}{t}\left(dG(t)\frac{h}{t+a},\frac{h}{t+a}\right)_E \\ &= \int_0^\infty t\left(dG(t)\frac{h}{t+a},\frac{h}{t+a}\right)_E + 2a\int_0^\infty \left(dG(t)\frac{h}{t+a},\frac{h}{t+a}\right)_E \\ &\quad + a^2\int_0^\infty \frac{(dG(t)h,h)_E}{t(t+a)^2}. \end{aligned} \tag{9.178}$$

Since $\operatorname{Re}\mathbb{A}$ is a non-negative self-adjoint bi-extension of $\dot{A}$ in the model system, then we can apply Theorem 9.6.6 to get (9.117). Then first two integrals in (9.178) converge for a fixed a because of (9.117) and equality

$$\int_0^\infty \left(dG(t)\frac{h}{t+a},\frac{h}{t+a}\right)_E = \int_0^\infty d\left(E(t)\varphi,\varphi\right), \quad \varphi \in \mathfrak{N}_{-a}.$$

Therefore the divergence of integral in (9.170) completely depends on divergence of the last integral in (9.178).

Now we can prove the first part of the theorem. Let Θ be our L-system with A_K that is a quasi-kernel for $\operatorname{Re}\mathbb{A}$, and the impedance function $V_\Theta(z)$. Without loss of generality we can consider Θ as a minimal system, otherwise we would take the principal part of Θ that is minimal and has the same impedance function (see Theorem 6.6.1 and Remark 6.6.3). Furthermore, $V_\Theta(z)$ can be realized as an impedance function of the model L-system Θ_1 constructed in the proof of Theorem 7.1.5. Some of the elements of Θ_1 were already described above during the proof of the second part of the theorem. If the L-system Θ_1 is not minimal, we consider its principal part $\Theta_{1,0}$ that is described by (6.93) and has the same impedance function as Θ_1. Since both Θ and $\Theta_{1,0}$ share the same impedance function $V_\Theta(z)$ they also have the same transfer function $W_\Theta(z)$ and thus we can apply Theorem 6.6.10 on bi-unitary equivalence. According to this theorem the quasi-kernel operator A_0 of $\Theta_{1,0}$ is unitarily equivalent to the quasi-kernel A_K in Θ. Consequently, property (9.175) of A_K gets transferred by the unitary equivalence mapping to the corresponding property of A_0 making it, by Proposition 9.5.2, the Kreĭn-von

Neumann self-adjoint extension of the corresponding symmetric operator $\dot{A}_0$ of $\Theta_{1,0}$. But this (see Theorem 6.6.1) implies that the quasi-kernel operator A of Θ_1 (defined by (6.69) and (9.172)) is also the Kreĭn-von Neumann self-adjoint extension and hence has property (9.175) that causes (9.177). Using (9.177) in conjunction with (9.178) we obtain (9.170). That proves the theorem. □

9.9 Realization of inverse Stieltjes functions

A scalar version of the following definition can be found in [159].

Definition 9.9.1. We will call an operator-valued Herglotz-Nevanlinna function $V(z)$ in a finite-dimensional Hilbert space E an **inverse Stieltjes** if $V(z)$ is holomorphic in $\mathrm{Ext}[0,+\infty)$ and

$$\frac{\mathrm{Im}[V(z)/z]}{\mathrm{Im}\, z} \geq 0. \tag{9.179}$$

Combining (9.179) with (6.7) we obtain

$$\sum_{k,l=1}^{n} \left(\frac{V(z_k)/z_k - V(\bar{z}_l)/\bar{z}_l}{z_k - \bar{z}_l} h_k, h_l \right)_E \geq 0,$$

for an arbitrary sequence $\{z_k\}$ $(k = 1, \ldots, n)$ of $(\mathrm{Im}\, z_k > 0)$ complex numbers and a sequence of vectors $\{h_k\}$ in E. It can be shown (see [159]) that every inverse Stieltjes function $V(z)$ in a finite-dimensional Hilbert space E admits the integral representation

$$V(z) = \alpha + z\beta + \int_0^\infty \left(\frac{1}{t-z} - \frac{1}{t} \right) dG(t), \tag{9.180}$$

where $\alpha \leq 0$, $\beta \geq 0$, and $G(t)$ is a non-decreasing on $[0,+\infty)$ operator-valued function such that

$$\int_0^\infty \frac{(dG(t)h,h)}{t+t^2} < \infty, \quad \forall h \in E.$$

The following definition provides the description of all realizable inverse Stieltjes operator-valued functions.

Definition 9.9.2. An operator-valued inverse Stieltjes function $V(z)$ in a finite-dimensional Hilbert space E is a member of the **class** $S^{-1}(R)$ if in the representation (9.180) we have

$$\text{i)} \qquad \beta = 0,$$
$$\text{ii)} \qquad \alpha h = 0,$$

for all $h \in E$ with

$$\int_0^\infty (dG(t)h,h)_E < \infty.$$

In what follows we will, however, be mostly interested in the following subclass of $S^{-1}(R)$.

Definition 9.9.3. An inverse Stieltjes function $V(z) \in S^{-1}(R)$ is a member of the **class** $S_0^{-1}(R)$ if

$$\int_0^\infty (dG(t)h, h)_E = \infty,$$

for all $h \in E$, $h \neq 0$.

It is not hard to see that $S_0^{-1}(R)$ is the analogue of the class $N_0(R)$ introduced in Section 7.1 and of the class $S_0(R)$ discussed in Section 9.8.

A $(*)$-extensions $\mathbb{A}$ of an operator $T \in \Lambda(\dot{A})$ is called **accumulative** if

$$(\mathrm{Re}\,\mathbb{A}f, f) \leq (\dot{A}^* f, f) + (f, \dot{A}^* f), \quad f \in \mathcal{H}_+. \tag{9.181}$$

An L-system Θ of the form (6.31)-(6.36) is called **accumulative** if its operator $\mathbb{A}$ is accumulative, i.e., satisfies (9.181). It is easy to see that if an L-system is accumulative, then (9.181) implies that the operator $\dot{A}$ of the system is non-negative and both operators T and T^* are accretive.

The following statement is the direct realization theorem for the functions of the class $S_0^{-1}(R)$.

Theorem 9.9.4. *Let Θ be an accumulative L-system of the form* (6.31)–(6.36) *with an invertible channel operator K and $\overline{\mathrm{Dom}(\dot{A})} = \mathcal{H}$. Then its impedance function $V_\Theta(z)$ of the form* (6.47) *belongs to the class $S_0^{-1}(R)$.*

Proof. First we will show that $V_\Theta(z)$ is an inverse Stieltjes function. Let $\{z_k\}$ $(k = 1, \ldots, n)$ be a sequence of non-real $(z_k \neq \bar{z}_k)$ complex numbers and φ_k $(z_k \neq \bar{z}_k)$ is a sequence of elements of $\mathfrak{N}_{z_k}$, the defect subspace of the operator $\dot{A}$. Then for every k there exists $h_k \in E$ such that

$$\varphi_k = z_k(\mathrm{Re}\,\mathbb{A} - z_k I)^{-1} K h_k, \qquad (k = 1, \ldots, n). \tag{9.182}$$

Taking into account that $\dot{A}^* \varphi_k = z_k \varphi_k$, formula (9.182), and letting $\varphi = \sum_{k=1}^n \varphi_k$ we get

$$\begin{aligned}
&(\dot{A}^*\varphi, \varphi) + (\varphi, \dot{A}^*\varphi) - (\mathrm{Re}\,\mathbb{A}\varphi, \varphi) \\
&= \sum_{k,l=1}^n \Big[(\dot{A}^*\varphi_k, \varphi_l) + (\varphi_k, \dot{A}^*\varphi_l) - (\mathrm{Re}\,\mathbb{A}\varphi_k, \varphi_l)\Big] \\
&= \sum_{k,l=1}^n ([-\mathrm{Re}\,\mathbb{A} + z_k + \bar{z}_l]\varphi_k, \varphi_l) \\
&= \sum_{k,l=1}^n \Bigg(\frac{(\mathrm{Re}\,\mathbb{A} - \bar{z}_l I)^{-1}(\bar{z}_l(\mathrm{Re}\,\mathbb{A} - \bar{z}_l I) - z_k(\mathrm{Re}\,\mathbb{A} - z_k I))(\mathrm{Re}\,\mathbb{A} - z_k I)^{-1}}{z_k \bar{z}_l (z_k - \bar{z}_l)} \\
&\qquad \times K h_k, K h_l\Bigg)
\end{aligned}$$

$$= \sum_{k,l=1}^{n} \left(\frac{\bar{z}_l K^*(\operatorname{Re}\mathbb{A} - z_k I)^{-1} K - z_k K^*(\operatorname{Re}\mathbb{A} - z_l I)^{-1} K}{z_k \bar{z}_l (z_k - \bar{z}_l)} h_k, h_l \right)$$
$$= \sum_{k,l=1}^{n} \left(\frac{\bar{z}_l V_\Theta(z_k) - z_k V_\Theta(\bar{z}_l)}{z_k z_l (z_k - \bar{z}_l)} h_k, h_l \right) \geq 0.$$

The last line can be re-written as

$$\sum_{k,l=1}^{n} \left(\frac{V_\Theta(z_k)/z_k - V_\Theta(\bar{z}_l)/\bar{z}_l}{z_k - \bar{z}_l} h_k, h_l \right) \geq 0. \tag{9.183}$$

Letting in (9.183) $n = 1$, $z_1 = z$, and $h_1 = h$ we get

$$\left(\frac{V_\Theta(z)/z - V_\Theta(\bar{z})/\bar{z}}{z - \bar{z}} h, h \right) \geq 0, \tag{9.184}$$

which means

$$\frac{\operatorname{Im}\left(V_\Theta(z)/z\right)}{\operatorname{Im} z} \geq 0,$$

and therefore $V_\Theta(z)/z$ is a Herglotz-Nevanlinna function. In Theorem 6.4.3 we have shown that $V_\Theta(z) \in N(R)$. Applying (9.179) we conclude that $V_\Theta(z)$ is an inverse Stieltjes function.

Now we will show that $V_\Theta(z)$ belongs to $S^{-1}(R)$. As any inverse Stieltjes function $V_\Theta(z)$ has its integral representation (9.180) where $\alpha \leq 0$, $\beta \geq 0$, and

$$\int_0^\infty \frac{(dG(t)h, h)}{t + t^2} < \infty, \quad \forall h \in E.$$

In a neighborhood of zero the expression $(t + t^2)$ is equivalent to the $(t + t^3)$ and in a neighborhood of the point at infinity

$$\frac{1}{t + t^3} < \frac{1}{t + t^2}.$$

Hence,

$$\int_0^\infty \frac{(dG(t)h, h)}{t + t^3} < \infty, \quad \forall h \in E.$$

Furthermore,

$$V_\Theta(z) = \alpha + z\beta + \int_0^\infty \left(\frac{1}{t - z} - \frac{t}{1 + t^2} + \frac{t}{1 + t^2} - \frac{1}{t} \right) dG(t)$$
$$= \left(\alpha - \int_0^\infty \frac{dG(t)}{t + t^3} \right) + z\beta + \int_0^\infty \left(\frac{1}{t - z} - \frac{t}{1 + t^2} \right) dG(t).$$

On the other hand, as it was shown in Section 6.4, a Herglotz-Nevanlinna function can be realized if and only if it belongs to the class $N(R)$ and hence in representation (6.52) $F = 0$ and

$$Qh = \int_{-\infty}^{+\infty} \frac{t}{1+t^2} dG(t)h,$$

for all $h \in E$ such that (6.54) holds. Considering this and the uniqueness of the function $G(t)$ we obtain

$$\left(\alpha - \int_0^{\infty} \frac{dG(t)}{t+t^3}\right) f = \int_0^{+\infty} \frac{t}{1+t^2} dG(t) f, \tag{9.185}$$

for all $f \in E$ such that $\int_{-\infty}^{+\infty} (dG(t)f, f)_E < \infty$. Solving (9.185) for α we get

$$\alpha f = \int_0^{\infty} \frac{1}{t}\, dG(t) f, \tag{9.186}$$

for the same selection of f. The left-hand side of (9.186) is non-positive but the right-hand side is non-negative. This means that $\alpha = 0$ and $V_\Theta(z) \in S^{-1}(R)$.

The proof of the fact that $V_\Theta(z) \in S_0^{-1}(R)$ is similar to the proof of Theorem 7.1.4 and follows from Theorem 6.2.10. □

The inverse realization theorem can be stated and proved for the class $S_0^{-1}(R)$ as follows.

Theorem 9.9.5. *Let an operator-valued function $V(z)$ belong to the class $S_0^{-1}(R)$. Then $V(z)$ can be realized as an impedance function of an accumulative minimal L-system Θ of the form* (6.31)–(6.36) *with an invertible channel operator K, a non-negative densely-defined symmetric operator $\dot{A}$ and $J = I$.*

Proof. The class $S_0^{-1}(R)$ is a subclass of $N_0(R)$ and hence it is realizable by a minimal L-system Θ with a densely-defined symmetric operator $\dot{A}$ and $J = I$. Thus all we have to show is that the L-system Θ we have constructed in the proof of Theorem 7.1.5 is an accumulative L-system, i.e., satisfying the condition (9.181), that is

$$(\mathrm{Re}\,\mathbb{A}f, f) \leq (\dot{A}^* f, f) + (f, \dot{A}^* f), \qquad f \in \mathcal{H}_+.$$

Since the L-system Θ is minimal then the operator $\dot{A}$ is prime. Applying Lemma 6.6.4 provides us with (6.94) which implies

$$\underset{z \neq \bar{z}}{c.l.s.}\, \mathfrak{N}_z = \mathcal{H}, \quad z \neq \bar{z}. \tag{9.187}$$

In the proof of Theorem 9.9.4 we have shown that

$$(\mathrm{Re}\,\mathbb{A}\varphi, \varphi) \leq (\dot{A}^*\varphi, \varphi) + (\varphi, \dot{A}^*\varphi), \quad \varphi = \sum_{k=1}^{n} \varphi_k, \quad \varphi_k \in \mathfrak{N}_{z_k}, \tag{9.188}$$

is equivalent to (9.183), where z_k are defined by (9.182). Combining (9.187) and (9.188) we get property (9.181) and conclude that Θ is an accumulative L-system. $\square$

Now we define a subclass of the class $S_0^{-1}(R)$.

Definition 9.9.6. An operator-valued Stieltjes function $V(z)$ of the class $S_0^{-1}(R)$ is said to be a member of the **class** $S_{0,F}^{-1}(R)$ if

$$\int_0^\infty \frac{t}{t^2+1}\,(dG(t)h,h)_E = \infty, \tag{9.189}$$

for all non-zero $h \in E$.

Theorem 9.9.7. *Let Θ be an accumulative L-system of the form (6.31)–(6.36) with an invertible channel operator K and a symmetric densely-defined operator $\dot{A}$. If Friedreichs extension A_F is a quasi-kernel for $\operatorname{Re}\mathbb{A}$, then the impedance $V_\Theta(z)$ of the form (6.47) belongs to the class $S_{0,F}^{-1}(R)$.*

Conversely, if $V(z) \in S_{0,F}^{-1}(R)$, then it can be realized as an impedance of an accumulative L-system Θ of the form (6.31)–(6.36) with $\operatorname{Re}\mathbb{A}$ containing A_F as a quasi-kernel and a preassigned direction operator J for which $I + iV(-i)J$ is invertible.

Proof. Following the framework of the proof of Theorem 9.8.14, we begin with the proof of the second part. First we use realization Theorems 6.5.1, 7.1.5, and 9.9.5 to construct a minimal model L-system Θ whose impedance function is $V(z)$. Then we will show that (9.170) is equivalent to the fact that self-adjoint operator A defined in (6.69), that we constructed in these theorems to be a quasi-kernel for $\operatorname{Re}\mathbb{A}$, coincides with A_F, that is the Friedreichs extension of the symmetric operator $\dot{A}$ of the form (6.76). Let $L^2_G(E)$ be a model space constructed in the proof or theorem (6.5.1). Let also $E(s)$ be the orthoprojection operator in $L^2_G(E)$ defined by (9.171). Then for the operator A defined in 6.69 in the proof of Theorem 6.5.1 we have

$$A = \int_0^\infty t\,dE(t),$$

and $E(t)$ is the spectral function of operator A. As we have shown in the proof of Theorem 9.8.14 the relations (9.173) and (9.174) hold. The equality $A = A_F$ holds (see Proposition 9.5.2) if for all $\varphi \in \mathfrak{N}_{-a}$

$$\int_0^\infty t\,(dE(t)\varphi,\varphi)_E = \infty, \tag{9.190}$$

where $\mathfrak{N}_{-a}$ is the deficiency subspace of the operator $\dot{A}$ corresponding to the point $(-a)$, $(a > 0)$. But according to Theorem 6.5.1 we have $\mathfrak{N}_{-a}$ described by (9.176). Taking into account (9.189) we have, for all $h \in E$,

$$\int_0^\infty s(dE(s)\varphi,\varphi)_{L^2_G(E)} = \int_0^\infty sd\left(E(s)\frac{h}{t+a},\frac{h}{t+a}\right)_{L^2_G(E)} = \int_0^\infty \frac{s\,(dG(s)h,h)_E}{(s+a)^2}.$$

Hence the operator $A = A_F$ iff

$$\int\limits_0^\infty \frac{t\,(dG(t)h,h)_E}{(t+a)^2} = \infty, \quad \forall h \in E,\ h \neq 0. \tag{9.191}$$

Let us transform (9.189)

$$\begin{aligned}\int_0^\infty \frac{t}{t^2+1}(dG(t)h,h)_E &= \int_0^\infty \frac{t(t+a)^2}{t^2+1}\left(dG(t)\frac{h}{t+a},\frac{h}{t+a}\right)_E \\ = \int_0^\infty \frac{t}{(t+a)^2}\cdot\frac{t^2}{t^2+1}&\left(dG(t)\frac{h}{t+a},\frac{h}{t+a}\right)_E \\ + 2a\int_0^\infty \frac{t^2}{(t+a)^2(t^2+1)}&\left(dG(t)\frac{h}{t+a},\frac{h}{t+a}\right)_E \\ + a^2\int_0^\infty \frac{1}{t^2+1}\cdot\frac{t\,(dG(t)h,h)_E}{(t+a)^2}&.\end{aligned} \tag{9.192}$$

Consider the obvious inequality

$$\frac{t^2}{(t+a)^2(t^2+1)} - \frac{1}{t^2+1} = \frac{t^2-(t+a)^2}{(t+a)^2(t^2+1)} = \frac{(2t+a)(-a)}{(t+a)^2(t^2+1)} < 0.$$

Taking into account this inequality and the fact that the integral

$$\int_0^\infty \frac{(dG(t)h,h)_E}{t^2+1}$$

converges for all $h \in E$, we conclude that the second integral in (9.192) is convergent. Let us denote this integral as Q. Then using (9.192) and obvious estimates we obtain

$$\begin{aligned}\int_0^\infty \frac{t}{t^2+1}(dG(t)h,h)_E \leq \int_0^\infty &\frac{t}{(t+a)^2}\left(dG(t)\frac{h}{t+a},\frac{h}{t+a}\right)_E \\ &+ 2aQ + a^2\int_0^\infty \frac{t\,(dG(t)h,h)_E}{(t+a)^2},\end{aligned}$$

or

$$\int_0^\infty \frac{t}{t^2+1}(dG(t)h,h)_E \leq (a^2+1)\int_0^\infty \frac{t}{(t+a)^2}\left(dG(t)\frac{h}{t+a},\frac{h}{t+a}\right)_E + 2aQ.$$

Since $V(z) \in S_0^{-1}(R)$, then (9.186) holds and the integral on the left diverges causing the integral on the right-hand to side diverge as well. Thus $A = A_F$.

Now we can prove the first part of the theorem. Let Θ be our L-system with A_F that is a quasi-kernel for $\mathrm{Re}\,\mathbb{A}$, and the impedance function $V_\Theta(z)$. Then $V_\Theta(z)$ can be realized as an impedance function of the model L-system Θ_1 constructed in the proof of Theorem 7.1.5. Repeating the argument of the second part of the proof of Theorem 9.8.14 with A_K replaced by A_F we conclude that the quasi-kernel operator A of Θ_1 (defined by (6.69)) is the Friedreichs self-adjoint extension and hence has property (9.190) that in turn causes (9.191) for any $a > 0$. Let $a = 1$, then by (9.191),

$$\infty = \int\limits_0^\infty \frac{t\,(dG(t)h,h)_E}{(t+1)^2} \leq \int\limits_0^\infty \frac{t\,(dG(t)h,h)_E}{t^2+1}, \quad \forall h \in E,\ h \neq 0,$$

and hence the integral on the right diverges and (9.189) holds. This completes the proof. □

Chapter 10

L-systems with Schrödinger operator

In this chapter we apply some of the previous results to the problems related to the Schrödinger operator. First we give a complete characterization of all $(*)$-extensions of ordinary differential operators. Then we use it to provide a thorough description of all L-systems with a Schrödinger operator T_h. Moreover, we describe the class of scalar Stieltjes/(inverse Stieltjes)-like functions that can be realized as impedance functions of L-systems with a Schrödinger operator T_h. The formulas that restore an L-system uniquely from a given Stieltjes/(inverse Stieltjes)-like function as the impedance function of this L-system are derived. These formulas allow us to solve the inverse problem and find the exact value of the parameter h in the definition of T_h as well as a real parameter μ that appears in the construction of the elements of the L-system being realized. A detailed study of these formulas shows the dynamics of the restored parameters h and μ in terms of the changing free term in the integral representation of a realizable function. We also provide a full description of accretive, sectorial, and extremal boundary value problems for a Schrödinger operator T_h on the half-line in terms of the boundary parameter h.

10.1 $(*)$-extensions of ordinary differential operators

In this section we will give a complete description of all $(*)$-extensions of ordinary differential operators. First we note that for the case of a densely-defined symmetric operator $\dot{A}$ the set of formulas (4.60) can be re-written (see p. 110). Let $\mathcal{U}$ be an isometric operator from the defect subspace $\mathfrak{N}_i$ of the symmetric operator $\dot{A}$ onto the defect subspace $\mathfrak{N}_{-i}$. Then the formulas below establish a one-to one correspondence between all $(*)$-extensions $\mathbb{A}$ of an operator T and all isometries $\mathcal{U}$,

$$\mathbb{A}f = \dot{A}^*f + i\mathcal{R}^{-1}(\mathcal{U} - I)\varphi, \quad \mathbb{A}^*f = \dot{A}^*f + i\mathcal{R}^{-1}(\mathcal{U} - I)\psi, \qquad f \in \mathcal{H}_+, \quad (10.1)$$

where $\varphi, \psi \in \mathfrak{N}_i$ are uniquely determined from the conditions

$$f - (\mathcal{U} + I)\varphi \in \mathrm{Dom}(T), \quad f - (\mathcal{U} + I)\psi \in \mathrm{Dom}(T^*),$$

and $\mathcal{R} \in [\mathcal{H}_-, \mathcal{H}_+]$ is the Riesz-Berezansky operator of the triplet $\mathcal{H}_+ \subset \mathcal{H} \subset \mathcal{H}_-$. If the symmetric operator $\dot{A}$ has deficiency indices (n, n), then formulas (10.1) can be rewritten in the form

$$\mathbb{A}f = \dot{A}^* f + \sum_{k=1}^{n} \Delta_k(f)\chi_k, \quad \mathbb{A}^* f = \dot{A}^* f + \sum_{k=1}^{n} \delta_k(f)\chi_k, \tag{10.2}$$

where $\{\chi_j\}_1^n \in \mathcal{H}_-$ is a basis in the subspace $\mathcal{R}^{-1}(\mathcal{U} - I)\mathfrak{N}_i$, and $\{\Delta_k\}_1^n$, $\{\delta_k\}_1^n$, are bounded linear functionals on $\mathcal{H}_+$ with the properties

$$\Delta_k(f) = 0, \quad \forall f \in \mathrm{Dom}(T), \quad \delta_k(f) = 0, \quad \forall f \in \mathrm{Dom}(T^*). \tag{10.3}$$

We will give a description of all $(*)$-extensions of a non-symmetric differential operator on a half-line. Consider the following self-adjoint quasi-differential expression on $L_2[a, +\infty]$,

$$l(y) = (-1)^n (p_0 y^{(n)}) + (-1)^{n-1}(p_1 y^{(n-1)}) + \ldots + p_n y,$$

where $\frac{1}{p_0(x)}, p_1(x), \ldots, p_n(x)$ are locally summable functions on $[a, +\infty)$. Denote by D^* the set of functions $y \in L_2[a, +\infty)$ for which the quasi-derivatives [207]

$$\begin{aligned} y^{[k]} &= \frac{d^k y}{dx^k}, \; k = 0, 1 \ldots, n-1, \quad y^{[n]} = p_0 \frac{d^n y}{dx^n}, \\ y^{[n+k]} &= p_k \frac{d^{n-k} y}{dx^{n-k}} - \frac{d}{dx}\left(y^{[n+k-1]}\right), \; k = 1, 2 \ldots, n, \end{aligned}$$

are locally absolutely continuous and $y^{[2k]} = l(y)$ belongs to $L_2[a, +\infty)$. Consider the symmetric operator

$$\begin{cases} \dot{A}y = l(y), \\ y^{[k-1]}(a) = 0, \end{cases} \quad y \in D^*, \; k = 1, 2, \ldots, 2n,$$

and suppose that this operator has deficiency indices (n, n). Consider the operators

$$\begin{cases} Ty = l(y), \\ y \in D^*, \\ \sum\limits_{k=1}^{2n} \upsilon_{jk} y^{[k-1]}(a) = 0, \end{cases} \quad \begin{cases} T^*y = l(y), \\ y \in D^*, \\ \sum\limits_{k=1}^{2n} \upsilon_{*jk} y^{[k-1]}(a) = 0, \end{cases} \quad (j = 1, 2, \ldots, n). \tag{10.4}$$

Suppose also that $\mathrm{Dom}(\dot{A}) = \mathrm{Dom}(T) \cap \mathrm{Dom}(T^*)$ and $\rho(T) \neq \emptyset$. It is well known that

$$\mathrm{Dom}(\dot{A}^*) = D^*, \quad \dot{A}^* y = l(y), \quad y \in D^*.$$

Consider $\mathcal{H}_+ = \mathrm{Dom}(\dot{A}^*) = D^*$ with the inner product

$$(y, z)_+ = (y, z)_{L_2[a,+\infty)} + (l(y), l(y))_{L_2[a,+\infty)},$$

and construct the rigged Hilbert space $\mathcal{H}_+ \subset L_2[a, +\infty) \subset \mathcal{H}_-$. Let $\{\upsilon_j\}_i^n$ and $\{\upsilon_{*j}\}_i^n$ be the set of elements in $\mathcal{H}_-$ generating the functionals

$$(f, \upsilon_j) = \sum_{k=1}^{2n} \upsilon_{jk} f^{[k-1]}(a), \quad (f, \upsilon_{*j}) = \sum_{k=1}^{2n} \upsilon_{*jk} f^{[k-1]}(a), \quad (j = 1, \dots, n).$$

Let $\{\chi_j\}_i^n \in \mathcal{H}_-$ be the set of elements linearly independent with $\{\upsilon_j\}_i^n$ and generating the functionals

$$(f, \chi_j) = \sum_{k=1}^{2n} \chi_{jk} f^{[k-1]}(a), \quad (j = 1, 2, \dots, n).$$

We also suppose that the operator

$$\begin{cases} \widehat{A}y = l(y), \\ (y, \chi_j) = 0, \end{cases} \quad y \in \mathcal{H}_+, \quad (j = 1, 2, \dots, n),$$

is self-adjoint in $L_2[a, +\infty)$.

Theorem 10.1.1. *Let T be a differential operator of the form (10.4). Then the formulas*

$$\mathbb{A}y = l(y) + \sum_{k,j=1}^{n} (y, \upsilon_j) c_{kj} \chi_k, \quad \mathbb{A}^* y = l(y) + \sum_{k,j=1}^{n} (y, \upsilon_{*j}) d_{kj} \chi_k, \tag{10.5}$$

establish a one-to-one correspondence between the set of (*)-extensions $\mathbb{A}$ of the operator T and the set of matrices $\mathcal{C} = \|c_{jk}\|$ and $\mathcal{D} = \|d_{kj}\|$ uniquely determined by the matrices $\mathcal{U} = \|\upsilon_{jk}\|$, $\mathcal{U}_* = \|\upsilon_{*jk}\|$, $\mathcal{V}_* = \|\chi_{jk}\|$ by the relation

$$\mathcal{U}_*^* \mathcal{D}^* \mathcal{V} - \mathcal{V}^* \mathcal{C} \mathcal{U} = \begin{bmatrix} 0 & I_{\nearrow} \\ -I_{\nearrow} & 0 \end{bmatrix}, \quad \left(\pm I_{\nearrow} := \begin{bmatrix} 0 \dots \pm 1 \\ \pm 1 \dots 0 \end{bmatrix} \right). \tag{10.6}$$

Proof. Let $\mathbb{A}$ be a (*)-extension of an operator T of the form (10.4). Then, as we mentioned above, $\mathbb{A}$ has the form (10.2) with the property (10.3). Therefore

$$\Delta_k(f) = \sum_{j=1}^{n} (f, \upsilon_j) c_{kj}, \quad \delta_k(f) = \sum_{j=1}^{n} (f, \upsilon_{*j}) d_{kj},$$

and the matrices $C = \|c_{kj}\|$ and $D = \|d_{kj}\|$ are invertible. It follows from (10.2) that

$$(\dot{A}^* f, g) - (f, \dot{A}^* g) = \sum_{k,j=1}^{n} (\upsilon_{*j}, g) \overline{d_{kj}} (f, \chi_k) - \sum_{k,j=1}^{n} (f, \upsilon_j) c_{kj} (\chi_k, g). \tag{10.7}$$

Applying the Lagrange formula for differential operators and taking into account that the deficiency indices of the operator $\dot{A}$ are (n,n), we get, using (10.7) with $g=f$, that

$$
\begin{aligned}
-\sum_k [f^{[k-1]}(a)^{[2n-k]}(a) - f^{[2n-k]}(a) f^{[k-1]}(a)] \\
= \sum_{l,m} \Big(\sum_{k,j} \overline{d_{kj}}\,\overline{\upsilon_{*jl}}\,\chi_{km}\Big) \overline{f^{[l-1]}(a)} f^{[m-1]}(a) \\
- \sum_{l,m} \Big(\sum_{k,j} c_{kj}\upsilon_{jl}\overline{\chi_{km}}\Big) f^{[l-1]}(a) \overline{f^{[m-1]}(a)}.
\end{aligned} \tag{10.8}
$$

Hence, we obtain (10.6). Going back to (10.6) we get (10.8), and thus (10.5) determines a $(*)$-extension of an operator T of the form (10.4). □

10.2 Canonical L-systems with Schrödinger operator

We begin this section by applying Theorem 10.1.1 in the context of a non-self-adjoint Schrödinger operator. Let $\mathcal{H} = L_2[a,+\infty)$ and $l(y) = -y'' + q(x)y$, where q is a real locally summable function. Suppose that the symmetric operator

$$
\begin{cases} \dot{A}y = -y'' + q(x)y, \\ y(a) = y'(a) = 0, \end{cases} \tag{10.9}
$$

has deficiency indices (1,1). Let D^* be the set of functions locally absolutely continuous together with their first derivatives such that $l(y) \in L_2[a,+\infty)$. Consider $\mathcal{H}_+ = \mathrm{Dom}(\dot{A}^*) = D^*$ with the scalar product

$$
(y,z)_+ = \int_a^\infty \Big(y(x)\overline{z(x)} + l(y)\overline{l(z)} \Big)\, dx, \qquad y,\, z \in D^*.
$$

Let $\mathcal{H}_+ \subset L_2[a,+\infty) \subset \mathcal{H}_-$ be the corresponding triplet of Hilbert spaces. Consider elements $\upsilon, \upsilon_*, \chi \in \mathcal{H}_-$, generating functionals

$$
\begin{aligned}
&(f,\chi) = \frac{(\mathrm{Im}\,h)^{\frac{1}{2}}}{|\mu - h|}[\mu f(a) - f'(a)], \qquad \mathrm{Im}\,h > 0,\ \mathrm{Im}\,\mu = 0, \\
&(f,\upsilon) = h f(a) - f'(a), \qquad (f,\upsilon_*) = \overline{h} f(a) - f'(a),
\end{aligned}
$$

and operators

$$
\begin{cases} Ty = l(y) = -y'' + q(x)y, \\ (y,\upsilon) = hy(a) - y'(a) = 0, \end{cases} \qquad \begin{cases} T^*y = l(y) = -y'' + q(x)y, \\ (y,\upsilon_*) = \overline{h}y(a) - y'(a) = 0, \end{cases} \tag{10.10}
$$

$$
\begin{cases} \widehat{A}y = l(y) = -y'' + q(x)y, \\ (y,\chi) = \mu y(a) - y'(a) = 0, \end{cases}, \quad \mathrm{Im}\,\mu = 0.
$$

It is well known [3] that $\widehat{A} = \widehat{A^*}$. By Theorem 10.1.1, the $(*)$ - extension $\mathbb{A}$ of an operator T of the form (10.10) can be represented as

$$\mathbb{A}y = -y'' + q(x)y + c(y, \upsilon)\chi, \quad \mathbb{A}^*y = -y'' + q(x)y + d(y, \upsilon_*)\chi,$$

where the numbers c and d satisfy the relation (10.6). According to (10.4) and (10.10), the matrices $\mathcal{U}$, $\mathcal{U}_*$ and $\mathcal{V}$ have the form

$$\mathcal{U} = \begin{bmatrix} h, & -1 \end{bmatrix}, \quad \mathcal{U}_* = \begin{bmatrix} \overline{h}, & -1 \end{bmatrix}, \quad \mathcal{V} = \begin{bmatrix} \frac{(\operatorname{Im} h)^{1/2}}{|\mu-h|}\mu, & \frac{-(\operatorname{Im} h)^{1/2}}{|\mu-h|} \end{bmatrix}.$$

Therefore,

$$\begin{aligned}\overline{d}\,\mathcal{U}_*^*\,\mathcal{V} - c\mathcal{V}^*\mathcal{U} &= \overline{d}\begin{bmatrix} h \\ -1 \end{bmatrix}\begin{bmatrix} \frac{(\operatorname{Im} h)^{1/2}}{|\mu-h|}\mu & -\frac{(\operatorname{Im} h)^{1/2}}{|\mu-h|} \end{bmatrix} - c\begin{bmatrix} \frac{(\operatorname{Im} h)^{1/2}}{|\mu-h|}\mu \\ -\frac{(\operatorname{Im} h)^{1/2}}{|\mu-h|} \end{bmatrix}\begin{bmatrix} h, & -1 \end{bmatrix} \\ &= \begin{bmatrix} 0 & 1 \\ -1 & 0 \end{bmatrix}.\end{aligned}$$

Solving this matrix equation with respect to c and d we get

$$c = \frac{|\mu - h|}{(\mu - h)(\operatorname{Im} h)^{\frac{1}{2}}}, \qquad d = \frac{|\mu - h|}{(\mu - \overline{h})(\operatorname{Im} h)^{\frac{1}{2}}}.$$

Consider now the elements $\mu\delta(x-a) + \delta'(x-a)$ and $\overline{h}\delta(x-a) + \delta'(x-a)$ in $\mathcal{H}_-$, and generating functionals

$$\begin{aligned}(f, \mu\delta(x-a) + \delta'(x-a)) &= \mu f(a) - f'(a), \\ (f, \overline{h}\delta(x-a) + \delta'(x-a)) &= hf(a) - f'(a),\end{aligned}$$

where $\delta(x-a)$ and $\delta'(x-a)$ are the delta-function and the derivative of the delta-function at the point a. Thus, we have proved the following.

Theorem 10.2.1. *The set of all $(*)$-extensions $\mathbb{A}$ of a non-self-adjoint Schrödinger operator T_h of the form* (10.10) *can be represented in the form*

$$\begin{aligned}\mathbb{A}y &= -y'' + q(x)y - \frac{1}{\mu - h}\,[y'(a) - hy(a)]\,[\mu\delta(x-a) + \delta'(x-a)], \\ \mathbb{A}^*y &= -y'' + q(x)y - \frac{1}{\mu - \overline{h}}\,[y'(a) - \overline{h}y(a)]\,[\mu\delta(x-a) + \delta'(x-a)].\end{aligned} \tag{10.11}$$

Moreover, the formulas (10.11) *establish a one-to-one correspondence between the set of all $(*)$-extensions of a Schrödinger operator T_h of the form* (10.10) *and all real numbers $\mu \in [-\infty, +\infty]$.*

Consider the symmetric operator $\dot{A}$ of the form (10.9) with defect indices (1,1), generated by the differential operation $l(y) = -y'' + q(x)y$. Let $\varphi_k(x,\lambda)$, $(k = 1,2)$ be the solutions of the following Cauchy problems:

$$\begin{cases} l(\varphi_1) = \lambda\varphi_1 \\ \varphi_1(a,\lambda) = 0 \\ \varphi_1'(a,\lambda) = 1 \end{cases}, \qquad \begin{cases} l(\varphi_2) = \lambda\varphi_2 \\ \varphi_2(a,\lambda) = -1 \\ \varphi_2'(a,\lambda) = 0 \end{cases}. \tag{10.12}$$

It is well known [3] that there exists a function $m_\infty(\lambda)$ (note that $-m_\infty(\lambda)$ is a Herglotz-Nevanlinna function called the **Weyl-Titchmarsh function**) for which

$$\varphi(x,\lambda) = \varphi_2(x,\lambda) + m_\infty(\lambda)\varphi_1(x,\lambda) \tag{10.13}$$

belongs to $L_2[a,+\infty)$.

Consider an L-system with a $(*)$-extension of a non-self-adjoint Schrödinger operator described in Section 10.1 as a state-space operator. One can easily check that the $(*)$-extension

$$\mathbb{A}y = -y'' + q(x)y - \frac{1}{\mu - h}\,[y'(a) - hy(a)]\,[\mu\delta(x-a) + \delta'(x-a)], \quad \operatorname{Im} h > 0,$$

of the non-self-adjoint Schrödinger operator T_h of the form (10.10) satisfies the condition

$$\operatorname{Im}\mathbb{A} = \frac{\mathbb{A} - \mathbb{A}^*}{2i} = (.,g)g, \tag{10.14}$$

where

$$g = \frac{(\operatorname{Im} h)^{\frac{1}{2}}}{|\mu - h|}\,[\mu\delta(x-a) + \delta'(x-a)], \tag{10.15}$$

and $\delta(x-a), \delta'(x)$ are the delta-function and its derivative at the point a. Moreover,

$$(y,g) = \frac{(\operatorname{Im} h)^{\frac{1}{2}}}{|\mu - h|}\,[\mu y(a) - y'(a)], \tag{10.16}$$

where $y \in \mathcal{H}_+$, $g \in \mathcal{H}_-$, and $\mathcal{H}_+ \subset L_2(a,+\infty) \subset \mathcal{H}_-$ is the triplet of Hilbert spaces as discussed in Theorem 10.2.1. Let $E = \mathbb{C}$, then we define

$$Kc = cg, \quad K^*y = (y,g), \quad y \in \mathcal{H}_+ \quad c \in \mathbb{C}, \tag{10.17}$$

and obtain $\operatorname{Im}\mathbb{A} = KK^*$. Applying (6.36) the array

$$\Theta = \begin{pmatrix} \mathbb{A} & K & 1 \\ \mathcal{H}_+ \subset L_2[a,+\infty) \subset \mathcal{H}_- & & \mathbb{C} \end{pmatrix} \tag{10.18}$$

is a canonical L-system with state-space operator $\mathbb{A}$ of the form (10.11), the direction operator $J = 1$, and the channel operator K of the form (10.17). Indeed, the following theorem applies.

Theorem 10.2.2. *The system* Θ *of the form*

$$\begin{cases} (T_h - zI)x = \beta[\mu\delta(x-a) + \delta'(x-a)]\varphi_-, \\ \varphi_+ = \varphi_- - 2i\beta[\mu x(a) - x'(a)], \end{cases} \qquad \varphi_\pm \in \mathbb{C}, \tag{10.19}$$

with $\beta > 0$ *and* T_h $(\operatorname{Im} h > 0)$ *defined by* (10.10), *is a scattering L-system if and only if* $\beta = \frac{(\operatorname{Im} h)^{\frac{1}{2}}}{|\mu - h|}$. *The transfer function* $W_\Theta(z)$ *of this system is given by the formula*

$$W_\Theta(z) = \frac{\mu - h}{\mu - \bar{h}}\,\frac{m_\infty(z) + \bar{h}}{m_\infty(z) + h}. \tag{10.20}$$

Moreover, the function

$$V_\Theta(z) = \frac{(m_\infty(z) + \mu)\operatorname{Im} h}{(\mu - \operatorname{Re} h)\, m_\infty(z) + \mu\operatorname{Re} h - |h|^2}, \tag{10.21}$$

is the transfer function of the impedance system

$$\begin{cases} (A_\xi - zI)x = \beta[\mu\delta(x-a) + \delta'(x-a)]\psi_-, \\ \psi_+ = \beta[\mu x(a) - x'(a)]), \end{cases} \qquad \psi_\pm \in \mathbb{C},$$

where A_ξ *has the form*

$$\begin{cases} A_\xi y = -y'' + q(x)y \\ \xi y(a) = y'(a) \end{cases},$$

and

$$\xi = \frac{\mu\operatorname{Re} h - |h|^2}{\mu - \operatorname{Re} h}. \tag{10.22}$$

Proof. Let Θ of the form (10.19) be an L-system with $J = I$. Then by Definition 6.3.4 there is a $(*)$-extension $\mathbb{A}$ of T_h such that $\operatorname{Im}\mathbb{A} = K_\beta K_\beta^*$, where

$$K_\beta c = c\beta[\mu\delta(x-a) + \delta'(x-a)], \quad K_\beta^* y = \beta[\mu y(a) - y'(a)], \quad y \in \mathcal{H}_+,\ c \in \mathbb{C}. \tag{10.23}$$

All the $(*)$-extensions of T_h, however, are described by (10.11). Comparing formula (10.23) with formulas (10.14)–(10.17), applying uniqueness Theorem 4.3.9, and performing straightforward calculations, we obtain

$$|\beta|^2 = \frac{(\operatorname{Im} h)}{|\mu - h|^2}.$$

Since β is real and positive we have $\beta = \frac{(\operatorname{Im} h)^{\frac{1}{2}}}{|\mu - h|}$. Formulas (10.20) and (10.21) are obtained by direct computations. The remaining statement of the theorem follows from the fact that A_ξ with ξ described by (10.22) is the quasi-kernel of $\operatorname{Re}\mathbb{A}$ that corresponds to system (10.19) and the definition of the impedance system.

Now let $\beta = \frac{(\operatorname{Im} h)^{\frac{1}{2}}}{|\mu - h|}$. We need to show that (10.19) is an L-system. As we already mentioned, all the $(*)$-extensions of T_h are described by (10.11). Hence, (10.14) and (10.15) imply that $\operatorname{Im}\mathbb{A} = KK^*$ only if β has the above value. Consequently, (10.19) is an L-system only for that β. □

One can also see that (10.21) implies

$$\operatorname{Im} V_\Theta(i) = -\operatorname{Im} h\, \frac{|\mu - h|^2}{(\mu - \operatorname{Re} h)^2} \frac{\operatorname{Im} m_\infty(i)}{\left| m_\infty(i) + \frac{\mu \operatorname{Re} h - |h|^2}{\mu - \operatorname{Re} h}\right|^2}. \tag{10.24}$$

Now we can give a good illustration to Theorems 8.2.1 and 8.2.2 of Section 8.2 for the case when a J-unitary factor $B = -1$. Since Theorem 10.2.1 establishes a one-to-one correspondence between all $(*)$-extensions $\mathbb{A}$ of the form (10.11) and all real numbers $\mu \in [-\infty, +\infty]$, we can parameterize all $\mathbb{A}$'s via μ's and label them accordingly. We set then

$$\begin{aligned} \mathbb{A}_\mu y &= -y'' + qy - \frac{1}{\mu - h}[y'(a) - hy(a)]\,[\mu\delta(x-a) + \delta'(x-a)], \\ \mathbb{A}_\mu^* y &= -y'' + qy - \frac{1}{\mu - \bar{h}}[y'(a) - \bar{h}y(a)]\,[\mu\delta(x-a) + \delta'(x-a)], \end{aligned} \tag{10.25}$$

and

$$\begin{aligned} \mathbb{A}_\xi y &= -y'' + qy - \frac{1}{\xi - h}[y'(a) - hy(a)]\,[\xi\delta(x-a) + \delta'(x-a)], \\ \mathbb{A}_\xi^* y &= -y'' + qy - \frac{1}{\xi - \bar{h}}[y'(a) - \bar{h}y(a)]\,[\xi\delta(x-a) + \delta'(x-a)], \end{aligned} \tag{10.26}$$

with ξ defined by (10.22). Let

$$\Theta_\mu = \begin{pmatrix} & \mathbb{A}_\mu & & K_\mu & 1 \\ \mathcal{H}_+ \subset L_2[a, +\infty) \subset \mathcal{H}_- & & & & \mathbb{C} \end{pmatrix}, \tag{10.27}$$

and

$$\Theta_\xi = \begin{pmatrix} & \mathbb{A}_\xi & & K_\xi & 1 \\ \mathcal{H}_+ \subset L_2[a, +\infty) \subset \mathcal{H}_- & & & & \mathbb{C} \end{pmatrix}, \tag{10.28}$$

be the corresponding L-systems, where K_μ and K_ξ are parameterized and labeled according to (10.16)-(10.17). From relation (10.20) it follows that

$$W_{\Theta_\mu}(z) = \frac{\mu - h}{\mu - \bar{h}} \frac{m_\infty(z) + \bar{h}}{m_\infty(z) + h}, \qquad W_{\Theta_\xi}(z) = \frac{\xi - h}{\xi - \bar{h}} \frac{m_\infty(z) + \bar{h}}{m_\infty(z) + h}. \tag{10.29}$$

Because

$$\xi = \frac{\mu \operatorname{Re} h - |h|^2}{\mu - \operatorname{Re} h},$$

one obtains

$$\begin{aligned}\frac{\xi-h}{\xi-\bar{h}}&=\frac{\frac{\mu\mathrm{Re}\,h-|h|^2}{\mu-\mathrm{Re}\,h}-h}{\frac{\mu\mathrm{Re}\,h-|h|^2}{\mu-\mathrm{Re}\,h}-\bar{h}}=\frac{\mu\mathrm{Re}\,h-|h|^2-\mu h+\mathrm{Re}\,hh}{\mu\mathrm{Re}\,h-|h|^2-\mu\bar{h}+\mathrm{Re}\,h\bar{h}}\\ &=\frac{\mu(\mathrm{Re}\,h-h)-h(\bar{h}-\mathrm{Re}\,h)}{\mu(\mathrm{Re}\,h-\bar{h})-\bar{h}(\bar{h}-\mathrm{Re}\,h)}=\frac{-i\mu\mathrm{Im}\,h+ih\mathrm{Im}\,h}{i\mu\mathrm{Im}\,h-i\bar{h}\mathrm{Im}\,h}\\ &=\frac{-(\mu-h)\mathrm{Im}\,h}{(\mu-\bar{h})\mathrm{Im}\,h}=-\frac{\mu-h}{\mu-\bar{h}}.\end{aligned}$$

Therefore, from (10.29) we get

$$W_{\Theta_\mu}(z)=-W_{\Theta_\xi}(z). \tag{10.30}$$

The relations (10.30) and (6.48) imply

$$\begin{aligned}V_{\Theta_\mu}(z)&=K_\mu^*(\mathrm{Re}\,\mathbb{A}_\mu-zI)^{-1}K_\mu=i[W_{\Theta_\mu}(z)-I]\,[W_{\Theta_\mu}(z)+I]^{-1}\\ &=i[-W_{\Theta_\xi}(z)-I][-W_{\Theta_\xi}(z)+I]^{-1}=-V_{\Theta_\xi}^{-1}(z).\end{aligned}$$

Finally,

$$V_{\Theta_\mu}(z)=-V_{\Theta_\xi}^{-1}(z). \tag{10.31}$$

10.3 Accretive and sectorial boundary problems for a Schrödinger operator

Suppose that the symmetric operator $\dot{A}$ of the form (10.9) with deficiency indices (1,1) is non-negative.

Theorem 10.3.1. *A Schrödinger operator T_h, ($\mathrm{Im}\,h>0$) of the form (10.10) is accretive if and only if the function*

$$V_h(z)=-i\frac{1-[(m_\infty(z)+\bar{h})/(m_\infty(z)+h)][(m_\infty(-1)+h)/(m_\infty(-1)+\bar{h})]}{1+[(m_\infty(z)+\bar{h})/(m_\infty(z)+h)][(m_\infty(-1)+h)/(m_\infty(-1)+\bar{h})]} \tag{10.32}$$

is holomorphic in $\mathrm{Ext}[0,+\infty)$, $V_h(-0)\neq 0$, $V_h(-\infty)\neq 0$, *and*

$$1+V_h(-0)\,V_h(-\infty)\geq 0. \tag{10.33}$$

Moreover, an accretive operator T_h of the form (10.10) is α-sectorial if and only if inequality (10.33) is strict. In this case the exact value of the angle α is

$$\alpha=arccot\,\frac{1+V_h(-0)V_h(-\infty)}{|V_h(-\infty)-V_h(-0)|}. \tag{10.34}$$

Proof. Taking into account that (-1) is a point of regular type for the operator $\dot{A}$ of the form (10.9), we can show that (-1) is a regular point for the operator T_h of the form (10.10). Indeed, assume that (-1) is an eigenvalue for T_h. Then for some non-zero $f \in \mathrm{Dom}(T_h)$ we have $\mathbb{A}f = -f$, or

$$\mathrm{Re}\,\mathbb{A}f + i(f,g)g = -f,$$

where g is defined by (10.15). This implies

$$(\mathrm{Re}\,\mathbb{A}f, f) + i|(f,g)|^2 = -(f,f).$$

The right-hand side of the above equation is real and hence $(f,g) = 0$. Using (10.15) and (10.10) yields the system

$$\begin{cases} hf(a) - f'(a) = 0, \\ \mu f(a) - f'(a) = 0. \end{cases}$$

Taking into account that $\mathrm{Im}\,h \neq 0$ and μ is real, the system has only the solution $f(a) = f'(a) = 0$, which means that $f \in \mathrm{Dom}(\dot{A})$. But this contradicts the fact that (-1) is a point of regular type for the operator $\dot{A}$. Consequently, (-1) is also a point of regular type for the operator T_h. Indeed, $(I+T_h)^{-1}$ can not be unbounded since $(I+\dot{A})^{-1}$ is bounded and $\mathrm{Ran}(I+T_h)$ is different from $\mathrm{Ran}(I+\dot{A})$ no more than by one dimension. Thus, all we need to show is that $\mathrm{Ran}(I+T_h) = \mathcal{H}$. Assuming the contrary, we get $\mathrm{Ran}(I+T_h) = \mathrm{Ran}(I+\dot{A})$. Now if we take a vector $f \in \mathrm{Dom}(T_h)$ such that $f \notin \mathrm{Dom}(\dot{A})$, then the vector $(T_h + I)f = h \in \mathrm{Ran}(I+T_h) = \mathrm{Ran}(I+\dot{A})$ can be represented as

$$h = (I+\dot{A})f_0 = (I+T_h)f, \quad f_0 \in \mathrm{Dom}(\dot{A}).$$

Hence $(I+T_h)(f - f_0) = 0$, which contradicts the existence of $(I+T_h)^{-1}$. This proves that (-1) is a regular point for the operator T_h.

Consider the linear-fractional transformation

$$S_h = (I - T_h)(I+T_h)^{-1},$$

of the operator T_h. According to Lemma 9.5.12, the operator S_h is an α-co-sectorial contraction if and only if T_h is an α-sectorial operator. Without loss of generality, we can assume that operator T_h (and consequently S_h) is prime. If T_h is not prime then, since (-1) is its regular point, we can apply Lemma 6.6.9 and conclude that the non-negative operator $\dot{A}$ of the form (10.9) is not prime as well. Then there is an invariant subspace $\mathcal{H}_2$ of $\mathcal{H}$ such that $\mathcal{H} = \mathcal{H}_1 \oplus \mathcal{H}_2$ and

$$\dot{A}_2 = \dot{A}\restriction \mathcal{H}_2, \quad \dot{A} = \dot{A}_1 \oplus \dot{A}_2,$$

where $\dot{A}_2$ is a non-negative self-adjoint operator in $\mathcal{H}_2$. Similarly we can represent $T_h = T_{h,1} \oplus \dot{A}_2$. Hence, the operator S_h will also split into $S_h = S_{h,1} \oplus S_{h,2}$, where $S_{h,2}$ is a self-adjoint contraction.

Applying (9.83), (9.84), (8.30) together with Lemma 9.5.12 and Theorem 9.4.2, we have a proof of the statement of the theorem. □

Consider a non-negative Schrödinger operator $\dot{A}$ of the form (10.9) and T_h defined by (10.10). According to Theorem 10.2.1 the set of all $(*)$-extensions $\mathbb{A}$ of $T_h \supset \dot{A}$ is given by (10.11). We can use (10.11) to describe all self-adjoint $(*)$-extensions $\mathbb{A}$ of T_h when $h = \bar{h}$. For a real ξ we have

$$\begin{aligned}\mathbb{A}y &= -y'' + q(x)y - \frac{1}{\mu - \xi}\,[y'(a) - \xi y(a)]\,[\mu\delta(x-a) + \delta'(x-a)]\\ &= -y'' + q(x)y - \frac{\xi}{\mu - \xi}\,[\frac{1}{\xi}y'(a) - y(a)]\,[\mu\delta(x-a) + \delta'(x-a)]\\ &= -y'' + q(x)y - \frac{1}{\mu/\xi - 1}\,[\frac{1}{\xi}y'(a) - y(a)]\,[\mu\delta(x-a) + \delta'(x-a)].\end{aligned}$$

Setting $h = 1/\xi$ we obtain the formula for self-adjoint $(*)$-extensions of $T_h \supset \dot{A}$

$$\mathbb{A}\,y = -y'' + q(x)y - \frac{1}{h\mu - 1}\,[hy'(a) - y(a)]\,[\mu\delta(x-a) + \delta'(x-a)]. \tag{10.35}$$

Let $\varphi_k(x,\lambda), (k = 1,2)$ be the solutions of Cauchy problems defined in (10.12). As we have already mentioned in (10.13)

$$\varphi(x,\lambda) = \varphi_2(x,\lambda) + m_\infty(\lambda)\varphi_1(x,\lambda)$$

belongs to $L_2[a,+\infty)$. Consider an equation

$$(\mathbb{A} - \lambda I)\psi(x,\lambda) = \chi, \tag{10.36}$$

where $\chi = \mu\delta(x-a) + \delta'(x-a)$ and $\mathbb{A}$ is defined by (10.35). It is known from Chapter 4 that its solution $\psi(x,\lambda)$ belongs to $\mathfrak{N}_\lambda$, the deficiency subspace of the operator $\dot{A}$. Thus we have

$$\psi(x,\lambda) = g(\lambda)\varphi(x,\lambda),$$

where $\varphi(x,\lambda)$ is defined by (10.13). It is easy to see that $l(\psi) = \lambda\psi$. Furthermore,

$$\begin{aligned}(\psi, \upsilon) &= \psi(a,\lambda) - h\psi'(a,\lambda) = g(\lambda)[\varphi(a,\lambda) - h\varphi'(a,\lambda)]\\ &= g(\lambda)[-1 - h\,m_\infty(\lambda)],\end{aligned}$$

where $\upsilon = \delta(x-a) + h\delta'(x-a)$. Using (10.11) and (10.36) we have

$$\frac{1}{h\mu - 1}g(\lambda)[-1 - h\,m_\infty(\lambda)] = 1.$$

This implies

$$g(\lambda) = \frac{1 - h\mu}{1 + h\,m_\infty(\lambda)},$$

and hence

$$((\mathbb{A} - \lambda I)^{-1}\chi, \chi) = \frac{1 - h\mu}{1 + h\,m_\infty(\lambda)}[-\mu - m_\infty(\lambda)]. \tag{10.37}$$

Let us set in (10.37) $h = 0$ and $\mu = -m_\infty(-0)$ (assuming that $m_\infty(-0) < \infty$). Then we get

$$Q_F(\lambda) = ((\mathbb{A} - \lambda I)^{-1}\chi, \chi) = m_\infty(-0) - m_\infty(\lambda). \tag{10.38}$$

It is easy to see that the function Q_F of the form (10.38) satisfies all the conditions to be a Q-function that corresponds to the Friedrichs extension $A_F y = -y'' + q(x)y$, $y(a) = 0$ of $\dot{A}$. Thus we can apply Theorem 9.5.18 to Q_F and obtain that $Q_F(x) \to -\infty$ when $x \to -\infty$ on the real axis. Then (10.38) implies that for a real x,

$$m_\infty(x) \to +\infty \text{ when } x \to -\infty. \tag{10.39}$$

Theorem 10.3.2. *A non-negative Schrödinger operator $\dot{A}$ of the form* (10.9) *admits non-self-adjoint accretive* (α*-sectorial,* $0 < \alpha < \pi/2$) *extensions T_h of the form* (10.10) *if and only if $m_\infty(-0) < \infty$.*

Proof. Let $m_\infty(-0) < 0$. We write $c = m_\infty(-1)$ and $b = m_\infty(-0)$. First, we find the values of h ($\operatorname{Im} h > 0$) for which the inequality (10.33) holds. Using (10.32) and the above notations we get

$$V_h(-0) = \frac{(b-c)\operatorname{Im} h}{cb + (c+b)\operatorname{Re} h + h\,\bar{h}}. \tag{10.40}$$

Taking into account the properties of the Q_F-function and (10.32) we have

$$V_h(-\infty) = \frac{\operatorname{Im} h}{c + \operatorname{Re} h}. \tag{10.41}$$

Furthermore,

$$1 + V_h(-0)V_h(-\infty) = 1 + \frac{(\operatorname{Im} h)^2 (b-c)}{[cb + (c+b)\operatorname{Re} h + h\,\bar{h}](c + \operatorname{Re} h)}. \tag{10.42}$$

Setting $h = x + iy$ and using (10.42) we obtain

$$1 + V_h(-0)V_h(-\infty) = \frac{[cb + (c+b)x + x^2 + y^2](c+x) + (b-c)y^2}{[cb + (c+b)x + x^2 + y^2](c+x)}. \tag{10.43}$$

We introduce functions $f(x,y)$ and $g(x,y)$ to denote numerator and denominator of (10.43), respectively

$$\begin{aligned} f(x,y) &= x^2(b + 2c + x) + y^2(b + x) + (2bc + c^2)x + bc^2, \\ g(x,y) &= x^2(x + b + 2c) + y^2(x + c) + (2bc + c^2)x + bc^2. \end{aligned} \tag{10.44}$$

Let us find out when $f(x,y) \geq 0$. If $x = -b$, then

$$f(-b,y) = 2b^2c - 2b^2c - bc^2 + bc^2 = 0,$$

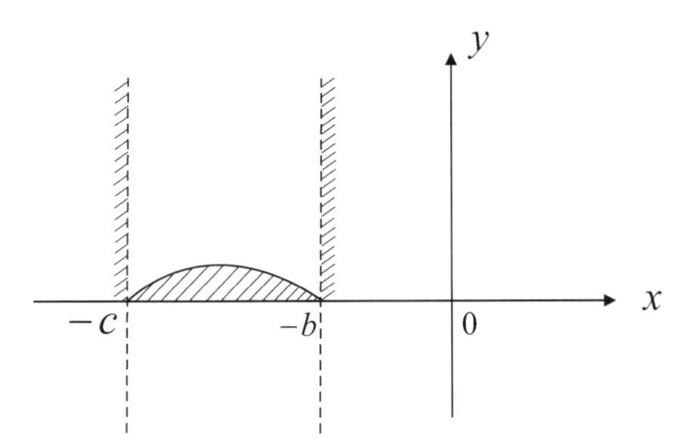

Figure 10.1: $f(x,y)/g(x,y) \geq 0$ domain

for $y > 0$. Clearly, for $x > -b$ we have that $y^2(b+x) > 0$ while for $x < -b$ that $y^2(b+x) < 0$. Consider the function

$$f(x) = x^3 + (b+2c)x^2 + (2bc+c^2)x + bc^2.$$

Using the derivative to find the intervals where $f(x)$ is monotone we get that it is increasing on $(-\infty,-c) \cup (-\frac{2b+c}{3},+\infty)$. Besides, since

$$f(-c) = 0, \quad f\left(-\frac{2b+c}{3}\right) < 0, \quad -c < -\frac{2b+c}{3} < -b,$$

$$f(x,y) = f(x) + y^2(b+x),$$

then for any $x < -b$, $f(x,y) < 0$ for all $y > 0$ and for any $x > -b$, $f(x,y) > 0$ for all $y > 0$. Similar study can be applied to the function $g(x,y)$. As the result we obtain that for any $x > -b$, $g(x,y) > 0$ for all $y > 0$ and for any $x < -c$, $g(x,y) < 0$ for all $y > 0$, since $g(x,y) = f(x) + y^2(x+c)$.

Now let us describe the behavior of $g(x,y)$ for $-c < x < -b$. Evidently, the domain where $g(x,y) > 0$ for $-c < x < -b$ follows from the inequality

$$y^2 > -x^2 - (b+c)x - bc.$$

The roots of the quadratic expression on the right are $x_1 = -c$ and $x_2 = -b$. Hence the domain satisfying the inequality $f(x,y)/g(x,y) \geq 0$ has the form presented in Figure 10.1. We are going to check whether $V_h(z)$ is holomorphic on $\mathrm{Ext}[0,+\infty)$ if our parameter h belongs to this domain. First we will show that $V_h(z)$ is not holomorphic on $\mathrm{Ext}[0,+\infty)$ if h belongs to the domain shown in Figure 10.2 below. In order to do that it is enough to prove that there is a $z \in \mathrm{Ext}[0,+\infty)$ such that

$$1 + \frac{m_\infty(z) + \bar{h}}{m_\infty(z) + h} \cdot \frac{c+h}{c+\bar{h}} = 0, \tag{10.45}$$

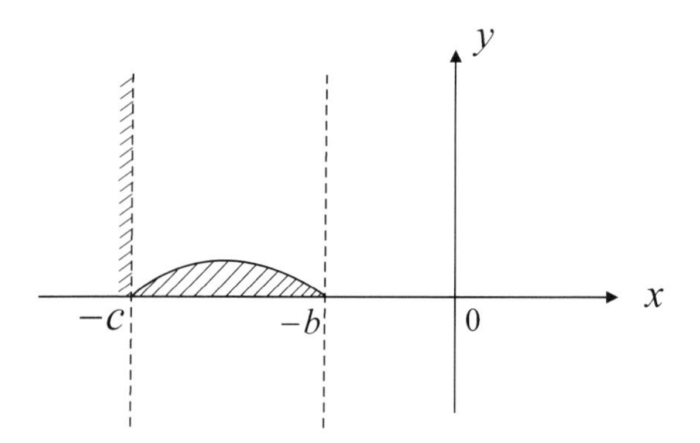

Figure 10.2: Domain for $h = x + iy$

for h in the domain in Figure 10.2. Since $h = x + iy$, then equation (10.45) is equivalent to $m_\infty(z) = (-x^2 - y^2 - cx)/(c + x)$ and hence

$$\begin{aligned} Q_F(z) &= m_\infty(-0) - m_\infty(z) = b - \frac{-x^2 - y^2 - cx}{c + x} \\ &= \frac{x^2 + (c + b)x + cb + y^2}{c + x}. \end{aligned} \tag{10.46}$$

By a direct check one confirms that the fraction

$$\frac{x^2 + (c + b)x + cb + y^2}{c + x},$$

considered in the shaded region in Figure 10.1, is negative if and only if a point (x, y) belongs to that shaded region in Figure 10.2. Using Theorem 9.5.18, (10.38), and (10.39) we get that the Q_F-function is negative on the left real semi-axis, and hence equation (10.46) is solvable for z. Therefore, the function $V_h(z)$ is holomorphic on $\mathrm{Ext}[0, +\infty)$ only if h is in the domain (shaded region $y > 0$, $x > -b = -m_\infty(-0)$) depicted in Figure 10.3, and inequality (10.33) holds for h from that domain as well. Thus, according to Theorem 10.3.1, any h from the domain in Figure 10.3 generates an accretive extension T_h of the operator $\dot{A}$. Using (10.40) and (10.41) we obtain

$$\begin{aligned} \frac{1 + V_h(-0)V_h(-\infty)}{V_h(-\infty) - V_h(-0)} &= \frac{[cb + (b + c)x + x^2 + y^2](c + x) + (b - c)y^2}{y([cb + (b + c)x + x^2 + y^2] - (b - c)(c + x))} \\ &= \frac{b + x}{y} = \frac{m_\infty(-0) + x}{y}. \end{aligned}$$

Therefore, we have shown that if for any given h, $(\mathrm{Im}\, h > 0)$ an operator T_h of the form (10.10) is α-sectorial, then the exact value of the angle α is shown

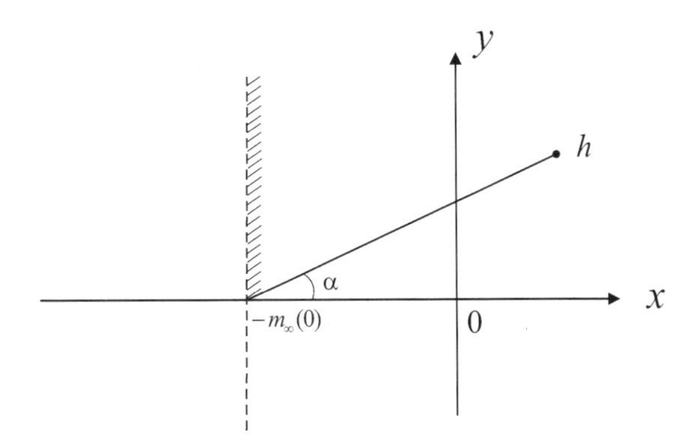

Figure 10.3: Value for the angle α

in Figure 10.3. Conversely, let for some h, ($\operatorname{Im} h > 0$) the operator T_h of the form (10.10) be α-sectorial for some angle α, ($0 < \alpha < \pi/2$). We will show that in this case $m_\infty(-0) < \infty$. Assume the contrary, i.e., $m_\infty(-0) = \infty$. Then by Theorem 10.3.1 the value of sectorial angle α is found via (10.34). Consequently, our assumption and (10.32) imply that $V_h(-0) = V_h(-\infty)$. But then (10.34) yields that $\alpha = arccot(\infty) = 0$ which is impossible. □

The following theorem immediately follows from the reasoning in the proof above.

Theorem 10.3.3. *If an accretive Schrödinger operator T_h, ($\operatorname{Im} h > 0$) of the form (10.10) is α-sectorial, then*

$$\alpha = \operatorname{arccot}\left(\frac{\operatorname{Re} h + m_\infty(-0)}{\operatorname{Im} h}\right). \tag{10.47}$$

Conversely, if h, ($\operatorname{Im} h > 0$) is such that $\operatorname{Re} h + m_\infty(-0) > 0$, then operator T_h of the form (10.10) is α-sectorial and α is determined by (10.47).

It is easy to see that $T_h^* = T_{\bar h}$ and hence all the accretive and α-sectorial non-self-adjoint operators T_h are described by only such values of parameter h for which $\operatorname{Re} h \geq -m_\infty(-0)$. Also, since a numeric range of an α-sectorial operator T_h lies in the sector $|\arg\zeta| \leq \alpha$, then the spectrum of T_h also belongs to this sector. Therefore, formula (10.47) allows one to "control" the spectrum with the help of a boundary parameter h.

Now let us find the value of the parameter h for which operator T_h of the form (10.10) generates a Kreĭn-von Neumann extension (see Section 9.5) of operator $\dot A$ of the form (10.9). Let $h \neq \bar h$ belong to the line $\operatorname{Re} h = -m_\infty(-0)$ (see Figure 10.3). Then the operator $(I - T_h)(I + T_h)^{-1}$ is a contraction according to Lemma

9.5.12. Taking into account (9.84) we get

$$(I - T_h)(I + T_h)^{-1} = \frac{1}{2}(S_M + S_\mu) + \frac{1}{2}\zeta(h)(S_M - S_\mu), \tag{10.48}$$

where S_M and S_μ are the extreme sc-extensions of $\dot{S} = (I - \dot{A})(I + \dot{A})^{-1}$ given by (9.37) and (9.38), and $|\zeta(h)| \leq 1$ for all h's in the domain in Figure 10.3. Since the left-hand side of (10.48) (for non-real h that belongs to the line $\operatorname{Re} h = -m_\infty(-0)$) is a contraction but not a α-co-sectorial contraction for any α, $(0 < \alpha < \pi/2)$, then (see Theorem 9.2.7) $|\zeta(h)| = 1$, for all $h \in (\operatorname{Re} h = -m_\infty(-0))$. Besides, when h moves along the above line, $\zeta(h)$ traces a unit circle with points ± 1 removed. By virtue of Theorem 8.2.4, the function $V(z)$ of the form (9.83) constructed for the operator $(I - T_h)(I + T_h)^{-1}$ possesses a property that $V(1+) = V_h(-0)$, $V(-1-) = V_h(-\infty)$, where $V_h(-0)$ and $V_h(-\infty)$ are defined by (10.40) and (10.41). Taking into account (10.40) and (10.41), for $h \to -m_\infty(-0)$ along the line $\operatorname{Re} h = -m_\infty(-0)$ we have that $\zeta(h) \to 1$. Since the right-hand side of (10.48) coincides with soft extension S_M, then for $h = -m_\infty(-0)$ the operator T_h generates a Kreĭn-von Neumann extension A_K of $\dot{A}$. Therefore the set of all self-adjoint extensions of a non-negative operator $\dot{A}$ of the form (10.9) is defined by operators T_h of the form (10.10) under the condition that h is real and $h \geq -m_\infty(-0)$. Consequently, the set of all accretive and α-sectorial boundary value problems for a Schrödinger operator is defined by a parameter h from the domain depicted in Figure 10.3. Moreover, if:

1. h belongs to the real line, then we obtain all accretive self-adjoint boundary value problems;
2. h does not belong to a line $\operatorname{Re} h = -m_\infty(-0)$ and $h \neq \bar{h}$, then we obtain all α-sectorial boundary value problems and the exact value of the angle α is given by (10.47);
3. $h \neq \bar{h}$ lies on the line $\operatorname{Re} h = -m_\infty(-0)$, then we obtain all accretive boundary value problems that are not α-sectorial for any α, $(0 < \alpha < \pi/2)$.

Example. Consider a non-negative symmetric Schrödinger (Bessel type) operator with deficiency indices $(1, 1)$:

$$\dot{A}y = -y'' + \frac{\nu^2 - (1/4)}{x^2}y, \quad \nu = \frac{k+1}{2},\ k = 0, 1, 2, \dots,$$
$$y'(1) = y(1) = 0,$$

in the Hilbert space $\mathcal{H} = L^2(1, +\infty)$. It was shown in [53] that

$$m_{\infty,\nu}(-0) = \nu - \frac{1}{2}.$$

The Friedrichs extension A_F of $\dot{A}$ in this case has the form

$$A_F y = -y'' + \frac{\nu^2 - (1/4)}{x^2}y, \quad y(1) = 0.$$

The Kreĭn-von Neumann extension is the boundary value problem

$$A_K\, y = -y'' + \frac{\nu^2 - (1/4)}{x^2} y, \quad y'(1) + (\nu - (1/2))\, y(1) = 0.$$

When $\nu = \frac{1}{2}$, the Kreĭn-von Neumann extension is simply the Neuman boundary value problem

$$A_K\, y = -y'', \quad y'(1) = 0.$$

This example shows that the Kreĭn-von Neumann extension may be not only the Neuman but also a mixed boundary value problem.

10.4 Functional model for symmetric operator with deficiency indices (1,1)

In this section we present material of auxiliary nature dealing with a functional model for a symmetric operator with deficiency indices (1,1). Let $\mathcal{H}$ be a Hilbert space and $\dot{A}$ be a closed densely-defined symmetric operator with deficiency indices (1,1). According to von Neumann formulas (1.13) every self-adjoint extension A_α of $\dot{A}$ can be represented in the form

$$\begin{aligned} \mathrm{Dom}(A_\alpha) &= \mathrm{Dom}(\dot{A}) + c(\varphi - e^{2i\alpha}\psi), \quad \varphi \in \mathfrak{N}_i,\ \psi \in \mathfrak{N}_{-i}, \\ A_\alpha &= \dot{A}h + c(i\varphi + ie^{2i\alpha}\psi), \quad h \in \mathrm{Dom}(\dot{A}),\ c \in \mathbb{C}, \end{aligned} \tag{10.49}$$

where $\|\varphi\| = \|\psi\| = 1$ and $\alpha \in [0, \pi)$. As usual we set $\mathcal{H}_+ = \mathrm{Dom}(\dot{A}^*)$ and construct a rigged Hilbert triplet $\mathcal{H}_+ \subset \mathcal{H} \subset \mathcal{H}_-$. Let us now consider a Hilbert space $\mathcal{H}_{+,\alpha} = \mathrm{Dom}(A_\alpha)$ equipped with the scalar product

$$(f, g)_{+,\alpha} = (f, g) + (A_\alpha f, A_\alpha g), \qquad (f, g \in \mathcal{H}),$$

where A_α is some self-adjoint extension of $\dot{A}$ in $\mathcal{H}$ of the form (10.49). Following the steps of Sections 2.1 and 6.5 we construct another rigged Hilbert space $\mathcal{H}_{+,\alpha} \subset \mathcal{H} \subset \mathcal{H}_{-,\alpha}$ with the Riesz-Berezansky operator $\mathsf{R} \in [\mathcal{H}_{-,\alpha}, \mathcal{H}_{+,\alpha}]$ such that

$$(f, g) = (f, \mathsf{R}g)_{+,\alpha}, \quad \|g\|_{-,\alpha} = \|\mathsf{R}g\|_{+,\alpha}, \qquad (\forall f \in \mathcal{H}_{+,\alpha},\ g \in \mathcal{H}), \tag{10.50}$$

and

$$(f, g)_{-,\alpha} = (\mathsf{R}f, \mathsf{R}g)_{+,\alpha}, \qquad (\forall f, g \in \mathcal{H}_{-,\alpha}).$$

Lemma 10.4.1. *Let $\dot{A}$ be a densely-defined symmetric operator in Hilbert space $\mathcal{H}$ with deficiency indices $(1, 1)$ and A_α be its self-adjoint extension in $\mathcal{H}$. If an element $g \in \mathcal{H}_{-,\alpha}$ is such that $g \notin \mathcal{H}$, then the operator*

$$\hat{A}_\alpha f = A_\alpha f, \quad \mathrm{Dom}(\hat{A}_\alpha) = \{f \in \mathrm{Dom}(A_\alpha) \mid (f, g) = 0\}, \tag{10.51}$$

is densely defined in $\mathcal{H}$.

Proof. Assuming that $\overline{\mathrm{Dom}(\hat{A}_\alpha)} \neq \mathcal{H}$, we have the existence of a vector $\phi \in \mathcal{H}$ such that $(f, \phi) = 0$ for all $f \in \mathrm{Dom}(\hat{A}_\alpha)$. Taking in to account (10.49), (10.50), and (10.51) we get that, for all $f \in \mathrm{Dom}(\hat{A}_\alpha)$,

$$(f, \mathsf{R}g)_{+,\alpha} = 0 \text{ and } (f, \mathsf{R}\phi)_{+,\alpha} = 0.$$

Therefore the vectors $\mathsf{R}g$ and $\mathsf{R}\phi$ are proportional and hence g and ϕ are proportional as well. This contradicts the fact that $g \notin \mathcal{H}$. □

Lemma 10.4.2. *Let $\dot{A}$ be a closed, densely-defined, symmetric operator in Hilbert space $\mathcal{H}$ with deficiency indices $(1,1)$. Then $\mathcal{H}$ can be decomposed into the orthogonal sum*

$$\mathcal{H} = \mathcal{H}_0 \oplus \mathcal{H}_1, \tag{10.52}$$

where both $\mathcal{H}_0$ and $\mathcal{H}_1$ are invariant under the resolvents R_z^α of self-adjoint extensions A_α of $\dot{A}$ for all $\alpha \in [0,\pi)$. Moreover, all the resolvents R_z^α coincide on $\mathcal{H}_1$.

Proof. Let $\mathcal{H}_0^\alpha$ be the subspace of $\mathcal{H}$ obtained by forming the closure of all finite linear combinations of vectors of the form $R_z^\alpha \varphi$, $\varphi \in \mathfrak{N}_i$ for all non-real z. This subspace is evidently invariant under R_z^α because of the resolvent identity

$$R_z^\alpha - R_\lambda^\alpha = (z - \lambda) R_z^\alpha R_\lambda^\alpha.$$

Since the adjoint of R_z^α is $R_{\bar{z}}^\alpha$, then $\mathcal{H}_0^{\alpha\perp}$ is invariant under $R_{\bar{z}}^\alpha$ as well. Since the vector $-zR_z^\alpha\varphi$, $(\varphi \in \mathfrak{N}_i)$ converges weakly to φ as z goes to infinity along the imaginary axis, then φ belongs to $\mathcal{H}_0^{\alpha\perp}$. Let $f \in \mathcal{H}_0^{\alpha\perp}$. Then $g = R_z^\alpha f \in \mathcal{H}_0^{\alpha\perp}$, $(\mathrm{Im}\, z \neq 0)$ and

$$(\varphi, R_z^\alpha f) = 0, \tag{10.53}$$

for $\varphi \in \mathcal{H}_0^\alpha$ and $R_z^\alpha f \in \mathcal{H}_0^{\alpha\perp}$. It is easy to see that the vector $g = R_z^\alpha f$ does not belong to $\ker(\dot{A}^* \mp iI)$. Let us assume the contrary, i.e., $\dot{A}^* g = \pm ig$. Since $g \in \mathrm{Dom}(A_\alpha)$, then we get $A_\alpha g = \pm ig$, which is impossible for a self-adjoint operator A_α. It follows from the relation $g = R_z^\alpha f$ that $(A_\alpha - zI)g = f$, and

$$A_\alpha g = zg + f. \tag{10.54}$$

Consider the space $\mathcal{H}_+ = \mathrm{Dom}(\dot{A}^*)$. Then (10.53) and (10.54) yield $(\varphi, g)_+ = 0$. Indeed,

$$\begin{aligned}(\varphi, g)_+ &= (\varphi, g) + (\dot{A}^*\varphi, \dot{A}^* g) = (\varphi, g) + i(\varphi, A_\alpha g)\\ &= (\varphi, g) + i(\varphi, zg + f) = 0.\end{aligned}$$

Since we have proved that $g = R_z^\alpha f$ does not belong to $\ker(\dot{A}^* \mp iI)$, then taking into account (2.9) we get that $g = R_z^\alpha f \in \mathrm{Dom}(\dot{A})$. Therefore (10.54) implies

$$A_\alpha g = A_\beta g = \dot{A} g = zg + f, \qquad (\forall \alpha, \beta \in [0,\pi)).$$

Thus all the family of resolvents R_z^α coincide on $\mathcal{H}_0^{\alpha\perp}$ and for any $\beta \in [0,\pi)$ vectors $R_z^\beta \varphi$ are orthogonal to $\mathcal{H}_0^{\alpha\perp}$. Since both α and β are arbitrary values from $[0,\pi)$, then these spaces must coincide. □

It was shown in Lemma 6.6.6 that if $\dot{A}$ is a densely-defined, prime, symmetric operator in a Hilbert space $\mathcal{H}$ with deficiency indices $(1,1)$, then

$$\underset{z\neq\bar{z}}{c.l.s.}\, R_z^\alpha \varphi = \mathcal{H}, \quad (\varphi \in \mathfrak{N}_i). \tag{10.55}$$

Let us now consider a self-adjoint operator in $L_2^\mu(\mathbb{R})$ defined by

$$Bx(t) = tx(t), \quad \mathrm{Dom}(B) = \{x(t) \in L_2^\mu(\mathbb{R}) \mid tx(t) \in L_2^\mu(\mathbb{R})\}, \tag{10.56}$$

where the measure μ (non-decreasing, non-negative function $\mu(t)$ on $\mathbb{R}$) satisfies the conditions

$$\int_{\mathbb{R}} d\mu(t) = \infty, \qquad \int_{\mathbb{R}} \frac{1}{1+t^2}\, d\mu(t) = 1. \tag{10.57}$$

Consider the operator

$$\dot{B}x(t) = tx(t), \quad \mathrm{Dom}(\dot{B}) = \{x(t),\, tx(t) \in L_2^\mu(\mathbb{R}) \mid \int_{\mathbb{R}} x(t)\mu(t) = 0\}. \tag{10.58}$$

Utilizing Theorems 6.5.1 and 6.6.7 we conclude that the operator $\dot{B}$ of the form (10.58) is closed, prime, and symmetric. Obviously, the deficiency indices of this operator is $(1,1)$. It was also shown in the proof of Theorem 6.5.1 that

$$\ker(\dot{B}^* \mp iI) = \left\{\frac{c}{t \mp i},\ c \in \mathbb{C}\right\}.$$

Consider vectors

$$\varphi = \frac{1}{t-i}, \qquad \psi = \frac{1}{t+i}.$$

Obviously,

$$\|\varphi\|^2_{L_2^\mu(\mathbb{R})} = \|\psi\|^2_{L_2^\mu(\mathbb{R})} = \int_{\mathbb{R}} \frac{1}{1+t^2}\, d\mu(t) = 1,$$

and

$$\mathcal{H}_{+,0} = \mathrm{Dom}(B) = \mathrm{Dom}(\dot{B}) + c\left(\frac{1}{t-1} - \frac{1}{t+i}\right), \tag{10.59}$$

$$Bx(t) = tu(t) + c\left(\frac{i}{t-1} + \frac{i}{t+i}\right) = tu(t) + \frac{2cit}{1+t^2} = t\left(u(t) + \frac{2ic}{1+t^2}\right), \tag{10.60}$$

where $u(t) \in \mathrm{Dom}(\dot{B})$ and $x(t) \in \mathrm{Dom}(B)$. It follows from (10.59)-(10.60) that any vector $x(t) \in \mathcal{H}_{+,0}$ can be represented in the form

$$x(t) = u(t) + \frac{2ic}{1+t^2}, \qquad (c \in \mathbb{C}). \tag{10.61}$$

Taking into account (10.61) and $\int_{\mathbb{R}} u(t)d\mu(t) = 0$, we have

$$(x,1) = \int_{\mathbb{R}} \left(u(t) + \frac{2ic}{1+t^2}\right) d\mu(t) = \int_{\mathbb{R}} \frac{2ic}{1+t^2} d\mu(t),\ (z \in \mathrm{Dom}(\dot{B}),\ c \in \mathbb{C}). \tag{10.62}$$

The relation (10.62) above shows that the functional $(x,1)=\int_{\mathbb{R}}x(t)d\mu(t)$ is continuous on $\mathcal{H}_{+,0}$ and therefore $x(t)\equiv 1$ belongs to $\mathcal{H}_{-,0}$. Hence we can apply Lemma 10.4.1 and relation (10.58) to conclude that symmetric operator $\dot{B}$ is densely defined.

Theorem 10.4.3. *Let $\dot{A}$ be a closed densely-defined, symmetric, prime operator with deficiency indices* $(1,1)$ *in a Hilbert space $\mathcal{H}$, and let A be its self-adjoint extension. Then the pair $(\dot{A},A)$ is unitary equivalent to the pair $(\dot{B},B)$ of the form* (10.58), (10.56) *in some Hilbert space $L_2^\mu(\mathbb{R})$, where the measure $\mu(t)$ satisfies* (10.57).

Proof. Let $\varphi\in\ker(\dot{A}^*-iI)$, $\|\varphi\|=1$, and A have the resolution of identity $E(t)$. Consider an operator

$$Ux(t)=\int_{\mathbb{R}}x(t)(t-i)dE(t)\varphi. \tag{10.63}$$

We are going to show that U is an isometric map of $L_2^\mu(\mathbb{R})$ onto $\mathcal{H}$ where

$$d\mu(t)=(1+t^2)d(E(t)\varphi,\varphi)_{\mathcal{H}}. \tag{10.64}$$

It follows from (10.64) that

$$\begin{aligned}(Ux,Uy)_{\mathcal{H}}&=\int_{\mathbb{R}}x(t)\overline{y(t)}(t^2+1)d(E(t)\varphi,\varphi)_{\mathcal{H}}=\int_{\mathbb{R}}x(t)\overline{y(t)}d\mu(t)\\&=\big(x(t),y(t)\big)_{L_2^\mu(\mathbb{R})}.\end{aligned}$$

Let $\phi=\frac{1}{t-i}$. Clearly, $\phi\in L_2^\mu(\mathbb{R})$ and $\|\phi\|_{L_2^\mu(\mathbb{R})}=1$. It is also easy to see that

$$R_z(B)\phi=(B-zI)^{-1}\left(\frac{1}{t-i}\right)=\frac{1}{(t-z)(t-i)}. \tag{10.65}$$

Combining (10.63) and (10.65) we obtain

$$UR_z(B)\phi=\int_{\mathbb{R}}\frac{(t-i)}{(t-z)(t-i)}dE(t)\,\varphi=\int_{\mathbb{R}}\frac{dE(t)}{t-z}\,\varphi=R_z(A)\varphi. \tag{10.66}$$

Since the operator $\dot{A}$ is a closed densely-defined, symmetric, prime operator with deficiency indices $(1,1)$, then we can apply Lemma 6.6.6 or Lemma 10.4.2 to obtain relation (10.55) for some α such that $A_\alpha=A$ and $R_z^\alpha(A_\alpha)=R_z(A)$. Consequently, (10.66) implies that the closed linear span of all vectors of the form $R_z(B)\phi$ coincides with $L_2^\mu(\mathbb{R})$. Using the resolvent identity and relation (10.66) we have $UR_z(B)R_\lambda(B)\phi=R_z(A)R_\lambda(A)\varphi$, and hence

$$UR_z(B)U^{-1}R_\lambda(A)\varphi=R_z(A)R_\lambda(A)\varphi.$$

Thus

$$UR_z(B)U^{-1}=R_z(A). \tag{10.67}$$

Applying (10.63), we get

$$U\phi = U\left(\frac{1}{t-i}\right) = \int_{\mathbb{R}} \frac{1}{t-i}(t-i)dE(t)\varphi = \int_{\mathbb{R}} dE(t)\varphi = \varphi. \tag{10.68}$$

The latter means that $U\ker(\dot{B}^* - iI) = \ker(\dot{A}^* - iI)$. Since

$$\frac{1}{(t+i)(t-i)} = \frac{i}{2}\left(\frac{1}{t+i} - \frac{1}{t-i}\right),$$

then

$$R_{-i}(B)\phi + \frac{i}{2}\phi = \left(\frac{i}{2}\right)\frac{1}{t+i} \in \ker(\dot{B}^* + iI).$$

On the other hand, applying (10.67) and (10.68) we obtain

$$U\left(\frac{i}{2}\right)\frac{1}{t+i} = UR_{-i}(B)\phi + U\frac{i}{2}\phi = R_{-i}(A)\varphi + \frac{i}{2}\varphi. \tag{10.69}$$

By direct calculation one checks that the right-hand side of (10.69) belongs to $\ker(\dot{A}^* + iI)$. Therefore, operator U maps $\ker(\dot{B}^* + iI)$ onto $\ker(\dot{A}^* + iI)$. As a result we get

$$\dot{A} = U\dot{B}U^{-1}, \qquad A = UBU^{-1}.$$

This completes the proof. □

A result analogous to Theorem 10.4.3 can be proved in a similar way for the case of deficiency indices (n, n).

10.5 Accretive (∗)-extensions of a Schrödinger operator

Suppose that the symmetric operator $\dot{A}$ of the form

$$\begin{cases} \dot{A}y = -y'' + q(x)y, \\ y(a) = y'(a) = 0, \end{cases}$$

(see also (10.9)) with deficiency indices (1,1) is non-negative, i.e., $(\dot{A}f, f) \geq 0$ for all $f \in \mathrm{Dom}(\dot{A})$). It was shown in Theorem 10.3.3 that the Schrödinger operator T_h of the form

$$\begin{cases} T_h y = -y'' + q(x)y \\ hy(a) - y'(a) = 0 \end{cases}, \quad \begin{cases} T_h^* y = -y'' + q(x)y \\ \overline{h}y(a) - y'(a) = 0 \end{cases}, \tag{10.70}$$

(see also (10.10)) is accretive if and only if

$$\mathrm{Re}\, h \geq -m_\infty(-0). \tag{10.71}$$

For real h such that $h \geq -m_\infty(-0)$ we get a description of all non-negative self-adjoint extensions of an operator $\dot{A}$. For $h = -m_\infty(-0)$ the corresponding operator

$$\begin{cases} A_K y = -y'' + q(x)y, \\ y'(a) + m_\infty(-0)y(a) = 0, \end{cases} \tag{10.72}$$

is the Kreĭn-von Neumann extension of $\dot{A}$ and for $h = +\infty$ the corresponding operator

$$\begin{cases} A_F y = -y'' + q(x)y, \\ y(a) = 0, \end{cases} \tag{10.73}$$

is the Friedrichs extension of $\dot{A}$. It follows from (10.71), (10.72) and (10.73) that a non-negative operator $\dot{A}$ of the form (10.10) admits non-self-adjoint accretive extensions if and only if $m_\infty(-0) < \infty$.

Theorem 10.5.1. *Let T_h ($\operatorname{Im} h > 0$) be an accretive Schrödinger operator of the form (10.70). Then for all real μ satisfying the inequality*

$$\mu \geq \frac{(\operatorname{Im} h)^2}{m_\infty(-0) + \operatorname{Re} h} + \operatorname{Re} h, \tag{10.74}$$

the operators

$$\begin{aligned} \mathbb{A}y &= -y'' + q(x)y + \frac{1}{\mu - h}\,[hy(a) - y'(a)]\,[\mu\delta(x-a) + \delta'(x-a)], \\ \mathbb{A}^*y &= -y'' + q(x)y + \frac{1}{\mu - \bar{h}}\,[\bar{h}y(a) - y'(a)]\,[\mu\delta(x-a) + \delta'(x-a)] \end{aligned} \tag{10.75}$$

define the set of all accretive $(*)$-extensions $\mathbb{A}$ of the operator T_h. The operator T_h has a unique accretive $(*)$-extension $\mathbb{A}$ if and only if

$$\operatorname{Re} h = -m_\infty(-0).$$

In this case this unique $(*)$-extension has the form

$$\begin{aligned} \mathbb{A}y &= -y'' + q(x)y + [hy(a) - y'(a)]\,\delta(x-a), \\ \mathbb{A}^*y &= -y'' + q(x)y + [\bar{h}y(a) - y'(a)]\,\delta(x-a). \end{aligned} \tag{10.76}$$

Proof. It follows from Theorem 10.2.1 that the set of all $(*)$-extensions of the Schrödinger operator T_h ($\operatorname{Im} h > 0$) of the form (10.70) can be described by the formula (10.75). Suppose that $\mathbb{A}$ of the form (10.75) is an accretive $(*)$ - extension of the accretive Schrödinger operator T_h of the form (10.70). Then

$$\mu + m_\infty(-0) \geq 0.$$

Let $\xi < 0$. By the direct computations using (10.20) one obtains

$$\begin{aligned} \operatorname{Im} W_\Theta(\xi) = &\frac{-2[\mu + m_\infty(\xi)\operatorname{Im} h]}{|\mu - h|^2 |m_\infty(\xi) + h|^2} \\ &\times \left[-(\operatorname{Re} h)^2 + (\mu - m_\infty(\xi))\operatorname{Re} h + \mu m_\infty(\xi) - (\operatorname{Im} h)^2\right]. \end{aligned} \tag{10.77}$$

Taking into account (see (10.39)) that $m_\infty(\xi) \to +\infty$ monotonically as $\xi \to -\infty$, we get that $\mu + m_\infty(\xi) > 0$ for $\xi < 0$. Using (6.47) and (6.48) yields

$$V_\Theta(z) = i[W_\Theta(z)+1]^{-1}\,[W_\Theta(z)-1] = i - 2i[W_\Theta(z)+1]^{-1}.$$

Therefore,

$$\mathrm{Re}\, V_\Theta(z) = -2[W_\Theta^*(z)+1]^{-1}\,\frac{W_\Theta(z)-W_\Theta^*(z)}{2i}\,[W_\Theta(z)+1]^{-1}$$

and for $\xi < 0$,

$$V_\Theta(\xi) = -2[W_\Theta^*(\xi)+1]^{-1}\,\mathrm{Im}\, W_\Theta(\xi)\,[W_\Theta(\xi)+1]^{-1}. \tag{10.78}$$

Comparing (10.78) and (6.47) we obtain $\mathrm{Im}\, W_\Theta(\xi) \le 0$ for $\xi < 0$. It follows from (10.77) that

$$-(\mathrm{Re}\, h)^2 - (\mu - m_\infty(\xi))\mathrm{Re}\, h + \mu m_\infty(\xi) - (\mathrm{Im}\, h)^2 \ge 0$$

for all $\xi < 0$. Taking the limit $\xi \to -0$ we get

$$-(\mathrm{Re}\, h)^2 - (\mu - m_\infty(-0))\,\mathrm{Re}\, h + \mu m_\infty(-0) - (\mathrm{Im}\, h)^2 \ge 0,$$

which is equivalent to (10.74). Suppose now that the inequality (10.74) holds. We will show that the operator $\mathbb{A}$ of the form (10.75) is accretive. Because $m_\infty(-0) < m_\infty(\xi)$ for all $\xi < 0$, it is easy to verify that the inequality

$$\begin{aligned} &-(\mathrm{Re}\, h)^2 - (\mu - m_\infty(-0))\,\mathrm{Re}\, h + \mu m_\infty(-0) - (\mathrm{Im}\, h)^2 \\ &\le -(\mathrm{Re}\, h)^2 - (\mu - m_\infty(\xi))\,\mathrm{Re}\, h + \mu m_\infty(\xi) - (\mathrm{Im}\, h)^2 \end{aligned} \tag{10.79}$$

holds for

$$\mu \ge \mathrm{Re}\, h. \tag{10.80}$$

Using (10.74) one concludes that (10.80) and the inequality (10.74) are equivalent to the left-hand side of (10.79) being non-negative, i.e.,

$$\mathrm{Im}\, W_\Theta(\xi) \le 0, \quad \xi < 0.$$

Therefore, applying (6.47) we get $V_\Theta(\xi) \ge 0$ for $\xi < 0$. Since $V_\Theta(z)$ is a Herglotz-Nevanlinna function, and it is non-negative on the negative axis, then $V_\Theta(z)$ is a Stieltjes function. Therefore, it follows from Theorem 9.8.2 that $\mathbb{A}$ is accretive. When $\mathrm{Re}\, h = -m_\infty(-0)$ then (10.74) implies that $\mu = \infty$. Taking the limit $\mu \to +\infty$ in (10.75) we get that T_h has only one accretive (*)-extension $\mathbb{A}$ if and only if $\mathrm{Re}\, h = -m_\infty(-0)$. This unique (*)-extension $\mathbb{A}$ has the form (10.76). $\square$

10.6 Stieltjes functions and L-systems with accretive Schrödinger operator

Next, we consider a scattering L-system of the form

$$\Theta = \begin{pmatrix} \mathbb{A} & K & 1 \\ \mathcal{H}_+ \subset L_2[a,+\infty) \subset \mathcal{H}_- & & \mathbb{C} \end{pmatrix},$$

where $\mathbb{A}$ is defined by (10.11), and suppose that the operator

$$\begin{cases} T_h y = -y'' + q(x)y, \\ y'(a) = hy(a), \end{cases} \tag{10.81}$$

is accretive and the symmetric operator

$$\begin{cases} \dot{A}y = -y'' + q(x)y, \\ y'(a) = y(a) = 0, \end{cases} \tag{10.82}$$

in $L_2[a,+\infty)$ has defect indices $(1,1)$. It was shown in Theorem 10.3.3 that the operator T_h is accretive if and only if $\operatorname{Re} h \geq -m_\infty(-0)$. Assume that $\operatorname{Re} h = -m_\infty(-0)$. Then, it follows from Theorem 10.5.1 that the $(*)$-extension (10.11) of the operator T_h ($\operatorname{Re} h = -m_\infty(-0)$) is accretive if and only if $\mu = \infty$ (see (10.74)). From the relations (10.11) for $\mu = \infty$ we get (10.76) and

$$\begin{aligned} \operatorname{Re} \mathbb{A} &= -y'' + q(x)y + [m_\infty(-0)y(a) + y'(a)]\delta(x-a), \\ \operatorname{Im} \mathbb{A} &= (\operatorname{Im} h)y(a)\delta(x-a) = (y,g)g, \quad \operatorname{Im} h > 0. \end{aligned} \tag{10.83}$$

Here

$$g = (\operatorname{Im} h)^{\frac{1}{2}}\delta(x-a), \tag{10.84}$$

and the operator $\operatorname{Re} \mathbb{A}$ of the form (10.83) is non-negative, i.e., $(\operatorname{Re} \mathbb{A}y, y) \geq 0$ for all $y \in \mathcal{H}_+$. Moreover, according to Remark 9.8.6,

$$V_\Theta(z)\phi = K^*(\operatorname{Re} \mathbb{A} - zI)^{-1}K\phi = \int_0^\infty \frac{d\sigma(t)}{t-z}\phi, \quad \phi \in \mathbb{C}.$$

So, we can associate $V_\Theta(z)$ with the Stieltjes function

$$V_\Theta(z) = \int_0^\infty \frac{d\sigma(t)}{t-z}, \tag{10.85}$$

where

$$\int_0^\infty d\sigma(t) = \infty, \quad \int_0^\infty \frac{d\sigma(t)}{1+t} < \infty. \tag{10.86}$$

Since $\operatorname{Re}\mathbb{A} \supset A_K$ (see (10.83)), where

$$\begin{cases} A_K y = -y'' + q(x)y, \\ y'(a) + m_\infty(-0)y(a) = 0, \end{cases} \tag{10.87}$$

is the Kreĭn-von Neumann extension of a non-negative operator $\dot{A}$ of the form (10.82).

Theorem 10.6.1. *Let Θ be a minimal L-system of the form (10.18), where $\mathbb{A}$ is a $(*)$-extension of the form (10.76) of the accretive Schrödinger operator T_h with $\operatorname{Re} h = -m_\infty(-0)$. Then the spectral measure in the representation*

$$V_\Theta(z) = \int_0^\infty \frac{d\sigma(t)}{t-z} \tag{10.88}$$

satisfies the relation

$$\int_0^\infty \frac{d\sigma(t)}{1+t^2} = \operatorname{Im} h \left(\sup_{y \in \operatorname{Dom}(A_K)} \frac{|y(a)|}{\left(\int_a^\infty (|y(x)|^2 + |l(y)|^2)\, dx\right)^{\frac{1}{2}}} \right)^2. \tag{10.89}$$

Proof. Consider the Hilbert space $L_2^\sigma[0,+\infty)$, where $\sigma(t)$ is the function from the representation (10.85) and the operator

$$\dot{\Lambda}_\sigma f = tf(t),$$
$$\operatorname{Dom}(\dot{\Lambda}_\sigma) = \left\{ f \in L_2^\sigma[0,+\infty) \mid tf(t) \in L_2^\sigma[0,+\infty),\ \int_0^\infty f(t)d\sigma(t) = 0 \right\}.$$

According to Section 10.4, the operator $\dot{\Lambda}_\sigma$ is a prime, symmetric operator with deficiency indices $(1,1)$ and

$$\Lambda_\sigma f = tf(t), \qquad \operatorname{Dom}(\Lambda_\sigma) = \{ f \in L_2^\sigma[0,+\infty) \mid tf(t) \in L_2^\sigma[0,+\infty) \},$$

is its self-adjoint extension. It was shown in Section 9.8 that Stieltjes function (10.88) can be realized as an impedance function $V_{\vec{\Theta}}(z)$ of some L-system of the form

$$\Theta_\Lambda = \begin{pmatrix} \mathbf{\Lambda} & K^\sigma & 1 \\ \mathcal{H}_+^\sigma \subset L_2^\sigma[0,+\infty) \subset \mathcal{H}_-^\sigma & & \mathbb{C} \end{pmatrix},$$

where

$$\mathbf{\Lambda} = \operatorname{Re}\mathbf{\Lambda} + i(\cdot,\vec{1})\vec{1}, \quad (\vec{1} \in \mathcal{H}_-^\sigma), \tag{10.90}$$

and $\operatorname{Re}\mathbf{\Lambda} \supset \Lambda_\sigma$. Therefore,

$$V_\Theta(z) = \int_0^\infty \frac{d\sigma(t)}{t-z} = (K^\sigma)^*(\operatorname{Re}\mathbf{\Lambda} - zI)^{-1}K^\sigma = \left((\operatorname{Re}\mathbf{\Lambda} - zI)^{-1}\vec{1},\vec{1}\right)$$
$$= V_{\Theta_\Lambda}(z).$$

Notice that we may assume that the vector $\vec{1}$ is such that

$$(\operatorname{Re}\mathbf{\Lambda} - zI)^{-1}\vec{1} = \frac{1}{t-z}. \tag{10.91}$$

Indeed, since $(\operatorname{Re}\mathbf{\Lambda} - zI)^{-1}\vec{1} = \frac{\xi}{t-z}$, $\xi \in \mathbb{C}$, we get

$$\operatorname{Im} V_\Theta(i) = \int_0^\infty \frac{d\sigma(t)}{t^2+1} = \left\|(\operatorname{Re}\mathbf{\Lambda} - iI)^{-1}\vec{1}\right\|^2 = |\xi|^2 \int_0^\infty \frac{d\sigma(t)}{t^2+1}.$$

So, $|\xi|^2 = 1$. It follows from (6.44), (6.47) and (6.48) that

$$\begin{aligned} W_\Theta(z) &= [I - iV_\Theta(z)]\,[I + iV_\Theta(z)]^{-1} = [I - iV_{\vec{\Theta}}(z)]\,[I + iV_{\Theta_\Lambda}(z)]^{-1} \\ &= W_{\Theta_\Lambda}(z). \end{aligned}$$

By Theorem 6.6.10 there exists a triple of isometric operators (U_+, U, U_-) that maps the triplet $\mathcal{H}_+ \subset L_2[a,+\infty) \subset \mathcal{H}_-$ isometrically onto $\mathcal{H}_+^\sigma \subset L_2^\sigma[0,+\infty) \subset \mathcal{H}_-^\sigma$ with $\mathbf{\Lambda} = U_- \mathbb{A} U_+^{-1}$. Therefore,

$$\operatorname{Im}\mathbf{\Lambda} = U_- \operatorname{Im}\mathbb{A}U_+^{-1}.$$

It follows from (10.84) and (10.90) that $U_- g = \vec{1}$, $\quad g = (\operatorname{Im} h)^{\frac{1}{2}}\delta(x-a)$ and $\|g\|_{\mathcal{H}_-}^2 = \left\|\vec{1}\right\|_{\mathcal{H}_-^\sigma}^2$. On the other hand, because $\operatorname{Dom}(A_K)$ is a subspace in $\mathcal{H}_+$ and dense in $L_2[a,+\infty)$, we can construct a new triple of Hilbert spaces, writing $\operatorname{Dom}(A_K) = \mathcal{H}_+^K$,

$$\mathcal{H}_+^K \subset \mathcal{H}_+ \subset L_2[a,+\infty) \begin{array}{l} \subset \mathcal{H}_-^K \\ \subset \mathcal{H}_-. \end{array}$$

In the same manner, writing $\operatorname{Dom}(\Lambda_\sigma) = \vec{\mathfrak{H}}_+$, we get

$$\vec{\mathfrak{H}}_+ \subset \mathcal{H}_+^\sigma \subset L_2^\sigma[0,+\infty) \begin{array}{l} \subset \vec{\mathfrak{H}}_- \\ \subset \mathcal{H}_-^\sigma \end{array}. \tag{10.92}$$

Since the operators A_K and Λ_σ are unitarily equivalent and $U_+ \restriction \mathcal{H}_+^K = U_+^K$ maps $\mathcal{H}_+^K$ onto $\vec{\mathfrak{H}}_+$, there exists a triple of operators (U_+^K, U, U_-^K) that maps the triplet of Hilbert spaces $\mathcal{H}_+^K \subset L_2[a,+\infty) \subset \mathcal{H}_-^K$ onto the triplet of Hilbert spaces

$$\vec{\mathfrak{H}}_+ \subset L_2^\sigma[0,+\infty) \subset \vec{\mathfrak{H}}_-. \tag{10.93}$$

It was shown in Theorem 6.5.2 that

$$\operatorname{Dom}(\Lambda_\sigma) = \operatorname{Dom}(\dot{\Lambda}_\sigma) + \frac{c}{1+t^2}, \quad c \in \mathbb{C}.$$

Consider the linear, continuous functional of the form $(f, \vec{1})$, $(f \in \mathcal{H}_+^\sigma)$ on $\vec{\mathcal{H}}_+$. Due to the embedding (10.92), we can consider this functional on $\vec{\mathfrak{H}}_+$ and it can be represented in the form

$$(f, \vec{1}) = (f, \widehat{\vec{1}}), \quad f \in \vec{\mathfrak{H}}_+,\ \widehat{\vec{1}} \in \vec{\mathfrak{H}}_-,\ \vec{1} \in \mathcal{H}_-^\sigma. \tag{10.94}$$

Let R be the Riesz-Berezansky operator in the triple of Hilbert spaces (10.93) which maps $\vec{\mathfrak{H}}_-$ isometrically onto $\vec{\mathfrak{H}}_+$. As we showed before in the proof of Theorem 6.5.1,

$$\mathsf{R}\, c = \frac{c}{1+t^2}, \quad c \in \mathbb{C}.$$

Therefore,

$$\mathsf{R}\,1 = \frac{1}{1+t^2},$$

and 1 belongs to $\vec{\mathfrak{H}}_-$. We will now show that

$$(f, \vec{1}) = (f, 1), \quad f \in \vec{\mathfrak{H}}_+, \quad 1 \in \vec{\mathfrak{H}}_-, \quad \vec{1} \in \mathcal{H}_-^\sigma. \tag{10.95}$$

Taking into account (10.91), (10.94), and properties of the resolvent of self-adjoint bi-extensions of a symmetric operator, we get

$$\begin{aligned}
&\left((\Lambda_\sigma - \bar{z}I)^{-1}u, \vec{1}\right) = \left((\Lambda_\sigma - \bar{z}I)^{-1}u, \widehat{\vec{1}}\right),\\
&\left(u, (\mathrm{Re}\,\mathbf{\Lambda} - zI)^{-1}\vec{1}\right) = \left(u, (\Lambda_\sigma - zI)^{-1}\widehat{\vec{1}}\right),\\
&\left(u, \frac{1}{t-z}\right) = \left(u, (\Lambda_\sigma - zI)^{-1}c\right),\\
&\left(u, \frac{1}{t-z}\right) = \bar{c}\left(u, \frac{1}{t-z}\right), \qquad u \in \mathcal{H}^\sigma = L_2^\sigma[0,+\infty)),\ c \in \mathbb{C}.
\end{aligned}$$

Therefore, $c = 1$ and $\widehat{\vec{1}} = 1$. It follows from (10.95) that

$$\begin{aligned}
&\sup_{f \in \vec{\mathfrak{H}}_+} \frac{|(f, \vec{1})|}{\|f\|_{\vec{\mathfrak{H}}_+}} = \sup_{f \in \vec{\mathfrak{H}}_+} \frac{|(f, 1)|}{\|f\|_{\vec{\mathfrak{H}}_+}} = \|1\|_{\vec{\mathfrak{H}}_-} = (\|1\|^2_{\vec{\mathfrak{H}}_-})^{\frac{1}{2}}\\
&= ((1,1)_{\vec{\mathfrak{H}}_-})^{\frac{1}{2}} = ((\mathsf{R}1, 1))^{\frac{1}{2}} = \left((\frac{1}{1+t^2}, 1)\right)^{\frac{1}{2}} = \left(\int_0^\infty \frac{d\sigma}{1+t^2}\right)^{\frac{1}{2}}.
\end{aligned}$$

On the other hand, for $y \in \mathcal{H}_+^K$, $g \in \mathcal{H}_-$,

$$(y, g) = (U^*Uy, g) = (U_+^K y, (U^*)^* g) = (U_+^K y, U_- g) = (U_+^K y, \vec{1}).$$

Therefore,

$$\sup_{y\in\mathcal{H}_+^K}\frac{|(y,g)|}{\|y\|_{\mathcal{H}_+^K}}=\sup_{y\in\mathcal{H}_+^K}\frac{|(U_+^K y,\vec{1})|}{\|y\|_{\mathcal{H}_+^K}}$$

$$=\sup_{y\in\mathcal{H}_+^K}\frac{|(U_+^K y,\vec{1})|}{\|U_+^K y\|_{\mathfrak{H}_+}}=\sup_{f\in\mathfrak{H}_+}\frac{|(f,\vec{1})|}{\|f\|_{\mathfrak{H}_+}}=\left(\int_0^\infty\frac{d\sigma}{1+t^2}\right)^{\frac{1}{2}}.$$

Taking into account (10.84) we get (10.89). □

Theorem 10.6.2. *Let Θ be an L-system of the form* (10.18)*, where $\mathbb{A}$ is a* $(*)$*-extension of the form* (10.11) *of the accretive Schrödinger operator T_h of the form* (10.81)*. Then its impedance function $V_\Theta(z)$ is a Stieltjes function if and only if*

$$\begin{cases}\mu\ge\dfrac{(\operatorname{Im}h)^2}{m_\infty(-0)+\operatorname{Re}h}+\operatorname{Re}h,\\ \operatorname{Re}h\ge -m_\infty(-0).\end{cases}\tag{10.96}$$

Proof. The proof of this theorem follows from Theorems 10.3.3 and 10.5.1 that T_h is accretive if and only if $\operatorname{Re}h\ge -m_\infty(-0)$, as well as from Theorem 9.8.2 stating that $V_\Theta(\lambda)$ is a Stieltjes function if and only if $\mathbb{A}$ is accretive. □

We can summarize this part of the section with the following general result.

Theorem 10.6.3. *An L-system Θ of the form*

$$\begin{cases}(T_h-zI)x=\dfrac{(\operatorname{Im}h)^{\frac{1}{2}}}{|\mu-h|}[\mu\delta(x-a)+\delta'(x-a)]\varphi_-,\\ \varphi_+=\varphi_- -2i\dfrac{(\operatorname{Im}h)^{\frac{1}{2}}}{|\mu-h|}[\mu x(a)-x'(a)],\end{cases}\quad \varphi_\pm\in\mathbb{C},\tag{10.97}$$

with an accretive Schrödinger operator T_h of the form (10.81) *has the Stieltjes impedance function $V_\Theta(z)$ if and only if relation* (10.96) *holds.*

Proof. The proof immediately follows from Theorems 10.2.2 and 10.6.2. □

Now let us consider applications related to the subclass S^{α_1,α_2} of scalar Stieltjes functions. Let again Θ be an L-system of the form (10.18), where $\mathbb{A}$ is a $(*)$-extension (10.11) of the accretive Schrödinger operator T_h (10.81). As we know from Theorem 10.3.3 and formula (10.47), the operator T_h is α-sectorial if and only if $\operatorname{Re}h>-m_\infty(-0)$ and is accretive but not α-sectorial for any $\alpha\in(0,\pi/2)$ if and only if $\operatorname{Re}h=-m_\infty(-0)$. It also follows from Theorem 10.5.1 that the operator $\mathbb{A}$ of Θ is accretive if and only if (10.74) holds. Using (10.21) we can write the impedance function $V_\Theta(z)$ in the form

$$V_\Theta(z)=\frac{(m_\infty(z)+\mu)\operatorname{Im}h}{(\mu-\operatorname{Re}h)(m_\infty(z)+\operatorname{Re}h)-(\operatorname{Im}h)^2}.\tag{10.98}$$

Consider our system Θ with $\mu = +\infty$. Then in (10.98) we obtain

$$V_\Theta(z) = \frac{\operatorname{Im} h}{m_\infty(z) + h}.$$

Then in this case

$$\lim_{x \to -\infty} V_\Theta(x) = \lim_{x \to -\infty} \frac{\operatorname{Im} h}{m_\infty(x) + h} = 0, \tag{10.99}$$

since $m_\infty(x) \to +\infty$ as $x \to -\infty$. Also,

$$\lim_{x \to -0} V_\Theta(x) = \frac{\operatorname{Im} h}{m_\infty(-0) + h}.$$

Assuming that T_h is α-sectorial and hence $\operatorname{Re} h > -m_\infty(-0)$, we use (9.150) and obtain

$$\lim_{x \to -\infty} V_\Theta(x) = 0 = \tan 0 = \tan \alpha_1, \quad \lim_{x \to -0} V_\Theta(x) = \frac{\operatorname{Im} h}{m_\infty(-0) + h} = \tan \alpha_2.$$

On the other hand since T_h is α-sectorial, then via Theorem 10.3.3 we have that

$$\tan \alpha = \tan \alpha_2 = \frac{\operatorname{Im} h}{m_\infty(-0) + h},$$

and hence, by Corollary 9.8.5, $V_\Theta(z)$ belongs to the class $S^{0,\alpha}$.

Let now $\mu \neq +\infty$ and satisfies inequality (10.74). Then

$$\begin{aligned} \lim_{x \to -\infty} V_\Theta(x) &= \lim_{x \to -\infty} \frac{(m_\infty(x) + \mu) \operatorname{Im} h}{(\mu - \operatorname{Re} h)(m_\infty(x) + \operatorname{Re} h) - (\operatorname{Im} h)^2} \\ &= \frac{\operatorname{Im} h}{\mu - \operatorname{Re} h} = \tan \alpha_1, \end{aligned} \tag{10.100}$$

and

$$\lim_{x \to -0} V_\Theta(x) = \frac{(m_\infty(-0) + \mu) \operatorname{Im} h}{(\mu - \operatorname{Re} h)(m_\infty(-0) + \operatorname{Re} h) - (\operatorname{Im} h)^2} = \tan \alpha_2. \tag{10.101}$$

Therefore, in this case $V_\Theta(z) \in S^{\alpha_1, \alpha_2}$.

Theorem 10.6.4. *Let Θ be an L-system of the form (10.18), where $\mathbb{A}$ is a $(*)$-extension of an α-sectorial operator T_h with the exact angle of sectoriality $\alpha \in (0, \pi/2)$. Then $\mathbb{A}$ is an α-sectorial $(*)$-extension of T_h (with the same angle of sectoriality) if and only if $\mu = +\infty$ in (10.11).*

Proof. It follows from (10.99)-(10.101) that in this case $V_\Theta(z) \in S^{0,\alpha}$ if and only if $\mu = +\infty$. Thus using Corollary 9.8.5 for the function $V_\Theta(z)$ we obtain that $\mathbb{A}$ is α-sectorial $(*)$-extension of T_h. □

We note that if T_h is α-sectorial with the exact angle of sectoriality α, then it admits only one α-sectorial $(*)$-extension $\mathbb{A}$ with the same angle of sectoriality α. Consequently, $\mu = +\infty$ and $\mathbb{A}$ has the form (10.76) as it was explained in the proof of Theorem 10.5.1.

Theorem 10.6.5. *Let Θ be an L-system of the form* (10.18)*, where $\mathbb{A}$ is a $(*)$-extension of an α-sectorial operator T_h with the exact angle of sectoriality $\alpha \in (0, \pi/2)$. Then $\mathbb{A}$ is accretive but not α-sectorial for any $\alpha \in (0, \pi/2)$ a $(*)$-extension of T_h if and only if in* (10.11)

$$\mu = \mu_0 = \frac{(\operatorname{Im} h)^2}{m_\infty(-0) + \operatorname{Re} h} + \operatorname{Re} h. \tag{10.102}$$

Proof. Let $V_\Theta(z)$ be the impedance function of our system Θ. If in (10.100) we set $\mu = \mu_0$ where μ_0 is given by (10.102), then

$$\begin{aligned}\lim_{x \to -\infty} V_\Theta(x) &= \frac{\operatorname{Im} h}{\mu_0 - \operatorname{Re} h} = \frac{m_\infty(-0) + \operatorname{Re} h}{\operatorname{Im} h} = \frac{1}{\tan \alpha} \\ &= \tan\left(\frac{\pi}{2} - \alpha\right) = \tan \alpha_1,\end{aligned} \tag{10.103}$$

where $\alpha_1 = \frac{\pi}{2} - \alpha$. On the other hand, using (10.101) with $\mu = \mu_0$ we obtain

$$\begin{aligned}\lim_{x \to -0} V_\Theta(x) &= \frac{\operatorname{Im} h \left(m_\infty(-0) + \frac{(\operatorname{Im} h)^2}{m_\infty(-0) + \operatorname{Re} h} \right)}{\frac{(\operatorname{Im} h)^2}{m_\infty(-0) + \operatorname{Re} h}(m_\infty(-0) + \operatorname{Re} h) - (\operatorname{Im} h)^2} = \infty \\ &= \tan \frac{\pi}{2} = \tan \alpha_2.\end{aligned} \tag{10.104}$$

Hence, (10.103) and (10.104) yield that $V_\Theta(z) \in S^{\frac{\pi}{2} - \alpha, \frac{\pi}{2}}$. If we assume that $\mathbb{A}$ is a β-sectorial $(*)$-extension, then by Theorem 9.8.7,

$$\tan \beta \leq \tan \alpha_2 + 2\sqrt{\tan \alpha_1 (\tan \alpha_2 - \tan \alpha_1)} = \infty.$$

Therefore, $\mathbb{A}$ is accretive but not β-sectorial for any $\beta \in (0, \pi/2)$. □

Note that it follows from the above theorem that any α-sectorial operator T_h with the exact angle of sectoriality $\alpha \in (0, \pi/2)$ admits only one accretive $(*)$-extension $\mathbb{A}$. This extension takes form (10.11) with $\mu = \mu_0$ where μ_0 is given by (10.102).

10.7 Inverse Stieltjes functions and systems with Schrödinger operator

Consider an L-system of the form (10.18), with $\dot{A}$ and $\dot{A}^*$ given by (10.9) and $\mu = \operatorname{Re} h = -m_\infty(-0)$. Then we obtain the following scattering L-system:

$$\Theta = \begin{pmatrix} \mathbb{A} & K & 1 \\ \mathcal{H}_+ \subset L_2[a, +\infty) \subset \mathcal{H}_- & & \mathbb{C} \end{pmatrix}, \tag{10.105}$$

where

$$\begin{aligned}\mathbb{A}y &= -y'' + q(x)y - \frac{i}{\operatorname{Im} h}\,[y'(a) - hy(a)]\,[\delta'(x-a) - m_\infty(-0)\delta(x-a)],\\ \mathbb{A}^* y &= -y'' + q(x)y + \frac{i}{\operatorname{Im} h}\,[y'(a) - \overline{h}y(a)]\,[\delta'(x-a) - m_\infty(-0)\delta(x-a)],\end{aligned} \tag{10.106}$$

and

$$\begin{aligned}\operatorname{Re}\mathbb{A} &= -y'' + q(x)y - y(a)[\delta'(x-a) - m_\infty(-0)\delta(x-a)],\\ \operatorname{Im}\mathbb{A} &= \frac{1}{\operatorname{Im} h}\,[-y'(a) - m_\infty(-0)y(a)]\,[\delta'(x-a) - m_\infty(-0)\delta(x-a)]\\ &= \frac{1}{\operatorname{Im} h}(y,g)g,\end{aligned} \tag{10.107}$$

where

$$g = \frac{1}{(\operatorname{Im} h)^{\frac{1}{2}}}[\delta'(x-a) - m_\infty(-0)\delta(x-a)].$$

Using relation (10.20) we get that the transfer function $W_\Theta(z)$ of an L-system Θ of the form (10.105) has the representation

$$W_\Theta(z) = I - 2iK^*(\mathbb{A} - zI)^{-1}K = -\frac{m_\infty(z) + \overline{h}}{m_\infty(z) + h}. \tag{10.108}$$

Denote by Θ_K an L-system of the form (10.18) with (10.76) and by Θ_F an L-system of the form (10.105), (10.106). Then it follows from (10.108) that $W_{\Theta_F}(z) = -W_{\Theta_K}(z)$. Moreover (see (6.48)),

$$\begin{aligned}V_{\Theta_F}(z) &= i[W_{\Theta_F}(z) - I][W_{\Theta_F}(z) + I]^{-1} = -i[W_{\Theta_K}(z) + I][-W_{\Theta_K}(z) + I]^{-1}\\ &= i[W_{\Theta_K}(z) + I][W_{\Theta_K}(z) - I]^{-1}\\ &= -\{i[W_{\Theta_K}(z) - I][W_{\Theta_K}(z) + I]^{-1}\}^{-1} = -V_{\Theta_K}^{-1}(z).\end{aligned}$$

Since $V_{\Theta_K}(z)$ is a Stieltjes function with representation

$$V_{\Theta_K}(z) = \int_0^\infty \frac{d\sigma(t)}{t - z},$$

$-V_{\Theta_K}^{-1}(z)$ is the inverse Stieltjes function and has the representation

$$V_{\Theta_F}(z) = -V_{\Theta_K}^{-1}(z) = \alpha + \int_0^\infty \left(\frac{1}{t-z} - \frac{1}{t}\right) d\tau(t),$$

where $\alpha \le 0$, $\tau(t)$ is a nondecreasing function satisfying the relations

$$\int_0^\infty d\tau(t) = \infty, \quad \int_0^\infty \frac{d\tau(t)}{t^2 + t} < \infty.$$

It follows from (10.107) that $\operatorname{Re}\mathbb{A} \supset A_F$, where

$$\begin{cases} A_F y = -y'' + q(x)y, \\ y(a) = 0 \end{cases}$$

is the Friedrichs extension of a non-negative operator $\dot{A}$ of the form (6.44). Taking into account calculations similar to those in connection with Theorem 10.6.1, we get

Theorem 10.7.1. *Let Θ be a minimal L-system of the form* (10.105), *where $\mathbb{A}$ is a $(*)$-extension given by* (10.106) *of the accretive Schrödinger operator T_h ($\mu = \operatorname{Re} h = -m_\infty(-0)$). Then the spectral measure $\tau(t)$ in the representation*

$$V_{\Theta_F}(z) = V_\Theta(z) = \alpha + \int_0^\infty \left(\frac{1}{t-z} - \frac{1}{t}\right) d\tau(t), \tag{10.109}$$

satisfies the relation

$$\int_0^\infty \frac{d\tau(t)}{1+t^2} = \frac{1}{\operatorname{Im} h} \left(\sup_{y \in \operatorname{Dom}(A_F)} \frac{|y'(a)|}{\left(\int_a^\infty (|y(x)|^2 + |l(y)|^2)\,dx\right)^{\frac{1}{2}}} \right)^2. \tag{10.110}$$

As a consequence of Theorems 10.6.1 and 10.7.1, it follows that the spectral measures $\sigma(t)$ and $\tau(t)$ in the representations (10.88) and (10.109), which correspond to the Kreĭn-von Neumann extension A_K and the Friedrichs extension A_F, satisfy the relation

$$\int_0^\infty \frac{d\sigma(t)}{1+t^2} \int_0^\infty \frac{d\tau(t)}{1+t^2} = \left(\sup_{y \in \operatorname{Dom}(A_K)} \frac{|y(a)|}{\left(\int_a^\infty (|y(x)|^2 + |l(y)|^2)\,dx\right)^{\frac{1}{2}}} \right)^2 \times \left(\sup_{y \in \operatorname{Dom}(A_F)} \frac{|y'(a)|}{\left(\int_a^\infty (|y(x)|^2 + |l(y)|^2)\,dx\right)^{\frac{1}{2}}} \right)^2.$$

Remark 10.7.2. It follows from relations (10.89) and (10.110) that

$$\int_a^\infty \left(|y(x)|^2 + |l(y)|^2\right) dx \geq \frac{(\operatorname{Im} h)\,|y(a)|^2}{\int_0^\infty \frac{d\sigma(t)}{1+t^2}}, \qquad y'(a) + m_\infty(-0)y(a) = 0,$$

$$\int_a^\infty \left(|y(x)|^2 + |l(y)|^2\right) dx \geq \frac{|y'(a)|^2}{(\operatorname{Im} h)\int_0^\infty \frac{d\sigma(t)}{1+t^2}}, \qquad y(a) = 0.$$

As we have already mentioned in Section 10.5, any Stieltjes function $V_\Theta(z)$ has a parametric representation of the type (9.141)

$$V_\Theta(z) = \gamma + \int_0^\infty \frac{d\sigma(t)}{t-z}, \quad (\gamma \geq 0), \tag{10.111}$$

where $\sigma(t)$ is a non-decreasing function for which

$$\int_0^\infty \frac{d\sigma(t)}{t+1} < \infty. \tag{10.112}$$

In addition to (10.112), since the maximal symmetric part of T_h has dense domain in $L_2[a,+\infty)$, one concludes that

$$\int_0^\infty d\sigma(t) = \infty. \tag{10.113}$$

Thus, the function $V_\Theta(z)$ has the parametric representation (10.111), (10.112), (10.113) if and only if the inequalities (10.96) hold.

Theorem 10.7.3. *Let Θ be an L-system of the form* (10.105), *where $\mathbb{A}$ is a $(*)$-extension of the form* (10.11) *of an accretive Schrödinger operator T_h of the form* (10.81). *Then its impedance function $V_\Theta(z)$ is an inverse Stieltjes function if and only if*

$$-m_\infty(-0) \leq \mu \leq \operatorname{Re} h. \tag{10.114}$$

Proof. Consider the $(*)$-extensions $\mathbb{A}_\mu$ and $\mathbb{A}_\xi$ of the accretive Schrödinger operator T_h defined by (10.25) and (10.26), respectively, with

$$\xi = \frac{\mu \operatorname{Re} h - |h|^2}{\mu - \operatorname{Re} h}. \tag{10.115}$$

Let

$$\Theta_\mu = \begin{pmatrix} \mathbb{A}_\mu & K_\mu & 1 \\ \mathcal{H}_+ \subset L_2[a,+\infty) \subset \mathcal{H}_- & & \mathbb{C} \end{pmatrix} \quad \text{and} \quad \Theta_\xi = \begin{pmatrix} \mathbb{A}_\xi & K_\xi & 1 \\ \mathcal{H}_+ \subset L_2[a,+\infty) \subset \mathcal{H}_- & & \mathbb{C} \end{pmatrix},$$

be the corresponding L-systems described in (10.27) and (10.28). Applying (10.29) and taking into account (10.115) we have

$$W_{\Theta_\mu}(z) = -W_{\Theta_\xi}(z). \tag{10.116}$$

The relations (10.116) and (6.48) imply

$$\begin{aligned} V_{\Theta_\mu}(z) &= K_\mu^*(\operatorname{Re} \mathbb{A}_\mu - zI)^{-1} K_\mu = i[W_{\Theta_\mu}(z) - I]\,[W_{\Theta_\mu}(z) + I]^{-1} \\ &= i[-W_{\Theta_\xi}(z) - I][-W_{\Theta_\xi}(z) + I]^{-1} = -V_{\Theta_\xi}^{-1}(z), \end{aligned}$$

and $V_{\Theta_\mu}(z) = -V_{\Theta_\xi}^{-1}(z)$. As was shown in Theorem 10.6.2, the function $V_{\Theta_\xi}(z)$ is a Stieltjes function if and only if

$$\xi \geq \frac{(\mathrm{Im}\, h)^2}{m_\infty(-0) + \mathrm{Re}\, h} + \mathrm{Re}\, h.$$

Therefore,

$$\xi = \frac{\mu \mathrm{Re}\, h - |h|^2}{\mu - \mathrm{Re}\, h} \geq \frac{(\mathrm{Im}\, h)^2}{m_\infty(-0) + \mathrm{Re}\, h} + \mathrm{Re}\, h. \tag{10.117}$$

Furthermore,

$$\frac{\mu \mathrm{Re}\, h - |h|^2}{\mu - \mathrm{Re}\, h} = \frac{\mu \mathrm{Re}\, h - (\mathrm{Im}\, h)^2 - (\mathrm{Re}\, h)^2}{\mu - \mathrm{Re}\, h} = -\frac{(\mathrm{Im}\, h)^2}{\mu - \mathrm{Re}\, h} + \mathrm{Re}\, h.$$

It follows from (10.117) that

$$-\frac{(\mathrm{Im}\, h)^2}{\mu - \mathrm{Re}\, h} + \mathrm{Re}\, h \geq \frac{(\mathrm{Im}\, h)^2}{m_\infty(-0) + \mathrm{Re}\, h} + \mathrm{Re}\, h,$$

and

$$\frac{1}{\mathrm{Re}\, h - \mu} - \frac{1}{m_\infty(-0) + \mathrm{Re}\, h} \geq 0.$$

The last inequality is equivalent to (10.114). When μ satisfies (10.114), then $V_{\Theta_\xi}(z)$ is a Stieltjes function and therefore $V_{\Theta_\mu}(z)$ is an inverse Stieltjes function. □

Theorem 10.7.4. *Let*

$$\Theta = \begin{pmatrix} \mathbb{A}_\mu & K_\mu & 1 \\ \mathcal{H}_+ \subset L_2[a,+\infty) \subset \mathcal{H}_- & & \mathbb{C} \end{pmatrix},$$

be a minimal scattering L-system, where $\mathbb{A}_\mu$ is a $(*)$-extension of the form (10.25) of the accretive Schrödinger operator T_h and

$$\mu \geq \frac{(\mathrm{Im}\, h)^2}{m_\infty(-0) + \mathrm{Re}\, h} + \mathrm{Re}\, h.$$

Then the spectral measure $\sigma_\xi(t)$ in the representation

$$V_{\Theta_\mu}(z) = \gamma + \int_0^\infty \frac{d\sigma_\xi(t)}{t - z}, \qquad \xi = \frac{\mu \mathrm{Re}\, h - |h|^2}{\mu - \mathrm{Re}\, h} \tag{10.118}$$

satisfies the relation

$$\int_0^\infty \frac{d\sigma_\xi(t)}{1 + t^2} = \frac{\mathrm{Im}\, h}{|\mu - h|^2} \left(\sup_{y \in \mathrm{Dom}(\widehat{A}_\xi)} \frac{|\mu y(a) - y'(a)|}{\left(\int_a^\infty \left(|y(x)|^2 + |l(y)|^2\right) dx \right)^{\frac{1}{2}}} \right)^2, \tag{10.119}$$

where

$$\begin{cases} \widehat{A}_\xi y = -y'' + qy, \\ y'(a) = \xi y(a), \end{cases} \qquad \xi = \frac{\mu \mathrm{Re}\, h - |h|^2}{\mu - \mathrm{Re}\, h}, \tag{10.120}$$

is a non-negative self-adjoint extension of the non-negative operator $\dot{A}$ of the form (10.82).

Proof. It is easy to check that when $\mu \in \left[\frac{(\mathrm{Im}\, h)^2}{m_\infty(-0)+\mathrm{Re}\, h} + \mathrm{Re}\, h, +\infty\right)$ then

$$\xi = \frac{\mu \mathrm{Re}\, h - |h|^2}{\mu - \mathrm{Re}\, h} \in [-m_\infty(-0), \mathrm{Re}\, h).$$

Therefore, $\widehat{A}_\xi$ of the form (10.120) is a non-negative self-adjoint extension of the non-negative minimal operator $\dot{A}$ of the form (10.82). From (10.114) it follows that

$$\mathrm{Re}\, \mathbb{A}_\mu = -y'' + qy + \left[\frac{\mu \mathrm{Re}\, h - |h|^2}{|\mu - h|^2} y(a) - \frac{\mu - \mathrm{Re}\, h}{|\mu - h|^2} y'(a)\right] \\ \times [\mu \delta(x-a) + \delta'(x-a)],$$

and that

$$\mathrm{Re}\, \mathbb{A}_\mu \supset \widehat{A}_\xi, \quad \xi = \frac{\mu \mathrm{Re}\, h - |h|^2}{\mu - \mathrm{Re}\, h}.$$

Taking into account that $\mathrm{Im}\, \mathbb{A}_\mu = (\cdot, g)g$, where g is defined by (10.15), and following the proof of Theorem 10.6.1, one obtains relation (10.119). □

Theorem 10.7.5. *Let*

$$\Theta_\xi = \begin{pmatrix} \mathbb{A}_\xi & K_\xi & 1 \\ \mathcal{H}_+ \subset L_2[a, +\infty) \subset \mathcal{H}_- & & \mathbb{C} \end{pmatrix},$$

be a minimal L-system, where $\mathbb{A}_\xi$ is a $(*)$-extension of the form (10.26) of the accretive Schrödinger operator T_h and $-m_\infty(-0) \le \xi \le \mathrm{Re}\, h$. Then the spectral measure $\tau_\mu(t)$ in the representation

$$V_{\Theta_\xi}(z) = \alpha + \int_0^\infty \left(\frac{1}{t-z} - \frac{1}{t}\right) d\tau_\mu(t), \quad (\alpha \le 0),$$

satisfies the relation

$$\int_0^\infty \frac{d\tau_\mu(t)}{1+t^2} = \frac{\mathrm{Im}\, h}{|\xi - h|^2} \left(\sup_{y \in \mathrm{Dom}(\widehat{A}_\mu)} \frac{|\xi y(a) - y'(a)|}{\left(\int_a^\infty (|y(x)|^2 + |l(y)|^2)\, dx\right)^{\frac{1}{2}}} \right)^2, \tag{10.121}$$

where

$$\xi = \frac{\mu \mathrm{Re}\, h - |h|^2}{\mu - \mathrm{Re}\, h}, \qquad \mu > \mathrm{Re}\, h, \tag{10.122}$$

and

$$\begin{cases} \widehat{A}_\mu y = -y'' + qy, \\ y'(a) = \mu y(a), \end{cases}$$

is a non-negative self-adjoint extension of the non-negative operator $\dot{A}$ of the form (10.82).

Proof. It was established in the proof of Theorem 10.7.3 that the inequalities

$$\mu \geq \frac{(\mathrm{Im}\, h)^2}{m_\infty(-0) + \mathrm{Re}\, h} + \mathrm{Re}\, h \tag{10.123}$$

and

$$-m_\infty(-0) \leq \xi \leq \mathrm{Re}\, h, \quad \mu > \mathrm{Re}\, h,$$

where

$$\xi = \frac{\mu \mathrm{Re}\, h - |h|^2}{\mu - \mathrm{Re}\, h},$$

are equivalent. Consider two $(*)$-extensions $\mathbb{A}_\mu$ and $\mathbb{A}_\xi$ of the accretive Schrödinger operator T_h. As μ satisfies (10.123), the operator-valued function $V_{\Theta_\xi}(z)$ is an inverse Stieltjes function and $V_{\Theta_\mu}(z)$ is a Stieltjes function by Theorem 10.7.3, and $V_{\Theta_\mu}(z) = -V^{-1}_{\Theta_\xi}(z)$. It follows from (10.26) that

$$\begin{aligned} \mathrm{Re}\, \mathbb{A}_\xi &= -y'' + qy \\ &+ \left[\frac{\xi \mathrm{Re}\, h - |h|^2}{|\xi - h|^2} y(a) - \frac{\xi - \mathrm{Re}\, h}{|\xi - h|^2} y'(a)\right] [\xi\delta(x-a) + \delta'(x-a)]. \end{aligned}$$

Using (10.122) we get

$$\mu = \frac{\xi \mathrm{Re}\, h - |h|^2}{\xi - \mathrm{Re}\, h}.$$

Therefore, $\mathrm{Re}\, \mathbb{A}_\xi \supset \widehat{A}_\mu$. Taking into account that $\mathrm{Im}\, \mathbb{A}_\xi = (\cdot, g)g$, where

$$g = \frac{(\mathrm{Im}\, h)^{\frac{1}{2}}}{|\xi - h|} [\xi\delta(x-a) + \delta'(x-a)],$$

and following the proof of Theorem 10.7.4, one will obtain relation (10.121). □

As a consequence of Theorems 10.7.4 and 10.7.5 we get the following: Let

$$\Theta_\mu = \begin{pmatrix} \mathbb{A}_\mu & K_\mu & 1 \\ \mathcal{H}_+ \subset L_2[a,+\infty) \subset \mathcal{H}_- & & \mathbb{C} \end{pmatrix} \quad \text{and} \quad \Theta_\xi = \begin{pmatrix} \mathbb{A}_\xi & K_\xi & 1 \\ \mathcal{H}_+ \subset L_2[a,+\infty) \subset \mathcal{H}_- & & \mathbb{C} \end{pmatrix},$$

where ξ is related to μ by (10.115), be L-systems described in (10.27) and (10.28), where $\mathbb{A}_\mu$ and $\mathbb{A}_\xi$ are $(*)$-extensions of the form (10.25), (10.26) of the accretive Schrödinger operator T_h and

$$\mu \geq \frac{(\operatorname{Im} h)^2}{m_\infty(-0)+\operatorname{Re} h} + \operatorname{Re} h.$$

Then the spectral measures in the representations

$$V_{\Theta_\mu}(z) = \gamma + \int_0^\infty \frac{d\sigma_\xi(t)}{t-z} \quad \text{and} \quad V_{\Theta_\xi}(z) = \alpha + \int_0^\infty \left(\frac{1}{t-z} - \frac{1}{t}\right) d\tau_\mu(t),$$

satisfy the relation

$$\begin{aligned}\int_0^\infty \frac{d\sigma_\xi(t)}{1+t^2}\int_0^\infty \frac{d\tau_\mu(t)}{1+t^2} = &\left(\sup_{y\in \operatorname{Dom}(\widehat{A}_\xi)} \frac{\left|\frac{\mu}{\mu-h}y(a) - \frac{1}{\mu-h}y'(a)\right|}{\left(\int_a^\infty \left(|y(x)|^2+|l(y)|^2\right)dx\right)^{\frac{1}{2}}}\right)^2 \\ &\times \left(\sup_{y\in \operatorname{Dom}(\widehat{A}_\mu)} \frac{\left|\frac{\mu\operatorname{Re} h - h^2}{\mu-h}y(a) - \frac{\mu-\operatorname{Re} h}{\mu-h}y'(a)\right|}{\left(\int_a^\infty \left(|y(x)|^2+|l(y)|^2\right)dx\right)^{\frac{1}{2}}}\right)^2.\end{aligned} \tag{10.124}$$

The proof of relation (10.124) follows from Theorems 10.7.4 and 10.7.5 by multiplying both parts of the relations (10.119) and (10.121), from the fact that

$$\xi = \frac{\mu\operatorname{Re} h - |h|^2}{\mu - h},$$

and

$$\begin{aligned}\xi - h &= \frac{\mu\operatorname{Re} h - |h|^2}{\mu - \operatorname{Re} h} - h = \frac{\mu\operatorname{Re} h - |h|^2 - \mu h - h\operatorname{Re} h}{\mu - \operatorname{Re} h} \\ &= \frac{\mu(\operatorname{Re} h - h) - h(\overline{h} - \operatorname{Re} h)}{\mu - \operatorname{Re} h} = \frac{-i\mu\operatorname{Im} h + ih\operatorname{Im} h}{\mu - \operatorname{Re} h} = i\operatorname{Im} h\frac{h-\mu}{\mu-\operatorname{Re} h}.\end{aligned}$$

Remark 10.7.6. Using (10.24), we can rewrite relations (10.119) and (10.110) in the following way:

$$\begin{aligned}\sup_{y'(a)=\xi y(a)} \frac{|y(a)|^2}{\int_a^\infty \left(|y(x)|^2+|l(y)|^2\right)dx} &= -\frac{\operatorname{Im} m_\infty(i)}{|m_\infty(i)+\xi|^2}, \\ \sup_{y(a)=0} \frac{|y'(a)|^2}{\int_a^\infty \left(|y(x)|^2+|l(y)|^2\right)dx} &= -\operatorname{Im} m_\infty(i).\end{aligned}$$

These relations allow us to obtain some new inequalities (see [53], [135]). For instance, performing calculations for $m_\infty(i)$ of a symmetric Schrödinger operator $\dot{A}$ of the form (10.9) with $q(x) = 0$ in $L_2[0,+\infty)$ we obtain the sharp inequality

$$|y'(0)| \le 2^{-\frac{1}{4}} \left(\int_0^\infty |y(x)|^2\, dx + \int_0^\infty |y''(x)|^2\, dx \right)^{\frac{1}{2}}, \tag{10.125}$$

where $y(x)$, $y'(x)$ are absolutely continuous on any interval $[0,b] \subset [0,+\infty)$, $y(0) = 0$, and $y'' \in L_2[0,+\infty)$.

Remark 10.7.7. One can prove that

$$\sup_{\sin\alpha y'(a)+\cos\alpha y(a)=0} \frac{|y'(a)|^2}{\int_a^\infty (|y(x)|^2 + |l(y)|^2)\, dx} = \frac{1}{l(y_0)|_{x=a} + \tan\alpha\, (l(y_0))'\,|_{x=a}},$$

where $y_0(x)$ is the unique solution of the boundary value problem

$$\begin{cases} l\,(l(y)) + y = 0, \\ y(a) = -\tan\alpha,\ y'(a) = 1, \\ y,\, l(y) \in D^*. \end{cases}$$

We conclude this section with the following general result.

Theorem 10.7.8. *An L-system Θ of the form*

$$\begin{cases} (T_h - zI)x = \dfrac{(\operatorname{Im} h)^{\frac{1}{2}}}{|\mu - h|}[\mu\delta(x-a) + \delta'(x-a)]\varphi_-, \\ \varphi_+ = \varphi_- - 2i\dfrac{(\operatorname{Im} h)^{\frac{1}{2}}}{|\mu - h|}[\mu x(a) - x'(a)], \end{cases} \qquad \varphi_\pm \in \mathbb{C},$$

with an accretive Schrödinger operator T_h of the form (10.81) *has the inverse Stieltjes impedance function $V_\Theta(z)$ if and only if relation* (10.114) *holds.*

Proof. The proof immediately follows from Theorems 10.2.2 and 10.7.3. □

10.8 Stieltjes-like functions and inverse spectral problems for systems with Schrödinger operator

In this section we are going to use the realization results for Stieltjes functions developed in Section 9.8 to obtain the solution of inverse spectral problem for L-systems with a Schrödinger operator of the form (10.81) in $L_2[a,+\infty)$ with non-self-adjoint boundary conditions

$$\begin{cases} T_h y = -y'' + q(x)y, \\ y'(a) = hy(a), \end{cases} \qquad \left(q(x) = \overline{q(x)},\ \operatorname{Im} h \neq 0\right). \tag{10.126}$$

In particular, we will show that if a non-decreasing function $\sigma(t)$ is the spectral function of non-negative self-adjoint boundary value problem

$$\begin{cases} A_\theta y = -y'' + q(x)y, \\ y'(a) = \theta y(a), \end{cases}$$

and satisfies conditions

$$\int_0^\infty d\sigma(t) = \infty, \quad \int_0^\infty \frac{d\sigma(t)}{1+t} < \infty,$$

then, for every $\gamma \geq 0$, a Stieltjes function

$$V(z) = \gamma + \int_0^\infty \frac{d\sigma(t)}{t-z},$$

can be realized in a unique way as the impedance function $V_\Theta(z)$ of an accretive L-system Θ with some Schrödinger operator T_h.

Let $\mathcal{H} = L_2[a,+\infty)$ and $l(y) = -y''+q(x)y$ where q is a real locally summable function. Consider a symmetric operator with defect indices $(1,1)$,

$$\begin{cases} \dot{B}y = -y'' + q(x)y, \\ y'(a) = y(a) = 0, \end{cases} \tag{10.127}$$

together with its non-negative self-adjoint extension of the form

$$\begin{cases} B_\theta y = -y'' + q(x)y, \\ y'(a) = \theta y(a). \end{cases} \tag{10.128}$$

A non-decreasing function $\sigma(t)$ defined on $(-\infty,+\infty)$ is called the **distribution function of an operator pair** $(\dot{B}, B_\theta)$, where B_θ is a self-adjoint extension of symmetric operator $\dot{B}$ with deficiency indices (1,1) in a Hilbert space $\mathcal{H}$, if the formulas

$$\varphi(t) = Uf(x), \quad f(x) = U^{-1}\varphi(t), \tag{10.129}$$

establish a one-to-one isometric correspondence U between $L_2^\sigma(\mathbb{R})$ and $\mathcal{H}$. Moreover, this correspondence is such that the operator B_θ is unitarily equivalent to the operator

$$\Lambda_\sigma\varphi(t) = t\varphi(t), \quad (\varphi(t) \in L_2^\sigma(\mathbb{R})) \tag{10.130}$$

in $L_2^\sigma(\mathbb{R})$ while symmetric operator $\dot{B}$ in (10.127) is unitarily equivalent to the symmetric operator

$$\dot{\Lambda}_\sigma\varphi(t) = t\varphi(t), \quad \operatorname{Dom}(\dot{\Lambda}_\sigma) = \left\{ \varphi(t),\, t\varphi(t) \in L_2^\sigma(\mathbb{R}) \mid \int_{-\infty}^{+\infty} \varphi(t)d\sigma(t) = 0 \right\}. \tag{10.131}$$

We are going to introduce a class of **Stieltjes-like functions** structure similar to that of $S_0(R)$ of Section 9.8 but dealing with scalar functions only.

Definition 10.8.1. A scalar function $V(z)$ is said to be a member of the **class** $SL_0(R)$ if it admits the integral representation

$$V(z) = \gamma + \int_0^\infty \frac{d\sigma(t)}{t-z}, \qquad (\gamma \in (-\infty, +\infty)), \tag{10.132}$$

where the non-decreasing function $\sigma(t)$ satisfies the conditions

$$\int_0^\infty d\sigma(t) = \infty, \quad \int_0^\infty \frac{d\sigma(t)}{1+t} < \infty. \tag{10.133}$$

Consider the following subclasses of $SL_0(R)$.

Definition 10.8.2. A scalar function $V(z) \in SL_0(R)$ belongs to the **class** $SL_0(R,K)$ if

$$\int_0^\infty \frac{d\sigma(t)}{t} = \infty.$$

Definition 10.8.3. A scalar function $V(z) \in SL_0(R)$ belongs to the **class** $SL_{01}(R,K)$ if

$$\int_0^\infty \frac{d\sigma(t)}{t} < \infty.$$

Consider an operator $\dot{\Lambda}_\sigma$ of the form (10.131) in $L_2[0,+\infty)$. Let T^σ be a quasi-self-adjoint extension of $\dot{\Lambda}_\sigma$ and let $\mathcal{H}_+^\sigma = \mathrm{Dom}(\dot{\Lambda}_\sigma^*)$. Then the following theorem describes the realization of the class $SL_0(R)$.

Theorem 10.8.4. *Let $V(z) \in SL_0(R)$. Then it can be realized by a minimal model L-system*

$$\Theta_\Lambda = \begin{pmatrix} \mathbf{\Lambda} & K^\sigma & 1 \\ \mathcal{H}_+^\sigma \subset L_2^\sigma[0,+\infty) \subset \mathcal{H}_-^\sigma & & \mathbb{C} \end{pmatrix}, \tag{10.134}$$

where $\mathbf{\Lambda} = \mathrm{Re}\,\mathbf{\Lambda} + iK^\sigma(K^\sigma)^*$ is a $(*)$-extension of an operator T^σ such that $\mathbf{\Lambda} \supset T^\sigma \supset \dot{\Lambda}_\sigma$, $\dot{\Lambda}_\sigma$ is defined via (10.131), and $K^\sigma c = c \cdot \vec{1}$, $(K^\sigma)^* x = (x, \vec{1})$, $c \in \mathbb{C}$, $\vec{1} \in \mathcal{H}_-^\sigma$, $x(t) \in \mathcal{H}_+^\sigma$.

Proof. We start by applying the general realization Theorem 6.5.2 to a Herglotz-Nevanlinna function $V(z)$ and obtain a rigged L-system of the form (10.134) such that $V(z) = V_{\Theta_\Lambda}(z)$. Following the steps for construction of the model L-system described in Theorem 6.5.2, we note that all the components of the system are described as in the statement of the theorem. Also, the real part $\mathrm{Re}\,\mathbf{\Lambda}$ is a self-adjoint bi-extension of $\dot{\Lambda}_\sigma$ that has a quasi-kernel Λ_σ of the form (10.130). According to Theorem 6.6.7, operator $\dot{\Lambda}_\sigma$ of the form (10.131) is a prime operator. Thus, L-system Θ_Λ of the form (10.134) is minimal. □

In addition we can observe that for the system Θ_Λ of the form (10.134), the function $\eta(\lambda) \equiv 1$ belongs to $\mathcal{H}_-^\sigma$. To confirm this we need to show that $(x, 1)$

defines a continuous linear functional for every $x \in \mathcal{H}_+^\sigma$. It was shown in Theorem 6.5.2 that

$$\mathcal{H}_+^\sigma = \operatorname{Dom}(\dot{\Lambda}_\sigma) \dotplus \left\{\frac{c_1}{1+t^2}\right\} \dotplus \left\{\frac{c_2 t}{1+t^2}\right\}, \quad c_1, c_2 \in \mathbb{C}. \tag{10.135}$$

Consequently, every vector $x \in \mathcal{H}_+^\sigma$ has three components $x = x_1 + x_2 + x_3$ according to the decomposition (10.135) above. Obviously, $(x_1, 1)$ and $(x_2, 1)$ yield convergent integrals while $(x_3, 1)$ boils down to

$$\int_0^\infty \frac{t}{1+t^2}\, d\sigma(t).$$

To see the convergence of the above integral we notice that

$$\frac{t}{1+t^2} = \frac{t-1}{(1+t^2)(t+1)} + \frac{1}{1+t} \le \frac{1}{1+t^2} + \frac{1}{1+t}.$$

The integrals taken of the last two expressions on the right side converge due to (6.52) and (10.133), and hence so does the integral of the left side. Thus, $(x, 1)$ defines a continuous linear functional for every $x \in \mathcal{H}_+^\sigma$, and hence $1 \in \mathcal{H}_-^\sigma$. The state space of the L-system Θ_Λ is $\mathcal{H}_+^\sigma \subset L_2^\sigma[0, +\infty) \subset \mathcal{H}_-^\sigma$, where $\mathcal{H}_+^\sigma = \operatorname{Dom}\big(\dot{\Lambda}_\sigma^*\big)$.

At this point we are ready to state and prove the main realization result of this section.

Theorem 10.8.5. *Let $V(z) \in SL_0(R)$ and the function $\sigma(t)$ be the distribution function of an operator pair $(\dot{B}, B_\theta)$ of the form* (10.127), (10.128). *Then there exist a unique Schrödinger operator T_h* ($\operatorname{Im} h > 0$) *of the form* (10.126), *an operator $\mathbb{A}$ given by* (10.11), *an operator K as in* (10.17), *and a minimal L-system*

$$\Theta = \begin{pmatrix} \mathbb{A} & K & 1 \\ \mathcal{H}_+ \subset L_2[a, +\infty) \subset \mathcal{H}_- & & \mathbb{C} \end{pmatrix}, \tag{10.136}$$

of the form (10.18) *such that $V(z) = V_\Theta(z)$.*

Proof. It follows from the definition of the distribution function above that there is an operator U defined in (10.129) establishing a one-to-one isometric correspondence between $L_2^\sigma[0, +\infty)$ and $L_2[a, +\infty)$ while providing for unitary equivalence between the operator B_θ and the operator of multiplication by an independent variable Λ_σ of the form (10.130).

Let us consider the L-system Θ_Λ of the form (10.134) constructed in the proof of Theorem 10.8.4. Applying Theorem 6.6.10 on unitary equivalence to the isometry U defined in (10.129) we obtain a triplet of isometric operators U_+, U, and U_-, where

$$U_+ = U \upharpoonright \mathcal{H}_+^\sigma, \quad U_-^* = U_+^*.$$

This triplet of isometric operators maps the rigged Hilbert space $\mathcal{H}_+^\sigma \subset L_2^\sigma[0, +\infty) \subset \mathcal{H}_-^\sigma$ into another rigged triplet $\mathcal{H}_+ \subset L_2^\sigma[a, +\infty) \subset \mathcal{H}_-$. Moreover, U_+ is an isometry from $\mathcal{H}_+^\sigma = \mathrm{Dom}(\dot{\Lambda}_\sigma^*)$ onto $\mathcal{H}_+ = \mathrm{Dom}(\dot{B}^*)$, and $U_-^* = U_+^*$ is an isometry from $\mathcal{H}_+^\sigma$ onto $\mathcal{H}_-$. This is true since the operator U provides unitary equivalence between the symmetric operators $\dot{B}$ and $\dot{\Lambda}_\sigma$.

Now we construct an L-system

$$\Theta = \begin{pmatrix} \mathbb{A} & K & 1 \\ \mathcal{H}_+ \subset L_2[a,+\infty) \subset \mathcal{H}_- & & \mathbb{C} \end{pmatrix},$$

where $K = U_- K^\sigma$ and $\mathbb{A} = U_- \mathbf{\Lambda} U_+^{-1}$ is a $(*)$-extension of operator $T = UT^\sigma U^{-1}$ such that $\mathbb{A} \supset T \supset \dot{B}$. This system is clearly minimal due to the unitary equivalence between the operator $\dot{B}$ and prime operator $\dot{\Lambda}_\sigma$. The real part $\mathrm{Re}\,\mathbb{A}$ contains the quasi-kernel B_θ. This construction of $\mathbb{A}$ is unique due to Theorem 4.3.9 on the uniqueness of a $(*)$-extension for a given quasi-kernel. On the other hand, all $(*)$-extensions based on a pair $(\dot{B}, B_\theta)$ must take the form (10.11) for some values of parameters h and μ. Consequently, our function $V(z)$ is realized by the L-system Θ given by (10.136) and

$$V(z) = V_{\Theta_\Lambda}(z) = V_\Theta(z).$$ □

Theorem 10.8.6. *The operator T_h in Theorem 10.8.5 is accretive if and only if*

$$\gamma^2 + \gamma \int_0^\infty \frac{d\sigma(t)}{t} + 1 \geq 0. \tag{10.137}$$

The operator T_h is α-sectorial for some $\alpha \in (0, \pi/2)$ if and only if the inequality (10.137) is strict. In this case the exact value of angle α can be calculated by the formula

$$\tan\alpha = \frac{\int_0^\infty \frac{d\sigma(t)}{t}}{\gamma^2 + \gamma \int_0^\infty \frac{d\sigma(t)}{t} + 1}. \tag{10.138}$$

Proof. It was shown in Theorem 10.3.1 that for the L-system Θ in (10.136), described in the previous theorem, the operator T_h is accretive if and only if the function $V_h(z)$ of the form (10.32) is holomorphic in $\mathrm{Ext}[0,+\infty)$ and satisfies the inequality (10.33), i.e.,

$$1 + V_h(-0)\, V_h(-\infty) \geq 0.$$

Here $W_\Theta(z)$ is the transfer function of (10.136). According to Theorem 10.3.1, the operator T_h is α-sectorial for some $\alpha \in (0, \pi/2)$ if and only if the inequality (10.33) is strict while the exact value of angle α can be calculated by the formula (10.34). Using Theorem 10.8.5 and equation (5.41) yields

$$W_\Theta(z) = (I - iV(z)J)(I + iV(z)J)^{-1}.$$

By direct calculations one obtains

$$W_\Theta(-1) = \frac{1 - i\left[\gamma + \int_0^\infty \frac{d\sigma(t)}{t+1}\right]}{1 + i\left[\gamma + \int_0^\infty \frac{d\sigma(t)}{t+1}\right]}, \quad W_\Theta^{-1}(-1) = \frac{1 + i\left[\gamma + \int_0^\infty \frac{d\sigma(t)}{t+1}\right]}{1 - i\left[\gamma + \int_0^\infty \frac{d\sigma(t)}{t+1}\right]}.$$

Using the notations

$$a=\int_0^\infty \frac{d\sigma(t)}{t+1} \quad \text{and} \quad b=\int_0^\infty \frac{d\sigma(t)}{t},$$

and performing straightforward calculations we obtain

$$V_h(-0)=\frac{a-b}{1+ab} \quad \text{and} \quad V_h(-\infty)=\frac{a-\gamma}{1+a\gamma}. \tag{10.139}$$

Substituting (10.139) into (10.34) and doing the necessary steps we get

$$\cot\alpha=\frac{1+b\gamma}{b-\gamma}=\frac{\gamma^2+\gamma\int_0^\infty \frac{d\sigma(t)}{t}+1}{\int_0^\infty \frac{d\sigma(t)}{t}}. \tag{10.140}$$

Taking into account that $b-\gamma>0$, we combine (10.33), (10.34) with (10.140) and this completes the proof of the theorem. □

Corollary 10.8.7. *The operator T_h in Theorem 10.8.5 is accretive if and only if*

$$1+V(-0)\,V(-\infty)\geq 0. \tag{10.141}$$

The operator T_h is α-sectorial for some $\alpha\in(0,\pi/2)$ if and only if the inequality (10.141) is strict. In this case the exact value of angle α can be calculated by the formula

$$\tan\alpha=\frac{V(-\infty)-V(-0)}{1+V(-0)\,V(-\infty)}. \tag{10.142}$$

Proof. Taking into account that

$$V(-0)=\gamma+\int_0^\infty \frac{d\sigma(t)}{t},$$

and $V_\Theta(-\infty)=\gamma$, we use (10.137) and (10.138) to obtain (10.141) and (10.142). □

We note that for the remainder of this chapter the accretive operator T_h is extremal if and only if it is not α-sectorial for any $\alpha\in(0,\pi/2)$ (see Remark 9.5.17). It also directly follows from Theorem 9.8.2 that an L-system Θ of the form (10.136) is accretive if and only if the function $V(z)$ belongs to the class $S_0(R)$.

Now let us consider $V(z)\in SL_0(R)$ satisfying the conditions of Theorem 10.8.5. According to Theorem 10.8.5 there exists a minimal L-system Θ given by (10.136) with unique Schrödinger operator T_h ($\operatorname{Im} h>0$) of the form (10.126). In the remainder of the section we will derive the formulas for calculation of the boundary parameter h in T_h as well as a real parameter μ that is used in construction (10.11) of the operator $\mathbb{A}$ of the L-system Θ realization. An elaborate investigation of these formulas will show the dynamics of the restored parameters

h and μ in terms of the changing free term γ from the integral representation (10.132) of the function $V(z)$.

We consider two major cases.

Case 1. In the first case we assume that $\int_0^\infty \frac{d\sigma(t)}{t} < \infty$. This means that our function $V(z)$ belongs to the class $SL_{01}(R,K)$. In what follows we write

$$b = \int_0^\infty \frac{d\sigma(t)}{t} \quad \text{and} \quad m = m_\infty(-0).$$

Suppose that $b \geq 2$. Then the quadratic inequality (10.137) implies that, for all γ such that

$$\gamma \in \left(-\infty, \frac{-b - \sqrt{b^2 - 4}}{2}\right] \cup \left[\frac{-b + \sqrt{b^2 - 4}}{2}, +\infty\right), \tag{10.143}$$

the restored operator T_h is accretive. Clearly, this operator is extremal accretive if

$$\gamma = \frac{-b \pm \sqrt{b^2 - 4}}{2}.$$

In particular if $b = 2$, then $\gamma = -1$ and the function

$$V(z) = -1 + \int_0^\infty \frac{d\sigma(t)}{t - z}$$

is realized using an extremal accretive T_h.

Now suppose that $0 < b < 2$. For every $\gamma \in (-\infty, +\infty)$ the restored operator T_h will be accretive and α-sectorial for some $\alpha \in (0, \pi/2)$. Consider a function $V(z)$ defined by (10.132). Conducting realizations of $V(z)$ by operators T_h for different values of $\gamma \in (-\infty, +\infty)$ we notice that the operator T_h with the largest angle of sectorialilty occurs when $\gamma = -(b/2)$ and the angle is found according to the formula

$$\alpha = \arctan \frac{b}{1 - b^2/4}.$$

This follows from formula (10.138), the fact that $\gamma^2 + \gamma b + 1 > 0$ for all γ, and the formula

$$\gamma^2 + \gamma b + 1 = \left(\gamma + \frac{b}{2}\right)^2 + \left(1 - \frac{b^2}{4}\right).$$

Now we will focus on the description of the parameter h in the restored operator T_h. It was shown in Theorem 10.7.4 that the quasi-kernel $\hat{A}$ of the realizing L-system Θ from Theorem 10.8.5 takes a form of (10.120) or more specifically

$$\left\{ \begin{array}{l} \widehat{A}y = -y'' + qy \\ y'(a) = \xi y(a) \end{array} \right., \quad \xi = \frac{\mu \mathrm{Re}\, h - |h|^2}{\mu - \mathrm{Re}\, h}.$$

On the other hand, since $\sigma(t)$ is also the spectral function of the operator pair $(\dot{B}, B_\theta)$, we can conclude that $\hat{A}$ is equal to the operator B_θ of the form (10.128). This connection allows us to obtain

$$\theta = \xi = \frac{\mu \mathrm{Re}\, h - |h|^2}{\mu - \mathrm{Re}\, h}. \tag{10.144}$$

Assuming that

$$h = u + iv,$$

we will use (10.144) to derive the formulas for u and v in terms of γ. First, to eliminate parameter μ, we notice that (10.20) and (5.41) imply

$$W_\Theta(\lambda) = \frac{\mu - h}{\mu - \overline{h}}\,\frac{m_\infty(\lambda) + \overline{h}}{m_\infty(\lambda) + h} = \frac{1 - iV(z)}{1 + iV(z)}.$$

Passing to the limit when $z \to -\infty$ and taking into account that $V(-\infty) = \gamma$ and utilizing (10.39), we obtain

$$\frac{\mu - h}{\mu - \overline{h}} = \frac{1 - i\gamma}{1 + i\gamma}. \tag{10.145}$$

Let us write

$$p = \frac{1 - i\gamma}{1 + i\gamma}. \tag{10.146}$$

Solving (10.145) for μ and using (10.146) yields

$$\mu = \frac{h - p\bar{h}}{1 - p}.$$

Substituting this value into (10.144) after simplification produces

$$\frac{u + iv - p(u - iv)u - (u^2 + v^2)(1 - p)}{u + iv - p(u - iv) - u(1 - p)} = \theta.$$

After straightforward calculations aiming to represent the numerator and denominator of the last equation in standard form, one obtains the relation

$$u - \gamma v = \theta. \tag{10.147}$$

It was shown in Theorem 10.3.3 that the α-sectorialilty of the operator T_h and (10.34) lead to

$$\tan \alpha = \frac{\mathrm{Im}\, h}{\mathrm{Re}\, h + m_\infty(-0)} = \frac{v}{u + m_\infty(-0)}. \tag{10.148}$$

Combining (10.147) and (10.148) one obtains

$$u - \gamma(u \tan \alpha + m_\infty(-0) \tan \alpha) = \theta,$$

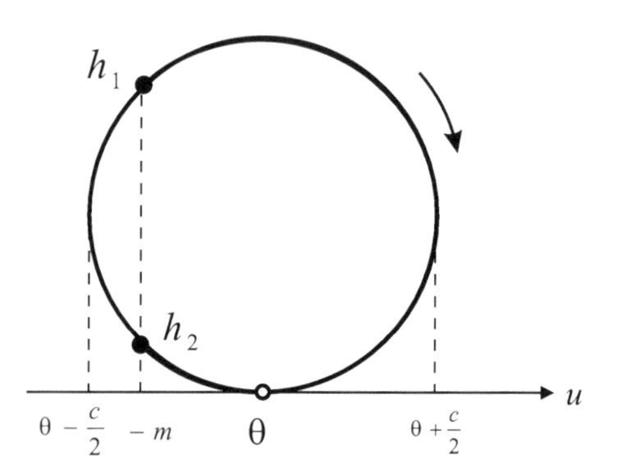

Figure 10.4: $b > 2$

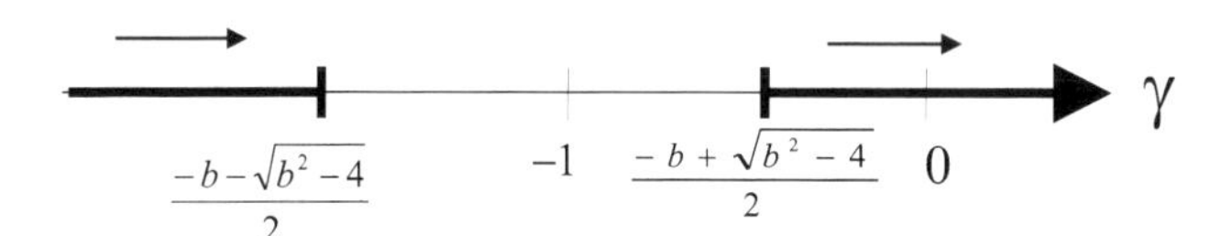

Figure 10.5: γ interval

or

$$u = \frac{\theta + \gamma m_\infty(-0)\tan\alpha}{1 - \gamma \tan\alpha}.$$

But $\tan\alpha$ is also determined by (10.138). Direct substitution of

$$\tan\alpha = \frac{b}{1 + \gamma(\gamma + b)}$$

into the above equation yields

$$u = \theta + \frac{\left[\theta + m_\infty(-0)\right] b\gamma}{1 + \gamma^2}.$$

Using the short notation $m = m_\infty(-0)$ and finalizing calculations we get

$$h = u + iv, \quad u = \theta + \frac{\gamma[\theta + m]b}{1 + \gamma^2}, \quad v = \frac{[\theta + m]b}{1 + \gamma^2}. \tag{10.149}$$

At this point we can use (10.149) to provide analytical and graphical interpretation of the parameter h in the restored operator T_h. Let

$$c = (\theta + m)b.$$

Again we consider three subcases.

Subcase 1. $b > 2$ Using basic algebra we transform (10.149) into

$$(u-\theta)^2 + \left(v - \frac{c}{2}\right)^2 = \frac{c^2}{4}. \tag{10.150}$$

Since in this case the parameter γ belongs to the interval in (10.143), we can see that h traces the highlighted part of the circle in Figure 10.4 as γ moves in the direction from $-\infty$ towards $+\infty$. We also notice that the removed point $(\theta, 0)$ corresponds to the value of $\gamma = \pm\infty$ while the points h_1 and h_2 correspond to the values $\gamma_1 = \frac{-b-\sqrt{b^2-4}}{2}$ and $\gamma_2 = \frac{-b+\sqrt{b^2-4}}{2}$, respectively.

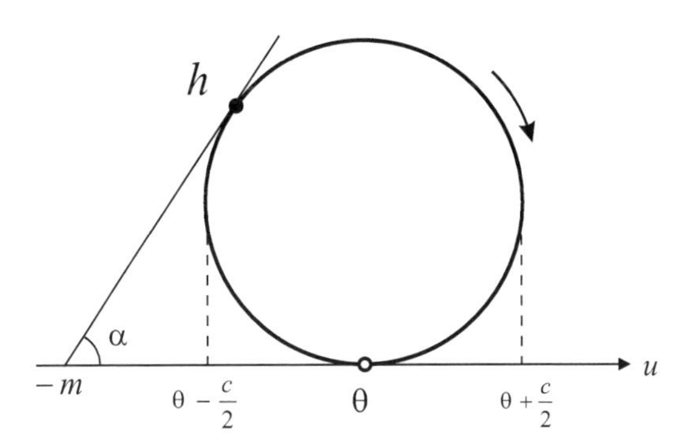

Figure 10.6: $b < 2$

Subcase 2. $b < 2$ For every $\gamma \in (-\infty, +\infty)$ the restored operator T_h will be accretive and α-sectorial for some $\alpha \in (0, \pi/2)$. As we have mentioned above, the operator T_h achieves the largest angle of sectorialilty when $\gamma = -\frac{b}{2}$. In this particular case (10.149) becomes

$$h = u + iv, \quad u = \frac{\theta(4-b^2) - 2b^2 m}{4+b^2}, \quad v = \frac{4(\theta+m)b}{4+b^2}. \tag{10.151}$$

The value of h from (10.151) is marked in Figure 10.6.

Subcase 3. $b = 2$ The behavior of parameter h in this case is depicted in Figure 10.7. It shows that in this case the function $V(z)$ can be realized using an extremal accretive T_h when $\gamma = -1$. The value of the parameter h according to (10.149) then becomes

$$h = u + iv, \quad u = -m, \quad v = \theta + m.$$

Clockwise direction of the circle again corresponds to the change of γ from $-\infty$ to $+\infty$ and the marked value of h occurs when $\gamma = -1$.

Now we consider the second case.

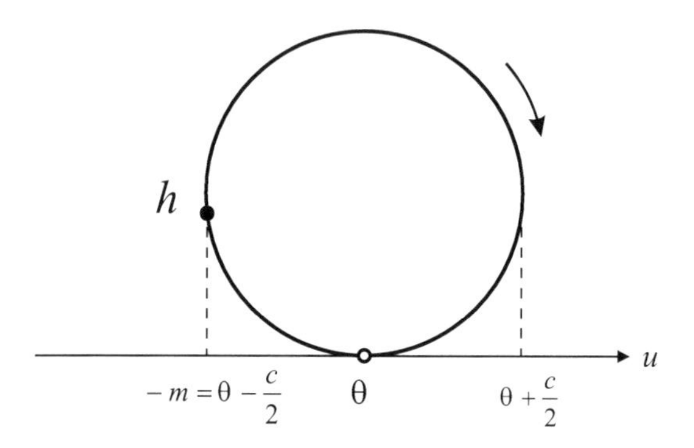

Figure 10.7: $b = 2$

Case 2. Here we assume that $\int_0^\infty \frac{d\sigma(t)}{t} = \infty$. This means that our function $V(z)$ belongs to the class $SL_0(R, K)$ and $b = \infty$. According to Theorem 10.8.6 and formulas (10.137) and (10.138), the restored operator T_h is accretive if and only if $\gamma \geq 0$ and α-sectorial if and only if $\gamma > 0$. It directly follows from (10.138) that the exact value of the angle α is then found via

$$\tan\alpha = \frac{1}{\gamma}.$$

The latter implies that the restored operator T_h is extremal if $\gamma = 0$. This means that a function $V(z) \in SL_0(R, K)$ is realized by an L-system with an extremal operator T_h if and only if

$$V(z) = \int_0^\infty \frac{d\sigma(t)}{t - z}.$$

On the other hand, since $\gamma \geq 0$, the function $V(z)$ is a Stieltjes function of the class $S_0^K(R)$. Applying Theorem 9.8.14 we conclude that $V(z)$ admits realization by an accretive L-system θ of the form (6.36) with $\operatorname{Re}\mathbb{A}$ containing the Kreĭn-von Neumann extension A_K as a quasi-kernel. Here A_K is defined by (10.87). This yields

$$\theta = -m_\infty(-0) = -m.$$

As in the beginning of the previous case we derive the formulas for u and v, where $h = u + iv$. Using (10.144) and (10.147) leads to

$$\begin{cases} \theta = \frac{\mu u - (u^2+v^2)}{\mu - u}, \\ u = \theta + \gamma v. \end{cases} \tag{10.152}$$

Solving this system for u and v leads to

$$u = \frac{\theta + \mu\gamma^2}{1+\gamma^2}, \quad v = \frac{(\mu - \theta)\gamma}{1+\gamma^2}. \tag{10.153}$$

Combining (10.152) and (10.153) gives

$$u = \frac{-m + \mu\gamma^2}{1+\gamma^2}, \quad v = \frac{(m+\mu)\gamma}{1+\gamma^2}. \tag{10.154}$$

To proceed, we first notice that our function $V(z)$ satisfies the conditions of Theorem 10.7.4. Indeed, the inequality

$$\mu \geq \frac{(\operatorname{Im} h)^2}{m_\infty(-0) + \operatorname{Re} h} + \operatorname{Re} h,$$

turns into

$$\mu = \frac{v^2}{u-m} + u,$$

if one uses $\theta = -m$ and the first equation in (10.152). Applying Theorem 10.7.4 yields

$$\int_0^\infty \frac{d\sigma(t)}{1+t^2} = \frac{\operatorname{Im} h}{|\mu - h|^2} \left(\sup_{y \in \operatorname{Dom}(A_K)} \frac{|\mu y(a) - y'(a)|}{\left(\int_a^\infty (|y(x)|^2 + |l(y)|^2)\, dx \right)^{\frac{1}{2}}} \right)^2.$$

Taking into account that $\mu y(a) - y'(a) = (\mu + m) y(a)$ and setting

$$c^{1/2} = \sup_{y \in \operatorname{Dom}(A_K)} \frac{|y(a)|}{\left(\int_a^\infty (|y(x)|^2 + |l(y)|^2)\, dx \right)^{\frac{1}{2}}}, \tag{10.155}$$

we obtain

$$\frac{\operatorname{Im} h}{|\mu - h|^2} (\mu + m)^2 c = \int_0^\infty \frac{d\sigma(t)}{1+t^2}. \tag{10.156}$$

Considering that $\operatorname{Im} h = v$ and combining (10.156) with (10.154) we use straightforward calculations to get

$$\mu = -m + \left(\frac{1}{\gamma c} \right) \int_0^\infty \frac{d\sigma(t)}{1+t^2}.$$

Let

$$\xi = \frac{1}{c} \int_0^\infty \frac{d\sigma(t)}{1+t^2}. \tag{10.157}$$

Then the last relation becomes

$$\mu = -m + \frac{\xi}{\gamma}. \tag{10.158}$$

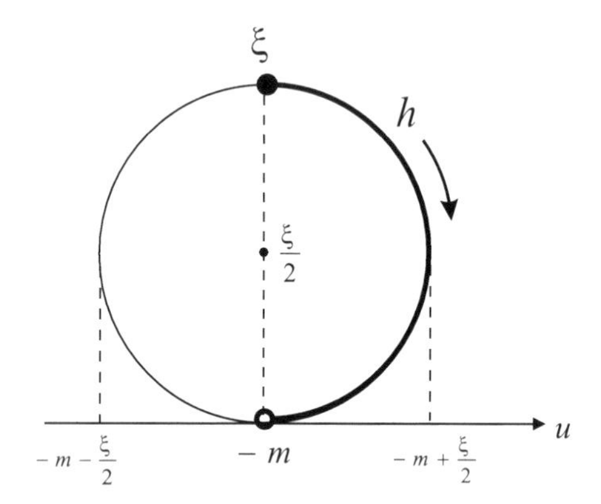

Figure 10.8: $b = \infty$

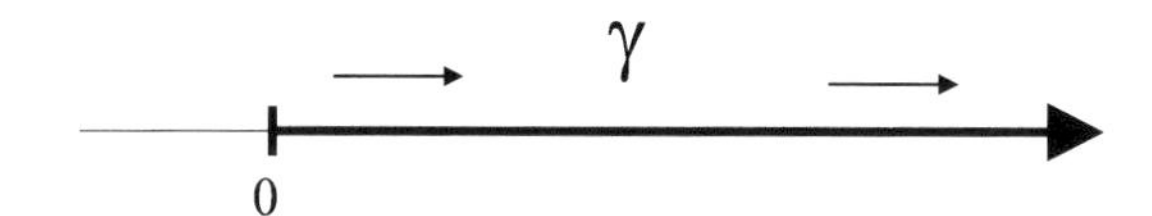

Figure 10.9: $\gamma \geq 0$

Applying (10.158) on (10.154) yields

$$u = -m + \frac{\gamma\xi}{1+\gamma^2}, \quad v = \frac{\xi}{1+\gamma^2}, \qquad \gamma \geq 0. \tag{10.159}$$

Following the previous case approach we transform (10.159) into

$$(u+m)^2 + \left(v - \frac{\xi}{2}\right)^2 = \frac{\xi^2}{4}.$$

The connection between the parameters γ and h in the accretive restored operator T_h is depicted in Figures 10.8 and 10.9. As we can see h traces the highlighted part of the circle clockwise in Figure 10.8 as γ moves from 0 towards $+\infty$.

As we mentioned earlier the restored operator T_h is extremal if $\gamma = 0$. In this case formulas (10.159) become

$$u = -m, \quad v = \xi, \qquad \gamma = 0, \tag{10.160}$$

where ξ is defined by (10.157).

Now once we have described all the possible outcomes for the restored accretive operator T_h, we can concentrate on the state-space operator $\mathbb{A}$ of the L-system (10.136). We recall that $\mathbb{A}$ is defined by formulas (10.11) and besides the parameter

h above contains also parameter μ. We will obtain the behavior of μ in terms of the components of our function $V(z)$ the same way we treated the parameter h. As before we consider two major cases dividing them into subcases when necessary.

Case 1. Assume that $b = \int_0^\infty \frac{d\sigma(t)}{t} < \infty$. In this case our function $V(z)$ belongs to the class $SL_{01}(R, K)$. First we will obtain the representation of μ in terms of u and v, where $h = u + iv$. We recall that

$$\mu = \frac{h - p\bar{h}}{1 - p},$$

where p is defined by (10.146). By direct computations we derive that

$$p = \frac{1 - \gamma^2}{1 + \gamma^2} - \frac{2\gamma}{1 + \gamma^2} i, \quad 1 - p = \frac{2\gamma^2}{1 + \gamma^2} + \frac{2\gamma}{1 + \gamma^2} i,$$

and

$$h - p\bar{h} = \left(\frac{2\gamma^2}{1 + \gamma^2} u + \frac{2\gamma}{1 + \gamma^2} v\right) + \left(\frac{2}{1 + \gamma^2} v + \frac{2\gamma}{1 + \gamma^2} u\right) i.$$

Plugging the last two equations into the formula for μ above and simplifying we obtain

$$\mu = u + \frac{1}{\gamma} v. \tag{10.161}$$

We recall that during the present case u and v parts of h are described by the formulas (10.149).

Once again we elaborate in three subcases.

Subcase 1. $b > 2$ As we have shown this above, the formulas (10.149) can be transformed into an equation of the circle (10.150). In this case the parameter γ belongs to the interval in (10.143), the accretive operator T_h corresponds to the values of h shown in the bold part of the circle in Figure 10.4 as γ moves from $-\infty$ towards $+\infty$. Substituting the expressions for u and v from (10.149) into (10.161) and simplifying we get

$$\mu = \theta + \frac{(\theta + m)b}{\gamma}.$$

The connection between values of γ and μ is depicted in Figure 10.10. We note that $\mu = 0$ when $\gamma = -\frac{(\theta+m)b}{\theta}$. Also, the endpoints

$$\gamma_1 = \frac{-b - \sqrt{b^2 - 4}}{2} \quad \text{and} \quad \gamma_2 = \frac{-b + \sqrt{b^2 - 4}}{2},$$

of γ-interval (10.143) are responsible for the μ-values

$$\mu_1 = \theta + \frac{(\theta + m)b}{\gamma_1} \quad \text{and} \quad \mu_2 = \theta + \frac{(\theta + m)b}{\gamma_2}.$$

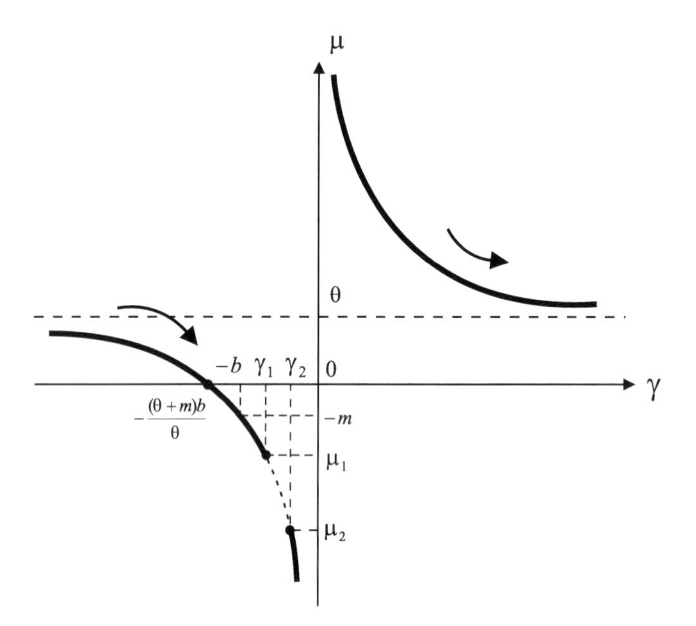

Figure 10.10: $b > 2$

The values of μ that are acceptable parameters of operator $\mathbb{A}$ of the restored L-system make the bold part of the hyperbola in Figure 10.10. It follows from Theorem 9.8.2 that the operator $\mathbb{A}$ of the form (10.11) is accretive if and only if $\gamma \geq 0$ and thus μ sweeps the right branch on the hyperbola. We note that Figure 10.10 shows the case when $-m < 0$, $\theta > 0$, and $\theta > -m$. Other possible cases, such as $(-m < 0,\ \theta < 0,\ \theta > -m)$, $(-m < 0,\ \theta = 0)$, and $(m = 0,\ \theta > 0)$ require corresponding adjustments to the graph shown in Figure 10.10.

Subcase 2. $b < 2$ For every $\gamma \in (-\infty, +\infty)$ the restored operator T_h will be accretive and α-sectorial for some $\alpha \in (0, \pi/2)$. As we have mentioned above, the operator T_h achieves the largest angle of sectorialilty when $\gamma = -\frac{b}{2}$. In this particular case (10.149) becomes

$$h = u + iv, \quad u = \frac{\theta(4 - b^2) - 2b^2 m}{4 + b^2}, \quad v = \frac{4(\theta + m)b}{4 + b^2}.$$

Substituting $\gamma = -b/2$ into (10.161) we obtain

$$\mu = -(\theta + 2m). \tag{10.162}$$

This value of μ from (10.162) is marked in Figure 10.11. The corresponding operator $\mathbb{A}$ of the realized L-system is based on these values of parameters h and μ.

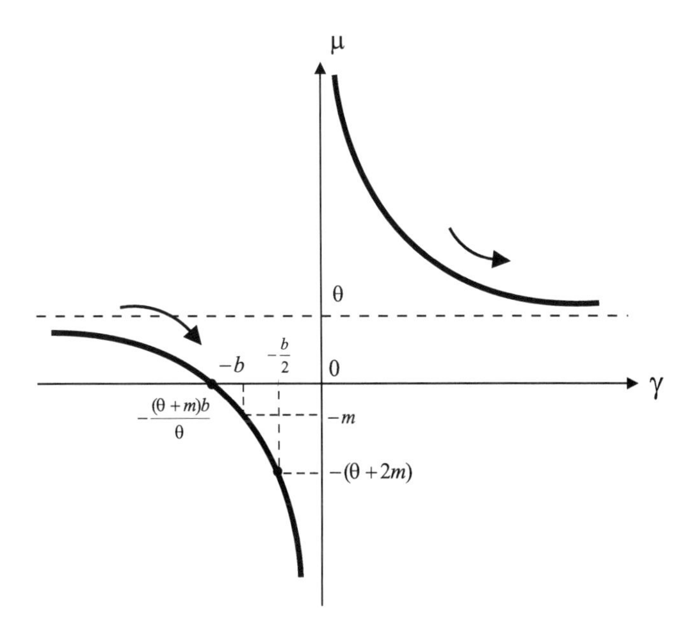

Figure 10.11: $b < 2$ and $b = 2$

Subcase 3. $b = 2$ The behavior of parameter μ in this case is also shown in Figure 10.11. It was shown above that in this case the function $V(z)$ can be realized using an extremal accretive T_h when $\gamma = -1$. The values of the parameters h and μ then become

$$h = u + iv, \quad u = -m, \quad v = \theta + m, \quad \mu = -(\theta + 2m).$$

The value of μ above is marked on the left branch of the hyperbola and occurs when $\gamma = -1 = -b/2$.

Case 2. Again we assume that $\int_0^\infty \frac{d\sigma(t)}{t} = \infty$. Hence $V(z) \in SL_0(R, K)$ and $b = \infty$. As we mentioned above the restored operator T_h is accretive if and only if $\gamma \geq 0$ and α-sectorial if and only if $\gamma > 0$. It is extremal if $\gamma = 0$. The values of u, v, and μ were already calculated and are given in (10.159) and (10.158), respectively. That is

$$u = -m + \frac{\gamma\xi}{1+\gamma^2}, \quad v = \frac{\xi}{1+\gamma^2}, \quad \mu = -m + \frac{\xi}{\gamma}, \qquad \gamma \geq 0,$$

where ξ is defined in (10.157). Figure 10.12 gives a graphical representation of this case. Only the right-hand bold branch of the hyperbola shows the values of μ in the case $b = \infty$. If $m = 0$, then

$$\mu = \frac{\xi}{\gamma}$$

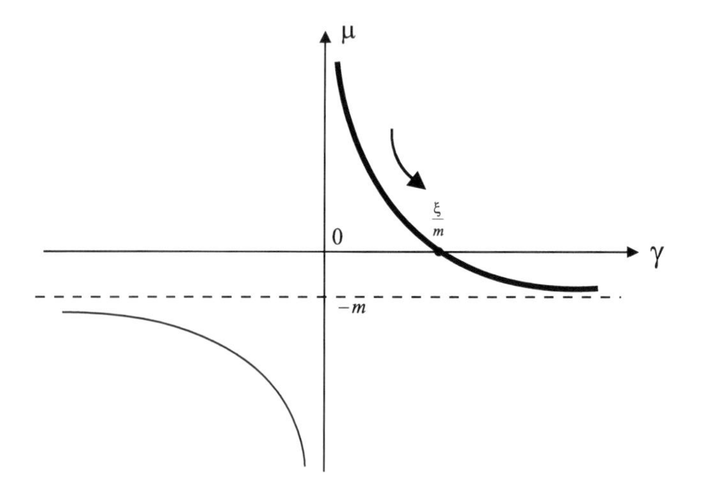

Figure 10.12: $b = \infty$

and the graph should be adjusted accordingly.

In the case when $\gamma = 0$ and T_h is extremal we have

$$u = -m, \quad v = \xi, \quad \mu = \infty, \quad h = -m + i\xi,$$

and according to (10.11) we have

$$\mathbb{A} = -y'' + q(x)y + [(-m + i\xi)y(a) - y'(a)]\delta(x - a), \tag{10.163}$$

which is the state-space operator of the realized L-system.

Example. Consider a function

$$V(z) = \frac{i}{\sqrt{z}}.$$

A direct check confirms that $V(z)$ is a Stieltjes function. Let us also consider a function $\sigma(\lambda)$ such that $\sigma(\lambda) = 0$ for $\lambda \leq 0$ and

$$\sigma(\lambda) = C + \lim_{y \to 0} \frac{1}{\pi} \int_0^\lambda \operatorname{Im} \left(\frac{i}{\sqrt{x + iy}} \right) dx, \quad \text{for} \quad \lambda > 0.$$

One can verify [207] that

$$\sigma'(\lambda) = \frac{1}{\pi\sqrt{\lambda}} \quad \text{for} \quad \lambda > 0.$$

By direct calculations we have that

$$V(z) = \int_0^\infty \frac{d\sigma(t)}{t - z} = \frac{i}{\sqrt{z}},$$

and that

$$\int_0^\infty \frac{d\sigma(t)}{t} = \int_0^\infty \frac{dt}{\pi t^{3/2}} = \infty.$$

It is also clear that the constant term in the integral representation (9.141) is zero, i.e., $\gamma = 0$.

Let us consider an operator pair $(\dot{A},\, A_K)$, where

$$\begin{cases} \dot{A}y = -y'', \\ y(0) = y'(0) = 0, \end{cases} \tag{10.164}$$

is a symmetric operator and

$$\begin{cases} A_K\, y = -y'', \\ y'(0) = 0, \end{cases}$$

is a self-adjoint one. Let us assume that $\sigma(t)$ satisfies our definition of spectral distribution function of the pair $(\dot{A},\, A_K)$ given in Section 10.8 (we will prove it a bit later). Operating under this assumption, we proceed to restore parameters h and μ and apply formulas (10.159) for the values $\gamma = 0$ and $m = -\theta = 0$. This yields $u = 0$. To obtain v we first find the value of

$$\int_0^\infty \frac{d\sigma(t)}{1+t^2} = \frac{1}{\sqrt{2}},$$

and then use formula (10.155) to get the value of c. This yields $c = 1/\sqrt{2}$. Consequently,

$$\xi = \frac{1}{c}\int_0^\infty \frac{d\sigma(t)}{1+t^2} = 1,$$

and hence $h = vi = i$. From (10.158) we have that $\mu = \infty$ and (10.163) becomes

$$\mathbb{A}\, y = -y'' + [iy(0) - y'(0)]\delta(x). \tag{10.165}$$

The operator T_h in this case is

$$\begin{cases} T_h y = -y'', \\ y'(0) = iy(0). \end{cases}$$

The channel vector g of the form (10.15) then equals $g = \delta(x)$, satisfying

$$\operatorname{Im}\mathbb{A} = \frac{\mathbb{A} - \mathbb{A}^*}{2i} = KK^* = (.,g)g,$$

and channel operator $Kc = cg$, $(c \in \mathbb{C})$ with

$$K^*y = (y,g) = y(0). \tag{10.166}$$

The real part of $\mathbb{A}$,

$$\mathrm{Re}\,\mathbb{A}\,y = -y'' - y'(0)\delta(x),$$

contains the self-adjoint quasi-kernel

$$\begin{cases} \widehat{A}y = -y'', \\ y'(0) = 0. \end{cases}$$

The L-system with Schrödinger operator of the form (10.18) that realizes $V(z)$ can now be written as

$$\Theta = \begin{pmatrix} \mathbb{A} & K & 1 \\ \mathcal{H}_+ \subset L_2[0,+\infty) \subset \mathcal{H}_- & & \mathbb{C} \end{pmatrix},$$

where $\mathbb{A}$ and K are defined above via (10.165) and (10.166), respectively.

Now we can back up our assumption on $\sigma(t)$ to be the spectral distribution function of the pair $(\dot{A},\, A_K)$. Indeed, calculating the function $V_\Theta(z)$ for the L-system Θ above directly via formula (10.21) with $\mu = \infty$ and comparing the result to $V(z)$ gives the exact value of $h = i$. It is known that operator $\dot{A}$ in (10.164) is prime. On the other hand, by Theorem 10.8.4, $V(z)$ can be realized as an impedance function of a model system Θ_Λ of the form (10.134), whose symmetric operator $\dot{\Lambda}_\sigma$ is also prime (see Theorem 6.6.7). Hence, we can apply Theorem 6.6.10 on bi-unitary equivalence that provides, with the unitary mapping U, for a definition of the spectral distribution function. Thus we confirm that $\sigma(t)$ is the spectral distribution function of the pair $(\dot{A},\, A_K)$.

All the derivations above can be repeated for a Stieltjes-like function

$$V(z) = \gamma + \frac{i}{\sqrt{z}}, \qquad -\infty < \gamma < +\infty, \quad \gamma \neq 0$$

with very minor changes. In this case the restored values for h and μ are described as follows:

$$h = u + iv, \quad u = \frac{\gamma}{1+\gamma^2}, \quad v = \frac{1}{1+\gamma^2}, \quad \mu = \frac{1}{\gamma}.$$

The dynamics of changing h according to changing γ is depicted in Figure 10.8 where the circle has its center at the point $i/2$ and radius of $1/2$. The behavior of μ is described by a hyperbola $\mu = 1/\gamma$ (see Figure 10.12 with $\theta = 0$). In the case when $\gamma > 0$ our function becomes Stieltjes and the restored L-system Θ is accretive. The operators $\mathbb{A}$ and K of the restored L-system are given according to the formulas (10.11) and (10.17), respectively.

10.9 Inverse Stieltjes-like functions and inverse spectral problems for systems with Schrödinger operator

In this section we are going to use the realization technique and results developed for inverse Stieltjes functions in Section 9.9 to obtain a solution of the inverse spectral problem for L-systems with Schrödinger operator of the form (10.126).

Definition 10.9.1. A scalar Herglotz-Nevanlinna function $V(z)$ is called an **inverse Stieltjes-like function** if it has an integral representation

$$V(z) = \alpha + \int_0^\infty \left(\frac{1}{t-z} - \frac{1}{t}\right) d\tau(t), \quad \int_0^\infty \frac{d\tau(t)}{t+t^2} < \infty, \tag{10.167}$$

similar to (9.180) but with an arbitrary (not necessarily non-positive) constant α.

We are going to introduce a new class of realizable scalar inverse Stieltjes-like functions whose structure is similar to that of $S_0^{-1}(R)$ of Section 9.9.

Definition 10.9.2. An inverse Stieltjes-like function $V(z)$ belongs to the *class* $SL_0^{-1}(R)$ if it admits an integral representation

$$V(z) = \alpha + \int_0^\infty \left(\frac{1}{t-z} - \frac{1}{t}\right) d\tau(t), \tag{10.168}$$

where non-decreasing function $\tau(t)$ satisfies the conditions

$$\int_0^\infty d\tau(t) = \infty, \quad \int_0^\infty \frac{d\tau(t)}{t+t^2} < \infty. \tag{10.169}$$

Consider the following subclasses of $SL_0^{-1}(R)$.

Definition 10.9.3. A function $V(z) \in SL_0^{-1}(R)$ is a member of the class $SL_0^{-1}(R,F)$ if

$$\int_0^\infty \frac{d\tau(t)}{t} = \infty. \tag{10.170}$$

Definition 10.9.4. A function $V(z) \in SL_0^{-1}(R)$ is a member of the *class* $SL_{01}^{-1}(R,F)$ if

$$\int_0^\infty \frac{d\tau(t)}{t} < \infty. \tag{10.171}$$

The following theorem describes the realization of the class $SL_0^{-1}(R)$.

Theorem 10.9.5. *Let $V(z) \in SL_0^{-1}(R)$. Then it can be realized as an impedance function of a minimal L-system.*

Proof. We start by applying the general realization Theorem 6.5.2 to a Herglotz-Nevanlinna function $V(z)$ and obtain an L-system

$$\Theta_\Lambda = \begin{pmatrix} \mathbf{\Lambda} & K^\tau & 1 \\ \mathcal{H}_+^\tau \subset L_2^\tau[0,+\infty) \subset \mathcal{H}_-^\tau & & \mathbb{C} \end{pmatrix}, \tag{10.172}$$

such that $V(z) = V_{\Theta_\Lambda}(z)$. Following the steps for construction of the model L-system described in Theorem 6.5.2, we note that

$$\mathbf{\Lambda} = \operatorname{Re}\mathbf{\Lambda} + iK^\tau(K^\tau)^*$$

is a $(*)$-extension of an operator T^τ such that $\mathbf{\Lambda} \supset T^\tau \supset \dot{\Lambda}_\tau$ where $\dot{\Lambda}_\tau$ is defined in (10.131). The real part $\operatorname{Re}\mathbf{\Lambda}$ is a self-adjoint bi-extension of $\dot{\Lambda}_\tau$ that has a quasi-kernel Λ_τ of the form (10.130). It was also shown in Section 9.9 that the operator $\mathbf{\Lambda}$ possess the accumulative property (9.181). The operator K^τ in the above L-system is defined by

$$K^\tau c = c \cdot \vec{1}, \quad (K^\tau)^* x = (x, \vec{1}) \quad c \in \mathbb{C}, \quad \vec{1} \in \mathcal{H}_-^\tau,\ x(t) \in \mathcal{H}_+^\tau.$$

In addition we can observe that the function $\eta(\lambda) \equiv 1$ belongs to $\mathcal{H}_-^\tau$. To confirm this we need to show that $(x, 1)$ defines a continuous linear functional for every $x \in \mathcal{H}_+^\tau$. It was shown in Theorem 6.5.2 that

$$\mathcal{H}_+^\tau = \operatorname{Dom}(\dot{\Lambda}_\tau) \dotplus \left\{\frac{c_1}{1+t^2}\right\} \dotplus \left\{\frac{c_2 t}{1+t^2}\right\}, \qquad c_1, c_2 \in \mathbb{C}. \tag{10.173}$$

Consequently, every vector $x \in \mathcal{H}_+^\tau$ has three components $x = x_1 + x_2 + x_3$ according to the decomposition (10.173) above. Obviously, $(x_1, 1)$ and $(x_2, 1)$ yield convergent integrals while $(x_3, 1)$ boils down to

$$\int_0^\infty \frac{t}{1+t^2}\, d\tau(t).$$

The convergence of the latter is guaranteed by the definition of a inverse Stieltjes-like function. The state-space of the L-system Θ_Λ is $\mathcal{H}_+^\tau \subset L_2^\tau[0, +\infty) \subset \mathcal{H}_-^\tau$, where $\mathcal{H}_+^\tau = \operatorname{Dom}(\dot{\Lambda}_\tau^*)$. According to Theorem 6.6.7 the operator $\dot{\Lambda}_\tau$ of the form (10.131) is a prime operator. Thus, L-system Θ_Λ of the form (10.172) is minimal. □

At this point we are ready to state and prove the main realization result of this section.

Theorem 10.9.6. *Let $V(z) \in SL_0^{-1}(R)$ and the function $\tau(t)$ be the distribution function of a pair $(\dot{B}, B_\theta)$ of the form* (10.127), (10.128). *Then there exist a unique Schrödinger operator T_h $(\operatorname{Im} h > 0)$ of the form* (10.126), *an operator $\mathbb{A}$ given by* (10.11), *an operator K as in* (10.17), *and an L-system*

$$\Theta = \begin{pmatrix} \mathbb{A} & K & 1 \\ \mathcal{H}_+ \subset L_2[a, +\infty) \subset \mathcal{H}_- & & \mathbb{C} \end{pmatrix}, \tag{10.174}$$

of the form (10.18) *such that $V(z) = V_\Theta(z)$.*

Proof. It follows from the definition of the distribution function above that there is an operator U defined in (10.129) establishing a one-to-one isometric correspondence between $L_2^\tau[0, +\infty)$ and $L_2[a, +\infty)$ while providing for unitary equivalence between the operator B_θ and the operator of multiplication by independent variable Λ_τ of the form (10.130).

Let us consider the L-system Θ_Λ of the form (10.172) constructed in the proof of Theorem 10.9.5. Applying Theorem 6.6.10 on unitary equivalence to the isometry U defined in (10.129) we obtain a triplet of isometric operators U_+, U, and U_-, where

$$U_+ = U\upharpoonright \mathcal{H}_+^\tau, \quad U_-^* = U_+^*.$$

This triplet of isometric operators will map the rigged Hilbert space $\mathcal{H}_+^\tau \subset L_2^\tau[0,+\infty) \subset \mathcal{H}_-^\tau$ into the triplet $\mathcal{H}_+ \subset L_2^\tau[a,+\infty) \subset \mathcal{H}_-$. Moreover, U_+ is an isometry from $\mathcal{H}_+^\tau = \mathrm{Dom}(\dot{\Lambda}_\tau^*)$ onto $\mathcal{H}_+ = \mathrm{Dom}(\dot{B}^*)$, and $U_-^* = U_+^*$ is an isometry from $\mathcal{H}_+^\tau$ onto $\mathcal{H}_-$. This is true since the operator U provides the unitary equivalence between the symmetric operators $\dot{B}$ and $\dot{\Lambda}_\tau$.

Now we construct an L-system

$$\Theta = \begin{pmatrix} \mathbb{A} & K & 1 \\ \mathcal{H}_+ \subset L_2[a,+\infty) \subset \mathcal{H}_- & & \mathbb{C} \end{pmatrix},$$

where $K = U_- K^\tau$ and $\mathbb{A} = U_- \mathbf{\Lambda} U_+^{-1}$ is a $(*)$-extension of operator $T = UT^\tau U^{-1}$ such that $\mathbb{A} \supset T \supset \dot{B}$. The real part $\mathrm{Re}\,\mathbb{A}$ contains the quasi-kernel B_θ. This construction of $\mathbb{A}$ is unique due to Theorem 4.3.9 on the uniqueness of a $(*)$-extension for a given quasi-kernel. On the other hand, all $(*)$-extensions based on a pair $(\dot{B}, B_\theta)$ must take form (10.11) for some values of parameters h and μ. Consequently, our function $V(z)$ is realized by the L-system Θ of the form (10.174) and

$$V(z) = V_{\Theta_\Lambda}(z) = V_\Theta(z). \qquad \square$$

The theorem below gives the criteria for the operator T_h of the realizing L-system to be accretive.

Theorem 10.9.7. *Let $V(z) \in SL_0^{-1}(R)$ satisfy the conditions of Theorem* 10.9.6. *Then the operator T_h in the conclusion of Theorem* 10.9.6 *is accretive if and only if*

$$\alpha^2 - \alpha \int_0^\infty \frac{d\tau(t)}{t} + 1 \geq 0. \tag{10.175}$$

The operator T_h is ϕ-sectorial for some $\phi \in (0, \pi/2)$ if and only if the inequality (10.175) *is strict. In this case the exact value of angle ϕ can be calculated by the formula*

$$\tan\phi = \frac{\int_0^\infty \frac{d\tau(t)}{t}}{\alpha^2 - \alpha \int_0^\infty \frac{d\tau(t)}{t} + 1}. \tag{10.176}$$

Proof. It was shown in Theorem 10.3.1 that for the L-system Θ in (10.174) described in the previous theorem the operator T_h is accretive if and only if the function $V_h(z)$ of the form (10.32) is holomorphic in $\mathrm{Ext}[0,+\infty)$ and satisfies the inequality (10.33). Here $W_\Theta(z)$ is the transfer function of (10.174). According to Theorem 10.3.1, the operator T_h is ϕ-sectorial for some $\phi \in (0, \pi/2)$ if and only if the inequality (10.33) is strict while the exact value of angle ϕ can be calculated by the formula (10.34), i.e.,

$$\cot\phi = \frac{1 + V_h(-0)\, V_h(-\infty)}{|V_h(-\infty) - V_h(-0)|}. \tag{10.177}$$

According to Theorem 10.9.6 and equation (6.48)

$$W_\Theta(z) = (I - iV(z)J)(I + iV(z)J)^{-1}.$$

By direct calculations one obtains

$$W_\Theta(-1) = \frac{1 - i\left[\alpha - \int_0^\infty \frac{d\tau(t)}{t+t^2}\right]}{1 + i\left[\alpha - \int_0^\infty \frac{d\tau(t)}{t+t^2}\right]},\ W_\Theta^{-1}(-1) = \frac{1 + i\left[\alpha - \int_0^\infty \frac{d\tau(t)}{t+t^2}\right]}{1 - i\left[\alpha - \int_0^\infty \frac{d\tau(t)}{t+t^2}\right]}.$$

Using the notations

$$c = \alpha - \int_0^\infty \frac{d\tau(t)}{t+t^2} \quad \text{and} \quad d = \alpha - \int_0^\infty \frac{d\tau(t)}{t},$$

and performing straightforward calculations we obtain

$$W_\Theta(-1) = \frac{1 - i\,c}{1 + i\,c}, \quad W_\Theta(-\infty) = \frac{1 - i\,d}{1 + i\,d},$$

and

$$V_h(-0) = \frac{c - \alpha}{1 + c\alpha} \quad \text{and} \quad V_h(-\infty) = \frac{c - d}{1 + c\,d}. \tag{10.178}$$

Substituting (10.178) into (10.177) and performing the necessary steps we get

$$\cot\phi = \frac{1 + \alpha\,d}{\alpha - d} = \frac{\alpha^2 - \alpha\int_0^\infty \frac{d\tau(t)}{t} + 1}{\int_0^\infty \frac{d\tau(t)}{t}}. \tag{10.179}$$

Taking into account that $\alpha - d > 0$ we combine (10.175), (10.177) with (10.179) and this completes the proof of the theorem. □

Now let us consider $V(z) \in SL_0^{-1}(R)$ satisfying the conditions of Theorem 10.9.6. According to Theorem 10.9.6 there exists a minimal L-system Θ of the form (10.174) with unique Schrödinger operator T_h ($\operatorname{Im} h > 0$) of the form (10.126). For the rest of this section we will derive the formulas for calculation of the boundary parameter h in T_h as well as a real parameter μ that is used in construction (10.11) of the operator $\mathbb{A}$ of the realizing L-system Θ. An elaborate investigation of these formulas is going to show the dynamics of the restored parameters h and μ in terms of the changing free term α from the integral representation (10.168) of the function $V(z)$.

Below we will derive the formulas for calculation of the boundary parameter h in the restored Schrödinger operator T_h of the form (10.126). We consider two major cases.

Case 1. In the first case we assume that $\int_0^\infty \frac{d\tau(t)}{t} < \infty$. This means that our function $V(z)$ belongs to the class $SL_{01}^{-1}(R,F)$. In what follows we write

$$b = \int_0^\infty \frac{d\tau(t)}{t} \quad \text{and} \quad m = m_\infty(-0).$$

Suppose that $b \geq 2$. Then the quadratic inequality (10.175) implies that, for all α such that

$$\alpha \in \left(-\infty, \frac{b-\sqrt{b^2-4}}{2}\right] \cup \left[\frac{b+\sqrt{b^2-4}}{2}, +\infty\right), \tag{10.180}$$

the restored operator T_h is accretive. Clearly, this operator is extremal accretive if

$$\alpha = \frac{b \pm \sqrt{b^2-4}}{2}.$$

In particular if $b = 2$, then $\alpha = 1$ and the function

$$V(z) = 1 + \int_0^\infty \left(\frac{1}{t-z} - \frac{1}{t}\right) d\tau(t)$$

is realized using an extremal accretive T_h.

Now suppose that $0 < b < 2$. Then for every $\alpha \in (-\infty, +\infty)$ the restored operator T_h will be accretive and ϕ-sectorial for some $\phi \in (0, \pi/2)$. Consider a function $V(z)$ defined by (10.168). Conducting realizations of $V(z)$ by operators T_h for different values of $\alpha \in (-\infty, +\infty)$ we notice that the operator T_h with the largest angle of sectoriality occurs when $\alpha = b/2$ and is found according to the formula

$$\phi = \arctan \frac{b}{1 - b^2/4}.$$

This follows from the formula (10.176), the fact that $\alpha^2 - \alpha b + 1 > 0$ for all α, and the formula

$$\alpha^2 - \alpha b + 1 = \left(\alpha - \frac{b}{2}\right)^2 + \left(1 - \frac{b^2}{4}\right).$$

Now we will focus on the description of the parameter h in the restored operator T_h. It was shown in Theorem 10.7.5 that the quasi-kernel $\hat{A}$ of the realizing L-system Θ from Theorem 10.9.5 takes the form

$$\begin{cases} \widehat{A}y = -y'' + q(x)y, \\ y'(a) = \eta y(a), \end{cases} \qquad \eta = \frac{\mu \operatorname{Re} h - |h|^2}{\mu - \operatorname{Re} h}.$$

On the other hand, since $\tau(t)$ is also the distribution function of the non-negative self-adjoint operator, we can conclude that $\hat{A}$ equals the operator B_θ of the form (10.128). This connection allows us to obtain

$$\theta = \eta = \frac{\mu \operatorname{Re} h - |h|^2}{\mu - \operatorname{Re} h}. \tag{10.181}$$

Assuming that

$$h = u + iv,$$

we will use (10.181) to derive the formulas for u and v in terms of γ. First, to eliminate parameter μ, we notice that (10.20) and (5.41) imply

$$W_\Theta(\lambda) = \frac{\mu - h}{\mu - \overline{h}} \frac{m_\infty(\lambda) + \overline{h}}{m_\infty(\lambda) + h} = \frac{1 - iV(z)}{1 + iV(z)}. \tag{10.182}$$

Passing to the limit in (10.182) when $\lambda \to -\infty$ and taking into account that $V(-\infty) = \alpha - b$ and $m_\infty(-\infty) = \infty$ we obtain

$$\frac{\mu - h}{\mu - \overline{h}} = \frac{1 - i(\alpha - b)}{1 + i(\alpha - b)}. \tag{10.183}$$

Let us set

$$p = \frac{1 - i(\alpha - b)}{1 + i(\alpha - b)}. \tag{10.184}$$

Solving (10.183) for μ and using (10.184) yields

$$\mu = \frac{h - p\bar{h}}{1 - p}.$$

Substituting this value of μ in (10.181) and simplifying we obtain

$$\frac{u + iv - p(u - iv)u - (u^2 + v^2)(1 - p)}{u + iv - p(u - iv) - u(1 - p)} = \theta.$$

After straightforward calculations aiming to represent the numerator and denominator of the last equation in standard form, one obtains the relation

$$u - (\alpha - b)\, v = \theta. \tag{10.185}$$

It was shown in Theorem 10.3.3 that the ϕ-sectorialilty of the operator T_h and (10.34) lead to

$$\tan\phi = \frac{\operatorname{Im} h}{\operatorname{Re} h + m_\infty(-0)} = \frac{v}{u + m_\infty(-0)}. \tag{10.186}$$

Combining (10.185) and (10.186) one obtains

$$u - (\alpha - b)(u \tan\phi + m_\infty(-0)\, \tan\phi) = \theta,$$

or

$$u = \frac{\theta + (\alpha - b)m_\infty(-0)\, \tan\phi}{1 - (\alpha - b)\, \tan\phi}.$$

But $\tan\phi$ is also determined by (10.176). Direct substitution of

$$\tan\phi = \frac{b}{1 + \alpha(\alpha - b)}$$

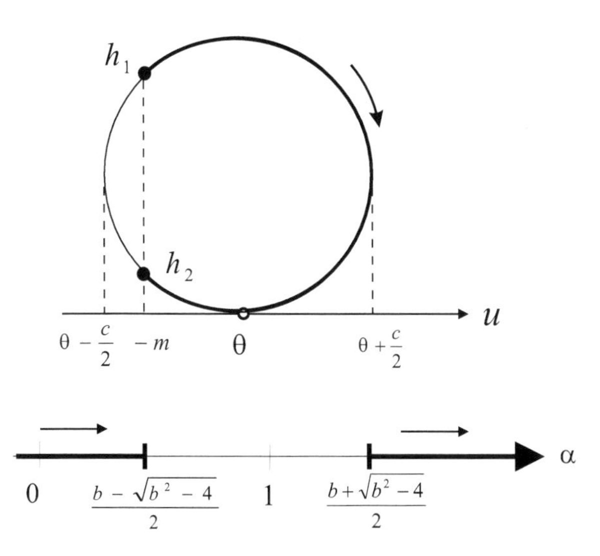

Figure 10.13: $b > 2$

into the above equation yields

$$u = \theta + \frac{\left[\theta + m_\infty(-0)\right] b(\alpha - b)}{1 + (\alpha - b)^2}.$$

Using the short notation and finalizing calculations we get

$$h = u + iv, \quad u = \theta + \frac{(\alpha - b)[\theta + m]b}{1 + (\alpha - b)^2}, \quad v = \frac{[\theta + m]b}{1 + (\alpha - b)^2}. \tag{10.187}$$

At this point we can use (10.187) to provide analytical and graphical interpretation of the parameter h in the restored operator T_h. Let

$$c = (\theta + m)b.$$

Again we consider three subcases.

Subcase 1. $b > 2$ Using basic algebra we transform (10.187) into

$$(u - \theta)^2 + \left(v - \frac{c}{2}\right)^2 = \frac{c^2}{4}. \tag{10.188}$$

Since in this case the parameter α belongs to the interval in (10.180), we can see that h traces the highlighted part of the circle in Figure 10.13 as α moves from $-\infty$ towards $+\infty$. We also notice that the removed point $(\theta, 0)$ corresponds to the value of $\alpha = \pm\infty$ while the points h_1 and h_2 correspond to the values $\alpha_1 = \frac{b-\sqrt{b^2-4}}{2}$ and $\alpha_2 = \frac{b+\sqrt{b^2-4}}{2}$, respectively (see Figure 10.13).

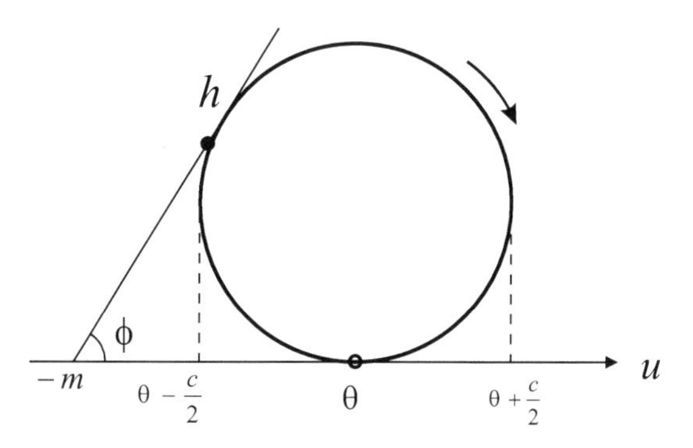

Figure 10.14: $b < 2$

Subcase 2. $b < 2$ For every $\alpha \in (-\infty, +\infty)$ the restored operator T_h will be accretive and ϕ-sectorial for some $\phi \in (0, \pi/2)$. As we have mentioned above, the operator T_h achieves the largest angle of sectorialilty when $\alpha = \frac{b}{2}$. In this particular case (10.187) becomes

$$h = u + iv, \quad u = \theta - \frac{2(\theta + m)b^2}{4 + b^2}, \quad v = \frac{4(\theta + m)b}{4 + b^2}. \tag{10.189}$$

The value of h from (10.189) is marked in Figure 10.14.

Subcase 3. $b = 2$ The behavior of parameter h in this case is depicted in Figure 10.15. It shows that in this case the function $V(z)$ can be realized using an extremal accretive T_h when $\alpha = 1$. The value of the parameter h according to (10.187) then becomes

$$h = u + iv, \quad u = -m, \quad v = \theta + m.$$

Clockwise direction of the circle again corresponds to the change of α from $-\infty$ to $+\infty$ and the marked value of h occurs when $\alpha = 1$.

Now we consider the second case.

Case 2. Here we assume that $\int_0^\infty \frac{d\tau(t)}{t} = \infty$. This means that our function $V(z)$ belongs to the class $SL_0^{-1}(R, F)$ and $b = \infty$. According to Theorem 10.9.7 and formulas (10.175) and (10.176), the restored operator T_h is accretive if and only if $\alpha \leq 0$ and ϕ-sectorial if and only if $\alpha < 0$. It directly follows from (10.176) that the exact value of the angle ϕ is then found from

$$\tan \phi = -\frac{1}{\alpha}. \tag{10.190}$$

The latter implies that the restored operator T_h is extremal if $\alpha = 0$. This means that a function $V(z) \in SL_0^{-1}(R, F)$ is realized by an L-system with an extremal

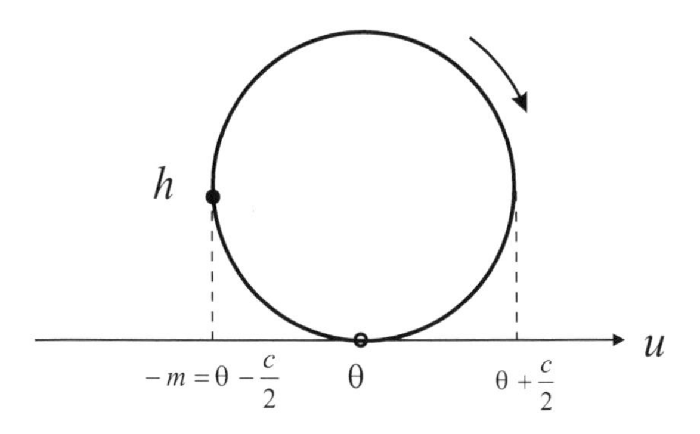

Figure 10.15: $b = 2$

operator T_h if and only if

$$V(z) = \int_0^\infty \left(\frac{1}{t-z} - \frac{1}{t}\right) d\tau(t).$$

On the other hand, since $\alpha \le 0$, the function $V(z)$ is an inverse Stieltjes function of the class $S_{0,F}^{-1}(R)$. Applying realization Theorem 9.9.7, we conclude that $V(z)$ admits realization by an accumulative L-system Θ of the form (6.36) with $\mathrm{Re}\,\mathbb{A}$ containing the Friedrichs extension A_F as a quasi-kernel. Here A_F is defined by (10.73). This yields

$$\theta = \frac{\mu u - (u^2+v^2)}{\mu - u} = \infty,$$

and hence $\mu = u$. As at the beginning of the previous case, we derive the formulas for u and v, where $h = u + iv$. Assuming that $\alpha \neq 0$ and using (10.186) and (10.190) one obtains

$$u = \mu, \quad v = -\frac{u+m}{\alpha}. \tag{10.191}$$

To proceed, we first notice that our function $V(z)$ satisfies the conditions of Theorem 10.7.5. Indeed, since $V(z) \in S_0^{-1}(R)$, we can use Theorem 10.7.3 and obtain inequality (10.114) that is required for Theorem 10.7.5 to hold. Applying Theorem 10.7.5 yields

$$\int\limits_0^\infty \frac{d\tau(t)}{1+t^2} = \frac{\mathrm{Im}\, h}{|\mu - h|^2}\left(\sup_{y\in\mathrm{Dom}(A_F)} \frac{|\mu y(a) - y'(a)|}{\left(\int\limits_a^\infty \left(|y(x)|^2 + |l(y)|^2\right) dx\right)^{\frac{1}{2}}}\right)^2.$$

Taking into account that for the case of A_F we have $y(a) = 0$ and hence $|\mu y(a) -$

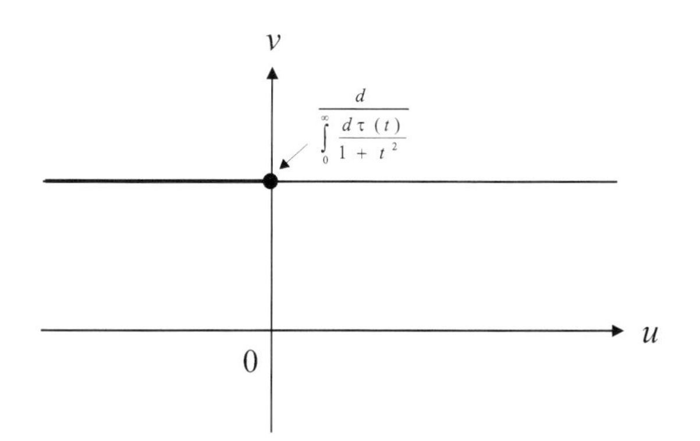

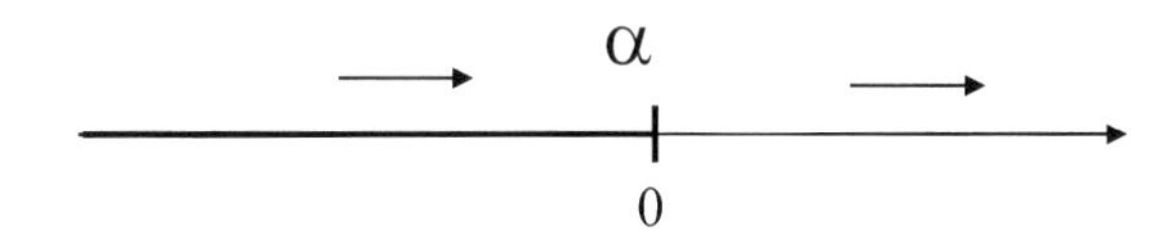

Figure 10.16: $b = \infty$

$y'(a)| = |y'(a)|$ we set

$$d^{1/2} = \sup_{y \in \mathrm{Dom}(A_F)} \frac{|y'(a)|}{\left(\int\limits_a^{\infty} \left(|y(x)|^2 + |l(y)|^2\right) dx\right)^{\frac{1}{2}}}, \tag{10.192}$$

and obtain

$$\frac{\operatorname{Im} h}{|\mu - h|^2}\, d = \int\limits_0^{\infty} \frac{d\tau(t)}{1+t^2}. \tag{10.193}$$

Since $\operatorname{Im} h = v$ and $\mu = u$ (see (10.191)), we solve (10.193) for v to get

$$v = \frac{d}{\int_0^\infty \frac{d\tau(t)}{1+t^2}}.$$

Consequently, equations (10.191) describing $h = u + iv$ take the form

$$u = -m + \frac{\alpha\, d}{\int_0^\infty \frac{d\tau(t)}{1+t^2}}, \quad v = \frac{d}{\int_0^\infty \frac{d\tau(t)}{1+t^2}}. \tag{10.194}$$

The equations (10.194) above provide parametrical equations of the straight horizontal line shown in Figure 10.16. The connection between the parameters α and h in the accretive restored operator T_h is depicted in bold.

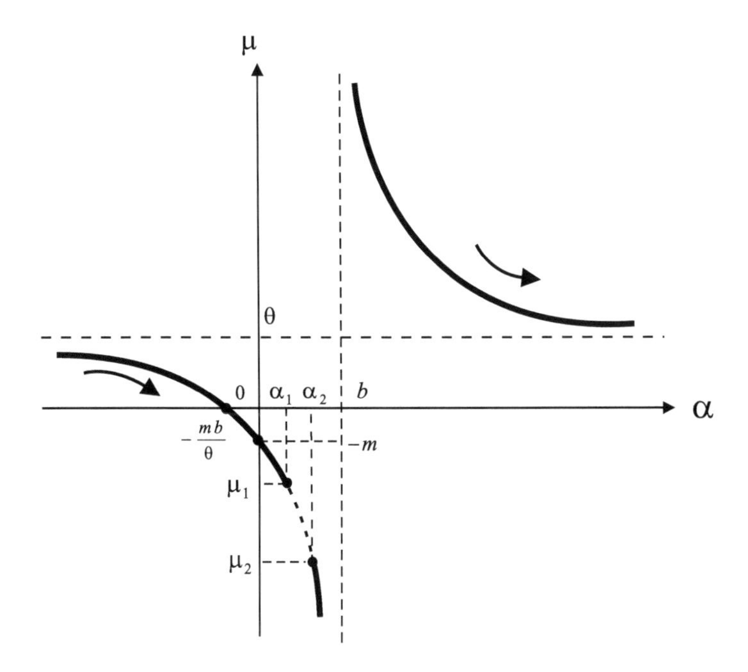

Figure 10.17: $b > 2$

As we mentioned earlier, the restored operator T_h is extremal if $\alpha = 0$. In this case formulas (10.194) become

$$u = -m, \quad v = \frac{d}{\int_0^\infty \frac{d\tau(t)}{1+t^2}}.$$

Now once we have described all possible outcomes for the restored accretive operator T_h, we can concentrate on the main operator $\mathbb{A}$ of the L-system (10.174). We recall that $\mathbb{A}$ is defined by formulas (10.11) and besides the parameter h above contains also parameter μ. We will obtain the behavior of μ in terms of the components of our function $V(z)$ in the same way as we treated the parameter h. As before we consider two major cases dividing them into subcases when necessary.

Case 1. Assume that $b = \int_0^\infty \frac{d\tau(t)}{t} < \infty$. In this case our function $V(z)$ belongs to the class $SL_{01}^{-1}(R,F)$. First we will obtain the representation of μ in terms of u and v, where $h = u + iv$. We recall that

$$\mu = \frac{h - p\bar{h}}{1 - p},$$

where p is defined by (10.184). By direct computations we derive that

$$p = \frac{1-(\alpha-b)^2}{1+(\alpha-b)^2} - \frac{2(\alpha-b)}{1+(\alpha-b)^2}i, \quad 1 - p = \frac{2(\alpha-b)^2}{1+(\alpha-b)^2} + \frac{2(\alpha-b)}{1+(\alpha-b)^2}i,$$

and

$$h - p\bar{h} = \left(\frac{2(\alpha - b)^2}{1 + (\alpha - b)^2} u + \frac{2(\alpha - b)}{1 + (\alpha - b)^2} v \right) + \left(\frac{2}{1 + (\alpha - b)^2} v + \frac{2(\alpha - b)}{1 + (\alpha - b)^2} u \right) i.$$

Plugging the last two equations into the formula for μ above and simplifying we obtain

$$\mu = u + \frac{v}{\alpha - b}. \tag{10.195}$$

We recall that during the present case u and v parts of h are described by the formulas (10.187).

Once again we elaborate in three subcases.

Subcase 1. $b > 2$ As we have shown before, the formulas (10.187) can be transformed into an equation of the circle (10.188). In this case the parameter α belongs to the interval in (10.180), the accretive operator T_h corresponds to the values of h shown in the bold part of the circle in Figure 10.13 as α moves from $-\infty$ towards $+\infty$. Substituting the expressions for u and v from (10.187) into (10.195) and simplifying we get

$$\mu = \theta + \frac{(\theta + m)b}{\alpha - b}.$$

The connection between values of α and μ is depicted in Figure 10.17.

We note that $\mu = 0$ when $\alpha = -\frac{mb}{\theta}$. Also, the endpoints

$$\alpha_1 = \frac{b - \sqrt{b^2 - 4}}{2} \quad \text{and} \quad \alpha_2 = \frac{b + \sqrt{b^2 - 4}}{2},$$

of α-interval (10.180) are responsible for the μ-values

$$\mu_1 = \theta + \frac{(\theta + m)b}{\alpha_1} \quad \text{and} \quad \mu_2 = \theta + \frac{(\theta + m)b}{\alpha_2}.$$

The values of μ that are acceptable parameters of operator $\mathbb{A}$ of the restored L-system with an accretive operator T_h make the bold part of the hyperbola in Figure 10.17. It follows from Theorems 9.9.4 and 9.9.5 that the operator $\mathbb{A}$ of the form (10.11) is accumulative if and only if $\alpha \leq 0$ and thus μ belongs to the part of the left branch on the hyperbola when $\alpha \in (-\infty, 0]$. We note that Figure 10.17 shows the case when $-m < 0$, $\theta > 0$, and $\theta > -m$. Other possible cases, such as ($-m < 0$, $\theta < 0$, $\theta > -m$), ($-m < 0$, $\theta = 0$), and ($m = 0$, $\theta > 0$) require corresponding adjustments to the graph shown in Figure 10.17.

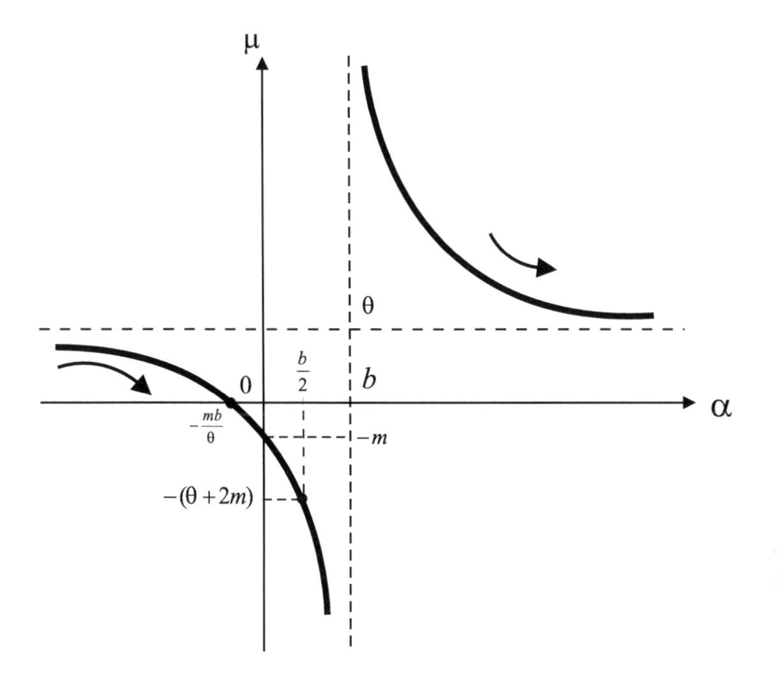

Figure 10.18: $b < 2$ and $b = 2$

Subcase 2. $b < 2$ For every $\alpha \in (-\infty, +\infty)$ the restored operator T_h is accretive and ϕ-sectorial for some $\phi \in (0, \pi/2)$. As we have mentioned above, the operator T_h achieves the largest angle of sectorialilty when $\alpha = \frac{b}{2}$. In this particular case (10.187) becomes (10.189). Substituting $\alpha = b/2$ and (10.189) into (10.195) we obtain

$$\mu = -(\theta + 2m). \tag{10.196}$$

This value of μ from (10.196) is marked in Figure 10.18. The corresponding operator $\mathbb{A}$ of the realizing L-system is based on these values of parameters h and μ.

Subcase 3. $b = 2$ The behavior of parameter μ in this case is also shown in Figure 10.18. It was shown above that in this case the function $V(z)$ can be realized using an extremal accretive operator T_h when $\alpha = 1$. The values of the parameters h and μ then become

$$h = u + iv, \quad u = -m, \quad v = \theta + m, \quad \mu = -(\theta + 2m).$$

The value of μ above is marked on the left branch of the hyperbola and occurs when $\alpha = 1 = b/2$.

Case 2. Again we assume that $\int_0^\infty \frac{d\tau(t)}{t} = \infty$. Hence $V(z) \in SL_0^{-1}(R, F)$ and $b = \infty$. As we mentioned above the restored operator T_h is accretive if and only if

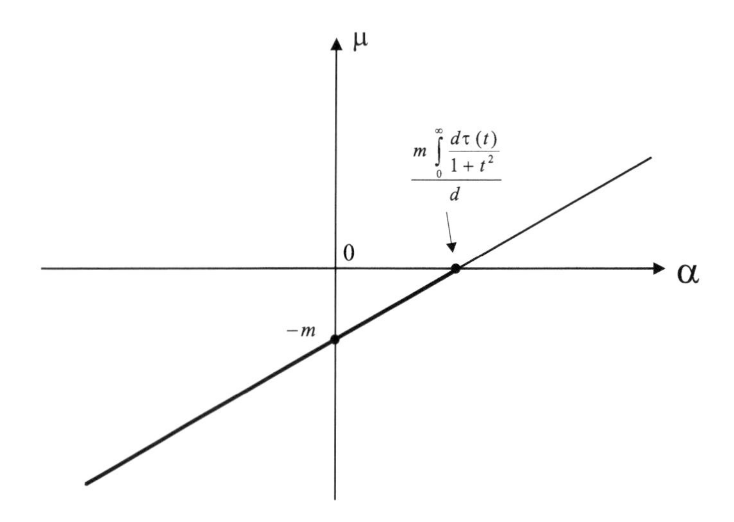

Figure 10.19: $b=\infty$

$\alpha\le 0$ and ϕ-sectorial if and only if $\alpha<0$. It is extremal if $\alpha=0$. The values of u and v, were already calculated and are given in (10.194). In particular, the value for μ is given by

$$\mu=u=-m+\frac{\alpha\, d}{\int_0^\infty \frac{d\tau(t)}{1+t^2}}. \tag{10.197}$$

where d is defined in (10.192). Figure 10.19 gives a graphical representation of this case. The left bold part of the line corresponds to the values of μ that yield an accumulative realizing L-system. If $m=0$, then the line passes through the origin and the graph should be adjusted accordingly. In the case when $\alpha=0$ and T_h is extremal we have $\mu=m$.

Example. Consider a function

$$V(z)=i\sqrt{z}.$$

A direct check confirms that $V(z)$ is an inverse Stieltjes function. Let us also consider a function $\tau(\lambda)$ such that $\tau(\lambda)=0$ for $\lambda\le 0$ and

$$\tau(\lambda)=C+\lim_{y\to 0}\frac{1}{\pi}\int_0^\lambda \mathrm{Im}\left(i\sqrt{x+iy}\right)\,dx.$$

One can verify (see also [207]) that

$$\tau'(\lambda)=\frac{1}{\pi}\sqrt{\lambda}\quad\text{for}\quad \lambda>0.$$

By direct calculations we have that

$$V(z)=\int_0^\infty\left(\frac{1}{t-z}-\frac{1}{t}\right)d\tau(t)=i\sqrt{z},$$

and that

$$\int_0^\infty\frac{d\tau(t)}{t}=\int_0^\infty\frac{dt}{\pi\sqrt{t}}=\infty.$$

It is also clear that the constant term in the integral representation (10.167) is zero, i.e., $\alpha=0$.

Let us consider an operator pair $(\dot{A},\,A_F)$ with a symmetric operator of the form (10.9) given by

$$\begin{cases}\dot{A}y=-y'',\\ y(0)=y'(0)=0,\end{cases}\tag{10.198}$$

and a self-adjoint operator

$$\begin{cases}A_F\,y=-y'',\\ y(0)=0.\end{cases}$$

We assume that $\tau(t)$ satisfies our definition of distribution function of the pair $(\dot{A},\,A_F)$ given in Section 10.8 (we are going to show it later). Operating under this assumption, we proceed to restore parameters h and μ and apply formulas (10.194) for the values $\alpha=0$ and $m=m_\infty(-0)=0$. This yields $u=0$. To obtain v we first find the value of

$$\int_0^\infty\frac{d\tau(t)}{1+t^2}=\frac{1}{\sqrt{2}},$$

and then use formula (10.192) to get the value of d. This yields $d=1/\sqrt{2}$. Consequently,

$$v=\frac{d}{\int_0^\infty\frac{d\tau(t)}{1+t^2}}=1,$$

and hence $h=vi=i$. From (10.197) we have that $\mu=0$ and (10.11) becomes

$$\mathbb{A}\,y=-y''-[iy(0)+y'(0)]\delta'(x).\tag{10.199}$$

The operator T_h in this case is

$$\begin{cases}T_hy=-y'',\\ y'(0)=iy(0).\end{cases}$$

The channel vector g of the form (10.15) then equals $g=\delta'(x)$, satisfying

$$\operatorname{Im}\mathbb{A}=\frac{\mathbb{A}-\mathbb{A}^*}{2i}=KK^*=(.,g)g,$$

and channel operator $Kc=cg$, $(c\in\mathbb{C})$ with

$$K^*y=(y,g)=-y'(0).\tag{10.200}$$

The real part of $\mathbb{A}$,

$$\operatorname{Re} \mathbb{A}\, y = -y'' - y(0)\delta'(x),$$

contains the self-adjoint quasi-kernel

$$\begin{cases} \widehat{A}y = -y'', \\ y(0) = 0. \end{cases}$$

An L-system with Schrödinger operator of the form (10.18) that realizes $V(z)$ can now be written as

$$\Theta = \begin{pmatrix} \mathbb{A} & K & 1 \\ \mathcal{H}_+ \subset L_2[0,+\infty) \subset \mathcal{H}_- & & \mathbb{C} \end{pmatrix},$$

where $\mathbb{A}$ and K are defined above via (10.199) and (10.200), respectively. Now we can back up our assumption on $\tau(t)$ to be the spectral distribution function of the pair $(\dot{A},\, A_F)$. Indeed, calculating the function $V_\Theta(z)$ for the L-system Θ above directly via formula (10.21) with $\mu = 0$ and comparing the result to $V(z)$ gives the exact value of $h = i$. It is known that operator $\dot{A}$ in (10.198) is prime. On the other hand, we know that $V(z)$ can be realized as an impedance function of a model system Θ_Λ of the form (10.134), whose symmetric operator $\dot{\Lambda}_\tau$ is also prime (see Section 10.4). Hence, we can apply Theorem 6.6.10 on bi-unitary equivalence that, with the unitary mapping U, provides for the definition of a spectral distribution function. Thus we confirm that $\tau(t)$ is the spectral distribution function of the pair $(\dot{A},\, A_F)$.

All the derivations above can be repeated for an inverse Stieltjes-like function

$$V(z) = \alpha + i\sqrt{z}, \qquad -\infty < \alpha < +\infty,$$

with very minor changes. In this case the restored values for h and μ are described as follows:

$$h = u + iv, \quad u = \alpha, \quad v = 1, \quad \mu = \alpha.$$

The dynamics of changing h according to changing α is depicted in Figure 10.16 where the horizontal line has a y-intercept of 1. The behavior of μ is described by a sloped line $\mu = \alpha$ (see Figure 10.19 with $m = 0$). In the case when $\alpha \le 0$ our function becomes inverse Stieltjes and the restored L-system Θ is accretive. The operators $\mathbb{A}$ and K of the restored L-system are given according to formulas (10.11) and (10.17), respectively.

Chapter 11

Non-self-adjoint Jacobi Matrices and System Interpolation

In this chapter we consider a new type of solutions of the Nevanlinna-Pick interpolation problem for the class of scalar Herglotz-Nevanlinna functions, the so-called, explicit system solutions that are impedance functions of the Livšic canonical systems. The conditions of the existence and uniqueness of solutions are presented in terms of interpolation data. We derive an exact formula for the angle of sectoriality of the corresponding state-space operator in the explicit system solution. The criterion for this operator to be accretive but not α-sectorial for any angle $\alpha \in (0, \pi/2)$ is obtained in terms of interpolation data and the classic Pick matrices. We find conditions on interpolation data when the explicit system solution is generated by the Livšic canonical dissipative system whose state-space operator is a non-self-adjoint, prime dissipative Jacobi matrix with a rank-one imaginary part. These results are based on a new model for prime, bounded, dissipative operators with rank-one imaginary part. In addition, an inverse spectral problem for finite non-self-adjoint Jacobi matrices with rank-one imaginary part is solved. We show that any finite sequence of non-real numbers in the open upper half-plane is the set of eigenvalues (counting multiplicity) of some dissipative non-self-adjoint Jacobi matrix with rank-one imaginary part. The algorithm of reconstruction of the unique Jacobi matrix from its non-real eigenvalues is presented.

11.1 Systems with Jacobi matrices

The three-diagonal matrices of the form

$$\mathcal{J} = \begin{pmatrix} b_1 & a_1 & 0 & 0 & \cdot & \cdot & \cdot \\ a_1 & b_2 & a_2 & 0 & \cdot & \cdot & \cdot \\ 0 & a_2 & b_3 & a_3 & \cdot & \cdot & \cdot \\ \cdot & \cdot & \cdot & \cdot & \cdot & \cdot & \cdot \\ \cdot & \cdot & \cdot & \cdot & \cdot & \cdot & a_{n-1} \\ \cdot & \cdot & \cdot & \cdot & 0 & a_{n-1} & b_n \end{pmatrix}, \tag{11.1}$$

and

$$\mathcal{J} = \begin{pmatrix} b_1 & a_1 & 0 & 0 & 0 & \cdot & \cdot \\ a_1 & b_2 & a_2 & 0 & 0 & \cdot & \cdot \\ 0 & a_2 & b_3 & a_3 & 0 & \cdot & \cdot \\ \cdot & \cdot & \cdot & \cdot & \cdot & \cdot & \cdot \end{pmatrix}, \tag{11.2}$$

where $a_k > 0$, and b_k are real numbers for all $k \geq 1$ are called **self-adjoint Jacobi matrices** [2]. We call the matrices of the form (11.1) or (11.2), with $a_k > 0$ for $k \geq 1$, and b_k are real numbers for $k \geq 2$, and $\operatorname{Im} b_1 > 0$, **dissipative finite** (respect., **semi-infinite**) Jacobi matrices.

Let a linear space $\mathbb{C}^n$ of columns be equipped with the usual inner product $(x, y) = \sum\limits_{k=1}^{n} x_k \overline{y_k}$ and let $l_2(\mathbb{N})$ be the Hilbert space of square summable sequences $x = \{x_1, x_2, \ldots, x_k, \ldots\}$ which we consider as semi-infinite vector-columns with the inner product given by

$$(x, y) = \sum_{k=1}^{\infty} x_k \overline{y_k}.$$

By $\{\delta_k\}$ we denote the canonical orthonormal basis in $\mathbb{C}^n$ ($l_2(\mathbb{N})$). An $(n \times n)$ complex Jacobi matrix $\mathcal{J}$ determines a linear operator in the Hilbert space $\mathbb{C}^n$ by means of the matrix product $\mathcal{J} \cdot x$. For the semi-infinite case we also assume that

$$\sup_k \{|a_k| + |b_k|\} < \infty. \tag{11.3}$$

Then this condition is necessary and sufficient for boundedness of the Jacobi operator in $l_2(\mathbb{N})$ defined as $\mathcal{J} \cdot x$, where $\mathcal{J}$ is a semi-infinite complex Jacobi matrix. The complex semi-infinite Jacobi matrix determines a compact operator in $l_2(\mathbb{N})$ if and only if

$$\lim_{k\to\infty} b_k = \lim_{k\to\infty} a_k = 0.$$

Suppose that (11.3) is fulfilled. In this case

$$||\mathcal{J}|| \leq 3 \max\left\{\sup_k\{|a_k|\},\ \sup_k\{|b_k|\}\right).$$

Because

$$\left(\mathcal{J}^k\delta_1\right)_{k+1} = a_k a_{k-1}\dots a_1, \qquad \left(\mathcal{J}^k\delta_1\right)_m = 0, \quad m \geq k+2, \tag{11.4}$$

and $a_k \neq 0$, the vectors $\delta_1, \mathcal{J}\delta_1, \dots, \mathcal{J}^k\delta_1, \dots$ are linearly independent.

Let A be a self-adjoint operator in a separable Hilbert space $\mathcal{H}$ and let $E(t)$ be its resolution of identity. We say that the operator A has a **simple spectrum** if there exists a nonzero vector g in $\mathcal{H}$ such that the linear span of all vectors of the form $E(\Delta)g$, where Δ runs through all intervals of $\mathbb{R}$, is dense in $\mathcal{H}$ [3].

In general, a vector $g \in \mathcal{H}$ is called a **cyclic vector** for a densely defined operator A if

$$\text{c.l.s.}\{g, Ag, \dots, A^k g, \dots\} = \mathcal{H}.$$

A cyclic vector g is called **normalized** if $\|g\| = 1$. In particular, the vector δ_1 above is a cyclic vector for the operator $\mathcal{J}$ in $l_2(\mathbb{N})$. The next two theorems are well known [239].

Theorem 11.1.1. *If A is a self-adjoint operator with a simple spectrum in a Hilbert space $\mathcal{H}$, then there exists a cyclic vector for A.*

Theorem 11.1.2. *If A is a self-adjoint operator with a simple spectrum in a Hilbert space $\mathcal{H}$, then there exists an orthonormal basis in which the matrix of the operator A is a self-adjoint Jacobi matrix.*

Proof. Let $\chi \in \mathcal{H}$, $||\chi|| = 1$ be a cyclic vector for A. Define

$$\mathcal{H}_k = \text{c.l.s.}\{\chi, A\chi, \dots, A^{k-1}\chi\}, \quad k \geq 1.$$

Then $\dim \mathcal{H}_k \leq k$ for all $k \geq 1$. Suppose that n is a minimal natural number such that $\dim \mathcal{H}_{n+1} < n+1$. Since ξ is a cyclic vector for A, we get $\mathcal{H}_{n+1} = \mathcal{H}_n$. Hence $\dim \mathcal{H} = n$. Clearly, $\dim \mathcal{H} = \infty$ if and only if $\dim \mathcal{H}_k = k$ for all natural k. Suppose $\dim \mathcal{H} = \infty$. Set

$$\mathcal{N}_1 = \mathcal{H}_1 = \{\lambda\chi, \quad \lambda \in \mathbb{C}\}, \quad \mathcal{N}_k = \mathcal{H}_k \ominus \mathcal{H}_{k-1}, \quad k \geq 2.$$

Then $\dim \mathcal{N}_k = 1$. We can find the system of vectors $\{\chi_k\}_{k\geq 1}$ such that

$$\chi_1 = \chi \in \mathcal{N}_1, \quad \chi_k \in \mathcal{N}_k, \quad ||\chi_k|| = 1, \quad (\chi_{k+1}, A\chi_k) > 0, \quad k \geq 1.$$

The system $\{\chi_k\}_{k\geq 1}$ forms an orthonormal basis in $\mathcal{H}$. This system in fact is obtained from the system $\{A^{k-1}\chi\}_{k\geq 1}$ by the Gram-Schmidt orthonormalization procedure. Because $A\mathcal{H}_k \subset \mathcal{H}_{k+1}$, we get

$$A\chi_k \perp \chi_j \quad \text{for all} \quad j \geq k+2.$$

The self-adjointness of A yields $A\chi_k \perp \chi_j$ for $j \leq k-2$ $(k \geq 3)$. Hence, with respect to the orthonormal basis $\{\chi_k\}_{k\geq 1}$ the matrix of the operator A takes the three-diagonal form. Since $(A\chi_k, \chi_{k+1}) > 0$ for $k \geq 1$, this matrix is a self-adjoint Jacobi matrix of the form (11.2).

The above construction is valid also in the case $\dim \mathcal{H} < \infty$. □

Lemma 11.1.3. *Let T be a dissipative operator with a rank-one imaginary part and let g be a vector in $\mathcal{H}$ such that*

$$2\,\mathrm{Im}\,Th = (h, g)g, \quad h \in \mathcal{H}.$$

Then T is prime if and only if the vector g is cyclic for the real part $\mathrm{Re}\,T$.

Proof. Suppose that T is prime. Then (5.7) holds. Let us prove that g is a cyclic vector for $\mathrm{Re}\,T$. Let

$$\mathcal{H}' = \mathrm{c.\,l.\,s.}\,\{(\mathrm{Re}\,T)^n g,\ n = 0, 1, \dots\} \neq \mathcal{H}.$$

Then $\mathcal{H}'$ and $\mathcal{H}'' = \mathcal{H} \ominus \mathcal{H}'$ are invariant with respect to $\mathrm{Re}\,T$. Since $\mathcal{H}'' \subset \mathrm{Ker}\,\mathrm{Im}\,T$, it follows that

$$\mathrm{Re}\,T \restriction \mathcal{H}'' = T \restriction \mathcal{H}'' = T^* \restriction \mathcal{H}'',$$

$T^{*n}\mathcal{H}'' \subset \mathcal{H}$ and $\mathrm{Im}\,TT^{*n} \restriction \mathcal{H}'' = 0$ for all $n = 0, 1, \dots$. Now from (5.7) we obtain that $\mathcal{H}'' = \{0\}$, i.e., g is a cyclic vector for $\mathrm{Re}\,T$.

Conversely, suppose that the vector g is cyclic for $\mathrm{Re}\,T$, i.e., $\mathcal{H}' = \mathcal{H}$. Let the subspace $\mathcal{H}_s$ be defined by the right-hand side of (5.7). Then

$$T \restriction (\mathcal{H} \ominus \mathcal{H}_s) = \mathrm{Re}\,T \restriction (\mathcal{H} \ominus \mathcal{H}_s),$$

and $\mathcal{H} \ominus \mathcal{H}_s$ as well as $\mathcal{H}_s$ reduces $\mathrm{Re}\,T$. Because $g \in \mathcal{H}_s$, we get that $\mathcal{H}' \subset \mathcal{H}_s$. It follows that $\mathcal{H}_s = \mathcal{H}$, i.e., T is a prime operator. □

Theorem 11.1.4. *Let $\mathcal{H}$ be a separable Hilbert space and let T be a bounded, prime, dissipative operator in $\mathcal{H}$ with a rank-one imaginary part. Then there exists an orthonormal basis in $\mathcal{H}$ in which the matrix of the operator T is a bounded dissipative Jacobi matrix.*

Proof. The operator T takes the form $\mathrm{Re}\,T + i\mathrm{Im}\,T$ and $\mathrm{Im}\,T = l(\cdot, \chi)\chi$, where $\chi \in \mathrm{Ran}(\mathrm{Im}\,T)$, $\|\chi\| = 1$, $l > 0$. Since T is a bounded prime operator, then, by Theorem 5.2.1, χ is a cyclic vector for T. Consequently, χ is a cyclic vector for bounded self-adjoint operator $\mathrm{Re}\,T$. By Stone's Theorem 11.1.2, there exists an orthonormal basis $\{\chi_k\}_{k\geq 1}$, $\chi_1 = \chi$ in $\mathcal{H}$ such that the matrix of $\mathrm{Re}\,T$ is a self-adjoint Jacobi matrix. It follows from $(T\chi, \chi) = (\mathrm{Re}\,T\chi, \chi) + il$ that the matrix of T with respect to $\{\chi_k\}_{k\geq 1}$ is a dissipative Jacobi matrix. □

Let $\mathcal{J}^*$ be the adjoint matrix to a Jacobi matrix $\mathcal{J}$ and let

$$\mathrm{Re}\,\mathcal{J} = \frac{1}{2}(\mathcal{J} + \mathcal{J}^*), \quad \mathrm{Im}\,\mathcal{J} = \frac{1}{2i}(\mathcal{J} - \mathcal{J}^*),$$

be the Hermitian components of $\mathcal{J}$. One has for the $(n \times n)$ case

$$\mathrm{Re}\,\mathcal{J} = \begin{pmatrix} \mathrm{Re}\,b_1 & a_1 & 0 & 0 & 0 & \cdot & \cdot & \cdot \\ a_1 & b_2 & a_2 & 0 & 0 & \cdot & \cdot & \cdot \\ 0 & a_2 & b_3 & a_3 & 0 & \cdot & \cdot & \cdot \\ \cdot & \cdot & \cdot & \cdot & \cdot & \cdot & \cdot & \cdot \\ \cdot & \cdot & \cdot & \cdot & \cdot & \cdot & \cdot & a_{n-1} \\ \cdot & \cdot & \cdot & \cdot & \cdot & 0 & a_{n-1} & b_n \end{pmatrix},$$

$$\operatorname{Im}\mathcal{J} = \begin{pmatrix} \operatorname{Im} b_1 & 0 & 0 & \cdot & \cdot & \cdot & 0 \\ 0 & 0 & 0 & \cdot & \cdot & \cdot & 0 \\ 0 & 0 & 0 & \cdot & \cdot & \cdot & 0 \\ \cdot & \cdot & \cdot & \cdot & \cdot & \cdot & \cdot \\ \cdot & \cdot & \cdot & \cdot & \cdot & \cdot & \cdot \\ 0 & 0 & 0 & \cdot & \cdot & \cdot & 0 \end{pmatrix},$$

and for the semi-infinite case

$$\operatorname{Re}\mathcal{J} = \begin{pmatrix} \operatorname{Re} b_1 & a_1 & 0 & 0 & 0 & \cdot & \cdot & \cdot \\ a_1 & b_2 & a_2 & 0 & 0 & \cdot & \cdot & \cdot \\ 0 & a_2 & b_3 & a_3 & 0 & \cdot & \cdot & \cdot \\ \cdot & \cdot & \cdot & \cdot & \cdot & \cdot & \cdot & \cdot \end{pmatrix},$$

$$\operatorname{Im}\mathcal{J} = \begin{pmatrix} \operatorname{Im} b_1 & 0 & 0 & 0 & 0 & \cdot & \cdot & \cdot \\ 0 & 0 & 0 & 0 & 0 & \cdot & \cdot & \cdot \\ 0 & 0 & 0 & 0 & 0 & \cdot & \cdot & \cdot \\ \cdot & \cdot & \cdot & \cdot & \cdot & \cdot & \cdot & \cdot \end{pmatrix}.$$

In addition, for every $x \in \mathbb{C}^n$ $(l_2(\mathbb{N}))$,

$$\operatorname{Im}\mathcal{J}x = (x, g)\, g = \mathcal{K}\mathcal{K}^*,$$

where

$$g = \sqrt{\operatorname{Im} b_1}\,\delta_1, \quad \mathcal{K}c = cg, \quad \mathcal{K}^*x = (x, g), \quad c \in \mathbb{C}. \tag{11.5}$$

The Livšic canonical system of the form (5.6) with finite Jacobi matrix $\mathcal{J}$ as a state-space operator

$$\Delta = \begin{pmatrix} \mathcal{J} & \mathcal{K} & 1 \\ \mathbb{C}^n & & \mathbb{C} \end{pmatrix}, \tag{11.6}$$

and, respectively with semi-infinite Jacobi matrix as a state-space operator,

$$\Delta = \begin{pmatrix} \mathcal{J} & \mathcal{K} & 1 \\ l_2(\mathbb{N}) & & \mathbb{C} \end{pmatrix} \tag{11.7}$$

is called the **Livšic system in Jacobi form**.

11.2 The Stone theorem and its generalizations

Let A be a bounded self-adjoint operator with simple spectrum in a separable Hilbert space $\mathcal{H}$ and let χ be a normalized ($\|\chi\| = 1$) cyclic vector for A.

Definition 11.2.1. The function

$$m(z) = ((A - zI)^{-1}\chi, \chi), \quad z \in \rho(A),$$

is called the **Weyl function** (or the m**-function**) of A.

The Weyl function of A is a Herglotz-Nevanlinna function holomorphic on $\mathbb{C} \setminus [-||A||, ||A||]$. It has the Taylor expansion

$$m(z) = -\sum_{n=1}^{\infty} (A^{n-1}\chi, \chi) z^{-n}, \quad |z| > ||A||.$$

In particular

$$\lim_{z \to \infty} zm(z) = -1.$$

Let $E(t)$ be a resolution of identity for A. Then the Weyl function $m(z)$ admits the integral representation

$$m(z) = \int_{\mathbb{R}} \frac{d\sigma(t)}{t - z},$$

where $\sigma(t) = (E(t)\chi, \chi)$ and $d\sigma$ is a probability measure with a compact support $\operatorname{supp}(d\sigma) \subseteq [-||A||, ||A||]$. It follows from the definition of the Weyl function and the Hilbert identity that

$$\frac{m(z) - m(\bar{\zeta})}{z - \bar{\zeta}} = ((A - zI)^{-1}\chi, (A - \zeta I)^{-1}\chi), \quad z, \zeta \in \rho(A). \tag{11.8}$$

Theorem 11.2.2. *Let A_1 and A_2 be two bounded self-adjoint operators in Hilbert spaces $\mathcal{H}_1$ and $\mathcal{H}_2$, respectively. Let χ_1 and χ_2 be normalized cyclic vectors for A_1 and A_2. If the corresponding Weyl functions $m_1(z)$ and $m_2(z)$ coincide in a neighborhood of infinity, then there exists a unitary operator $U \in [\mathcal{H}_1, \mathcal{H}_2]$ such that*

$$UA_1 = A_2 U \quad \text{and} \quad U\chi_1 = \chi_2.$$

Proof. Let $m_1(z) = m_2(z)$ for all z in a neighborhood G of infinity. From (11.8) we get that

$$((A_1 - zI)^{-1}\chi_1, (A_1 - \zeta I)^{-1}\chi_1)_{\mathcal{H}_1} = ((A_2 - zI)^{-1}\chi_2, (A_2 - \zeta I)^{-1}\chi_2)_{\mathcal{H}_2}$$

for $z, \zeta \in G$. Since χ_1 and χ_2 are cyclic vectors for A_1 and A_2, respectively, we get the relations

$$c.l.s.\{(A_1 - zI)^{-1}\chi_1,\ z \in G\} = \mathcal{H}_1, \quad c.l.s.\{(A_2 - zI)^{-1}\chi_2,\ z \in G\} = \mathcal{H}_2.$$

Let

$$\mathcal{L}_1 = \operatorname{span}\{(A_1 - zI)^{-1}\chi_1,\ z \in G\}, \quad \mathcal{L}_2 = \operatorname{span}\{(A_2 - zI)^{-1}\chi_2,\ z \in G\}.$$

Define on $\mathcal{L}_1$ a linear operator U as

$$U\left(\sum c_k (A_1 - z_k I)^{-1}\chi_1\right) = \left(\sum c_k (A_2 - z_k I)^{-1}\chi_2\right).$$

Then $U\mathcal{L}_1 = \mathcal{L}_2$ and U is an isometry. Because $\mathcal{L}_1$ and $\mathcal{L}_2$ are dense in $\mathcal{H}_1$ and $\mathcal{H}_2$, respectively, the operator U has unitary continuation on $\mathcal{H}_1$. Since

$$zU(A_1 - zI)^{-1}\chi_1 = z(A_2 - zI)^{-1}\chi_2,$$

for all $z \in G$, we get

$$U\chi_1 = -U(\lim_{z\to\infty} z(A_1 - zI)^{-1}\chi_1) = -\lim_{z\to\infty} z(A_2 - zI)^{-1}\chi_2 = \chi_2.$$

Furthermore,

$$\begin{aligned} &UA_1\left(\sum c_k(A_1 - z_kI)^{-1}\chi_1\right) = U\left(\sum\left(c_k\chi_1 + z_k(A_1 - z_kI)^{-1}\chi_1\right)\right) \\ &= \sum\left(c_k\chi_2 + z_k(A_2 - z_kI)^{-1}\chi_2\right) = A_2\left(\sum c_k(A_2 - z_kI)^{-1}\chi_2\right) \\ &= A_2U\left(\sum c_k(A_1 - z_kI)^{-1}\chi_1\right). \end{aligned}$$

Hence, $UA_1 = A_2U$. □

Definition 11.2.3. A Herglotz-Nevanliina function $m(z)$ belongs to the **class** $\mathbf{N}^0$ if $m(z)$ is holomorphic in the neighborhood of infinity and

$$\lim_{z\to\infty} zm(z) = -1.$$

A function $m(z) \in \mathbf{N}^0$ has the integral representation

$$m(z) = \int_{\mathbb{R}} \frac{d\sigma(t)}{t - z},$$

where $d\sigma$ is a probability measure with compact support. The asymptotic expansion

$$m(z) = -z^{-1} - b_1z^{-2} - (b_1^2 + a_1^2)z^{-3} + o(z^{-3}), \quad z \to \infty,$$

holds, where

$$b_1 = \int_{\mathbb{R}} t\,d\sigma(t), \quad a_1^2 + b_1^2 = \int_{\mathbb{R}} t^2 d\sigma(t).$$

We can recover the numbers b_1 and a_1^2 from the function $m(z)$,

$$b_1 = -\lim_{z\to\infty} z^2(m(z) + z^{-1}), \quad a_1^2 = -\lim_{z\to\infty} z^3(m(z) + z^{-1} + b_1z^{-2}) - b_1^2.$$

Notice that if $\operatorname{supp}(d\sigma) \subset [-C, C]$, then $|b_1| \le C$ and $|a_1| \le C$. In the following we assume that $a_1 > 0$. Define the function $m_1(z)$ via the formula

$$m_1(z) = \frac{1}{a_1^2}\left(b_1 - z - m^{-1}(z)\right).$$

Clearly, $-m^{-1}(z)$ is a Herglotz-Nevanlinna function and

$$-m^{-1}(z) = z - b_1 - a_1^2z^{-1} + o(z^{-1}), \quad z \to \infty.$$

If $\operatorname{supp}(d\sigma) \subset [-C, C]$, then $m(z) > 0$ for $z < -C$ and $m(z) < 0$ for $z > C$. Hence $-m^{-1}(z)$ is holomorphic on $\mathbb{C} \setminus [-C, C]$. It follows that m_1 is Herglotz-Nevanlinna function, holomorphic on $\mathbb{C} \setminus [-C, C]$ and $\lim_{z\to\infty} zm_1(z) = -1$. Thus, m_1 belongs to the class $\mathbf{N}^0$.

Repeating such a procedure, we obtain finite or infinite sequences of real numbers $\{b_k\}_{k\geq 1}$ and positive numbers $\{a_k\}_{k\geq 1}$. Moreover, $|b_k| \leq C$ and $|a_k| \leq C$ for all k. The process terminates on the N-th step if and only if $a_N = 0$, i.e., $m(z)$ is a rational function of the form

$$m(z) = \sum_{n=1}^{N} \frac{c_k}{\mu_k - z},$$

where $\mu_1, \ldots, \mu_N$ are distinct real numbers and $c_1, \ldots, c_N$ are positive numbers such that $\sum_{n=1}^{N} c_k = 1$. Thus we have obtained the following representation of $m(z)$ by means of the continued fraction

$$m(z) = \cfrac{-1}{z - b_1 + \cfrac{-a_1^2}{z - b_2 + \cfrac{-a_2^2}{z - b_3 + \ldots + \cfrac{-a_{n-1}^2}{z - b_n + \ldots}}}}. \tag{11.9}$$

Using the sequences $\{b_k\}$ and $\{a_k\}$, we construct a Jacobi matrix $\mathcal{J}$ of the form (11.1) or (11.2).

Theorem 11.2.4. *Let $m(z) \in \mathbf{N}^0$ and let $\{b_k\}$ and $\{a_k\}$ be corresponding parameters of $m(z)$. Then the Weyl function*

$$m_{\mathcal{J}}(z) := ((\mathcal{J} - zI)^{-1}\delta_1, \delta_1),$$

where $\mathcal{J}$ is a self-adjoint Jacobi matrix of the form (11.1) *or* (11.2)*, coincides with $m(z)$.*

Proof. Recall that the Schur-Frobenius formula

$$P_{\mathcal{H}_1}(A - zI)^{-1} \upharpoonright \mathcal{H}_1 = \left(-zI + A_{11} - A_{12}(A_{22} - zI)^{-1}A_{21}\right)^{-1}, \tag{11.10}$$

written for

$$A = \begin{pmatrix} A_{11} & A_{12} \\ A_{21} & A_{22} \end{pmatrix} : \begin{matrix} \mathcal{H}_1 \\ \oplus \\ \mathcal{H}_2 \end{matrix} \to \begin{matrix} \mathcal{H}_1 \\ \oplus \\ \mathcal{H}_2 \end{matrix}$$

holds for $z \in \rho(A) \cap \rho(A_{22})$. Here $P_{\mathcal{H}_1}$ is the orthogonal projection in $\mathcal{H}$ onto $\mathcal{H}_1$. Applying (11.10) to $\mathcal{J}$ gives

$$\frac{1}{a_1^2}\left(b_1 - z - m_{\mathcal{J}}^{-1}(z)\right) = m_{\mathcal{J}_1}(z),$$

where $m_{\mathcal{J}_1}(z)$ is the Weyl function of the Jacobi matrix $\mathcal{J}_1$ obtained from $\mathcal{J}$ by crossing out the first row and the first column. It follows that the parameters corresponding to $m_{\mathcal{J}}(z)$ are the sequences $\{b_k\}$ and $\{a_k\}$. Hence, the continued fraction expansion of $m_{\mathcal{J}}(z)$ is the same as for $m(z)$ (see (11.9)). Therefore, $m_{\mathcal{J}}(z) = m(z)$. □

Theorem 11.2.2 and Theorem 11.2.4 yield the Stone theorem.

Theorem 11.2.5. *Let A be a bounded self-adjoint operator with simple spectrum. Then A is unitarily equivalent to the self-adjoint operator given by a Jacobi matrix of the form* (11.1) *or* (11.2).

The following is a generalization of the Stone Theorem.

Theorem 11.2.6. *Let $\mathcal{H}$ be separable Hilbert space and T be a bounded prime dissipative operator in $\mathcal{H}$ with a rank-one imaginary. Then T is unitarily equivalent to the operator given by a Jacobi matrix of the form* (11.1) *or* (11.2).

Proof. Let $g \in \mathcal{H}$ be such that $2\,\mathrm{Im}\,Th = (h,g)g$, $h \in \mathcal{H}$. According to Theorem 5.1.2, there is the Livšic canonical system

$$\Theta = \begin{pmatrix} T & K & 1 \\ \mathcal{H} & & \mathbb{C} \end{pmatrix},$$

where $Kc = c(g/\sqrt{2})$, $c \in \mathbb{C}$. Let $W_\Theta(z)$ be the transfer function of the system Θ of the form (5.17). Then [1]

$$W_\Theta(z) = 1 - i\left((T - zI)^{-1}g, g\right), \quad z \in \rho(T).$$

It follows that

$$\lim_{z\to\infty} z(W_\Theta(z) - 1) = i\|g\|^2.$$

Let $c = \|g\|^2$. Then

$$V_\Theta(z) = i\,\frac{W_\Theta(z) - 1}{W_\Theta(z) + 1} = \frac{1}{2}\left((\mathrm{Re}\,T - zI)^{-1}g, g\right),$$

We set

$$m(z) = \frac{2}{\|g\|^2}\,V_\Theta(z).$$

Then $m(z)$ is the Weyl function of $\mathrm{Re}\,T$. Hence there exists a unique Jacobi matrix

$$H = \begin{pmatrix} b & a_1 & 0 & 0 & 0 & \cdot & \cdot & \cdot \\ a_1 & b_2 & a_2 & 0 & 0 & \cdot & \cdot & \cdot \\ 0 & a_2 & b_3 & a_3 & 0 & \cdot & \cdot & \cdot \\ \cdot & \cdot & \cdot & \cdot & \cdot & \cdot & \cdot & \cdot \end{pmatrix},$$

[1] When we deal with the Livšic canonical system whose input-output space is $\mathbb{C}$, the transfer and impedance operator-functions $W_\Theta(z)$ and $V_\Theta(z)$ applied to an element $c \in \mathbb{C}$ can be considered as scalar functions multiplied by a scalar c, i.e., $W_\Theta(z)c = W_\Theta(z)\cdot c$ and $V_\Theta(z)c = V_\Theta(z)\cdot c$.

with real entries $b, b_2, b_3, \ldots$ and positive entries $a_1, a_2, \ldots$ such that

$$m(z) = \left((H - zI)^{-1}\delta_1, \delta_1\right).$$

Note that (11.3) holds because H defines a bounded operator in $l_2(\mathbb{N})$. Moreover, the entries of H can be found by means of the continued fraction expansion

$$m(z) = \cfrac{-1}{z - b + \cfrac{-a_1^2}{z - b_2 + \cfrac{-a_2^2}{z - b_3 + \ldots + \cfrac{-a_{n-1}^2}{z - b_n + \ldots}}}}.$$

Let

$$\begin{aligned} \mathcal{J} &= \begin{pmatrix} b + i\|g\|^2/2 & a_1 & 0 & 0 & 0 & \cdot & \cdot & \cdot \\ a_1 & b_2 & a_2 & 0 & 0 & \cdot & \cdot & \cdot \\ 0 & a_2 & b_3 & a_3 & 0 & \cdot & \cdot & \cdot \\ \cdot & \cdot & \cdot & \cdot & \cdot & \cdot & \cdot & \cdot \end{pmatrix} \\ &= H + i \begin{pmatrix} \|g\|^2/2 & 0 & 0 & 0 & \cdot & \cdot & \cdot \\ 0 & 0 & 0 & 0 & \cdot & \cdot & \cdot \\ 0 & 0 & 0 & 0 & \cdot & \cdot & \cdot \\ \cdot & \cdot & \cdot & \cdot & \cdot & \cdot & \cdot \end{pmatrix}. \end{aligned} \tag{11.11}$$

Suppose $\Theta_{\mathcal{J}}$ is the Livšic canonical system constructed via Theorem 5.1.2 with $\mathcal{J}$ as the state-space operator. Then the transfer function of $\Theta_{\mathcal{J}}$ is

$$W_{\Theta_{\mathcal{J}}}(z) = 1 - i\|g\|^2 \left((\mathcal{J} - zI)^{-1}\delta_1, \delta_1\right).$$

Since $\operatorname{Re}\mathcal{J} = H$ and $m(z) = \left((H - zI)^{-1}\delta_1, \delta_1\right)$, we get that $W_{\Theta_{\mathcal{J}}}(z) = W_{\Theta}(z)$. Because the matrix $\mathcal{J}$ is prime, we can apply Theorem 5.4.3 and conclude that the operator T is unitarily equivalent to $\mathcal{J}$. □

An application of the Schur-Frobenius formula (11.10) to $\mathcal{J}$ given by (11.11) shows that the entries of $\mathcal{J}$ can be also found using the continued fraction expansion

$$M(z) = \cfrac{-1}{z - b_1 + \cfrac{-a_1^2}{z - b_2 + \cfrac{-a_2^2}{z - b_3 + \ldots + \cfrac{-a_{n-1}^2}{z - b_n + \ldots}}}},$$

where

$$M(z) = ((\mathcal{J} - zI)^{-1}\delta_1, \delta_1) = \frac{i}{\beta}(W(z) - 1),$$

and $\beta = \lim\limits_{z\to\infty} (iz(1 - W(z)))$.

Theorem 11.2.7. *Any Herglotz-Nevanlinna function of the form*

$$V(z)=\int_a^b \frac{1}{t-z}\,d\sigma(t),$$

such that $\sigma(t)$ is a non-negative, non-decreasing function on finite interval $[a,b]$, can be realized in the form

$$V(z)=i\frac{W_\Delta(z)-1}{W_\Delta(z)+1},$$

where $W_\Delta(z)$ is the transfer function of the Livšic canonical system in Jacobi form (11.6), (11.7).

Proof. Consider an auxiliary system Θ of the form (5.37) constructed in Theorem 5.5.1,

$$\Theta=\begin{pmatrix} \mathcal{A} & K & 1\\ L_2([a,b],d\sigma) & & \mathbb{C}\end{pmatrix}, \tag{11.12}$$

where the state-space operator $\mathcal{A}$ is given by

$$(\mathcal{A}f)(t)=tf(t)+2i\int_a^b f(t)d\sigma(t), \tag{11.13}$$

and

$$Kc=ch,\ c\in\mathbb{C},\quad K^*f=(f,h)h,\ f\in L_2([a,b],d\sigma),\quad h=h(t)=1.$$

Applying Theorem 5.5.1 we have

$$V_\Theta(z)=K^*(\mathrm{Re}\,\mathcal{A}-zI)^{-1}K=V(z),$$

and

$$V(z)=i\frac{W_\Theta(z)-1}{W_\Theta(z)+1} \tag{11.14}$$

where $W_\Theta(z)$ is a transfer function of an auxiliary system Θ of the form (11.12). The system in Jacobi form (11.6), (11.7) is a minimal system since the vector δ_1 is a cyclic vector (see (5.8)) for Jacobi matrix $\mathcal{J}$ (11.1), (11.2) and $\mathcal{J}$ is a prime operator. It is established in Theorem 11.2.6 that any bounded, dissipative, prime operator A with a rank-one imaginary part is unitarily equivalent to a Jacobi matrix of the form (11.1), (11.2). The auxiliary model operator $\mathcal{A}$ (11.13) is prime, since vector $h=h(t)=1$ is a cyclic vector for $(\mathrm{Re}\,\mathcal{A}f)(t)=tf(t)$ in $L_2([a,b],d\sigma)$ and therefore this vector is cyclic for $\mathcal{A}$. By the above mentioned theorem, auxiliary model operator $\mathcal{A}$ is unitarily equivalent to the Jacobi matrix $\mathcal{J}$. Therefore, there exists a unitary operator $\mathcal{U}$ from $L_2([a,b],d\sigma)$ onto $\mathbb{C}^n$ (respectively $l_2(\mathbb{N})$) such that

$$\mathcal{U}\mathcal{A}\mathcal{U}^{-1}=\mathcal{J},$$

Obviously, $(\mathcal{U}h)(t) = e^{i\phi}g$ for $h = h(t) = 1$, $g = \sqrt{\text{Im } b_1}\,\delta_1$, and thus $\mathcal{U}Kc = \mathcal{K}_\phi c = ce^{i\phi}g$, $c \in \mathbb{C}$. Considering a new unitary operator $U = e^{-i\phi}\mathcal{U}$ we get that $U\mathcal{A}U^{-1} = \mathcal{J}$ and $Uh = g$. Hence systems Θ and Δ of the form

$$\Delta = \begin{pmatrix} \mathcal{J} & \mathcal{K} & 1 \\ \mathbb{C}^n & & \mathbb{C} \end{pmatrix}, \tag{11.15}$$

or of the form (respectively with semi-infinite Jacobi matrix as a state-space operator)

$$\Delta = \begin{pmatrix} \mathcal{J} & \mathcal{K} & 1 \\ l_2(\mathbb{N}) & & \mathbb{C} \end{pmatrix},$$

are unitarily equivalent and have the same transfer functions, i.e., $W_\Delta(z) = W_\Theta(z)$. The representation (11.14) completes the proof of the theorem. □

11.3 Inverse spectral problems for finite dissipative Jacobi matrices

Let $\mathcal{J}$ be a dissipative $(n \times n)$ Jacobi matrix of the form (11.1) or (11.2). Then the corresponding operator in $\mathbb{C}^n$ is a prime dissipative operator with rank-one imaginary part. Therefore, the matrix $\mathcal{J}$ has only non-real eigenvalues with positive imaginary parts. The next theorem establishes that the n arbitrary non-real numbers, counting algebraic multiplicity taken from the open upper half-plane, determine uniquely some dissipative $(n \times n)$ Jacobi matrix with rank-one imaginary part.

Theorem 11.3.1. *Suppose that $z_1, \dots, z_n$ are not necessarily distinct complex numbers with positive imaginary parts. Then there exists a unique $(n \times n)$ dissipative Jacobi matrix whose eigenvalues (counting algebraic multiplicity) coincide with $\{z_k\}_{k=1}^n$.*

Proof. Let

$$W(z) = \prod_{k=1}^{n} \frac{z - \bar{z}_k}{z - z_k}.$$

Then

$$\lim_{z\to\infty} z(W(z) - 1) = 2i \sum_{k=1}^{n} \text{Im } z_k.$$

Let $c = \sum_{k=1}^n \text{Im } z_k$ and define

$$m(z) = \frac{i}{c}\,\frac{W(z) - 1}{W(z) + 1}. \tag{11.16}$$

The Herglotz-Nevanlinna function $m(z)$ has the expansion in the neighborhood of infinity

$$m(z) \sim -\frac{1}{z} - \frac{b}{z^2} - \frac{b^2 + a_1^2}{z^3} + O\left(\frac{1}{z^4}\right),$$

and determines a probability measure supported at n points. By Theorem 11.2.4 there exists a unique self-adjoint $(n \times n)$ Jacobi matrix

$$H = \begin{pmatrix} b & a_1 & 0 & 0 & \cdot & \cdot & \cdot \\ a_1 & b_2 & a_2 & 0 & \cdot & \cdot & \cdot \\ 0 & a_2 & b_3 & a_3 & \cdot & \cdot & \cdot \\ \cdot & \cdot & \cdot & \cdot & \cdot & \cdot & \cdot \\ \cdot & \cdot & \cdot & \cdot & \cdot & \cdot & a_{n-1} \\ \cdot & \cdot & \cdot & \cdot & 0 & a_{n-1} & b_n \end{pmatrix},$$

such that $m(z) = \left((H - zI)^{-1}\delta_1, \delta_1\right)$. Let

$$\mathcal{J} = \begin{pmatrix} b + ic & a_1 & 0 & 0 & \cdot & \cdot & \cdot \\ a_1 & b_2 & a_2 & 0 & \cdot & \cdot & \cdot \\ 0 & a_2 & b_3 & a_3 & \cdot & \cdot & \cdot \\ \cdot & \cdot & \cdot & \cdot & \cdot & \cdot & \cdot \\ \cdot & \cdot & \cdot & \cdot & \cdot & \cdot & a_{n-1} \\ \cdot & \cdot & \cdot & \cdot & 0 & a_{n-1} & b_n \end{pmatrix} = H + i \begin{pmatrix} c & 0 & 0 & 0 & \cdot & \cdot & \cdot \\ 0 & 0 & 0 & 0 & \cdot & \cdot & \cdot \\ 0 & 0 & 0 & 0 & \cdot & \cdot & \cdot \\ \cdot & \cdot & \cdot & \cdot & \cdot & \cdot & \cdot \\ \cdot & \cdot & \cdot & \cdot & \cdot & \cdot & \cdot \\ \cdot & \cdot & \cdot & \cdot & 0 & 0 & 0 \end{pmatrix}.$$

Then

$$(\operatorname{Im} \mathcal{J})\, x = (x, g)\, g, \quad x \in \mathbb{C}^n,$$

where $g = \sqrt{c}\,\delta_1 \in \mathbb{C}^n$. By Theorem 5.1.2, operator $\mathcal{J}$ can be included in the system Δ of the form (11.6) with the channel operator $\mathcal{K}$ defined by (11.5) using $g = \sqrt{c}\,\delta_1$. Then the transfer function $W_\Delta(z)$ has a form

$$W_\Delta(z) = 1 - 2i\, c\left((\mathcal{J} - zI)^{-1}\delta_1, \delta_1\right),$$

while the impedance function is

$$V_\Delta(z) = i\,\frac{W_\Delta(z) - 1}{W_\Delta(z) + 1} = c\,\left((H - zI)^{-1}\delta_1, \delta_1\right) = c\, m(z).$$

From (11.16) we get

$$V_\Delta(z) = i\,\frac{W(z) - 1}{W(z) + 1}.$$

Therefore, $W_\Delta(z) = W(z)$. Hence, the eigenvalues of $\mathcal{J}$ coincide with $\{z_k\}$, counting algebraic multiplicity. □

Example. Let us construct a (3×3) dissipative Jacobi matrix with eigenvalues $z_1 = i$ of multiplicity 2 and $z_2 = 2i$ (of multiplicity 1). Then the transfer function $W_\Delta(z)$

of the system Δ of the form (11.6) with $\mathcal{J}$ as a state-space operator (constructed in the above proof) is

$$W_\Delta(z) = \left(\frac{z+i}{z-i}\right)^2 \frac{z+2i}{z-2i}.$$

Then $\lim_{z\to\infty} z(W_\Delta(z)-1) = 4i$. Note that $2\operatorname{Im} z_1 + \operatorname{Im} z_2 = 4$. Let

$$m(z) = \frac{i}{4}\frac{W_\Delta(z)-1}{W_\Delta(z)+1} = \frac{i}{4}\frac{(z+i)^2(z+2i)-(z-i)^2(z-2i)}{(z+i)^2(z+2i)+(z-i)^2(z-2i)} = \frac{-2z^2+1}{2z^3-10z}.$$

Then

$$m(z) = \cfrac{-1}{z - \cfrac{9/2}{z - \cfrac{1/2}{z}}}.$$

It follows that $b_1 = 4i$, $b_2 = b_3 = 0$, $a_1 = \sqrt{9/2}$, $a_2 = \sqrt{1/2}$ and the Jacobi matrix $\mathcal{J}$ takes the form

$$\mathcal{J} = \begin{pmatrix} 4i & \frac{3}{\sqrt{2}} & 0 \\ \frac{3}{\sqrt{2}} & 0 & \frac{1}{\sqrt{2}} \\ 0 & \frac{1}{\sqrt{2}} & 0 \end{pmatrix}.$$

Example. In order to construct a dissipative $(n\times n)$ Jacobi matrix with eigenvalue $z_0 = x_0 + iy_0$ of algebraic multiplicity n $(y_0 > 0)$, it is sufficient to construct an $(n \times n)$ Jacobi matrix $\mathcal{J}_n$ with eigenvalue $z_0 = i$ of algebraic multiplicity n. The transfer function $W_\Delta(z)$ of the system Δ of the form (11.6) with $\mathcal{J}_n$ as a state-space operator (constructed in the proof of Theorem 11.3.1) is

$$W_\Delta(z) = \left(\frac{z+i}{z-i}\right)^n.$$

The corresponding Weyl function of the real part of $\mathcal{J}_n$ is

$$m_n(z) = \frac{i}{n}\frac{(z+i)^n-(z-i)^n}{(z+i)^n+(z-i)^n}.$$

Because $\operatorname{trace}\mathcal{J}_n = in$, we get $\operatorname{Im} b_1 = n$. Expanding the function $m_n(z)$ via continued fractions we obtain the real part of $\mathcal{J}_n$. In particular,

$$m_2(z) = \cfrac{-1}{z - \cfrac{1}{z}}, \quad m_3(z) = \cfrac{-1}{z - \cfrac{8/3}{z - \cfrac{1/3}{z}}},$$

$$m_4(z) = \cfrac{-1}{z + \cfrac{-5}{z + \cfrac{-4/5}{z + \cfrac{-1/5}{z}}}}, \quad m_5(z) = \cfrac{-1}{z + \cfrac{-8}{z + \cfrac{-7/5}{z + \cfrac{-16/35}{z + \cfrac{-1/7}{z}}}}}.$$

Therefore,

$$\mathcal{J}_2 = \begin{pmatrix} 2i & 1 \\ 1 & 0 \end{pmatrix}, \quad \mathcal{J}_3 = \begin{pmatrix} 3i & \frac{2\sqrt{2}}{\sqrt{3}} & 0 \\ \frac{2\sqrt{2}}{\sqrt{3}} & 0 & \frac{1}{\sqrt{3}} \\ 0 & \frac{1}{\sqrt{3}} & 0 \end{pmatrix},$$

$$\mathcal{J}_4 = \begin{pmatrix} 4i & \sqrt{5} & 0 & 0 \\ \sqrt{5} & 0 & \frac{2}{\sqrt{5}} & 0 \\ 0 & \frac{2}{\sqrt{5}} & 0 & \frac{1}{\sqrt{5}} \\ 0 & 0 & \frac{1}{\sqrt{5}} & 0 \end{pmatrix}, \quad \mathcal{J}_5 = \begin{pmatrix} 5i & \frac{2\sqrt{2}}{\sqrt{5}} & 0 & 0 & 0 \\ \frac{2\sqrt{2}}{\sqrt{5}} & 0 & \frac{\sqrt{7}}{\sqrt{5}} & 0 & 0 \\ 0 & \frac{\sqrt{7}}{\sqrt{5}} & 0 & \frac{4}{\sqrt{35}} & 0 \\ 0 & 0 & \frac{4}{\sqrt{35}} & 0 & \frac{1}{\sqrt{7}} \\ 0 & 0 & 0 & \frac{1}{\sqrt{7}} & 0 \end{pmatrix}.$$

Note that the eigenvalues of the real part of $\mathcal{J}_n$ are solutions of the equations

$$\left(\frac{z+i}{z-i}\right)^n = -1,$$

and given by $\lambda_k = -\cot\left(\frac{\pi+2\pi k}{2n}\right), \quad k = 0, 1 \dots, n-1.$

11.4 Reconstruction of a dissipative Jacobi matrix from its triangular form

Let $z_1, z_2, \dots, z_n$ be (not necessarily distinct) complex numbers with positive imaginary parts. According to Theorem 11.3.1 there exists a unique dissipative $(n \times n)$ Jacobi matrix with one-dimensional imaginary part,

$$\mathcal{J} = \begin{pmatrix} b_1 & a_1 & 0 & 0 & \cdot & \cdot & \cdot \\ a_1 & b_2 & a_2 & 0 & \cdot & \cdot & \cdot \\ 0 & a_2 & b_3 & a_3 & \cdot & \cdot & \cdot \\ \cdot & \cdot & \cdot & \cdot & \cdot & \cdot & \cdot \\ \cdot & \cdot & \cdot & \cdot & \cdot & \cdot & a_{n-1} \\ \cdot & \cdot & \cdot & \cdot & 0 & a_{n-1} & b_n \end{pmatrix},$$

whose eigenvalues coincide with $\{z_k\}_{k=1}^n$ counting algebraic multiplicity. On the other hand, by Theorem 5.6.4, such a matrix is unitarily equivalent to the triangular matrix of the form

$$\vec{J} = \begin{pmatrix} z_1 & i\beta_1\beta_2 & \cdot & \cdot & \cdot & i\beta_1\beta_n \\ 0 & z_2 & \cdot & \cdot & \cdot & i\beta_2\beta_n \\ \cdot & \cdot & \cdot & \cdot & \cdot & \cdot \\ 0 & 0 & \cdot & \cdot & \cdot & z_n \end{pmatrix}, \tag{11.17}$$

where

$$z_k = \alpha_k + i\frac{\beta_k^2}{2}, \quad \operatorname{Im} z_k = \frac{\beta_k^2}{2}, \quad \beta_k > 0, \quad k = 1, \dots, n. \tag{11.18}$$

It follows from (11.17) that

$$2\,\mathrm{Im}\,\vec{J} = \begin{pmatrix} \beta_1^2 & \beta_1\beta_2 & \cdot & \cdot & \cdot & \beta_1\beta_n \\ \beta_1\beta_2 & \beta_2^2 & \cdot & \cdot & \cdot & \beta_2\beta_n \\ \cdot & \cdot & \cdot & \cdot & \cdot & \cdot \\ \beta_1\beta_n & \beta_2\beta_n & \cdot & \cdot & \cdot & \beta_n^2 \end{pmatrix},$$

and that $2\,\mathrm{Im}\,\vec{J}\vec{x} = (\vec{x},\vec{g})\,\vec{g}, \quad \vec{x} \in \mathbb{C}^n$, where

$$\vec{g} = \begin{pmatrix} \beta_1 \\ \beta_2 \\ \cdot \\ \cdot \\ \cdot \\ \beta_n \end{pmatrix}. \tag{11.19}$$

Let U be a unitary matrix

$$U = \begin{pmatrix} u_{11} & u_{12} & \cdot & \cdot & \cdot & u_{1n} \\ u_{21} & u_{22} & \cdot & \cdot & \cdot & u_{2n} \\ \cdot & \cdot & \cdot & \cdot & \cdot & \cdot \\ u_{n1} & u_{n2} & \cdot & \cdot & \cdot & u_{nn} \end{pmatrix},$$

such that

$$U\,\mathcal{J} = \vec{J}U, \quad Ug = \vec{g}, \tag{11.20}$$

where $g = \sqrt{2\mathrm{Im}\,b_1}\,\delta_1$. Next we present an algorithm which allows us to find the Jacobi matrix $\mathcal{J}$ and the unitary matrix U satisfying (11.20). It follows from (11.20) that

$$U\,\mathcal{J}^k g = (\vec{J})^k\vec{g}, \quad k = 1,2,\dots,n. \tag{11.21}$$

Since U is a unitary matrix, we have $\|g\|^2 = \|\vec{g}\|^2$, and taking into account (11.5) and (11.19) we obtain

$$\|g\|^2 = 2\mathrm{Im}\ b_1 = \beta_1^2 + \beta_2^2 + \ldots + \beta_n^2 = \|\vec{g}\|^2.$$

Thus

$$\mathrm{Im}\ b_1 = \sum_{k=1}^{n} \mathrm{Im}\ z_k. \tag{11.22}$$

Since $Ug = \vec{g}$, we have

$$u_{k1} = \frac{\sqrt{\mathrm{Im}\ z_k}}{\sqrt{\sum\limits_{j=1}^{n} \mathrm{Im}\ z_j}}, \quad k = 1,2,\dots,n. \tag{11.23}$$

Relations (11.20) and (11.21) yield

$$\left(\vec{J}\vec{g},\vec{g}\right) = (U\mathcal{J}g, Ug) = (\mathcal{J}g, g)\,.$$

Taking into account (11.1), (11.5), (11.17) and (11.19), we get

$$\mathcal{J}g = \begin{pmatrix} b_1\sqrt{2\,\mathrm{Im}\,b_1} \\ a_1\sqrt{2\,\mathrm{Im}\,b_1} \\ 0 \\ \cdot \\ \cdot \\ \cdot \\ 0 \end{pmatrix}, \quad \vec{J}\vec{g} = \begin{pmatrix} \left(z_1 + i\sum\limits_{k=2}^{n}\beta_k^2\right)\beta_1 \\ \left(z_2 + i\sum\limits_{k=3}^{n}\beta_k^2\right)\beta_2 \\ \cdot \\ \cdot \\ \cdot \\ z_n\beta_n \end{pmatrix}. \tag{11.24}$$

Therefore, $(\mathcal{J}g, g) = 2\,b_1\,\mathrm{Im}\,b_1$ and

$$\left(\vec{J}\vec{g},\vec{g}\right) = \sum_{j=1}^{n}\left(z_j + i\sum_{k=j+1}^{n}\beta_k^2\right)\beta_j^2.$$

Since $(\mathcal{J}g, g) = \left(\vec{J}\vec{g},\vec{g}\right)$, we get from (11.18) and (11.22) that

$$b_1 = \frac{\sum\limits_{j=1}^{n}\left(z_j + i\sum\limits_{k=j+1}^{n}\beta_k^2\right)\mathrm{Im}\,z_j}{\sum\limits_{k=1}^{n}\mathrm{Im}\,z_k}, \qquad \mathrm{Re}\,b_1 = \frac{\sum\limits_{k=1}^{n}\mathrm{Re}\,z_k\,\mathrm{Im}\,z_k}{\sum\limits_{k=1}^{n}\mathrm{Im}\,z_k}.$$

Since U is unitary and $U\mathcal{J}g = \vec{J}\vec{g}$, we get $\|\mathcal{J}g\|^2 = \left\|\vec{J}\vec{g}\right\|^2$. This equality and (11.24) yield

$$|b_1|^2\,(2\,\mathrm{Im}\,b_1) + a_1^2(2\,\mathrm{Im}\,b_1) = \sum_{j=1}^{n}\left|z_j + i\sum_{k=j+1}^{n}\beta_k^2\right|^2\beta_j^2,$$

and hence

$$a_1 = \left(\frac{\sum\limits_{j=1}^{n}\left|z_j + i\sum\limits_{k=j+1}^{n}\beta_k^2\right|^2\mathrm{Im}\,z_j}{\sum\limits_{k=1}^{n}\mathrm{Im}\,z_k} - \frac{\left|\sum\limits_{j=1}^{n}\left(z_j + i\sum\limits_{k=j+1}^{n}\beta_k^2\right)\mathrm{Im}\,z_j\right|^2}{\left(\sum\limits_{k=1}^{n}\mathrm{Im}\,z_k\right)^2}\right)^{1/2}.$$

Recall that by (11.4),

$$(\mathcal{J}^{m-1}g)_m = a_{m-1}a_{m-2}\cdots a_1\sqrt{2\,\mathrm{Im}\, b_1},$$

and

$$(\mathcal{J}^{m-1}g)_{m+1} = (\mathcal{J}^{m-1}g)_{m+2} = \cdots = (\mathcal{J}^{m-1}g)_n = 0,$$

for $m = 1, \ldots, n$. Let $m \geq 2$. Suppose that $(\mathcal{J}^{m-1}g)_k$ are already known, where $k = 1, 2 \ldots, m-1$. Then from (11.1) we obtain

$$\mathcal{J}^m g = \begin{pmatrix} b_1(\mathcal{J}^{m-1}g)_1 + a_1(\mathcal{J}^{m-1}g)_2 \\ a_1(\mathcal{J}^{m-1}g)_1 + b_2(\mathcal{J}^{m-1}g)_2 + a_2(\mathcal{J}^{m-1}g)_3 \\ a_2(\mathcal{J}^{m-1}g)_2 + b_3(\mathcal{J}^{m-1}g)_3 + a_3(\mathcal{J}^{m-1}g)_4 \\ \vdots \\ a_{m-1}(\mathcal{J}^{m-1}g)_{m-1} + b_m(\mathcal{J}^{m-1}g)_m \\ a_m(\mathcal{J}^{m-1}g)_m \\ 0 \\ \vdots \\ 0 \end{pmatrix}.$$

Consequently

$$\begin{aligned}(\mathcal{J}^m g, \mathcal{J}^{m-1}g) &= \sum_{j=1}^{n} (\mathcal{J}^m g)_j \,\overline{(\mathcal{J}^{m-1}g)_j} \\ &= \sum_{j=1}^{m-1} (\mathcal{J}^m g)_j \,\overline{(\mathcal{J}^{m-1}g)_j} + \left(a_{m-1}(\mathcal{J}^{m-1}g)_{m-1} + b_m(\mathcal{J}^{m-1}g)_m\right) \overline{(\mathcal{J}^{m-1}g)_m}.\end{aligned}$$

From

$$(\mathcal{J}^m g, \mathcal{J}^{m-1}g) = \left((\vec{J})^m \vec{g}, (\vec{J})^{m-1}\vec{g}\right),$$

and $(\mathcal{J}^{m-1}g)_m = a_{m-1}a_{m-2}\cdots a_1\sqrt{2\,\mathrm{Im}\, b_1}$ we get a linear equation with respect to b_m:

$$\begin{aligned}\sum_{j=1}^{m-1} (\mathcal{J}^m g)_j \,\overline{(\mathcal{J}^{m-1}g)_j} + \left(a_{m-1}(\mathcal{J}^{m-1}g)_{m-1} + b_m(\mathcal{J}^{m-1}g)_m\right) \overline{(\mathcal{J}^{m-1}g)_m} \\ = \left((\vec{J})^m \vec{g}, (\vec{J})^{m-1}\vec{g}\right).\end{aligned}$$

This equation can be solved since the coefficient $(\mathcal{J}^{m-1}g)_m$ is non-zero by (11.4). In order to find a_m we use the relation $\|\mathcal{J}^m g\|^2 = \left\|(\vec{J})^m \vec{g}\right\|^2$ which takes the form

$$\begin{aligned}\sum_{j=1}^{m-1} |(\mathcal{J}^m g)_j|^2 + \left|a_{m-1}(\mathcal{J}^{m-1}g)_{m-1} + b_m(\mathcal{J}^{m-1}g)_m\right|^2 \\ + a_m^2(\mathcal{J}^{m-1}g)_m^2 = \left\|(\vec{J})^m \vec{g}\right\|^2.\end{aligned}$$

Solving for a_m^2 we can find a_m.

According to (11.23), the elements $\{u_{k1}\}_{k=1}^n$ are expressed by means of $z_1, z_2, \dots, z_n$. The equality $U\mathcal{J}g = \vec{J}\vec{g}$ gives the following linear system with respect to the second column $\{u_{k2}\}_{k=1}^n$ of the matrix U:

$$\begin{cases} u_{11}(\mathcal{J}g)_1 + u_{12}(\mathcal{J}g)_2 = (\vec{J}\vec{g})_1, \\ u_{21}(\mathcal{J}g)_1 + u_{22}(\mathcal{J}g)_2 = (\vec{J}\vec{g})_2, \\ \dots \quad \dots \\ u_{n1}(\mathcal{J}g)_1 + u_{n2}(\mathcal{J}g)_2 = (\vec{J}\vec{g})_n. \end{cases}$$

Since $(\mathcal{J}g)_2 = a_1\sqrt{2\,\mathrm{Im}\, b_1} \neq 0$, one can find $\{u_{k2}\}_{k=1}^n$. By induction, the equality $U\mathcal{J}^{m-1}g = (\vec{J})^{m-1}\vec{g}$ enables us to find $\{u_{k\,m}\}_{k=1}^n$ for $m \leq n$. Finally, the relation $\mathcal{J} = U^{-1}\vec{J}U$, and the already known entries $\{a_k\}_{k=1}^{n-1}$ and $\{b_k\}_{k=1}^{n-1}$ allow us to find the entry b_n.

In conclusion, we provide formulas for the reconstruction of a (2×2) dissipative Jacobi matrix $\mathcal{J}$ with a rank-one imaginary part from its eigenvalues z_1, z_2, and the corresponding unitary matrix U:

$$\mathcal{J} = \begin{pmatrix} \frac{\mathrm{Re}\, z_1 \mathrm{Im}\, z_1 + \mathrm{Re}\, z_2 \mathrm{Im}\, z_2}{\mathrm{Im}\, z_1 + \mathrm{Im}\, z_2} + i(\mathrm{Im}\, z_1 + \mathrm{Im}\, z_2) & \sqrt{\omega\,\mathrm{Im}\, z_1\,\mathrm{Im}\, z_2} \\ \sqrt{\omega\,\mathrm{Im}\, z_1\,\mathrm{Im}\, z_2} & \frac{\mathrm{Re}\, z_1\,\mathrm{Im}\, z_2 + \mathrm{Re}\, z_2\,\mathrm{Im}\, z_1}{\mathrm{Im}\, z_1 + \mathrm{Im}\, z_2} \end{pmatrix},$$

where $\omega = \left(\left(\frac{\mathrm{Re}\, z_1 - \mathrm{Re}\, z_2}{\mathrm{Im}\, z_1 + \mathrm{Im}\, z_2}\right)^2 + 1\right)$ and

$$U = \begin{pmatrix} \sqrt{\frac{\mathrm{Im}\, z_1}{\mathrm{Im}\, z_1 + \mathrm{Im}\, z_2}} & \sqrt{\frac{\mathrm{Im}\, z_2}{\mathrm{Im}\, z_1 + \mathrm{Im}\, z_2}} \\ \sqrt{\frac{\mathrm{Im}\, z_2}{\mathrm{Im}\, z_1 + \mathrm{Im}\, z_2}} & -\sqrt{\frac{\mathrm{Im}\, z_1}{\mathrm{Im}\, z_1 + \mathrm{Im}\, z_2}} \end{pmatrix}.$$

The presented algorithm of reconstruction of the unique dissipative non-self-adjoint Jacobi matrix, with a rank-one imaginary part having given non-real numbers as its eigenvalues, allows us to recover the set of tri-diagonal matrices with the same non-real eigenvalues.

11.5 System Interpolation and Sectorial Operators

We begin this section with the following definition of interpolation systems.

Definition 11.5.1. Let Θ be the Livšic canonical system of the form (5.6) and $V_\Theta(z)$ be its impedance function defined by (5.28),(5.30). Then Θ is called an **interpolation system (solution)** in the **Nevanlinna-Pick interpolation problem** for the data $\{z_\ell \in \mathbb{C}_+,\ \ell = 1, \dots m\}$ and $\{v_\ell \in [E, E],\ \mathrm{Im}\, v_\ell \geq 0,\ \ell = 1, \dots m\}$ if

$$V_\Theta(z_\ell) = v_\ell, \quad \ell = 1, \dots m. \tag{11.25}$$

We will be interested in constructing, when possible, the Livšic canonical scattering system with given interpolation data whose state-space operator is α-sectorial. We will also focus on systems with the state-space $\mathcal{H}$ of minimal dimension. The following uniqueness result explains the importance of this last requirement.

Theorem 11.5.2. *Let $\{z_\ell,\ \ell = 1,\dots m\}$ and $\{v_\ell,\ \ell = 1,\dots m\}$ be the interpolation data (as in Definition 11.5.1). Let Θ_1 and Θ_2 be two minimal scattering interpolation systems for this data such that*

$$\Theta_1 = \begin{pmatrix} T_1 & K_1 & I \\ \mathcal{H}_1 & & E \end{pmatrix}, \qquad \Theta_2 = \begin{pmatrix} T_2 & K_2 & I \\ \mathcal{H}_2 & & E \end{pmatrix},$$

$\dim(E) < \infty$, $\mathcal{H}_1$ and $\mathcal{H}_2$ are finite dimensional (of dimensions n_1 and n_2 respectively), and $\ker(K_1) = \ker(K_2) = \{0\}$. Assume further that

$$\begin{aligned} &\text{c.l.s.}\{\operatorname{Ran}((T_1 - z_\ell I)^{-1}K_1), \quad \ell = 1,\dots,m\} = \mathcal{H}_1, \\ &\text{c.l.s.}\{\operatorname{Ran}((T_2 - z_\ell I)^{-1}K_2), \quad \ell = 1,\dots,m\} = \mathcal{H}_2, \end{aligned} \tag{11.26}$$

and that

$$|z_\ell| > \max\,(\|T_1\|, \|T_2\|), \quad \ell = 1,\dots m. \tag{11.27}$$

Then, the two systems are unitarily equivalent, $n_1 = n_2$, and

$$V_{\Theta_1}(z) = V_{\Theta_2}(z), \quad z \in \rho(T_1) \cap \rho(T_2).$$

Proof. Condition (11.27) forces the points z_ℓ to be in the resolvent sets of the operators T_1 and T_2. Therefore the interpolation conditions $V_{\Theta_1}(z_\ell) = V_{\Theta_2}(z_\ell)$ for $\ell = 1,\dots m$ can be rewritten as

$$K_1^*(T_1 - z_\ell I)^{-1}K_1 = K_2^*(T_2 - z_\ell I)^{-1}K_2, \quad \ell = 1,\dots m, \tag{11.28}$$

that is, $W_{\Theta_1}(z_\ell) = W_{\Theta_2}(z_\ell)$, $\ell = 1,\dots m$. Moreover, we claim that

$$((T_1 - z_\ell I)^{-1}K_1\phi, (T_1 - z_j I)^{-1}K_1\phi)_{\mathcal{H}_1} = ((T_2 - z_\ell I)^{-1}K_2\phi, (T_2 - z_j I)^{-1}K_2\phi)_{\mathcal{H}_2} \tag{11.29}$$

for any $\phi \in E$ and all $\ell, j = 1,\dots m$. Indeed, taking into account that

$$\operatorname{Im} T_1 = K_1K_1^*, \qquad \operatorname{Im} T_2 = K_2K_2^*,$$

we have that

$$\begin{aligned} &(T_1 - z_\ell I)^{-1} - (T_1^* - \bar{z}_j I)^{-1} \\ &= (T_1^* - \bar{z}_j I)^{-1}\left((T_1^* - \bar{z}_j I) - (T_1 - z_\ell I)\right)(T_1 - z_\ell)^{-1} \\ &= -2i(T_1^* - \bar{z}_j)^{-1}K_1K_1^*(T_1 - z_\ell I)^{-1} + (z_\ell - \bar{z}_j)(T_1^* - \bar{z}_j)^{-1}(T_1 - z_\ell I)^{-1}, \end{aligned} \tag{11.30}$$

and the same relation holds for T_2. Equations (11.28) and (11.30) lead to

$$K_1^*(T_1^* - \bar{z}_j I)^{-1}(T_1 - z_\ell)^{-1}K_1 = K_2^*(T_2^* - \bar{z}_j I)^{-1}(T_2 - z_\ell)^{-1}K_2, \tag{11.31}$$

and hence to (11.29). Due to (11.31), we can define an operator U via

$$U(T_1 - z_\ell I)^{-1}K_1\phi = (T_2 - z_\ell I)^{-1}K_2\phi, \quad (\phi \in E, \quad \ell = 1, \dots, m). \tag{11.32}$$

This operator is isometric and in fact unitary, because of the range conditions (11.26). Rewriting (11.28) as

$$K_1^*U^*U(T_1 - z_j I)^{-1}K_1\phi = K_2^*(T_2 - z_j I)^{-1}K_2\phi,$$

and taking into account the definition of U, we obtain that

$$K_1^*U^*U(T_1 - z_j I)^{-1}K_1\phi = K_2^*(T_2 - z_j I)^{-1}K_2\phi.$$

The range conditions (11.26) then leads to $K_1^*U^* = K_2^*$. We now show that $UT_1 = T_2U$. Condition (11.27) allows us to write the power expansions

$$(T_1 - z_j I)^{-1} = -\frac{I}{z_j} - \frac{T_1}{z_j^2} - \frac{T_1^2}{z_j^3} - \cdots,$$

and

$$(T_2 - z_j I)^{-1} = -\frac{I}{z_j} - \frac{T_2}{z_j^2} - \frac{T_2^2}{z_j^3} - \cdots,$$

for $j = 1, \dots m$. The definition (11.32) of U gives us

$$U\left(-\frac{I}{z_j} - \frac{T_1}{z_j^2} - \frac{T_1^2}{z_j^3} - \cdots\right)K_1 = \left(-\frac{I}{z_j} - \frac{T_2}{z_j^2} - \frac{T_2^2}{z_j^3} - \cdots\right)K_2,$$

and therefore

$$-\frac{UK_1}{z_j} + U\left(-\frac{T_1}{z_j^2} - \frac{T_1^2}{z_j^3} - \cdots\right)K_1 = -\frac{K_2}{z_j} + \left(-\frac{T_2}{z_j^2} - \frac{T_2^2}{z_j^3} - \cdots\right)K_2.$$

Since $K_2 = UK_1$, we obtain

$$U\frac{T_1}{z_j}\left(-\frac{I}{z_j} - \frac{T_1}{z_j^2} - \frac{T_1^2}{z_j^3} - \cdots\right)K_1 = \frac{T_2}{z_j}\left(-\frac{I}{z_j} - \frac{T_2}{z_j^2} - \frac{T_2^2}{z_j^3} - \cdots\right)K_2,$$

so that $UT_1(T_1 - z_j I)^{-1}K_1 = T_2(T_2 - z_j I)^{-1}K_2$, for $j = 1, \cdots m$. This last equation can be rewritten as

$$UT_1U^{-1}U(T_1 - z_j I)^{-1}K_1 = T_2(T_2 - z_j I)^{-1}K_2, \quad j = 1, \cdots m.$$

The range conditions (11.26) and the definition of U lead then to

$$UT_1U^{-1} = T_2,$$

which concludes the proof. □

Now we focus on bounded α-sectorial operators whose definition was given in Section 9.5. It follows from (9.104) that when the bounded operator T is α-sectorial, the exact value of the angle α can be computed via the formula

$$\tan\alpha = \sup_{\xi\in\mathrm{Dom}(T)} \frac{|\mathrm{Im}\,(T\xi,\xi)|}{\mathrm{Re}\,(T\xi,\xi)}. \tag{11.33}$$

When the state space is finite dimensional, the angle α is given by another formula:

Theorem 11.5.3. *Let*

$$\Theta = \begin{pmatrix} T & K & I \\ \mathcal{H} & & E \end{pmatrix},$$

be the Livšic canonical system with accretive state-space operator T (with both $\mathcal{H}$ and E finite dimensional) and invertible channel operator K. Then T is α-sectorial if and only if the limit

$$\|V_\Theta(-0)\| = \lim_{x\in\mathbb{R}^-,x\to 0} \|V_\Theta(x)\|$$

is finite. In this case, the angle α is given by the formula

$$\tan\alpha = \|V_\Theta(-0)\|.$$

Proof. Assume that the operator T is α–sectorial. It follows from (11.33) that

$$\begin{aligned}
\tan\alpha &= \sup_{\xi\in\mathrm{Ran}(\mathrm{Re}\,T)} \frac{((\mathrm{Im}\,T)\xi,\xi)}{((\mathrm{Re}\,T)\xi,\xi)} \\
&= \sup_{f\in\mathrm{Ran}(\mathrm{Re}\,T)} \frac{((\mathrm{Im}\,T)(\mathrm{Re}\,T)^{-1/2}f,(\mathrm{Re}\,T)^{-1/2}f)}{||f||^2} \\
&= \sup_{f\in\mathrm{Ran}(\mathrm{Re}\,T)} \frac{||K^*(\mathrm{Re}\,T)^{-1/2}f||^2}{||f||^2} \\
&= ||K^*(\mathrm{Re}\,T)^{-1/2}||^2 = ||(\mathrm{Re}\,T)^{-1/2}K||^2.
\end{aligned}$$

On the other hand

$$(V_\Theta(-0)g,g)_E = \lim_{x\uparrow 0}((\mathrm{Re}\,T - xI)^{-1}Kg,Kg) = ||(\mathrm{Re}\,T)^{-1/2}Kg||^2, \quad g\in E.$$

Hence, $||V_\Theta(-0)|| = ||(\mathrm{Re}\,T)^{-1/2}K||^2 = \tan\alpha$.

If $||V_\Theta(-0)|| < \infty$, then $\mathrm{Ran}(\mathrm{Im}\,T) = \mathrm{Ran}(K) \subseteq \mathrm{Ran}(\mathrm{Re}\,T)$. Therefore T is an α-sectorial operator and, as before, $\tan\alpha = ||V_\Theta(-0)||$. □

Obviously, the operator T is accretive but not α-sectorial for any $\alpha\in(0,\pi/2)$ if and only if

$$\lim_{x\to 0^-} ||V_\Theta(x)|| = \infty.$$

Theorem 11.5.4. *Let*

$$\Theta = \begin{pmatrix} T & K & I \\ \mathcal{H} & & E \end{pmatrix},$$

be the Livšic canonical system with a bounded state-space operator T. If the operator T is α-sectorial, then for every set of non-real points $z_1, \dots, z_p \in \mathbb{C}_+$ and every set of vectors $h_1, \dots, h_p \in E$, the following inequalities are valid:

$$\sum_{k,\ell=1}^{p} \left(\frac{V_\theta(z_k) - V_\theta^*(z_\ell)}{z_k - \bar{z}_\ell} h_k, h_\ell \right)_E \geq 0, \tag{11.34}$$

$$\sum_{k,\ell=1}^{p} \left(\frac{z_k V_\theta(z_k) - \bar{z}_\ell V_\theta^*(z_\ell)}{z_k - \bar{z}_\ell} h_k, h_\ell \right)_E \geq (\cot \alpha) \sum_{k,\ell=1}^{p} (V_\theta^*(z_\ell) V_\theta(z_k) h_k, h_\ell)_E. \tag{11.35}$$

Proof. Writing $V_\theta(z) = K^*(\mathrm{Re}\, T - zI)^{-1} K$ we obtain

$$\begin{aligned} \frac{V_\theta(z_k) - V_\theta^*(z_\ell)}{z_k - \bar{z}_\ell} &= \frac{1}{z_k - \bar{z}_\ell} \left(K^*(\mathrm{Re}\, T - z_k I)^{-1} K - K^*(\mathrm{Re}\, T - \bar{z}_\ell I)^{-1} K \right) \\ &= K^*((\mathrm{Re}\, T - z_\ell I)^{-1})^* (\mathrm{Re}\, T - z_k I)^{-1} K. \end{aligned}$$

We set

$$\xi_k = (\mathrm{Re}\, T - z_k I)^{-1} K h_k, \quad (k = 1, 2, \dots, n), \tag{11.36}$$

and

$$\xi = \sum_{k=1}^{n} \xi_k. \tag{11.37}$$

Then

$$\begin{aligned} &\sum_{k,\ell=1}^{n} \left(\frac{V_\theta(z_k) - V_\theta^*(z_\ell)}{z_k - \bar{z}_\ell} h_k, h_\ell \right)_E \\ &= \sum_{k,\ell=1}^{n} (K^*(\mathrm{Re}\, T - \bar{z}_\ell I)^{-1} (\mathrm{Re}\, T - z_k I)^{-1} K h_k, h_\ell)_E \\ &= \sum_{k,\ell=1}^{n} (\xi_k, \xi_\ell)_{\mathcal{H}} = (\xi, \xi)_{\mathcal{H}} \geq 0. \end{aligned}$$

Since the operator T is α–sectorial, then

$$\cot \alpha \cdot |(\mathrm{Im}\, T\, \xi, \xi)| \leq (\mathrm{Re}\, T\, \xi, \xi).$$

It follows from (11.36) and (11.37) that

$$\sum_{k,\ell=1}^{n} (\mathrm{Re}\, T \xi_k, \xi_\ell) \geq \cot \alpha \cdot \sum_{k,\ell} (\mathrm{Im}\, T \xi_k, \xi_\ell).$$

Since $\operatorname{Im} T = KK^*$, we get

$$\sum_{k,\ell=1} (\operatorname{Re} T(\operatorname{Re} T - z_k I)^{-1} K h_k, (\operatorname{Re} T - z_\ell I)^{-1} K h_\ell)$$
$$\geq \cot\alpha \cdot \sum_{k,\ell} (KK^*(\operatorname{Re} T - z_k I)^{-1} K h_k, (\operatorname{Re} T - z_\ell I)^{-1} K h_\ell),$$

which leads to

$$\sum_{k,\ell=1} (K^*(\operatorname{Re} T - \bar{z}_\ell I)^{-1} \operatorname{Re} T(\operatorname{Re} T - z_k I)^{-1} K h_k, h_\ell)_E$$
$$\geq \cot\alpha \cdot \sum_{k,\ell=1} (K^*(\operatorname{Re} T - \bar{z}_\ell I)^{-1} KK^*(\operatorname{Re} T - z_k I)^{-1} K h_k, h_\ell)_E.$$

Since

$$K^*(\operatorname{Re} T - \bar{z}_\ell I)^{-1} \operatorname{Re} T(\operatorname{Re} T - z_k I)^{-1} K$$
$$= K^* \frac{\bar{z}_\ell(\operatorname{Re} T - \bar{z}_\ell I)^{-1} - z_k K^*(\operatorname{Re} T - z_k I)^{-1}}{\bar{z}_\ell - z_k} K,$$

we obtain (11.35). □

11.6 The Livšic interpolation systems in the Pick form

In this section we consider the Nevanlinna-Pick interpolation problem in the class **N** of scalar Herglotz-Nevanlinna functions. We set

$$\mathbb{P} = \left(\frac{v_k - \bar{v}_j}{z_k - \bar{z}_j} \right)_{k,j=1,\dots m}, \qquad \mathbb{Q} = \left(\frac{z_k v_k - \bar{z}_j \bar{v}_j}{z_k - \bar{z}_j} \right)_{k,j=1,\dots m}. \tag{11.38}$$

Theorem 11.6.1. *Let $\{z_k,\ k = 1,\dots,m\} \in \mathbb{C}_+$ and $\{v_k,\ k = 1,\dots,m\} \in \mathbb{C}_+$ be system interpolation data for which the matrix $\mathbb{P}$ in (11.38) is strictly positive.*[2] *Then there exists a scattering interpolation system Θ with the state space of dimension m that is a solution of the interpolation problem for this data. The state-space operator of Θ is α-sectorial if and only if*

$$\left(\frac{z_k v_k - \bar{z}_j \bar{v}_j}{z_k - \bar{z}_j} \right)_{k,j=1,\dots m} \geq \cot\alpha \ (v_k \bar{v}_j)_{k,j=1,\dots m}. \tag{11.39}$$

Proof. Consider the space $\mathbb{C}^m$ with the inner product

$$(\xi, \eta)_{\mathbb{C}^m_{\mathbb{P}}} = \eta^* \mathbb{P} \xi.$$

[2] A matrix is called strictly positive if it is non-negative and invertible.

It is easily seen that the operator $\vec{A} = \mathbb{P}^{-1}\mathbb{Q}$ is self–adjoint with respect to this inner product. Set

$$\vec{g} = \mathbb{P}^{-1}\varphi, \qquad \varphi = \begin{pmatrix} \bar{v}_1 \\ \vdots \\ \bar{v}_m \end{pmatrix}. \tag{11.40}$$

The operators

$$\begin{aligned} \vec{T} &= \vec{A} + i(\cdot\,, \vec{g})_{\mathbb{C}^m_{\mathbb{P}}} \vec{g}, \quad \vec{A} = \mathbb{P}^{-1}\mathbb{Q}, & (11.41) \\ \vec{K}c &= c \cdot \vec{g} & (11.42) \end{aligned}$$

satisfy $\operatorname{Im} \vec{T} = \vec{K}\vec{K}^*$ and therefore define the Livšic canonical scattering system

$$\vec{\Theta} = \begin{pmatrix} \vec{T} & \vec{K} & 1 \\ \mathbb{C}^m_{\mathbb{P}} & & \mathbb{C} \end{pmatrix}.$$

Let

$$V_{\vec{\Theta}}(z) = K^*(\vec{A} - zI)^{-1}K = \vec{K}^*(\operatorname{Re} \vec{T} - zI)^{-1}\vec{K}. \tag{11.43}$$

We are going to check that $V_{\vec{\Theta}}(z_k) = v_k, \quad k = 1, \dots m$. We have

$$(\mathbb{Q} - z_k\mathbb{P})_{j,k} = \frac{z_k v_k - \bar{z}_j \bar{v}_j}{z_k - \bar{z}_j} - z_k \frac{v_k - \bar{v}_j}{z_k - \bar{z}_j} = \bar{v}_j. \tag{11.44}$$

It follows from (11.44) that the k-th column of the matrix $\mathbb{Q} - z_k\mathbb{P}$ is equal to φ defined in (11.40). Hence, $(\vec{A} - z_k I_m)e_k = g$ (where e_k denotes the $(m \times 1)$ column vector whose all entries are equal to 0, besides the k–th one, which is equal to 1). Therefore,

$$\begin{aligned} V_{\vec{\Theta}}(z_k) &= ((\vec{A} - z_k I_m)^{-1} g, g)_{\mathbb{C}^m_{\mathbb{P}}} = (e_k, g)_{\mathbb{C}^m_{\mathbb{P}}} = (\mathbb{P}e_k, \mathbb{P}^{-1}g)_{\mathbb{C}^m_{\mathbb{P}}} \\ &= (\mathbb{P}e_k, \mathbb{P}^{-1}g)_{\mathbb{C}^m} = (e_k, g)_{\mathbb{C}^m} = v_k. \end{aligned}$$

Hence we proved that the Livšic scattering system of the form (11.41)-(11.42) is the interpolation system for the Nevanlinna-Pick interpolation problem. If the operator $\vec{T}$ is α-sectorial, it follows from Theorem 11.5.4 that (11.39) holds.

Conversely, let us suppose that (11.39) holds. We prove that the operator $\vec{T}$ is α-sectorial. Condition (11.39) implies that for an arbitrary vector $\xi = (\xi_j) \in \mathbb{C}^m$, we have the inequality

$$\cot \alpha \sum_{k,j=1}^{m} v_k \bar{v}_j \xi_k \xi_j^* \le \sum_{k,j=1}^{m} \frac{z_k v_k - \bar{z}_j \bar{v}_j}{z_k - \bar{z}_j} \xi_k \xi_j^*,$$

i.e.,

$$\cot \alpha \left(\sum_1^m v_k \xi_k \right) \left(\sum_1^m v_j \xi_j \right)^* \le (\mathbb{Q}\xi, \xi)_{\mathbb{C}^m}.$$

This inequality can be rewritten as

$$\cot\alpha \left(\sum_1^m v_k\xi_k\right)\left(\sum_1^m v_j\xi_j\right)^* \le (\vec{A}\xi,\xi)_{\mathbb{C}^m_{\mathbb{P}}}. \tag{11.45}$$

Besides, we have

$$(\xi,\vec{g})_{\mathbb{C}^m_{\mathbb{Q}}} = (\mathbb{P}\xi,\mathbb{P}^{-1}\varphi)_{\mathbb{C}^m} = (\xi,\varphi)_{\mathbb{C}^m}. \tag{11.46}$$

It follows from (11.45) and (11.46) that $\cot\alpha\ |(\xi,\vec{g})_{\mathbb{C}^m_{\mathbb{P}}}|^2 \le (\vec{A}\xi,\xi)_{\mathbb{C}^m_{\mathbb{P}}}$, or

$$\cot\alpha\ |\mathrm{Im}\ (\vec{T}\xi,\xi)_{\mathbb{C}^m_{\mathbb{P}}}| \le \mathrm{Re}\ (\vec{T}\xi,\xi)_{\mathbb{C}^m_{\mathbb{P}}}.$$

Thus the operator $\vec{T}$ is α-sectorial. □

The system

$$\vec{\Theta} = \begin{pmatrix} \vec{T} & \vec{K} & 1 \\ \mathbb{C}^m_{\mathbb{P}} & & \mathbb{C} \end{pmatrix} \tag{11.47}$$

with operators $\vec{T}$ and $\vec{K}$ defined by (11.41) and (11.42) constructed in the proof of Theorem 11.6.1, we call the **Livšic system in the Pick form**.

Theorem 11.6.2. *Let $z_1,\dots,z_m \in \mathbb{C}_+$ and $v_1,\dots,v_m \in \mathbb{C}_+$ be the interpolation data. If the Livšic canonical scattering system*

$$\Theta = \begin{pmatrix} T & K & 1 \\ \mathcal{H} & & \mathbb{C} \end{pmatrix}, \qquad \mathrm{Im}\ T = KK^*$$

is the interpolation system for the Nevanlinna-Pick interpolation problem with this data and $\dim\mathcal{H} = m$, z_k are not eigenvalues of the state-space operator T, and

$$\mathrm{c.l.s.}\{(T - z_kI)^{-1}K\mathbb{C},\quad k = 1,\dots,m\} = \mathcal{H}, \tag{11.48}$$

then the matrix $\mathbb{P}$ is strictly positive. The given system is unitarily equivalent to the system $\vec{\Theta}$ of the form (11.47) *and*

$$\mathrm{c.l.s.}\{\mathrm{Ran}((\vec{T} - z_kI)^{-1}\vec{K}),\quad k = 1,\dots,m\} = \mathbb{C}^m_{\mathbb{P}}.$$

Proof. Set $V_\Theta(z) = K^*(\mathrm{ReT} - zI)^{-1}K$. Then,

$$\begin{aligned} \frac{V_\Theta(z_k) - V_\Theta^*(z_j)}{z_k - \bar{z}_j} &= K^*(\mathrm{ReT} - \bar{z}_jI)^{-1}(\mathrm{ReT} - z_kI)^{-1}K \\ &= K^*(\mathrm{ReT} - \bar{z}_jI)^{-1}KK^{-1}(K^{-1})^*K^*(\mathrm{ReT} - z_kI)^{-1}K \\ &= x^2V_\Theta(z_j)^*V_\Theta(z_k), \end{aligned}$$

where we have set $K^{-1}(K^{-1})^* = x^2I$, (K^{-1} denotes the inverse of K on its range). Since the function $V_\Theta(z)$ solves the interpolation problem associated to the interpolation data, we have

$$\frac{v_k - \bar{v}_j}{z_k - \bar{z}_j} = x^2\bar{v}_jv_k. \tag{11.49}$$

Now we are going to show that the matrix $\mathbb{P}$ is nonsingular. Let $\xi = (\xi_j) \in \mathbb{C}^m$ be in the kernel of $\mathbb{P}$. Then,

$$\sum_{k=1}^{m} \frac{v_k - \bar{v}_j}{z_k - \bar{z}_j}\xi_k = 0, \qquad j = 1, \dots m.$$

Equation (11.49) then implies that

$$\sum_{k=1}^{m} x^2 \bar{v}_j v_k \xi_k = 0, \qquad j = 1, \dots m,$$

and hence $\sum_{k=1}^{m} v_k \xi_k = 0$, i.e., $\sum_{k=1}^{m} K^*(\mathrm{Re}\, T - z_k I)^{-1} K \xi_k = 0$. Thus, we have

$$\sum_{k=1}^{m} (\mathrm{Re}\, T - z_k I)^{-1} K \xi_k = 0. \tag{11.50}$$

The operator T has a one-dimensional imaginary part. Therefore T can be represented in the form

$$T = \mathrm{Re}\, T + i\,(\,\cdot\,, g)\; g. \tag{11.51}$$

It follows from (11.48) that the vectors

$$x_k = 2(T - z_k I)^{-1} g, \qquad k = 1, \dots m,$$

are linearly independent. Hence (11.51) yields that

$$x_k = (2 - i(x_k, g))(\mathrm{Re}\, T - z_k I)^{-1} g.$$

Set

$$\omega(z) = 1 - 2i((T - zI)^{-1} g, g).$$

Then we have

$$x_k = (1 + \omega(z_k))(\mathrm{Re}\, T - z_k I)^{-1} g.$$

Since the z_k are not the eigenvalues of T, we have $|\omega(z_k)| > 1$. We may assume that $K(1) = g$ and therefore (11.50) implies that

$$\sum_{k=1}^{m} \frac{\xi_k}{1 + \omega(z_k)} x_k = 0.$$

Since the vectors x_k are linearly independent, we obtain that $\frac{\xi_k}{1+\omega(z_k)} = 0$, and therefore $\xi_k = 0$, that is, $\mathbb{P}$ is nonsingular. Therefore we can consider the Livšic system $\vec{\Theta}$ in the Pick form (11.47). Since $\vec{\Theta}$ is also an interpolation system we have (11.25). Thus

$$V_{\vec{\Theta}}(z_k) = V_\Theta(z_k) = v_k, \qquad k = 1, \dots m$$

and $K^*(\mathrm{Re}\, T - z_k I)^{-1} K = \vec{K}^*(\mathrm{Re}\, \vec{T} - z_k I)^{-1} \vec{K}$ for $k = 1, \ldots m$. Hence we obtain that

$$W_\Theta(z_k) = W_{\vec{\Theta}}(z_k), \qquad k = 1, \ldots m,$$

that is

$$K^*(T - z_k I)^{-1} K = \vec{K}^*(\vec{T} - z_k I)^{-1} \vec{K}, \qquad k = 1, \ldots m.$$

The operator U defined by

$$U(T - z_k I)^{-1} K = (\vec{T} - z_k I)^{-1} \vec{K} \tag{11.52}$$

is isometric from $\mathcal{H}$ onto $\mathbb{C}_{\mathbb{P}^m}$ and satisfies $\vec{K} = UK$. We leave it to the reader to check that (11.52) implies that $UT = \vec{T}U$ and therefore Θ and $\vec{\Theta}$ are unitarily equivalent. Hence $V_\Theta(z) = V_{\vec{\Theta}}(z)$ and the theorem is proved. □

Theorem 11.6.3. *Let z_k and v_k be the interpolation data with z_k, $v_k \in \mathbb{C}_+$, $k = 1, \ldots, m$, for which the matrix $\mathbb{P}$ of the form* (11.38) *is strictly positive. Let also $\vec{\Theta}$ be the Livšic system in the Pick form* (11.47) *corresponding to this data. Assume that z_{k_j}, $j = 1, \ldots, p$ are eigenvalues of $\vec{T}$. Then,*

$$v_{k_1} = \cdots = v_{k_p} = i. \tag{11.53}$$

Proof. By Theorem 11.6.1 there exists the Livšic system $\vec{\Theta}$ in the Pick form (11.47) that is the interpolation system for the data. Thus,

$$V_{\vec{\Theta}}(z_k) = v_k, \qquad k = 1, \ldots m,$$

and $V_{\vec{\Theta}}(z_k) = \vec{K}^*(\mathrm{Re}\, \vec{T} - z_k I)^{-1} \vec{K}$. Recall that

$$W_{\vec{\Theta}}(z) = i \frac{V_{\vec{\Theta}}(z) - 1}{V_{\vec{\Theta}}(z) + 1}. \tag{11.54}$$

It is proved in [91] that the spectrum of the operator $\vec{T}$ coincides with the set of singular points of $W_{\vec{\Theta}}(z)$. Therefore the z_{k_i} are poles of $W_{\vec{\Theta}}(z)$ and $\lim_{z \to z_{k_j}} W_{\vec{\Theta}}(z) = \infty$. Hence

$$v_{k_j} = \lim_{z \to z_{k_j}} V_{\vec{\Theta}}(z) = i,$$

due to (11.54). □

Below we have the converse statement.

Theorem 11.6.4. *Let $z_1, \ldots, z_m$ and $v_1, \ldots, v_m$ be the system interpolation data with z_k, $v_k \in \mathbb{C}_+$, $k = 1, \ldots, m$. Suppose that the matrix $\mathbb{P}$ of the form* (11.38) *is strictly positive and* (11.53) *holds. Let also $\vec{\Theta}$ be the Livšic system in the Pick form* (11.47) *for this data. Then $z_{k_1}, \ldots, z_{k_p}$ are eigenvalues of the state-space operator $\vec{T}$ defined by* (11.41).

Proof. If the matrix $\mathbb{P}$ is strictly positive, then Theorem 11.6.1 implies that the interpolation Nevanlinna-Pick problem is solvable with $\vec{\Theta}$, the Livšic system in the Pick form (11.47). Thus, $V_{\vec{\Theta}}(z_{k_j}) = i$ for $j = 1, \dots p$. Since

$$W_{\vec{\Theta}}(z) = \frac{1 - iV_{\vec{\Theta}}(z)}{1 + iV_{\vec{\Theta}(z)}},$$

we get

$$\lim_{z \to z_{k_j}} W_{\vec{\Theta}}(z) = \lim_{z \to z_{k_j}} \frac{1 - iV_{\vec{\Theta}}(z)}{1 + iV_{\vec{\Theta}}(z)} = \infty, \qquad j = 1, \dots p.$$

This means that the z_{k_j} are the poles of $W_{\vec{\Theta}}(z)$. Therefore, they are eigenvalues of $\vec{T}$. □

Theorem 11.6.5. *Let z_k, $k = 1, \dots m$ and $v_k = i$, $k = 1, \dots m$ be the interpolation data with z_k, $v_k \in \mathbb{C}_+$. Then there exists the Livšic scattering system solution of the corresponding system interpolation problem with state-space of dimension m. This system solution is unique up to a unitary equivalence.*

Proof. Set $W(z) = \prod_{k=1}^{m} \frac{z - z_k^*}{z - z_k}$ and

$$T_\Delta = \begin{pmatrix} z_1 & i\sqrt{2\mathrm{Im}\ z_1}\sqrt{2\mathrm{Im}\ z_2} & \cdots & i\sqrt{2\mathrm{Im}\ z_1}\sqrt{2\mathrm{Im}\ z_m} \\ 0 & z_2 & \cdots & i\sqrt{2\mathrm{Im}\ z_2}\sqrt{2\mathrm{Im}\ z_m} \\ \cdot & \cdot & \cdot & \cdot \\ 0 & 0 & \cdots & z_m \end{pmatrix}. \tag{11.55}$$

Then, $\mathrm{Im}\, T_\Delta = (\cdot, g_\Delta) g_\Delta$ with

$$g_\Delta = \begin{pmatrix} \sqrt{\mathrm{Im}\ z_1} \\ \sqrt{\mathrm{Im}\ z_2} \\ \vdots \\ \sqrt{\mathrm{Im}\ z_m} \end{pmatrix}.$$

Consider the Livšic system

$$\Theta_\Delta = \begin{pmatrix} T_\Delta & K_\Delta & 1 \\ \mathbb{C}^m & & \mathbb{C} \end{pmatrix}, \tag{11.56}$$

where $K_\Delta(c) = c \cdot g_\Delta$, $c \in \mathbb{C}$. The transfer function of the system Θ_Δ is

$$W_{\Theta_\Delta}(z) = 1 - 2i((T_\Delta - zI)^{-1} g_\Delta, g_\Delta) = \prod_{k=1}^{m} \frac{z - \bar{z}_k}{z - z_k}. \tag{11.57}$$

The impedance function

$$V_{\Theta_\Delta}(z) = i \frac{W_{\Theta_\Delta}(z) - 1}{W_{\Theta_\Delta}(z) + 1}, \tag{11.58}$$

clearly satisfies the interpolation conditions

$$V_{\Theta_\Delta}(z_k) = i, \qquad k = 1, \dots m.$$

Therefore the Livšic scattering system (11.56) is the interpolation system for the stated Nevanlinna-Pick interpolation problem.

Let the system

$$\Theta = \begin{pmatrix} T & K & 1 \\ \mathcal{H} & & \mathbb{C} \end{pmatrix}, \qquad \dim \mathcal{H} = m, \qquad \operatorname{Im} T = KK^*,$$

be the interpolation system in the Nevanlinna-Pick interpolation problem for the data $z_k, k = 1, \dots m$ and $v_k = i, k = 1, \dots m$. Without loss of generality, we can assume that $\operatorname{Im} T = (\cdot, g)g$ and $K\,1 = g$. Then, $K^*x = (x, g)$. Define $V_\Theta(z)$ and $W_\Theta(z)$ as above, with this choice of T and K. Obviously (since Θ is the interpolation system), $V_\Theta(z_k) = i$ for $k = 1, \dots m$ and

$$\lim_{z \to z_{k_j}} \frac{1 - iV_\Theta(z)}{1 + iV_\Theta(z)} = \infty, \qquad j = 1, \dots m.$$

Therefore, each point z_k is a pole of $W_\Theta(z)$. Since $\dim \mathcal{H} = m$, then according to (5.46) (see also Section 5.6), z_k are exactly the points of the spectrum of T and

$$W_\Theta(z) = \prod_{k=1}^{m} \frac{z - \bar{z}_k}{z - z_k} = W_{\Theta_\Delta}(z).$$

By Theorem 5.4.3 on unitary equivalence of systems, the systems Θ and Θ_Δ are unitary equivalent and therefore $V_\Theta(z) = V_{\Theta_\Delta}(z)$. □

Corollary 11.6.6. *Let z_k, $k = 1, \dots, m$, be arbitrary points in the open upper half–plane. There exists a unique matrix U such that $U^*\mathbb{P}U = I_m$,*

$$U \left(\frac{2i}{z_k - \bar{z}_j} \right)^{-1} \left(i \left(\frac{z_k + \bar{z}_j}{z_k - \bar{z}_j} \right) + i\Pi \right) = T_\Delta U,$$

and

$$U \left(\frac{2i}{z_k - \bar{z}_j} \right)^{-1} \begin{pmatrix} -i \\ -i \\ \vdots \\ -i \end{pmatrix} = \begin{pmatrix} \sqrt{\operatorname{Im} z_1} \\ \vdots \\ \sqrt{\operatorname{Im} z_m} \end{pmatrix}, \tag{11.59}$$

where

$$\Pi = \begin{pmatrix} 1 & 1 & \cdots & 1 \\ 1 & 1 & \cdots & 1 \\ \vdots & & & \vdots \\ 1 & 1 & \cdots & 1 \end{pmatrix},$$

and T_Δ is of the form (11.55).

Proof. The matrix $\mathbb{P} = \left(\frac{2i}{z_k - \bar{z}_j}\right)$ is strictly positive and the function $V_{\Theta_\Delta}(z)$ defined by (11.58) satisfies $V_{\Theta_\Delta}(z_k) = i$ for $k = 1, \dots, m$. It has been shown that the system $\vec{\Theta}$ of the form (11.47) also generates a solution of this Nevanlinna-Pick interpolation problem, of the form

$$V_{\vec{\Theta}}(z) = ((\vec{A} - zI)^{-1}\vec{g}, \vec{g})_{\mathbb{C}^m_{\mathbb{P}}},$$

where

$$\vec{A} = \mathrm{Re}\,\vec{T} = \mathbb{P}^{-1}\left(\frac{z_k + \bar{z}_j}{z_k - \bar{z}_j}\right), \qquad \vec{g} = \mathbb{P}^{-1}\begin{pmatrix} -i \\ \vdots \\ -i \end{pmatrix}.$$

Since for $c \in \mathbb{C}$

$$\vec{K}^*(\mathrm{Re}\,\vec{T} - zI)^{-1}\vec{K}c = V_{\vec{\Theta}}(z) \cdot c,$$

we get that $V_{\vec{\Theta}}(z_k) = i$ for $k = 1, \dots, m$ and $W_{\Theta_\Delta}(z) = W_{\vec{\Theta}}(z)$, where $W_{\Theta_\Delta}(z)$ has the form (11.57) and $W_{\vec{\Theta}}(z)$ has the form

$$W_{\vec{\Theta}}(z) = I - 2i\vec{K}^*(\vec{T} - zI)^{-1}\vec{K}.$$

There exits thus a unique unitary mapping U from $\mathbb{C}^m_{\mathbb{P}}$ onto $\mathbb{C}^m$ such that

$$U\vec{T} = T_\Delta U, \qquad U\vec{K} = K_\Delta. \tag{11.60}$$

The fact that U is an isometry imlies that

$$U^*\mathbb{P}U = I_m. \tag{11.61}$$

Taking into account (11.55), (11.56), (11.41)–(11.42) we get that relations (11.60) and (11.61) coincide with (11.59). □

Theorem 11.6.7. *Let z_k and v_k be the interpolation data with z_k, $v_k \in \mathbb{C}_+$, $k = 1, \dots, m$. Assume that the matrix $\mathbb{P}$ of the form (11.38) is strictly positive and the matrix $\mathbb{Q}$ of the form (11.38) is non-negative. Let also $\vec{\Theta}$ be the Livšic system in the Pick form (11.47) for this data. Then the state-space operator $\vec{T}$ of $\vec{\Theta}$ defined by (11.41) is accretive. Moreover, $\vec{T}$ is accretive but not α-sectorial for any $\alpha \in (0, \pi/2)$ if and only if* $\det \mathbb{Q} = 0$ *and for the unitary operator U on $\mathbb{C}^m_{\mathbb{P}}$ for which*

$$U\vec{A}U^* = U\mathbb{P}^{-1}\mathbb{Q}U^* = \mathrm{diag}\,(0, \dots 0, \lambda_1, \dots, \lambda_q), \tag{11.62}$$

(with $\lambda_i > 0$, p elements equal to 0 and $p + q = m$) at least one from the first p coordinates of the vector $U\vec{g}$ with respect to the orthonormal basis of eigenvectors of the self-adjoint operator $\vec{A} = \mathbb{P}^{-1}\mathbb{Q}$ is not equal to 0.

Proof. The accretiveness of $\vec{T}$ immediately follows from (11.41) and positivity of $\mathbb{P}$ and $\mathbb{Q}$. Set

$$V_{\vec{\Theta}}(z) = ((\mathbb{P}^{-1}\mathbb{Q} - zI)^{-1}\vec{g}, \vec{g})_{\mathbb{C}^m_{\mathbb{P}}},$$

and let T of the form (11.41) be accretive but not α-sectorial for any $\alpha \in (0, \pi/2)$ operator. From Theorem 11.5.3 we have that

$$\lim_{x \to 0^-} \|V_{\vec{\Theta}}(x)\| = \lim_{x \to 0^-} ((\mathbb{P}^{-1}\mathbb{Q} + xI)^{-1}\vec{g}, \vec{g})_{\mathbb{C}^m_{\mathbb{P}}} = \infty. \tag{11.63}$$

Assume that $\mathbb{Q}$ is invertible. Then,

$$\lim_{x \to 0^-} (\mathbb{P}^{-1}\mathbb{Q} + xI)^{-1} = \mathbb{Q}^{-1}\mathbb{P},$$

and the limit (11.63) would then be finite. Therefore, $\det \mathbb{Q} = 0$. Since the operator $\vec{A}$ is self-adjoint, there exists a unitary operator U on $\mathbb{C}^m_{\mathbb{P}}$ for which (11.62) holds. Suppose that all the p first coordinates of the vector $U\vec{g}$ with respect to this orthonormal basis are equal to 0, i.e.,

$$U\vec{g} = \begin{pmatrix} 0 \\ \vdots \\ 0 \\ \xi_1 \\ \vdots \\ \xi_q \end{pmatrix},$$

with $\xi_i \neq 0$. We have

$$\begin{aligned}
\|V_{\vec{\Theta}}(-0)\| &= \lim_{\varepsilon \to 0}((\mathbb{P}^{-1}\mathbb{Q} + \varepsilon I)^{-1}\vec{g}, \vec{g})_{\mathbb{C}^m_{\mathbb{P}}} \lim_{\varepsilon \to 0}(U(\mathbb{P}^{-1}\mathbb{Q} + \varepsilon I)^{-1}U^*U\vec{g}, U\vec{g})_{\mathbb{C}^m_{\mathbb{P}}} \\
&= \lim_{\varepsilon \to 0}((U(\mathbb{P}^{-1}\mathbb{Q} + \varepsilon I)U^*)^{-1}U\vec{g}, U\vec{g})_{\mathbb{C}^m_{\mathbb{P}}} \\
&= \lim_{\varepsilon \to 0}\left(\begin{pmatrix} \frac{1}{\varepsilon} & 0 & 0 & & \\ & \ddots & & & \\ 0 & \frac{1}{\varepsilon} & 0 & 0 & \\ & & & \frac{1}{\lambda_1 + \varepsilon} & \\ & & & & \frac{1}{\lambda_q + \varepsilon} \end{pmatrix} \begin{pmatrix} 0 \\ \vdots \\ 0 \\ \xi_1 \\ \vdots \\ \xi_q \end{pmatrix}, \begin{pmatrix} 0 \\ \vdots \\ 0 \\ \xi_1 \\ \vdots \\ \xi_q \end{pmatrix} \right)_{\mathbb{C}^m} \\
&= \lim_{\varepsilon \to 0} \sum_1^q \frac{1}{\lambda_k + \varepsilon} |\xi_k|^2 \neq \infty,
\end{aligned}$$

since $\vec{g} \neq 0$. This contradiction gives the proof of the necessity. Conversely, assume $\det \mathbb{Q} = 0$. With the notation already used above, assume that there is $\eta \neq 0$ such that

$$U\vec{g} = \begin{pmatrix} \vdots \\ \eta \\ \vdots \\ \xi_1 \\ \vdots \\ \xi_q \end{pmatrix}.$$

We have

$$\begin{aligned}
|\lim_{\varepsilon\to 0} V_{\vec{\Theta}}(-\varepsilon)| &= \lim_{\varepsilon\to 0}((\mathbb{P}^{-1}\mathbb{Q}+\varepsilon I)^{-1}\vec{g},\vec{g})_{\mathbb{C}^m_{\mathbb{P}}} \\
&= ((U\mathbb{P}^{-1}\mathbb{Q}U^*+\varepsilon I)^{-1}U\vec{g},U\vec{g})_{\mathbb{C}^m_{\mathbb{P}}} \\
&= \lim_{\varepsilon\to 0}\left(\operatorname{diag}\left(\frac{1}{\varepsilon},\dots,\frac{1}{\varepsilon},\frac{1}{\lambda_1+\varepsilon},\dots,\frac{1}{\lambda_q+\varepsilon}\right)\begin{pmatrix}\vdots\\ \eta\\ \vdots\\ \xi_1\\ \vdots\\ \xi_q\end{pmatrix},\begin{pmatrix}\vdots\\ \eta\\ \vdots\\ \xi_1\\ \vdots\\ \xi_q\end{pmatrix}\right)_{\mathbb{C}^m} \\
&\geq \lim_{\varepsilon\to 0}\frac{|\eta|^2}{\varepsilon}=\infty.
\end{aligned}$$

Applying Theorem 11.5.3 we see that $\vec{T}$ is accretive but not α-sectorial for any $\alpha\in(0,\pi/2)$. □

Theorem 11.6.8. *Let $\{z_k\}$ and $\{v_k\}$ be the interpolation data with $z_k, v_k\in\mathbb{C}_+$, $k=1,\dots,m$. Suppose that matrices $\mathbb{Q}$ and $\mathbb{P}$ of the form (11.38) are strictly positive. Then, the state-space operator $\vec{T}$ of $\vec{\Theta}$ defined in (11.41) is α-sectorial and the angle α is given by*

$$\tan\alpha=(\mathbb{Q}^{-1}\varphi,\varphi)_{\mathbb{C}^m},\qquad \varphi=\begin{pmatrix}\bar{v}_1\\ \vdots\\ \bar{v}_m\end{pmatrix}. \tag{11.64}$$

Proof. By Theorem 11.5.3,

$$\begin{aligned}
\tan\alpha &= \|V_\Theta(-0)\|=\lim_{x\to 0^-}((\mathbb{P}^{-1}\mathbb{Q}+xI)^{-1}g,g)_{\mathbb{C}^m_{\mathbb{P}}} \\
&= (\mathbb{Q}^{-1}\mathbb{P}g,g)_{\mathbb{C}^m_{\mathbb{P}}}=(\mathbb{Q}^{-1}\mathbb{P}g,\mathbb{P}g)_{\mathbb{C}^m}
\end{aligned}$$

which allows us to conclude, since $g=\mathbb{P}^{-1}\varphi$. □

Corollary 11.6.9. *Let $\{z_k\}$ and $\{v_k\}$ be an interpolation data with $z_k, v_k\in\mathbb{C}_+$, $k=1,\dots,m$ for which the matrices $\mathbb{P}$ and $\mathbb{Q}$ are strictly positive. Then,*

$$\mathbb{Q}\geq\frac{1}{(\mathbb{Q}^{-1}\varphi,\varphi)_{\mathbb{C}^m}}\varphi^*\varphi.$$

Proof. The result follows from Theorems 11.6.1, 11.6.8 and inequality (11.39). □

Theorem 11.6.10. *Let $\{z_k\}$ and $\{v_k\}$ be the interpolation data with $z_k, v_k\in\mathbb{C}_+$, $k=1,\dots,m$. Suppose also that the corresponding matrix $\mathbb{Q}$ of the form (11.38) is strictly positive. Assume that the Livšic canonical scattering system*

$$\Theta=\begin{pmatrix}T & K & 1\\ \mathcal{H} & & \mathbb{C}\end{pmatrix}$$

is the interpolation system in the Nevanlinna-Pick interpolation problem with $\dim \mathcal{H} = m$ *for which the* $z_k \in \rho(T)$ *and such that*

$$\text{c.l.s.}\{(T - z_k I)^{-1} K\mathbb{C}, \quad k = 1, \dots, m\} = \mathcal{H}.$$

Then, the state-space operator T *is* α*-sectorial and the value of the angle* α *is given by formula* (11.64).

Proof. Theorem 11.6.2 implies that the corresponding matrix $\mathbb{P}$ is strictly positive. The same theorem yields that Θ is unitarily equivalent to the system (11.41)-(11.42). Theorem 11.6.8 implies then that the main operator T is α-sectorial and that the angle α is given by (11.64). □

11.7 The Nevanlinna-Pick rational interpolation with distinct poles

In this section we consider the Nevanlinna-Pick interpolation problem in the class $\mathbf{N}^r$ of rational Herglotz-Nevanlinna functions $V(z) \in \mathbf{N}$ with n distinct poles and $V(\infty) = 0$. It is formulated as follows:

> *Given* n *points* $z_1, \dots, z_n \in \mathbb{C}_+$ *and* n *not necessarily distinct points* $v_1, \dots, v_n \in \mathbb{C}_+$, *find a function* $V(z) \in \mathbf{N}^r$ *with* n *distinct poles such that* $V(z_k) = v_k$, $k = 1, \dots, n$.

Let

$$\mathbb{P} = \left(\frac{v_k - \bar{v}_j}{z_k - \bar{z}_j}\right)_{k,j=1,\dots n} \quad \text{and} \quad \mathbb{Q} = \left(\frac{z_k v_k - \bar{z}_j \bar{v}_j}{z_k - \bar{z}_j}\right)_{k,j=1,\dots n} \tag{11.65}$$

be the Pick matrices corresponding to the interpolation data.

Theorem 11.7.1. *Let*

$$\Theta = \begin{pmatrix} T & K & 1 \\ \mathcal{H} & & \mathbb{C} \end{pmatrix}, \tag{11.66}$$

be the Livsic canonical system with n*-dimensional state space* $\mathcal{H}$ *and channel operator* $Kc = cg$, $c \in \mathbb{C}$, $g \in \mathcal{H}$, $K^*x = (x, g)$, $x \in \mathcal{H}$. *Then the following statements are equivalent:*

(i) *the vector* g *is a cyclic vector for* $\operatorname{Re} T$;

(ii) *for every distinct* n *numbers* $z_1, \dots, z_n$ *from* $\rho(\operatorname{Re} T)$ *the vectors*

$$(\operatorname{Re} T - z_1 I)^{-1} g, \dots, (\operatorname{Re} T - z_n I)^{-1} g,$$

are linearly independent;

(iii) *for every distinct non-real n numbers $z_1, \dots, z_n$ from the open (lower) half-plane and for numbers $v_k = ((\mathrm{Re}\, T - z_k I)^{-1} g, g)$, $k = 1, \dots, n$ the corresponding Pick matrix $\mathbb{P}$ given by (11.65) is strictly positive.*

Proof. (i)⇒(ii): Let the vectors $g, \mathrm{Re}\, Tg, \dots, \mathrm{Re}\, T^{n-1} g$ be linearly independent. Suppose that the vectors

$$h_1 = (\mathrm{Re}\, T - z_1 I)^{-1} g, \ \dots, \ h_n = (\mathrm{Re}\, T - z_n I)^{-1} g,$$

are not linearly independent. Let then $h_1, \dots, h_m$, $m \le n-1$ be a maximal linearly independent subsystem. Then the vector h_n is a linear combination of $h_1 \dots, h_m$, i.e.,

$$h_n = \sum_{k=1}^{m} \lambda_k^{(0)} h_k.$$

Since $T(T - \xi I)^{-1} \varphi = \varphi + \xi (T - \xi I)^{-1} \varphi$ for every $\varphi \in \mathcal{H}$ and every $\xi \in \rho(\mathrm{Re}\, T)$, we get

$$g + z_n h_n = \sum_{k=1}^{m} \left(\lambda_k^{(0)} g + \lambda_k^{(0)} z_k h_k \right).$$

It follows that

$$\left(1 - \sum_{k=1}^{m} \lambda_k^{(0)} \right) g = \sum_{k=1}^{m} (z_k - z_n) \lambda_k^{(0)} h_k.$$

Hence $(1 - \sum_{k=1}^{m} \lambda_k^{(0)}) \neq 0$ and, therefore, $g = \sum_{k=1}^{m} \lambda_k^{(1)} h_k$. Since

$$\mathrm{Re}\, Tg = \sum_{k=1}^{m} \lambda_k^{(1)} (g + z_k h_k),$$

we get that

$$\mathrm{Re}\, Tg = \sum_{k=1}^{m} \lambda_k^{(2)} h_k.$$

Continuing this process we conclude that the vectors $g, \mathrm{Re}\, Tg, \dots, \mathrm{Re}\, T^{n-1} g$ are linear combinations of $h_1, \dots, h_m$ that yields a contradiction. Thus,

$$(\mathrm{Re}\, T - z_1 I)^{-1} g, \ \dots, \ (\mathrm{Re}\, T - z_n I)^{-1} g,$$

are linearly independent.

(ii)⇒(i): Let the vectors $(\mathrm{Re}\, T - z_1 I)^{-1} g, \dots, (\mathrm{Re}\, T - z_n I)^{-1} g$ be linearly independent. Because the subspace

$$\mathcal{H}^0 := c.l.s\{g, \mathrm{Re}\, Tg, \dots, \mathrm{Re}\, T^{n-1} g\},$$

reduces $\operatorname{Re} T$, the vectors $(\operatorname{Re} T - z_1 I)^{-1}g, \ldots, (\operatorname{Re} T - z_n I)^{-1}g$ lie in $\mathcal{H}^0$. Therefore, $\mathcal{H}^0 = \mathcal{H}$.
(ii) $\Longleftrightarrow$ (iii). Let $\vec{\lambda} = (\lambda_1, \ldots, \lambda_n) \in \mathbb{C}^n$. Since

$$\left\| \sum_{k=1}^{n} \lambda_k (T - z_k I)^{-1} g \right\|^2 = \sum_{k,j=1}^{n} \lambda_k \bar{\lambda}_j ((T - z_k I)^{-1} g, (T - z_j I)^{-1} g)$$

$$= \sum_{k,j=1}^{n} \lambda_k \bar{\lambda}_j \frac{((T - z_k I)^{-1} g, g) - ((T - z_j^* I)^{-1} g, g)}{z_k - z_j^*}$$

$$= \sum_{k,j=1}^{n} \lambda_k \bar{\lambda}_j \frac{v_k - v_j^*}{z_k - z_j^*} = \vec{\lambda}^* \mathbb{P} \vec{\lambda},$$

we get that $\mathbb{P}$ is non-negative. This matrix is invertible if and only if the vectors $(\operatorname{Re} T - z_1 I)^{-1}g, \ldots, (\operatorname{Re} T - z_n I)^{-1}g$ are linearly independent. □

Lemma 11.7.2. *Any rational function $V(z)$ from the class $\mathbf{N}^r$ of the form*

$$V(z) = \sum_{k=1}^{n} \frac{a_k}{t_k - z}, \tag{11.67}$$

where $t_1, \ldots, t_n$ are distinct real numbers and $a_k > 0$, $k = 1, 2, \ldots, n$, can be represented in the form

$$V(z) = ((\operatorname{Re} T - zI)^{-1} g, g) = i \frac{W_\Theta(z) - 1}{W_\Theta(z) + 1},$$

where $W_\Theta(z)$ is a transfer function of the minimal system of the form

$$\Theta = \begin{pmatrix} T & K & 1 \\ \mathbb{C}^n & & \mathbb{C} \end{pmatrix}, \tag{11.68}$$

g is a cyclic vector for $\operatorname{Re} T$, and $\operatorname{Im} T = (\cdot, g)g$.

Proof. Let

$$\delta_k = (\underbrace{0 \ldots 01}_{k} 0 \ldots 0), \quad k = 1, \ldots, n,$$

be an orthonormal system of vectors from $\mathbb{C}^n$. Define a self-adjoint operator $\operatorname{Re} T$ in $\mathbb{C}^n$ as

$$\operatorname{Re} T \delta_k = t_k \delta_k, \quad k = 1, \ldots, n,$$

and let $g = \sum_{k=1}^{n} \sqrt{a_k} \delta_k$. Then

$$((\operatorname{Re} T - zI)^{-1} g, g) = \sum_{k=1}^{n} \frac{a_k}{t_k - z}, \quad z \neq t_k, \ k = 1, \ldots, n.$$

It follows that $V(z) = ((\mathrm{Re}\, T - zI)^{-1} g, g)$. Moreover, the vectors g, $\mathrm{Re}\, Tg$, ..., $\mathrm{Re}\, T^{n-1} g$ are linearly independent. Consider the system Θ of the form (11.68) where

$$T = \mathrm{Re}\, T + i(.,g)g, \quad Kc = cg,\ c \in \mathbb{C}, \quad K^* x = (x,g), \quad x \in \mathbb{C}^n.$$

It was shown in Lemma 11.1.3 that vector g is cyclic for the real part $\mathrm{Re}\, T$ of operator T if and only if it is cyclic for the operator T. Therefore operator T is prime and the system Θ as a result of that is minimal. □

Denote by $\{z_k,\ k = 1,\ldots,n\}$ and $\{v_k,\ k = 1,\ldots,n\}$ interpolation data on the open upper half-plane $\mathbb{C}_+$, and let the matrices $\mathbb{P}$ and $\mathbb{Q}$ be defined by (11.65). Let also

$$\vec{\Theta} = \begin{pmatrix} \vec{T} & \vec{K} & 1 \\ \mathbb{C}^n_{\mathbb{P}} & & \mathbb{C} \end{pmatrix}, \tag{11.69}$$

be the system Livšic system in the Pick form (written for this data) defined by (11.47) with operators $\vec{T}$ and $\vec{K}$ defined by (11.41) and (11.42), respectively.

Theorem 11.7.3. *Let $\{z_k,\ k = 1,\ldots,n\} \in \mathbb{C}_+$ and $\{v_k,\ k = 1,\ldots,n\} \in \mathbb{C}_+$ be interpolation data. The system $\vec{\Theta}$ in the Pick form* (11.69) *is a minimal interpolation system in the Nevanlinna-Pick interpolation problem and function*

$$V_{\vec{\Theta}}(z) = \left((\mathrm{Re}\, \vec{T} - zI)^{-1}\chi, \chi\right)$$

is a rational function from the class $\mathbf{N}^r$ with n distinct poles and representation (11.67).

Proof. We need to show that the function

$$V_{\vec{\Theta}}(z) = \left((\mathrm{Re}\, \vec{T} - zI)^{-1}\chi, \chi\right)_{\mathbb{C}^n_{\mathbb{P}}} = ((\mathbb{Q} - z\mathbb{P})^{-1}\varphi, \varphi)_{\mathbb{C}^n}, \quad z \in \rho(\mathrm{Re}\, \vec{T}) \tag{11.70}$$

satisfies conditions $V_{\vec{\Theta}}(z_k) = v_k$, $k = 1,\ldots,n$. We have that

$$(\mathbb{Q} - z_k\mathbb{P})_{j,k} = \frac{z_k v_k - \bar{z}_j \bar{v}_j}{z_k - \bar{z}_j} - z_k \frac{v_k - \bar{v}_j}{z_k - \bar{z}_j} = \bar{v}_j.$$

Therefore the k-th column of the matrix $\mathbb{Q} - z_k\mathbb{P}$ is equal to the vector φ of the form (11.40) . Hence, $(\mathrm{Re}\, \vec{T} - z_k I)\delta_k = \chi$ (where δ_k denotes the $(n \times 1)$ column vector whose entries are all equal to 0, except the k-th one, which is equal to 1). Thus,

$$\begin{aligned} V(z_k) &= ((\mathrm{Re}\, \vec{T} - z_k I)^{-1}\chi, \chi)_{\mathbb{C}^n_{\mathbb{P}}} = (\delta_k, \chi)_{\mathbb{C}^n_{\mathbb{P}}} \\ &= (\mathbb{P}\delta_k, \mathbb{P}^{-1}\chi)_{\mathbb{C}^n_{\mathbb{P}}} = (\mathbb{P}\delta_k, \mathbb{P}^{-1}\chi)_{\mathbb{C}^n} = (\delta_k, \chi)_{\mathbb{C}^n} = v_k. \end{aligned}$$

Theorem 11.7.1 implies that vector χ is cyclic for operator $\mathrm{Re}\, \vec{T}$ and this operator has n distinct eigenvalues. Therefore function $V_{\vec{\Theta}}(z)$ is rational with n poles and belongs to the class $\mathbf{N}^r$. □

Theorem 11.7.4. *The Nevanlinna-Pick interpolation problem in the class* $\mathbf{N}^r$ *has a solution if and only if the Pick matrix* $\mathbb{P}$ (11.65) *is strictly positive. Moreover, if the matrix* $\mathbb{P}$ *is strictly positive, then the Nevanlinna-Pick interpolation problem has the unique solution* $V_{\vec{\Theta}}(z)$ *of the form* (11.70) *that is represented by the minimal Livsic interpolation system in the Pick form* (11.69).

Proof. Suppose that function $V(z)$ is a solution of Nevanlinna-Pick interpolation problem in the class $\mathbf{N}^r$. By Lemma 11.7.2 function $V(z)$ has a representation

$$V(z) = ((\mathrm{Re}\,T - zI)^{-1}g, g) = i\frac{W_\Theta(z) - 1}{W_\Theta(z) + 1}, \tag{11.71}$$

where $W_\Theta(z)$ is a transfer function of the minimal system of the form

$$\Theta = \begin{pmatrix} T & K & 1 \\ \mathbb{C}^n & & \mathbb{C} \end{pmatrix}, \tag{11.72}$$

and $Kc = cg,\ K^*x = (x, g),\ x, g \in \mathbb{C}^n$, g is a cyclic vector for $\mathrm{Re}\,T$. It follows from Theorem 11.7.1 that the Pick matrix $\mathbb{P}$ is invertible and non-negative. If matrix $\mathbb{P}$ is invertible and non-negative, then by Theorem 11.7.3 the function $V_{\vec{\Theta}}(z)$ of the form (11.70) is a solution of the Nevanlinna-Pick interpolation problem in the class $\mathbf{N}^r$ and is represented by the minimal interpolation system $\vec{\Theta}$ of the Livsic type in the Pick form (11.69).

Now we will establish the solution uniqueness of the Nevanlinna-Pick interpolation problem in the class $\mathbf{N}^r$. Suppose that in addition to the solution $V_{\vec{\Theta}}$ (11.70), (11.69) there is another solution $V(z)$. As we mentioned at the beginning of the proof, this solution can be represented by (11.71)

$$V(z) = V_\Theta(z) = ((\mathrm{Re}\,T - zI)^{-1}g, g),$$

and is generated by the system Θ of the form (11.72). Then from the equality $V(z_k) = V_\Theta(z_k) = V_{\vec{\Theta}}(z_k),\quad k = 1, \dots, n$ follows

$$\begin{aligned}
\left((\mathrm{Re}\,\vec{T} - z_k I)^{-1}\chi, \chi\right)_{\mathbb{C}^n_{\mathbb{P}}} &= \left((\mathrm{Re}\,T - z_k I)^{-1}g, g\right)_{\mathbb{C}^n} = v_k, \\
\left((\mathrm{Re}\,\vec{T} - \bar{z}_k I)^{-1}\chi, \chi\right)_{\mathbb{C}^n_{\mathbb{P}}} &= \left((\mathrm{Re}\,T - \bar{z}_k I)^{-1}g, g\right)_{\mathbb{C}^n} = \bar{v}_k, \quad k = 1, \dots, n,
\end{aligned}$$

and

$$\begin{aligned}
&\left((\mathrm{Re}\,\vec{T} - z_k I)^{-1}\chi, \chi\right)_{\mathbb{C}^n_{\mathbb{P}}} - \left((\mathrm{Re}\,\vec{T} - z_m^* I)^{-1}\chi, \chi\right)_{\mathbb{C}^n_{\mathbb{P}}} \\
&\quad = \left((\mathrm{Re}\,T - z_k I)^{-1}g, g\right)_{\mathbb{C}^n} - \left((\mathrm{Re}\,T - z_m^* I)^{-1}g, g\right)_{\mathbb{C}^n}.
\end{aligned} \tag{11.73}$$

Applying the Hilbert identity for resolvents, we get

$$\begin{aligned}
&((\mathrm{Re}\,\vec{T} - z_k I)^{-1}\chi, (\mathrm{Re}\,\vec{T} - z_m I)^{-1}\chi)_{\mathbb{C}^n_{\mathbb{P}}} \\
&\quad = \left((\mathrm{Re}\,T - z_k I)^{-1}g, (T - z_m I)^{-1}g\right)_{\mathbb{C}^n}.
\end{aligned} \tag{11.74}$$

Let the operator $U : \mathbb{C}^n_{\mathbb{P}} \to \mathbb{C}^n$ be defined as

$$U(\mathrm{Re}\,\vec{T} - z_k I)^{-1}\chi = (\mathrm{Re}\,T - z_k I)^{-1} g, \quad k = 1, \ldots n.$$

Then from Theorem 11.7.1 and relation (11.74) we get that U is a unitary operator. From (11.73) we obtain

$$\left(U(\mathrm{Re}\,\vec{T} - z_k I)^{-1}\chi, g\right)_{\mathbb{C}^n} = (\mathrm{Re}\,T - z_k I)^{-1} g, g)_{\mathbb{C}^n} = ((\mathrm{Re}\,\vec{T} - z_k I)^{-1}\chi, \chi)_{\mathbb{C}^n_{\mathbb{P}}},$$

for $k = 1, \ldots, n$, and hence $(Uh, g)_{\mathbb{C}^n} = (h, \chi)_{\mathbb{C}^n_{\mathbb{P}}}$ for all $h \in \mathbb{C}^n_{\mathbb{P}}$. Therefore $U\chi = g$. Furthermore

$$\begin{aligned}
(\mathrm{Re}\,T)U(\mathrm{Re}\,\vec{T} - z_k I)^{-1}\chi &= (\mathrm{Re}\,T)(\mathrm{Re}\,T - z_k I)^{-1} g \\
&= g + z_k(\mathrm{Re}\,T - z_k I)^{-1} g = U\chi + z_k U(\mathrm{Re}\,\vec{T} - z_k I)^{-1}\chi \\
&= U(\mathrm{Re}\,\vec{T})(\mathrm{Re}\,\vec{T} - z_k I)^{-1}\chi.
\end{aligned}$$

Finally, $(\mathrm{Re}\,T)Uh = U\mathrm{Re}\,\vec{T}h$ for all $h \in \mathbb{C}^n_{\mathbb{P}}$, i.e., the operator $\mathrm{Re}\,T$ is unitarily equivalent to the operator $\mathrm{Re}\,\vec{T}$ and moreover, $V_\Theta(z) = V_{\vec{\Theta}}(z)$ for all $z \in \rho(\mathrm{Re}\,T) = \rho(\mathrm{Re}\,\vec{T})$. □

Theorem 11.7.5. *Let $z_1, \ldots, z_n \in \mathbb{C}_+$ be n distinct points, and $v_1, \ldots, v_n \in \mathbb{C}_+$ be n not necessarily distinct points for which the Pick matrix $\mathbb{P}$ (11.65) is strictly positive. Then the Livšič canonical system $\vec{\Theta}$ in the Pick form (11.69) corresponding to the given data is unitarily equivalent to the system Δ of the form (11.15) with some $(n \times n)$ Jacobi matrix $\mathcal{J}$ (11.1) as a state-space operator of the system.*

Proof. Consider interpolation system $\vec{\Theta}$ of the form (11.69). By Theorem 11.7.4 the system $\vec{\Theta}$ is minimal and operator $\vec{T}$ is prime. Therefore by Theorem 5.6.4 this operator has only non-real eigenvalues in the upper half-plane. By Theorem 11.2.6 operator $\vec{T}$ is unitarily equivalent to an operator given by a Jacobi matrix $\mathcal{J}$ of the form (11.1) that can be obtained by the algorithm of reconstruction (see Section 11.4) from non-real eigenvalues of the matrix $\vec{T}$. Consequently, we obtain that the system $\vec{\Theta}$ is unitarily equivalent to some system Δ of the form (11.15). □

It follows from this theorem that the system Δ with the Jacobi matrix $\mathcal{J}$ as a state-space operator is a system solution of the Nevanlinna-Pick interpolation problem in the class $\mathbf{N}^r$ as well.

11.8 Examples

Example. Consider the following interpolation data:

$$z_1 = 2i, \quad z_2 = 3i, \quad v_1 = \frac{2i}{5}, \quad v_2 = \frac{3i}{10}.$$

We will find an interpolation system $\vec{\Theta}$ in the Pick form whose impedance function is the solution of the Nevanlinna-Pick interpolation problem for the given interpolation data. We will also show that $\vec{\Theta}$ is unitarily equivalent to the system Δ in Jacobi form with a non-self-adjoint Jacobi matrix (with a rank-one imaginary part) as a state-space operator of the system. We have

$$\mathbb{P} = \begin{pmatrix} \frac{1}{5} & \frac{7}{50} \\ \frac{7}{50} & \frac{1}{10} \end{pmatrix}, \qquad \mathbb{Q} = \begin{pmatrix} 0 & \frac{i}{50} \\ -\frac{i}{50} & 0 \end{pmatrix},$$

and thus

$$\operatorname{Re} \vec{T} = \mathbb{P}^{-1}\mathbb{Q} = \begin{pmatrix} 7i & 5i \\ -10i & -7i \end{pmatrix}.$$

Furthermore, we have

$$\chi = \mathbb{P}^{-1} \begin{pmatrix} v_1^* \\ v_2^* \end{pmatrix} = \begin{pmatrix} -5i \\ -10i \end{pmatrix}, \qquad \vec{T} = \begin{pmatrix} -5i & \frac{7}{2}i \\ -6i & -4i \end{pmatrix}.$$

Thus the corresponding interpolation system in the Pick form is equal to

$$\vec{\Theta} = \begin{pmatrix} \begin{pmatrix} -5i & \frac{7}{2}i \\ -6i & -4i \end{pmatrix} & Kc = c \cdot \begin{pmatrix} 5i \\ -10i \end{pmatrix} & J = 1 \\ \mathbb{C}^2 \begin{pmatrix} \frac{1}{5} & \frac{7}{50} \\ \frac{7}{50} & \frac{1}{10} \end{pmatrix} & & \mathbb{C} \end{pmatrix}.$$

Applying the algorithm of reconstruction we get from the system $\vec{\Theta}$ the Livsič canonical system Δ with a non-self-adjoint Jacobi matrix as a state-space operator of the system. This system has the form

$$\Delta = \begin{pmatrix} \begin{pmatrix} i & 1 \\ 1 & 0 \end{pmatrix} & Kc = c \cdot \begin{pmatrix} 1 \\ 0 \end{pmatrix} & J = 1 \\ \mathbb{C}^2 & & \mathbb{C} \end{pmatrix}.$$

where the state-space operator is the Jacobi matrix $\mathcal{J} = \begin{pmatrix} i & 1 \\ 1 & 0 \end{pmatrix}$ and

$$\operatorname{Re} \mathcal{J} = \begin{pmatrix} 0 & 1 \\ 1 & 0 \end{pmatrix}, \qquad g = \begin{pmatrix} 1 \\ 0 \end{pmatrix}.$$

The corresponding impedance function that is the system solution is

$$V_\Delta(z) = ((\operatorname{Re} \mathcal{J} - zI)^{-1} g, g) = \frac{z}{1 - z^2}.$$

Chapter 12

Non-canonical Systems

In Chapter 6 we described the class $N(R)$ of Herglotz-Nevanlinna functions in a finite-dimensional Hilbert space that can be realized as impedance functions of canonical L-systems. Since the class $N(R)$ is substantially narrower than the set of all Herglotz-Nevanlinna functions, the problem of the general realization, or description of a new non-canonical type of systems that realize an arbitrary Herglotz-Nevanlinna function, remain open.

In the present chapter we introduce the notion of a **non-canonical impedance system** (**NCI-system**) as well as a **non-canonical L-system** (**NCL-system**). We are going to show that any Herglotz-Nevanlinna function in finite-dimensional Hilbert space E can be realized as a transfer function of an NCI-system of the form

$$\begin{cases} (\mathbb{D} - zF_+)x = K\varphi_-, \\ \varphi_+ = K^*x, \end{cases} \tag{12.1}$$

where $\mathbb{D}$ and F_+ are self-adjoint operators acting from $\mathcal{H}_+$ into $\mathcal{H}_-$ and in addition F_+ is an orthogonal projector in $\mathcal{H}_+$ and $\mathcal{H}$. In this case, the associated **transfer function of the NCI-system** is given by

$$V(z) = K^*(\mathbb{D} - zF_+)^{-1}K, \tag{12.2}$$

and every Herglotz-Nevanlinna function can be represented in the form (12.2). We also consider the following **non-canonical L-systems** or **NCL-systems**

$$\begin{cases} (\mathbb{A} - zF_+)x = KJ\varphi_-, \\ \varphi_+ = \varphi_- - 2iK^*x. \end{cases} \tag{12.3}$$

This system can be expressed via an array similar to the one of an L-system in (6.36), that is

$$\Theta_{F_+} = \begin{pmatrix} \mathbb{A} & F_+ & K & J \\ \mathcal{H}_+ \subset \mathcal{H} \subset \mathcal{H}_- & & & E \end{pmatrix}. \tag{12.4}$$

The additional ingredient in (12.4) is the operator F_+ which is an orthogonal projection in $\mathcal{H}_+$ and $\mathcal{H}$. The corresponding **transfer function of an NCL-system** is

$$W_{\Theta_{F_+}}(z) = I - 2iK^*(\mathbb{A} - zF_+)^{-1}KJ,$$

and the **impedance function of an NCL-system** is

$$V_{\Theta_{F_+}}(z) = K^*(\mathrm{Re}\,\mathbb{A} - zF_+)^{-1}K = i[W_{\Theta_{F_+}}(z) + I]^{-1}[W_{\Theta_{F_+}}(z) - I]J. \tag{12.5}$$

We will show that, after some restrictions on the linear coefficient L in the integral representation

$$V(z) = Q + zL + \int_{-\infty}^{+\infty} \left(\frac{1}{t-z} - \frac{t}{1+t^2}\right) dG(t), \tag{12.6}$$

any Herglotz-Nevanlinna function $V(z)$ is realizable as the impedance function of an NCL-system of the form (12.3).

We begin this chapter with a special case. In the following two sections, we give realization results for Herglotz-Nevanlinna functions of the form

$$V(z) = Q + Lz + \int_{-\infty}^{+\infty} \frac{dG(t)}{t-z}, \tag{12.7}$$

where $Q = Q^*$, $L \geq 0$, and $G(t)$ is a non-decreasing function on $\mathbb{R}$ such that

$$\int_{-\infty}^{+\infty} (dG(t)x, x)_E < \infty, \qquad \text{for all } x \in E. \tag{12.8}$$

In particular, it is shown that each Herglotz-Nevanlinna function $V(z)$ of the form (12.7) satisfying (12.8) gives rise to an F**-system** of the form

$$\Theta_F = \begin{pmatrix} M & F & K & J \\ \mathcal{H} & & & E \end{pmatrix}, \tag{12.9}$$

where the state-space operator M is now allowed to be unbounded. Moreover, the additional ingredient in (12.9) is the operator F which is an orthogonal projection in $\mathcal{H}$. The system determined by (12.9) has properties that are analogous to those of canonical Livšic systems (5.6) in Chapter 5. The function $V_{\Theta_F}(z)$ is obtained from (12.9) via an F-resolvent in the form

$$V_{\Theta_F}(z) = K^*(D - zF)^{-1}K, \quad z \in \mathbb{C} \setminus \mathbb{R},$$

where $D = \mathrm{Re}\,M$. The transfer function $W_{\Theta_F}(z)$ associated to the F-system (12.9) is defined by

$$W_{\Theta_F}(z) = 1 - 2iK^*(M - zF)^{-1}KJ,$$

for all $z \in \mathbb{C}$ for which $(M - zF)^{-1}$ exists, bounded, and defined on entire $\mathcal{H}$. Again the functions $V_{\Theta_F}(z)$ and $W_{\Theta_F}(z)$ are related via (12.5).

12.1 F-systems: definition and basic properties

Let $\mathcal{H}$ be a Hilbert space with inner product $(\cdot,\cdot)$. Let A be a closed linear operator in $\mathcal{H}$ and let F be an orthogonal projection in $\mathcal{H}$. Associated to the pair (A,F) is the **resolvent set** $\rho(A,F)$, i.e., the set of all $z \in \mathbb{C}$ for which $A - zF$ is boundedly invertible in $\mathcal{H}$ and $(A-zF)^{-1}$ is defined on entire $\mathcal{H}$. The corresponding **resolvent operator** is defined as $(A - zF)^{-1}$, $z \in \rho(A,F)$. When the resolvent set $\rho(A,F)$ is non-empty, the following analog of the resolvent identity holds:

$$(A - zF)^{-1} - (A - wF)^{-1} = (z - w)(A - wF)^{-1}F(A - zF)^{-1}, \qquad (12.10)$$

$z, w \in \rho(A,F)$. Note that if A is self-adjoint, then $\rho(A,F)$ is symmetric with respect to the real axis. In the sequel an important role is played by the class of densely-defined closed linear operators A in $\mathcal{H}$ which are of the form

$$A = D + iR, \quad D = D^*, \quad R = R^* \in [\mathcal{H},\mathcal{H}]. \qquad (12.11)$$

Here D, and hence A, may be unbounded, but R is bounded. Therefore $A^* = D^* - iR^*$, which implies $2D = A + A^* = 2\mathrm{Re}\,A$ and $2iR = A - A^* = 2i\mathrm{Im}\,A$, so that

$$D = \mathrm{Re}\,A, \quad R = \mathrm{Im}\,A,$$

are the real part and the imaginary part of A, respectively.

Lemma 12.1.1. *Let $A = D + iR$ be a closed linear operator in the Hilbert space $\mathcal{H}$ of the form* (12.11), *so that $D = D^*$ and $R = R^* \in [\mathcal{H},\mathcal{H}]$. Let F be an orthogonal projection in $\mathcal{H}$. Assume that $R \geq 0$ or that $R \leq 0$. Then:*

(i) $\ker(A - zF) = \ker(A^* - \bar{z}F) = \ker(A) \cap \ker(F)$ *for every $z \in \mathbb{C}_-$ or for every $z \in \mathbb{C}_+$, respectively;*

(ii) *if $z \in \rho(A,F)$ for some $z \in \mathbb{C}_-$ or for some $z \in \mathbb{C}_+$, then $z \in \rho(A,F)$ for every $z \in \mathbb{C}_-$ or for every $z \in \mathbb{C}_+$, respectively.*

Moreover, if A is self-adjoint or equivalently $R = 0$, then the statements (i) *and* (ii) *hold for all $z \in \mathbb{C}_- \cup \mathbb{C}_+$.*

Proof. It suffices to give a proof for the case $R \geq 0$. The case $R \leq 0$ can be dealt with in an analogous manner.

(i) Clearly, $\ker(A) \cap \ker(F) \subset \ker(A - zF)$ for any $z \in \mathbb{C}$. To see the converse, assume that $x \in \ker(A - zF)$, $z \in \mathbb{C}_-$. Then

$$0 = ((A - zF)x, x) = (Dx, x) + i(Rx, x) - z(Fx, x),$$

and, since $z \in \mathbb{C}_-$, one concludes that $(Rx, x) = (Fx, x) = 0$. Therefore, $Rx = Fx = 0$ and consequently $Ax = 0$, so that $x \in \ker(A) \cap \ker(F)$. Similarly one proves $\ker(A^* - \bar{z}F) = \ker(A) \cap \ker(F)$.

(ii) Assume that $z_0 \in \rho(A,F)$ for some $z_0 \in \mathbb{C}_-$, so that $\ker(A - z_0F) = \{0\}$ and $\mathrm{Ran}(A - z_0F) = \mathcal{H}$. It follows from (i) that $\ker(A - zF) = \{0\}$ for every

$z \in \mathbb{C}_-$. It remains to show that $\operatorname{Ran}(A - zF) = \mathcal{H}$ for every $z \in \mathbb{C}_-$. Again by (i) $\ker(A^* - \bar{z}F) = \{0\}$, and thus it is enough to prove that $\operatorname{Ran}(A - zF)$ is closed for every $z \in \mathbb{C}_-$; or equivalently, that for every $z \in \mathbb{C}_-$ there is a positive constant $c(z)$ such that $\|(A - zF)x\| \geq c(z)\|x\|$ for all $x \in \operatorname{Dom}(A)$. Assume the converse, that there is $z \in \mathbb{C}_-$ such that $(A - zF)x_n \to 0$ for some sequence (x_n), $\|x_n\| = 1$, $x_n \in \operatorname{Dom}(A)$. Then

$$((A - zF)x_n, x_n) = (Dx_n, x_n) + i(Rx_n, x_n) - z(Fx_n, x_n) \to 0, \tag{12.12}$$

implies that $(Fx_n, x_n) = \|Fx_n\|^2 \to 0$. Hence $Fx_n \to 0$, $Ax_n \to 0$, and consequently $(A - z_0F)x_n \to 0$, a contradiction to the assumption $z_0 \in \rho(A, F)$.

It is clear that if A is self-adjoint, then the assertions (i) and (ii) are true for every $z \in \mathbb{C} \setminus \mathbb{R}$. □

Corollary 12.1.2. *Let $A = D + iR$ and F be as in Lemma* 12.1.1, *and assume that $R = R^* \in [\mathcal{H}, \mathcal{H}]$ satisfies $R \geq 0$ or $R \leq 0$, respectively. Then:*

(i) $\ker(A) \subset \ker(\operatorname{Re} A)$*;*

(ii) $\rho(\operatorname{Re} A, F) \neq \emptyset$ *implies* $\rho(A, F) \neq \emptyset$.

Proof. (i) Assume that $Ax = 0$ for some $x \in \operatorname{Dom}(A)$. Then $((\operatorname{Re} A + iR)x, x) = 0$, which implies $(\operatorname{Re} Ax, x) = (Rx, x) = 0$. Consequently, $Rx = 0$ and hence also $\operatorname{Re} Ax = 0$.

(ii) Assume that $\rho(\operatorname{Re} A, F) \neq \emptyset$. Then, by Lemma 12.1.1, $\mathbb{C}_+ \cup \mathbb{C}_- \subset \rho(\operatorname{Re} A, F)$. With $z \in \mathbb{C}_-$ part (i) and Lemma 12.1.1 imply $\ker(A - zF) = \ker(A) \cap \ker(F) \subset \ker(\operatorname{Re} A) \cap \ker(F) = \{0\}$. To see that $z \in \rho(A, F)$ consider a sequence (x_n), $\|x_n\| = 1$, with $(A - zF)x_n \to 0$ as in the proof of Lemma 12.1.1. Then (12.12) shows that $Fx_n \to 0$ and $R^{\frac{1}{2}}x_n \to 0$. Therefore also $Rx_n \to 0$ and $\operatorname{Re} Ax_n \to 0$. Consequently, $(\operatorname{Re} A - zF)x_n \to 0$ and this implies (ii). □

Let M be a closed linear operator in $\mathcal{H}$ of the form (12.11) and let F be an orthogonal projection in $\mathcal{H}$, $K \in [E, \mathcal{H}]$, and J be a bounded, self-adjoint, and unitary operator in E. Let also $\operatorname{Im} M = KJK^*$ and $L^2_{[0,\tau_0]}(E)$ be the Hilbert space of E-valued functions equipped with an inner product

$$(\varphi, \psi)_{L^2_{[0,\tau_0]}(E)} = \int_0^{\tau_0} (\varphi, \psi)_E \, dt, \quad \Big(\varphi(t),\, \psi(t) \in L^2_{[0,\tau_0]}(E)\Big). \tag{12.13}$$

Consider the system of equations

$$\begin{cases} iF\frac{d\chi}{dt} + M\chi(t) = KJ\psi_-(t), \\ \chi(0) = x \in \mathcal{H}, \\ \psi_+ = \psi_- - 2iK^*\chi(t). \end{cases} \tag{12.14}$$

We have the following lemma.

Lemma 12.1.3. *If for a given continuous $\psi_-(t) \in L^2_{[0,\tau_0]}(E)$ we have $\chi(t) \in \mathcal{H}$ and $\psi_+(t) \in L^2_{[0,\tau_0]}(E)$ satisfy (12.14), then a system of the form (12.14) satisfies the metric conservation law*

$$2\|F\chi(\tau)\|^2 - 2\|F\chi(0)\|^2 = \int_0^\tau (J\psi_-, \psi_-)_E\, dt - \int_0^\tau (J\psi_+, \psi_+)_E\, dt, \tag{12.15}$$

$\tau \in [0, \tau_0]$.

Proof. Following the algebraic steps of the proof of Lemma 5.1.1 and taking into account that F is a bounded operator and $\operatorname{Im} M = KJK^*$, we obtain

$$\frac{d}{dt}(F\chi(t), F\chi(t)) = \frac{1}{2}(J\psi_-(t), \psi_-(t))_E - \frac{1}{2}(J\psi_+(t), \psi_+(t))_E.$$

Taking into account that $\psi_\pm(t) \in L^2_{[0,\tau_0]}(E)$, we integrate both sides from 0 to $\tau \in [0, \tau_0]$ and multiply by 2 to obtain (12.15). $\square$

Given an input vector $\psi_- = \varphi_- e^{izt} \in E$, we seek solutions to the system (12.14) as an output vector $\psi_+ = \varphi_+ e^{izt} \in E$ and a state-space vector $\chi(t) = xe^{izt} \in \mathcal{H}$. Substituting the expressions for $\psi_\pm(t)$ and $\chi(t)$ allows us to cancel exponential terms and convert the system (12.14) to the stationary form

$$\begin{cases} (M - zF)x = KJ\varphi_-, \\ \varphi_+ = \varphi_- - 2iK^*x, \end{cases} \quad z \in \rho(M, F). \tag{12.16}$$

Following the canonical case we can re-write (12.16) as an array and have the following definition.

Definition 12.1.4. Let $\mathcal{H}$ and E be Hilbert spaces with $\dim E < \infty$, and let M, F, K, and J be linear operators. The array

$$\Theta_F = \begin{pmatrix} M & F & K & J \\ \mathcal{H} & & & E \end{pmatrix}, \tag{12.17}$$

is called an F**-system** if:

(i) M is of the form (12.11);

(ii) $J = J^* = J^{-1} \in [E, E]$;

(iii) $\operatorname{Im} M = KJK^*$, where $K \in [E, \mathcal{H}]$;

(iv) F is an orthogonal projection in $\mathcal{H}$;

(v) the resolvent sets $\rho(\operatorname{Re} M, F)$ and $\rho(M, F)$ are nonempty.

The system (12.17) is called a **scattering** F**-system** if in (ii) $J = I$, in which case the state-space operator M in (12.17) is dissipative: $\operatorname{Im} M \geq 0$, as follows from (iii). Moreover, in this case the first condition $\rho(\operatorname{Re} M, F) \neq \emptyset$ in (v) implies the second condition $\rho(M, F) \neq \emptyset$ in (v), as follows from Corollary 12.1.2.

To each F-system in Definition 12.1.4 one can associate the **impedance function**

$$V_{\Theta_F}(z) = K^*(\operatorname{Re} M - zF)^{-1}K, \quad z \in \rho(\operatorname{Re} M, F). \tag{12.18}$$

Lemma 12.1.5. *Let Θ_F be an F-system of the form (12.17). Then for all $z, w \in \rho(\operatorname{Re} M, F)$, the function $V_{\Theta_F}(z)$ in (12.18) satisfies*

$$V_{\Theta_F}(z) - V_{\Theta_F}(\bar{w})^* = (z - \bar{w})K^*(\operatorname{Re} M - zF)^{-1}F(\operatorname{Re} M - \bar{w}F)^{-1}K. \tag{12.19}$$

If $\rho(\operatorname{Re} M, F)$ is nonempty, then $\mathbb{C}_+ \cup \mathbb{C}_- \subset \rho(\operatorname{Re} M, F)$ and $V_{\Theta_F}(z)$, $z \in \mathbb{C} \setminus \mathbb{R}$, is a Herglotz-Nevanlinna function.

Proof. For each $z, w \in \rho(\operatorname{Re} M, F)$ the resolvent identity (12.10) reads

$$(\operatorname{Re} M - zF)^{-1} - (\operatorname{Re} M - \bar{w}F)^{-1} = (z - \bar{w})(\operatorname{Re} M - zF)^{-1}F(\operatorname{Re} M - \bar{w}F)^{-1}.$$

The definition (12.18) implies (12.19). If $\rho(\operatorname{Re} M, F)$ is nonempty, then according to Lemma 12.1.1, $\mathbb{C} \setminus \mathbb{R} \subset \rho(\operatorname{Re} M, F)$. Clearly, $V_{\Theta_F}(z)^* = V_{\Theta_F}(\bar{z})$ when $z \in \rho(\operatorname{Re} M, F)$. Moreover, it follows from (12.19) that $V_{\Theta_F}(z)$ is (locally) holomorphic on $\rho(\operatorname{Re} M, F)$ and has a non-negative imaginary in the upper half-plane, so that $V_{\Theta_F}(z)$ is an operator-valued Herglotz-Nevanlinna function. □

To each F-system in Definition 12.1.4 one can also associate the **transfer function**

$$W_{\Theta_F}(z) = I - 2iK^*(M - zF)^{-1}KJ, \quad z \in \rho(M, F). \tag{12.20}$$

Lemma 12.1.6. *Let Θ_F be an F-system of the form (12.17). Then for all $z, w \in \rho(M, F)$, the function $W_\Theta(z)$ in (12.20) satisfies*

$$W_{\Theta_F}(z)JW^*_{\Theta_F}(w) - J = 2i(\bar{w} - z)K^*(M - zF)^{-1}F(M^* - \bar{w}F)^{-1}K,$$
$$W^*_{\Theta_F}(w)JW_{\Theta_F}(z) - J = 2i(\bar{w} - z)JK^*(M^* - \bar{w}F)^{-1}F(M - zF)^{-1}KJ.$$

In particular, if $z \in \rho(M, F)$, then $W_{\Theta_F}(z)$ is J-unitary when $z \in \mathbb{R}$, J-expansive when $z \in \mathbb{C}_+$, and J-contractive when $z \in \mathbb{C}_-$.

Proof. The property (iii) in Definition 12.1.4 shows that for all $z, w \in \rho(M, F)$,

$$\begin{aligned}
&(M - zF)^{-1} - (M^* - \bar{w}F)^{-1} \\
&= (M - zF)^{-1}[(M^* - \bar{w}F) - (M - zF)](M^* - \bar{w}F)^{-1} \\
&= (z - \bar{w})(M - zF)^{-1}F(M^* - \bar{w}F)^{-1} - 2i(M - zF)^{-1}KJK^*(M^* - \bar{w}F)^{-1}.
\end{aligned}$$

This identity together with (12.20) implies that

$$\begin{aligned}
&W_{\Theta_F}(z)JW^*_{\Theta_F}(w) - J \\
&\qquad = [I - 2iK^*(M - zF)^{-1}KJ]J[I + 2iJK^*(M^* - \bar{w}F)^{-1}K] - J \\
&\qquad = 2i(\bar{w} - z)K^*(M - zF)^{-1}F(M^* - \bar{w}F)^{-1}K.
\end{aligned}$$

This proves the first equality. Likewise, by using

$$(M - zF)^{-1} - (M^* - \bar{w}F)^{-1} = (z - \bar{w})(M^* - \bar{w}F)^{-1}F(M^* - zF)^{-1} \\ - 2i(M^* - \bar{w}F)^{-1}KJK^*(M - zF)^{-1},$$

one proves the second identity. □

The impedance function $V_{\Theta_F}(z)$ defined in (12.18) and the transfer function $W_{\Theta_F}(z)$ defined in (12.20) are closely connected.

Lemma 12.1.7. *Let Θ_F be an F-system of the form (12.17). Then for all $z \in \rho(\mathrm{Re}\, M, F) \cap \rho(M, F)$ the operators $I + iV_{\Theta_F}(z)J$ and $I + W_{\Theta_F}(z)$ are bounded and invertible. Moreover,*

$$\begin{aligned} V_{\Theta_F}(z) &= i[W_{\Theta_F}(z) - I][W_{\Theta_F}(z) + I]^{-1}J \\ &= i[W_{\Theta_F}(z) + I]^{-1}[W_{\Theta_F}(z) - I]J, \end{aligned} \tag{12.21}$$

and

$$\begin{aligned} W_{\Theta_F}(z) &= [I - iV_{\Theta_F}(z)J][I + iV_{\Theta_F}(z)J]^{-1} \\ &= [I + iV_{\Theta_F}(z)J]^{-1}[I - iV_{\Theta_F}(z)J]. \end{aligned} \tag{12.22}$$

Proof. The following identity with $z \in \rho(M, F) \cap \rho(\mathrm{Re}\, M, F)$

$$(\mathrm{Re}\, M - zF)^{-1} - (M - zF)^{-1} = i(M - zF)^{-1}\mathrm{Im}\, M(\mathrm{Re}\, M - zF)^{-1},$$

leads to

$$K^*(\mathrm{Re}\, M - zF)^{-1}K - K^*(M - zF)^{-1}K \\ = iK^*(M - zF)^{-1}KJK^*(\mathrm{Re}\, M - zF)^{-1}K.$$

Now in view of (12.20) and (12.18)

$$2V_{\Theta_F}(z) + i(I - W_{\Theta_F}(z))J = (I - W_{\Theta_F}(z))V_{\Theta_F}(z),$$

or equivalently,

$$[I + W_{\Theta_F}(z)][I + iV_{\Theta_F}(z)J] = 2I. \tag{12.23}$$

Similarly, the identity

$$(\mathrm{Re}\, M - zF)^{-1} - (M - zF)^{-1} = i(\mathrm{Re}\, M - zF)^{-1}\mathrm{Im}\, M(M - zF)^{-1}$$

with $z \in \rho(M, F) \cap \rho(\mathrm{Re}\, M, F)$ leads to

$$[I + iV_{\Theta_F}(z)J][I + W_{\Theta_F}(z)] = 2I. \tag{12.24}$$

The equalities (12.23) and (12.24) show that the operators are boundedly invertible and consequently one obtains (12.21) and (12.22). □

When M is a bounded operator and $F = I$, the F-system in Definition 12.1.4 reduces to the Livsič canonical system studied in Chapter 5. The present situation with M, generally speaking, unbounded and F an orthogonal projection is of interest as it is phrased in pure Hilbert space terminology; it does not require any rigged Hilbert space triplets as in Chapter 6.

If the system (12.17) is a scattering system ($J = I$), then it follows from (12.22) that $W_{\Theta_F}(z)$ is a Schur function; it is holomorphic on $\mathbb{C}_-$ and its values are contractions. In general, when (12.17) is not a scattering system, the domain of holomorphy of $W_{\Theta_F}(z)$ need not be $\mathbb{C}_-$.

12.2 Multiplication theorems for F-systems

Consider the two F-systems Θ_{F_1} and Θ_{F_2} of the form (12.17), defined by

$$\Theta_{F_1} = \begin{pmatrix} M_1 \; F_1 & K_1 & J \\ \mathcal{H}_1 & & E \end{pmatrix}, \tag{12.25}$$

and

$$\Theta_{F_2} = \begin{pmatrix} M_2 \; F_2 & K_2 & J \\ \mathcal{H}_2 & & E \end{pmatrix}. \tag{12.26}$$

Define the Hilbert space $\mathcal{H}$ by

$$\mathcal{H} = \mathcal{H}_1 \oplus \mathcal{H}_2, \tag{12.27}$$

and let P_j be the orthoprojections from $\mathcal{H}$ onto $\mathcal{H}_j$, $j = 1, 2$. Define the operators M, F, and K by

$$M = M_1P_1 + M_2P_2 + 2iK_1JK_2^*P_2, \quad F = F_1P_1 + F_2P_2, \quad K = K_1 + K_2. \tag{12.28}$$

Theorem 12.2.1. *Let Θ_{F_1} be the F_1-system in* (12.25) *and let Θ_{F_2} be the F_2-system in* (12.26). *Then the aggregate*

$$\Theta = \begin{pmatrix} M \; F & K & J \\ \mathcal{H} & & E \end{pmatrix}, \tag{12.29}$$

with $\mathcal{H}$, M, F, and K, defined by (12.27) *and* (12.28), *is an F-system.*

Proof. Taking adjoints in (12.28) gives

$$M^* = M_1^*P_1 + M_2^*P_2 - 2iK_2^*JK_1^*P_1, \quad K^* = K_1^*P_1 + K_2^*P_2,$$

and therefore,

$$\begin{aligned} M - M^* &= (M_1 - M_1^*)P_1 + (M_2 - M_2^*)P_2 + 2iK_1JK_2^*P_2 + 2iK_2JK_1^*P_1 \\ &= 2iK_1JK_1^*P_1 + 2iK_2JK_2^*P_2 + 2iK_1JK_2^*P_2 + 2iK_2JK_1^*P_1 \\ &= 2i(K_1 + K_2)J(K_1^*P_1 + K_2^*P_2) = 2iKJK^*. \end{aligned}$$

Clearly, F is an orthoprojection. Hence, the aggregate (12.29) with $\mathcal{H}$, M, F, and K, defined by (12.27) and (12.28), is an F-system. □

The F-system Θ in (12.29) is called the **coupling** of the F_1-system Θ_{F_1} and the F_2-system Θ_{F_2}; it is denoted by

$$\Theta = \Theta_{F_1}\Theta_{F_2}.$$

Theorem 12.2.2. *Let the F-system Θ be the coupling of the F_1-system Θ_{F_1} and the F_2-system Θ_{F_2}. Then the associated transfer functions satisfy*

$$W_\Theta(z) = W_{\Theta_{F_1}}(z)W_{\Theta_{F_2}}(z), \quad z \in \rho(M_1, F_1) \cap \rho(M_2, F_2). \tag{12.30}$$

Proof. Let $z \in \rho(M_1, F_1) \cap \rho(M_2, F_2)$. Observe that

$$\begin{aligned} M - zF &= M_1P_1 + M_2P_2 + 2iK_1JK_2^*P_2 - z(F_1P_1 + F_2P_2) \\ &= (M_1 - zF_1)P_1 + (M_2 - zF_2)P_2 + 2iK_1JK_2^*P_2. \end{aligned}$$

Therefore,

$$\begin{aligned} &(M - zF)\left[(M_1 - zF_1)^{-1}P_1 + (M_2 - zF_2)^{-1}P_2\right. \\ &\qquad\qquad \left.-2i(M_1 - zF_1)^{-1}K_1JK_2^*(M_2 - zF_2)^{-1}P_2\right] \\ &= P_1 - 2iK_1JK_2^*(M_2 - zF_2)^{-1}P_2 + P_2 + 2iK_1JK_2^*(M_2 - zF_2)^{-1}P_2 \\ &= P_1 + P_2 = I. \end{aligned}$$

A similar result is obtained when the factors are multiplied in the reverse order. Hence, it follows that $z \in \rho(M, F)$ and that

$$\begin{aligned} (M - zF)^{-1} =& (M_1 - zF_1)^{-1}P_1 + (M_2 - zF_2)^{-1}P_2 \\ &- 2i(M_1 - zF_1)^{-1}K_1JK_2^*(M_2 - zF_2)^{-1}P_2. \end{aligned}$$

Furthermore, (12.30) follows from

$$\begin{aligned} W_\Theta(z) &= I - 2iK^*(M - zF)^{-1}KJ \\ &= I - 2i(K_1^*P_1 + K_2^*P_2)\left[(M_1 - zF_1)^{-1}P_1 + (M_2 - zF_2)^{-1}P_2\right. \\ &\quad \left.-2i(M_1 - zF_1)^{-1}K_1JK_2^*(M_2 - zF_2)^{-1}P_2\right](K_1 + K_2)J \\ &= I - 2iK_1^*(M_1 - zF_1)^{-1}K_1J - 2iK_2^*(M_2 - zF_2)^{-1}K_2J \\ &\quad + (2i)^2K_1^*(M_1 - zF_1)^{-1}K_1JK_2^*(M_2 - zF_2)^{-1}K_2J \\ &= [I - 2iK_1^*(M_1 - zF_1)^{-1}K_1J][I - 2iK_2^*(M_2 - zF_2)^{-1}K_2J] \\ &= W_{\Theta_{F_1}}(z)W_{\Theta_{F_2}}(z). \end{aligned}$$

□

When $F_1 = I$ and $F_2 = I$, the above results reduce to the multiplication Theorem 7.3.2 for L-systems.

Let Θ_F be an F-system of the form (12.17). Assume that $\mathcal{H}_1$ is a closed linear subspace of $\mathcal{H}$, which is invariant with respect to M and F. Let P_1 be the orthogonal projection from $\mathcal{H}$ onto $\mathcal{H}_1$. Consider the aggregate Θ_{F_1} defined by

$$\Theta_{F_1} = \begin{pmatrix} M_1 & F_1 & K_1 & J \\ \mathcal{H}_1 & & & E \end{pmatrix}, \tag{12.31}$$

where

$$M_1 = M\upharpoonright \mathcal{H}_1, \quad K_1 = P_1 K, \quad F_1 = F\upharpoonright \mathcal{H}_1. \tag{12.32}$$

Since $\mathcal{H}_1$ is invariant under M and F, $M_1 \in [\mathcal{H}_1, \mathcal{H}_1]$, and F_1 is an orthoprojection. Moreover, $\operatorname{Im} M_1 = K_1 J K_1^*$, and hence (12.31) is an F_1-system. The F_1-system Θ_{F_1} is called the **projection** of Θ_F onto the invariant subspace $\mathcal{H}_1$, and is denoted by:

$$\Theta_{F_1} = \operatorname{pr}_{\mathcal{H}_1} \Theta_F.$$

Let P_2 be the orthogonal projection from $\mathcal{H}$ onto $\mathcal{H}_2 = \mathcal{H} \ominus \mathcal{H}_1$. Define the aggregate Θ_{F_2}

$$\Theta_{F_2} = \begin{pmatrix} M_2 & F_2 & K_2 & J \\ \mathcal{H} \ominus \mathcal{H}_1 & & & E \end{pmatrix}, \tag{12.33}$$

where

$$M_2 = P_2 M\upharpoonright \mathcal{H}_2, \quad K_2 = P_2 K, \quad F_2 = F\upharpoonright \mathcal{H}_2. \tag{12.34}$$

Since $\mathcal{H}_2 = \mathcal{H}\ominus\mathcal{H}_1$ is invariant under F_2, then F_2 is an orthoprojection. Moreover, $\operatorname{Im} M_2 = K_2 J K_2^*$, so that (12.33) is an F_2-system. The F_2-system Θ_{F_2} is called the projection of Θ_F onto the orthogonal complement $\mathcal{H}_2 = \mathcal{H} \ominus \mathcal{H}_1$ of the invariant subspace $\mathcal{H}_1$; it is denoted by $\Theta_{F_2} = \operatorname{pr}_{\mathcal{H}\ominus\mathcal{H}_1} \Theta_F$.

Theorem 12.2.3. *Let Θ_F be an F-system of the form* (12.17)*. Assume that $\mathcal{H}_1$ is a closed linear subspace of $\mathcal{H}$, which is invariant under M and F. Then Θ_F is the coupling of the F_1-system* (12.31) *and the F_2-system* (12.33)*, i.e., the projections of Θ_F onto the invariant subspace $\mathcal{H}_1$ and its orthogonal complement $\mathcal{H}_2 = \mathcal{H}\ominus\mathcal{H}_1$, respectively. Moreover,*

$$W_\Theta(z) = W_{\Theta_{F_1}}(z) W_{\Theta_{F_2}}(z). \tag{12.35}$$

Proof. Consider the operators defined in (12.32) and (12.34). Since $\mathcal{H}_1$ is invariant under M and F, it is clear that $\mathcal{H}_2 = \mathcal{H} \ominus \mathcal{H}_1$ is invariant under M^* and F. Therefore one obtains

$$\begin{aligned} M &= P_1 M P_1 + P_2 M P_2 + P_1 M P_2 + P_2 M P_1 = P_1 M P_1 + P_2 M P_2 + P_1 M P_2 \\ &= M_1 P_1 + M_2 P_2 + P_1 M P_2 = M_1 P_1 + M_2 P_2 + 2i P_1 \operatorname{Im} M P_2 \\ &= M_1 P_1 + M_2 P_2 + 2i P_1 K J K^* P_2 = M_1 P_1 + M_2 P_2 + 2i K_1 J K_2^*, \end{aligned}$$

where the identities $K_1^* = K^* P_1$ and $K_2^* = K^* P_2$ have been used. Moreover, it is clear that $F = F_1 P_1 + F_2 P_2$ and $K = K_1 + K_2$. Hence, Θ_F has the factorization $\Theta_{F_1} \Theta_{F_2}$, where Θ_{F_1} and Θ_{F_2} are given by (12.31) and (12.33). The identity (12.35) is now a consequence of Theorem 12.2.2. □

Theorem 12.2.4. *Each constant J-unitary operator W on a finite-dimensional Hilbert space E can be realized as a transfer function of some F-system of the form* (12.17).

Proof. Assume that (-1) belongs to the resolvent set of the J-unitary operator W, and define

$$Q = i[W - I][W + I]^{-1}J,$$

so that

$$Q^* = -iJ[W^* + I]^{-1}[W^* - I].$$

The operator Q is selfadjoint, since $W^*JW - J = 0$ implies

$$\begin{aligned}Q - Q^* &= i[W - I][W + I]^{-1}J + iJ[W^* + I]^{-1}[W^* - I] \\ &= iJ[W^* + I]^{-1}\{(W^* + I)J(W - I) + (W^* - I)J(W + I)\}[W + I]^{-1}J \\ &= 2iJ[W^* + I]^{-1}(W^*JW - J)[W + I]^{-1}J = 0.\end{aligned}$$

Now assume also that $1 \in \rho(W)$, so that Q is invertible. Define the Hilbert space $\mathcal{H}$ by $\mathcal{H} = E$ and let $K : E \to E = \mathcal{H}$ be any bounded and boundedly invertible operator. Define the operator M by

$$M = KQ^{-1}(I + iQJ)K^*,$$

so that $\operatorname{Im} M = KJK^*$. Obviously, the aggregate

$$\Theta_0 = \begin{pmatrix} KQ^{-1}(I + iQJ)K^* \ 0 & K & J \\ & E & E \end{pmatrix},$$

is an F-system with $F = 0$. By means of $I + iQJ = 2[W + I]^{-1}$, one obtains

$$\begin{aligned}W_{\Theta_0}(z) &= I - 2iK^*(M - zF)^{-1}KJ = I - 2iK^*M^{-1}KJ \\ &= I - 2iK^*(K^*)^{-1}(I + iQJ)^{-1}QK^{-1}KJ = I - 2i(I + iQJ)^{-1}QJ \\ &= (I + iQJ)^{-1}[(I + iQJ) - 2iQJ] = (I + iQJ)^{-1}(I - iQJ) = W.\end{aligned}$$

Hence, the theorem has been shown for the case that $\pm 1 \in \rho(W)$. Now consider the case of an arbitrary J-unitary operator W acting in a Hilbert space E. Since E is finite-dimensional, it is easy to see that W can be represented in the form

$$W = W_1 W_2,$$

where W_j is a J-unitary operator in E and $\pm 1 \in \rho(W_j)$, $j = 1, 2$. By the previous case, each of the operators W_1 and W_2 can be realized as transfer operator-valued functions of two F-systems Θ_{F_1} and Θ_{F_2}, respectively, i.e.,

$$W_{\Theta_{F_1}}(z) = W_1, \quad W_{\Theta_{F_2}}(z) = W_2.$$

Consider the coupling $\Theta_F = \Theta_{F_1}\Theta_{F_2}$ of these operator systems, and apply the multiplication theorem. Then

$$W_{\Theta_F}(z) = W_{\Theta_{F_1}}(z)W_{\Theta_{F_2}}(z) = W_1 W_2 = W.$$

This completes the proof. □

Theorem 12.2.2 and Theorem 12.2.4 imply the following result.

Corollary 12.2.5. *Let Θ_F be an F-system of the form (12.17). Let U and V be constant J-unitary operators on E. Then there exists an F-system $\widetilde{\Theta}_F$, such that its transfer function takes the form*

$$W_{\widetilde{\Theta}_F}(z) = UW_{\Theta_F}(z)V.$$

Example. As an instructive example, consider $\mathcal{H} = L^2_{[0,a]}$ with $a > 1$, $E = \mathbb{C}$, and the operators $D \in [\mathcal{H}, \mathcal{H}]$, $F \in [\mathcal{H}, \mathcal{H}]$, and $K \in [E, \mathcal{H}]$ defined by

$$\begin{aligned} &(Df)(t) = t\,f(t), \quad (Ff)(t) = \chi_{[0,1]}(t), \quad f \in L^2_{[0,a]}, \\ &(K\xi)(t) = \xi, \quad t \in [0,a], \quad \xi \in \mathbb{C} = E. \end{aligned}$$

Here $\chi_{[0,1]}(t)$ denotes the characteristic function of the closed interval $[0,1]$. Define the operator M by

$$(Mf)(t) = (Df)(t) + i(KK^*f)(t) = tf(t) + i\int_0^a f(t)dt.$$

Then $\operatorname{Im} M = KJK^*$ with $J = 1$ and the aggregate

$$\Theta_F = \begin{pmatrix} M & F & K & I \\ L^2_{[0,a]} & & & \mathbb{C} \end{pmatrix}$$

is an F-system. Since $\operatorname{Re} M = D$, the function $V_{\Theta_F}(z)$ in (12.18) is given by

$$\begin{aligned} V_\Theta(z) &= K^*(\operatorname{Re} M - zF)^{-1}K = K^*(D - zF)^{-1}K \\ &= ((D - zF)^{-1}\mathbf{1}, \mathbf{1})_{L^2_{[0,a]}} = \int_0^a \frac{1}{t - z\chi_{[0,1]}(t)}\,dt \\ &= \left(\int_0^1 \frac{dt}{t-z} + \int_1^a \frac{dt}{t}\right) = \left[\log\left(1 - \frac{1}{z}\right) + \log(a)\right]. \end{aligned}$$

By the formula (12.22) we obtain the transfer function

$$W_\Theta(z) = \frac{1 - i[\log(1 - 1/z) + \log(a)]}{1 + i[\log(1 - 1/z) + \log(a)]}.$$

12.3 Realizations in the case of a compactly supported measure

Let A be a self-adjoint operator in a Hilbert space $\mathcal{H}$, and let $\mathfrak{L}$ be a closed linear subspace of $\mathcal{H}$. We recall that when A is bounded, the subspace $\mathfrak{L}$ is said to be invariant under A if $A\mathfrak{L} \subset \mathfrak{L}$. When A is possibly unbounded, the subspace $\mathfrak{L}$ is said to be invariant under A if for some point $z \in \mathbb{C} \setminus \mathbb{R}$,

$$(A - zI)^{-1}\mathfrak{L} \subset \mathfrak{L}. \tag{12.36}$$

Lemma 12.3.1. *Let $\mathfrak{L}$ be a subspace of $\mathcal{H}$. If (12.36) holds for some $z \in \mathbb{C} \setminus \mathbb{R}$, then $\mathfrak{L}$ reduces $A = A^*$:*

$$A = A_1 \oplus A_2, \quad A_j = A_j^* \text{ in } \mathcal{H}_j,\ j = 1,2, \quad \mathcal{H}_1 = \mathfrak{L}, \quad \mathcal{H}_2 = \mathcal{H} \ominus \mathfrak{L}. \tag{12.37}$$

Proof. By means of the resolvent operator of A one can describe the graph of A for each $z \in \rho(A)$ as follows:

$$A = \{\, \{(A - zI)^{-1}h, (I + z(A - zI)^{-1})h\} : h \in \mathcal{H} \,\}.$$

By assumption $(A - zI)^{-1}\mathfrak{L} \subset \mathfrak{L}$, and therefore also $(A - \bar{z}I)^{-1}\mathfrak{L}^\perp \subset \mathfrak{L}^\perp$. This gives rise to the following restrictions of A in $\mathcal{H}_1$ and $\mathcal{H}_2$, respectively:

$$\begin{aligned} A_1 &= \{\, \{(A - zI)^{-1}h, (I + z(A - zI)^{-1})h\} : h \in \mathfrak{L} \,\}, \\ A_2 &= \{\, \{(A - \bar{z}I)^{-1}h, (I + \bar{z}(A - \bar{z}I)^{-1})h\} : h \in \mathcal{H} \ominus \mathfrak{L} \,\}. \end{aligned}$$

Now A_j as a restriction of A is symmetric, say, with defect numbers (n_j^+, n_j^-) in $\mathcal{H}_j$, $j = 1,2$. Since $A_1 \oplus A_2 = A$ and A is self-adjoint one concludes that $(n_1^+ + n_2^+, n_1^- + n_2^-) = (0,0)$. Consequently, A_1 and A_2 are (graphs of) self-adjoint operators. □

Lemma 12.3.2. *Let D be a bounded self-adjoint operator in a Hilbert space $\mathcal{H}$ and let F be an orthogonal projector in $\mathcal{H}$ with $\dim \ker(F) < \infty$. Then the resolvent set $\rho(D,F)$ is non-empty if and only if*

$$\ker(D) \cap \ker(F) = \{0\}, \tag{12.38}$$

in which case $\mathbb{C} \setminus \mathbb{R} \subset \rho(D,F)$.

Proof. Let $\mathcal{H} = \operatorname{Ran}(F) \oplus \ker(F)$ and let $D = (D_{ij})_{i,j=1}^2$ and $F = (F_{ij})_{i,j=1}^2$ be decomposed accordingly, so that $F_{11} = I$ and $F_{ij} = 0$ otherwise. Then $z \in \rho(D,F)$ if and only if the operator

$$D_{22} - D_{12}^*(D_{11} - z)^{-1}D_{12}$$

is boundedly invertible. Since $\dim \ker(F) < \infty$, this is equivalent to

$$\ker\left(D_{22} - D_{12}^*(D_{11} - z)^{-1}D_{12}\right) = \{0\}.$$

Assume that $D_{22}f = D_{12}^*(D_{11} - z)^{-1}D_{12}f$ for some $f = (I - F)g$. Then the resolvent identity applied to $(D_{11} - zI)^{-1}$ yields

$$(z - \bar{z})\left(D_{12}^*(D_{11} - \bar{z}I)^{-1}(D_{11} - zI)^{-1}D_{12}f, f\right)_{\ker(F)} = 0.$$

For $z \in \mathbb{C} \setminus \mathbb{R}$, this implies $(D_{11} - zI)^{-1}D_{12}f = 0$ and thus $D_{12}f = 0$. But then also $D_{22}f = 0$, i.e., $f \in \ker(D) \cap \ker(F)$. The converse statement is obvious. □

Lemma 12.3.3. *Let D be a bounded self-adjoint operator and F be an orthogonal projection with* $\dim\ker(F) < \infty$ *in a Hilbert space $\mathcal{H}$. Let also $\rho(D,F) \neq \emptyset$ and K be a bounded operator from a finite-dimensional Hilbert space E into $\mathcal{H}$. Then there exists an interval $[a,b] \subset \mathbb{R}$, such that $\mathbb{C} \setminus [a,b] \subset \rho(D,F)$, and the function $V(z) = K^*(D - zF)^{-1}K$, $z \in \rho(D,F)$ admits the integral representation*

$$V(z) = Q + zL + \int_a^b \frac{dG(t)}{t-z}, \quad Q = Q^*, \quad L \geq 0, \tag{12.39}$$

where Q is a self-adjoint operator, L is a non-negative operator, and $G(t)$ is a non-decreasing operator-valued function in E.

Proof. Define the subspaces $\mathcal{H}_1$, $\mathcal{H}_2$, and $\mathcal{H}_3$ by

$$\mathcal{H}_1 = F\mathcal{H}, \quad \mathcal{H}_2 = P(I-F)\mathcal{H}, \quad \mathcal{H}_3 = (I-P)(I-F)\mathcal{H},$$

where P denotes the orthogonal projection onto $\ker((I-F)D{\restriction}(I-F)\mathcal{H})$, so that $\mathcal{H} = \mathcal{H}_1 \oplus \mathcal{H}_2 \oplus \mathcal{H}_3$. This gives rise to the following orthogonal operator matrix representations for D and F:

$$D = \begin{pmatrix} D_{11} & D_{12} & D_{13} \\ D_{12}^* & D_{22} & 0 \\ D_{13}^* & 0 & 0 \end{pmatrix}, \quad F = \begin{pmatrix} I & 0 & 0 \\ 0 & 0 & 0 \\ 0 & 0 & 0 \end{pmatrix}.$$

Since $\dim\ker(F) < \infty$ and $\rho(D,F) \neq \emptyset$, we can apply Lemma 12.3.2 to obtain (12.38). The latter implies that D_{22} is invertible and therefore the following inverse exists and is bounded:

$$\begin{aligned} H_{11}(z)^{-1} &= \begin{pmatrix} D_{11} - zI & D_{12} \\ D_{12}^* & D_{22} \end{pmatrix}^{-1} \\ &= \begin{pmatrix} 0 & 0 \\ 0 & D_{22}^{-1} \end{pmatrix} + \begin{pmatrix} I \\ -D_{22}^{-1}D_{12}^* \end{pmatrix} (\hat{D} - zI)^{-1} \begin{pmatrix} I & -D_{12}D_{22}^{-1} \end{pmatrix}, \end{aligned} \tag{12.40}$$

where $\hat{D}$ denotes the Schur complement of $(D_{ij})_{i,j=1}^2$,

$$\hat{D} = D_{11} - D_{12}D_{22}^{-1}D_{21}, \quad \hat{D} = \hat{D}^*. \tag{12.41}$$

By assumption $(H_{ij}(z))_{i,j=1}^2 := (D - zF)$ is boundedly invertible for $z \in \rho(D,F)$. The invertibility of $H_{11}(z)$ in (12.40) now implies that also

$$U(z) = H_{21}(z)H_{11}^{-1}(z)H_{12}(z) = D_{13}^*(\hat{D} - zI)^{-1}D_{13}, \tag{12.42}$$

has a bounded inverse for $z \in \rho(D,F)$. The boundedness of $\hat{D}$ in (12.41) implies that $(\hat{D} - z)^{-1}$ is holomorphic outside a compact interval $[a,b]$ of the real line. Since the operator $U(z)$ in (12.42) has a bounded inverse, $\ker(D_{13}) = \{0\}$. Moreover, the inverse of $U(z)$ is holomorphic on $\mathbb{C} \setminus [a,b]$. In fact, if $\mathcal{H}_1$ and $\hat{D} = (\hat{D}_{ij})_{i,j=1}^2$ are decomposed orthogonally according to $\operatorname{Ran}(D_{13})$, one can write

$$U(z)^{-1} = D_{13}^{-1}(T(z) - z)D_{13}^{-*},$$

where

$$T(z) = \hat{D}_{11} - \hat{D}_{12}(\hat{D}_{22} - z)^{-1}\hat{D}_{12}^*.$$

The function $T(z)$ is holomorphic on $\mathbb{C} \setminus [a, b]$, since the numerical ranges satisfy the obvious inclusion $W(\hat{D}_{22}) \subset W(\hat{D})$. Now the analog of the inverse formula in (12.40), when applied to $(D - zF) = (H_{ij}(z))_{i,j=1}^2$, implies that also $V(z)$ is holomorphic on $\mathbb{C} \setminus [a, b]$, i.e., $\mathbb{C} \setminus [a, b] \subset \rho(V)$. Therefore, by Stieltjes' inversion formula the measure in the Nevanlinna integral representation of $V(z)$ is supported on $[a, b]$, in which case this integral representation reduces to (12.39). □

It was shown in Lemma 12.1.5 that the function $V_{\Theta_F}(z)$ associated with the F-system Θ_F in (12.17) is a Herglotz-Nevanlinna function, assuming that $\rho(\operatorname{Re} M, F) \neq \emptyset$. However, more can be said when $\operatorname{Re} M$ and F have a "bounded non-commuting part" in the sense to be specified in the next theorem.

Theorem 12.3.4. *Let Θ_F be an F-system of the form* (12.17) *such that* $\dim \ker(F) < \infty$. *Assume that $\rho(\operatorname{Re} M, F)$ is nonempty, and let $V_{\Theta_F}(z)$ be its impedance function defined by* (12.18). *Then the following statements hold:*

(i) *If there is an orthogonal projection P commuting with F and $(\operatorname{Re} M - zI)^{-1}$ for some $z \in \mathbb{C} \setminus \mathbb{R}$, such that $\operatorname{Ran}(P) \subset \operatorname{Ran}(F)$ and the restriction $\operatorname{Re} M \restriction \ker(P)$ is bounded, then $V_{\Theta_F}(z)$ is an operator-valued Herglotz-Nevanlinna function with the integral representation*

$$V_{\Theta_F}(z) = Q + Lz + \int_{-\infty}^{+\infty} \frac{dG(t)}{t - z}, \qquad z \in \rho(M, F), \tag{12.43}$$

where $Q = Q^*$, $L \geq 0$, and $G(t)$ is a non-decreasing operator-valued function on $\mathbb{R}$ satisfying (12.8).

(ii) *If, in particular, $F(\operatorname{Re} M - zI)^{-1} = (\operatorname{Re} M - zI)^{-1}F$ for some $z \in \mathbb{C} \setminus \mathbb{R}$, then the integral representation of $V(z)$ is of the form* (12.43) *with $L = 0$.*

Proof. Let P be an orthogonal projection satisfying the assumptions in (i). Then, by Lemma 12.3.1, $\operatorname{Re} M$ admits an orthogonal decomposition of the form (12.37), where $M_2 = \operatorname{Re} M \restriction \ker(P)$ is bounded by the assumptions. Decomposing $K = K_1 \oplus K_2$ according to $\mathcal{H} = \operatorname{Ran}(P) \oplus \ker(P)$ one obtains

$$V_{\Theta_F}(z) = K_1^*(M_1 - zI)^{-1}K_1 + K_2^*(M_2 - zF_2)^{-1}K_2, \tag{12.44}$$

where $F_2 = (I - P)F$ is an orthogonal projection in $\ker(P)$. Now $\dim \ker(F_2) < \infty$ and according to Lemma 12.3.3 the second summand in (12.44) has the integral representation

$$K_2^*(M_2 - zF_2)^{-1}K_2 = Q_2 + L_2 z + \int_a^b \frac{dG_2(t)}{t - z}, \qquad a, b \in \mathbb{R}. \tag{12.45}$$

Clearly, the first summand in (12.44) has the integral representation

$$K_1^*(M_1 - zI)^{-1}K_1 = \int_{-\infty}^{+\infty} \frac{dG_1(t)}{t-z}, \quad \int_{-\infty}^{+\infty} (dG_1(t)x, x) < \infty, \quad x \in E. \tag{12.46}$$

Hence, (12.43) follows from (12.45) and (12.46) with $Q = Q_2$, $L = L_2$, and $G(t) = G_1(t) + G_2(t)$.

To prove (ii) observe that the decomposition $\operatorname{Re} M = M_1 \oplus M_2$ for $\operatorname{Re} M$ in Lemma 12.3.1 gives (12.44) with $F_2 = (I - F)F = 0$. Since $\rho(\operatorname{Re} M, F) \neq \emptyset$, the operator M_2 must be invertible, and therefore

$$(\operatorname{Re} M - zF)^{-1} = (M_1 - zI)^{-1} \oplus M_2^{-1},$$

and

$$V_{\Theta_F}(z) = K_1^*(M_1 - zI)^{-1}K_1 + K_2^* M_2^{-1} K_2.$$

Now one obtains (12.43) with $Q = K_2^* M_2^{-1} K_2$ and $L = 0$. □

In the remaining part of this section the converse to Theorem 12.3.4 will be established.

Lemma 12.3.5. *Let Q and $L \geq 0$ be self-adjoint operators in E, $\dim E < \infty$. Then the following representations hold:*

$$Q = K_1^*(D_1 - zF_1)^{-1}K_1, \quad z \in \rho(D_1, F_1)\,(= \mathbb{C}), \tag{12.47}$$

$$zL = K_2^*(D_2 - zF_2)^{-1}K_2, \quad z \in \rho(D_2, F_2)\,(= \mathbb{C}), \tag{12.48}$$

where K_j is an injective operator from E into a Hilbert space $\mathcal{H}_j$, $\dim \mathcal{H}_j < \infty$, D_j is a self-adjoint operator in $\mathcal{H}_j$, and F_j is an orthogonal projection in $\mathcal{H}_j$, $j = 1, 2$.

Proof. The representation (12.47) is straightforward to show if Q is invertible and one can take $\mathcal{H}_1 = E$, $K = I$, $D = Q^{-1}$, $F = 0$. In the general case write $Q = Q_1 + Q_2$ with two invertible self-adjoint operators Q_1 and Q_2. Let $\mathcal{H}_1 = E \oplus E$ and introduce in $\mathcal{H}_1$ the operators

$$K_1 = \begin{pmatrix} I \\ I \end{pmatrix}, \quad D_1 = \begin{pmatrix} Q_1^{-1} & 0 \\ 0 & Q_2^{-1} \end{pmatrix}, \quad F_1 = 0.$$

Then K_1 is an injective operator from E into $\mathcal{H}_1 = E \oplus E$, D_1 is self-adjoint and invertible, and $F_1 = 0$ is an orthogonal projection in $\mathcal{H}_1$. Clearly, $\rho(D_1, F_1) = \mathbb{C}$ and it is straightforward to check that $K_1^*(D_1 - zI)^{-1}K_1 = Q_1 + Q_2 = Q$. This proves (12.47).

Now consider the function zL. Let $\mathcal{H}_2$ be the Hilbert space $E \oplus E$ and define in $\mathcal{H}_2$ the operators

$$K_2 = \begin{pmatrix} L^{\frac{1}{2}} \\ kI \end{pmatrix}, \quad D_2 = \begin{pmatrix} 0 & iI \\ -iI & 0 \end{pmatrix}, \quad F_2 = \begin{pmatrix} 0 & 0 \\ 0 & I \end{pmatrix},$$

where $k \in \mathbb{R}\backslash\{0\}$. Then K_2 is an injective operator from E into $\mathcal{H}_2$, D_2 is self-adjoint and F_2 is an orthogonal projection in $\mathcal{H}_2$. Moreover, $\rho(D_2, F_2) = \mathbb{C}$, and it is easy to check that

$$K_2^*(D_2 - zF_2)^{-1}K_2 = zL + iL^{\frac{1}{2}}k - ikL^{\frac{1}{2}} = zL. \qquad \square$$

Theorem 12.3.6. *Each Herglotz-Nevanlinna function $V(z)$ in a Hilbert space E, $n = \dim E < \infty$, of the form*

$$V(z) = Q + zL + \int_{-\infty}^{+\infty} \frac{dG(t)}{t - z}$$

where $Q = Q^*$, $L \geq 0$, and $G(t)$ is a non-decreasing operator-valued function on $\mathbb{R}$ satisfying (12.8), can be represented in the form

$$V(z) = K^*(D - zF)^{-1}K, \quad z \in \mathbb{C} \setminus \mathbb{R} \subset \rho(D, F), \tag{12.49}$$

where K is an operator from E into $\mathcal{H}$, D is a self-adjoint operator in a Hilbert space $\mathcal{H}$, and F is an orthogonal projection in $\mathcal{H}$ with $\dim \ker(F) < \infty$. *Moreover, if $L = 0$, then the operators D and F can be selected such that $(D - zI)^{-1}$, $z \in \mathbb{C} \setminus \mathbb{R}$, and F commute.*

Proof. Consider the Hilbert space $\mathcal{H}_3 = L^2_G(E)$ of measurable vector functions $f = f(t)$ with

$$\|f\|^2 = \int_{-\infty}^{+\infty} (dG(t)f(t), f(t)) < \infty.$$

Define the vector functions $\omega_j(t) \equiv e_j$, $j = 1, \ldots, n$, where $\{e_j\}$, $j = 1, \ldots, n$, is an orthonormal basis in E. The integrability property (12.8) guarantees that $\omega_j \in \mathcal{H}_3$. Let $K_3 : E \to \mathcal{H}_3$ be the linear operator determined by $K_3 e_j = \omega_j(t)$, $j = 1, \ldots, n$. Let $D_3 = t$ be the multiplication operator by the independent variable in $\mathcal{H}_3$, and let $F_3 = I$. Then

$$K_3^*(D_3 - zF_3)^{-1}K_3 = \int_{\mathbb{R}} \frac{dG(t)}{t - z}.$$

Now let the Hilbert space $\mathcal{H}$ be defined by $\mathcal{H} = \mathcal{H}_1 \oplus \mathcal{H}_2 \oplus \mathcal{H}_3$, where $\mathcal{H}_1$ and $\mathcal{H}_2$ are as in Lemma 12.3.5. In addition, using the operators introduced in the proof of Lemma 12.3.5, define

$$K = \begin{pmatrix} K_1 \\ K_2 \\ K_3 \end{pmatrix}, \quad D = \begin{pmatrix} D_1 & 0 & 0 \\ 0 & D_2 & 0 \\ 0 & 0 & D_3 \end{pmatrix}, \quad F = \begin{pmatrix} F_1 & 0 & 0 \\ 0 & F_2 & 0 \\ 0 & 0 & F_3 \end{pmatrix}. \tag{12.50}$$

Then K is an operator from E into $\mathcal{H}$, D is a self-adjoint operator and F is an orthogonal projection in $\mathcal{H}$. Moreover, it is easy to check that (12.49) is satisfied. Observe, that if $L = 0$, then the definitions of K, D, and F can be reduced to $K^* = (K_1^*, K_3^*)$, $D = D_1 \oplus D_3$, and $F = F_1 \oplus F_3$, in which case the commutativity of D and F follows from $F_1 = 0$, $F_3 = I$. $\square$

The next result shows that each Nevanlinna function of the form (12.43) can be realized via an F-system (12.17) with $J = I$, i.e., by means of a scattering F-system.

Theorem 12.3.7. *For each Herglotz-Nevanlinna function acting on a Hilbert space E, $\dim E < \infty$, of the form*

$$V(z) = Q + zL + \int_{-\infty}^{+\infty} \frac{dG(t)}{t - z},$$

where $Q = Q^*$, $L \geq 0$, and $G(t)$ satisfies (12.8), there is a scattering F-system Θ_F of the form (12.17) with $\dim\ker(F) < \infty$, whose transfer function $W_{\Theta_F}(z)$ is defined and holomorphic on $\mathbb{C}_-$ and such that

$$V(z) = V_{\Theta_F}(z) = i[W_{\Theta_F}(z) - I][W_{\Theta_F}(z) + I]^{-1}.$$

Proof. By Theorem 12.3.6 the function $V(z)$ can be represented via an F-resolvent in the form (12.49). Let $\mathcal{H}$, K, D, and F be as in Theorem 12.3.6. Define in $\mathcal{H}$ a dissipative operator M by $M = D + iKK^*$. Then $D = \operatorname{Re} M$ and by construction $\mathbb{C} \setminus \mathbb{R} \subset \rho(D, F)$. Now Corollary 12.1.2 shows that $\rho(M, F) \neq \emptyset$ and hence $\mathbb{C}_- \subset \rho(M, F)$, cf. Lemma 12.1.1. This gives rise to an F-system Θ_F satisfying all the properties required in Definition 12.1.4. Moreover, the transfer function $W_{\Theta_F}(z)$ associated to Θ_F in (12.20) is holomorphic on $\mathbb{C}_-$. It remains to apply Lemma 12.1.7 to $V_{\Theta_F}(z) = V(z)$ and $W_{\Theta_F}(z)$ with $z \in \mathbb{C}_- \subset \rho(M, F) \cap \rho(D, F)$. □

12.4 Definitions of NCI-systems and NCL-systems

Now we can give the proper definitions for NCI-systems and NCL-systems.

Definition 12.4.1. Let $\dot{A}$ be a closed symmetric operator in a Hilbert space $\mathcal{H}$ and let $\mathcal{H}_+ \subset \mathcal{H} \subset \mathcal{H}_-$ be the rigged Hilbert space associated with $\dot{A}$. The system of equations

$$\begin{cases} (\mathbb{D} - zF_+)x = K\varphi_-, \\ \varphi_+ = K^*x, \end{cases} \tag{12.51}$$

where E is a finite-dimensional Hilbert space is called a **non-canonical impedance system** or **NCI-system** if:

(i) $\mathbb{D} \in [\mathcal{H}_+, \mathcal{H}_-]$ is a t-self-adjoint bi-extension of a symmetric operator $\dot{A}$;

(ii) $K \in [E, \mathcal{H}_-]$;

(iii) F_+ is an orthogonal projection in $\mathcal{H}_+$ and $\mathcal{H}$;

(iv) the set $\rho(\mathbb{D}, F_+, K)$ of all points $z \in \mathbb{C}$ where $(\mathbb{D} - zF_+)^{-1}$ exists on $\mathcal{H} + \operatorname{Ran}(K)$ and $(-, \cdot)$-continuous is open.

Let $T \in \Lambda(\dot{A})$, $\mathbb{A} \in [\mathcal{H}_+, \mathcal{H}_-]$ be a $(*)$-extension of T, K be a bounded linear operator from a finite-dimensional Hilbert space E into $\mathcal{H}_-$, $K^* \in [\mathcal{H}_+, E]$, $J = J^* = J^{-1} \in [E, E]$, and $\operatorname{Im}\mathbb{A} = KJK^*$. Let also F_+ be an orthogonal projection in $\mathcal{H}_+$ and $\mathcal{H}$. Consider the system of equations

$$\begin{cases} iF_+\frac{d\chi}{dt} + \mathbb{A}\chi(t) = KJ\psi_-(t), \\ \chi(0) = x \in \mathcal{H}_+, \\ \psi_+ = \psi_- - 2iK^*\chi(t). \end{cases} \tag{12.52}$$

Let $L^2_{[0,\tau_0]}(E)$ be the Hilbert space of E-valued functions equipped with an inner product (12.13). We have the following lemma.

Lemma 12.4.2. *If for a given continuous $\psi_-(t) \in L^2_{[0,\tau_0]}(E)$ we have that $\chi(t) \in \mathcal{H}_+$ and $\psi_+(t) \in L^2_{[0,\tau_0]}(E)$ satisfy (12.52), then a system of the form (12.52) satisfies the metric conservation law*

$$2\|F_+\chi(\tau)\|^2 - 2\|F_+\chi(0)\|^2 = \int_0^\tau (J\psi_-, \psi_-)_E\, dt - \int_0^\tau (J\psi_+, \psi_+)_E\, dt, \tag{12.53}$$
$\tau \in [0, \tau_0]$.

The proof of Lemma 12.4.2 follows from Lemmas 6.3.3 and 12.1.3.

Given an input vector $\psi_- = \varphi_- e^{izt} \in E$, we seek solutions to the system (12.52) as an output vector $\psi_+ = \varphi_+ e^{izt} \in E$ and a state-space vector $\chi(t) = xe^{izt} \in \mathcal{H}_+$. Substituting the expressions for $\psi_\pm(t)$ and $\chi(t)$ allows us to cancel exponential terms and convert the system (12.52) to the stationary form

$$\begin{cases} (\mathbb{A} - zF_+)x = KJ\varphi_-, \\ \varphi_+ = \varphi_- - 2iK^*x, \end{cases} \qquad z \in \rho(\mathbb{A}, F_+, K), \tag{12.54}$$

where $\rho(\mathbb{A}, F_+, K)$ is defined below. Following the canonical case we can re-write (12.54) as an array and have the following definition.

Definition 12.4.3. Let $\dot{A}$ be a closed symmetric operator in a Hilbert space $\mathcal{H}$ and let $\mathcal{H}_+ \subset \mathcal{H} \subset \mathcal{H}_-$ be the rigged Hilbert space associated with $\dot{A}$. The array

$$\Theta = \Theta_{F_+} = \begin{pmatrix} \mathbb{A} & F_+ & K & J \\ \mathcal{H}_+ \subset \mathcal{H} \subset \mathcal{H}_- & & & E \end{pmatrix}, \tag{12.55}$$

where E is a finite-dimensional Hilbert space is called a **non-canonical L-system** or **NCL-system** if:

(i) $\mathbb{A} \in [\mathcal{H}_+, \mathcal{H}_-]$ is a $(*)$-extension of $T \in \Lambda(\dot{A})$;

(ii) $J = J^* = J^{-1} : E \to E$;

(iii) $\mathbb{A} - \mathbb{A}^* = 2iKJK^*$, where $K \in [E, \mathcal{H}_-]$;

(iv) F_+ is an orthogonal projection in $\mathcal{H}_+$ and $\mathcal{H}$;

(v) the set $\rho(\mathbb{A}, F_+, K)$ of all points $z \in \mathbb{C}$, where $(\mathbb{A} - zF_+)^{-1}$ exists on $\mathcal{H} + \mathrm{Ran}(K)$ and $(-,\cdot)$-continuous, is open;

(vi) the set $\rho(\mathrm{Re}\,\mathbb{A}, F_+, K)$ of all points $z \in \mathbb{C}$, where $(\mathrm{Re}\,\mathbb{A} - zF_+)^{-1}$ exists on $\mathcal{H} + \mathrm{Ran}(K)$ and $(-,\cdot)$-continuous, and the set $\rho(\mathbb{A}, F_+, K) \cap \rho(\mathrm{Re}\,\mathbb{A}, F_+, K)$ are both open;

(vii) if $z \in \rho(\mathbb{A}, F_+, K)$, then $\bar{z} \in \rho(\mathbb{A}^*, F_+, K)$; if $z \in \rho(\mathrm{Re}\,\mathbb{A}, F_+, K)$, then $\bar{z} \in \rho(\mathrm{Re}\,\mathbb{A}, F_+, K)$.

We say that Θ_{F_+} is a **scattering NCL-system** if $J = I$. In this case the state-space operator $\mathbb{A}$ in (12.55) is dissipative: $\mathrm{Im}\,\mathbb{A} \geq 0$. It is easy to see that Theorem 4.3.10 implies that any L-system can be considered as an NCL-system with $F_+ = I$.

To each NCL-system in Definition 12.4.3 one can associate a **transfer function**, via

$$W_{\Theta_{F_+}}(z) = I - 2iK^*(\mathbb{A} - zF_+)^{-1}KJ, \quad z \in \rho(\mathbb{A}, F_+, K). \tag{12.56}$$

Lemma 12.4.4. *Let Θ_{F_+} be an NCL-system of the form* (12.55). *Then for all $z, w \in \rho(\mathbb{A}, F_+, K)$,*

$$\begin{aligned} W_{\Theta_{F_+}}(z)JW^*_{\Theta_{F_+}}(w) - J &= 2i(\bar{w} - z)K^*(\mathbb{A} - zF_+)^{-1}F_+(\mathbb{A}^* - \bar{w}F_+)^{-1}K, \\ W^*_{\Theta_{F_+}}(w)JW_{\Theta_{F_+}}(z) - J &= 2i(\bar{w} - z)JK^*(\mathbb{A}^* - \bar{w}F_+)^{-1}F_+(\mathbb{A} - zF_+)^{-1}KJ. \end{aligned}$$

Proof. By the properties (iii) and (vi) in Definition 12.4.3 one has, for all $z, w \in \rho(\mathbb{A}, F_+, K)$,

$$\begin{aligned} (\mathbb{A} - &zF_+)^{-1} - (\mathbb{A}^* - \bar{w}F_+)^{-1} \\ &= (\mathbb{A} - zF_+)^{-1}[(\mathbb{A}^* - \bar{w}F_+) - (\mathbb{A} - zF_+)](\mathbb{A}^* - \bar{w}F_+)^{-1} \\ &= (z - \bar{w})(\mathbb{A} - zF_+)^{-1}F_+(\mathbb{A}^* - \bar{w}F_+)^{-1} \\ &\quad - 2i(\mathbb{A} - zF_+)^{-1}KJK^*(\mathbb{A}^* - \bar{w}F_+)^{-1}. \end{aligned}$$

This identity together with (12.56) implies that

$$\begin{aligned} W_{\Theta_{F_+}}&(z)JW^*_{\Theta_{F_+}}(w) - J \\ &= [I - 2iK^*(\mathbb{A} - zF_+)^{-1}KJ]J[I + 2iJK^*(\mathbb{A}^* - \bar{w}F_+)^{-1}K] - J \\ &= 2i(\bar{w} - z)K^*(\mathbb{A} - zF_+)^{-1}F_+(\mathbb{A}^* - \bar{w}F_+)^{-1}K. \end{aligned}$$

This proves the first equality. Likewise one proves the second identity by using

$$\begin{aligned} (\mathbb{A} - zF_+)^{-1} - (\mathbb{A}^* - \bar{w}F_+)^{-1} = (z - \bar{w})&(\mathbb{A}^* - \bar{w}F_+)^{-1}F_+(\mathbb{A}^* - zF_+)^{-1} \\ &- 2i(\mathbb{A}^* - \bar{w}F_+)^{-1}KJK^*(\mathbb{A} - zF_+)^{-1}. \end{aligned}$$

This completes the proof. $\square$

Lemma 12.4.4 shows that the transfer function $W_{\Theta_{F_+}}(z)$ in (12.56) associated to an NCL-system of the form (12.55) is J-unitary on the real axis, J-expansive in the upper half-plane, and J-contractive in the lower half-plane with $z \in \rho(\mathbb{A}, F_+, K)$.

There is another function that one can associate to each NCL-system Θ_{F_+} of the form (12.55). This is the **impedance function** of the form

$$V_{\Theta_{F_+}}(z) = K^*(\mathrm{Re}\,\mathbb{A} - zF_+)^{-1}K, \quad z \in \rho(\mathrm{Re}\,\mathbb{A}, F_+, K), \tag{12.57}$$

where $\rho(\mathrm{Re}\,\mathbb{A}, F_+, K)$ is defined above. Clearly, $\rho(\mathrm{Re}\,\mathbb{A}, F_+, K)$ is symmetric with respect to the real axis.

Theorem 12.4.5. *Let Θ_{F_+} be an NCL-system of the form (12.55) and let $W_{\Theta_{F_+}}(z)$ and $V_{\Theta_{F_+}}(z)$ be defined by (12.56) and (12.57), respectively. Then for all $z, w \in \rho(\mathrm{Re}\,\mathbb{A}, F_+, K)$, the impedance function $V_{\Theta_{F_+}}(z)$ satisfies*

$$V_{\Theta_{F_+}}(z) - V_{\Theta_{F_+}}(w)^* = (z - \bar{w})K^*(\mathrm{Re}\,\mathbb{A} - zF_+)^{-1}F_+(\mathrm{Re}\,\mathbb{A} - \bar{w}F_+)^{-1}K, \tag{12.58}$$

$V_{\Theta_{F_+}}(z)$ is a Herglotz-Nevanlinna function, and for each

$$z \in \rho(\mathrm{Re}\,\mathbb{A}, F_+, K) \cap \rho(\mathbb{A}, F_+, K)$$

the operators $I + iV_{\Theta_{F_+}}(z)J$ and $I + W_{\Theta_{F_+}}(z)$ are invertible. Moreover,

$$V_{\Theta_{F_+}}(z) = i[W_{\Theta_{F_+}}(z) + I]^{-1}[W_{\Theta_{F_+}}(z) - I]J \tag{12.59}$$

and

$$W_{\Theta_{F_+}}(z) = [I + iV_{\Theta_{F_+}}(z)J]^{-1}[I - iV_{\Theta_{F_+}}(z)J]. \tag{12.60}$$

Proof. For each $z, w \in \rho(\mathrm{Re}\,\mathbb{A}, F_+, K)$ one has

$$(\mathrm{Re}\,\mathbb{A} - zF_+)^{-1} - (\mathrm{Re}\,\mathbb{A} - \bar{w}F_+)^{-1} = (z - \bar{w})(\mathrm{Re}\,\mathbb{A} - zF_+)^{-1}F_+(\mathrm{Re}\,\mathbb{A} - \bar{w}F_+)^{-1}.$$

In view of (12.57) this relation implies (12.58). Clearly,

$$V_{\Theta_{F_+}}(z)^* = V_{\Theta_{F_+}}(\bar{z}).$$

Moreover, it follows from (12.58) and the Definition 12.4.3 that $V_{\Theta_{F_+}}(z)$ is an operator-valued Herglotz-Nevanlinna function. The following identity for $z \in \rho(\mathbb{A}, F_+, K) \cap \rho(\mathrm{Re}\,\mathbb{A}, F_+, K)$,

$$(\mathrm{Re}\,\mathbb{A} - zF_+)^{-1} - (\mathbb{A} - zF_+)^{-1} = i(\mathbb{A} - zF_+)^{-1}\mathrm{Im}\,\mathbb{A}(\mathrm{Re}\,\mathbb{A} - zF_+)^{-1},$$

leads to

$$\begin{aligned} K^*(\mathrm{Re}\,\mathbb{A} - zF_+)^{-1}K - &K^*(\mathbb{A} - zF_+)^{-1}K \\ &= iK^*(\mathbb{A} - zF_+)^{-1}KJK^*(\mathrm{Re}\,\mathbb{A} - zF_+)^{-1}K. \end{aligned}$$

Now in view of (12.56) and (12.57), we have

$$2V_{\Theta_{F_+}}(z) + i(I - W_{\Theta_{F_+}}(z))J = (I - W_{\Theta_{F_+}}(z))V_{\Theta_{F_+}}(z),$$

or equivalently, that

$$[I + W_{\Theta_{F_+}}(z)][I + iV_{\Theta_{F_+}}(z)J] = 2I. \tag{12.61}$$

Similarly, the identity

$$(\mathrm{Re}\,\mathbb{A} - zF_+)^{-1} - (\mathbb{A} - zF_+)^{-1} = i(\mathrm{Re}\,\mathbb{A} - zF_+)^{-1}\mathrm{Im}\,\mathbb{A}(\mathbb{A} - zF_+)^{-1},$$

with $z \in \rho(\mathbb{A}, F_+, K) \cap \rho(\mathrm{Re}\,\mathbb{A}, F_+, K)$ leads to

$$[I + iV_{\Theta_{F_+}}(z)J][I + W_{\Theta_{F_+}}(z)] = 2I. \tag{12.62}$$

The equalities (12.61) and (12.62) show that the operators are invertible and consequently one obtains (12.59) and (12.60). □

12.5 NCI realizations of Herglotz-Nevanlinna functions

The realization of Herglotz-Nevanlinna functions has been obtained for various subclasses. In this section earlier realizations are combined to present a general realization of an arbitrary Herglotz-Nevanlinna function by an NCI-system.

Lemma 12.5.1. *Let Q be a self-adjoint operator in a finite-dimensional Hilbert space E. Then $V(z) \equiv Q$ admits a representation of the form*

$$V(z) = K^*(D - zF_+)^{-1}K, \quad z \in \rho(D, F),$$

where K is an invertible mapping from E into a Hilbert space $\mathcal{H}$, D is a bounded self-adjoint operator in $\mathcal{H}$, and F_+ is an orthogonal projection in $\mathcal{H}$ whose kernel $\ker(F_+)$ *is finite-dimensional.*

Proof. First assume that Q is invertible. Let $\mathcal{H} = E$, let K be any invertible mapping from E onto $\mathcal{H}$, and let $D = KQ^{-1}K^*$. Then D is a bounded self-adjoint operator in $\mathcal{H}$. Clearly, $V(z) = K^*(D - zF)^{-1}K$ with $F = 0$, an orthogonal projection in $\mathcal{H}$. In the general case, Q can be written as the sum of two invertible self-adjoint operators $Q = Q^{(1)} + Q^{(2)}$ (for example, $Q^{(1)} = Q - \varepsilon I$ and $Q^{(2)} = \varepsilon I$, where ε is a real number), so that

$$Q^{(1)} = K^{(1)*}(D^{(1)} - zF^{(1)})^{-1}K^{(1)}, \quad Q^{(2)} = K^{(2)*}(D^{(2)} - zF^{(2)})^{-1}K^{(2)},$$

where $K^{(i)}$ is an operator from E into a Hilbert space $\mathcal{H}^{(i)} = E$, $D^{(i)}$ is a bounded self-adjoint operator in $\mathcal{H}^{(i)}$, and $F^{(i)} = 0$ is an orthogonal projection in $\mathcal{H}^{(i)}$,

$i = 1, 2$. (Note that since $K^{(i)}$ is an arbitrary operator from E into $\mathcal{H}^{(i)} = E$ it may as well be chosen as $K^{(i)} = I$). Define

$$\mathcal{H} = \mathcal{H}^{(1)} \oplus \mathcal{H}^{(2)},\ K = \begin{pmatrix} K^{(1)} \\ K^{(2)} \end{pmatrix},\ D = \begin{pmatrix} D^{(1)} & 0 \\ 0 & D^{(2)} \end{pmatrix},\ F_+ = \begin{pmatrix} F^{(1)} & 0 \\ 0 & F^{(2)} \end{pmatrix}.$$

Then K is an operator from E into the Hilbert space $\mathcal{H}$, D is a bounded self-adjoint operator, and $F_+ = 0$ is an orthogonal projection in $\mathcal{H}$. Moreover,

$$\begin{aligned} Q = Q^{(1)} + Q^{(2)} &= K^{(1)*}(D^{(1)} - zF^{(1)})^{-1}K^{(1)} + K^{(2)*}(D^{(2)} - zF^{(2)})^{-1}K^{(2)} \\ &= K^*(D - zF_+)^{-1}K, \end{aligned}$$

which proves the lemma. □

Herglotz-Nevanlinna functions of the form (12.6) which belong to the class $N(R)$ can be realized using Theorem 6.5.4. By means of Lemma 12.5.1 these realizations can be extended to Herglotz-Nevanlinna functions of the form (12.6) with $L = 0$.

Theorem 12.5.2. *Let $V(z)$ be a Herglotz-Nevanlinna function in a finite-dimensional Hilbert space E, with the integral representation*

$$V(z) = Q + \int_{\mathbb{R}} \left(\frac{1}{t-z} - \frac{t}{1+t^2} \right) dG(t), \tag{12.63}$$

where $Q = Q^*$ and $G(t)$ is a non-decreasing operator-valued function on $\mathbb{R}$. Then $V(z)$ admits a realization of the form

$$V(z) = K^*(\mathbb{D} - zF_+)^{-1}K, \quad z \in \mathbb{C} \setminus \mathbb{R} \subset \rho(\mathbb{D}, F_+, K), \tag{12.64}$$

where $\mathbb{D} \in [\mathcal{H}_+, \mathcal{H}_-]$ is a t-self-adjoint bi-extension of a symmetric operator $\dot{A}$, $\mathcal{H}_+ \subset \mathcal{H} \subset \mathcal{H}_-$ is a rigged Hilbert space, F_+ is an orthogonal projection in $\mathcal{H}_+$ and $\mathcal{H}$, K is an operator from E into $\mathcal{H}_-$, $K^* \in [\mathcal{H}_+, E]$. Moreover, the operators $\mathbb{D}$ and F_+ can be selected such that the following commutativity condition holds:

$$F_-\mathbb{D} = \mathbb{D}F_+, \qquad F_- = \mathcal{R}^{-1}F_+\mathcal{R} \in [\mathcal{H}_-, \mathcal{H}_-], \tag{12.65}$$

where $\mathcal{R}$ is the Riesz-Berezansky operator defined in (2.1).

Proof. According to Theorem 6.5.4 each operator-valued Herglotz-Nevanlinna function of the form (12.63) admits a realization as the impedance function of an L-system of the form (6.36), i.e.,

$$V(z) = K^*(\mathrm{Re}\,\mathbb{A} - zI)^{-1}K = i[W_\Theta(z) + I]^{-1}[W_\Theta(z) - I],$$

where $W_\Theta(z)$ is the transfer function (6.44), if and only if the following condition holds:

$$Qf = \int_{\mathbb{R}} \frac{t}{1+t^2}\, dG(t)f, \tag{12.66}$$

for every vector $f \in E$, such that

$$\int_{\mathbb{R}} (dG(t)f, f)_E < \infty. \tag{12.67}$$

To prove the existence of the representation (12.64) for Herglotz-Nevanlinna functions $V(z)$ which do not satisfy the condition (12.66), the realization result in Lemma 12.5.1 will be used. Denote by E_1 the subspace of vectors $f \in E$ with the property (12.67) and let $E_2 = E \ominus E_1$, so that $E = E_1 \oplus E_2$. Rewrite Q in the block matrix form

$$Q = \begin{pmatrix} Q_{11} & Q_{12} \\ Q_{21} & Q_{22} \end{pmatrix}, \quad Q_{ij} = P_{E_i} Q \upharpoonright E_j, \quad j = 1, 2,$$

and let $G(t) = (G_{ij}(t))_{i,j=1}^2$ be decomposed accordingly. Observe, that by (12.66), (12.67) the integrals

$$S_{11} := \int_{\mathbb{R}} \frac{t}{1+t^2}\, dG_{11}(t), \quad S_{12} := \int_{\mathbb{R}} \frac{t}{1+t^2}\, dG_{12}(t),$$

are convergent in the strong topology. Let the self-adjoint operator S be defined by

$$S = \begin{pmatrix} S_{11} & S_{12} \\ S_{12}^* & C \end{pmatrix}, \tag{12.68}$$

where $C = C^*$ is arbitrary. Now rewrite $V(z) = V_1(z) + V_2(z)$ with

$$V_1(z) = Q - S, \quad V_2(z) = S + \int_{\mathbb{R}} \left(\frac{1}{t-z} - \frac{t}{1+t^2} \right) dG(t). \tag{12.69}$$

Clearly, for every $f \in E_1$ the equality

$$Sf = \int_{\mathbb{R}} \frac{t}{1+t^2}\, dG(t) f,$$

holds. Consequently, $V_2(z) \in N(R)$ and thus by Theorem 6.5.2 admits the representation

$$V_2(z) = K_2^* (\operatorname{Re} \mathbb{A}^{(2)} - zI)^{-1} K_2,$$

where $K_2 : E \to \mathcal{H}_{-2}$, $K_2^* : \mathcal{H}_{+2} \to E$ with $\mathcal{H}_{+2} \subset \mathcal{H}_2 \subset \mathcal{H}_{-2}$ a rigged Hilbert space, and where $\operatorname{Re} \mathbb{A}^{(2)}$ is a t-self-adjoint bi-extension of a symmetric operator $\dot{A}_2$. The operator K_2 has the properties

$$\begin{aligned} &\operatorname{Ran}(K_2) \subset \operatorname{Ran}(\mathbb{A}^{(2)} - zI), && \operatorname{Ran}(K_2) \subset \operatorname{Ran}(\operatorname{Re} \mathbb{A}^{(2)} - zI), \\ &(\mathbb{A}^{(2)} - zI)^{-1} K_2 \in [E, \mathcal{H}_+], && (\operatorname{Re} \mathbb{A}^{(2)} - zI)^{-1} K_2 \in [E, \mathcal{H}_+]; \end{aligned} \tag{12.70}$$

for further details, see the proof of Theorems 6.5.1–6.5.2. Now, by Lemma 12.5.1 the function $V_1(z)$ admits the representation

$$V_1(z) = K_1^* (D_1 - zF_{+,1})^{-1} K_1,$$

where $D_1 = D_1^*$ and $F_{+,1} = 0$ are acting on a finite-dimensional Hilbert space $\mathcal{H}_1 = E \oplus E$. Recall from Lemma 12.5.1 that

$$D_1 = \begin{pmatrix} D_1^{(1)} & 0 \\ 0 & D_1^{(2)} \end{pmatrix}, \quad K_1 = \begin{pmatrix} K_1^{(1)} \\ K_1^{(2)} \end{pmatrix},$$

where $K_1^{(i)} : E \to E$, $i = 1, 2$, and $D_1^{(1)}$, $D_1^{(2)}$ are defined by means of the decomposition of $Q - S$ into the sum of two invertible self-adjoint operators

$$Q - S = (Q^{(1)} - S^{(1)}) + (Q^{(2)} - S^{(2)}).$$

Then

$$D_1^{(i)} = K_1^{(i)*}(Q^{(i)} - S^{(i)})^{-1} K_1^{(i)}, \quad i = 1, 2.$$

To obtain the realization (12.64) for $V(z)$ in (12.63), introduce the triplet of Hilbert spaces

$$\mathcal{H}_+^{(1)} := E \oplus E \oplus \mathcal{H}_{+2} \subset E \oplus E \oplus \mathcal{H}_2 \subset E \oplus E \oplus \mathcal{H}_{-2} := \mathcal{H}_-^{(1)}, \tag{12.71}$$

i.e., a rigged Hilbert space corresponding to the block representation of symmetric operator $D_1 \oplus \dot{A}_2$ in $\mathcal{H}^{(1)} := \mathcal{H}_1 \oplus \mathcal{H}_2$ (where $\mathcal{H}_1 = E \oplus E$). Also introduce the operators

$$\mathbb{D} = \begin{pmatrix} D_1^{(1)} & 0 & 0 \\ 0 & D_1^{(2)} & 0 \\ 0 & 0 & \operatorname{Re}\mathbb{A}^{(2)} \end{pmatrix}, \quad F_+ = \begin{pmatrix} 0 & 0 & 0 \\ 0 & 0 & 0 \\ 0 & 0 & I \end{pmatrix}, \quad K = \begin{pmatrix} K_1^{(1)} \\ K_1^{(2)} \\ K_2 \end{pmatrix}. \tag{12.72}$$

It is straightforward to check that

$$\begin{aligned} V(z) = V_1(z) + V_2(z) &= K_1^{(1)*}(D_1^{(1)} - zF_{+,1})^{-1} K_1^{(1)} \\ &\quad + K_1^{(2)*}(D_1^{(2)} - zF_{+,1})^{-1} K_1^{(2)} + K_2^*(\operatorname{Re}\mathbb{A}^{(2)} - zI)^{-1} K_2 \\ &= K^*(\mathbb{D} - zF_+)^{-1} K. \end{aligned} \tag{12.73}$$

By the construction, $\dot{A}_2 \subset \hat{A}_2 = \hat{A}_2^* \subset \operatorname{Re}\mathbb{A}^{(2)}$, where $\hat{A}_2$ is the quasi-kernel of $\operatorname{Re}\mathbb{A}^{(2)}$ and $\dot{A}_2$ is a symmetric operator in $\mathcal{H}_2$. Moreover, $\mathbb{D}$ as an operator in $[\mathcal{H}_+^{(1)}, \mathcal{H}_-^{(1)}]$ is self-adjoint, i.e., $\mathbb{D} = \mathbb{D}^*$, and since

$$\hat{D} = \begin{pmatrix} D_1 & 0 \\ 0 & \hat{A}_2 \end{pmatrix} \subset \begin{pmatrix} D_1 & 0 \\ 0 & \operatorname{Re}\mathbb{A}^{(2)} \end{pmatrix} = \mathbb{D},$$

and $\dot{A} = D_1 \oplus \dot{A}_2 \subset \hat{D}$, the operator $\mathbb{D}$ is a t-self-adjoint bi-extension of the symmetric operator $\dot{A}$ in $\mathcal{H}_1 \oplus \mathcal{H}_2$. It is easy to see that with operators in (12.72) one obtains the representation (12.64) for $V(z)$ in (12.63) and the system constructed with these operators satisfies the Definition 12.4.1 of a NCI-system. Finally, from (12.72) one obtains $F_-\mathbb{D} = \mathbb{D}F_+$, where F_+ and F_- are connected as in (12.65). This completes the proof of the theorem. □

The general impedance realization result for Herglotz-Nevanlinna functions of the form (12.7) is going to be built on Theorem 12.5.2 and the following representation for the linear term in integral representation (12.6).

Lemma 12.5.3. *Let L be a non-negative operator in a finite-dimensional Hilbert space E. Then it admits a realization of the form*

$$zL = z\hat{K}^* P\hat{K} = K_3^*(D_3 - zF_3)^{-1}K_3, \tag{12.74}$$

where D_3 is a self-adjoint operator in a Hilbert space $\mathcal{H}_3$, P is the orthogonal projection onto $\operatorname{Ran}(L)$*, and K_3 is an invertible operator from E into $\mathcal{H}_3$.*

Proof. Since $L \geq 0$, there is a unique non-negative square root $L^{1/2} \geq 0$ of L with

$$\ker(L^{1/2}) = \ker(L), \quad \operatorname{Ran}(L^{1/2}) = \operatorname{Ran}(L).$$

Define the operator $\hat{K}$ in E by

$$\hat{K}u = \begin{cases} u, & u \in \ker(L), \\ L^{1/2}u, & u \in \operatorname{Ran}(L). \end{cases} \tag{12.75}$$

Then $\hat{K}$ is invertible and $L^{1/2} = P\hat{K}$, where P denotes the orthogonal projection onto $\operatorname{Ran}(L)$. Define

$$\mathcal{H}_3 = E \oplus E, \quad K_3 = \begin{pmatrix} P\hat{K} \\ \hat{K} \end{pmatrix}, \quad D_3 = \begin{pmatrix} 0 & iI \\ -iI & 0 \end{pmatrix}, \quad F_{+,3} = \begin{pmatrix} 0 & 0 \\ 0 & I \end{pmatrix}. \tag{12.76}$$

Then K_3 is an invertible operator from E into $\mathcal{H}_3$, D_3 is a bounded self-adjoint operator, and $F_{+,3}$ is an orthogonal projection in $\mathcal{H}_3$. Moreover,

$$V_3(z) = zL = z\hat{K}^* P\hat{K} = K_3^*(D_3 - zF_{+,3})^{-1}K_3.$$

This completes the proof. □

The general realization result for Herglotz-Nevanlinna functions of the form (12.7) is now being obtained by combining the earlier realizations.

Theorem 12.5.4. *Let $V(z)$ be an operator-valued Herglotz-Nevanlinna function in a finite-dimensional Hilbert space E with the integral representation*

$$V(z) = Q + zL + \int_{\mathbb{R}} \left(\frac{1}{t-z} - \frac{t}{t^2+1} \right) dG(t), \tag{12.77}$$

where $Q = Q^*$, $L \geq 0$, and $G(t)$ is a non-decreasing, non-negative operator-valued function on $\mathbb{R}$. Then $V(z)$ admits a realization of the form

$$V(z) = K^*(\mathbb{D} - zF_+)^{-1}K \tag{12.78}$$

where $\mathbb{D} \in [\mathcal{H}_+, \mathcal{H}_-]$ is a t-self-adjoint bi-extension in a rigged Hilbert space $\mathcal{H}_+ \subset \mathcal{H} \subset \mathcal{H}_-$, F_+ is an orthogonal projection in $\mathcal{H}_+$ and $\mathcal{H}$, and $K \in [E, \mathcal{H}_-]$.

Proof. Let us define the functions

$$V_1(z) = Q + \int_{\mathbb{R}} \left(\frac{1}{t-z} - \frac{t}{1+t^2} \right) dG(t), \quad V_2(z) = zL.$$

According to Theorem 12.5.2 the function $V_1(z)$ has a representation

$$V_1(z) = K_1^*(\mathbb{D}_1 - zF_{+,1})^{-1}K_1,$$

where $\mathbb{D}_1$, K_1 and $F_{+,1}$ are given by the formula (12.72). We recall that $\mathbb{D}_1$ is a t-self-adjoint bi-extension in a rigged Hilbert space $\mathcal{H}_-^{(1)} \subset \mathcal{H}^{(1)} \subset \mathcal{H}_+^{(1)}$ given by (12.71), $F_{+,1}$ is an orthogonal projection in $\mathcal{H}_+^{(1)}$, and K_1 is an invertible mapping from E into $\mathcal{H}_-^{(1)}$. According to Lemma 12.5.3 the functions $V_2(z)$ has a realization of the form (12.74) with components $\mathcal{H}_3$, D_3, K_3 and $F_{+,3}$ described by (12.76).

Now the final result follows by introducing the rigged Hilbert space $\mathcal{H}_3 \oplus \mathcal{H}_+^{(1)} \subset \mathcal{H}_3 \oplus \mathcal{H}^{(1)} \subset \mathcal{H}_3 \oplus \mathcal{H}_-^{(1)}$ and the operators

$$\mathbb{D} = \begin{pmatrix} D_3 & 0 \\ 0 & \mathbb{D}_1 \end{pmatrix} \in [\mathcal{H}_3 \oplus \mathcal{H}_+^{(1)}, \mathcal{H}_3 \oplus \mathcal{H}_-^{(1)}], \quad F_+ = \begin{pmatrix} F_{+,3} & 0 \\ 0 & F_{+,1} \end{pmatrix}, \quad K = \begin{pmatrix} K_3 \\ K_1 \end{pmatrix}.$$

It is straightforward to check that with these operators one obtains the representation (12.78) for $V(z)$ in (12.77) and the system constructed with these operators satisfies the Definition 12.4.1 of a NCI-system. □

For the sake of clarity an extended version for the NCI-realization in the proof of Theorem 12.5.4 is provided. The rigged Hilbert space used is

$$E \oplus E \oplus E \oplus E \oplus \mathcal{H}_{+2} \subset E \oplus E \oplus E \oplus E \oplus \mathcal{H}_2 \subset E \oplus E \oplus E \oplus E \oplus \mathcal{H}_{-2}, \tag{12.79}$$

and the operators are given by

$$\begin{aligned} \mathbb{D} &= \begin{pmatrix} 0 & iI & 0 & 0 & 0 \\ -iI & 0 & 0 & 0 & 0 \\ 0 & 0 & D_1^{(1)} & 0 & 0 \\ 0 & 0 & 0 & D_1^{(2)} & 0 \\ 0 & 0 & 0 & 0 & \operatorname{Re}\mathbb{A}^{(2)} \end{pmatrix}, \\ F_+ &= \begin{pmatrix} 0 & 0 & 0 & 0 & 0 \\ 0 & I & 0 & 0 & 0 \\ 0 & 0 & 0 & 0 & 0 \\ 0 & 0 & 0 & 0 & 0 \\ 0 & 0 & 0 & 0 & I \end{pmatrix}, \quad K = \begin{pmatrix} P\hat{K} \\ \hat{K} \\ K_1^{(1)} \\ K_1^{(2)} \\ K_2 \end{pmatrix}. \end{aligned} \tag{12.80}$$

All the operators in (12.80) are defined above.

12.6 Realization by NCL-systems

In the NCI-realization results of Theorem 12.5.2 and Theorem 12.5.4, the realizations are presented in terms of the operators in (12.64) and (12.78), respectively. It remains to identify the Herglotz-Nevanlinna functions as impedance functions of appropriate NCL-systems.

Theorem 12.6.1. *Let $V(z)$ be a Herglotz-Nevanlinna function acting on a finite-dimensional Hilbert space E with the integral representation*

$$V(z) = Q + \int_{\mathbb{R}} \left(\frac{1}{t-z} - \frac{t}{1+t^2} \right) dG(t),$$

where $Q = Q^*$ and $G(t)$ is a non-decreasing, non-negative operator-valued function on $\mathbb{R}$. Then the function $V(z)$ can be realized as the impedance function of a scattering NCL-system Θ_{F_+} defined in (12.55), that is

$$V(z) = i[W_{\Theta_{F_+}}(z) + I]^{-1}[W_{\Theta_{F_+}}(z) - I], \tag{12.81}$$

where $W_{\Theta_{F_+}}(z)$ is the transfer function of Θ_{F_+}.

Proof. By Theorem 12.5.2 the function $V(z)$ can be represented in the form $V(z) = K^*(\mathbb{D} - zF_+)^{-1}K$, where K, $\mathbb{D}$, and F_+ are as in (12.72) corresponding to the decomposition

$$V(z) = V_1(z) + V_2(z),$$

where

$$V_1(z) = Q - G, \quad V_2(z) = G + \int_{\mathbb{R}} \left(\frac{1}{t-z} - \frac{t}{1+t^2} \right) dG(t),$$

with a self-adjoint operator G of the form (12.68). With the notation used in the proof of Theorem 12.5.2 one may rewrite $V_1(z)$ and $V_2(z)$ as in (12.73) with

$$D_1^{(1)} = (Q - G - \varepsilon I)^{-1}, \quad D_1^{(2)} = (\varepsilon I)^{-1}, \quad K_1^{(1)} = \lambda I_E, \quad K_1^{(2)} = I_E, \tag{12.82}$$

$\operatorname{Re}\mathbb{A}^{(2)} \in [\mathcal{H}_{+2}, \mathcal{H}_{-2}]$, $\mathbb{A}^{(2)}$ is a $(*)$-extension of an operator $T_2 \in \Lambda(\dot{A}_2)$ for which $(-i) \in \rho(T_2)$, cf. Section 6.5. The remaining operators are defined in (12.72) and $\varepsilon, \lambda > 0$.

Recall that K_2 and the resolvents $(\mathbb{A}^{(2)} - zI)^{-1}$, $(\operatorname{Re}\mathbb{A}^{(2)} - zI)^{-1}$ satisfy the properties (12.70). To construct an NCL-system of the form (12.55) we introduce an operator $\mathbb{A}$ by

$$\mathbb{A} = \mathbb{D} + iKK^* \in [\mathcal{H}_+, \mathcal{H}_-],$$

where K, $\mathbb{D}$, and F_+ are defined in (12.72). Then the block-matrix form of $\mathbb{A}$ is

$$\mathbb{A} = \begin{pmatrix} D_1^{(1)} + i\lambda^2 I & i\lambda I & i\lambda K_2^* \\ i\lambda I & D_1^{(2)} + iI & iK_2^* \\ i\lambda K_2 & iK_2 & \mathbb{A}^{(2)} \end{pmatrix}.$$

Let

$$\Theta_{F_+} = \begin{pmatrix} \mathbb{A} & F_+ & K & I \\ \mathcal{H}_+ \subset \mathcal{H} \subset \mathcal{H}_- & & & E \end{pmatrix},$$

where the rigged Hilbert triplet $\mathcal{H}_+ \subset \mathcal{H} \subset \mathcal{H}_-$ is defined in (12.71), i.e.,

$$E \oplus E \oplus \mathcal{H}_{+2} \subset E \oplus E \oplus \mathcal{H}_2 \subset E \oplus E \oplus \mathcal{H}_{-2}.$$

It remains to show that all the properties in Definition 12.4.3 are satisfied. For this purpose, consider the equation

$$(\mathbb{A} - zF_+)x = (\mathbb{D} + iKK^*)x - zF_+x = Kg, \quad g \in E,$$

or

$$\begin{pmatrix} D_1^{(1)} + i\lambda^2 I & i\lambda I & i\lambda K_2^* \\ i\lambda I & D_1^{(2)} + iI & iK_2^* \\ i\lambda K_2 & iK_2 & \mathbb{A}^{(2)} - zI \end{pmatrix} \begin{pmatrix} x_1 \\ x_2 \\ x_3 \end{pmatrix} = \begin{pmatrix} \lambda g \\ g \\ K_2 g \end{pmatrix}.$$

Using the decomposition of the operators and taking into account that $\mathbb{A}^{(2)} = \operatorname{Re}\mathbb{A}^{(2)} + iK_2K_2^*$, we can re-write this equation in form of the system

$$\begin{cases} D_1^{(1)}x_1 + i\lambda^2 Ix_1 + i\lambda Ix_2 + i\lambda K_2^* x_3 = \lambda g, \\ D_1^{(2)}x_2 + i\lambda Ix_1 + iIx_2 + iK_2^*x_3 = g, \\ (\mathbb{A}^{(2)} - zI)x_3 + i\lambda K_2 x_1 + iK_2 x_2 = K_2 g, \end{cases} \tag{12.83}$$

or

$$\begin{cases} \frac{1}{\lambda} D_1^{(1)}x_1 + i\lambda Ix_1 + iIx_2 + iK_2^* x_3 = g, \\ D_1^{(2)}x_2 + i\lambda Ix_1 + iIx_2 + iK_2^*x_3 = g, \\ (\mathbb{A}^{(2)} - zI)x_3 + i\lambda K_2 x_1 + iK_2 x_2 = K_2 g. \end{cases}$$

In a neighborhood of $(-i)$ the resolvent $(\mathbb{A}^{(2)} - zI)^{-1}$ is well defined so that by (12.70) the third equation in (12.83) can be solved for x_3:

$$x_3 = (\mathbb{A}^{(2)} - zI)^{-1}K_2 g - i(\mathbb{A}^{(2)} - zI)^{-1}K_2(\lambda x_1 + x_2). \tag{12.84}$$

Substituting (12.84) into the first line of the system yields

$$\frac{1}{\lambda}D_1^{(1)}x_1 + iI(\lambda x_1 + x_2) + K_2^*(\mathbb{A}^{(2)} - zI)^{-1}K_2(\lambda x_1 + x_2) \\ = g - iK_2^*(\mathbb{A}^{(2)} - zI)^{-1}K_2 g,$$

Denoting the right-hand side by C and using (12.56) we get

$$C = g - iK_2^*(\mathbb{A}^{(2)} - zI)^{-1}K_2 g = \frac{1}{2}\left[I + W_{\Theta_2}(z)\right] g.$$

Then

$$\frac{1}{\lambda}D_1^{(1)}x_1 + iI(\lambda x_1 + x_2) + K_2^*(\mathbb{A}^{(2)} - zI)^{-1}K_2(\lambda x_1 + x_2) = C.$$

Multiplying both sides by $2i$ and using (12.56) one more time yields

$$\frac{2i}{\lambda}D_1^{(1)}x_1 - [I + W_{\Theta_2}(z)]\,(\lambda x_1 + x_2) = 2iC.$$

Writing for further convenience $B = [I + W_{\Theta_2}(z)]$ we obtain

$$\frac{2i}{\lambda}D_1^{(1)}x_1 - \lambda Bx_1 - Bx_2 = 2iC$$

or

$$\frac{2i}{\lambda}D_1^{(1)}x_1 - \lambda Bx_1 - 2iC = Bx_2. \tag{12.85}$$

Now we subtract the second equation of the system from the first and obtain $D_1^{(1)}x_1 = \lambda D_1^{(2)}x_2$, or

$$\lambda(D_1^{(1)})^{-1}D_1^{(2)}x_2 = x_1. \tag{12.86}$$

Applying (12.86) to (12.85) we get

$$2iD_1^{(2)}x_2 - B\lambda^2(D_1^{(1)})^{-1}D_1^{(2)}x_2 - Bx_2 = 2iC,$$

and using (12.82)

$$\frac{2i}{\varepsilon}Ix_2 - B[\lambda^2(Q - G - \varepsilon I)\frac{1}{\varepsilon} + I]x_2 = 2iC,$$

or

$$\Big(2iI - [I + W_{\Theta_2}(z)]\,[\lambda^2(Q - G) + \varepsilon(1 - \lambda^2)I]\Big)x_2 = 2i\varepsilon C. \tag{12.87}$$

Choosing λ and ε sufficiently small the operator on the left-hand side of (12.87) can be made invertible for $z = -i$. Using an invertibility criteria from [89] we deduce that (12.87) is also invertible in a neighborhood of $(-i)$. Consequently, the system (12.83) has a unique solution and $(\mathbb{A} - zF_+)^{-1}K$ is well defined in a neighborhood of $(-i)$.

In order to show that the remaining properties in Definition 12.4.3 are satisfied we need to present an operator $T \in \Lambda(\dot{A})$ such that $\mathbb{A}$ is a $(*)$-extension of T. To construct T we note first that $(\mathbb{A} - zF_+)\mathcal{H}_+ \supset \mathcal{H}$ for some z in a neighborhood of $(-i)$. This can be confirmed by considering the equation

$$(\mathbb{A} - zF_+)x = g, \quad x \in \mathcal{H}_+, \tag{12.88}$$

and showing that it has a unique solution for every $g \in \mathcal{H}$. The procedure then is reduced to solving the system (12.83) with an arbitrary right-hand side $g \in \mathcal{H}$. Following the steps for solving (12.83) we conclude that the system (12.88) has a unique solution. Similarly one shows that $(\mathbb{A}^* - zF_+)\mathcal{H}_+ \supset \mathcal{H}$. Using the technique developed in Section 4.5 we can conclude that operators $(\mathbb{A} + iF_+)^{-1}$ and $(\mathbb{A}^* - iF_+)^{-1}$ are $(-,\cdot)$-continuous. Define

$$\begin{aligned} T &= \mathbb{A}, \quad \mathrm{Dom}(T) = (\mathbb{A} + iF_+)^{-1}\mathcal{H}, \\ T_1 &= \mathbb{A}^*, \quad \mathrm{Dom}(T_1) = (\mathbb{A}^* - iF_+)^{-1}\mathcal{H}. \end{aligned} \tag{12.89}$$

One can see that both $\mathrm{Dom}(T)$ and $\mathrm{Dom}(T_1)$ are dense in $\mathcal{H}$ while operator T is closed in $\mathcal{H}$. Indeed, assuming that there is a vector $\phi \in \mathcal{H}$ that is $(\cdot)$-orthogonal to $\mathrm{Dom}(T)$ and representing $\phi = (\mathbb{A}^* - iF_+)\psi$, we can immediately get $\phi = 0$. It is also easy to see that $T_1 = T^*$. Thus, operator T defined by (12.89) fits the definition of a $(*)$-extension for operator $\mathbb{A}$. Property (vi) of Definition 12.4.3 follows from Theorem 12.5.4 and the fact that $\mathrm{Re}\,\mathbb{A} = \mathbb{D}$.

Consequently all the properties for an NCL-system Θ in Definition 12.4.3 are fulfilled with the operators and spaces defined above. □

Now we present the principal result of this section.

Theorem 12.6.2. *Let $V(z)$ be an operator-valued Herglotz-Nevanlinna function in a finite-dimensional Hilbert space E with the integral representation*

$$V(z) = Q + zL + \int_{\mathbb{R}} \left(\frac{1}{t-z} - \frac{t}{t^2+1} \right) dG(t), \tag{12.90}$$

where $Q = Q^*$, $L \geq 0$ is an invertible operator, and $G(t)$ is a non-decreasing, non-negative operator-valued function on $\mathbb{R}$. Then $V(z)$ can be realized as the impedance function of a scattering NCL-system Θ_{F_+} of the form (12.55).

Proof. Decompose the function $V(z)$ as follows:

$$V_1(z) = Q + \int_{\mathbb{R}} \left(\frac{1}{t-z} - \frac{t}{t^2+1} \right) dG(t) \quad \text{and} \quad V_2(z) = zL,$$

and use the earlier realizations for each of these functions. By Theorem 12.6.1 the function $V_1(z)$ can be represented by

$$V_1(z) = i[W_{\Theta_{F_{1,+}}}(z) + I]^{-1}[W_{\Theta_{F_{1,+}}}(z) - I],$$

where $W_{\Theta_{F_{1,+}}}(z)$ is an operator-valued transfer function of some scattering NCL-system,

$$W_{\Theta_{F_{1,+}}}(z) = I - 2iK_1^*(\mathbb{A}_1 - zF_{1,+})^{-1}K_1, \tag{12.91}$$

$\mathbb{A}_1 = \mathbb{D}_1 + iK_1K_1^*$ maps $\mathcal{H}_{+1}$ continuously into $\mathcal{H}_{-1}$, $\mathbb{D}_1$ is a self-adjoint bi-extension, and $\mathbb{D}_1 \in [\mathcal{H}_{+1}, \mathcal{H}_{-1}]$, $K_1 \in [E, \mathcal{H}_{-1}]$.

Following the proof of Theorem 12.5.4, the function $V_2(z)$ can be represented in the form

$$V_2(z) = K_2^*(D_2 - zF_{2,+})^{-1}K_2,$$

where

$$D_2 = \begin{pmatrix} 0 & iI \\ -iI & 0 \end{pmatrix}, \quad F_{2,+} = \begin{pmatrix} 0 & 0 \\ 0 & I \end{pmatrix}, \quad K_2 = \begin{pmatrix} P\widehat{K} \\ \widehat{K} \end{pmatrix}, \tag{12.92}$$

and P and $\widehat{K}$ are as in (12.75), so that K_2 is an operator from E into $\mathcal{H}_2 = E \oplus E$. Introduce the triplet $\mathcal{H}_{+1} \oplus \mathcal{H}_2 \subset \mathcal{H}_1 \oplus \mathcal{H}_2 \subset \mathcal{H}_{-1} \oplus \mathcal{H}_2$, and consider the operator

$$\mathbb{A} = \mathbb{D} + iKK^*, \tag{12.93}$$

from $\mathcal{H}_{+1} \oplus \mathcal{H}_2$ into $\mathcal{H}_{-1} \oplus \mathcal{H}_2$ given by the block form

$$\begin{aligned}\mathbb{A} &= \begin{pmatrix} \mathbb{D}_1 & 0 \\ 0 & D_2 \end{pmatrix} + i \begin{pmatrix} K_1 \\ K_2 \end{pmatrix} \begin{pmatrix} K_1^* & K_2^* \end{pmatrix} \\ &= \begin{pmatrix} \mathbb{A}_1 & iK_1K_2^* \\ iK_2K_1^* & \mathbb{A}_2 \end{pmatrix}.\end{aligned} \tag{12.94}$$

Here $\mathbb{A}_2 = D_2 + iK_2K_2^*$. It will be shown that the equation

$$(\mathbb{A} - zF_+)x = Kh, \quad h \in E, \tag{12.95}$$

with

$$F_+ = \begin{pmatrix} F_{1,+} & 0 \\ 0 & F_{2,+} \end{pmatrix}, \quad K = \begin{pmatrix} K_1 \\ K_2 \end{pmatrix},$$

has always a unique solution $x \in \mathcal{H}_{+1} \oplus \mathcal{H}_2$ and

$$(\mathbb{A} - zF_+)^{-1}K \in [E, \mathcal{H}_{+1} \oplus \mathcal{H}_2].$$

Taking into account (12.94), the equation (12.95) can be written as the system

$$\begin{cases} (\mathbb{A}_1 - zF_{1,+})x_1 + iK_1K_2^*x_2 = K_1h, \\ (\mathbb{A}_2 - zF_{2,+})x_2 + iK_2K_1^*x_1 = K_2h, \end{cases} \tag{12.96}$$

where

$$\mathbb{A}_1 = \mathbb{D}_1 + iK_1K_1^*, \quad \mathbb{A}_2 = D_2 + iK_2K_2^*.$$

By Theorem 12.6.1 it follows that $(\mathbb{A}_1 - zF_{1,+})^{-1}K_1 \in [E, \mathcal{H}_{+1}]$. Therefore, the first equation in (12.96) gives

$$x_1 = (\mathbb{A}_1 - zF_{1,+})^{-1}K_1h - i(\mathbb{A}_1 - zF_{1,+})^{-1}K_1K_2^*x_2. \tag{12.97}$$

Now substituting x_1 in the second equation in (12.96) yields

$$\begin{aligned}(\mathbb{A}_2 - zF_{2,+})x_2 + K_2K_1^*(\mathbb{A}_1 - zF_{1,+})^{-1}K_1K_2^*x_2 \\ = K_2h - iK_2K_1^*(\mathbb{A}_1 - zF_{1,+})^{-1}K_1h.\end{aligned} \tag{12.98}$$

Taking into account (12.91), (12.92), and (12.94) the identity (12.98) leads to

$$\begin{aligned}&\left(2iI - (D_2 - zF_+^{(2)})^{-1}K_2[I + W_{\Theta_{F_{1,+}}}(z)]K_2^*\right)x_2 \\ &\quad = 2i(D_2 - zF_+^{(2)})^{-1}\left(K_2h - iK_2K_1^*(\mathbb{A}_1 - zF_+^{(1)})^{-1}K_1h\right).\end{aligned}$$

It will be shown that the operator-function on the left-hand side, in front of x_2, is invertible. First by straightforward calculations one obtains

$$(D_2 - zF_+^{(2)})^{-1} = \begin{pmatrix} zI & iI \\ -iI & 0 \end{pmatrix} \in [E \oplus E, E \oplus E].$$

The operator-function $M(z)$ defined by

$$M(z) = I + W_{\Theta_{F_{1,+}}}(z) \in [E, E]$$

is invertible according to Theorem 12.4.5. It follows from (12.92) that

$$K_2 M(z) = \begin{pmatrix} P\widehat{K} \\ \widehat{K} \end{pmatrix} M(z) = \begin{pmatrix} L^{1/2}M(z) \\ \widehat{K}M(z) \end{pmatrix} \in [E \oplus E, E \oplus E],$$

and that

$$K_2 M(z) K_2^* = \begin{pmatrix} L^{1/2}M(z)L^{1/2} & L^{1/2}M(z)\widehat{K} \\ \widehat{K}M(z)L^{1/2} & \widehat{K}M(z)\widehat{K} \end{pmatrix} \in [E \oplus E, E \oplus E].$$

For any 2×2 block-matrix

$$Z = \begin{pmatrix} a & b \\ c & d \end{pmatrix}$$

with entries in E, define the operator-function

$$N(z) = 2iI - (D_2 - zF_+^{(2)})^{-1} \begin{pmatrix} a & b \\ c & d \end{pmatrix} = i \begin{pmatrix} 2 + zai - c & zbi - d \\ a & b + 2 \end{pmatrix}.$$

Since the operator $L > 0$ is invertible, therefore $\ker(L) = \{0\}$ and $\hat{K} = L^{1/2}$. Now choose

$$Z = \begin{pmatrix} L^{1/2}M(z)L^{1/2} & L^{1/2}M(z)L^{1/2} \\ L^{1/2}M(z)L^{1/2} & L^{1/2}M(z)L^{1/2} \end{pmatrix} = \begin{pmatrix} A_0 & A_0 \\ A_0 & A_0 \end{pmatrix},$$

where $A_0 = A_0(z) = L^{1/2}M(z)L^{1/2}$. Note that the operator-function A_0 is invertible and that $A_0^{-1} = L^{-1/2}M(z)^{-1}L^{-1/2}$. With this choice of Z one obtains

$$N = N(z) = i \begin{pmatrix} 2I + ziA_0 - A_0 & ziA_0 - A_0 \\ A_0 & A_0 + 2I \end{pmatrix}.$$

To investigate the invertibility of N consider the system

$$\begin{pmatrix} 2I + ziA_0 - A_0 & ziA_0 - A_0 \\ A_0 & A_0 + 2I \end{pmatrix} \begin{pmatrix} x_1 \\ x_2 \end{pmatrix} = \begin{pmatrix} 0 \\ 0 \end{pmatrix},$$

or

$$\begin{cases} (2I + ziA_0 - A_0)x_1 + (ziA_0 - A_0)x_2 = 0, \\ A_0x_1 + (A_0 + 2)x_2 = 0. \end{cases}$$

Solving the second equation for x_1 yields

$$\begin{cases} 2x_1 + ziA_0x_1 - A_0x_1 + ziA_0x_2 - A_0x_2 = 0, \\ x_1 = -x_2 - 2A_0^{-1}x_2. \end{cases}$$

Substituting x_1 into the first equation gives $(2A_0^{-1} + zi)x_2 = 0$, or equivalently,

$$A_0 x_2 = \frac{2i}{z} x_2. \tag{12.99}$$

Recall that

$$A_0 = A_0(z) = L^{1/2} M(z) L^{1/2} = L^{1/2}[I + W_{\Theta_1}(z)]L^{1/2}.$$

For every z in the lower half-plane $W_{\Theta_1}(z)$ is a contraction (see (6.46)) and thus $\|A_0(z)\| \leq 2\|L\|$. This means that for every z (Im $z < 0$) the norm of the left-hand side of (12.99) is bounded while the norm of the right-hand side can be made unboundedly large by letting $z \to 0$ along the imaginary axis. This leads to a conclusion that $x_2 = 0$ and then also $x_1 = 0$. Hence, $N = N(z)$ is invertible. Consequently,

$$2iI - (D_2 - zF_+^{(2)})^{-1} K_2[I + W_{\Theta_1}(z)]K_2^*$$

is invertible and x_2 depends continuously on $h \in E$ in (12.98), while (12.97) shows that x_1 depends continuously on $h \in E$.

Now we will follow the steps taken in the proof of Theorem 12.6.1 to show that the remaining properties in Definition 12.4.3 are satisfied. We introduce an operator $T \in \Lambda(\dot{A})$ such that $\mathbb{A}$ is a $(*)$-extension of T. To construct T we note first that $(\mathbb{A} - zF_+)\mathcal{H}_+ \supset \mathcal{H}$ for some z in a neighborhood of $(-i)$. This can be confirmed by considering the equation

$$(\mathbb{A} - zF_+)x = g, \quad x \in \mathcal{H}_+, \tag{12.100}$$

and showing that it has a unique solution for every $g \in \mathcal{H}$. The procedure then is reduced to solving the system (12.96) with an arbitrary right-hand side $g \in \mathcal{H}$. Inspecting the steps of solving (12.96) we conclude that the system (12.100) has a unique solution. Similarly one shows that $(\mathbb{A}^* - zF_+)\mathcal{H}_+ \supset \mathcal{H}$. Once again relying on Section 4.5 we can conclude that operators $(\mathbb{A} + iF_+)^{-1}$ and $(\mathbb{A}^* - iF_+)^{-1}$ are $(-,\cdot)$-continuous and define

$$\begin{aligned} T &= \mathbb{A}, \quad \mathrm{Dom}(T) = (\mathbb{A} + iF_+)^{-1}\mathcal{H}, \\ T_1 &= \mathbb{A}^*, \quad \mathrm{Dom}(T_1) = (\mathbb{A}^* - iF_+)^{-1}\mathcal{H}. \end{aligned} \tag{12.101}$$

Using arguments similar to the proof of Theorem 12.6.1, we note that both $\mathrm{Dom}(T)$ and $\mathrm{Dom}(T_1)$ are dense in $\mathcal{H}$ while operator T is closed in $\mathcal{H}$. It is also easy to see that $T_1 = T^*$. Thus, operator T defined by (12.101) fits the construction of $(*)$-extension $\mathbb{A}$. Property (vi) of Definition 12.4.3 follows from Theorem 12.5.4 and the fact that $\mathrm{Re}\,\mathbb{A} = \mathbb{D}$. Therefore, the array

$$\Theta_{F_+} = \begin{pmatrix} \mathbb{A} & K \quad F_+ & I \\ \mathcal{H}_{+1} \oplus \mathcal{H}_2 \subset \mathcal{H}_1 \oplus \mathcal{H}_2 \subset \mathcal{H}_{-1} \oplus \mathcal{H}_2 & & E \end{pmatrix}$$

is an NCL-system and $V(z)$ admits the realizations

$$V(z) = K^*(\mathbb{D} - zF_+)^{-1}K = i[W_{\Theta_{F_+}}(z) + I]^{-1}[W_{\Theta_{F_+}}(z) - I].$$

This completes the proof. □

12.7 Minimal NCL-realization

Recall from Section 6.6 that a symmetric operator $\dot{A}$ in a Hilbert space $\mathcal{H}$ is called a *prime operator* if there exists no reducing non-trivial invariant subspace on which it induces a self-adjoint operator. An NCL-system of the form (12.55) is called F_+-**minimal** if there are no nontrivial reducing invariant subspaces $\mathcal{H}^1 = \overline{\mathcal{H}^1_+}$, ($\mathcal{H}^1_+$ is a (+)-subspace of $\text{Ran}(F_+)$) of $\mathcal{H}$ where the symmetric operator $\dot{A}$ induces a self-adjoint operator. Here the closure is taken with respect to the $(\cdot)$-metric. In the case that $F_+ = I$ this definition coincides with the definition of minimality for L-systems given in Section 6.5.

Theorem 12.7.1. *Let the operator-valued Herglotz-Nevanlinna function $V(z)$ be realized as the impedance function of a scattering NCL-system Θ_{F_+} of the form* (12.55) *Then this NCL-system can be reduced to an F_+-minimal NCL-system that also realizes $V(z)$.*

Proof. Let the operator-valued Herglotz-Nevanlinna function $V(z)$ be realized in the form (12.81) by an NCL-system of the type (12.55). Assume that its symmetric operator $\dot{A}$ has a reducing invariant subspace $\mathcal{H}^1 = \overline{\mathcal{H}^1_+}$, ($\mathcal{H}^1_+$ is a (+)-subspace of $\text{Ran}(F_+)$) on which it generates a self-adjoint operator A_1. Then there is the $(\cdot,\cdot)$-orthogonal decomposition

$$\mathcal{H} = \mathcal{H}^0 \oplus \mathcal{H}^1, \quad \dot{A} = \dot{A}_0 \oplus A_1, \tag{12.102}$$

where $\dot{A}_0$ is an operator induced by $\dot{A}$ on $\mathcal{H}^0$. The identity (12.102) shows that the adjoint of $\dot{A}$ in $\mathcal{H}$ admits the orthogonal decomposition $\dot{A}^* = \dot{A}_0^* \oplus A_1$. Now consider operators $T \supset \dot{A}$ and $T^* \supset \dot{A}$ as in the definition of the system Θ_{F_+}. It is easy to see that both T and T^* admit the $(\cdot,\cdot)$-orthogonal decompositions

$$T = T_0 \oplus A_1, \quad \text{and} \quad T^* = T_0^* \oplus A_1,$$

where $T_0 \supset \dot{A}_0$ and $T_0^* \supset \dot{A}_0$. Since $T \in \Lambda(\dot{A})$, the identity

$$\dot{A}_0 \oplus A_1 = T \cap T^* = (T_0 \cap T_0^*) \oplus A_1,$$

holds and $(-i)$ is a regular point of $T = T_0 \oplus A_1$ or, equivalently, $(-i)$ is a regular point of T_0. (Here we use the notation $T \cap T^*$ to denote the maximal symmetric part of the operators T and T^*.) The above relation shows that $T_0 \in \Lambda(\dot{A}_0)$. Clearly,

$$\mathcal{H}_+ = \mathcal{H}^0_+ \oplus \mathcal{H}^1_+ = \text{Dom}(A_0^*) \oplus \text{Dom}(A_1).$$

This decomposition remains valid in the sense of (+)-orthogonality. Indeed, if $f_0 \in \mathcal{H}^0_+$ and $f_1 \in \mathcal{H}^1_+ = \text{Dom}(A_1)$, then by considering the adjoint of $\dot{A} : \mathcal{H}_0 (= \text{dom}\,\dot{A}) \to \mathcal{H}$ as a mapping from $\mathcal{H}$ into $\mathcal{H}_0$ one obtains

$$(f_0, f_1)_+ = (f_0, f_1) + (\dot{A}^* f_0, \dot{A}^* f_1) = (f_0, f_1) + (\dot{A}_0^* f_0, A_1 f_1) = 0 + 0 = 0.$$

Consequently, the inclusions $\mathcal{H}_+ \subset \mathcal{H} \subset \mathcal{H}_-$ can be rewritten in the decomposed forms

$$\begin{aligned}&\mathcal{H}_+^0 \oplus \mathcal{H}_+^1 \subset \mathcal{H}^0 \oplus \mathcal{H}^1 \subset \mathcal{H}_-^0 \oplus \mathcal{H}_-^1\\ &= \mathcal{H}_+^0 \oplus \mathrm{Dom}(A_1) \subset \mathcal{H}^0 \oplus \mathcal{H}^1 \subset \mathcal{H}_-^0 \oplus \mathcal{H}_-^1.\end{aligned}$$

Now let $\mathbb{A} \in [\mathcal{H}_+, \mathcal{H}_-]$ be the $(*)$-extension of $\dot{A}$ in the definition of the system Θ_{F_+}. Then $\mathbb{A}$ admits the decomposition $\mathbb{A} = \mathbb{A}_0 \oplus A_1$ and $\mathbb{A}^* = \mathbb{A}_0^* \oplus A_1$. Since A_1 is self-adjoint in $\mathcal{H}^1$, $\mathbb{A}_0$ is a $(*)$-extension of T_0. Moreover,

$$\begin{aligned}\frac{\mathbb{A} - \mathbb{A}^*}{2i} &= \frac{(\mathbb{A}_0 \oplus A_1) - (\mathbb{A}_0^* \oplus A_1)}{2i} = \frac{\mathbb{A}_0 - \mathbb{A}_0^*}{2i} \oplus \frac{A_1 - A_1}{2i}\\ &= \frac{\mathbb{A}_0 - \mathbb{A}_0^*}{2i} \oplus O,\end{aligned} \tag{12.103}$$

where O stands for the zero operator. Then (12.103) implies that

$$KJK^* = K_0JK_0^* \oplus O.$$

Let P_+^0 be the orthogonal projection operator of $\mathcal{H}_+$ onto $\mathcal{H}_+^0$ and set $K = K_0 \oplus O$. Then $K^* = K_0^* P_+^0$, since for all $f \in E$, $g \in \mathcal{H}_+$ one has

$$\begin{aligned}(Kf, g) &= (K_0 f, g) = (K_0 f, g_0 + g_1) = (K_0 f, g_0) + (K_0 f, g_1)\\ &= (K_0 f, g_0) = (f, K_0^* g_0) = (f, K_0^* P_+^0 g).\end{aligned}$$

Since $\mathcal{H}_+^1$ is a closed subspace of $\mathrm{Ran}(F_+)$, P_+^0 commutes with F_+ and therefore $F_+^0 := F_+ P_+^0$ defines an orthogonal projection in $\mathcal{H}_+^0$. Now, let $h \in E$, $z \in \rho(\mathbb{A}, F_+, K)$, and $\phi = \phi_0 + \phi_1 \in \mathcal{H}_+ = \mathcal{H}_+^0 \oplus \mathcal{H}_+^1$ be such that

$$(\mathbb{A} - zF_+)\phi = Kh.$$

Since $K = K_0 \oplus O$, the previous identity is equivalent to

$$(\mathbb{A}_0 \oplus A_1 - zF_+)(\phi_0 + \phi_1) = (K_0 \oplus O)h.$$

Since $F_+\phi_1 = \phi_1$ and P_+^0 commutes with F_+, this yields

$$(\mathbb{A}_0 - zF_+^0)\phi_0 = K_0 h, \qquad (A_1 - zI)\phi_1 = 0.$$

It follows from the previous equations that $z \in \rho(A_1)$ because $z \in \rho(\mathbb{A}, F_+, K)$. Thus, $\rho(\mathbb{A}, F_+, K) \subset \rho(\mathbb{A}_0, F_+^0, K_0)$ and hence $\phi_0 = (\mathbb{A}_0 - zF_+^0)^{-1}K_0 h$. On the other hand, $\phi_0 = \phi = (\mathbb{A} - zF_+)^{-1}Kh$ and therefore for all $h \in E$ one obtains $(\mathbb{A} - zF_+)^{-1}Kh = (\mathbb{A}_0 - zF_+^0)^{-1}K_0 h$ and

$$K^*(\mathbb{A} - zF_+)^{-1}Kh = K_0^*(\mathbb{A}_0 - zF_+^0)^{-1}K_0 h.$$

This means that the transfer functions of the system Θ_{F_+} in (12.55) and of the system

$$\Theta^0_{F_+} = \begin{pmatrix} \mathbb{A}_0 & F^0_+ & K_0 & J \\ \mathcal{H}^0_+ \subset \mathcal{H}^0 \subset \mathcal{H}^0_- & & & E \end{pmatrix}$$

coincide. Therefore, the system Θ_{F_+} in (12.55) can be reduced to an F_+-minimal system of the same form such that the corresponding transfer and, consequently, impedance functions coincide. This completes the proof of the theorem. □

The definition of minimality can be extended to NCI-systems in the same manner. Moreover, an NCL-system of the form (12.3)

$$\begin{cases} (\mathbb{A} - zF_+)x = KJ\varphi_-, \\ \varphi_+ = \varphi_- - 2iK^*x, \end{cases}$$

and a NCI-system of the form (12.1)

$$\begin{cases} (\operatorname{Re}\mathbb{A} - zF_+)x = K\varphi_-, \\ \varphi_+ = K^*x, \end{cases}$$

where $\operatorname{Re}\mathbb{A}$ is the real part of $\mathbb{A}$, are minimal (or non-minimal) simultaneously.

For the NCI-systems constructed in Section 12.4 the minimality can be characterized as follows.

Theorem 12.7.2. *The realization of the operator-valued Herglotz-Nevanlinna function $V(z)$ constructed in Theorem 12.5.4 is minimal if and only if the symmetric part $\dot{A}_2$ of $\operatorname{Re}\mathbb{A}^{(2)}$ defined by (12.80) is prime.*

Proof. Assume that the system constructed in Theorem 12.5.4 is not minimal. Let $\mathcal{H}^1$ (with $\mathcal{H}^1_+ \subset \operatorname{Ran}(F_+)$) be a reducing invariant subspace from Theorem 12.7.1 on which $\dot{A}$ generates a self-adjoint operator A_1. Then $\mathbb{D} = \mathbb{D}_0 \oplus A_1$ and it follows from the block representations of $\mathbb{D}$ and F_+ in (12.80) that $\mathcal{H}^1$ is necessarily a subspace of $\mathcal{H}_2$ in (12.79) while $\mathcal{H}^1_+$ is a subspace of $\mathcal{H}_{+2}$. To see this let us describe $\operatorname{Ran}(F_+)$ first. According to (12.79)

$$\mathcal{H}_+ \subset \mathcal{H} \subset \mathcal{H}_- = E^4 \oplus \mathcal{H}_{+2} \subset E^4 \oplus \mathcal{H}_2 \subset E^4 \oplus \mathcal{H}_{-2},$$

where $E^4 = E \oplus E \oplus E \oplus E$. Hence every vector $x \in \mathcal{H}_+$ can be written as

$$x = \begin{pmatrix} x_1 \\ x_2 \\ x_3 \\ x_4 \\ x_5 \end{pmatrix}, \text{ where } x_1, x_2, x_3, x_4 \in E, x_5 \in \mathcal{H}_{+2}.$$

By (12.80),

$$F_+x = \begin{pmatrix} 0 \\ x_2 \\ 0 \\ 0 \\ x_5 \end{pmatrix}, \quad \text{and} \quad \mathbb{D}(F_+x) = \begin{pmatrix} ix_2 \\ 0 \\ 0 \\ 0 \\ \operatorname{Re}\mathbb{A}^{(2)}x_5 \end{pmatrix}.$$

This means that $x \in \mathcal{H}^1_+ \subset \mathrm{Ran}(F_+)$ only if $x_2 = 0$. Therefore the only possibility for a reducing invariant subspace $\mathcal{H}^1$ to exist, is to be a subspace of $\mathcal{H}_2$ while $\mathcal{H}^1_+$ is a subspace of $\mathcal{H}_{+2}$. This proves the claim $\mathcal{H}^1_+ \subset \mathcal{H}_{+2}$. Consequently, $\mathcal{H}^1$ is a reducing invariant subspace for the symmetric operator A_2, in which case the operator A_2 is not prime.

Conversely, if the symmetric operator A_2 is not prime, then a reducing invariant subspace on which A_2 generates a self-adjoint operator is automatically a reducing invariant subspace for the operator $\dot{A}$ which belongs to $\mathrm{Ran}(F_+)$. This completes the proof. □

Finally, Theorem 12.7.2 implies that a realization of an arbitrary operator-valued Herglotz-Nevanlinna function in Theorem 12.5.4 can be provided by a minimal NCI-system.

12.8 Examples and non-canonical system interpolation

Example. Consider the Herglotz-Nevanlinna function

$$V(z) = 1 + z - i\tanh\left(\frac{i}{2}zl\right), \quad z \in \mathbb{C} \setminus \mathbb{R}, \tag{12.104}$$

where $l > 0$. We are going to construct an NCL-system Θ_{F_+} whose impedance function coincides with $V(z)$. Let the differential operator T_2 in $\mathcal{H}_2 = L^2_{[0,l]}$ be given by

$$\begin{aligned} T_2 x &= \frac{1}{i}\frac{dx}{dt}, \\ \mathrm{Dom}(T_2) &= \{\, x(t) \in \mathcal{H}_2 : x(t) - \text{abs. cont.},\ x'(t) \in \mathcal{H}_2,\ x(0) = 0 \,\}, \end{aligned}$$

with the adjoint

$$\begin{aligned} T_2^* x &= \frac{1}{i}\frac{dx}{dt}, \\ \mathrm{Dom}(T_2^*) &= \{\, x(t) \in \mathcal{H}_2 : x(t) - \text{abs. cont.}, x'(t) \in \mathcal{H}_2,\ x(l) = 0 \,\}. \end{aligned}$$

Let $\dot{A}_2$ be the symmetric operator defined by

$$\begin{aligned} \dot{A}_2 x &= \frac{1}{i}\frac{dx}{dt}, \\ \mathrm{Dom}(\dot{A}_2) &= \{\, x(t) \in \mathcal{H}_2 : x(t) - \text{abs. cont.}, x'(t) \in \mathcal{H}_2,\ x(0) = x(l) = 0 \,\}, \end{aligned} \tag{12.105}$$

with the adjoint

$$\dot{A}_2^* x = \frac{1}{i}\frac{dx}{dt}, \quad \mathrm{Dom}(\dot{A}_2^*) = \{\, x(t) \in \mathcal{H}_2 : x(t) - \text{abs. cont.}, x'(t) \in \mathcal{H}_2 \,\}.$$

Then $\mathcal{H}_+ = \mathrm{Dom}(\dot{A}_2^*) = W_2^1$ is a Sobolev space with the scalar product

$$(x,y)_+ = \int_0^l x(t)\overline{y(t)}\,dt + \int_0^l x'(t)\overline{y'(t)}\,dt.$$

Now consider the rigged Hilbert space $W_2^1 \subset L^2_{[0,l]} \subset (W_2^1)_-$ and the operators

$$\mathbb{A}_2 x = \frac{1}{i}\frac{dx}{dt} + ix(0)\left[\delta(x-l) - \delta(x)\right],\ \mathbb{A}_2^* x = \frac{1}{i}\frac{dx}{dt} + ix(l)\left[\delta(x-l) - \delta(x)\right],$$

where $x(t) \in W_2^1$ and $\delta(x)$, $\delta(x-l)$ are delta-functions in $(W_2^1)_-$. Define the operator K_2 by

$$K_2 c = c \cdot \frac{1}{\sqrt{2}}[\delta(x-l) - \delta(x)], \quad c \in \mathbb{C}^1,$$

so that

$$K_2^* x = \left(x, \frac{1}{\sqrt{2}}[\delta(x-l) - \delta(x)]\right) = \frac{1}{\sqrt{2}}[x(l) - x(0)], \quad x(t) \in W_2^1.$$

Let $D_1 = K_1 Q_1^{-1} K_1^* = 1$, where $Q = 1$, and $K_1 = 1$, $K_1 : \mathbb{C} \to \mathbb{C}$. Following (12.76) define

$$\mathcal{H}_3 = \mathbb{C} \oplus \mathbb{C}, \quad K_3 = \begin{pmatrix} 1 \\ 1 \end{pmatrix}, \quad D_3 = \begin{pmatrix} 0 & iI \\ -iI & 0 \end{pmatrix}, \quad F_{+,3} = \begin{pmatrix} 0 & 0 \\ 0 & I \end{pmatrix}.$$

Now the corresponding NCL-system can be constructed. According to (12.80) one has

$$\mathbb{D} = \begin{pmatrix} 0 & i & 0 & 0 \\ -i & 0 & 0 & 0 \\ 0 & 0 & 1 & 0 \\ 0 & 0 & 0 & \mathrm{Re}\,\mathbb{A}_2 \end{pmatrix}, \quad F_+ = \begin{pmatrix} 0 & 0 & 0 & 0 \\ 0 & 1 & 0 & 0 \\ 0 & 0 & 0 & 0 \\ 0 & 0 & 0 & I \end{pmatrix}, \quad K = \begin{pmatrix} 1 \\ 1 \\ 1 \\ K_2 \end{pmatrix},$$

and it follows from (12.93)-(12.94) that

$$\mathbb{A} = \mathbb{D} + iKK^* = \begin{pmatrix} i & 2i & i & iK_2^* \\ 0 & i & i & iK_2^* \\ i & i & 1+i & iK_2^* \\ iK_2 & iK_2 & iK_2 & \mathbb{A}_2 \end{pmatrix}.$$

Consequently, the corresponding NCL-system is given by

$$\Theta_{F_+} = \begin{pmatrix} \mathbb{A} & K & F_+ & I \\ \mathbb{C}^3 \oplus W_2^1 \subset \mathbb{C}^3 \oplus L^2_{[0,l]} \subset \mathbb{C}^3 \oplus (W_2^1)_- & & & \mathbb{C} \end{pmatrix}, \tag{12.106}$$

where $\mathbb{C}^3 = \mathbb{C} \oplus \mathbb{C} \oplus \mathbb{C}$ and all the operators are described above. Since the symmetric operator $\dot{A}_2$ defined in (12.105) is prime, then according to Theorem

12.7.2, the system (12.106) is prime. Thus, $V(z)$ in (12.104) are the impedance functions of the F_+-minimal NCL-system (12.106). The transfer function of this system is

$$W_{\Theta_{F_+}}(z) = \frac{2 - i(1+e^{izl})(z+1)}{2e^{izl} + i(1+e^{izl})(z+1)} = \frac{1 - i - zi - \tanh\left(\frac{i}{2}zl\right)}{1 + i + zi + \tanh\left(\frac{i}{2}zl\right)}.$$

Example. Consider the Herglotz-Nevanlinna function

$$V(z) = \begin{pmatrix} 1 & 0 \\ 0 & 1 \end{pmatrix} + z\begin{pmatrix} 1 & 0 \\ 0 & 1 \end{pmatrix} + \begin{pmatrix} -i\tanh(\pi i z) & 0 \\ 0 & \frac{1-z}{z^2-z-1} \end{pmatrix}. \tag{12.107}$$

An explicit NCL-system Θ_{F_+} will be constructed so that $V(z) \equiv V_{\Theta_{F_+}}(z)$. Let T_{21} be a differential operator $\mathcal{H}_2 = L^2_{[0,2\pi]}$ given by

$$T_{21}x = \frac{1}{i}\frac{dx}{dt},$$
$$\mathrm{Dom}(T_{21}) = \{\, x(t) \in \mathcal{H}_2 : x(t) - \text{abs. cont.}, x'(t) \in \mathcal{H}_2,\, x(0) = 0 \,\},$$

with the adjoint

$$T^*_{21}x = \frac{1}{i}\frac{dx}{dt},$$
$$\mathrm{Dom}(T^*_{21}) = \{\, x(t) \in \mathcal{H}_2 : x(t) - \text{abs. cont.}, x'(t) \in \mathcal{H}_2,\, x(2\pi) = 0 \,\}.$$

Let $\dot{A}_{21}$ be the symmetric operator defined by

$$\dot{A}_{21}x = \frac{1}{i}\frac{dx}{dt},$$
$$\mathrm{Dom}(\dot{A}_{21}) = \{\, x(t) \in \mathcal{H}_2 : x(t) - \text{abs. cont.}, x'(t) \in \mathcal{H}_2,\, x(0) = x(2\pi) = 0 \,\}, \tag{12.108}$$

with the adjoint

$$\dot{A}^*_{21}x = \frac{1}{i}\frac{dx}{dt}, \quad \mathrm{Dom}(\dot{A}^*_{21}) = \{\, x(t) \in \mathcal{H}_2 : x(t) - \text{abs. cont.}, x'(t) \in \mathcal{H}_2 \,\}.$$

Then $\mathcal{H}_+ = \mathrm{Dom}(\dot{A}^*_{21}) = W^1_2$ is a Sobolev space with the scalar product

$$(x,y)_+ = \int_0^{2\pi} x(t)\overline{y(t)}\,dt + \int_0^{2\pi} x'(t)\overline{y'(t)}\,dt.$$

Consider the rigged Hilbert space $W^1_2 \subset L^2_{[0,2\pi]} \subset (W^1_2)_-$ and the operators

$$\mathbb{A}_{21}x = \frac{1}{i}\frac{dx}{dt} + ix(0)\left[\delta(x-2\pi) - \delta(x)\right],$$
$$\mathbb{A}^*_{21}x = \frac{1}{i}\frac{dx}{dt} + ix(2\pi)\left[\delta(x-2\pi) - \delta(x)\right],$$

where $x(t) \in W_2^1$ and $\delta(x)$, $\delta(x-2\pi)$ are delta-functions in $(W_2^1)_-$. Define the operator K_{21} by

$$K_{21}c = c \cdot \frac{1}{\sqrt{2}}[\delta(x-2\pi)-\delta(x)], \quad c \in \mathbb{C}^1,$$

so that

$$K_{21}^* x = \left(x, \frac{1}{\sqrt{2}}[\delta(x-2\pi)-\delta(x)]\right) = \frac{1}{\sqrt{2}}\,[x(2\pi)-x(0)],$$

for $x(t) \in W_2^1$. Define

$$T_{22} = \begin{pmatrix} i & i \\ -i & 1 \end{pmatrix} \text{ and } K_{22} = \begin{pmatrix} 1 \\ 0 \end{pmatrix}, \tag{12.109}$$

and set

$$\mathbb{A}_2 = \begin{pmatrix} \mathbb{A}_{21} & 0 \\ 0 & T_{22} \end{pmatrix} \text{ and } K_2 = \begin{pmatrix} K_{21} & 0 \\ 0 & K_{22} \end{pmatrix}. \tag{12.110}$$

Now let $D_1 = K_1 Q^{-1} K_1^* = I_2$, where $Q = I_2$, and $K_1 = I_2$, $K_1 : \mathbb{C}^2 \to \mathbb{C}^2$. Following (12.76) define

$$\mathcal{H}_3 = \mathbb{C}^4, \quad K_3 = \begin{pmatrix} 1 \\ 1 \\ 1 \\ 1 \end{pmatrix}, \quad D_3 = \begin{pmatrix} 0 & 0 & i & 0 \\ 0 & 0 & 0 & i \\ -i & 0 & 0 & 0 \\ 0 & -i & 0 & 0 \end{pmatrix}, \quad F_{+,3} = \begin{pmatrix} 0 & 0 & 0 & 0 \\ 0 & 0 & 0 & 0 \\ 0 & 0 & 1 & 0 \\ 0 & 0 & 0 & 1 \end{pmatrix}.$$

Now the corresponding NCL-system will be constructed. According to (12.80) one has

$$\mathbb{D} = \begin{pmatrix} D_3 & \vdots & 0 \\ \cdots & \cdots & \cdots \\ 0 & D_1 & 0 \\ \cdots & \cdots & \cdots \\ 0 & \vdots & \operatorname{Re}\mathbb{A}_2 \end{pmatrix}, \quad F_+ = \begin{pmatrix} F_{+,3} & \vdots & 0 \\ \cdots & \cdots & \cdots \\ 0 & 0 & 0 \\ \cdots & \cdots & \cdots \\ 0 & \vdots & I \end{pmatrix}, \quad K = \begin{pmatrix} K_3 \\ \cdots \\ K_1 \\ \cdots \\ K_2 \end{pmatrix}, \tag{12.111}$$

and it follows from (12.93)-(12.94) that

$$\mathbb{A} = \mathbb{D} + iKK^*. \tag{12.112}$$

Consequently, the corresponding NCL-system is given by

$$\Theta_{F_+} = \begin{pmatrix} \mathbb{A} & K & F_+ & I \\ \mathbb{C}^6 \oplus W_2^1 \subset \mathbb{C}^6 \oplus L^2_{[0,2\pi]} \subset \mathbb{C}^6 \oplus (W_2^1)_- & & & \mathbb{C}^2 \end{pmatrix}, \tag{12.113}$$

where all the operators are described above. The transfer function of this system is given by

$$W_{\Theta_{F_+}}(z) = \begin{pmatrix} \frac{1-i-zi-\tanh(\pi i z)}{1+i+zi+\tanh(\pi i z)} & 0 \\ 0 & \frac{z^3+iz^2-(3+i)z-i}{-z^3+iz^2+(3-i)z-i} \end{pmatrix}.$$

It is easy to see that the maximal symmetric part of the operator T_{22} in (12.109) is a non-densely-defined operator

$$\dot{A}_{22} = \begin{pmatrix} 0 & i \\ -i & 1 \end{pmatrix}, \quad \mathrm{Dom}(\dot{A}_{22}) = \left\{ \begin{pmatrix} 0 \\ c \end{pmatrix} : c \in \mathbb{C} \right\}. \tag{12.114}$$

Consequently, the symmetric operator $\dot{A}_2$ defined by $\mathbb{A}_2$ in (12.110), $\mathbb{D}$ in (12.111), and $\mathbb{A}$ in (12.112) is given by

$$\dot{A}_2 = \begin{pmatrix} \frac{1}{i}\frac{dx}{dt} & 0 & 0 \\ 0 & 0 & i \\ 0 & -i & 1 \end{pmatrix},$$

$$\mathrm{Dom}(\dot{A}_2) = \left\{ \begin{pmatrix} x(t) \\ 0 \\ c \end{pmatrix} : x(t), x'(t) \in \mathcal{H}_2,\ x(0) = x(2\pi) = 0,\ c \in \mathbb{C} \right\}.$$

Hence, this operator $\dot{A}_2$ does not have nontrivial reducing invariant subspaces on which it induces self-adjoint operators. Thus, the NCL-system in (12.113) is an F_+-minimal realization of the function $V(z)$ in (12.107).

Example. Let us consider the interpolation data

$$z_1 = i, \quad z_2 = 2i, \quad v_1 = 1 + i, \quad v_2 = 1 + \frac{i}{2}. \tag{12.115}$$

We construct Pick matrices $\mathbb{P}$ and $\mathbb{Q}$ of the form (11.38) and get

$$\mathbb{P} = \begin{pmatrix} 1 & \frac{1}{2} \\ \frac{1}{2} & \frac{1}{4} \end{pmatrix}, \qquad \mathbb{Q} = \begin{pmatrix} 1 & 1 \\ 1 & 1 \end{pmatrix}.$$

Clearly $\det(P) = 0$ and $\mathbb{P}$ is not invertible. Thus we can not apply Theorem 11.6.1 to deduce the existence of a canonical interpolation system that is a solution to the Nevanlinna-Pick interpolation problem for the data (12.115). Moreover, such a canonical system does not exist. To show that we assume the contrary, i.e., that there is a system of the form (11.66) with state-space $\mathcal{H}$, $\dim \mathcal{H} = 2$, and the state-space operator T having one-dimensional imaginary part $\mathrm{Im}\, T = (\cdot, g)g$. Without loss of generality we can assume that T is prime. Then, by Lemma 11.1.3, $\mathrm{Re}\, T$ has a cyclic vector g. Consequently, we can apply Theorem 11.7.1 that gives us invertibility of $\mathbb{P}$. Thus, we have arrived at a contradiction.

Let φ be a vector in $\mathbb{C}^2$ of the form (11.40), i.e., $\varphi = \begin{pmatrix} \bar{v}_1 \\ \bar{v}_2 \end{pmatrix} = \begin{pmatrix} 1 - i \\ 1 - \frac{i}{2} \end{pmatrix}$. Consider the function

$$V(z) = ((\mathbb{Q} - z\mathbb{P})^{-1}\varphi, \varphi). \tag{12.116}$$

It was shown in the proof of Theorem 11.6.1 that $V(z_k) = v_k$, $(k = 1, 2)$. Taking into account that

$$\mathbb{Q} - z\mathbb{P} = \begin{pmatrix} 1 - z & 1 - \frac{z}{2} \\ 1 - \frac{z}{2} & 1 - \frac{z}{4} \end{pmatrix},$$

we compute $V(z)$ using (12.116) to obtain

$$V(z) = ((\mathbb{Q} - z\mathbb{P})^{-1}\varphi, \varphi) = 1 - \frac{1}{z}.$$

According to Theorem 12.3.6 the function $V(z)$ can be realized as the impedance function of a non-canonical scattering F-system Θ_F of the form (12.17). The algorithm of reconstruction of Θ_F is described in the proof of Theorem 12.3.6 while the elements of Θ_F are given by (12.50). This yields

$$K = \begin{pmatrix} 1 \\ 1 \end{pmatrix}, \quad K^* = (1,\ 1), \qquad D = \begin{pmatrix} 1 & 0 \\ 0 & 1 \end{pmatrix}, \quad F = \begin{pmatrix} 0 & 0 \\ 0 & 1 \end{pmatrix},$$

and $M = D + iKK^* = \begin{pmatrix} 1+i & i \\ i & i \end{pmatrix}$. Finally,

$$\Theta_F = \begin{pmatrix} M & F & K & 1 \\ \mathbb{C}^2 & & & \mathbb{C} \end{pmatrix}, \tag{12.117}$$

and

$$W_{\Theta_F}(z) = I - 2iK^*(M - zF)^{-1}K = \frac{z - iz + i}{z + iz - i}.$$

Therefore we have shown that even though the canonical system interpolation solution to the Nevanlinna-Pick problem for the data (12.115) does not exist, there is a non-canonical system solution provided by the system Θ_F in (12.117) and its impedance function $V_{\Theta_F}(z) = 1 - 1/z$.

Example. Consider the interpolation data

$$z_1 = i, \quad z_2 = 2i, \quad v_1 = 2i, \quad v_2 = \frac{5i}{2}, \tag{12.118}$$

and construct Pick matrices $\mathbb{P}$ and $\mathbb{Q}$ of the form (11.38)

$$\mathbb{P} = \begin{pmatrix} 2 & \frac{3}{2} \\ \frac{3}{2} & \frac{5}{4} \end{pmatrix}, \qquad \mathbb{Q} = \begin{pmatrix} 0 & -i \\ i & 0 \end{pmatrix}. \tag{12.119}$$

Unlike Example 12.8, $\det(P) \neq 0$ and $\mathbb{P}$ is invertible. Hence we can apply Theorem 11.6.1 which provides us with a canonical system $\vec{\Theta}$ of the form (11.47) with the state space $\mathbb{C}^2_{\mathbb{P}}$ of dimension 2 that is a solution to the Nevanlinna-Pick interpolation problem for the data (12.118). Applying (11.40) we get

$$\varphi = \begin{pmatrix} \bar{v}_1 \\ \bar{v}_2 \end{pmatrix} = \begin{pmatrix} -2i \\ -\frac{5i}{2} \end{pmatrix}, \quad \vec{g} = \mathbb{P}^{-1}\varphi = \begin{pmatrix} 5 & -6 \\ -6 & 8 \end{pmatrix}\varphi = \begin{pmatrix} 5i \\ -8i \end{pmatrix}. \tag{12.120}$$

The state-space and channel operators $\vec{T}$ and $\vec{K}$ of $\vec{\Theta}$ are defined by (11.41) and (11.42) using $\mathbb{P}$, $\mathbb{Q}$, and $\vec{g}$ from (12.119)–(12.120) as

$$\vec{T} = \mathbb{P}^{-1}\mathbb{Q} + i(\cdot\,, \vec{g})_{\mathbb{C}^2_{\mathbb{P}}}\vec{g}, \quad \vec{K}c = c \cdot \vec{g}, \quad c \in \mathbb{C}.$$

Furthermore,

$$V_{\vec{\Theta}}(z) = ((\mathbb{Q} - z\mathbb{P})^{-1}\varphi, \varphi) = -\frac{10z}{z^2 - 4}, \tag{12.121}$$

is the impedance function of the canonical interpolation system solution

$$\vec{\Theta} = \begin{pmatrix} \vec{T} & \vec{K} & I \\ \mathbb{C}^2_{\mathbb{P}} & & \mathbb{C} \end{pmatrix}. \tag{12.122}$$

On the other hand, it is easy to see that our interpolation data (12.118) also satisfies another function

$$V(z) = z - \frac{1}{z}. \tag{12.123}$$

Clearly, the function $V(z)$ does not belong to the class $N(R)$ and hence can not be realized as an impedance function of any canonical L-system. However, $V(z)$ can be realized as an impedance function of a non-canonical scattering F-system Θ_F of the form (12.17). We follow the algorithm of reconstruction described in the proof of Theorem 12.3.6 and obtain the elements of Θ_F via (12.50). This yields

$$K = \begin{pmatrix} 1 \\ 1 \\ 1 \end{pmatrix}, \quad K^* = (1,\ 1,\ 1), \quad D = \begin{pmatrix} 0 & i & 0 \\ -i & 0 & 0 \\ 0 & 0 & 0 \end{pmatrix}, \quad F = \begin{pmatrix} 0 & 0 & 0 \\ 0 & 1 & 0 \\ 0 & 0 & 1 \end{pmatrix},$$

and $M = D + iKK^* = \begin{pmatrix} i & 2i & i \\ 0 & i & i \\ i & i & i \end{pmatrix}$. Finally,

$$\Theta_F = \begin{pmatrix} M & F & K & 1 \\ \mathbb{C}^3 & & & \mathbb{C} \end{pmatrix}, \tag{12.124}$$

and

$$W_{\Theta_F}(z) = \frac{z^2 + iz - 1}{-z^2 + iz - 1}, \qquad V_{\Theta_F}(z) = z - 1/z.$$

Therefore we have shown that there are two system solutions to the Nevanlinna-Pick problem for the data (12.118): one canonical and one non-canonical. The canonical one is given by the system $\vec{\Theta}$ of the form (12.122) and has the impedance function $V_{\vec{\Theta}}(z) = -10z/(z^2 - 4)$ in (12.121). The non-canonical system solution is provided by the system Θ_F in (12.124) and its impedance function $V_{\Theta_F}(z) = z - 1/z$.

Notes and Comments

Chapter 1

Theorem 1.1.2 for the case of a densely-defined symmetric operator belongs to von Neumann and for a non-densely-defined operator to Krasnoselskiĭ [169]. Theorems from Sections 1.1–1.4 were stated and proved by von Neumann. Definition of the aperture of two linear manifolds in Hilbert space due to Sz.-Nagy [242] and Kreĭn–Krasnoselskiĭ [170] (see also [3]). Formula (1.21) was used by Kreĭn–Krasnoselskiĭ–Milman [171] to determine the aperture of two linear manifolds in Banach space. Semi-deficiency subspaces and semi-deficiency indices have been introduced by Krasnoselskiĭ [169]. Theorem 1.5.3 belongs to Krasnoselskiĭ [169]. Symmetric and self-adjoint extensions of non-densely-defined symmetric operators were considered for the first time in [169]. Admissible isometric extensions were introduced and used by Krasnoselskiĭ [169] to obtain the analogues of the first and second von Neumann's formulas in order to get parametrization of symmetric and self-adjoint extensions. Theorems 1.7.4 and 1.7.5 belong to M. Naimark [206] and Krasnoselskiĭ [169]. The results of Section 1.7 were published in [169].

Chapter 2

Rigged Hilbert spaces were first studied by Leray [186] and Lax [184], but even earlier Krein [172] had studied the case when $\mathcal{H}_+$ is a Banach space and $\mathcal{H}_-$ is its dual. The theory of rigged Hilbert spaces (triplets of Hilbert spaces) used in the text was developed by Berezansky in [81]. Another approach to the rigged Hilbert spaces via so-called complimentary Hilbert spaces belongs to de Branges [85]. Theorem 2.1.1 is taken from the paper by Tsekanovskiĭ and Šmuljan [264]. The construction of an operator generated rigging was first carried out independently by Tsekanovskiĭ [247] and Šmuljan [264]. Theorem 2.1.2 is due to Douglas [112].

The analogues of von Neumann formulas (2.14) and (2.15) as well as Theorem 2.3.1 are contained in the papers by Arlinskiĭ and Tsekanovskiĭ [44], [45], and of Šmuljan [233]. Theorem 2.3.3 is due to Arlinskiĭ and Tsekanovskiĭ [44]. The decomposition of the class of all closed symmetric operators into the classes of regular and singular operators was found by Krasnoselskii [169]. Theorem 2.4.1

is taken from [264]. The concept of an O-operator was introduced and studied independently by Arlinskiĭ and Tsekanovskiĭ [45] and by Šmuljan [233]. Theorem 2.4.3 was first published in [264]. The notion of the minimal angle between two subspaces in a Hilbert space, investigations of its properties and results very close to Lemma 2.5.2 can be found in [171], [179], and [139]. Theorems 2.5.1–2.5.4 are due to Šmuljan [233], [264], while Theorems 2.5.5–2.5.6 are due to Arlinskiĭ, Tsekanovskiĭ, and Šmuljan [44], [264]. The definition of quasi-kernel belongs to Tsekanovskiĭ [254], [264].

Chapter 3

The concept of bi-extensions, including self-adjoint bi-extensions was introduced by Tsekanovskiĭ [251]. Most of the contents of Sections 3.1–3.2 belong to Tsekanovskiĭ and Šmuljan, and can be found in [264] and [265]. The content of Section 3.3 is due to Arlinskiĭ and Tsekanovskiĭ [44], [45]. The results of Section 3.4 were obtained for the first time by Tsekanovskiĭ [251], [252], [253] for densely-defined symmetric operators with finite and equal defect indices, and for a densely-defined symmetric operator with arbitrary and equal defect indices by Okunskiĭ and Tsekanovskiĭ, and published in [265]. Theorem 3.4.11 and the description of self-adjoint extensions of a symmetric densely-defined operator with finite equal deficiency indices in terms of abstract boundary conditions (3.40) belong to Tsekanovskiĭ [251]. Analogues to Theorem 3.4.11 result for quasi-self-adjoint extensions and the corresponding abstract boundary conditions can be found in [253]. The t-self-adjoint bi-extensions of a densely-defined symmetric operator have been used in the abstract problem of singular perturbation of a self-adjoint operator by Arlinskiĭ and Tsekanovskiĭ in [52].

Chapter 4

The results of this chapter are due to Arlinskiĭ, Tsekanovskiĭ, and Šmuljan [264], [265]. Lemma 4.1.4 and Theorems 4.1.9, 4.1.11 in Section 4.1 belong to Arlinskiĭ and Tsekanovskiĭ and can be found in [264]. Theorems 4.1.10, 4.1.12 belong to Arlinskiĭ, while Theorems 4.1.5 and 4.1.13 are due to Tsekanovskiĭ and Šmuljan [264], [265]. In Section 4.2, Theorems 4.2.5 and 4.2.8 for symmetric operators with dense domain and finite and equal defect indices and formulas (4.23) were obtained for the first time by Tsekanovskiĭ [254], [263] while Theorem 4.2.9 belongs to Arlinskiĭ [20]. In Section 4.3 the notion of $(*)$-extension belongs to Tsekanovskiĭ and Šmuljan [264], [265], [263]. Theorems 4.3.2, 4.3.3, 4.3.10, and 4.3.11 are due to Arlinskiĭ [20]. In the case of a densely-defined symmetric operator with equal defect indices, Theorems 4.3.2, 4.3.3, 4.3.5, and 4.3.7 via formulas (4.23) were obtained by Tsekanovskiĭ [254], [253]. Uniqueness Theorem 4.3.9 in the densely-defined case is due to Tsekanovskiĭ [254], [253], while the presented proof is due to Arlinskiĭ. In

Section 4.4 Theorem 4.4.2 belongs to Arlinskiĭ, Tsekanovskiĭ and Šmuljan [264], [265], and Theorem 4.4.3 is due to Arlinskiĭ [20]. The Uniqueness Theorem 4.4.6 for a densely-defined symmetric operator with equal defect indices was obtained by Tsekanovskiĭ [254], [253]; the presented proof belongs to Arlinskii. The results of Section 4.5 belong to Arlinskiĭ, Tsekanovskiĭ, and Šmuljan [254], [264], [265]. Theorems 4.5.1–4.5.6 belong to Arlinskiĭ, Tsekanovskiĭ and independently to Šmuljan [17], [231], [264], [265]. Theorem 4.5.11 was obtained by Šmuljan [231], [233]. Theorems 4.5.12–4.5.18 were originally set up for the densely-defined case with finite and equal defect indices by Tsekanovskiĭ [254], [264], [265].

Chapter 5

Canonical open systems (5.6), (5.4) with a bounded main (state-space) operator and its transfer functions (5.17) were introduced by Livšic [191]. The metric conservation law (5.2) belongs to Livšic and can be found in books by Livšic-Yantsevich [193] and by Zolotarev [272]. In Operator Theory [89], [193] the array (5.6) is called the operator colligation and the expression (5.17) is called the characteristic function of this colligation. The notion of the characteristic function of a bounded linear operator on a separable Hilbert space with nuclear imaginary part and its analytical properties without colligation approach was discovered and studied for the first time by Livšic [190]. Theorems 5.1.2 and 5.6 belong to Brodskiĭ [89], and Theorem 5.4 is due to Livšic [190]; the presented proof of this theorem is due to Brodskiĭ [89]. The geometrical theory of operator colligations was developed by Brodskiĭ [89] and we follow this approach and terminology (principal and excess subspaces, system projections, etc.) in our considerations. Coupling of two linear bounded operators originally belongs to Livšic [190], [91]. On its basis Livšic and Potapov set up for the first time the multiplication theorem [192] of the characteristic matrix-valued functions. The fact that the characteristic function of a bounded linear operator with nuclear imaginary part is a unitary invariant of this operator was discovered by Livšic [190]. The new approach to inverse problems in the theory of non-self-adjoint operators belongs to Livšic [190], [193], [188]. He described the class of matrix-valued functions that can be realized as characteristic functions of some bounded linear operator and set up connections with the spectrum of this operator. Theorems 5.4.1, 5.4.3, 5.5.3–5.5.8 are a colligation version of the Livšic approach and belong to M.Brodskiĭ [89]. Theorem 5.5.1 belongs to the authors and is a colligation version of a result implicitly contained in [91]. Triangular models of linear operators with nuclear imaginary part in Hilbert spaces were constructed by Livšic using factorizations of characteristic (transfer) functions and are presented in [188], [189], [190], [193], [91], [272]. Theorem 5.6.4 is a special case of a more general result established by Livšic. Functional models for contractions were obtained by B.Sz.-Nagy and Foias [243], by de Branges and Rovnyak [86], [87].

Chapter 6

Theorems 6.1.2–6.2.8 belong to Šmuljan [231], [232], [233], Theorem 6.2.10 is due to Arlinskiĭ [18]. The definition of a rigged canonical system in the case of a densely-defined symmetric part of a given non-self-adjoint state-space operator with finite and equal deficiency indices was introduced by Tsekanovskiĭ [254]. The definition of an L-system, an impedance system associated with an L-system, and simply an impedance system, belongs to the authors. The metric conservation law (6.34) for L-systems with unbounded operators is presented for the first time and its proof is based on bi-extension and (*)-extension theory developed in Chapters 3 and 4. Theorems 6.3.7 and 6.3.8 also belong to the authors and are being published for the first time. Class $N(R)$ of Herglotz-Nevanlinna functions were introduced by Belyi and Tsekanovskiĭ [74], [75], but in the implicit form this class was considered earlier by Arlinskiĭ [18], [19]. Matrix-valued functions of the class $N(R)$ were also treated by Alpay and Gohberg in [7]. The conservative state-space realizations of analytic functions that maps the right half-plane into itself were considered by Ball and Staffans [68], [69]. The realization problems with passive systems were treated by Arov and Nudelman [56], [57], [60]. Theorems 6.4.3–6.6.1 belong to Arlinskiĭ [18], Belyi and Tsekanovskiĭ [74], [75]. The proof of these theorems is based on the approach considered by Tsekanovskiĭ [254] and its further general development by Arlinskiĭ in [18], [19] regarding inverse problems of the theory of characteristic functions of unbounded operators. A description of the Hilbert space $L^2_G(\mathbb{C}^n)$ with matrix-valued function $G(t)$ has been given by Kac [158] (see also [61], [207]). In the case of $\dim E = \infty$ and operator-valued function $G(t)$ the characterization of $L^2_G(E)$ and the corresponding functional model for the symmetric operator were obtained by Gesztesy, Kalton, Makarov, and Tsekanovskiĭ [135] for the case of the dense domain and by S. Malamud and M. Malamud in [199]. The functional model for a symmetric operator and its primeness was obtained by Gesztesy and Tsekanovskiĭ in [137] for the case of finite and equal deficiency indices and in the general situation by S. Malamud and M. Malamud in [199]. Our proof of the primeness for a symmetric operator in Theorem 6.6.7 uses the approach of [199] to a finite-dimensional space E. Lemmas 6.6.4–6.6.9 belong to Tsekanovskiĭ [258], [260]. Theorem 6.6.10 and Corollary 6.6.12 are due to Arlinskiĭ and Tsekanovskiĭ [53].

Operator colligations (open systems) and their characteristic functions (transfer functions) related to unbounded quasi-self-adjoint state space operators, were defined and studied using different methods by Strauss [241], A. Kuzhel [181], Gubreev [144], [145], and Gubreev-Kovalenko [147]. Some connections between A. Kuzhel's approach and the (*)-extensions are established in [146]. In Zolotarev's papers [273] and [274] Kuzhel's approach is applied to the study of open systems with commuting unbounded operators. Triangular models for unbounded operators were constructed by A. Kuzhel [181], Arlinskiĭ [21], and Tsekanovskiĭ [259]. Functional models for unbounded dissipative and non-dissipative operators have been applied by Pavlov (see [214] and references therein), by Naboko [203], [204] for spectral analysis, and by Tikhonov [244].

Chapter 7

Subclasses $N_0(R)$, $N_1(R)$, and $N_{01}(R)$ were introduced by Belyi and Tsekanovskiĭ [77]. Theorems 7.1.4–7.1.11 are due to Belyi and Tsekanovskiĭ [77]. Classes $\Omega(R,J)$, $\Omega_0(R,J)$, $\Omega_1(R,J)$, $\Omega_{01}(R,J)$ are defined by Belyi and Tsekanovskiĭ [76]. In implicit form $\Omega_0(R,J)$ was considered in the inverse problem of characteristic functions of unbounded operators by Tsekanovskiĭ [254] and other classes by Arlinskiĭ [18]. Theorem 7.2.4 and Corollary 7.2.5 are due to Belyi and Tsekanovskiĭ [76]. Theorems 7.3.2–7.3.3, 7.3.5–7.3.8 belong to Belyi and Tsekanovskiĭ [76]. Coupling of two rigged canonical systems with transfer functions from the class $\Omega_0(R,J)$ was considered for the first time by Arlinskiĭ and Tsekanovskiĭ [46]. Theorem 7.3.4 is due to Arlinskiĭ and Tsekanovskiĭ [46]. Proposition 7.4.4, Theorem 7.4.5 belongs to Arlinskĭ and Kaplan [43]; Theorems 7.5.2 and 7.5.3 were established by Arlinskiĭ. The definition of the Q-function is due to Krein and Langer [174], [175], [176]. The definition of the Weyl-Titchmarsh function via boundary triplet (boundary value spaces) belong to Derkach and Malamud [100]. In [99], [102], [104] these authors established connections between Strauss characteristic functions [241] of quasi-self-adjoint (proper) extensions of a symmetric operator and the Weyl (Weyl-Titchmarsh) functions of boundary triplets.

Chapter 8

Lemma 8.1.1, Theorem 8.1.2 are due to Arlinskiĭ and Tsekanovskiĭ [46], [51]. An auxiliary system of the form (8.15) was introduced in [46] and [51]. Theorems 8.2.1–8.2.2 on the constant J-unitary factor belong to Arlinskiĭ and Tsekanovskiĭ [46], [51]. Theorem 8.2.3 is due to Arlinskiĭ, Belyi and Tsekanovskiĭ and is presented for the first time. Transform (8.31) was introduced in the scalar case by Donoghue [111]. Theorem 8.2.4 is due to Tsekanovskiĭ [255], [260]. Theorems 8.3.1 and 8.3.2 were discovered in the matrix-valued case for normalized Weyl-Titchmarsh functions by Makarov and Tsekanovskiĭ. The version of the theorems presented here belongs to Makarov and the authors. The Donoghue transform for normalized Herglotz-Nevanlinna matrix-functions was studied by Gesztesy and Tsekanovskiĭ [137]; the operator-valued version of the Donoghue transform has been considered and used in Krein's resolvent formula by Gesztesy, Makarov and Tsekanovskiĭ [134] and Gesztesy, Kalton, Makarov and Tsekanovskiĭ [135]. The results of Section 8.4 belong to Tsekanovskiĭ [247], [248], [249]. Definition of normalized canonical systems belongs to Tsekanovskiĭ [247], [248], [249]. Realization of $e^{\frac{il}{\lambda}}$ as transfer function belongs to Livšic, while realization of $e^{il\lambda}$ is due to Tsekanovskiĭ [247], [248], [249].

Chapter 9

The statements in Lemma 9.1.1 are due to Ando [13] and Krein–Ovcharenko [177]. Operator H_Ω presented in formula (9.8) was introduced by Krein [172], [173], the name shorted operator for H_Ω was first used by Anderson [10]. Properties of shorted operators were studied by Anderson [10], Anderson and Trapp [11], Anderson and Duffin [12], Fillmore and Williams[123], Kreĭn and Ovcharenko [177], Nishio and Ando [210], Pekarev [215], Shmul'yan [230]. Theorem 9.1.3 belongs to Crandall [94]. Theorems 9.1.4–9.1.5 belong to Arsene and Geondea [61], Davis, Kahan and Weinberger [96], Šmuljan and Yanovskaya [234]; different proofs are given by Foias and Frazho [125], Dritchel and Rovnyak [114], Kolmanovich and Malamud [165], Arlinskiĭ [37], Arlinskiĭ, Hassi, de Snoo [40]. The extremal sc-extensions A_μ and A_M, called rigid and soft contractive self-adjoint extensions of symmetric (Hermitian) contraction, were discovered by Krein [172], [173]; criterion (9.55) for uniqueness of *sc*-extensions, Proposition 9.2.6, and Theorem 9.2.4 belong to Krein [172]. Kreĭn's approach is presented in Akhiezer and Glazman [3] and also in A. Kuzhel and S. Kuzhel [182]. The matrix form of the rigid and soft extensions A_μ and A_M were obtained by Ando in [14] (see also [65], [165], [150]).

Definition of a sectorial operator belongs to Kato [163]; α-cosectorial contraction, class $C_{\mathfrak{H}}(\alpha)$ as well as qsc-extension were introduced by Arlinskiĭ and Tsekanovskiĭ [47], [49], [50], [257], [260], [25], [27]. Theorem 9.2.7 on parametrization of all α-cosectorial contractive extensions of a given symmetric contraction as well as Corollaries 9.2.14 and 9.2.15 belong to Arlinskiĭ and Tsekanovskiĭ [47], [49]. More general problem: extensions of the class $C_{\mathfrak{H}}(\beta)$ of a non-densely-defined $C_{\mathfrak{H}}(\alpha)$ sub-operator A ($\alpha \leq \beta < \pi/2$) have been studied by Arlinskiĭ in [28], [31], [34], [37] and Malamud in [196], [197], [198]. The definition of extremal qsc-extension of symmetric contraction as well as Proposition 9.2.9 are due to Arlinskiĭ and Tsekanovskiĭ [50]. Proposition 9.2.19 is a qsc-extended version of the corresponding proposition of Krein and Ovčarenko [177] obtained for sc-extensions of symmetric contraction. The condition (in terms of defect elements) for a non-densely-defined symmetric operator to be prime was considered by Langer and Textorius [183]. Theorem 9.3.1 is due to Brodskiĭ [89], Theorem 9.3.2, Propositions 9.3.3–9.3.6 are due to Arlinskiĭ, Hassi, de Snoo [40]. Theorems 9.4.1–9.4.2 belong to Derkach and Tsekanovskiĭ [108], [257], [258]. Proposition 9.5.2 and Theorem 9.5.4 were established by Krein [172], [173]. Theorem 9.5.7 belongs to Ando and Nishio [16]. Uniqueness of nonnegative self-adjoint extensions of the operator A_0 (9.100) in Example 9.5 was established using different methods by Gesztesy, Kalton, Makarov, and Tsekanovskiĭ in [135] and by Adamyan in [1]. The set of equivalent conditions on accretive operators in Section 9.5 belongs to Phillips [217], [218], [219]. Lemma 9.5.12 and Theorem 9.5.13 are due to Tsekanovskiĭ [255], [256]. The Phillips-Kato extension problem [217], [163] in the restricted sense consists of existence and description of maximal accretive and sectorial extensions T of a non-negative densely-defined operator $\dot{A}$ such that $\dot{A} \subset T \subset \dot{A}^*$. The solution of this problem was presented by Arlinskiĭ (Theorem 9.5.14) and Tsekanovskiĭ (Theorem 9.5.13) as well as in Theorem 9.2.7 [47], [49], [255], [256]. Theorems 9.6.1–9.6.3 and Lemma 9.6.4 were obtained by Tsekanovskiĭ [255], [260], [257], [258]; Corollary 9.6.2 and Theorem 9.6.6 are due to Okunskiĭ and Tsekanovskiĭ [211]. The results of Section 9.7 belong to Arlinskiĭ [23]. Theorem 9.8.2 was obtained by Derkach and Tsekanovskiĭ [107]. Theorems 9.8.11–9.8.14 on realization of Stieltjes functions and their connections with the Krein-von Neumann extensions of non-negative operators are due to Dovzhenko and Tsekanovskiĭ [113]. The L-system modifications are due to Belyi and Tsekanovskiĭ and proofs are being published for the first time. Definition of classes $S^{-1}(R)$ and $S_0^{-1}(R)$ as well as Theorems 9.9.4 and 9.9.5 were introduced in [73]. Results of Section 9.9 on realization of inverse Stieltjes functions and their connections with the Friedrichs extensions of non-negative operators belong to Belyi and Tsekanovskiĭ and are being published for the first time. Quasi-self-adjoint m-sectorial extensions were studied by Derkach, Malamud, and Tsekanovskiĭ [105], [106], and by Derkach and Malamud [104] via the characteristic function approach based on boundary triplets and corresponding Weyl functions. Unbounded m-sectorial operators and their characteristic functions were treated by Arlinskiĭ [36] by means of rigged Hilbert spaces.

Chapter 10

Theorem 10.1.1 belongs to Arlinskiĭ [22] and Theorem 10.2.1 is due to Tsekanovskiĭ [258], [260]. Theorem 10.2.2 on criterion for the rigged canonical system with Schrödinger operator to be an L-system belongs to the authors and is being published for the first time. Formulas (10.30)–(10.31) were established by Arlinskiĭ and Tsekanovskiĭ [53]. Theorems 10.3.1–10.3.2 are due to Tsekanovskiĭ [257], [258], [260]. Lemma 10.4.2 and Theorem 10.4.3 belong to Donoghue [111], for finite equal deficiency indices to Gesztesy and Tsekanovskiĭ [137], and for equal infinite deficiency indices to Gesztesy, Kalton, Makarov, and Tsekanovskiĭ [135]. Theorem 10.5.1 is due to Tsekanovskiĭ [260], Theorem 10.6.1 was obtained by Arlinskiĭ and Tsekanovskiĭ [53], Theorems 10.6.2–10.6.3 are due to Arlinskiĭ, Belyi and Tsekanovskiĭ, and their L-system version is being published for the first time. Inequality (10.125) was discovered by Gesztesy, Kalton, Makarov, and Tsekanovskiĭ in [135]; other inequalities of this nature were obtained by Arlinskiĭ and Tsekanovskiĭ in [53]. An elementary proof of (10.125) was offered by Kalton in private communications. Another proof of inequality (10.125) and other inequalities related to differential operators were established by Trigub in [246]. Theorems 10.7.1–10.7.5 belong to Arlinskiĭ and Tsekanovskiĭ [53], Theorem 10.7.8 is due to the authors and is presented for the first time. Stieltjes-like functions were introduced by Belyi and Tsekanovskiĭ [79]. Theorems 10.8.4–10.8.6 belong to Belyi and Tsekanovskiĭ [79] as well as formulas and dynamics of restoration of the whole L-system with Schrödinger operator based on its Stieltjes-like impedance function. Results of Section 10.9 belong to Belyi and Tsekanovskiĭ [80] as well as formulas and dynamics of restoration of the whole L-system with Schrödinger operator based on its impedance inverse Stieltjes-like function.

Chapter 11

Theorems 11.1.1–11.1.2 belong to Stone [239]. Lemma 11.1.3 is a reformulation of the definition of a completely non-self-adjoint linear bounded operator [190], [193], [89], [91]. Theorems 11.1.4–11.2.6 on non-self-adjoint strengthening of Stone's Theorem 11.1.1–11.1.2 were established by Arlinskiĭ and Tsekanovskiĭ [54]. Theorem 11.2.7 on realization of Herglotz-Nevanlinna function $v(z) = \int_a^b \frac{1}{t-z} d\sigma(t)$ as impedance function of some L-system with dissipative Jacobi matrix with one-dimensional imaginary part as state-space operator of the system belong to Arlinskiĭ and Tsekanovskiĭ and is being published for the first time. All the results on inverse spectral problems for non-self adjoint Jacobi matrices in Section 11.3 belong to Arlinskiĭ and Tsekanovskiĭ [54]. System interpolation was introduced and studied by Alpay and Tsekanovskiĭ [8]. Theorems 11.5.2–11.6.10 are due to Alpay and Tsekanovskiĭ [8]. The results of Section 11.7 on the Nevanlinna-Pick interpolation problem for a rational function with distinct poles and its connection with system interpolation and, in particular, with the Livšic system in the Pick form, belong to Arlinskiĭ and Tsekanovskiĭ and are presented for the first time.

Chapter 12

Non-canonical F-systems were introduced by Hassi, de Snoo and Tsekanovskiĭ in [151], [152], [153]. The metric conservation law of the forms (12.15), (12.53) for non-canonical systems is presented for the first time and generalizes its canonical versions described in Chapters 5 and 6. The results in Sections 12.1–12.3 belong to Hassi, de Snoo and Tsekanovskiĭ [151], [152], [153], [154]. The general realization Theorems 12.5.4 and 12.6.2 for Herglotz-Nevanlinna functions as well as other results in Sections 12.4–12.6 were established by Belyi, Hassi, de Snoo and Tsekanovskiĭ [72], [73], [74], [75], [77]. Conservative state-space realizations of analytic functions that map the right half-plane into itself were treated by Ball and Staffans in [68], [69] and Staffans in [235], [237]. The systems with operator pencils as state-space operator have been studied by Rutkas [222], [223], [224]. Realizations of operator-valued Herglotz-Nevanlinna functions and pairs by means of the Weyl families of boundary relations have been obtained by Derkach and Malamud in [99], [100], [101], [103] and by Derkach, Hassi, Malamud, and de Snoo in [97], [98].

Bibliography

[1] Adamyan, V.: Non-negative perturbations of non-negative self-adjoint operators. Methods of Functional Anal. and Topology, **13**, 103–109 (2007)

[2] Akhiezer, N.I.: The Classical Moment Problem. Oliver and Boyd, Edinburgh, (1965)

[3] Akhiezer, N.I., Glazman, I.M.: Theory of Linear Operators in Hilbert Space. Dover, New York, (1993)

[4] Alpay, D., Dijksma, A., Rovnyak, J., de Snoo, H.S.V.: Schur functions, operator colligations, and reproducing kernel Pontryagin spaces. Oper. Theory Adv. Appl., **96**, Birkhäuser Verlag, Basel, (1997)

[5] Albeverio, S., Kurasov, P.: Singular Pertrubations of Differential Operators, London Math. Soc. Lecture Notes, **271**, Cambridge Univesity Press, London, (2000)

[6] Alonso, S., Simon, B." The Birman-Kreĭn-Vishik theory of selfadjoint extensions of semibounded operators. J. Operator Theory, **4**, 251-270 (1980)

[7] Alpay, D., Gohberg, I.: Pairs of selfadjoint operators and their invariants. Algebra i Analiz, **16**, no. 1, 70–120, (2004); translation in St. Petersburg Math. J. **16**, no. 1, 59–104, (2005)

[8] Alpay, D., Tsekanovskiĭ, E.R.: Interpolation theory in sectorial Stieltjes classes and explicit system solutions. Lin. Alg. Appl., **314**, 91–136 (2000)

[9] Alpay, D., Tsekanovskiĭ, E.R.: Subclasses of Herglotz-Nevanlinna matrix-valued functions and linear systems. In: J. Du and S. Hu (ed) Dynamical systems and differential equations, An added volume to *Discrete and continuous dynamical systems*, 1–14 (2001)

[10] Anderson, W.N.: Shorted operators. SIAM J. Appl. Math., **20**, 520–525 (1971)

[11] Anderson, W.N., Trapp, G.E.: Shorted operators II. SIAM J. Appl. Math., **28**, 60–71 (1975)

[12] Anderson, W.N., Duffin, R.J.: Series and parallel addition of matrices. J. Math. Anal. Appl. **26**, 576–594 (1969)

[13] Ando, T.: Truncated moment problems for operators. Acta Sci. Mat., **31**, No.3-4, 319–334 (1970)

[14] Ando, T.: Topics on operator inequalities. Division of Applied Mathematics, Research Institute of Applied Electricity, Hokkaido University, Sapporo (1978)

[15] Ando, T.: de Branges spaces and analytic operator functions. Division of Applied Mathematics, Hokkaido University, Sapporo (1990)

[16] Ando, T., Nishio, K.: Positive Selfadjoint Extensions of Positive Symmetric Operators. Tohóku Math. J., **22**, 65–75 (1970)

[17] Arlinskiĭ, Yu.M.: On resolvents of generalized self-adjoint extensions of a Hermitian operator with a non-dense domain. Dopovidi Akad. Nauk Ukrain. RSR, Ser. A, 1–3 (1974)

[18] Arlinskiĭ, Yu.M.: On inverse problem of the theory of characteristic functions of unbounded operator colligations. Dopovidi Akad. Nauk Ukrain. RSR, Ser. A, No. 2, 105–109 (1976)

[19] Arlinskiĭ, Yu.M.: Bi-extensions of unbounded operators, characteristic operator-functions of generalized operator colligations and their applications. Candidate of Sciences Disseration (Russian), Donetsk State University, Donetsk, (1976)

[20] Arlinskiĭ, Yu.M.: Regular (*)-extensions of quasi-Hermitian operators in rigged Hilbert spaces. (Russian), Izv. Akad. Nauk Armyan. SSR, Ser.Mat., **14**, No. 4, 297–312 (1979)

[21] Arlinskiĭ, Yu.M.: A triangular model of an unbounded quasi-Hermitian operator with complete system of root subspaces. (Russian), Dokl.Akad. Nauk Ukraine, Ser.A, No. 11, 883–886 (1979)

[22] Arlinskiĭ, Yu.M.: On regular (*)-extensions and characteristic matrix valued functions of ordinary differential operators. Boundary value problems for differential operators, Kiev, 3–13, (1980)

[23] Arlinskiĭ, Yu.M.: On accretive (*)-extensions of a positive symmetric operator. (Russian), Dokl.Akad. Nauk Ukraine, Ser.A, No. 11, 3–5 (1980)

[24] Arlinskiĭ, Yu.M.: Contractive extensions and generalized resolvents of a dual pair of contractions. (Russian) Ukrain. Math.Journ., **37**, No. 2, 247–250 (1985)

[25] Arlinskiĭ, Yu.M.: A class of contractions in Hilbert space. Ukrain. Math. Journ., 39 , N6, 691–696 (1987)(Russian). English translation in: Ukrainian Mathematical Journal, **39**, No.6, 560–564 (1987)

[26] Arlinskiĭ, Yu.M.: Positive Spaces of Boundary Values and Sectorial Extensions of Nonnegative Operator. (Russian) Ukrainian Mat.J. **40**, No.1, 8–14 (1988)

[27] Arlinskiĭ, Yu.M.: Characteristic functions of operators of the class $C(\alpha)$. (Russian) Izv.Vyssh. Uchebn. Zaved. Mat., No. 2, 13–21 (1991)

[28] Arlinskiĭ, Yu.M.: A class of extensions of a $C(\alpha)$-suboperator. (Russian) Dokl. Akad.Nauk Ukraine, No.8, 12–16 (1992)

[29] Arlinskiĭ, Yu.M.: On Proper Accretive Extensions of Positive Linear Relations. Ukrainian Mat.J. **47**, No.6, 723-730 (1995)

[30] Arlinskiĭ, Yu.M.: Maximal Sectorial Extensions and Associated with them Closed Forms. (Russian) Ukrainian Mat.J. **48**, No.6, 723–739 (1996)

[31] Arlinskiĭ, Yu.M.: Extremal Extensions of Sectorial Linear Relations. Matematychnii Studii, **7**, No.1, 81–96 (1997)

[32] Arlinskiĭ, Yu.M.: On M-Accretive Extensions and Restrictions. Methods of Funct. Anal. and Topol., **4**, No.3, 1–26 (1998)

[33] Arlinskiĭ, Yu.M.: On Functions connected with sectorial Operators and their extensions. Int.Equat. Oper.Theory, **33**, No.2, 125–152 (1999)

[34] Arlinskiĭ, Yu.M.: On a class of contractions and their extensions. Journ. of Mathematical Sciences, **97**, No.5, 4391–4419 (1999)

[35] Arlinskiĭ, Yu.M.: Abstract Boundary Conditions for Maximal Sectorial Extensions of Sectorial Operators. Math. Nachr., **209**, 5–36 (2000)

[36] Arlinskiĭ, Yu.M.: Characteristic functions of maximal sectorial operators. Operator Theory: Advances and Applications, **124**, 89–108 (2001)

[37] Arlinskiĭ, Yu.M.: Extremal extensions of a $C(\alpha)$-suboperator and their representations. Operator Theory: Advances and Applications, **162**, 47–69 (2006)

[38] Arlinskiĭ, Yu.M., Belyi, S.V., Derkach, V.A., Tsekanovskiĭ, E.R.: On realization of the Krein-Langer class N_κ of matrix-valued functions in Hilbert spaces with indefinite metric. Mathematische Nachrichten, **241**, no. 10, 1380–1399 (2008)

[39] Arlinskiĭ, Yu.M., Derkach, V.A.: An inverse problem of the theory of characterstic functions in Π_κ. Ukr. Math. Journ., **31**, No.2, 115–122 (1979)

[40] Arlinskiĭ, Yu.M., Hassi, S., de Snoo, H.S.V.: Q-functions of quasiselfadjoint contractions. Operator Theory: Advances and Applications, **163**, 23–54 (2006)

[41] Arlinskiĭ, Yu.M., Hassi, S., de Snoo, H.S.V.: Parametrization of contractive block-operator matrices and passive disrete-time systems. Complex Analysis and Operator Theory, **1**, No.2, 211–233 (2007)

[42] Arlinskiĭ, Yu.M., Hassi, S., Sebestyen, Z., de Snoo, H.S.V.: On the class of extremal extensions of a nonnegative operators. Operator Theory: Advances and Applications, **127**, 41–81 (2001)

[43] Arlinskiĭ, Yu.M., Kaplan, V.L.: On selfajoint biextensions in rigged Hilbert spaces. (Russian), Functional Analysis, Ulyanovsk, Spektral Theory, No.29, 11–19 (1989)

[44] Arlinskiĭ, Yu.M., Tsekanovskiĭ, E.R.: Generalized self-adjoint extensions of Hermitian operators with non-dense domain. Dopovidi Akad. Nauk Ukrain. RSR, Ser. A, 1–3 (1973)

[45] Arlinskiĭ, Yu.M., E.R .Tsekanovskiĭ: The method of equipped spaces in the theory of extensions of Hermitian operators with a nondense domain of definition. Sibirsk. Mat. Zh., **15**, 597–610 (1974)

[46] Arlinskiĭ, Yu.M., Tsekanovskiĭ, E.R.: Regular (∗)-extension of unbounded operators, characteristic operator-functions and realization problems of transfer mappings of linear systems. Preprint, VINITI, -2867. -79, Dep.-72 (1979)

[47] Arlinskiĭ, Yu.M., Tsekanovskiĭ, E.R.: Nonselfadjoint contractive extensions of a Hermitian contraction and theorems of Krein. Russ. Math. Surv., **37:1**, 151–152 (1982)

[48] Arlinskiĭ, Yu.M., Tsekanovskiĭ, E.R.: On resolvents of m-accretive extensions of symmetric differential operator. (Russian), Math. Phys. Nonlin. Mech., No.1, 11–16 (1984)

[49] Arlinskiĭ, Yu.M., Tsekanovskiĭ, E.R.: Sectorial extensions of positive Hermitian operators and their resolvents. (Russian), Akad. Nauk. Armyan. SSR, Dokl., **79**, No.5, 199–203 (1984)

[50] Arlinskiĭ, Yu.M., Tsekanovskiĭ, E.R.: Quasi-selfajoint contractive extension of Hermitian contractions. Teor. Funk., Funk.Anal., (Russian) No.50, 9–16 (1988)

[51] Arlinskiĭ, Yu.M., Tsekanovskiĭ, E.R.: Constant J-unitary factor and operator-valued transfer functions. In: Dynamical systems and differential equations, Discrete Contin. Dyn. Syst., Wilmington, NC, 48–56 (2003)

[52] Arlinskiĭ, Yu.M., Tsekanovskiĭ, E.R.: Some remarks on singular perturbations of selfadjoint operators. Methods of Functional Analysis and Topology, **9**, No.4, 287–308 (2003)

[53] Arlinskiĭ, Yu.M., Tsekanovskiĭ, E.R.: Linear systems with Schrödinger operators and their transfer functions. Oper. Theory Adv. Appl., **149**, 47–77 (2004)

[54] Arlinskiĭ, Yu.M., Tsekanovskiĭ, E.R.: Non-selfadjoint Jacobi matrices with a rank-one imaginary part. J. Funct. Anal., **241**, 383–438 (2006)

[55] Arov, D.Z.: Passive linear steady-state dynamical systems. (Russian), Sibirsk. Mat. Zh., **20**, no. 2, 211–228, 457 (1979)

[56] Arov, D.Z.: Realization of a canonical system with a dissipative boundary condition at one end of the segment in terms of the coefficient of dynamical compliance. (Russian), Sibirsk. Mat. Z. **16**, no. 3, 440–463 (1975)

[57] Arov, D.Z.: Realization of matrix-valued functions according to Darlington. (Russian), Izv. Akad. Nauk SSSR Ser. Mat. **37**, 1299–1331 (1973)

[58] Arov, D.Z., Dym, H.: J-contractive matrix valued functions and related topics. Encyclopedia of Mathematics and its Applications, 116. Cambridge University Press, Cambridge, (2008)

[59] Arov, D., Grossman, L.: Scattering Matrices in the Theory of Unitary Extension of Isometric Operators, Math. Nachr., **157**, 105–135 (1992)

[60] Arov, D.Z., Nudelman, M.A.: Passive linear stationary dynamical scattering systems with continuous time. Integr. Equat. Oper. Th., **24**, 1–45 (1996)

[61] Arsene, Gr., Gheondea, A.: Completing matrix contractions. J. Operator Theory, **7**, 179–189 (1982)

[62] Ashbaugh, M.S., Gesztesy, F., Mitrea, M., Shterenberg, R., Teschl, G.: The Krein-von Neumann extension and its connection to an abstract buckling problem. Math. Nachr. **283**, No.2, 165-179 (2010)

[63] Atkinson, F.V.: Discrete and Continuous Boundary Problems. Academic Press, New York, (1964)

[64] Avdonin, S., Mikhaylov, S., Rybkin, A.: The boundary control approach to the Titchmarsh-Weyl m-function. Comm. Math. Phys., **275**, No.3, 791–803 (2007)

[65] Azizov, T.Ya., Iokhvidov, I.S.: Linear operators in spaces with indefinite metric, John Wiley and Sons, New-York, (1989)

[66] Ball, J.A., Gohberg, I., Rodman, L.: Interpolation of rational matrix functions. Birkhäuser Verlag, Basel, (1990)

[67] Ball, J.A., Cohen, N.: de Branges-Rovnyak operator models and systems theory: a survey. Topics in matrix and operator theory, Oper. Theory Adv. Appl., **50**, Birkhauser, Basel, 93-136 (1991)

[68] Ball, J.A., Staffans, O.J.: Conservative state-space realizations of dissipative system behaviors. Report No.37, Institute Mittag-Leffler, (2002/2003)

[69] Ball, J.A., Staffans, O.J.: Conservative state-space realizations of dissipative system behaviors. Integr. Equ. Oper. Theory (Online), Birkhäuser, DOI 10.1007/s00020-003-1356-3, (2005)

[70] Bart, H., Gohberg, I., Kaashoek, M.A.: Minimal Factorizations of Matrix and Operator Functions. Operator Theory: Advances and Applications, Vol. 1, Birkhäuser, Basel, (1979)

[71] Bart, H., Gohberg, I., Kaashoek, M.A., Ran, A.: Factorization of matrix and operator functions: the state space method. Operator Theory: Advances and Applications, **178**, Birkhauser Verlag, Basel, (2008)

[72] Belyi, S.V., Hassi, S., de Snoo, H.S.V., Tsekanovskiĭ, E.R.: A general realization theorem for matrix-valued Herglotz-Nevanlinna functions, Linear Algebra and Applications. vol. 419, 331–358 (2006)

[73] Belyi, S.V., Hassi, S., de Snoo, H.S.V., Tsekanovskiĭ, E.R.: On the realization of inverse of Stieltjes functions. Proceedings of MTNS-2002, University of Notre Dame, CD-ROM, 11p., (2002)

[74] Belyi, S.V., Tsekanovskiĭ, E.R.: Classes of operator R-functions and their realization by conservative systems. Sov. Math. Dokl. **44**, 692–696 (1992)

[75] Belyi, S.V., Tsekanovskiĭ, E.R.: Realization theorems for operator-valued R-functions. Operator theory: Advances and Applications, **98**, Birkhäuser Verlag Basel, 55–91 (1997)

[76] Belyi, S.V., Tsekanovskiĭ, E.R.: Multiplication Theorems for J-contractive operator-valued functions. Fields Institute Communications, **25**, 187–210 (2000)

[77] Belyi, S.V., Tsekanovskiĭ, E.R.: On classes of realizable operator-valued R-functions. Operator theory: Advances and Applications, **115**, Birkhäuser Verlag Basel, (2000), 85–112.

[78] Belyi, S.V., Menon, G., Tsekanovskiĭ, E.R.: On Krein's Formula in the Case of Non-densely Defined Symmetric Operators. J. Math. Anal. Appl., **264**(2), 598–616 (2001)

[79] Belyi, S.V., Tsekanovskiĭ, E.R.: Stieltjes like functions and inverse problems for systems with Schrödinger operator. Operators and Matrices, vol. 2, No.2, 265–296 (2008)

[80] Belyi, S.V., Tsekanovskiĭ, E.R.: Inverse Stieltjes like functions and inverse problems for systems with Schrödinger operator. Operator Theory: Advances and Applications, vol. 197, 21–49 (2009)

[81] Berezansky, Ju.M.: Expansion in eigenfunctions of self-adjoint operators. vol. 17, Transl. Math. Monographs, AMS, Providence, (1968)

[82] Berezin, F., Faddeev, L.: Remark on the Schrödinger equation with singular potential. Dokl. Akad. Nauk SSSR, **137**, 1011-1014 (1961)

[83] Birman, M.S.: On the selfadjoint extensions of positive definite operators. Mat.Sbornik, (Russian), **38** (1956), 431–450

[84] Bolotnikov, V.: Nevanlinna-Pick meromorphic interpolation: the degenerate case and minimal norm solutions. J. Math. Anal. Appl., **353**, No. 2, 642-651 (2009)

[85] de Branges, L.: Complementation in Krein spaces. Trans. Amer. Math. Soc., **305**, No. 1, 277–291 (1988)

[86] de Branges, L., Rovnyak, J.: Square summable power series. Holt, Rinehart and Winston, New York, (1966)

[87] de Branges, L., Rovnyak, J.: Appendix on square summable power series, Canonical models in quantum scattering theory. In: *Perturbation theory and its applications in quantum mechanics* (ed. C.H. Wilcox), New York, 295–392 (1966)

[88] de Branges, L., Rovnyak, J.: Canonical models in quantum scattering theory. Perturbation Theory and its Applications in Quantum Mechanics (Proc. Adv. Sem. Math. Res. Center, U.S. Army, Theoret. Chem. Inst., Univ. of Wisconsin, Madison, Wis.) Wiley, New York, 295–392 (1965)

[89] Brodskiĭ, M.S.: Triangular and Jordan Representations of Linear Operators. Amer. Math. Soc., Providence, RI, (1971)

[90] Brodskiĭ, V.M.: Certain theorems on colligations and their characteristic functions. (Russian) Funkcional. Anal. i Prilozen., **4** no. 3, 95–96 (1970)

[91] Brodskiĭ, M.S., Livšic, M.S.: Spectral analysis of non-selfadjoint operators and intermediate systems. Uspekhi Matem. Nauk, **XIII**, no. 1 (79), 3–84 (1958)

[92] Bruk, V.M.: On one Class of Boundary Value Problems with a Spectral Parameter in the Boundary Condition. (Russian), Mat. Sbornik, **100**, No.2, 210–216 (1976)

[93] Clark, S., Gesztesy, F., Zinchenko, M.: Borg-Marchenko-type uniqueness results for CMV operators. Skr. K. Nor. Vidensk. Selsk., No.1, 1-18, (2008)

[94] Crandall, M.G.: Norm preserving extensions of linear transformations in Hilbert space. Proc. Amer. Math. Soc. **21**, 335–340 (1969)

[95] Curtain, R., Zwart, H.: An introduction to infinite-dimensional linear systems theory. Texts in Applied Mathematics, **21**, Springer-Verlag, New York, (1995)

[96] Davis, C., Kahan, W.M., Weinberger, H.F.: Norm-preserving dilations and their applications to optimal error bound. Siam J. Numerical Anal., **19**, No.3, 445–469 (1982)

[97] Derkach, V.A., Hassi, S., Malamud, M.M., de Snoo, H.S.V.: Boundary relations and their Weyl families. Trans. Am. Math. Soc., **358**, No. 12, 5351–5400 (2006)

[98] Derkach, V.A., Hassi, S., Malamud, M.M., de Snoo, H.S.V.: Boundary relations and generalized resolvents of symmetric operators. Russ. J. Math. Phys., **16**, No.1, 17–60 (2009)

[99] Derkach, V.A., Malamud, M.M.: Weyl Function of Hermitian Operator and its Connection With the Characteristic Function. (Russian), Preprint 85-9, Fiz.-Tekhn. Inst. Akad. Nauk Ukraine, 1–50, (1985)

[100] V. A. Derkach, Malamud, M.M.: On the Weyl function and Hermitian operators with gaps. Sov. Math. Dokl. **35**, 393–398 (1987)

[101] Derkach, V.A., Malamud, M.M.: Generalized Resolvents and the Boundary Value Problems for Hermitian Operators with Gaps. J. Funct. Anal., **95**, No.1, 1–95 (1991)

[102] Derkach, V.A., Malamud, M.M.: Characteristic functions of almost solvable extensions of Hermitian operators. (Russian), Ukr. Mat. Zh. **44**, No.4, 435-459 (1992). English translation in Ukr. Math. J., **44**, No.4, 379-401 (1992)

[103] Derkach, V.A., Malamud, M.M.: The Extension Theory of Hermitian Operators and the Moment Problem. J. of Math. Sci., 73, No.2, 141-242 (1995)

[104] Derkach, V.A., Malamud, M.M.: Non-self-adjoint extensions of a Hermitian operator and their characteristic functions. J. Math. Sci., New York, **97**, No.5, 4461–4499 (1999)

[105] Derkach, V.A., Malamud, M.M., Tsekanovskiĭ, E.R.: Sectorial extensions of a positive operator and the characteristic function. Sov. Math. Dokl. **37**, 106-110 (1988).

[106] Derkach, V.A., Malamud, M.M., Tsekanovskiĭ, E.R.: Sectorial Extensions of Positive Operator. (Russian), Ukrainian Mat.J. **41**, No.2, 151-158 (1989)

[107] Derkach, V.A., Tsekanovskiĭ, E.R.: Characteristic operator-functions of accretive operator colligations. (Russian), Dokl. Akad. Nauk Ukrain. SSR Ser. A, No.8, 16–19 (1981)

[108] Derkach, V.A., Tsekanovskiĭ, E.R.: On characteristic function of quasi-hermitian contraction. Izvestia Vysshich Uchebnuch Zavedeni, Math., **6**, 46–51 (1987)

[109] Dewilde, P.: Input-output description of roomy systems. SIAM J. Control Optimization, **14**, No.4, 712736 (1976)

[110] Dewilde, P.: The geometry of Darlington synthesis. Infinite-dimensional systems theory and operator theory. Int. J. Appl. Math. Comput. Sci., **11**, No.6, 13791386 (2001)

[111] Donoghue, W.F.: On the perturbation of spectra. Commun. Pure Appl. Math. **18**, 559–579 (1965)

[112] Douglas, R.G.: On majorization, factorization and range inclusion of operators in Hilbert space. Proc. Amer. Math. Soc. **17**, 413–416 (1966)

[113] Dovzhenko, I., Tsekanovskiĭ, E.R.: Classes of Stieltjes operator-valued functions and their conservative realizations. Soviet Math.Dokl., **41**, no.2, 201–204 (1990)

[114] Dritschel, M., Rovnyak, J.: Extension theorems for contraction operators on Krein spaces. Oper. Theory Adv. Appl., **47**, Birkhäser, 221–305 (1990)

[115] Dubovoj, V.K.: Weyl families of operator colligations, and the open fields corresponding to them. (Russian) Teor. Funkcii Funkcional. Anal. i Prilozen. No. 14, 67–83 (1971)

[116] Dubovoj, V.K., Fritzsche, B., Kirstein, B.: Matricial Version of the Classical Schur Problem. Teubner-Texte zur Mathematik [Teubner Texts in Mathematics], B.G. Teubner Verlagsgesellschaft mbH, Stuttgart (1992)

[117] Dyukarev, Yu.M., Katsnelson, V.E.: Multiplicative and additive Stieltjes classes of analytic matrix-valued functions and interpolation problems connected with them. (Russian) Teor. Funktsii Funktsional. Anal. i Prilozhen. No.36, 13–27 (1981)

[118] Dunford, N., Schwartz, J.T.: Linear Operators Part II: Spectral Theory. Interscience, New York, (1988)

[119] Dym, H.: J contractive matrix functions, reproducing kernel Hilbert spaces and interpolation. Providence, RI, (1989)

[120] Efimov, A.V., Potapov, V.P.: J-expanding matrix-valued functions, and their role in the analytic theory of electrical circuits. (Russian) Uspehi Mat. Nauk, **28**, No.1 (169), 65–130 (1973)

[121] Feintuch, A.: Realization theory for symmetric systems. J. Math. Anal. Appl. **71**, No.1, 131–146 (1979)

[122] Feintuch, A.: Robust control theory in Hilbert space. Applied Mathematical Sciences, **130**, Springer-Verlag, New York, (1998)

[123] Fillmore, P.A., Williams, J.P.: On operator ranges. Advances in Math. **7**, 254-281 (1971)

[124] Foias, C., Özbay, H., Tannenbaum, A.: Robust control of infinite-dimensional systems. Frequency domain methods. Lecture Notes in Control and Information Sciences, **209**, Springer-Verlag London, Ltd., London, (1996)

[125] Foias, C., Frazho, A.: Redhefer products and the lifting of contractions on Hilbert space. Journ. Oper. Theory, **11**, No.1, 193–196 (1984)

[126] Foias, C., Frazho, A.: The commutant lifting approach to interpolation problems. Operator Theory: Advances and Applications, **44**. Birkhauser Verlag, Basel, (1990)

[127] Francis, B.: A course in H_∞ control theory. Lecture Notes in Control and Information Sciences, **88**. Springer-Verlag, Berlin, (1987)

[128] Friedrichs, K.: Spektraltheorrie Halbbeschrankter Operatoren, Math. Ann., **109**, 405–487 (1934)

[129] Fritzsche, B., Kirstein, B.: Representations of central matrix-valued Caratheodory functions in both non-degenerate and degenerate cases. Integral Equations Operator Theory, **50**, No. 3, 333-361 (2004)

[130] Fuhrmann, P.: Linear systems and operators in Hilbert space. McGraw-Hill International Book Co., New York, (1981)

[131] Fuhrmann, P.: Realization theory in Hilbert space for a class of transfer functions. J. Functional Analysis, **18**, 338–349 (1975)

[132] Gamelin, T.W.: Uniform algebras, Prentice-Hall, Inc., Englewood Cliffs, N.J., (1969)

[133] Gelfand, I.M., Levitan, B.M.: On the determination of a differential equation from its spectral function. (Russian) Izvestiya Akad. Nauk SSSR. Ser. Mat. **15**, (1951)

[134] Gesztesy, F., Makarov, K.A., Tsekanovskiĭ, E.R.: An addendum to Krein's formula, J. Math. Anal. Appl. **222**, 594–606 (1998)

[135] Gesztesy, F., Kalton, N.J., Makarov, K.A., Tsekanovskiĭ, E.R.: Some Applications of Operator-Valued Herglotz Functions, Operator Theory: Advances and Applications, **123**, Birkhäuser, Basel, 271–321 (2001)

[136] Gesztesy, F., Simon, B.: M-functions and inverse spectral analysis for finite and semifinite Jacobi matrices, J. Anal. Math., **73**, 267–297 (1997)

[137] Gesztesy, F., Tsekanovskiĭ, E.R.: On Matrix-Valued Herglotz Functions. Math. Nachr. **218**, 61-138 (2000)

[138] Ginzburg, Yu.: On J-contractive operator functions. (Russian) Dokl. Akad. Nauk SSSR, **117**, 171–173 (1957)

[139] Gohberg, I.P., Krein, M.G.: Introduction to the Theory of Linear Non-selfadjoint Operators. Nauka, Moscow, (1965)

[140] Goldstein, J.: Semigroups of linear operators and applications. Oxford Mathematical Monographs. The Clarendon Press, Oxford University Press, New York, (1985)

[141] Golinskiĭ, L.B.: On some generalization of the matricial Nevanlinna-Pick problem. (Russian) Izv. Akad. Nauk Armyan. SSR, **18**, No.3, 187–205 (1983)

[142] Gorbachuk, V.I., Gorbachuk, M.L.: Boundary value problems for differential-operator equations. Naukova Dumka, Kiev (1984)

[143] Gorbachuk, V.I., Gorbachuk M.L., Kochubei, A.N.: Extension Theory of Symmetric Operators and Boundary Value Problems. Ukrainian Mat.J., **41**, No.10, 1298-1313 (1989)

[144] Gubreev, G.M.: Characteristic matrix-valued functions of unbounded non-selfadjoint operators. (Russian) Teor. Funkcii Funkcional. Anal. i Prilozen. Vyp. 26, 12–21 (1976)

[145] Gubreev, G.M.: Factorization of characteristic operator functions. (Russian) Teor. Funkcii Funkcional. Anal. i Prilozen. Vyp. 28, 3–9 (1977)

[146] Gubreev, G.M.: Definition and fundamental properties of the characteristic function of a W-colligation. (Russian) Dokl. Akad. Nauk Ukrain. SSR Ser. A, No. 1, 3–6, (1978)

[147] Gubreev, G.M., Kovalenko, A.I.: On the equality criterion $D_A = D_{A^*}$ for dissipative operators. (Russian) Dokl. Akad. Nauk SSSR, **254**, No.5, 1044-1047 (1980)

[148] Han, Do Kong: A study of operators that realize certain classes of meromorphic functions. (Russian) Izv. Akad. Nauk Armjan. SSR Ser. Mat. 11 (1976)

[149] Hassi, S., Kaltenbäck, M., de Snoo, H.S.V.: Triplets of Hilbert spaces and Friedrichs extensions associated with the subclass N_1 of Nevanlinna functions. J. Operator Th. **37**, 155–181 (1997)

[150] Hassi, S., Malamud, M.M., de Snoo, H.S.V.: On Kreĭn extension theory of nonnegative operators. Math. Nach., **274-275**, 40–73 (2004)

[151] Hassi, S., de Snoo, H.S.V., Tsekanovskiĭ, E.R.: An addendum to the multiplication and factorization theorems of Brodskiĭ-Livšic-Potapov. Applicable Analysis, **77**, 125–133 (2001)

[152] Hassi, S., de Snoo, H.S.V., Tsekanovskiĭ, E.R.: Commutative and noncommutative representations of matrix-valued Herglotz-Nevanlinna functions. Applicable Analysis, **77**, 135–147 (2001)

[153] Hassi, S., de Snoo, H.S.V., Tsekanovskiĭ, E.R.: Realizations of Herglotz-Nevanlinna functions via F-colligations. Oper. Theory Adv. Appl., **132**, 183–198 (2002)

[154] Hassi, S., de Snoo, H.S.V., Tsekanovskiĭ, E.R.: The realization problem for Herglotz-Nevanlinna functions. In *Unsolved problems in mathematical systems and control theory*, (ed. V. Blondel and A. Megretski), Princeton University Press, 8–13 (2004)

[155] Helton, J.W.: Systems with infinite-dimensional state space: the Hilbert space approach. Proc. IEEE, **64**, No. 1, 145–160 (1976)

[156] Helton, J.W.: Discrete time systems, operator models, and scattering theory. J. Functional Analysis, **16**), 1538 (1974)

[157] Horn, R.A., Johnson, C.R.: Matrix Analysis. Cambridge University Press. Cambridge (1986)

[158] Kac I.S.: Hilbert spaces generated by monotonic Hermitian matrix-functions. Kharkov, State University, Proceedings Kharkov Math. Soc., **34**, 95–113 (1950)

[159] Kac, I.S., Kreĭn, M.G.: R-functions – analytic functions mapping the upper halfplane into itself. Amer. Math. Soc. Transl., (2), **103**, 1-18 (1974)

[160] Katsnelson, V.E., Kheifets, A.Ya., Yuditskii, P.M.: An abstract interpolation problem and the theory of extensions of isometric operators. (Russian) In: Operators in function spaces and problems in function theory, Naukova Dumka, Kiev, 8396, (1987)

[161] Kailath, T.: Linear systems. Prentice-Hall Information and System Sciences Series. Prentice-Hall, Inc., Englewood Cliffs, N.J., (1980)

[162] Kalman, R.E., Falb, P.L., Arbib, M.A.: Topics in mathematical system theory. McGraw-Hill Book Co., New York-Toronto, Ont.-London, (1969)

[163] Kato, T.: Perturbation Theory for Linear Operators. Springer-Verlag, (1966)

[164] Kochubeĭ, A.N.: On extensions of symmetric operators and symmetric binary relations. (Russian), Math. Zametki, **17**, No.1, 41- 48 (1975)

[165] Kolmanovich, V.U., Malamud, M.M.: Extensions of sectorial operators and dual pairs of contractions. (Russian) Manuscript No.4428-85, Deposited at VINITI, 1–57 (1985)

[166] Koosis, P.: Introduction to H_p Spaces. Cambridge University Press, (1980)

[167] Koshmanenko, V.D.: Singular quadratic forms in perturbations theory. Kluwer Academic Publishers, (1999)

[168] Kostrykin, V., Potthoff, J., Schrader, R.: Contraction semigroups on metric graphs. Analysis on graphs and its applications, Proc. Sympos. Pure Math., **77**, Amer. Math. Soc., Providence, RI, (2008)

[169] Krasnoselskiĭ, M.A.: On self-adjoint extensions of Hermitian operators. Ukrain. Mat. Zh., **1**, 21–38 (1949)

[170] Krasnoselskiĭ, M.A., Kreĭn, M.G.: Fundamental theorems on the extension of Hermitian operators and certain of their applications to the theory of orthogonal polynomials and the problem of moments. (Russian) Uspehi Matem. Nauk, **2**, No.3 (19), 60–106 (1947)

[171] Krasnoselskiĭ, M.A., Kreĭn, M.G., Milman, D.P.: On defect numbers of linear operators in Banach spaces. (Russian) Sb. Trudov Instituta Matematiki AN USSR, **11**, (1948)

[172] Kreĭn, M.G.: The Theory of Selfadjoint Extensions of Semibounded Hermitian Transformations and its Applications. I,(Russian) Mat.Sbornik **20**, No.3, 431-495 (1947)

[173] Kreĭn, M.G.: The Theory of Selfadjoint Extensions of Semibounded Hermitian Transformations and its Applications, II, (Russian) Mat.Sbornik **21**, No.3, 365-404 (1947)

[174] Kreĭn, M.G., Langer, H.: On Defect Subspaces and Generalized Resolvents of Hermitian Operator in the Space Π_κ. (Russian) Fuctional Analysis and Appl., **5**, No.2, 59-71 (1971)

[175] Kreĭn, M.G., Langer, H.: On Defect Subspaces and Generalized Resolvents of Hermitian Operator in the Space Π_κ. (Russian) Fuctional Analysis and Appl., **5**, No.3, 54-69 (1971)

[176] Kreĭn, M.G., Langer, H.: Über die Q - Function Eines Π - Hermiteschen Operators im Raum Π_κ. Acta Sci. Math. Szeged, **34**, 191-230 (1973)

[177] Kreĭn, M.G., Ovčarenko, I.E.: On Q-Functions and SC-Resolvents of Nondensely Defined Hermitian Contractions. Siberian Math. Zh., **18**, 728–746 (1977)

[178] Kreĭn, M.G., Ovčarenko, I.E.: Inverse problems for Q-functions and resolvent matrices of positive Hermitian operators. Sov. Math. Dokl. **19**, 1131–1134 (1978)

[179] Krein, S.G.: Linear equations in Banach space. Moscow, Nauka, (1971)

[180] Kuzhel, A.V.: Extensions of Hermitian operators. Mathematics today '87 (Russian), Vishcha Shkola, Kiev, (1987)

[181] Kuzhel, A.V.: Characteristic functions and models of nonself-adjoint operators. Kluwer Academic Publishers, Dordrecht, Boston, London, (1996)

[182] Kuzhel, A.V., Kuzhel, S.A.: Regular extensions of Hermitian operators. VSP, Netherlands, (1998)

[183] Langer, H., Textorius, B.: On generalized resolvents and Q-functions of symmetric linear relations (subspaces) in Hilbert space. Pacific J. Math. **72**, 135–165 (1977)

[184] Lax, P.: On Cauchy problem for hyperbolic equations and the differentiability of solutions of elliptic equations. Comm. Pure Appl. Math. **8**, 615–633 (1955)

[185] Lax, P., Phillips, R.: Scattering theory. Pure and Applied Mathematics. Vol. 26, Academic Press, New York-London, (1967)

[186] Leray, J.: Lectures on hyperbolic equations with variable coefficients, mimeographed notes. Inst. Adv. Studies, Princeton, NJ, (1952)

[187] Levitan, B.M.: Inverse Sturm-Liouville problems. Translated from the Russian by O. Efimov. VSP, Zeist, (1987)

[188] Livšic, M.S.: On an inverse problem of the theory of operators. (Russian) Dokl. Akad. Nauk SSSR, **97**, 399–402 (1954)

[189] Livšic, M.S.: On a class of linear operators in Hilbert space. Amer. Math. Soc. Transl. (2) **13**, 61–83 (1960)

[190] Livšic, M.S.: On a spectral decomposition of linear nonself-adjoint operator. Amer. Math. Soc. Transl. (2) **5**, 67–114 (1957)

[191] Livšic, M.S.: Operators, oscillations, waves. Moscow, Nauka, (1966)

[192] Livšic, M.S., Potapov, V.P.: A theorem on the multiplication of characteristic matrix-functions. Dokl. Akad. Nauk SSSR, **72**, 625–628 (1950)

[193] Livšic, M.S., Yantsevich, A.A.: Operator colligations in Hilbert spaces. Winston, (1979)

[194] Livšic, M.S., Kravitsky, N., Markus, A.S., Vinnikov, V.: Theory of commuting nonselfadjoint operators. Mathematics and its Applications. **332**, Kluwer Academic Publishers Group, Dordrecht, (1995)

[195] Lyantse, V., Storozh, O.: Methods of the theory of unbounded operators, Naukova Dumka, Kiev, (1983)

[196] Malamud, M.M.: On extensions of Hermitian and sectorial operators and dual pairs of contractions. (Russian), Sov. Math. Dokl. **39**, No.2, 253–254 (1989)

[197] Malamud, M.M.: On some classes of extensions of sectorial operators and dual pair of contractions. Operator Theory: Advances and Appl, **1**24, 401–448 (2001)

[198] Malamud, M.M.: Operator holes and extensions of sectorial operators and dual pair of contractions. Math. Nach., **2**79, 625–655 (2006)

[199] Malamud, M.M., Malamud, S.M.: Spectral theory of operator measures in Hilbert space. (English. Russian original) St. Petersbg. Math. J., **15**, No.3, 323–373 (2004); translation from Algebra Anal. 15, No.3, 1–77 (2003)

[200] Megrabjan, L.H.: Realization of certain classes of meromorphic functions in the theory of systems with discrete time. (Russian) Izv. Akad. Nauk Armjan. SSR Ser. Mat., **10**, No.6, 560580 (1975)

[201] Megrabjan, L.H.: Blaschke-Dzrbasjan type products in the upper half-plane. (Russian) Sibirsk. Mat. Zh., **20**, No.4, 807825 (1979)

[202] Milton, G.W., Guevara, F.V., Onofrei, D.: Complete Characterization And Synthesis Of The Response Function Of Elastodynamic Networks. arXiv:0911.1501v1, (2009)

[203] Naboko, S.N.: Absolutely continuous spectrum of a nondissipative operator and functional model. I. (English. Russian original) Journal of Mathematical Scince, **16**, No.3, 1109–1117 (1977); translation from Zap. LOMI, **65**, 90-102 (1981)

[204] Naboko, S.N.: Absolutely continuous spectrum of a nondissipative operator and functional model. II. (English. Russian original) Journal of Mathematical Scince, **34**, No.6, 2090–2101 (1986); translation from Zap. LOMI, **73**, 118-135 (1977)

[205] Naboko, S.N.: Nontangential boundary values of operator-valued R-functions in a half-plane. Leningrad Math. J., **1**, 1255–1278 (1990)

[206] Naimark, M.A.: Spectral functions of a symmetric operator. Izv. Akad. Nauk SSSR, **4**, 227–318 (1940)

[207] Naimark, M.A.: Linear Differential Operators II. F. Ungar Publ., New York, (1968)

[208] von Neumann, J.: Allgemeine Eigenwerttheorie hermitescher Funktionaloperatoren, *Math. Ann.* **102**, 49–131 (1929-30)

[209] Nikolski, N.: Operators, functions, and systems: an easy reading. Vol. 1-2, Mathematical Surveys and Monographs, **92**, American Mathematical Society, Providence, RI, (2002)

[210] Nishio, K., Ando, T.: Characterizations of operators derived from network connections. J. Math. Anal. Appl., **53**, 539–549 (1976)

[211] Okunskiĭ, M.D., Tsekanovskiĭ, E.R.: On the theory of generalized selfadjoint extensions of semibounded operators. (Russian) Funkcional. Anal. i Prilozen., **7**, No.3, 92–93 (1973)

[212] Partington, J.: Linear operators and linear systems. An analytical approach to control theory. London Mathematical Society Student Texts, **60**. Cambridge University Press, Cambridge, (2004)

[213] Pavlov, B. S.: Operator Extensions Theory and explicitly solvable models. Uspekhi Math. Nauk., **42**, 99–131, (1987)

[214] Pavlov, B. S.: Spectral analysis of a dissipative singular Schrödinger operator in terms of a functional model. (Russian) Partial differential equations 8, Itogi Nauki i Tekhniki. Ser. Sovrem. Probl. Mat. Fund. Napr., 65, VINITI, Moscow, 1991, 95163

[215] Pekarev, E.: Shorts of operators and some extremal problems. Acta Sci. Math. (Szeged), **56**, 147–163 (1992)

[216] Peller, V.: Hankel operators and their applications. Springer Monographs in Mathematics. Springer-Verlag, New York, (2003)

[217] Phillips, R.: Dissipative parabolic systems. Trans. Amer. Math. Soc., **86**, 109–173 (1957)

[218] Phillips, R.: Dissipative operators and hyperbolic systems of partial differential equations. Trans. Amer. Math. Soc., **90**, 192–254 (1959)

[219] Phillips, R.: On dissipative operators, Lectures in Differential Equations. **3**, 65–113 (1969)

[220] Potapov, V.P.: The multiplicative structure of J-contractive matrix functions. Amer. Math. Soc. Transl., (2), **15**, 131243 (1960)

[221] Riesz, F., Sz.-Nagy, B.: Lektsii po funktsionalnomu analizu. (Russian) [Lectures in functional analysis], Mir, Moscow, (1979)

[222] Rutkas, A., Radbel, N.: Linear operator pencils and noncanonical systems. (Russian) Teor. Func. Anal i Prilozhen., **17**, 3–14 (1973)

[223] Rutkas, A.: Characteristic function and a model of a linear operator pencil. (Russian) Teor. Func. Func. Anal. i Prilozhen., **45**, 98–111 (1986), [English Transl., J.Soviet Math., **48**, 451–464 (1990)]

[224] Rutkas, A.: On a function of an operator colligation. (Russian) Vestnik Harkov. Gos. Univ. No.113 Mat. i Meh. Vyp. **39**, 14–17, (1974)

[225] Rutkas, A., Čausovskii, D.: Indefinite operator colligations and wave functions of discrete structures. (Russian) Teor. Funkcii Funkcional. Anal. i Prilozen., Vyp. **23**, 93109 (1975)

[226] Sakhnovich, L.A.: The factorization of an operator-valued transmission function. (Russian) Dokl. Akad. Nauk SSSR, **226**, No.4, 781–784 (1976)

[227] Sakhnovich, L.A.: Interpolation theory and its applications. Mathematics and its Applications, **428**, Kluwer Academic Publishers, Dordrecht, (1997)

[228] Salamon, D.: Infinite-dimensional linear systems with unbounded control and observation: a functional analytic approach. Trans. Amer. Math. Soc. **300**, No.2, 383–431 (1987)

[229] Savchuk, A., Shkalikov, A.: Sturm-Liouville operators with singular potentials. Math. Notes, **66**, No. 6, 741–753 (1999)

[230] Šmuljan, Yu.L.: Hellinger's operator integral. Mat. Sb. **49**, No.4, 381–430 (1959)

[231] Šmuljan, Yu.L.: Extended resolvents and extended spectral functions of Hermitian operator. Math. USSR Sbornick, **13**, No.3, 435–450 (1971)

[232] Šmuljan, Yu.L.: On operator R-functions, Sib.Mat.J. **12**, 315–322 (1971)

[233] Šmuljan, Yu.L.: On closed Hermitian operators and their self-adjoint extensions. Math. USSR Sbornik, **22**, 151–166, (1974)

[234] Šmuljan, Yu.L., Yanovskaya, R.N.: Blocks of a contractive operator matrix. Izv. Vyssh. Uchebn. Zaved., Mat. **7**, 72–75 (1981)

[235] Staffans, O.J.: Well-posed linear systems. Cambridge University Press, Cambridge, (2005)

[236] Staffans, O.J.: Passive and conservative continuous time impedance and scattering systems, Part I: Well posed systems. Math. Control Signals Systems, **15**, 291–315 (2002)

[237] Staffans, O.J.: Passive and conservative infinite-dimensional impedance and scattering systems (from a personal point of view). In: Rosenthal, J., Gilliam, D.S. (eds) Mathematical Systems Theory in Biology, Communication, Computation, and Finance, IMA Volumes in Mathematics and its Applications, **134**, Springer-Verlag, New York, 375–413 (2002)

[238] Staffans, O.J., Weiss, G.: Transfer functions of regular linear systems, Part II: the system operator and the Lax-Phillips semigroup. Trans. Amer. Math. Soc., **354**, 3229–3262 (2002)

[239] Stone, M.: Linear transformations in Hilbert spaces and their applications in analysis. Amer. Math. Soc. Colloquium Publication, **15**, (1932)

[240] Storozh, O.: Extremal extensions of nonnegative operator and accretive boundary problems. Ukr. Math. J., **42**, No. 6, 758–760 (1990)

[241] Strauss, A.V.: Characteristic functions of linear operators. Izv. Akad. Nauk SSSR Ser. Math, **24**, No.1, 43-74 (1960)

[242] Sz.-Nagy, B.: Uber die Lage der Doppelgeraden von gewissen Flachen gegebener geometrischer Ordnung. (German) Comm. Math. Helv., **19**, 347–366 (1947)

[243] Sz.-Nagy, B., Foias, C.: Harmonic Analysis of Operators on Hilbert Space. North-Holland, Amsterdam, (1970)

[244] Tikhonov, A.S.: A functional model of the K-relation. J. Soviet Math., **65**, No.1, 1471–1474 (1993)

[245] Titchmarsh, E.C.: Eigenfunction Expansions Associated with Second-Order Differential Equations. Part I, 2nd ed., Oxford University Press, Oxford, (1962)

[246] Trigub, R.M.: On a comparison of linear differential operators. Math. Notes **82**, No.3-4, 380394 (2007)

[247] Tsekanovskiĭ, E.R.: Real and imaginary part of an unbounded operator. Soviet math. Dokl. **2**, 881–885 (1961)

[248] Tsekanovskiĭ, E.R.: Characteristic operator-functions of unbounded operators. Transactions of Mining Institute, Kharkov, vol. 11, 95–100, (1962)

[249] Tsekanovskiĭ, E.R.: Generalized extensions of non-symmetric operators. Mat. Sbornik, **68** (110), No.4, 527–548, (1965)

[250] Tsekanovskiĭ, E.R.: Description of the generalized extensions with rank one imaginary component of a differentiation operator without spectrum. (Russian), Dokl. Akad. Nauk SSSR, **176**, 1266–1269 (1967)

[251] Tsekanovskiĭ, E.R.: Generalized self-adjoint extensions of symmetric operators. Soviet Math. Dokl., **9**, 293–296 (1968)

[252] Tsekanovskiĭ, E.R.: Certain properties of generalized selfadjoint extensions of symmetric operators. (Russian) Mathematical Physics, **5**, 203–205 (1968)

[253] Tsekanovskiĭ, E.R.: The description and the uniqueness of generalized extensions of quasi-Hermitian operators. (Russian) Funkcional. Anal. i Prilozen., **3**, No.1, 95–96 (1969)

[254] Tsekanovskiĭ, E.R.: The theory of generalized extensions of unbounded linear operators. (Russian), Doctor of Science Thesis, Insitute of Low Temperatures, Ukrainian Academy of Sciences, Kharkov, (1970)

[255] Tsekanovskiĭ, E.R.: Non-self-adjoint accretive extensions of positive operators and theorems of Friedrichs-Krein-Phillips. Funct. Anal. Appl. **14**, 156–157 (1980)

[256] Tsekanovskiĭ, E.R.: Friedrichs and Krein extensions of positive operators and holomorphic contraction semigroups. Funct. Anal. Appl. **15**, 308–309 (1981)

[257] Tsekanovskiĭ, E.R.: Characteristic function and description of accretive and sectorial boundary value problems for ordinary differential operators. (Russian), Dokl. Akad. Nauk Ukrain. SSR Ser. A, No.6, 21–24, (1985)

[258] Tsekanovskiĭ, E.R.: Characteristic function and sectorial boundary value problems. Investigation on geometry and math. analysis, Novosibirsk, **7**, 180–185 (1987)

[259] Tsekanovskiĭ, E.R.: Triangular models of unbounded accretive operators and the regular factorization of their characteristic operator functions. (Russian), Dokl. Akad. Nauk SSSR **297**, No.3, 552–556 (1987); translation in Soviet Math. Dokl. **36**, No.3, 512–515 (1988)

[260] Tsekanovskiĭ, E.R.: Accretive Extensions and Problems on Stieltjes Operator-Valued Functions Relations. Operator Theory: Advan. and Appl., **59**, 328-347 (1992)

[261] Tsekanovskiĭ, V.E.: On classes of Stieltjes type operator-valued functions with gaps. Z. Anal. Anwendungen, **11**, No.2, 183–198 (1992)

[262] Tsekanovskiĭ, V.E., Tsekanovskiĭ, E.R.: Stieltjes operator-functions with the gaps and their realization by conservative systems. Recent advances in mathematical theory of systems, control, networks and signal processing, I, Kobe, Mita, Tokyo, 37–42 (1992)

[263] Tsekanovskiĭ, E.R., Šmuljan, Yu.L.: Questions in the theory of the extension of unbounded operators in rigged Hilbert spaces. (Russian) Mathematical analysis, Vol. 14, 59–100, (1977)

[264] Tsekanovskiĭ, E.R., Šmuljan, Yu.L.: The theory of bi-extensions of operators on rigged Hilbert spaces. Unbounded operator colligations and characteristic functions. Russ. Math. Surv., **32**, 73–131 (1977)

[265] Tsekanovskiĭ, E.R., Šmuljan, Yu.L.: Method of generalized functions in the theory of extensions of unbounded linear operators. Donetsk State Univeristy, Donetsk, (1973)

[266] Tucsnak, M., Weiss, G.: Observation and control for operator semigroups. Birkhäuser Advanced Texts. Birkhäuser Verlag, Basel, (2009)

[267] Vaksman, L., Livšic, M.S.: Open geometry, and operator colligations. (Russian) Ukrain. Geometr. Sb., No.15, 16–35 (1974)

[268] Vinnikov, V., Yuditskii, P.: Functional models for almost periodic Jacobi matrices and the Toda hierarchy. Mat. Fiz. Anal. Geom., **9**, No.2, 206219 (2002)

[269] Weyl, H.: Über gewöhnliche Differentialgleichungen mit Singularitäten und die zugehörigen Entwicklungen willkürlicher Funktionen. Math. Ann. **68**, 220–269 (1910).

[270] Yantsevich, A.A.: Operator J-colligations and associated open systems. (Russian), Teor. Funkcii Funkcional. Anal. i Prilozen., No.17, 215–220 (1973)

[271] Yosida, K.: Functional Analysis. Springer-Verlag New York, (1994)

[272] Zolotarev, V.A.: Analytic methods in spectral representations of non-selfadjoint and non-unitary operators. (Russian), Kharkov National University, Kharkov (2003)

[273] Zolotarev, V.A.: On commutative systems of nonselfadjoint unbounded linear operators. Journal of Mathematical Physics, Analysis, Geometry, **5**, No.3, 273–295 (2009)

[274] Zolotarev, V.A.: Properties of characteristic function of commutative system of nonselfadjoint unbounded linear operators. Journal of Mathematical Physics, Analysis, Geometry, **6**, No.2, 192–228 (2010)

Index